Hydrogen Sulfide in Plant Biology

Hydrogen Sulfide in Plant Biology

Past and Present

Edited by

Samiksha Singh

Vijay Pratap Singh

Durgesh Kumar Tripathi

Sheo Mohan Prasad

Devendra Kumar Chauhan

Nawal Kishore Dubey

ELSEVIER

ACADEMIC PRESS

An imprint of Elsevier

Academic Press is an imprint of Elsevier
125 London Wall, London EC2Y 5AS, United Kingdom
525 B Street, Suite 1650, San Diego, CA 92101, United States
50 Hampshire Street, 5th Floor, Cambridge, MA 02139, United States
The Boulevard, Langford Lane, Kidlington, Oxford OX5 1GB, United Kingdom

British Library Cataloguing-in-Publication Data
A catalogue record for this book is available from the British Library

Library of Congress Cataloging-in-Publication Data
A catalog record for this book is available from the Library of Congress

ISBN: 978-0-323-85862-5

For Information on all Academic Press publications
visit our website at https://www.elsevier.com/books-and-journals

Publisher: Charlotte Cockle
Editorial Project Manager: Lena Sparks
Production Project Manager: Punithavathy Govindaradjane
Cover Designer: Christian Bilbow

Typeset by MPS Limited, Chennai, India

Contents

List of contributors xv

1. **Hydrogen sulfide regulates temperature stress in plants** 1

Sajid Ali, Muhammad Akbar Anjum, Aamir Nawaz, Safina Naz, Hasan Sardar and Mahmood Ul Hasan

1.1 Introduction 1
1.2 Temperature stress 3
1.2.1 Chilling stress in plants 3
1.2.2 Chilling-induced effects on plants 3
1.2.3 High-temperature or heat stress in plants 7
1.2.4 High-temperature-induced effects on plants 7
1.2.5 H_2S-induced regulation of temperature stress in plants 10
1.2.6 Crosstalk of H_2S with other molecules during temperature stress 10
1.2.7 H_2S and chilling stress in plants 11
1.2.8 H_2S and high-temperature stress in plants 14
1.3 Conclusion and future perspectives 17
References 18

2. **Crosstalk of hydrogen sulfide with melatonin and nitric oxide in ripening of fruits** 25

Sajid Ali, Muhammad Akbar Anjum, Aamir Nawaz, Mahmood Ul Hasan, Safina Naz, Shaghef Ejaz and Sakeena Tul-Ain Haider

2.1 Introduction 25
2.2 Fruit ripening mechanism 27
2.3 H_2S, NO, and melatonin in regulation of fruit ripening 28
2.3.1 Crosstalk of H_2S with NO during ripening 28
2.3.2 Crosstalk of melatonin with NO 28
2.3.3 Crosstalk between H_2S, melatonin and NO 29
2.4 Sole role of H_2S in fruit ripening 29
2.4.1 H_2S and fruit ripening 29
2.4.2 Role of H_2S in delaying ripening and conserving quality during postharvest 30
2.5 Nitric oxide and fruit ripening 36

2.5.1 NO and postharvest fruit quality 37
2.5.2 NO and postharvest senescence reduction 37
2.5.3 NO and biochemical quality conservation 39
2.5.4 NO and chilling injury 39
2.5.5 NO and postharvest diseases 40
2.5.6 Action mechanism of NO 40
2.5.7 NO, intracellular energy, and sugar metabolism 41
2.6 Melatonin and fruit ripening 42
2.6.1 Role of melatonin in postharvest fruit quality 42
2.6.2 Melatonin and postharvest senescence 43
2.6.3 Melatonin, chilling injury, and disease reduction 43
2.6.4 Action mechanism of melatonin 46
2.7 Conclusion and future prospects 46
References 46

3. Role of hydrogen sulfide in alleviating oxidative stress in plants through induction of antioxidative defense mechanism, and modulations of physiological and biochemical components 55
Monika Patel and Asish Kumar Parida

3.1 Introduction 55
3.2 Biosynthesis of hydrogen sulfide in plant cells under adverse environmental conditions 58
3.3 Role of hydrogen sulfide as an emergent signaling molecule in plants 59
3.3.1 Interaction of H_2S with other signaling molecule to enhance antioxidative defense mechanism 61
3.4 Role of H_2S in acceleration of antioxidative mechanism in plants during abiotic stress condition 61
3.4.1 Function of hydrogen sulfide in mitigation of drought stress in plants 62
3.4.2 Hydrogen sulfide improves salinity tolerance in plants 62
3.4.3 Protective functions of H_2S in response to heavy metal and metalloid toxicity 63
3.4.4 Bioprotective nature of hydrogen sulfide on plant thermotolerance 65
3.4.5 Protective function of hydrogen sulfide under cold stress 66
3.5 Role of H_2S in ionic and osmotic homeostasis in plants under various abiotic stresses 66
3.6 Interaction of H_2S and physiological responses in plants imposed to various abiotic stresses 68
3.7 Conclusions and future prospective 70
Acknowledgments 76
References 76

4. Regulation of metal stress toxicity in plants by the hydrogen sulfide 87

Nimisha Amist and N.B. Singh

4.1 Introduction 87
4.2 H_2S in plants: metabolism and role in plant development 88
4.3 H_2S-induced protein persulfidation during oxidative stress 91
4.4 H_2S signaling triggered by heavy metal stress 92
4.5 H_2S-induced metal stress tolerance 93
4.6 Conclusions and future perspectives 96
References 97

5. Hydrogen sulfide and lateral root development in plants under stress 103

Muhammad Fasih Khalid, Sajjad Hussain, Anas Fadli, Faisal Shahzad, Muhammad Akbar Anjum, Iqra Zakir and Shakeel Ahmad

5.1 Introduction 103
5.2 Hydrogen sulfide and environmental stress 104
5.3 Signaling of hydrogen sulfide 105
5.4 Hydrogen sulfide and production of lateral roots 107
5.5 Conclusions 108
References 108

6. General view on H_2S and abiotic stress tolerance in plants 113

Savita Bhardwaj and Dhriti Kapoor

6.1 Introduction 113
6.2 Insight into H_2S metabolism 114
6.3 Protein persulfidation induced by H_2S due to oxidative stress 116
6.4 H_2S in abiotic stress tolerance 118
6.4.1 Ameliorative roles of H_2S during drought stress 119
6.4.2 Ameliorative roles of H_2S during salinity stress 120
6.4.3 Ameliorative roles of H_2S during temperature stress 121
6.4.4 Ameliorative roles of H_2S during heavy metal stress 123
6.5 H_2S and plant growth regulators 124
6.6 Conclusion 125
References 126

7. Role of H_2S in plants: a current update 133

Shristy Gautam, Savita Bhardwaj and Dhriti Kapoor

7.1 Introduction 133
7.2 Sulfide: an emerging signal molecule 135
7.3 H_2S synthesis in plants 137

7.4 Biological significance of H_2S 139
7.5 Crosstalk of H_2S with other signals 140
7.6 Conclusion 142
References 143

8. Hydrogen sulfide and nitric oxide crosstalk in plants under stress 149
Réka Szőllősi, Árpád Molnár, Gábor Feigl, Dóra Oláh and Zsuzsanna Kolbert

8.1 Introduction 149
8.2 Similarities and differences between NO and H_2S 149
8.2.1 Physicochemical aspects 149
8.2.2 Metabolic processes 150
8.2.3 Signaling pathways 152
8.3 NO and H_2S and their role in plants at physiological conditions 154
8.3.1 Seed dormancy/germination 155
8.3.2 Root organogenesis 155
8.3.3 Stomatal closure 155
8.3.4 Leaf senescence 156
8.3.5 Fruit ripening 156
8.4 Crosstalk between H_2S and NO in plants under stress conditions 156
8.4.1 Abiotic stress 156
8.4.2 Biotic stress 161
8.5 Conclusions and future perspectives 161
Acknowledgment 162
References 162

9. Gene regulation by H_2S in plants 171
Alireza Iranbakhsh, Zahra Oraghi Ardebili and Narges Oraghi Ardebili

9.1 Introduction 171
9.2 Multilayered transcriptional regulation 173
9.3 H_2S and epigenetics 174
9.3.1 Histone modification (histone methylation/demethylation or acetylation/deacetylation) by catalytic functions of histone acetyltransferases, histone deacetylases, methyltransferases, and demethylases 174
9.3.2 DNA cytosine methylation 176
9.3.3 Noncoding ribonucleic acid (RNA) regulation; fully discussed in the next title 176

9.4 H_2S and microRNA, and epigenetic feedback loop 176
9.4.1 MicroRNAs can influence both histone and DNA modifications through targeting the responsible enzymes 177
9.4.2 The epigenetic machinery controls gene transcriptions of microRNAs 177
9.4.3 MicroRNAs control H_2S biosynthesis 178
9.4.4 H_2S modulates microRNAs 179
9.5 H_2S, hormones, epigenetics, and transcriptional landscape 179
9.5.1 Phytohormones involves in H_2S production and homeostasis 179
9.5.2 H_2S contributes to the modulation of biosynthesis, signaling, and/or functions of phytohormones 180
9.6 H_2S, mitochondria, chloroplast, epigenetics, and gene regulation 181
9.7 H_2S and transcription factors 182
9.8 Ca^{2+}, Ca^{2+}/CaM, H_2S, and gene regulation 185
9.8.1 Ca^{2+}/CaM signaling regulates H_2S production and homeostasis 185
9.8.2 H_2S can effectively participate in the modulation of Ca^{2+} signaling 186
9.9 H_2S, mitogen-activated protein kinase, signal transduction, and transcriptional control 186
9.10 Crosstalk between NO and H_2S 187
9.11 Crosstalk between H_2O_2 and H_2S 189
9.12 H_2S and redox-based gene regulation 189
9.13 H_2S, cytoskeleton, and gene regulation 190
9.14 H_2S-mediated mechanisms of transcription reprogramming, concluding remarks, and future perspective 190
References 192

10. Hydrogen sulfide and reactive oxygen species crosstalk and acquisition of abiotic stress tolerance 201

Krishna Kumar Choudhary and Nivedita Chaudhary

10.1 Introduction 201
10.2 Role of hydrogen sulfide in plants 202
10.2.1 Generation of H_2S in plants 202
10.2.2 H_2S metabolism in plants 204
10.2.3 Importance of H_2S in plants as a signaling molecule 204
10.3 Hydrogen sulfide and reactive oxygen species crosstalk during abiotic stress in plants 205
10.3.1 H_2S and antioxidative defense mechanism 206
10.3.2 H_2S and reactive oxygen species 207
10.3.3 Cooperative interactions of H_2S and reactive oxygen species in plants 208
10.4 Conclusion and future perspectives 209

Acknowledgments 209
References 209

11. Regulation of salinity stress by hydrogen sulfide in plants 213

Owias Iqbal Dar, Kirpal Singh, Junaid Aslam, Sunil Sharma, Arvinder Kaur, Renu Bhardwaj and Anket Sharma

11.1 Introduction 213
11.2 Impact of salinity stress on plants 214
11.2.1 Impact of salinity on growth 214
11.2.2 Impact of salinity on photosynthesis 215
11.2.3 Impact of salinity on oxidative stress 215
11.3 Role of H_2S in plant biology 215
11.4 Role of H_2S in plants under salinity stress 217
11.4.1 Impact on photosynthesis 217
11.4.2 Impact on oxidative stress 218
11.4.3 Impact on antioxidative system 218
11.5 Role of H_2S in plant productivity under salinity stress 219
11.6 Conclusion 219
References 220

12. Regulation of drought stress by hydrogen sulfide in plants 229

Owias Iqbal Dar, Kirpal Singh, Sunil Sharma, Junaid Aslam, Arvinder Kaur, Renu Bhardwaj and Anket Sharma

12.1 Introduction 229
12.2 Impact of drought stress on plants 230
12.2.1 Impact on growth and photosynthesis 230
12.2.2 Impact on oxidative stress 231
12.3 Role of hydrogen sulfide in plant biology 231
12.4 Role of H_2S in plants under drought stress 233
12.4.1 Impact on photosynthesis 233
12.4.2 Impact on oxidative stress 234
12.4.3 Impact on antioxidative system 234
12.5 Role of H_2S in plant productivity under drought stress 235
12.6 Conclusion 236
References 237

13. Hydrogen sulfide and phytohormones crosstalk in plant defense against pathogen attack 243

Shivam Jasrotia, Raman Jasrotia and Anket Sharma

13.1 Introduction 243
13.2 Impact of pathogens on plant physiology 244
13.3 Effect of H_2S on plant morphology 245

13.4 Hydrogen sulfide–plant hormone crosstalk under pathogen attack 246
13.4.1 Crosstalk between hydrogen sulfide and abscisic acid 246
13.4.2 Crosstalk between hydrogen sulfide and salicylic acid 247
13.4.3 Crosstalk between hydrogen sulfide and auxins 248
13.4.4 Crosstalk between hydrogen sulfide and jasmonic acid 249
13.4.5 Crosstalk between hydrogen sulfide and ethylene 249
13.4.6 Crosstalk between hydrogen sulfide and gibberellic acid 250
13.5 Conclusion 250
References 251

14. Role of H_2S in plants against radiation stress 257
Shivam Jasrotia

14.1 Introduction-role of H_2S 257
14.2 Effect of radiation 258
14.3 Plant's response against radiation 260
14.4 H_2S helps in the amelioration of radiation stress by interacting with other compounds 260
14.5 H_2S itself is triggered by UV radiation 261
14.6 Antistress role of H_2S exhibited by regulating antioxidative enzyme activities and antioxidant contents 261
References 263

15. Hydrogen sulfide and phytohormones crosstalk in plant defense against abiotic stress 267
Jaspreet Kour, Kanika Khanna, Pooja Sharma, Arun Dev Singh, Indu Sharma, Priya Arora, Pardeep Kumar, Kamini Devi, Mohd Ibrahim, Puja Ohri, Bilal Ahmad Mir, Ashutosh Sharma and Renu Bhardwaj

15.1 Introduction 267
15.2 Generation of H_2S in plants 268
15.3 Role of H_2S in plants under abiotic stress 270
15.3.1 Salt stress 270
15.3.2 Water stress 275
15.3.3 Heat stress 276
15.3.4 Metal stress 276
15.4 Ameliorative role of phytohormones in plants under abiotic stress 284
15.5 Crosstalk of H_2S with phytohormones under various abiotic stresses 287
15.5.1 H_2S and auxin interaction 288
15.5.2 H_2S and ethylene interaction 289

15.5.3 H_2S and abscisic acid interaction 290
15.5.4 H_2S and gibberellic acid interaction 290
15.5.5 H_2S and jasmonic acid 291
15.5.6 An outline of crosstalk between H_2S and phytohormones 291
15.6 Conclusion and future perspectives 292
References 292

16. Implication of H_2S and Ca signaling in abiotic stress tolerance in plants 303

Shalini Dhiman, Palak Bakshi, Arun Dev Singh, Tamanna Bhardwaj, Neerja Sharma, Nitika Kapoor, Dhriti Kapoor, Ravdeep Kaur, Upma Arora, Anket Sharma, Bilal Ahmad Mir and Renu Bhardwaj

16.1 Introduction 303
16.2 Role of different signaling molecules in plants during abiotic stresses 305
16.2.1 Calcium 305
16.2.2 Reactive sulfur species 305
16.2.3 Reactive carbonyl species 306
16.2.4 Reactive oxygen species 307
16.2.5 Reactive nitrogen species 308
16.3 Hydrogen sulfide (H_2S) and Ca^{2+} signaling during different types of abiotic stress tolerance in plants 309
16.3.1 Salt stress 309
16.3.2 Hydrogen sulfide (H_2S) and Ca^{2+} signaling during temperature stress 311
16.3.3 Hydrogen sulfide (H_2S) and Ca^{2+} signaling during heavy metal stress 319
16.3.4 Hydrogen sulfide (H_2S) and Ca^{2+} signaling during drought stress 322
16.3.5 Hydrogen sulfide (H_2S) and Ca^{2+} signaling during flooding stress 325
16.3.6 Hydrogen sulfide (H_2S) and Ca^{2+} signaling during other stresses 327
16.4 Crosstalk between H_2S and Ca^{2+} signaling during plant responses to abiotic stress 328
16.5 Conclusion and future perspective 330
References 330

17. Hydrogen sulfide homeostasis in plants: An overview 341

Sunil K. Gupta, Naina Marwa, Ashutosh K. Pandey, Yun-Bing Zhang and Jiao-Lin Zhang

17.1 Introduction 341
17.1.1 Source of H_2S in plants 342
17.1.2 Role of H_2S in plants 343

17.2 H_2S effects on plant growth and development 344
17.3 H_2S effects on physiological homeostasis 345
17.3.1 On gas exchange parameter 345
17.3.2 H_2S effects on mineral nutrition 347
17.4 H_2S homeostasis during metal toxicity 348
17.5 Hydrogen sulfide effects and enzymatic homeostasis in plants 349
17.5.1 On membrane degradation 349
17.5.2 On reactive oxygen species defense and homeostasis 349
17.5.3 Antioxidant and antioxidant enzyme activity 350
17.6 Hydrogen sulfide and molecular homeostasis in plants 351
17.6.1 The molecular mechanism behind H_2S homeostasis 351
17.6.2 Biotechnological application of H_2S in plants 353
17.6.3 H_2S-mediated regulation of gene and protein expression 353
17.7 Conclusion and future prospects 355
Acknowledgments 355
References 355

Index 365

List of contributors

Shakeel Ahmad Department of Agronomy, Faculty of Agricultural Sciences and Technology, Bahauddin Zakariya University, Multan, Pakistan

Sajid Ali Department of Horticulture, Faculty of Agricultural Sciences and Technology, Bahauddin Zakariya University, Multan, Pakistan

Nimisha Amist Plant Physiology Laboratory, Department of Botany, University of Allahabad, Allahabad, India

Muhammad Akbar Anjum Department of Horticulture, Faculty of Agricultural Sciences and Technology, Bahauddin Zakariya University, Multan, Pakistan

Narges Oraghi Ardebili Department of Biology, Science and Research Branch, Islamic Azad University, Tehran, Iran

Zahra Oraghi Ardebili Department of Biology, Garmsar Branch, Islamic Azad University, Garmsar, Iran

Priya Arora Department of Botanical and Environmental Sciences, Guru Nanak Dev University, Amritsar, India

Upma Arora Department of Botany, Lyallpur Khalsa College, Jalandhar, India

Junaid Aslam Department of Botanical and Environmental Sciences, Guru Nanak Dev University, Amritsar, India

Palak Bakshi Department of Botanical and Environmental Sciences, Guru Nanak Dev University, Amritsar, India

Renu Bhardwaj Department of Botanical and Environmental Sciences, Guru Nanak Dev University, Amritsar, India

Savita Bhardwaj Department of Botany, School of Bioengineering and Biosciences, Lovely Professional University, Phagwara, India

Tamanna Bhardwaj Department of Botanical and Environmental Sciences, Guru Nanak Dev University, Amritsar, India

Nivedita Chaudhary Department of Environmental Science, School of Earth Sciences, Central University of Rajasthan, Kishangarh, Ajmer, India

Krishna Kumar Choudhary Botany Section, Mahila Mahavidyalaya, Banaras Hindu University, Varanasi, India

Owias Iqbal Dar Department of Zoology, Guru Nanak Dev University, Amritsar, India

Kamini Devi Department of Botanical and Environmental Sciences, Guru Nanak Dev University, Amritsar, India

Shalini Dhiman Department of Botanical and Environmental Sciences, Guru Nanak Dev University, Amritsar, India

Shaghef Ejaz Department of Horticulture, Faculty of Agricultural Sciences and Technology, Bahauddin Zakariya University, Multan, Pakistan

Anas Fadli Horticultural Sciences Department, Citrus Research and Education Centre, Institute of Food and Agricultural Sciences, University of Florida, Lake Alfred, FL, United States

Gábor Feigl Department of Plant Biology, University of Szeged, Szeged, Hungary

Shristy Gautam Department of Molecular Biology and Genetic Engineering, School of Bioengineering and Biosciences, Lovely Professional University, Phagwara, India

Sunil K. Gupta CAS Key Laboratory of Tropical Forest Ecology, Xishuangbanna Tropical Botanical Garden, Chinese Academy of Sciences, Yunnan, P.R. China; Plant Ecology & Climate Change Science Division, CSIR-National Botanical Research Institute, Rana Pratap Marg, Lucknow, India

Sakeena Tul-Ain Haider Department of Horticulture, Faculty of Agricultural Sciences and Technology, Bahauddin Zakariya University, Multan, Pakistan

Mahmood Ul Hasan Institute of Horticultural Sciences, University of Agriculture, Faisalabad, Pakistan

Sajjad Hussain Department of Horticulture, Faculty of Agricultural Sciences and Technology, Bahauddin Zakariya University, Multan, Pakistan

Mohd Ibrahim Department of Botanical and Environmental Sciences, Guru Nanak Dev University, Amritsar, India

Alireza Iranbakhsh Department of Biology, Science and Research Branch, Islamic Azad University, Tehran, Iran

Raman Jasrotia Department of Zoology, University of Jammu, Jammu, India

Shivam Jasrotia Government Degree College, Kathua, Jammu and Kashmir, India

Dhriti Kapoor Department of Botany, School of Bioengineering and Biosciences, Lovely Professional University, Phagwara, India

Nitika Kapoor PG Department of Botany, Hans Raj Mahila Maha Vidyalaya, Jalandhar, India

Arvinder Kaur Department of Zoology, Guru Nanak Dev University, Amritsar, India

Ravdeep Kaur Department of Botanical and Environmental Sciences, Guru Nanak Dev University, Amritsar, India

Muhammad Fasih Khalid Department of Horticulture, Faculty of Agricultural Sciences and Technology, Bahauddin Zakariya University, Multan, Pakistan; Horticultural Sciences Department, Citrus Research and Education Centre, Institute of Food and Agricultural Sciences, University of Florida, Lake Alfred, FL, United States

Kanika Khanna Department of Botanical and Environmental Sciences, Guru Nanak Dev University, Amritsar, India

Zsuzsanna Kolbert Department of Plant Biology, University of Szeged, Szeged, Hungary

Jaspreet Kour Department of Botanical and Environmental Sciences, Guru Nanak Dev University, Amritsar, India

Pardeep Kumar Department of Botanical and Environmental Sciences, Guru Nanak Dev University, Amritsar, India

Naina Marwa Plant Ecology & Climate Change Science Division, CSIR-National Botanical Research Institute, Rana Pratap Marg, Lucknow, India

Bilal Ahmad Mir Department of Botany, School of Life Sciences, Satellite Campus Kargil, University of Kashmir, Jammu and Kashmir, India

Árpád Molnár Department of Plant Biology, University of Szeged, Szeged, Hungary

Aamir Nawaz Department of Horticulture, Faculty of Agricultural Sciences and Technology, Bahauddin Zakariya University, Multan, Pakistan

Safina Naz Department of Horticulture, Faculty of Agricultural Sciences and Technology, Bahauddin Zakariya University, Multan, Pakistan

Puja Ohri Department of Zoology, Guru Nanak Dev University, Amritsar, India

Dóra Oláh Department of Plant Biology, University of Szeged, Szeged, Hungary

Ashutosh K. Pandey School of Agricultural Sciences, Sharda University, Greater Noida, India

Asish Kumar Parida Plant Omics Division, CSIR-Central Salt and Marine Chemicals Research Institute (CSIR-CSMCRI), Gijubhai Badheka Marg, Bhavnagar, India; Academy of Scientific and Innovative Research, Ghaziabad, India

Monika Patel Plant Omics Division, CSIR-Central Salt and Marine Chemicals Research Institute (CSIR-CSMCRI), Gijubhai Badheka Marg, Bhavnagar, India; Academy of Scientific and Innovative Research, Ghaziabad, India

Hasan Sardar Department of Horticulture, Faculty of Agricultural Sciences and Technology, Bahauddin Zakariya University, Multan, Pakistan

Faisal Shahzad Horticultural Sciences Department, Citrus Research and Education Centre, Institute of Food and Agricultural Sciences, University of Florida, Lake Alfred, FL, United States

Anket Sharma State Key Laboratory of Subtropical Silviculture, Zhejiang A&F University, Hangzhou, P.R. China

Ashutosh Sharma Department of Agriculture Sciences, DAV University, Jalandhar, India

Indu Sharma Sant Baba Bhag Singh University, Padhiana, Khiala, Jalandhar, India

Neerja Sharma Department of Botanical and Environmental Sciences, Guru Nanak Dev University, Amritsar, India

Pooja Sharma Department of Botanical and Environmental Sciences, Guru Nanak Dev University, Amritsar, India; Department of Microbiology, DAV University, Jalandhar, India

Sunil Sharma Department of Zoology, Guru Nanak Dev University, Amritsar, India

Arun Dev Singh Department of Botanical and Environmental Sciences, Guru Nanak Dev University, Amritsar, India

Kirpal Singh Department of Zoology, Guru Nanak Dev University, Amritsar, India

N.B. Singh Plant Physiology Laboratory, Department of Botany, University of Allahabad, Allahabad, India

Réka Szőllősi Department of Plant Biology, University of Szeged, Szeged, Hungary

Iqra Zakir Department of Agronomy, Faculty of Agricultural Sciences and Technology, Bahauddin Zakariya University, Multan, Pakistan

Jiao-Lin Zhang CAS Key Laboratory of Tropical Forest Ecology, Xishuangbanna Tropical Botanical Garden, Chinese Academy of Sciences, Yunnan, P.R. China

Yun-Bing Zhang CAS Key Laboratory of Tropical Forest Ecology, Xishuangbanna Tropical Botanical Garden, Chinese Academy of Sciences, Yunnan, P.R. China

Chapter 1

Hydrogen sulfide regulates temperature stress in plants

Sajid Ali[1,*], Muhammad Akbar Anjum[1], Aamir Nawaz[1], Safina Naz[1], Hasan Sardar[1] and Mahmood Ul Hasan[2]
[1]Department of Horticulture, Faculty of Agricultural Sciences and Technology, Bahauddin Zakariya University, Multan, Pakistan, [2]Institute of Horticultural Sciences, University of Agriculture, Faisalabad, Pakistan
**Corresponding author.*

1.1 Introduction

Plants are sessile and unable to move under different abiotic stress conditions. Abiotic stresses due to adverse change in environment, such as high and low temperatures, severely limit the vegetative growth and development of plants (Ohama et al., 2017; Sun et al., 2019). The response to temperature stress is a highly specific mechanism which involves various metabolic pathways, cellular signaling, and regulatory networks (Kotak et al., 2007). The capability of the plants to acclimatize to the prevailing environmental conditions is the leading determinant for coping with different stresses (Shah et al., 2013). A short period of temperature stress can prove highly detrimental to vegetative growth and yield, leading to severe economic losses of the crops (Shah et al., 2013).

The rapid population increase and climate change has negatively affected the status of global food security. These are the major constraints in securing food security in the world. In the present scenario of climate change, heat stress due to high temperature has become an imperative factor that limits crop production, subsequently affecting food security (Lobell et al., 2011). The average temperature of the ocean and land surface has increased by 0.85°C (Fahad et al., 2017). On average, a 0.2°C increase per decade has been projected from now onwards. The rising concentrations of different greenhouse gasses are becoming a leading factor affecting global warming (Friedlingstein et al., 2010; Fahad et al., 2017). Severe and prolonged heat stress substantially reduces the yield of crops by negatively influencing vegetative growth, physiology, and other metabolic cascades in the crop plants (Fahad et al., 2017). High-temperature-induced heat stress negatively affects

Hydrogen Sulfide in Plant Biology. DOI: https://doi.org/10.1016/B978-0-323-85862-5.00003-8

the vegetative growth and physiology of plants, ultimately leading to a substantial reduction in yield and quality of the produced commodity (Duc et al., 2018; Nemeskéri and Helyes, 2019). The plant growth is also significantly influenced by different biochemical, physiological, and morphological changes in response to high-temperature stress (Wahid et al., 2007; Fahad et al., 2017; Lamaoui et al., 2018). Due to climate change under the global warming scenario, high-temperature stress mitigation has now become an urgent issue in the world. Therefore the comprehensive understanding of various biochemical and physiological mechanisms under heat stress is critical for future sustainable production of various crops under high-temperature conditions (Zhou et al., 2015). High-temperature susceptibility normally varies with the developmental phases and genotypes of the crops (Wahid et al., 2007). Moreover, heat stress tolerance involves various regulations of physiological and molecular responses due to changes in the cell organelles (Wahid et al., 2007). The sexual reproduction of plants is highly sensitive to high temperature or heat stress, subsequently affecting flower production, number of flowers, pollen production, pollen viability, and fertilization (Sangu et al., 2015).

It has also been reported that different crops also experience extreme low-temperature stress in various parts of the world (Wang et al., 2016). Such exposure of the crops to chilling stress significantly hampers the growth of plants and therefore low temperature stress is considered extremely detrimental to sustainable crop production (Tommasini et al., 2008). The chilling stress poses certain negative impacts on stomata development and growth of leaves (Deng et al., 2012). Plants normally show specific physiological and molecular responses upon their exposure to chilling stress (Atkinson et al., 2013; Sewelam et al., 2014). Overall, low temperature or chilling exposure thermodynamically reduces various physiological kinetics and metabolic mechanisms occurring in plants (Hussain et al., 2016). The chilling stress severely decreases uniformity and germination rate, hampers vigor of seedlings and markedly delays ontogenic development of the exposed plants eventually leading to severe yield losses of the crops (Ruelland et al., 2009).

Several approaches have been reported for mitigation of temperature stress in plants. Among the used strategies, exogenous application of some suitable chemicals is also an effective method for the amelioration of temperature stress in plants. Hydrogen sulfide (H_2S) is the third major gaseous transmitter (Ali et al., 2019). It has been used in several studies for alleviation of different types of stresses including heat (Li et al., 2012), chilling (Ma et al., 2015), salt (Wang et al., 2012), drought (Jin et al., 2013), and heavy metal (He et al., 2018) stresses. Exogenous H_2S application has been found to enhance chilling tolerance in grapes (Fu et al., 2013), banana (Luo et al., 2015), and cucumber (Liu et al., 2020). H_2S showed promising results in the alleviation of various types of temperature stresses in a number of plants. Therefore the present chapter reviews possible H_2S-mediated roles and mechanisms in the alleviation of high- and low-temperature stress in different crop plants.

1.2 Temperature stress

Temperature stress is a type of abiotic stress that negatively impacts the growth and various other responses of plants. Temperature stress can be divided into two groups, such as high temperature or heat stress and chilling or freezing stress.

1.2.1 Chilling stress in plants

The unsuitable or suboptimal temperature generally evokes various negative effects on crop plants at any stage of growth and development. Such detrimental influences on crop plants normally depend upon the growth stages and magnitude of the chilling stress as well as the period of the stress. Such types of effects are generally elicited in the form of physiological, morphological, biochemical, and molecular changes in the plants. The chilling or cold stress leads to various detrimental effects in plants (Fig. 1.1).

1.2.2 Chilling-induced effects on plants

1.2.2.1 Seed germination

Certain crops are extremely susceptible to chilling stress, especially during the seed germination phase and the early period of seedlings development. Generally, every seed needs an optimal temperature and moisture in soil for rapid and uniform germination. Low temperature or chilling stress negatively and severely decreases germination and therefore significantly reduces seedling vigor (Kang and Saltveit, 2002; Wang et al., 2016). These negative

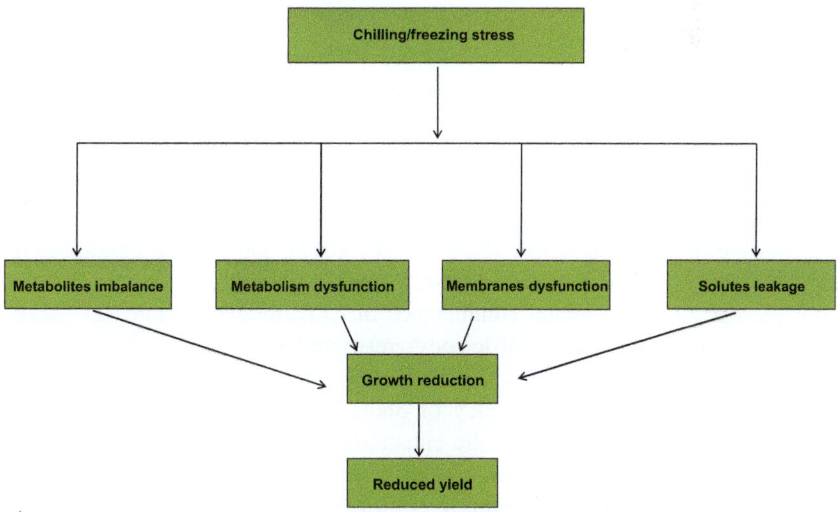

FIGURE 1.1 Low temperature (chilling/freezing)-induced effects on plants.

effects delay the development of plants (Ruelland et al., 2009). Overall, low temperature or chilling exposure thermodynamically reduces the various physiological kinetics and metabolic mechanisms occurring in plants (Ruelland et al., 2009). The chilling stress severely decreases uniformity and germination rate, hampers vigor of seedlings, and markedly delays ontogenic development of the exposed plants, eventually leading to severe yield losses of the crops (Oliver et al., 2007). Chilling stress condition induces disturbance in osmotic balance and impairs various metabolic activities of plants at cellular levels. The overproduction of reactive oxygen species (ROS) leads to alterations in protein structure, causes damage to membranes, decreases respiration, and impairs the production of cellular energy in the form of adenosine triphosphate (ATP), subsequently leading to reduced germination of seeds and seedling vigor (Hussain et al., 2018).

1.2.2.2 Growth and development

Chilling stress has also been known to cause a marked reduction in growth and development of the stressed plants. It has been observed that chilling stress imposes negative effects on the growth of numerous crops such as tomato (Starck et al., 2000), potato (Svensson et al., 2002), muskmelon (Wang et al., 2004), cotton (Zhao et al., 2012), and sugarcane (Zhu et al., 2013). Chilling stress may lead to the development of leaf necrotic lesions, delayed development, prolonged cell cycle, along with inhibited cells production, wilting and increased sensitivity to diseases and pathogens (Rymen et al., 2007; Hussain et al., 2018). Reduced cell division rate and elongation results in a small area of leaves (Jouyban et al., 2013) and reduced leaf initiation as well as lower leaf number during chilling conditions (Lukatkin et al., 2012; Jouyban et al., 2013; Hussain et al., 2018).

1.2.2.3 Gaseous exchange and photosynthesis

Gas exchange and photosynthesis is important for the synthesis of food reserves of crop plants. Chilling stress alters the normal photosynthesis rate as well as attributes of gas exchange in various crops. Chilling stress in combination with drought conditions severely influences photochemical activity, leaf gas exchange, CO_2 assimilation, electron transport, PS-II energetic pressure, and transpiration in sugarcane (Sales et al., 2013). During the conditions of chilling stress, the lower photosynthetic rate in plants has been ascribed to poor mesophyll and stomatal conductance of CO_2, reduced chloroplast development, decreased quantum efficiency, restricted metabolites transport, and reduced CO_2 assimilation yield (Hussain et al., 2018). It was noted that chilling exposure reduces the efficiency of photosynthesis and increases the process of photoinhibition owing to the enhanced thylakoid membrane excitation, and eventually leads to photosynthesis function impairment in tomato (Jung and Steffen, 1997). Exposure to chilling stress also reduces CO_2 fixation rate and limits the availability of $NADP^+$ for the acceptance of electrons

(Wise, 1995). Although chilling stress has some peculiar influences on gas exchange and photosynthesis function of plants, it is generally considered detrimental for the photosynthetic apparatus in plants and also it negatively impacts the process of light harvesting (Hussain et al., 2018).

1.2.2.4 Water and nutrient relations

Transpiration rate, leaf water potential, stomatal conductance, relative water content, and canopy temperatures are the major characteristics that describe the water relation status of plants (Farooq et al., 2009). Cold stress combined with drought significantly reduced the leaf water potential of sugarcane (Sales et al., 2013). Overall, cold sensitive plants showed symptoms of water stress during chilling temperature exposure that may result in lower leaf water potential and turgor pressure loss (Aroca et al., 2003). The stomatal control loss and reduced hydraulic conductivity decline in roots because of cold-induced plant wilting (Lee et al., 1993). The vapor pressure reduction during chilling stress may decrease root water uptake and transpiration rate (Aroca et al., 2003). Chilling stress led to water deficit in shoots of maize plants due to a significant decrease in water uptake rate in roots instead of leaf transpiration (Janowiak and Markowski, 1994). Therefore it may be said that chilling stress negatively affects the water relation of plants. Similar to water relations, chilling stress also influences the nutrient uptake in plants. It has been reported that chilling stress significantly impairs nutrient uptake in certain plants (Xu and Huang, 2006; Hussain et al., 2018). It has also been observed that the chilling temperature influences different microbiological and physicochemical mechanisms in soil that in turn may alter the plant nutrients relationships (Yan et al., 2012). Poor root development during chilling stress significantly reduces the uptake of various nutrients such as nitrogen, potassium, and phosphorous in certain plants (Yan et al., 2012). Low hydraulic conductance, reduced root growth, poor root branching, and thick axis of roots led to suppressed uptake of mineral nutrients in plants under chilling stress (Farooq et al., 2009). However, significant variations exist in plants depending upon the chilling exposure, stress period, plant species, and growth of plants under stress (Hussain et al., 2018).

1.2.2.5 Oxidative stress and osmotic balance

The chilling stress also significantly affects oxidative damage and the performance of plants. The imbalance between detoxification and production of ROS is expressed as oxidative stress. The overproduction of ROS results in membrane lipid peroxidation (Ali et al., 2019), membrane integrity loss (Ali et al., 2020a), protein oxidation, damage to biological macromolecules (DNA and RNA), and even leads to the death of cells (Apel and Hirt, 2004). Chilling stress causes oxidative damage due to excessive production of ROS in the plant cells (Shu et al., 2011). It has been observed that chilling stress has a detrimental impact on the membrane integrity and enzymatic activities, eventually resulting in tissue necrosis (Chinnusamy et al., 2007). Moreover, chilling-induced overproduction of ROS causes some serious damage to cells due to rapid reactions with the cellular structures (Sattler et al., 2000; Hussain et al., 2018). To combat, ROS plants have different antioxidative defense systems comprising enzymatic [ascorbate peroxidase (APX), catalase (CAT), POD, superoxide dismutase (SOD), etc.] and nonenzymatic [ascorbic acid (AsA), carotenoids, phenols, etc.] antioxidants (Gill and Tuteja, 2010).

The osmotic modification is a state of decreased osmotic potential due to solutes accumulation in response to an H_2O deficit condition (Zhang et al., 1999). During osmotic stress, certain plant species conserve water by the accumulation of different osmotically active ions such as organic acids, soluble sugars, proline, calcium, potassium, glycine betaine, and chloride ions. Osmoregulation helps to maintain the photosynthesis rate and vegetative growth and assimilate partitioning during various stresses (Subbarao et al., 2000). The chilling temperature disturbs osmotic adjustment and results in the excessive dehydration stress (Wang et al., 2016). It has been observed that in chilling sensitive tissues the concentration of proteins generally decreases normally due to lower synthesis that subsequently increases the levels of certain free amino acids, especially proline (Hussain et al., 2018).

1.2.2.6 *Yield attributes*

The unsuitable or suboptimal temperature generally evokes various negative effects on crop plants at any stage of growth and development subsequently leading to a reduction in yield of crops of economic importance in the world (Hussain et al., 2018). The exposure to chilling stress severely impaired the yield and production of barley (Rizza et al., 1994). The chilling temperature during the reproductive stage led to flower abortion and reduced pod setting percentage (Nayyar et al., 2005). In the temperate regions, a 30%–40% reduction in rice has been observed due to cold temperature conditions. The chilling stresses causes spikelet degeneration, panicle deformation, and significantly lower fertility of spikelet (Andaya and Mackill, 2003). Flower abscission, pollen sterility, reduced fruit setting, and markedly lower final yield have been reported in grain crops under low-temperature stress (Thakur et al., 2010). It has been noted that stem elongation was significantly reduced along with spikelet death and lower total biomass accumulation, eventually resulting in markedly less grain yield in wheat, under chilling stress (Whaley et al., 2004). In the same way, at the booting and jointing stage of wheat, productive tiller number on per plant basis was substantially reduced under chilling stress (Li et al., 2015). It was also observed that the chilling stress negatively influenced young spikes development (Thakur et al., 2010). It may be said that chilling stress has certain negative impacts on the yield attributes of various crops (Hussain et al., 2018).

1.2.2.7 Chilling stress and postharvest quality of fresh fruits and vegetables

In contrast to agronomic crops, fresh fruits and vegetables are living commodities that continue to respire even after harvest from their parent plants. Fresh fruits and vegetables have a significantly higher content of moisture and are highly perishable (Ali et al., 2019). Cold storage, alone or in combination with exogenous treatments, is generally used to maintain quality and to extend the postharvest shelf and/or storage life of fresh fruits and vegetables (Ali et al., 2018; Ali et al., 2020a,b). However, various tropical and subtropical fruit and vegetable crops are prone to a physiological disorder known as chilling injury (CI). The CI symptoms generally vary depending upon maturity stage and low-temperature exposure and commodity. Generally, CI symptoms can be categorized into two major groups: (1) metabolic or developmental symptoms, such as poor aroma and flavor or abnormal ripening, and (2) physiological indicators including discoloration, pitting, internal breakdown, and water soaking (Sevillano et al., 2009). In case of severe CI occurrence, substantial deterioration of quality takes place, subsequently leading to significant economic losses of fresh fruits and vegetables (Aghdam and Bodbodak, 2014). Hence, it may be concluded that low temperature not only causes various undesirable effects in crops during preharvest but it also leads to certain negative impacts on the storage potential and overall quality of fresh fruits and vegetables during postharvest storage.

1.2.3 High-temperature or heat stress in plants

Heat stress or high-temperature stress is a paramount challenge which deleteriously impacts the vegetative and reproductive phases of various plants (Ihsan et al., 2019). Heat or high-temperature stress refers to an extremely high-temperature duration above the optimal range needed for the normal growth of the plants. Heat stress has various detrimental effects on crops (Fig. 1.2). It negatively affects vegetative growth, pollen production, pollen fertility, and pollination in different plants (Asseng et al., 2011).

1.2.4 High-temperature-induced effects on plants

1.2.4.1 Seed germination and plant growth

Various plant responses such as seed germination, photosynthesis, respiration, growth, development, grain maturity, dry matter partitioning, and transpiration are significantly influenced in response to temperature (Zhao et al., 2013). Thus temperature is largely accountable for the modulation of seed germination and subsequent development of plants. Growth and development of plants normally needs a specific optimal temperature range and even a

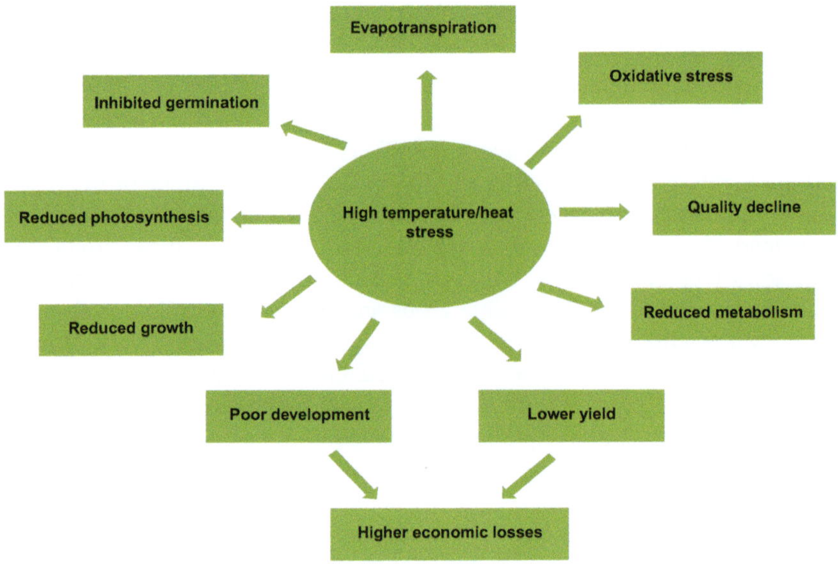

FIGURE 1.2 High temperature/heat-induced effects on plants.

slight variation during germination may negatively affect the final germination percentage, growth rate, and development of the plants (Ihsan et al., 2019). Heat or high-temperature stress influences various plants by altering certain pathways, such as primary and secondary metabolism. The reduced function of primary and secondary metabolites triggers certain crop plants to increase the defense-related mechanisms. Temperature above optimal range increases the biochemical reaction rate that in turn accelerates the development of crop plants that subsequently shortens the growing period (Fahad et al., 2016). Furthermore, temperature stress also results in early phenological stage occurrence, particularly booting (Ihsan et al., 2016). During the phase of vegetative growth, stem reserves accumulation and photosynthesis have a potential role in organs formation. Hence, an efficient growth and development process for higher yield requires an efficient strategy to target numerous development stages concurrently (Ihsan et al., 2019).

1.2.4.2 Gas exchange and photosynthesis

Gas exchange rate and photosynthesis play an important role with regard to the biosynthesis of food reserves in crop plants. Heat stress directly influences the gas exchange of plant leaves (Salvucci and Crafts-Brandner, 2004). A heat sensitive component exists in light- and dark-dependent photosynthesis reactions. The PS-II has been known as a heat sensitive photosynthesis component and is considered as a critical place for heat-induced damage (Allakhverdiev et al., 2008). It has been observed that a short period of mild high-temperature stress does not lead to any significant changes, however, markedly high temperature, especially above 45°C, greatly damages PS-II of plants leading to substantially reduced photosynthesis efficiency with a subsequent decline in the biosynthesis of food reserves (Haque et al., 2014; Ihsan et al., 2019).

1.2.4.3 Nutrient and water relations

Transpiration rate, leaf water potential, stomatal conductance, relative water content, and canopy temperatures are the major characteristics that describe the water relation status of the plants (Farooq et al., 2009). High-temperature or heat stress negatively affects the water and nutrient relations of the plants under stress conditions. The high temperature can leads to excessive transpiration and evaporation leading to a decline of water content in plants. The reduced water in the plants may lead to the decreased uptake and translocation of nutrients. Therefore plants should be provided with optimum water availability at critical phenological stages of the plants.

1.2.4.4 Osmotic balance and oxidative stress

The osmotic adjustment during changing temperature is a necessary mechanism of plants to protect them against high-temperature stress. The osmotic adjustment is a state of decreased osmotic potential due to solutes accumulation in response to conditions of H_2O deficit environment (Zhang et al., 1999). During osmotic stress conditions, certain plant species preserve the required water relations by accretion of various osmotically active ions including sugars, organic acid, proline, glycine betaine, calcium, potassium, and chloride ions. Osmotic modification efficiently helps to maintain the rate of photosynthesis and vegetative growth along with efficient partitioning (Subbarao et al., 2000). It has been observed that proline is an important stressor related to amino acid and the accumulation of proline has been found to be correlated with the tolerance of osmotic stress, involving excessive dehydration in response to heat or high-temperature stress (Ihsan et al., 2019). Heat or high-temperature stress also leads to oxidative damage by the increased production of ROS in plants. The overproduction of ROS is detrimental to lipids, proteins, DNA, and carbohydrates, ultimately causing a reduction in plant performance (Zlatev and Lidon, 2012). Therefore the reduction of the overproduction of ROS is an important strategy to increase the performance of plants under high-temperature or heat stress conditions.

1.2.4.5 Pollen production and pollination

Temperature significantly affects fertile pollen production, anther indehiscence, and pollen incompatibility (Ihsan et al., 2019). However, heat susceptibility normally varies with the developmental phases and genotypes of the crops (Wahid et al., 2007). The sexual reproduction in plants is highly sensitive to high temperature or heat stress, subsequently affecting flower production, number of flowers, pollen production, pollen viability, and fertilization (Sangu et al., 2015). It has been reported that vegetable crops such as tomatoes are highly heat sensitive. Heat stress detrimentally affects the male reproductive organs eventually leading to occasional heat-induced sterility (due to the desiccation of pollens) causing a severe reduction in the yield of tomatoes. Heat stress at pollination stage negatively affects pollen growth and hinders fertilization, eventually leading to severe flower drop without fruit setting. Moreover, heat stress causes blossom drop and blossom end rot in tomato. Heat stress also significantly reduces pollen production, pollen viability, fruit set, and flowers per inflorescence (Xu et al., 2017). Pollen grains are known to be highly sensitive to heat stress compared with stigma (Wassmann et al., 2009).

1.2.4.6 Yield attributes

A final high yield is a critical index of crop production. The high-temperature or heat stress markedly influences the yield of various crops. However, different crop species show a varied response to high-temperature stress. It has been observed that cereals have a narrow range of temperature stress at or during flowering time and therefore the high-temperature stress detrimentally affects quality of grains as compared with grains yield (Ihsan et al., 2019). It was noted that maize yield was reduced by 1% on per day basis when the daily mean temperature exceeded its optimum range in certain conditions (Lobell et al., 2011).

1.2.5 H_2S-induced regulation of temperature stress in plants

Several approaches have been reported for the alleviation of temperature stress in plants. The use of exogenous H_2S application is one of the effective ways of temperature stress amelioration in plants. H_2S regulates various high and/or chilling-associated responsive genes and helps in the alleviation of temperature stress in a variety of plant species. H_2S also crosstalks with certain other molecules in the regulation of high- and low-temperature stress. The detail is given below.

1.2.6 Crosstalk of H_2S with other molecules during temperature stress

It has been observed that H_2S interacts with various other molecules during a variety of stresses in plants. H_2S-induced closure of stomata has been reported due to its possible role in abscisic (ABA)-mediated signaling in *Vicia faba* guard cells (García-Mata and Lamattina, 2010). In addition, exogenous application of ABA enhanced L−cysteine desulfhydrase (LCD) activity that in turn resulted in increased endogenous H_2S accumulation and led to significantly higher tolerance to heat stress in tobacco cells (Li and Jin, 2016).

Moreover, positive roles of nitric oxide (NO) and H_2S have also been noted under abiotic stress. The exogenous treatment of H_2S markedly decreased NO concentration in guard cells, whereas its higher endogenous biosynthesis was found in root cells showing the regulation of NO concentration due to H_2S upstream (Bright et al., 2009; Zhang et al., 2009a). Increased indole acetic acid and NO concentrations were observed in sweet potatoes under exogenous H_2S application (Zhang et al., 2009b). Trehalose has been found to show enormous potential to conserve the membranes under water scarcity (Kim et al., 2018). In addition, it also acts as an imperative signaling molecule along with H_2S and provides help to acquire high-temperature tolerance. The association between trehalose and H_2S has been reported in maize to acquire heat tolerance (Li et al., 2014a). High-temperature exposure induced the upregulation of LCD enzyme in tobacco plants, which subsequently enhanced the biosynthesis of H_2S under high-temperature stress (Chen et al., 2016). The exposure of the plants to 35°C significantly enhanced the activity of LCD as compared with controls so that the response was even continued for 1 day after removal from the temperature stress. The higher biosynthesis of H_2S enhances the synthesis of nicotine, which results in jasmonic acid accumulation. The enhanced biosynthesis of H_2S was also noted in strawberry plants upon exposure to 42°C temperature (Christou et al., 2014). So, the concentration of H_2S changes upon exposure of stress (low or high temperature) and H_2S crosstalks with certain other molecules to develop better mechanisms in order to cope with undesirable temperature stresses.

1.2.7 H_2S and chilling stress in plants

1.2.7.1 H_2S regulates chilling stress responsive genes

It has been observed that H_2S regulates certain chilling stress-related genes in plants to ameliorate low-temperature stress. The exposure of *Vitis vinifera* to 7−15 μmol L^{-1} NaHS increased D-cysteine desulfhydrase (DCD) and LCD activities during 4°C cold stress (Fu et al., 2013). Furthermore, exposure of Bermuda grass seedlings to cold stress elevated the endogenous levels of H_2S (Shi et al., 2013). Ma et al. (2015) observed enhanced O-acetylserine thiol lyase (OAS-TL), L/DCD, and cyanoalanine synthase (CAS) expression involved in the biosynthesis of H_2S in *Lamiophlomis rotata* plants at higher altitudes. This evidence further supports the H_2S-induced function of plants in chilling or cold tolerance at higher altitude. H_2S donor NaHS exposed Bermuda grass plants showed osmotic adjustment and enhanced antioxidative enzymes such as CAT, SOD, glutathione reductase (GR), and glutathione peroxidase (GPX) activities induction along with significantly higher AsA-glutathione (AsA-GSH) level (Shi et al., 2013), as well as upregulation of *VvCBF3* and *VvICE1* genes expression (Fu et al., 2013). H_2S as an exogenous application decreased the production of ROS and reactive nitrogen species

(RNS) by increased antioxidative enzyme activities (Shi et al., 2015). The treatment of banana fruits exhibited the amelioration of chilling stress in response to H_2S due to lower malondialdehyde (MDA) and superoxide anion as well as H_2O_2 radicals. In addition, antioxidant enzyme activities and phenylalanine ammonia lyase (PAL)-induced biosynthesis of phenols was also enhanced under chilling stress in bananas (Luo et al., 2015). The mitogen-activated protein kinase (MAPK) plays important role in signal transduction and has been found to be involved in cold tolerance in *Arabidopsis* (Furuya et al., 2013). The role of MAPK in chilling stress tolerance was investigated by Du et al. (2017) in *Arabidopsis*. They reported that cold stress decreased the length of the roots which nevertheless became cold stress tolerant in response to H_2S treatment due to the upregulation of the *MAPK4* gene (Du et al., 2017).

1.2.7.2 H_2S-induced low-temperature/chilling tolerance

Low-temperature or cold stress consists of freezing and chilling stresses. The low-temperature stress generally results in oxidative and osmotic stresses. However, the plants can diminish low temperature-induced injuries by activating the antioxidative system and osmotic adjustments (Foyer and Noctor, 2011; Iqbal et al., 2016). It has been noted that H_2S treatments can mitigate low-temperature stress in plants (Table 1.1). NaHS exposure mitigated multiple stresses including freezing tolerance in Bermuda grass as indicated by increased survival percentage and lower leakage of electrolytes under the conditions of freezing stress (Shi et al., 2013). In addition, treatment of NaHS also ameliorated ROS-based damages to cells caused by freezing stress condition through the modulation of antioxidant such as GPX, CAT, and GR enzymes as well as nonenzymatic-based pool of GSH and redox state of the plants (Shi et al., 2013). The NaHS exposed grapes seedlings exhibited enhanced SOD activity, lower superoxide anion radical, MDA, and electrolyte leakage during 4°C cold stress. In contrast, control grapes seedlings had contrary impacts under same stress type (Fu et al., 2013). Under chilling stress, *Arabidopsis* seedlings exhibited overexpression of an LCD-associated gene upon treatment with NaHS and had higher H_2S endogenous levels along with significantly higher tolerance of chilling stress. On the other side, controls showed a knockdown of LCD and significantly reduced H_2S endogenous concentration and markedly lower tolerance to chilling stress. Moreover, upregulation of H_2S-related genes was observed under various biotic and abiotic stresses, with suppressed accumulation of ROS (Shi et al., 2015). The activities of CAS, L/DCD, and OAS-TL were found to be significantly higher in plants at high altitudes which subsequently resulted in markedly higher H_2S levels due to its (H_2S) increased biosynthesis (Ma et al., 2015). H_2S exogenous treatment suppressed ROS and RNS-induced damages due to higher antioxidant enzymes activities and reduced degradation of protein in *L. rotata*. These findings show that H_2S

TABLE 1.1 Effect of H_2S on low-temperature/chilling stress in plants.

Crop	Temperature	Concentration	Inferences	References
Arabidopsis	4°C for 6 h	ABA induced endogenous H_2S	Enhanced expression of *VvCBF3* and *VvICE1* and higher SOD activity, lower superoxide anion radical, MDA, and electrolyte leakage	Shi et al. (2015)
Bermuda grass	4°C for 14 days	500 μmol L^{-1} NaHS	Increased survival percentage and lower leakage of electrolytes, enhanced GPX, CAT, and GR enzymes activity increased GSH concentration	Shi et al. (2013)
Cucumber	5°C for 24 h	1 mmol L^{-1} NaHS	Reduced H_2O_2, MDA, electrolyte leakage, and increased reduced glutathione content, photosynthetic rate, and PS-II photochemical efficiency	Liu et al. (2020)
Grapes	4°C for 24 h	0.10 mmol L^{-1} NaHS	Higher SOD activity and enhanced *VvICE1* and *VvCBF3* genes expression	Fu et al. (2013)
Rye	2°C–4°C for 7 days	0.1 and 0.5 mmol L^{-1} NaHS	Increased GPX, CAT, and SOD activity along with increased anthocyanins and suppressed MDA content	Kolupaev et al. (2019)
Wheat	2°C–4°C for 7 days	0.1 and 0.5 mmol L^{-1} NaHS	Increased CAT, GPX, and SOD activity along with increased anthocyanins and suppressed MDA content	Kolupaev et al. (2019)
Banana	7°C for 14 days during postharvest	0.5 mmol L^{-1} NaHS	Higher skin lightness values, better peel firmness, and reduced MDA accumulation	Luo et al. (2015)
Hawthorn	1°C for 20 days during postharvest	1.5 mmol L^{-1} NaHS	Reduced chilling injury, MDA content, increased activities of DCD, LCD, CAT, SOD, APX, PAL, along with higher phenolics, anthocyanin contents, total flavonoid and ascorbic acid concentrations	Aghdam et al. (2018)

APX, ascorbate peroxidase; *CAT*, catalase; *DCD*, D-cysteine desulfhydrase; *GPX*, glutathione peroxidase; *GSH*, glutathione; *H*, hour; *H_2S*, hydrogen sulfide; H_2O_2, hydrogen peroxide; *LCD*, L–cysteine desulfhydrase; *MDA*, malondilaldehyde; *NaHS*, sodium hydrosulfide; *PAL*, phenylalanine ammonia lyase; *SA*, salicylic acid.

plays a critical function to induce tolerance against multiple stresses in *L. rotata* under alpine environment conditions (Ma et al., 2015). Fumigation of bananas with H_2S resulted in conserved firmness and lightness values as well as lower MDA accumulation in peel during chilling stress (Luo et al., 2015). Moreover, H_2S also upregulated SOD, CAT, GR, and APX, as well as PAL activities, that led to the higher accumulation of phenols, that in turn increased banana fruit antioxidant capacity and showed reduced superoxide anion and H_2O_2 concentrations during chilling stress (Table 1.1) (Luo et al., 2015). In another study, banana fruit fumigation with NaHS reduced MDA accumulation and inhibited membrane permeability. Moreover, H_2S noticeably conserved ATP content and higher Ca^{2+}-ATPase, H^+-ATPase, cytochrome C oxidase, and succinate dehydrogenase, which eventually mitigated chilling stress symptoms in bananas (Li et al., 2016a). The NaHS application considerably enhanced chilling tolerance in hawthorn fruits (Aghdam et al., 2018). Single NaHS foliar application reduced hydrogen peroxide, lowered photoinhibition, and reduced suppression of carbon dioxide assimilation rate in "Hass" avocado plants during high daylight and overnight frost stress conditions (Joshi et al., 2019). Exogenous NaHS treatment increased CAT, GPX, and SOD activity along with increased anthocyanins. In contrast, the level of MDA was suppressed in NaHS treatments. All these effects resulted in enhanced chilling resistance in rye and wheat seedlings (Kolupaev et al., 2019). Recently, NaHS treatment reduced H_2O_2, MDA, electrolyte leakage, and increased concentration of reduced GSH, photosynthetic rate, and PS-II photochemical efficiency in chilling-exposed cucumber seedlings. Overall, H_2S application in the form of NaHS markedly ameliorated chilling stress in seedlings of cucumber (Liu et al., 2020).

1.2.8 H_2S and high-temperature stress in plants

1.2.8.1 H_2S regulates high-temperature responsive genes

The extreme of temperature is one of the main restrictive causes that negatively affect the growth, development, and productivity of plants, and it has now become a leading global concern (Echevarría-Zomeño et al., 2016). The heat exposure-based damage includes aggregation and denaturation of proteins, damage to membranes owing to peroxidation of lipids, inhibition or inactivation of enzymes, and protein biosynthesis inhibition (Sehgal et al., 2016; Fahad et al., 2019). It has been observed that plants show numerous mechanisms to offer protection against temperature stress-induced damage, including MAPK upregulation and heat shock protein (HSP) accumulation under stressful conditions (Yu et al., 2016). Furthermore, an antioxidative defense system is normally activated with concurrent secondary metabolites adjustment and the accumulation of various osmolytes in plant cells under high-temperature stress (Ali et al., 2019). In addition, damage to proteins is repaired and cell homeostasis is preserved to ameliorate heat stress-induced detrimental effects (Li et al., 2015). High temperature negatively affects cell membranes. Trehalose has been found to show enormous potential to conserve the membranes even under water scarcity (Kim et al., 2018). It can effectively stabilize proteins and enzymes to ensure thermotolerance (Kaur and Thind, 2018). In addition, it also acts as an imperative signaling molecule along with H_2S and provides help to attain high-temperature tolerance. The association between trehalose and H_2S has been reported in maize to acquire heat tolerance (Li et al., 2014a). High-temperature exposure induces upregulation of LCD enzyme in tobacco plants that subsequently enhanced the biosynthesis of H_2S under high-temperature stress (Chen et al., 2016). The exposure of the plants to 35°C significantly enhanced the activity of LCD as compared with control plants. The higher biosynthesis of H_2S enhances the synthesis of nicotine that results in jasmonic acid accumulation. The enhanced biosynthesis of H_2S was also noted in strawberry plants upon exposure to 42°C temperature (Christou et al., 2014). The NaHS pretreatment led to the development of heat tolerance in maize (Li et al., 2013), wheat seedlings (Wu et al., 2013), and cell suspension culture of tobacco plants (Li et al., 2012). In the same way, NaHS pretreatment upregulated ascorbate, GSH and glutamate dehydrogenase-related genes expression and biosynthesis of these antioxidants in heat-stressed strawberry plants. Moreover, NaHS pretreatment also led to enhanced activities of CAT, APX, GR, and SOD enzymes in strawberry plants (Christou et al., 2014). These examples show that H_2S is quite capable of coordination to induce higher expression of heat stress-related genes and their associated pathways at molecular levels to induce high-temperature stress tolerance in a number of crops.

1.2.8.2 H_2S-induced high-temperature/heat tolerance

Under the current global warming scenario, high-temperature or heat stress has become a major abiotic stress around the globe. The high-temperature stress-induced mechanisms have received great attention in order to mitigate the detrimental impacts on crop plants (Asthir, 2015; Hemmati et al., 2015). Various strategies have been used to ameliorate heat or high-temperature-induced damage in plants. Exogenous treatments of H_2S are also among these strategies to alleviate the negative influences of temperature stress on plants (Li et al., 2016b). It has been observed that the pretreatment of strawberry roots with NaHS efficiently ameliorated leaf chlorophyll fluorescence reduction as well as a decline of relative water content and stomatal conductance at 42°C heat stress (Table 1.2). On the other side, H_2S application markedly inhibited electrolyte leakage and MDA production compared with control plants exposed to heat stress conditions (Christou et al., 2011). Moreover, the pretreatment of the plants with H_2S conserved the homeostasis of AsA/GSH as revealed by declined GSH and AsA redox pool disturbance and increased AsA/GSH-related gene expression that in turn led to its higher

TABLE 1.2 Effect of H_2S on high-temperature/heat stress in plants.

Crop	Temperature	Concentration	Inferences	References
Maize	48°C for 18 h	1.5 mmol L^{-1} NaHS	SA induced H_2S accumulation, increased seed germination and survival percentage and enhanced antioxidant capacity	Li et al. (2013)
Maize	47°C for 15 h	0.6 mmol L^{-1} NaHS	Increased antioxidant content, antioxidant enzymes activities, and higher nonenzymatic antioxidants concentration	Li et al. (2014b)
Maize	39°C for 5 days	0.5 mmol L^{-1} NaHS	Increased APX, CAT, GR, SOD, and GPX enzymes activities, higher glutathione and ascorbic acid. Increased glycine betaine and trehalose accumulation	Zhou et al. (2018)
Strawberry	42°C for 8 h	100 μmol L^{-1} NaHS	Ameliorated heat imposed phenotypic damages; enhanced antioxidants and HSP	Christou et al. (2014)
Tobacco	43°C for 7 days	50 μmol L^{-1} NaHS	Higher survival percentage, enhanced antioxidants, H_2S acted as a signal downstream of nitric oxide	Li and Gu (2016)
Wheat	38°C for 24 h	0.15 mmol L^{-1} NaHS	Reduced oxidative damage and upregulated antioxidant enzymes activities along with markedly inhibited heat stress	Min et al. (2016)

APX, ascorbate peroxidase; *CAT*, catalase; *H_2S*, hydrogen sulfide; *HSP*, heat shock protein; *H*, Hour; *DCD*, D-cysteine desulfhydrase; *GPX*, glutathione peroxidase; *NaHS*, sodium hydrosulfide; *SA*, salicylic acid.

biosynthesis 8 hours after the heat stress imposition. In addition, NaHs root pretreatment upregulated CAT, cAPX, GR, and MnSOD-associated gene expression and induced HSP70, HSP90, and HSP80 HSPs as well as aquaporins (Christou et al., 2014). These investigations showed that root H_2S pretreatment activates a well-organized heat shock-related defense pathway network at the level of transcription and efficiently saves strawberry plants from the detrimental impacts of high-temperature-induced damage (Christou et al., 2014). Treatment with NaHS also induced heat tolerance in maize seedlings due to an induction of enhanced GPX, CAT, GR, and SOD activities (Li et al., 2014b). Under the heat stress environment, antioxidant content, antioxidant enzyme activities, and the concentrations of non-enzymatic antioxidants were also enhanced in NaHS applied maize seedlings (Li et al., 2014b). Moreover, pretreatment with NaHS markedly enhanced the tobacco cell survival percentage during heat stress conditions and the ability of regrowth after imposition of heat environment was increased (Table 1.2). In addition, the increase in MDA concentration and electrolyte leakage was also suppressed in NaHS pretreated plants (Li et al., 2012). It has also been reported that the application of NaHS substantially enhanced germination and survival percentage of maize under heat stress conditions. The heat tolerance was possibly increased due to NaHS-induced lower MDA accumulation and electrolyte leakage and suppressed tissue vitality (Li et al., 2013). In addition, the pretreatment with NaHS improved P5CS activity and decreased proline dehydrogenase activity that subsequently improved endogenous proline contents in seedlings of maize (Li et al., 2013). The SA pretreatment-induced increased H_2S accumulation enhanced heat tolerance in maize seedlings (Table 1.2) (Li et al., 2015). Foliar H_2S application reduced oxidative damage and upregulated antioxidant enzyme activities along with markedly inhibited heat stress-induced negative effects on wheat plants (Min et al., 2016). H_2S application as a presoaking seed treatment significantly increased germination percentage, root length, sprout length, and fresh mass compared with controls. In addition, H_2S increased APX, CAT, GR, SOD, and GPX activities as well as GSH and AsA and increased glycine betaine and trehalose accumulation that in turn alleviated high-temperature stress in maize (Zhou et al., 2018). These reports suggest that pretreatment with NaHS can effectively increase heat tolerance in various plant species.

1.3 Conclusion and future perspectives

In conclusion, H_2S is a potential signaling molecule that plays an important role in numerous cellular mechanisms of plants. Its roles in germination, stomatal movement, root elongation, photosynthesis, and growth have also been reported. H_2S regulates various high-temperature and/or chilling-associated responsive genes and helps in the alleviation of temperature stress. H_2S crosstalks with certain other molecules in the regulation of high- and low-temperature stresses, lessens oxidative damage, and activates antioxidant enzymes, thereby it can help in sustainable crop production under undesirable temperature stress. However, biochemical, physiological, and molecular mechanisms behind the regulation of temperature stress need further comprehensive investigation. In addition, metabolomics, transcriptomics, and proteomics-based studies should be conducted to understand the possible H_2S-induced temperature stress mitigation in plants.

References

Aghdam, M.S., Bodbodak, S., 2014. Postharvest heat treatment for mitigation of chilling injury in fruits and vegetables. Food Bioprocess Technol. 7 (1), 37–53.

Aghdam, M.S., Mahmoudi, R., Razavi, F., Rabiei, V., Soleimani, A., 2018. Hydrogen sulfide treatment confers chilling tolerance in hawthorn fruit during cold storage by triggering endogenous H_2S accumulation, enhancing antioxidant enzymes activity and promoting phenols accumulation. Sci. Hortic. 238, 264–271.

Ali, S., Khan, A.S., Malik, A.U., Shaheen, T., Shahid, M., 2018. Pre-storage methionine treatment inhibits postharvest enzymatic browning of cold stored 'Gola' litchi fruit. Postharvest Biol. Technol. 140, 100–106.

Ali, S., Nawaz, A., Ejaz, S., Haider, S.T.A., Alam, M.W., Javed, H.U., 2019. Effects of hydrogen sulfide on postharvest physiology of fruits and vegetables: an overview. Sci. Hortic. 243, 290–299.

Ali, S., Khan, A.S., Anjum, M.A., Nawaz, A., Naz, S., Ejaz, S., et al., 2020a. Effect of postharvest oxalic acid application on enzymatic browning and quality of lotus (*Nelumbo nucifera* Gaertn.) root slices. Food Chem. 312, 126051.

Ali, S., Anjum, M.A., Nawaz, A., Naz, S., Hussain, S., Ejaz, S., et al., 2020b. Effect of pre-storage ascorbic acid and *Aloe vera* gel coating application on enzymatic browning and quality of lotus root slices. J. Food Biochem. e13136.

Allakhverdiev, S.I., Kreslavski, V.D., Klimov, V.V., Los, D.A., Carpentier, R., Mohanty, P., 2008. Heat stress. An overview of molecular responses in photosynthesis. Photosynth. Res. 98, 541–550.

Andaya, V.C., Mackill, D.J., 2003. Mapping of QTLs associated with cold tolerance during the vegetative stage in rice. J. Exp. Bot. 54, 2579–2585.

Apel, K., Hirt, H., 2004. Reactive oxygen species: metabolism, oxidative stress, and signal transduction. Annu. Rev. Plant Biol. 55, 373–399. Available from: https://doi.org/10.1146/annurev.arplant.55.031903.141701.

Aroca, R., Vernieri, P., Irigoyen, J.J., Sancher-Diaz, M., Tognoni, F., Pardossi, A., 2003. Involvement of abscisic acid in leaf and root of maize (*Zea mays* L.) in avoiding chilling-induced water stress. Plant Sci. 165, 671–679.

Asseng, S., Foster, I., Turner, N.C., 2011. The impact of temperature variability on wheat yields. Glob. Change Biol. 17, 997–1012.

Asthir, B., 2015. Mechanisms of heat tolerance in crop plants. Biol. Plant 59, 620–628.

Atkinson, N.J., Lilley, C.J., Urwin, P.E., 2013. Identification of genes involved in the response to simultaneous biotic and abiotic stress. Plant Physiol. 162, 2028–2041.

Bright, J., Hiscock, S.J., James, P.E., Hancock, J.T., 2009. Pollen generates nitric oxide and nitrite: a possible link to pollen-induced allergic responses. Plant Physiol. Biochem. 47, 49–55.

Chen, X., Chen, Q., Zhang, X., Li, R., Jia, Y., Ef, A., 2016. Hydrogen sulfide mediates nicotine biosynthesis in tobacco (*Nicotiana tabacum*) under high temperature conditions. Plant Physiol. Biochem. 104, 174–179.

Chinnusamy, V., Zhu, J., Zhu, J.K., 2007. Cold stress regulation of gene expression in plants. Trend. Plant Sci. 12, 444–451.

Christou, A., Manganaris, G., Papadopoulos, I., Fotopoulos, V., 2011. The importance of hydrogen sulfide as a systemic priming agent in strawberry plants grown under key abiotic stress factors. In: Proceedings of the 4th International Conference on Plant Abiotic Stress: From Systems Biology to Sustainable Agriculture, Limassol, 47.

Christou, A., Filippou, P., Manganaris, G.A., Fotopoulos, V., 2014. Sodium hydrosulfide induces systemic thermotolerance to strawberry plants through transcriptional regulation of heat shock proteins and aquaporin. BMC Plant Biol. 14, 42. Available from: https://doi.org/10.1186/1471-2229-14-42.

Deng, B., Du, W., Liu, C., Sun, W., Tian, S., Dong, H., 2012. Antioxidant response to drought, cold and nutrient stress in two ploidy levels of tobacco plants: low resource requirement confers polytolerance in polyploids? Plant Growth Regul. 66, 37–47.

Du, X., Jin, Z., Liu, D., Yang, G., Pei, Y., 2017. Hydrogen sulfide alleviates the cold stress through MPK4 in *Arabidopsis thaliana*. Plant Physiol. Biochem. 120, 112–119.

Duc, N.H., Csintalan, Z., Posta, K., 2018. Arbuscular mycorrhizal fungi mitigate negative effects of combined drought and heat stress on tomato plants. Plant Physiol. Biochem. 132, 297–307.

Echevarría-Zomeño, S., Fernández-Calvino, L., Castro-Sanz, A.B., López, J.A., Vázquez, J., Castellano, M.M., 2016. Dissecting the proteome dynamics of the early heat stress response leading to plant survival or death in Arabidopsis. Plant Cell Environ. 39, 1264–1278.

Fahad, S., Hussain, S., Saud, S., Hassan, S., Chauhan, B.S., Khan, F., et al., 2016. Responses of rapid viscoanalyzer profile and other rice grain qualities to exogenously applied plant growth regulators under high day and high night temperatures. PLoS One 11, e0159590.

Fahad, S., Bajwa, A.A., Nazir, U., Anjum, S.A., Farooq, A., Zohaib, A., et al., 2017. Crop production under drought and heat stress: plant responses and management options. Front. Plant Sci. 8, 1147. Available from: https://doi.org/10.3389/fpls.2017.01147.

Fahad, S., Adnan, M., Hassan, S., Saud, S., Hussain, S., Wu, C., et al., 2019. Rice responses and tolerance to high temperature. Advances in Rice Research for Abiotic Stress Tolerance. Woodhead Publishing, Sawston, Cambridge, pp. 201–224.

Farooq, M., Aziz, T., Wahid, A., Lee, D.J., Siddique, K.H.M., 2009. Chilling tolerance in maize: agronomic and physiological approaches. Crop Pasture Sci. 60, 501–516. Available from: https://doi.org/10.1071/CP08427.

Foyer, C.H., Noctor, G., 2011. Ascorbate and glutathione: the heart of the redox hub. Plant Physiol. 155, 2–18.

Friedlingstein, P., Houghton, R.A., Marland, G., Hackler, J., Boden, T.A., Conway, T.J., 2010. Update on CO_2 emissions. Nat. Geosci. 3, 811–812. Available from: https://doi.org/10.1038/ngeo1022.

Fu, P.N., Wang, W.J., Hou, L.X., Liu, X., 2013. Hydrogen sulfide is involved in the chilling stress response in *Vitis vinifera* L. Acta Soc. Bot. Pol. 82, 295–302.

Furuya, T., Matsuoka, D., Nanmori, T., 2013. Phosphorylation of *Arabidopsis thaliana*, MEKK1 via Ca^{2+} signaling as a part of the cold stress response. J. Plant Res. 126 (6), 833–840.

García-Mata, C., Lamattina, L., 2010. Hydrogen sulphide, a novel gasotransmitter involved in guard cell signalling. N. Phytol. 188, 977–984. Available from: https://doi.org/10.1111/j.1469-8137.2010.03465.x.

Gill, S.S., Tuteja, N., 2010. Reactive oxygen species and antioxidant machinery in abiotic stress tolerance in crop plants. Plant Physiol. Biochem. 48, 909–930.

Haque, M.S., Kjaer, K.H., Rosenqvist, E., Sharma, D.K., Ottosen, C.O., 2014. Heat stress and recovery of photosystem II efficiency in wheat (*Triticum aestivum* L.) cultivars acclimated to different growth temperatures. Environ. Exp. Bot. 99, 1–8.

He, H., Li, Y., He, L.F., 2018. The central role of hydrogen sulfide in plant responses to toxic metal stress. Ecotoxicol. Environ. Saf. 157, 403–408.

Hemmati, H., Gupta, D., Basu, C., 2015. Molecular physiology of heat stress responses in plant. In: Pandey, G.K. (Ed.), Elucidation of Abiotic Stress Signaling in Plants: Functional Genomics Gerspectives, Vol. 2. Springer, New York, pp. 109–142.

Hussain, S., Khan, F., Cao, W., Wu, L., Geng, M., 2016. Seed priming alters the production and detoxification of reactive oxygen intermediates in rice seedlings grown under sub-optimal temperature and nutrient supply. Front. Plant Sci. 7, 439.

Hussain, H.A., Hussain, S., Khaliq, A., Ashraf, U., Anjum, S.A., Men, S., et al., 2018. Chilling and drought stresses in crop plants: implications, cross talk, and potential management opportunities. Front. Plant Sci. 9, 393.

Ihsan, M.Z., El-Nakhlawy, F.S., Ismail, S.M., Fahad, S., Daur, I., 2016. Wheat phenological development and growth studies as affected by drought and late season high temperature stress under arid environment. Front. Plant Sci. 795, 1–14.

Ihsan, M.Z., Daur, I., Alghabari, F., Alzamanan, S., Rizwan, S., Ahmad, M., et al., 2019. Heat stress and plant development: role of sulphur metabolites and management strategies. Acta Agric. Scand. Sect. B—Soil Plant Sci. 69 (4), 332–342.

Iqbal, N., Nazar, R., Khan, N.A., 2016. Osmolytes and Plants Acclimation to Changing Environment: Emerging Omics Technologies. Springer, London.

Janowiak, F., Markowski, A., 1994. Changes in leaf water relations and injuries in maize seedlings induced by different chilling conditions. J. Agron. Crop. Sci. 172, 19–28.

Jin, Z., Xue, S., Luo, Y., Tian, B., Fang, H., Li, H., et al., 2013. Hydrogen sulfide interacting with abscisic acid in stomatal regulation responses to drought stress in Arabidopsis. Plant Physiol. Biochem. 62, 41–46.

Joshi, N.C., Yadav, D., Ratner, K., Kamara, I., Aviv-Sharon, E., Irihimovitch, V., et al., 2019. Sodium hydrosulfide priming improves the response of photosynthesis to overnight frost and day high light in avocado (*Persea americana* Mill, cv. 'Hass'). Physiol. Plant. Available from: https://doi.org/10.1111/ppl.13023.

Jouyban, Z., Hasanzade, R., Sharafi, S., 2013. Chilling stress in plants. Int. J. Agric. Crop. Sci. 5, 2961.

Jung, S., Steffen, K.L., 1997. Influence of photosynthetic photon flux densities before and during long-term chilling on xanthophyll cycle and chlorophyll fluorescence quenching in leaves of tomato (*Lycopersicon hirsutum*). Physiol. Plant 100, 958–966.

Kang, H.M., Saltveit, M.E., 2002. Chilling tolerance of maize cucumber and rice seedling leaves and roots are differentially affected by salicylic acid. Physiol. Plant 115, 571–576.

Kaur, A., Thind, S.K., 2018. Effect of exogenous trehalose in protection of cells and their membranes in wheat (*Triticum aestivum* L.) genotypes under heat stress. Curr. J. Appl. Sci. Technol. 30, 1–13.

Kim, S.X., Çamdere, G., Hu, X., Koshland, D., Tapia, H., 2018. Synergy between the small intrinsically disordered protein Hsp12 and trehalose sustain viability after severe desiccation. Elife 7, e38337.

Kolupaev, Y.E., Horielova, E.I., Yastreb, T.O., Ryabchun, N.I., Kirichenko, V.V., 2019. Stress-protective responses of wheat and rye seedlings whose chilling resistance was induced with a donor of hydrogen sulfide. Russian J. Plant Physiol. 66 (4), 540–547.

Kotak, S., Larkindale, J., Lee, U., Von Koskull-Döring, P., Vierling, E., Scharf, K.D., 2007. Complexity of the heat stress response in splants. Curr. Opin. Plant Biol. 10 (3), 310–316.

Lamaoui, M., Jemo, M., Datla, R., Bekkaoui, F., 2018. Heat and drought stresses in crops and approaches for their mitigation. Front. Chem. 6, 26. Available from: https://doi.org/10.3389/fchem.2018.00026.

Lee, T.M., Lur, H.S., Chu, C., 1993. Role of abscisic acid in chilling tolerance of rice (*Oryza sativa* L.) seedlings. Endogenous abscisic acid levels. Plant Cell Environ. 16, 481–490.

Li, Z.G., Jin, J.Z., 2016. Hydrogen sulfide partly mediates abscisic acid-induced heat tolerance in tobacco (*Nicotiana tabacum* L.) suspension cultured cells. Plant Cell Tissue Organ. Cult. 125, 207–214.

Li, Z.G., Gu, S.P., 2016. Hydrogen sulfide as a signal molecule in hematin-induced heat tolerance of tobacco cell suspension. Biol. Plant. 60 (3), 595–600.

Li, Z.G., Gong, M., Xie, H., Yang, L., Li J., 2012. Hydrogen sulfide donor sodium hydrosulfide-induced heat tolerance in tobacco (Nicotiana tabacum L.) suspension cultured cells and involvement of Ca^{2+} and calmodulin, Plant Sci. 185–189.

Li, Z.G., Ding, X.J., Du, P.F., 2013. Hydrogen sulfide donor sodium hydrosulfide-improved heat tolerance in maize and involvement of proline. J. Plant Physiol. 170, 741–747.

Li, Z.G., Luo, L.J., Zhu, L.P., 2014a. Involvement of trehalose in hydrogen sulfide donor sodium hydrosulfide-induced the acquisition of heat tolerance in maize *(Zea mays* L.) seedlings. Bot. Stud. 55 (1), 20.

Li, Z.G., Yi, X.Y., Li, Y.T., 2014b. Effect of pretreatment with hydrogen sulfide donor sodium hydrosulfide on heat tolerance in relation to antioxidant system in maize (*Zea mays*) seedlings. Biologia 69, 1001–1009. Available from: https://doi.org/10.2478/s11756-014-0396-2.

Li, Z.G., Xie, L.R., Li, X.J., 2015. Hydrogen sulfide acts as a downstream signal molecule in salicylic acid-induced heat tolerance in maize (*Zea mays* L.) seedlings. J. Plant Physiol. 177, 121–127.

Li, D., Limwachiranon, J., Li, L., Du, R., Luo, Z., 2016a. Involvement of energy metabolism to chilling tolerance induced by hydrogen sulfide in cold-stored banana fruit. Food Chem. 208, 272–278.

Li, Z.G., Min, X., Zhou, Z.H., 2016b. Hydrogen sulfide: a signal molecule in plant cross-adaptation. Front. Plant Sci. 7, 1621. Available from: https://doi.org/10.3389/fpls.2016.01621.

Liu, F., Zhang, X., Cai, B., Pan, D., Fu, X., Bi, H., et al., 2020. Physiological response and transcription profiling analysis reveal the role of glutathione in H_2S-induced chilling stress tolerance of cucumber seedlings. Plant Sci. 291, 110363.

Lobell, D.B., Bänziger, M., Magorokosho, C., Vivek, B., 2011. Nonlinear heat effects on African maize as evidenced by historical yield trials. Nat. Clim. Change. 1, 42–45.

Lukatkin, A.S., Brazaityte, A., Bobinas, C., Duchovskis, P., 2012. Chilling injury in chilling-sensitive plants: a review. Agriculture 99, 111–124.

Luo, Z., Li, D., Du, R., Mou, W., 2015. Hydrogen sulfide alleviates chilling injury of banana fruit by enhanced antioxidant system and proline content. Sci. Hortic. 183, 144–151.

Ma, L., Yang, L., Zhao, J., Wei, J., Kong, X., Wang, C., 2015. Comparative proteomic analysis reveals the role of hydrogen sulfide in the adaptation of the alpine plant Lamiophlomis rotate to altitude gradient in the Northern Tibetan Plateau. Planta 241, 887–906.

Min, Y., Qin, B.P., Ping, W., Li, M.L., Chen, L.L., Chen, L.T., et al., 2016. Foliar application of sodium hydrosulfide (NaHS), a hydrogen sulfide (H2S) donor, can protect seedlings against heat stress in wheat (*Triticum aestivum* L.). J. Int. Agric. 15 (12), 2745–2758.

Nayyar, H., Bains, T., Kumar, S., 2005. Low temperature induced floral abortion in chickpea: relationship to abscisic acid and cryoprotectants in reproductive organs. Environ. Exp. Bot. 53, 39–48. Available from: https://doi.org/10.1016/j.envexpbot.2004.02.011.

Nemeskéri, E., Helyes, L., 2019. Physiological responses of selected vegetable crop species to water stress. Agronomy 9, 447. Available from: https://doi.org/10.3390/agronomy9080447.

Ohama, N., Sato, H., Shinozaki, K., Yamaguchi-Shinozaki, K., 2017. Transcriptional regulatory network of plant heat stress response. Trends Plant Sci. 22 (1), 53–65.

Oliver, S.N., Dennis, E.S., Dolferus, R., 2007. ABA regulates apoplastic sugar transport and is a potential signal for cold-induced pollen sterility in rice. Plant Cell Physiol. 48, 1319–1330.

Rizza, F., Crosatti, C., Stanca, A.M., Cattivelli, L., 1994. Studies for assessing the influence of hardening on cold tolerance of barley genotypes. Euphytica 75, 131–138.

Ruelland, E., Vaultier, M.N., Zachowski, A., Hurry, V., 2009. Cold signaling and cold acclimation in plants. Adv. Bot. Res. 49, 35–150.

Rymen, B., Fiorani, F., Kartal, F., Vandepoele, K., Inzé, D., Beemster, G.T.S., 2007. Cold nights impair leaf growth and cell cycle progression in maize through transcriptional changes of cell cycle genes. Plant Physiol. 143, 1429–1438.

Sales, C.R.G., Ribeiro, R.V., Silveira, J.A.G., Machado, E.C., Martins, M.O., Lagôa, A.M.M.A., 2013. Superoxide dismutase and ascorbate peroxidase improve the recovery of photosynthesis in sugarcane plants subjected to water deficit and low substrate temperature. Plant Physiol. Biochem. 73, 326–336.

Salvucci, M.E., Crafts-Brandner, S.J., 2004. Inhibition of photosynthesis by heat stress: the activation state of Rubisco as a limiting factor in photosynthesis. Physiol. Plant 120, 179–186.

Sangu, E., Tibazarwa, F.I., Nyomora, A., Symonds, R.C., 2015. Expression of genes for the biosynthesis of compatible solutes during pollen development under heat stress in tomato (*Solanum lycopersicum*). J. Plant Physiol. 178, 10–16.

Sattler, U., Calsou, P., Boiteux, S., Salles, B., 2000. Detection of oxidative base dna damage by a new biochemical assay. Arch. Biochem. Biophys. 376, 26–33. Available from: https://doi.org/10.1006/abbi.2000.1701.

Sehgal, A., Sita, K., Nayyar, H., 2016. Heat stress in plants:sensing and defense mechanisms. J. Plant Sci. Res. 32, 195.

Sevillano, L., Sanchez-Ballesta, M.T., Romojaro, F., Flores, F.B., 2009. Physiological, hormonal and molecular mechanisms regulating chilling injury in horticultural species. Postharvest technologies applied to reduce its impact. J. Sci. Food Agric. 89, 555–573.

Sewelam, N., Oshima, Y., Mitsuda, N., Ohme-Takagi, M., 2014. A step towards understanding plant responses to multiple environmental stresses: a genome-wide study. Plant Cell Environ. 37, 2024–2035.

Shah, K., Singh, M., Rai, A.C., 2013. Effect of heat-shock induced oxidative stress is suppressed in BcZAT12 expressing drought tolerant tomato. Phytochemistry 95, 109–117.

Shi, H., Ye, T., Chan, Z., 2013. Exogenous application of hydrogen sulfide donor sodium hydrosulfide enhanced multiple abiotic stress tolerance in bermudagrass (*Cynodon dactylon* (L). Pers.). Plant Physiol. Biochem. 71, 226–234.

Shi, H., Ye, T., Han, N., Bian, H., Liu, X., Chan, Z., 2015. Hydrogen sulfide regulates abiotic stress tolerance and biotic stress resistance in Arabidopsis. J. Integr. Plant Biol. 57, 628–640.

Shu, D.F., Wang, L.Y., Duan, M., Deng, Y.S., Meng, Q.W., 2011. Antisense-mediated depletion of tomato chloroplast glutathione reductase enhances susceptibility to chilling stress. Plant Physiol. Biochem. 49, 1228–1237.

Starck, Z., Niemyska, B., Bogdan, J., Tawalbeh, R.N.A., 2000. Response of tomato plants to chilling stress in association with nutrient or phosphorus starvation. Plant Soil 226, 99–106.

Subbarao, G.V., Nam, N.H., Chauhan, Y.S., Johansen, C., 2000. Osmotic adjustment, water relations and carbohydrate remobilization in pigeonpea under water deficits. J. Plant Physiol. 157, 651–659.

Sun, M., Jiang, F., Zhou, R., Wen, J., Cui, S., Wang, W., et al., 2019. Respiratory burst oxidase homologue-dependent H_2O_2 is essential during heat stress memory in heat sensitive tomato. Sci. Hortic. 258, 108777.

Svensson, J., Ismail, A., Palva, E.T., Close, T.J., 2002. Dehydrins. In: Storey, K.B., Storey, J.B. (Eds.), Cell and Molecular Responses to Stress. Elsevier, Amsterdam, pp. 155–171.

Thakur, P., Kumar, S., Malik, J.A., Berger, J.D., Nayyar, H., 2010. An overview: cold stress effects on reproductive development in grain crops. Environ. Exp. Bot. 67, 429–443.

Tommasini, L., Svensson, J.T., Rodriguez, E.M., Wahid, A., Malatrasi, M., Kato, K., 2008. Dehydrin gene expression provides an indicator of low temperature and drought stress:

transcriptome-based analysis of Barley (*Hordeum vulgare* L.). Funct. Integr. Genom. 8, 387–405.

Wahid, A., Gelani, S., Ashraf, M., Foolad, M.R., 2007. Heat tolerance in plants: an overview. Environ. Exp. Bot. 61 (3), 199–223.

Wang, B., Vinocur, O., Altman, A.S., 2004. Role of plant heat-shock proteins and molecular chaperones in the abiotic stress response. Trends Plant Sci. 9, 244–252.

Wang, Y., Li, L., Cui, W., Xu, S., Shen, W., Wang, R., 2012. Hydrogen sulfide enhances alfalfa (*Medicago sativa*) tolerance against salinity during seed germination by nitric oxide pathway. Plant Soil. 351 (1-2), 107–119.

Wang, W., Chen, Q., Hussain, S., Mei, J., Dong, H., Peng, S., 2016. Pre-sowing seed treatments in direct-seeded early rice: consequences for emergence, seedling growth and associated metabolic events under chilling stress. Sci. Rep. 6, 19637. Available from: https://doi.org/10.1038/srep19637.

Wassmann, R., Jagadish, S., Sumfleth, K., Pathak, H., Howell, G., Ismail, A., et al., 2009. Regional vulnerability of climate change impacts on Asian rice production and scope for adaptation. Adv. Agron. 102, 91–133.

Whaley, J.M., Kirby, E.J.M., Spink, J.H., Foulkes, M.J., Sparkes, D.L., 2004. Frost damage to winter wheat in the UK: the effect of plant population density. Eur. J. Agron. 21, 105–115.

Wise, R.R., 1995. Chilling-enhanced photooxidation: the production, action and study of reactive oxygen species produced during chilling in the light. Photosyn. Res. 45, 79–97.

Wu, D.H., Li, Y.L., Xia, X., Pu, Z.P., Liao, J.M., Huang, K., et al., 2013. Hydrogen sulfide donor sodium hydrosulfide pretreatment improved multiple resistance abilities of wheat to high temperature and drought stress. J. Yunnan Norm. Univ. Nat. Sci. 33, 29–35.

Xu, Q., Huang, B., 2006. Seasonal changes in root metabolic activity and nitrogen uptake for two cultivars of creeping bentgrass. Hortic. Sci. 41, 822–826.

Xu, J., Wolters-Arts, M., Mariani, C., Huber, H., Rieu, I., 2017. Heat stress affects vegetative and reproductive performance and trait correlations in tomato (*Solanum lycopersicum*). Euphytica 213 (7), 156.

Yan, Q., Duan, Z., Jingdong, M., Xun, L., Fei, D., 2012. Effects of root-zone temperature and N, P, and K supplies on nutrient uptake of cucumber (*Cucumis sativus* L.) seedlings in hydroponics. Soil Sci. Plant Nutr. 58, 707–717.

Yu, L., Yan, J., Yang, Y., He, L., Zhu, W., 2016. Enhanced tolerance to chilling stress intomato by overexpression of a mitogen-activated protein kinase, SlMPK7. Plant Mol. Biol. Rep. 34 (1), 76–88.

Zhang, J., Nguyen, H.T., Blum, A., 1999. Genetic analysis of osmotic adjustment in crop plants. J. Exp. Bot. 50, 291–302.

Zhang, H., Tang, J., Liu, X.P., 2009a. Hydrogen sulfide promotes root organogenesis in *Ipomoea batatas*, *Salix matsudana* and *Glycine max*. J. Integr. Plant Biol. 51, 1086–1094.

Zhang, H., Ye, Y.K., Wang, S.H., 2009b. Hydrogen sulfide counteracts chlorophyll loss in sweet potato seedling leaves and alleviates oxidative damage against osmotic stress. Plant Growth Regul. 58, 243–250.

Zhao, J., Li, S., Jiang, T., Liu, Z., Zhang, W., 2012. Chilling stress- the key predisposing factor for causing *Alternaria alternata* infection and leading to cotton (*Gossypium hirsutum* L.) leaf senescence. PLoS One 7, e36126. Available from: https://doi.org/10.1371/journal.pone.0036126.

Zhao, J., Hartmann, H., Trumbore, S., Ziegler, W., Zhang, Y., 2013. High temperature causes negative whole-plant carbon balance under mild drought. New Phytol. 200, 330–339.

Zhou, R., Yu, X., Kjær, K.H., Rosenqvist, E., Ottosen, C.O., Wu, Z., 2015. Screening and validation of tomato genotypes under heat stress using Fv/Fm to reveal the physiological mechanism of heat tolerance. Environ. Exp. Bot. 118, 1–11.

Zhou, Z.H., Wang, Y., Ye, X.Y., Li, Z.G., 2018. Signaling molecule hydrogen sulfide improves seed germination and seedling growth of maize (Zea mays L.) under high temperature by inducing antioxidant system and osmolyte biosynthesis. Front. Plant Sci. 9, 1288.

Zhu, J.J., Li, Y.R., Liao, J.X., 2013. Involvement of anthocyanins in the resistance to chilling-induced oxidative stress in *Saccharum officinarum* L. leaves. Plant Physiol. Biochem. 73, 427–443.

Zlatev, Z., Lidon, F.C., 2012. An overview on drought induced changes in plant growth, water relations and photosynthesis. Emir. J. Food Agric. 24, 57–72.

Chapter 2

Crosstalk of hydrogen sulfide with melatonin and nitric oxide in ripening of fruits

Sajid Ali[1,*], Muhammad Akbar Anjum[1], Aamir Nawaz[1], Mahmood Ul Hasan[2], Safina Naz[1], Shaghef Ejaz[1] and Sakeena Tul-Ain Haider[1]
[1]Department of Horticulture, Faculty of Agricultural Sciences and Technology, Bahauddin Zakariya University, Multan, Pakistan, [2]Institute of Horticultural Sciences, University of Agriculture, Faisalabad, Pakistan
**Corresponding author.*

2.1 Introduction

Hydrogen sulfide (H_2S) is a potential signaling molecule, among several gasotransmitters, and it plays a critical role in multiple physiological developments in plants (Ali et al., 2019a). H_2S interacts with ethylene signaling and influences ripening physiology (Ziogas et al., 2018). H_2S affects ethylene biosynthesis and therefore regulates fruit ripening and delays senescence during postharvest storage (Hu et al., 2012; Ali et al., 2019a). The use of H_2S in the delay of the ripening process and quality conservation in fruits and vegetables is well documented (Ali et al., 2019a). Application of H_2S showed anti-senescence potential in kiwifruit and delayed its ripening (Li et al., 2017). H_2S probably inhibits cell wall loosening enzymes activities, such as polygalacturonase (PG), activity and suppresses fruit softening. During poststorage, H_2S maintains enzymatic antioxidants and alleviates oxidative damage, eventually resulting in delayed senescence and extended storage potential due to the inhibition of ripening and softening of fruits. The inhibition or enhancement of ripening normally depends upon its endogenous concentration (Ali et al., 2019a). So, it may be said that H_2S significantly influences the ripening and senescence of fruits. However, literature regarding the potential role of H_2S during preharvest fruit development and ripening is very limited.

Nitric oxide (NO) is a free radical gas with a highly reactive nature. It has been found to act as a multifunctional molecule during signaling cascades (Leshem et al., 1998; Hong et al., 2008; Hu et al., 2014a). NO has been

Hydrogen Sulfide in Plant Biology. DOI: https://doi.org/10.1016/B978-0-323-85862-5.00005-1

characterized as an ethylene-inhibiting regulator which delays fruit ripening. The ripening of fruits is closely associated with endogenous NO biosynthesis. Exogenous application of NO has also been found to regulate its endogenous concentration (Zhu et al., 2006). NO treatment in different climacteric fruits such as peaches, mangoes, and tomatoes showed increased shelf life with maintained quality due to the suppressed ripening phenomenon during postharvest storage (Chen et al., 2005; Hong et al., 2014; Eum et al., 2009). NO application suppressed the ripening of Japanese plums (Singh et al., 2009). NO has also been found to delay ripening of certain non-climacteric fruits. Its role in the regulation of fruit ripening is well documented. So, NO plays an important role in the regulation of fruit ripening and the suppression of senescence-induced quality loss of numerous fruits under postharvest storage (Mukherjee, 2019a). Crosstalk of NO with SA was observed in certain studies and it inhibited the ethylene production and delayed senescence. However, NO association with other phytohormones such as gibberellic acid, auxins, cytokinins, and polyamines is still unclear and there is still a gap to be filled with further scientific research (Mukherjee, 2019a).

Melatonin is a multifunctional biological signaling molecule. It has been found to influence a number of plant responses including germination, flowering, development, photosynthesis, and defense mechanisms (Wang et al., 2020). Its probable role in maturation, ripening, and postharvest senescence has also been reported (Xu et al., 2019). Melatonin has been found to regulate the ripening process of fruits during preharvest development, as noted in sweet cherry (Tijero et al., 2019). Several other reports also described melatonin regulating the ripening of various fruits (Meng et al., 2015; Sun et al., 2015; Aghdam and Fard, 2017). Its endogenous concentration was found to be influenced by changes in jasmonic acid, salicylic acid, and abscisic acid levels. It delayed the accumulation of anthocyanins and it also affected the endogenous content of cytokinins in sweet cherry fruits during ripening (Tijero et al., 2019). However, the mechanism of melatonin in fruit ripening regulation depends upon the type of fruits and the stage of ripening as well as its endogenous concentration. Tomatoes showed concentration dependent variability in endogenous levels of melatonin biosynthesis along with different stages of ripening, while its biosynthesis varied with changing genotypes and environmental conditions (Riga et al., 2014). In another study, the melatonin biosynthesis was found to be triggered in darkness and oxidative stress in different varieties of sweet cherry (Zhao et al., 2013). Higher accumulation of melatonin might assist in the climacteric rise of respiration in apple fruits (Lei et al., 2013). Its role in delaying ripening and conserving quality attributes during postharvest storage of fruits has also been reported (Wang et al., 2020). The role of NO has been reported in signaling related with H_2S and melatonin during fruit ripening (Mukherjee, 2019a). However, H_2S, NO, and melatonin are also important for the preharvest regulation of ripening and delaying ripening as well as senescence of fruits during postharvest storage. This chapter reviews the possible role of H_2S, NO, and melatonin in the regulation of ripening during preharvest and postharvest storage of different fruits.

2.2 Fruit ripening mechanism

The ripening of fruits is a well-coordinated, irreversible, and genetically controlled process that involves a number of biochemical, physiological, molecular, and organoleptic changes, eventually leading to soft and ripened fruits with edible quality characteristics (Prasanna et al., 2007). A wide range of physical and biochemical alterations, including chlorophyll degradation, increased ethylene production, and respiration rate, as well as biosynthesis of anthocyanins, carotenoids, aroma, and flavor compounds, take place during fruit ripening. In addition, activities of cell wall degradation based enzymes and textural changes also occur during the ripening of fruits (Brady, 1987; Prasanna et al., 2007).

The change of color takes place due to the dismantling of photosynthesis-related apparatus and chlorophyll degradation along with biosynthesis of carotenoids and anthocyanins and their subsequent buildup in vacuoles. The accumulation of xanthophyll esters, β-carotenoids, lycopene, and xanthophyll also correlates with fruit ripening, depending upon the type of fruits (Lizada, 1993; Prasanna et al., 2007). The increased aroma and flavor correlates with higher production of a number of diverse volatile organic compounds such as myrcene and ocimene (Lizada, 1993) as well as degradation of bitter compounds including tannins, flavonoids, and other associated compounds (Tucker and Grierson, 1987). The development of characteristic taste takes place owing to increased sweetness and decreased acidity and accumulation of certain organic acids and sugars leading to balanced acid/sugars blends (or ratio) (Lizada, 1993; Prasanna et al., 2007).

The softening of fruits takes place due to textural changes in response to upregulated activities of softening enzymes that lead to complete or partial degradation and solubilization of cellulose and pectin contents present in the cell wall polysaccharides (Tucker and Grierson, 1987; Razzaq et al., 2013) and hydrolysis of storage polysaccharides and starch (Selvaraj et al., 1989; Prasanna et al., 2007). Softening is a factor that limits the transport and storage potential of fruits. It involves a well-coordinated solubilization, disassembly, and de-esterification of polysaccharides in cell walls owing to the upregulation of various enzymes that have been found to be involved in fruit softening and rapid senescence (Brummell, 2006; Tucker et al., 2017; Razzaq et al., 2013).

Depending upon ethylene production, fruits have been grouped into two major classes: non-climacteric and climacteric (Pech et al., 2008; Cherian et al., 2014). Ethylene plays a significant role in the ripening of fruits. The excessive biosynthesis of ethylene takes place from its autocatalytic stimulation synthesis. The ACC oxidase and AAC synthase encodes certain genes that lead to the biosynthesis of ethylene during fruit ripening. So, ethylene plays a major part in the regulation of fruit ripening. Therefore in order to delay the ripening process, the suppression of ethylene biosynthesis is critical (Yang and Hoffman, 1984; Gray et al., 1992; Pech et al., 2008).

2.3 H_2S, NO, and melatonin in regulation of fruit ripening

2.3.1 Crosstalk of H_2S with NO during ripening

Significant insights have been reported regarding the crosstalk of H_2S with NO in the ripening of climacteric fruits. A well-established crosstalk has been reported between H_2S and NO regarding fruit ripening (Li et al., 2016). It has been observed that H_2S acts in an inhibitory and complementary manner regarding the signaling of NO. Negative and/or positive interaction has been noted in NO and H_2S depending upon their used concentrations (Li et al., 2012; Wang et al., 2012). The crosstalk between H_2S and NO exists in a synergistic manner regarding ethylene-dependent fruit ripening (Lisjak et al., 2010). A downstream H_2S position in ethylene−NO−H_2S signaling pathways has been reported by Liu et al. (2011). On the other hand, H_2S−ethylene interaction has been found to be antagonistic in the context of ripening in fruits. Negative and/or positive crosstalk exists in H_2S and NO. However, these types of interactions depend upon various physiological aspects. The exogenous combined application of NO and H_2S markedly enhanced antiripening influences in strawberry fruits (Chang et al., 2014). NO and H_2S increased ripening inhibition by conserving firmness and suppressing pectin methyl esterase (PME) enzyme activity. In addition, H_2S and NO application markedly decreased the impact of ethylene-dependent respiration rate and led to inhibited ripening in strawberry fruits. H_2S-dependent anti-senescence effects partly work via chloroplast-based biogenesis enhancement and thiol redox modulation cascade (Chen et al., 2011). Therefore H_2S and NO treatments have a synergistic effect regarding the reduction of senescence that in turn delays the ripening phenomenon during postharvest storage (Gong et al., 2018). Recently, it has been observed that H_2S and NO shows synergistic roles in the inhibition of cell wall metabolism and reduction of ethylene biosynthesis that subsequently lead to reduced ripening intensity in peach fruits during postharvest storage (Zhu et al., 2019). Similarly, exogenous NO treatment downregulated H_2S content and delayed ripening of peach fruits (Geng et al., 2019). However, the participation of reactive oxygen species (ROS) in H_2S and NO crosstalk pathways needs to be comprehensively studied in the near future (Mukherjee, 2019a).

2.3.2 Crosstalk of melatonin with NO

Melatonin and NO have emerged as molecules that play critical roles in the signaling pathways of various physiological cascades and fruit ripening. NO is a highly reactive biomolecule and brings quick responses regarding signaling (Singh et al., 2016). The physiological responses correlated with ripening of fruits involve reactive nitrogen species (RNS) and NO emerges from the nitrooxidative metabolism pathway. The role of RNS/ROS and NO has been well explained in signaling and it is found to be correlated with the ripening of fruits (Corpas et al., 2018). In this regard, variations in the endogenous concentration of melatonin have been investigated during the ripening of various climacteric fruits. Hence, potential formation of N-nitroso melatonin during different stages of fruit ripening cannot be overlooked (Mukherjee, 2019a). Melatonin was associated with the regulation of ripening in sweet cherry (Tijero et al., 2019). Its endogenous concentration was influenced by changes in salicylic acid, jasmonic acid, and abscisic acid. Melatonin delayed anthocyanins accumulation due to changes in endogenous content of cytokinins in sweet cherry fruits during ripening (Tijero et al., 2019). The variations in salicylic acid, jasmonic acid, cytokinins, and abscisic acid have been reported in melatonin-induced fruit ripening in cherries. Hence, changes in NO and H_2S in response to melatonin application may also be studied in various climacteric and non-climacteric fruits for further understanding and clarity.

2.3.3 Crosstalk between H_2S, melatonin and NO

The correlation of melatonin and NO has been investigated in plant systems in recent studies (Mukherjee, 2018, 2019b). Arnao and Hernandez-Ruiz (2019) reported the melatonin–NO interaction may be available in the redox system of plants, specifically in RNS and ROS homeostasis regulation mediated through the action of NO and H_2O_2 which appears to upregulate melatonin molecules in the signaling. Improvement in postharvest quality seems to be correlated with a change in the endogenous mechanism of melatonin–NO crosstalk. Delay in postharvest senescence of fruits due to the higher NO production being antagonistic to lower ethylene biosynthesis under the melatonin treatment could not be found to be effective as alone melatonin works internally. As previously stated by Liu et al. (2019a), pear fruits showed lower ethylene production under the application of melatonin but results were statistically non-significant for ethylene synthesis. Therefore fruit ripening might be attributed to an increase in NO-assisted development of RNS from nitrooxidative metabolism. Singh et al. (2016) reported that exo- and endogenous melatonin can produce N-nitroso melatonin and maintain RNS homeostasis but melatonin's role still needs to be further studied.

2.4 Sole role of H_2S in fruit ripening

2.4.1 H_2S and fruit ripening

H_2S has been considered a potential signaling molecule, among several gasotransmitters, playing a vital role in multiple physiological processes in plants (Ali et al., 2019a). Recent literature has reported that H_2S is being recognized as an endogenous regulator of certain plant biomolecules. H_2S acts as a signaling molecule of ethylene-mediator (Liu et al., 2011). In several plant systems, H_2S significantly affects ethylene production, inhibits fruit ripening, and delays senescence (Hu et al., 2012). Postharvest application of H_2S has been mostly implied through sodium hydrogen sulfide (NaHS) as a donor or a releasing agent of H_2S for extending shelf and storage life. During poststorage, H_2S has the ability to alter and help in maintaining enzymatic antioxidants. (Ni et al., 2016) reported that H_2S application significantly prevented oxidative burst and successful attainment of cellular homeostasis was noted. Ziogas et al. (2018) also studied the interaction of H_2S and ethylene signaling which induced fruit ripening with bursts of ROS and RNS signaling. Fruit ripening has been mediated through associated proteins in cysteine residues that operate via post-translational modification under the role of H_2S while acting antagonistically to ethylene. The application of H_2S showed antisenescence response in kiwifruit by delaying fruit ripening (Li et al., 2017). Inhibition of cell wall loosening enzyme such as PG enzyme activity was found to be responsible for H_2S function. Moreover, exogenous application of H_2S significantly improved CAT (catalase) and APX (ascorbate peroxidase), activities while PPO (polyphenol oxidase) and LOX (lipoxygenase) activities were inhibited in kiwifruits that resulted in the delay of senescence. The molecular study showed that ethylene expressing genes were found to be downregulated, representing the inhibition of ethylene biosynthesis in response to H_2S treatment (Ziogas et al., 2018). Similarly, the endogenous H_2S levels were modulated in mulberry fruits during storage after the application of exogenous NaHS application. Endogenous H_2S biosynthesis further declined with the onset of ripening while H_2S-related enzymes activities increased, including L-cysteine desulfhydrase (LCD) and D-cysteine desulfhydrase (DCD), respectively (Hu et al., 2014a). Ripening of fruits was closely associated with biosynthesis of carotenoids and degradation of chlorophyll content of H_2S-treated mulberry (Hu et al., 2014a). On the other hand, the application of H_2S also improved chilling tolerance in low-temperature-stored fruits. However, stress amelioration that is primarily linked with the integrity of cell wall might be affected by H_2S (exogenous or endogenous) while a delay in ripening could be due to a reduction in the biosynthesis of organic acids and other protein/sugar hydrolysis. Li et al. (2016) reported that H_2S-treated banana fruits exhibited lower ethylene production and chilling injury due to the reduction of ion leakage and MDA contents.

2.4.2 Role of H_2S in delaying ripening and conserving quality during postharvest

2.4.2.1 H_2S and color changes

Fruit color is the major indicator striking consumer choice in the market and determines purchasing decisions. Exogenous application of H_2S showed a bright color of strawberry fruits during storage (Hu et al., 2012). Hu et al. (2014a) reported similar results, where exogenous NaHS treatment significantly improved endogenous H_2S contents owing to increased DCD and LCD activities, that in turn maintained the red color of fruits by suppressing the degradation of anthocyanins in mulberry. H_2S application promisingly prevented chlorophyll breakdown and affected carotenoid accumulation that subsequently helped in maintaining the color of "Kyoho" grapes (Ni et al., 2016). Deshi et al. (2020) found that H_2S significantly suppressed the enzymatic activities of PPO and peroxidase (POD), increased anthocyanin content that assisted in retaining pericarp color, and inhibited the browning index of litchi fruits (Table 2.1).

2.4.2.2 H_2S and postharvest chilling injury

Fresh horticulture produce can last for more days on the consumer's shelf if stored in cold chain under lower temperature. Low-temperature storage delays postharvest senescence and maintains fruit quality but at the same time results in chilling damage (Aghdam et al., 2013; Hasan et al., 2019). Chilling injury symptoms clearly appear when fruits stay on the shelf for reconditioning after removal from low-temperature storage. Surface pitting and water soaked thin specks are specific symptoms of chilling injury. Aghdam and Bodbodak (2013) reported some indirect symptoms of chilling injury including ripening failure, higher ethylene biosynthesis and respiration rate, and rapid fruit senescence with lower shelf and storage life. Moreover, chilling injury also affects a few textural changes and alters sensory attributes with higher ion leakage and peroxidation related to the quick deterioration of quality. Postharvest application of H_2S is found to be promising for alleviating chilling injury during storage (Ali et al., 2019a). Luo et al. (2015) reported that banana fruits treated with exogenous NaHS application significantly improved the accumulation of proline content with reduced chilling injury due to increased levels of antioxidant activities under H_2S application. Similarly, banana fruits fumigated with H_2S showed lower loosening of cell wall integrity due to reduced malondialdehyde (MDA) contents. Zhi and Dong (2018) reported that NaHS fumigation alleviated chilling injury of "Lapins" and "Regina" cherries (Table 2.1). Moreover, NaHS-treated bananas exhibited less chilling injury as compared to untreated control fruits (Li et al., 2016). Furthermore, hawthorn fruits treated with NaHS showed lower surface pitting caused possibly by suppressed chilling injury incidence during storage (Aghdam et al., 2018a).

2.4.2.3 H_2S and postharvest oxidative stress

H_2S found to be closely associated with a burst of ROS produced during metabolic activities in each organism. During most of the postharvest trials, H_2O_2 and $O_2^{\bullet -}$ have been investigated in various fruits and vegetables to

TABLE 2.1 Impact of H_2S on quality of different fruits during storage.

Fruit	Concentration	Storage conditions	Inferences	References
Apple	0.4 mmol L^{-1}	Stored for 5 days at 25°C	Delayed senescence, conserved ascorbic acid, total flavonoid, phenolic, sugars, proteins, CAT, SOD, POD, APX, GR activities, along with lower H_2O_2 and superoxide anion contents	Zheng et al. (2016)
Banana	0.5 mmol L^{-1}	Stored for 14 days at 7°C + 6 days at 20°C	Reduced chilling injury, MDA, conserved firmness, increased proline content, total phenols, upregulated SOD, CAT, APX, PAL, and POD activity	Luo et al. (2015)
Banana	0.5 mmol L^{-1}	Stored for 14 days at 7°C + 6 days 20°C	Conserved hue angle, firmness, lessened MEL, MDA contents, ethylene biosynthesis, chilling injury, increased H^+-ATPase, CCO, Ca^{2+}-ATPase, and SDH activities	Li et al. (2016)
Banana	1 mmol L^{-1}	Stored for 6 days at 20°C	Reduced yellowing, fruit softening, total carotenoids, total proteins, MDA contents, superoxide anion, H_2O_2 production, and ethylene biosynthesis genes activities	Ge et al. (2017)
Grapes	1 mmol L^{-1}	Stored for 7 days at 25°C	Reduced weight loss, chlorophyll breakdown, MDA, H_2O_2 and superoxide anion radicals, conserved TA, SSC, increased CAT and APX activity	Ni et al. (2016)
Hawthorn	1.5 mmol L^{-1}	Stored for 20 days at 1°C	Reduced chilling injury, MDA content, increased activities of LCD, DCD, CAT, SOD, APX, PAL, phenolics, anthocyanin, and ascorbic acid content	Aghdam et al. (2018a)
Kiwifruit	1 mmol L^{-1}	Stored for 5 days at 20°C ± 0.5°C	Conserved sugar contents, ascorbic acid, chlorophyll, inhibited MDA concentration, H_2O_2 and superoxide anion production, conserved higher APX, POD, CAT, GR; while reduced LOX enzyme activity	Gao et al. (2013)

Kiwifruit	45 μmol L^{-1}	Stored for 105 days at 2°C	Presented greater ascorbic acid content, chlorophyll levels and conserved higher CAT, SOD, and POD enzymes activities	Yonggen et al. (2015)
Kiwifruit	1 mmol L^{-1}	Stored for 8 days at 25°C	Reduced ripening, MDA, H_2O_2, activity of PG enzyme, and ethylene biosynthesis genes expression	Li et al. (2017)
Mulberry	0.8 mmol L^{-1}	Stored for 6 days at 2°C ± 0.2°C	Increased DCD and LCD activities, reduced respiration burst, conserved higher TA, ascorbic acid content, CAT, SOD, POD, and DPPH scavenging activity	Hu et al. (2014a)
Strawberry	0.8 mmol L^{-1}	Stored for 8 days at 20°C ± 0.5°C	Reduced respiration, LOX, PG activity, disease index, increased firmness, and activity of CAT, APX, POD, and GR enzymes	Hu et al. (2012)
Strawberry	0.8 mmol L^{-1} + 5 μmol L^{-1} SNP	Stored for 4 days at 20°C ± 1°C	Reduced decay index, respiration activity, conserved color, showed higher chitinase, β-1,3-glucanase activity, and reduced PE and PG activities	Chang et al. (2014)

investigate the concentration of ROS (Ali et al., 2019a). The postharvest exogenous NaHS application showed lower H_2O_2 and $O_2^{\bullet-}$ production. On the other hand, the exogenous application of H_2S treatment showed markedly higher enzymatic activities of GR (glutathione reductase), APX, CAT, and POD enzymes. Similarly, banana fruits treated with NaHS had lower $O_2^{\bullet-}$ and H_2O_2 radicals and higher CAT, POD, superoxide dismutase (SOD), GR, and APX activities (Luo et al., 2015). Similarly, NaHS application on mulberry fruits showed higher nonenzymatic antioxidants in terms of anthocyanins, ascorbic acid contents, and total antioxidants, as well as enzymatic antioxidants including SOD, CAT, and POD. Moreover, NaHS-treated mulberry fruits also displayed lower production of H_2O_2 and O_2^- which reduced oxidative stress during cold storage (Hu et al., 2014a). Likewise, enzymatic antioxidants including SOD, POD, CAT, and APX were higher in grapes and kiwifruits treated with H_2S through the application of NaHS treatment and substantially reduced H_2O_2 and $O_2^{\bullet-}$ radicals during storage (Zhu et al., 2014; Ni et al., 2016). Hu et al. (2014b) tested the response of H_2S application through NaHS fumigation in fresh-cut pear fruits. H_2S treatment led to markedly lower LOX and PAL enzymes and increased POD, CAT, and APX activities while reduced contents of H_2O_2 and $O_2^{\bullet-}$ were observed with extended storage life. Gao et al. (2013) reported that fresh-cut kiwifruits subjected to NaHS treatment significantly increased enzymatic antioxidants including CAT, POD, SOD, GR, and APX activities during storage, while on the other side H_2O_2 and $O_2^{\bullet-}$ radicals were reduced (Table 2.1). Likewise, a similar response was observed in apples during storage (Zheng et al., 2016).

2.4.2.4 H_2S, ethylene production, and respiratory metabolism

Ethylene plays an important role in fruit ripening. Its excessive production leads to quick senescence and short storage potential of produce (Razzaq et al., 2013). Zheng et al. (2016) reported that NaHS application significantly reduced the ethylene biosynthesis in fresh-cut apples during storage (Zheng et al., 2016). Similarly, banana fruits treated with H_2S significantly inhibited the biosynthesis of ethylene with delayed senescence (Luo et al., 2015; Li et al., 2015). Grapes treated with H_2S showed alleviated senescence and lower rachis browning due to inhibition of ethylene production during storage (Ni et al., 2016). The used concentration of H_2S also affects its role in ethylene biosynthesis; the lower doses significantly reduce ethylene production while it is increased as a result of the application of higher doses, as observed in kiwifruits (Zhu et al., 2014). The NaHS-treated kiwifruits showed a markedly lower production of ethylene with downregulated gene expressions (Li et al., 2017). Postharvest storage of banana fruits was extended as a result of NaHS treatment due to the inhibition of ethylene biosynthesis with related lower gene expressions (Ge et al., 2017).

Fresh produce undergoes respiration continuously after harvest and dysfunction of cell membrane integrity during storage. It has been observed that the quality and life of harvested produce are closely linked with the rate of respiration (Hu et al., 2012). Hu et al. (2012) reported that strawberry fruits treated with exogenous NaHS treatment had a reduced rate of respiration (Table 2.1). In another study on spinach, the application of NaHS reduced the intensity of respiration by preserving the energy in terms of ATP through oxidative phosphorylation (Hu et al., 2015). Li et al. (2016) reported that H_2S-treated banana fruits showed a lower rate of respiration might be due to the higher energy metabolism and assisted in lower consumption of energy during storage that subsequently extended the marketable window with retained eating quality. Similarly, respiration rate was found to be lower in strawberry and mulberry fruits treated with exogenous NaHS (Chang et al., 2014; Hu et al., 2014a).

2.4.2.5 H_2S, membrane integrity, and lipid peroxidation

Cell membrane integrity is the major indicator and is regarded as the primary reason for the cause of different disorders in fresh horticulture produce during storage. Mostly, higher membrane leakage and lipid peroxidation result in higher water loss along with the development of surface pitting caused by chilling damage. Postharvest treatment of NaHS, a donor of H_2S, significantly controls ion leakage and reduces lipid peroxidation in fruits. Luo et al. (2015) and Li et al. (2016) studied the effect of H_2S fumigation on banana fruits and noted that treated fruits displayed reduced chilling injury due to substantial reduction in MDA contents and ion leakage (Table 2.1). Exogenous application of H_2S decreased MDA contents and ion leakage and decreased membrane permeability of grapes (Ni et al., 2016), kiwifruit (Yonggen et al., 2015), strawberry (Chang et al., 2014), and mulberry (Hu et al., 2014a) during storage. Likewise, a lower activity of LOX enzyme due to H_2S application in fresh-cut pear (Hu et al., 2014b), kiwifruit (Gao et al., 2013), and apple (Zheng et al., 2016) resulted in lower MDA contents and ion leakage during storage.

2.4.2.6 H_2S and postharvest diseases

H_2S is a sulfur-containing compound that has the ability to induce tolerance against biotic stresses (Bloem et al., 2004). In plants, resistance against pathogenicity was found to be directly associated with nutritional supplementation of sulfur and has drawn a concept of "sulfur-induced-resistance." Although plants have the capability to synthesize glucosinolates and alliins which are natural sulfur-containing compounds and work against several pathogenic diseases (Calderwood and Kopriva, 2014). The exogenous application of NaHS showed significant results in reducing fundal decay of sweet potatoes (fresh-cut) and emerged as a potential postharvest fungicide

(Tang et al., 2014). In-depth studies were also conducted to evaluate the response of H_2S against postharvest diseases, e.g., treated fresh-cut pear fruits exhibited significantly inhibited fungal decay during storage (Hu et al., 2014b). Likewise, decay-causing pathogens including *Aspergillus niger* and *Penicillium italicum* were found to be suppressed due to induced ROS burst and reduced activation of SOD and CAT activities with downregulated gene expression in H_2S-treated apple, kiwifruit, tomato, and citrus fruits (Fu et al., 2014). Strawberry fruits exposed to H_2S significantly reduced fungal decay due to the induction of higher chitinase and β-1,3-glucanase activities (Chang et al., 2014).

2.4.2.7 H_2S and postharvest fruit softening

The texture is a major indicator of quality assessment of fleshy fruits. However, the loss of texture in fleshy fruits leads to membrane loosening, disassembly of middle lamella and pectins, resulting in softening of tissues (Vicente et al., 2007). Different postharvest coating applications have been studied to reduce fruit softening. NaHS application significantly delayed postharvest senescence and exhibited minimum fruit softening in kiwifruits (fresh-cut) during storage (Gao et al., 2013). It has been reported that strawberry fruits treated with exogenous H_2S treatment reduced softening due to reduced activity of endo-β-1,4-glucanase, PG, and PME during storage (Hu et al., 2012; Chang et al., 2014). Hu et al. (2014a) showed that H_2S-treated mulberry fruits exhibited higher pectin content with delayed postharvest softening (Table 2.1).

2.5 Nitric oxide and fruit ripening

Fruit ripening is closely associated with endogenous levels of NO biosynthesis and its regulation in metabolic activities. NO has been characterized as an ethylene-inhibiting molecule which delays fruit ripening followed by senescence with extended shelf life. Exogenous application of NO significantly regulated its endogenous content and thereby retained cell wall integrity, reduced softening and ion leakage, and maintained firmness, along with markedly better quality of peaches (Zhu et al., 2006; Flores et al., 2008). NO treatment on different climacteric fruits such as peach, mango, and tomatoes showed increased shelf life with maintained quality (Chen et al., 2005; Hong et al., 2014; Eum et al., 2009). In addition, NO application also possesses the ability to delay fruit ripening of non-climacteric fruits. Likewise, Singh et al. (2009) reported that the damage associated with storage in a cool chain was found to be reduced in NO-fumigated Japanese plums. Similarly, NO treatment induced resistance against the development of pericarp browning by reducing PPO activity in longan fruits (Duan et al., 2007). Endogenous NO levels can also be modulated from some other exogenous treatments, such as methyl salicylate, which alleviates chilling injury associated with a surge of NO in cherry tomatoes

(Zhang et al., 2011). In most of the postharvest studies, sodium nitroprusside (SNP) has been employed as an appropriate NO donor while gaseous NO has also been employed as fumigation for large-scale application. NO application decreased the climacteric peak of respiration in bananas and delayed fruit ripening, while its higher levels significantly affected the shelf life of mangoes. Increased levels of endogenous NO were measured in mango fruits, resulting in reduced chilling injury and maintained quality (Ruan et al., 2015). Rodríguez-Ruiz et al. (2017) stated that pepper fruits exhibited higher ascorbic acid content during ripening as a result of higher activity of galactono-1,4-lactone dehydrogenase induced by NO. Likewise, it has been reported that levels of phytohormones and regulation of carbon metabolism mediated by NO are considered essential for ripening of fruits as studied in short root mutant of tomato (Bodanapu et al., 2016). Moreover, the storage life of persimmon fruits was extended in response to the application of NO treatment (Shahkoomahally et al., 2015). Crosstalk of NO with SA was observed in certain studies, where it inhibited the ethylene production and delayed senescence. However, NO association with other phytohormones such as GA, auxins, cytokinins, and certain polyamines is still unclear and there is still a gap to fill with further scientific research (Mukherjee, 2019a).

2.5.1 NO and postharvest fruit quality

In recent years, the trend has been of an increased use of NO as an exogenous application for extending shelf and storage life of numerous fruits and vegetables. The mode of application might be different in varying situations and availability; either fumigated with NO gas or dipping of fresh produce in NO-releasing SNP solution. Although, the immersion methodology is acceptable, the exposure to NO gas fumigation seems a better option for commercial-scale application. After confirmation of the role of NO in fruit ripening, several studies have been conducted and exogenous application of NO has been tested for delaying maturity and senescence in different climacteric and non-climacteric fruits (Table 2.2). In addition, NO application has also been reported to inhibit disease incidence and chilling injury during low-temperature storage (Singh et al., 2013).

2.5.2 NO and postharvest senescence reduction

Senescence is a well-established metabolic respiratory mechanism and is considered as a basic physiological process that occurs in living organism which leads to a release of energy for plant metabolism, subsequently resulting in stimulated ripening followed by shortened postharvest life. The climacteric fruits show a higher rate of respiration and ethylene biosynthesis even off-tree and show markedly higher indication of senescence (Zhang et al., 2019). Zaharah and Singh (2011b) reported that respiration rate and

TABLE 2.2 Impact of NO on quality of different fruits during storage.

Fruit	Concentration	Storage conditions	Inferences	References
Plum	NO 10 $\mu L\ L^{-1}$	Stored for 49 days at 0°C	Decreased respiration and ethylene production, decay and conserved hue angle, firmness, and TA	Singh et al. (2009)
Mango	NO 20 $\mu L\ L^{-1}$	Stored for 21 days at 13°C ± 1°C	Decreased respiration and ethylene production, ACO, ACS, ACO, endo-PG, PE, exo-PG, and EGase activities	Zaharah and Singh (2011b)
Tomato	SNP 1 $mmol\ L^{-1}$	Stored for 20 days at 25°C	Decreased ethylene generation, conserved TA and firmness, showed less increase in SSC, enhanced POD, SOD, and CAT activity	Lai et al. (2011)
Banana	NO 60 $\mu L\ L^{-1}$	Stored for 15 days at 7°C ± 1°C	Mitigated chilling injury, decreased ion leakage, MDA and superoxide anion along with enhanced APX, SOD, POD, and CAT activities	Wu et al. (2014)
Banana	SNP 0.05 $mmol\ L^{-1}$	Stored for 20 days at 7°C	Mitigated chilling injury, decreased ion leakage and increased energy content	Wang et al. (2015)
Blueberry	GSNO 1 $mmol\ L^{-1}$	Stored for 14 days at 4°C	Conserved better firmness, had higher citric acid, malic acid and ascorbic acid	Grozeff et al. (2017)
Kiwifruit	SNP 0.2 $mmol\ L^{-1}$	Stored for 13 days at 20°C	Suppressed increase in disease index, conserved ASA and TA content, and had less SSC, increased total phenols, total flavonoids, and lignin content	Zheng et al. (2017)
Peach	NO 10 $\mu L\ L^{-1}$	Stored for 24 days at 4°C ± 0.5°C	Decreased ethylene production, conserved higher firmness, and affected sugar contents	Han et al. (2018)
Jujube	NO 20 $\mu L\ L^{-1}$	Stored for 75 days at 0°C	Conserved higher firmness, reduced PME, PG, β-Gal, and cellulase activities	Zhao et al. (2019)
Pitaya	SNP 0.1 $mmol\ L^{-1}$	Stored for 8 days at 25°C	Reduced disease incidence, enhanced PPO and PAL activity, had higher total phenols, lignin, and total flavonoids	Hu et al. (2019)
Muskmelon	NO 60 $\mu L\ L^{-1}$	Stored for 10 days at 25°C	Decreased black spot disease, enhanced 4CL, PAL, POD, C4H, CHT, and GLU activities	Yan et al. (2019)
Apple	SNP 1 $mmol\ L^{-1}$	Stored for 18 days at 20°C	Abridged respiration metabolism, conserved higher firmness, TA, SSC, and ASA content	Chen et al. (2019)

ethylene production was delayed in mango fruits treated with an exogenous application of NO gas during storage. Likewise, in peaches and plums, NO fumigation was found to be promising to decrease respiration and ethylene production, and to increase storage life (Singh et al., 2009; Han et al., 2018). In contrast with NO fumigation, the application of dipping of different fruits in SNP solution also helps to reduce metabolic processes such as rate of respiration and ethylene biosynthesis (Lai et al., 2011; Shi et al., 2019). NO plays an antagonistic role with ethylene which is due to the enzymatic activity involved in ethylene biosynthesis and possible related gene expression (Table 2.2). Zaharah and Singh (2011b) showed that the enzymatic activity of ACC (1-aminocyclopropane-1-carboxylic acid oxidase), ACO (1-aminocyclopropane-1-carboxylic acid oxidase), and ACS (1-aminocyclopropane-1-carboxylic acid synthase) enzymes was decreased as a result of NO gas fumigation. Moreover, it has also been reported that SNP application downregulates the expression of genes (*LeACS2* and *LeACO1*) related to ethylene production (Lai et al., 2011).

2.5.3 NO and biochemical quality conservation

Conservation of biochemical quality attributes is considered most relevant in extending the shelf and storage life of fruits and vegetables. Most prominently, total soluble solids (TSS) and total titratable acidity are considered as indicators of good taste and eating quality (Ali et al., 2016). Zhang et al. (2019) reported that acidity decreases with the onset of ripening and storage life directly relates to changes in TSS, which might be due to the conversion of organic acids into sugars during the progression from ripening to senescence and leads to a taste deterioration of fruits. In tomato, immersion in SNP solution markedly sustained higher acidity and displayed lower to medium TSS during storage (Lai et al., 2011). In addition, mango fruits also exhibited similar results regarding TA and TSS during postharvest storage (Hu et al., 2014c). However, SNP also have the ability to preserve both TA and TSS values at higher percentages, as reported in apple (Chen et al., 2019). The effects of exogenous NO on SSC and TA during postharvest storage in fruits are mainly related to changes in respiration. The conservation of quality showed indirect relation with mass loss during storage and was found to be increased during shelf and storage associated with the respiration rate of fruits (Zhang and Jiang, 2019). NO treated table grapes and pitaya fruits showed lower mass loss and showed better quality as compared with non-treated controls (Zhang et al., 2019; Hu et al., 2019).

2.5.4 NO and chilling injury

Storage of fresh produce in a cold chain prolongs storage life (Aghdam and Bodbodak, 2013). However, cold storage of tropical and subtropical fruits leads to a disorder called "chilling injury," which limits the marketability of fresh produce (Hasan et al., 2019). As a result of chilling injury, watery spots appear on fruit skin due to disruption in cell membrane integrity (Nasef, 2018). The different enzymes play an important role in the destruction of cell membrane, including peroxidation of membrane fatty acids (Rui et al., 2010). Exogenous application of NO treatment showed a significant effect on the alleviation of chilling injury in tomato under 2°C ± 1°C of storage (Zhao et al., 2011). Similar results were noted in cucumber (Yang et al., 2011), banana (Wu et al., 2014; Wang et al., 2015) melon (Zhang et al., 2017), plums (Singh et al., 2009), and mango (Zaharah and Singh, 2011a).

2.5.5 NO and postharvest diseases

Fresh produce is susceptible to various types of diseases and fungal decay during storage. Along with other postharvest treatments, exogenous application of NO has been reported to control fruit diseases. Hu et al. (2014c) stated that SNP-treated mango fruits reduced expansion of anthracnose disease caused by *C. gloeosporioides*. Likewise, application of SNP leads to resistance against *B. cinerea* pathogen which causes gray mold in tomato during storage (Lai et al., 2011). Application of aqueous SNP induced disease resistance (*C. gloeosporioides*) in citrus fruits during storage at 20°C (Zhou et al., 2015). Similarly, exogenous application of SNP promisingly reduced *Botrytis* rot and decay in strawberry fruits stored at 2.5°C (Abdollahi et al., 2013). Additionally, black spot disease caused by *A. alternata* has been found to be reduced in response to fumigation with NO gas in muskmelon fruits during poststorage (Yan et al., 2019). During storage, fungal disease incidence is one of the most deleterious issues that occur in fresh produce due to preharvest and postharvest associated factors (Sun et al., 2017). Immersion of kiwifruits in SNP solution induced disease resistance by the activation of PAL and GLU enzymes (Table 2.2) (Zheng et al., 2017). Similarly, the application of SNP enhanced the activity of POD, PPO, PAL, CHI, and GLU enzymes involved in plant defense mechanisms and showed higher nonenzymatic antioxidants (total phenolics, flavonoids, and lignin) in pitaya fruits (Hu et al., 2019).

2.5.6 Action mechanism of NO

Ethylene is a well-accepted ripening hormone that regulates the senescence of climacteric (sharp ripening) fruits (Razzaq et al., 2013). Mukherjee (2019a) reported that fruit senescence was found to be delayed in response to exogenous NO application, which significantly inhibited the ethylene production through signal regulation. During ethylene biosynthesis, NO combines with ACC oxidase enzyme and develops ACC oxidase–NO complexes (binary) that progress into ACC chelates to form ACC–ACC oxidase–NO complexes (ternary), resulting in the decline of ethylene biosynthesis (Zhu et al., 2006). On the other hand, NO reacts with ACO, involved in ethylene biosynthesis pathway, and reduces ACO activity leading to decreased ethylene production (Zhu et al., 2006; Manjunatha et al., 2010). It has been studied that the application of NO treatment downregulates ACO-related genes in tomato fruit. However, some other genes were found to be unaffected in terms of their relation to the expression of ACS (Eum et al., 2009). Zhu et al. (2006) reported that activity of ACO was unaffected with NO application in peach fruits but decreased ACS activity (Zhu et al., 2006). NO was found to be related to the inhibition of ethylene production mediated through ACS posttranscriptional regulation. Lindermayr et al. (2006) reported that NO reduced the activity of MAT1 (methionine adenosyl transferase1) involved in the production of SAM (S-adenosyl methionine), the precursor of ethylene biosynthesis in *Arabidopsis*. Lastly, it has been observed that ethylene biosynthesis and respiration rate can be inhibited through exogenous NO treatment leading to the delay in fruit ripening and senescence.

The mechanism of intercellular ROS is widely established toward fruit ripening and senescence, and homeostasis needs to be balanced to maintain the end quality of the product (Zhang and Jiang, 2019). There are two systems (enzymatic and nonenzymatic) of ROS detoxification usually employed for its removal from cells. Among the enzymatic antioxidants, numerous enzymes including APX, CAT, SOD, POD, glutathione peroxidase, MDHAR (monodehydroascorbate reductase), and GR have been reported so far, while nonenzymatic antioxidants such as phenolic contents, anthocyanins, carotenoid and ascorbic acid content (vitamin C), glutathione, vitamin E, and lycopene contents have also been studied in different commodities (Mittler, 2002; Ali et al., 2016, 2019b). Preservation of antioxidants during storage determines the quality of fresh produce for the consumer's shelf. The exogenous application of NO led to markedly higher antioxidants in several crops. Lai et al. (2011) reported that SOD, POD, and CAT enzymes were enhanced in tomatoes treated with SNP treatment. Likewise, table grapes fumigated with NO showed upregulation of activities of APX, SOD, POD, CAT, and GR enzymes during storage. Moreover, molecular study also showed that genes, i.e., *VvSOD, VvCAT, VvPOD2*, and *VvGR* related to their respective enzymes, were highly expressed, decreased MDA contents, H_2O_2 and superoxide anion, eventually leading to the elimination of ROS along with maintained quality (Zhang et al., 2017). In addition, nonenzymatic antioxidants such as ascorbic acid, flavonoid, and phenolic content were found to be higher in SNP-treated kiwifruits (Zhang et al., 2017), followed by peach (Ma et al., 2019) treated with NO fumigation. Wu et al. (2014) revealed that activities of SOD, APX, CAT, and POD enzymes were upregulated in response to NO treatment in banana fruits during storage.

2.5.7 NO, intracellular energy, and sugar metabolism

Adequate supply of energy is a most critical aspect in maintaining quality during postharvest conditions (Aghdam et al., 2018b). The soluble sugar content is mainly composed of glucose, fructose, and sucrose, while sucrose molecules are found to be involved in the metabolism of ripening in fruits (Han et al., 2018). In addition to the conservation of antioxidative activities, NO also plays a major role in sugar metabolism. Shi et al. (2019) revealed that postharvest application of NO enhanced the activities of sugar accumulation enzymes such as SPS (sucrose phosphate synthase), SS (sucrose synthase), and NI (neutral invertase) followed by decrease in AI (acid invertase) enzyme activity. Apple fruits treated with SNP had higher activities of SPS, SS, NI, and Al (Chen et al., 2019). Fruits stored at low temperature are most vulnerable to chilling injury and its mechanism is quite similar to energy release due to cell degradation. This has been confirmed by Wang et al. (2015) who reported that chilling tolerance was found to increase as a result of increased energy in the form of ATP content in response to SNP.

2.6 Melatonin and fruit ripening

Several reports have shown that melatonin regulates ripening in grapes, tomato, and strawberry fruits (Meng et al., 2015; Sun et al., 2015; Aghdam and Fard, 2017). L-Tryptophan has been recognized as a precursor of melatonin because of its indoleamine nature (Galano et al., 2011). Recent studies have revealed the action mechanism of melatonin in fruits (Vitalini et al., 2011). The defense mechanism has been associated with its antioxidative properties during fruit ripening. The postharvest application of melatonin also showed similar effects, as observed with exogenous NO and H_2S prestorage treatments. Grapes treated with melatonin showed increased anthocyanin, flavonoid, proanthocyanidin, and phenol contents with maintained quality, altered carbohydrate metabolism, and increased ethylene production. Similarly, the treatment by melatonin of peaches stored at low temperature significantly prevented chilling damage and induced chilling tolerance, which might be due to stable ROS homeostasis that led to delayed ripening (Cao et al., 2018). The application of melatonin treatments seems to be correlated with an increase of endogenous H_2O_2 levels during storage of fruits. However, melatonin treatment was found to be responsible for increasing ascorbate−glutathione cycle and ascorbic acid synthesis. Tomatoes showed variability in endogenous levels of melatonin biosynthesis during different stages of ripening while its synthesis might vary with changing genotype and environmental exposure (Riga et al., 2014). In another study, the melatonin biosynthesis was triggered in darkness and oxidative stress in different varieties of sweet cherry fruits (Zhao et al., 2013). Higher accumulation of melatonin might assist in the sharp rise of respiration in apple fruits and affect the ripening physiology (Lei et al., 2013).

2.6.1 Role of melatonin in postharvest fruit quality

Initially, melatonin was considered to be a potential antioxidant compound, but subsequent studies showed its significance in the regulation of numerous physiological processes in plants, quite similar to the action of certain other plant hormones. Application of melatonin plays a critical role in multiple physiological processes under certain conditions. In addition, melatonin has also been reported to regulate postharvest physiology such as fruit ripening, senescence, chilling injury alleviation, and postharvest disease control (Table 2.3) (Zhang et al., 2020).

2.6.2 Melatonin and postharvest senescence

Exogenous application of melatonin can significantly inhibit the rate of respiration in climacteric fruits. It has been reported that peach and pear fruits treated with melatonin showed lower rates of respiration and water loss, and maintained TTS better than untreated control fruits during storage (Gao et al., 2016; Zhai et al., 2018). Liu et al. (2019a) stated that pear fruits treated with melatonin maintained significantly higher firmness, showed reduced ethylene biosynthesis, and exhibited lower expression of related genes (*PcPG*, *PcCel*, *PcACS1*, *PcACO1*, *PcACS2*, and *PcACO2*) under cold storage. Likewise, banana fruits treated with melatonin exhibited reduced ethylene production due to the downregulation of ethylene-related gene expression, such as *MaACO1* and *MaACS1*. The above studies showed that melatonin as a postharvest treatment might also act as a repressor of ethylene biosynthesis and plays a vital role in the regulation of ethylene balance (Zhai et al., 2018). Therefore, the impact of melatonin in the regulation of fruit ripening and senescence needs to be explored further.

2.6.3 Melatonin, chilling injury, and disease reduction

The role of exogenous melatonin application has also been studied in the alleviation of chilling injury and control of disease incidence in fruits during storage (Zhang et al., 2020). Gao et al. (2018) reported that peach fruits treated with melatonin showed decreased MDA contents and LOX activity, maintained higher linoleic acid and linolenic acid, and had lower stearic acid and palmitic acid, which in turn led to alleviation of chilling injury during low-temperature storage. Jannatizadeh (2019) stated that pomegranate arils treated with melatonin had lower chilling injury, ion leakage, MDA content, and H_2O_2 as well as PLD and LOX activities during ultracold storage. On the other hand, melatonin treatment stimulated resistance against *B. cinerea*, causing gray mold, and inhibited expansion of disease incidence on tomatoes under in vivo conditions. However, results were found to be statistically non-significant for the inhibition of *B. cinerea* spread (Li et al., 2019). In addition, melatonin-treated bananas showed delayed senescence and superficially reduced postharvest anthracnose disease caused by *Colletotrichum musae* might be due to increased mitogen-activated protein kinase, auxin, ethylene signaling pathways, multiple metabolites and disease resistant proteins (Li et al., 2019).

TABLE 2.3 Impact of melatonin on quality of different fruits during storage.

Fruits	Concentration	Storage conditions	Inferences	References
Peach	0.1 mmol L^{-1}	Stored for 7 days at 25°C–28°C	Reduced weight loss, respiration and MDA production, superoxide anion and H_2O_2, maintained better firmness, TSS and ASA, decreased LOX activity, increased SOD, POD, APX, and CAT activities	Gao et al. (2016)
Strawberry	0.1 mmol L^{-1}	Stored for 12 days at 4°C	Reduced decay incidence, increased activity of SOD and PAL, reduced APX and CAT activities, maintained higher phenols, flavonoids, energy content and unSFA/SFA ratio and affected pathway of GABA shunt	Aghdam and Fard (2017)
Strawberry	0.1, 1.0 and 10 mmol L^{-1}	Stored for 12 days at 4°C	Reduced decay incidence, weight loss, and MDA and H_2O_2, conserved firmness, TA, total phenolics and flavonoids, increased endogenous melatonin, upregulated FaTDC, FaT5H, FaSNAT, and FaASMT expression, the genes involved in biosynthesis of melatonin content	Liu et al. (2018)
Pear	0.1 mmol L^{-1}	Stored for 30 days cold storage at 4°C and for 3 days at 26°C	Reduced respiration, ethylene biosynthesis, maintained firmness, downregulated PcCel, PcPG, PcACS2, PcACS1, PcACO1, and PcACO2 expressions, also enhanced PcDHAR and PcPOD along with higher SOD, POD, and DHAR activities	Zhai et al. (2018)
Banana	0.05 mmol L^{-1}	Stored for 19 days at 25°C	Delayed senescence, lowered ethylene production, conserved higher starch contents and firmness, lowered soluble sugars, downregulated MaACO1 and MaACS1 expressions	Hu et al. (2017)
Peach	0.1 mmol L^{-1}	Stored for 28 days at 1°C	Mitigated CI, conserved increased firmness, lessened MDA production and LOX activity, enhanced unSFA/SFA ratio, total phenolics as well as G6PDH, SKDH, and PAL activities	Gao et al. (2018)
Litchi	0.4 mmol L^{-1}	Stored for 8 days at 25°C	Decreased skin browning, reduced MEL, MDA, superoxide anion and H_2O_2 concentration, enhanced SOD, GR, APX, and CAT activities along with higher total phenolics, flavonoids, anthocyanins, and melatonin endogenous concentration	Zhang et al. (2018)

Tomato	0.1 mmol L^{-1}	Stored for 6 days at 25°C ± 1°C	Enhanced decay resistance, CHI, PAL, GLU, SOD, PPO, and APX activities, reduced accumulation H_2O_2, and enhanced JA signaling pathways	Liu et al. (2019a,b)
Pomegranate	0.1 mmol L^{-1}	Stored for 120 days at 4°C	Lessened CI, reduced REL, MDA, and H_2O_2, increased APX, SOD, CAT, APX, GR, and PAL activities, along with higher phenols increase and reduced LOX and PLD activity	Jannatizadeh (2019)
Tomato	0.5 mmol L^{-1}	Stored for 28 days at 4°C ± 0.5°C	Lowered H_2O_2, increased melatonin accumulation, upregulated TDC, T5H, SNAT, and ASMT expressions involved in biosynthesis of melatonin, affected pathway of GABA shunt, showed increased accumulation of phenols, and had higher DPPH radical scavenging capacity	Sharafi et al. (2019)
Sweet cherry	0.1 mmol L^{-1}	Stored for 63 days at 0°C	Abridged decay, weight loss, respiration, conserved firmness, TSS, TA, and endogenous melatonin, decreased REL, MDA, superoxide anion, and H_2O_2, enhanced SOD, CAT, APX, DHAR, GR, and MDHAR activities	Wang et al. (2019)
Tomato	0.1 mmol L^{-1}	Stored for 28 days at 4°C	Alleviated CI index, lowered REL, reduced MDA, upregulated ZAT2/6/12 pathways, ADC, ARG1, ODC, and ARG 2 expressions, enhanced NOS, ODC, ADC, P5CS, OAT, and PDH activities and nitric oxide, polyamines, and proline content	Aghdam et al. (2019)
Plum	0.1 and 1 mmol L^{-1}	Stored for 40 days at 1°C	Reduced weight loss, decay, maintained higher firmness, TA, ASA, total phenol contents, and antioxidants activity	Bal (2019)
Pomegranate	0.1 mmol L^{-1}	Stored for 120 days at 4°C	Increased 6PGDH, G6PDH, AAO, GR, APX and PAL activity and ASA, GSH, NADPH accumulation, phenols, and anthocyanins	Aghdam et al. (2020)

2.6.4 Action mechanism of melatonin

Exogenous application of melatonin is positively correlated with biosynthesis of endogenous NO in fruits during storage. Aghdam et al. (2019) reported that melatonin application enhanced NO biosynthesis in tomato fruits and showed upregulation of *NOS1* gene expression. Similarly, higher accumulation of endogenous NO and relative enzymatic activities, such as NR and NOS, has been observed in pear in response to melatonin treatment (Liu et al., 2019a). It has been noted that melatonin reduces ethylene biosynthesis in pear fruits during their shelf life (Liu et al., 2019a). Liu et al. (2018) found that exogenous melatonin application on strawberry fruits increased endogenous melatonin content and antioxidant activity and decreased H_2O_2. Enzymatic antioxidants were also observed to be higher in melatonin-treated fruits as compared to the controls during storage. Wang et al. (2019) reported that melatonin-treated cherries exhibited higher enzyme activities, e.g., APX, CAT, SOD, dehydroascorbate reductase (DHAR), and GR, and reduced MDA content with maintained fruit quality. Likewise, APX and GR activities were also increased in melatonin-treated pomegranate fruits during cold storage (Aghdam et al., 2020). It has also been observed that melatonin-treated tomato fruits showed higher disease resistance, possibly due to jasmonate pathway activation (Liu et al., 2019b). Citrus fruits treated with exogenous melatonin treatment showed lower green mold disease while no positive impact on growth of *Penicillium digitatum* was observed (Lin et al., 2019).

2.7 Conclusion and future prospects

In conclusion, H_2S, NO, and melatonin play various vital roles in different stages of ripening in numerous fruits. H_2S, NO, and melatonin crosstalk either delays or stimulates ripening in fruits. In addition, these signaling molecules crosstalk with each other and influence senescence during preharvest and poststorage phases in a number of fruits. However, detailed investigations are still needed in order to understand the exact mechanisms of H_2S, NO, and melatonin-induced fruit ripening. Most of the work has been carried out during postharvest and literature regarding their effects on the growth, development, and ripening of fruit crops during preharvest is very limited. Proteomics and transcriptome-based studies have generally been overlooked and must be conducted for better understanding of the effects of H_2S, NO, and melatonin on fruits.

References

Abdollahi, R., Asghari, M., Esmaiili, M., Abdollahi, A., 2013. Postharvest nitric oxide treatment effectively reduced decays of selva strawberry fruit. Int. J. Agric. Crop. Sci. 6 (6), 353–355.

Aghdam, M.S., Bodbodak, S., 2013. Physiological and biochemical mechanisms regulating chilling tolerance in fruit and vegetables under postharvest salicylates and jasmonates treatments. Sci. Hortic. 156, 73–85.

Aghdam, M.S., Fard, J.R., 2017. Melatonin treatment attenuates postharvest decay and maintains nutritional quality of strawberry fruits (*Fragaria × anannasa* cv. Selva) by enhancing GABA shunt activity. Food Chem. 221, 1650–1657.

Aghdam, M.S., Sevillano, L., Flores, F.B., Bodbodak, S., 2013. Heat shock proteins as biochemical markers for postharvest chilling stress in fruit and vegetables. Sci. Hortic. 160, 54–64.

Aghdam, M.S., Mahmoudi, R., Razavi, F., Rabiei, V., Soleimani, A., 2018a. Hydrogen sulfide treatment confers chilling tolerance in hawthorn fruit during cold storage by triggering endogenous H_2S accumulation, enhancing antioxidant enzymes activity and promoting phenols accumulation. Sci. Hortic. 238, 264–271.

Aghdam, M.S., Jannatizadeh, A., Luo, Z., Paliyath, G., 2018b. Ensuring sufficient intracellular ATP supplying and friendly extracellular ATP signaling attenuates stresses, delays senescence and maintains quality in horticultural crops during postharvest life. Trends Food Sci. Technol. 76, 67–81.

Aghdam, M.S., Luo, Z., Jannatizadeh, A., Sheikh-Assadi, M., Sharafi, Y., Farmani, B., 2019. Employing exogenous melatonin applying confers chilling tolerance in tomato fruits by upregulating ZAT2/6/12 giving rise to promoting endogenous polyamines, proline, and nitric oxide accumulation by triggering arginine pathway activity. Food Chem. 275, 549–556.

Aghdam, M.S., Luo, Z., Li, L., Jannatizadeh, A., Fard, J.R., Pirzad, F., 2020. Melatonin treatment maintains nutraceutical properties of pomegranate fruits during cold storage. Food Chem. 303, 125385.

Ali, S., Khan, A.S., Malik, A.U., Shahid, M., 2016. Effect of controlled atmosphere storage on pericarp browning, bioactive compounds and antioxidant enzymes of litchi fruits. Food Chem. 206, 18–36.

Ali, S., Nawaz, A., Ejaz, S., Haider, S.T.A., Alam, M.W., Javed, H.U., 2019a. Effects of hydrogen sulfide on postharvest physiology of fruits and vegetables: an overview. Sci. Hortic. 243, 290–299.

Ali, S., Khan, A.S., Malik, A.U., Anjum, M.A., Nawaz, A., Shah, H.M.S., 2019b. Modified atmosphere packaging delays enzymatic browning and maintains quality of harvested litchi fruit during low temperature storage. Sci. Hortic. 254, 14–20.

Ali, S., Khan, A.S., Anjum, M.A., Nawaz, A., Naz, S., Ejaz, S., et al., 2019c. *Aloe vera* gel coating delays post-cut surface browning and maintains quality of cold stored lotus (*Nelumbo nucifera* Gaertn.) root slices. Sci. Hortic. 256, 108612.

Ali, S., Khan, A.S., Nawaz, A., Anjum, M.A., Naz, S., Ejaz, S., et al., 2019d. *Aloe vera* gel coating delays postharvest browning and maintains quality of harvested litchi fruit. Postharvest Biol. Technol. 157, 110960.

Arnao, M.B., Hernandez-Ruiz, J., 2019. Melatonin: a new plant hormone and/or a plant master regulator? Trends Plant Sci. 24 (1), 38–48.

Bal, E., 2019. Physicochemical changes in ‘Santa Rosa’ plum fruit treated with melatonin during cold storage. J. Food Measur. Charac. 13 (3), 1713–1720.

Bloem, E., Riemenschneider, A., Volker, J., Papenbrock, J., Schmidt, A., Salac, I., et al., 2004. Sulphur supply and infection with *Pyrenopeziza brassicae* influence L-cysteine desulphydrase activity in *Brassica napus* L. J. Exp. Bot. 55, 2305–2312.

Bodanapu, R., Gupta, S.K., Basha, P.O., Sakthivel, K., Sreelakshmi, Y., Sharma, R., 2016. Nitric oxide overproduction in tomato *shr* mutant shifts metabolic profiles and suppresses fruit growth and ripening. Front. Plant Sci. 7, 1714.

Brady, C.J., 1987. Fruit ripening. Annu. Rev. Plant Physiol. 38, 155–178.

Brummell, D.A., 2006. Cell wall disassembly in ripening fruit. Funct. Plant Biol. 33, 103–119.

Calderwood, A., Kopriva, S., 2014. Hydrogen sulfide in plants: from dissipation of excess sulfur to signaling molecule. Nitric Oxide 41, 72–78.

Cao, S., Shao, J., Shi, L., Xu, L., Shen, Z., Chen, W., et al., 2018. Melatonin increases chilling tolerance in postharvest peach fruit by alleviating oxidative damage. Sci. Rep. 8 (1), 1–11. Available from: https://doi.org/10.1038/s41598-018-19363-5.

Chang, Z., Jingying, S., Liqin, Z., Changle, L., Qingguo, W., 2014. Cooperative effects of hydrogen sulfide and nitric oxide on delaying softening and decay of strawberry. Int. J. Agric. Biol. Eng. 7, 114–122.

Chen, Y.F., Etheridge, N., Schaller, G.E., 2005. Ethylene signal transduction. Ann. Bot. 95 (6), 901–915.

Chen, J., Wu, F.H., Wang, W.H., Zheng, C.J., Lin, G.H., Dong, X.J., et al., 2011. Hydrogen sulphide enhances photosynthesis through promoting chloroplast biogenesis, photosynthetic enzyme expression, and thiol redox modification in *Spinacia oleracea* seedlings. J. Exp. Bot. 62 (13), 4481–4493.

Chen, Y., Ge, Y., Zhao, J., Wei, M., Li, C., Hou, J., 2019. Postharvest sodium nitroprusside treatment maintains storage quality of apple fruit by regulating sucrose metabolism. Postharvest Biol. Technol. 154, 115–120.

Cherian, S., Figueroa, C.R., Nair, H., 2014. 'Movers and shakers' in the regulation of fruit ripening: a cross-dissection of climacteric versus non-climacteric fruit. J. Exp. Bot. 65, 4705–4722.

Corpas, F.J., Freschi, L., Rodríguez-Ruiz, M., Mioto, P.T., González-Gordo, S., Palma, J.M., 2018. Nitro-oxidative metabolism during fruit ripening. J. Exp. Bot. 69 (14), 3449–3463.

Deshi, V., Siddiqui, M.W., Homa, F., Singh, J.P., 2020. Postharvest hydrogen sulfide infiltration modulates antioxidative metabolism and increases shelf life of litchi. Acta Physiol. Plant 42 (67), 1–9.

Duan, X., Su, X., You, Y., Qu, H., Li, Y., Jiang, Y., 2007. Effect of nitric oxide on pericarp browning of harvested longan fruit in relation to phenolic metabolism. Food Chem. 104 (2), 571–576.

Eum, H.L., Kim, H.B., Choi, S.B., Lee, S.K., 2009. Regulation of ethylene biosynthesis by nitric oxide in tomato (*Solanum lycopersicum* L.) fruit harvested at different ripening stages. Eur. Food Res. Technol. 228 (3), 331.

Flores, F.B., Sánchez-Bel, P., Valdenegro, M., Romojaro, F., Martínez-Madrid, M.C., Egea, M.I., 2008. Effects of a pretreatment with nitric oxide on peach (*Prunus persica* L.) storage at room temperature. Eur. Food Res. Technol. 227 (6), 1599.

Fu, L.H., Hu, K.D., Hu, L.Y., Li, Y.H., Hu, L.B., Yan, H., et al., 2014. An antifungal role of hydrogen sulfide on the postharvest pathogens *Aspergillus niger* and *Penicillium italicum*. PLoS One 9, e104206.

Galano, A., Tan, D.X., Reiter, R.J., 2011. Melatonin as a natural ally against oxidative stress: a physicochemical examination. J. Pineal Res. 51 (1), 1–16.

Gao, S.P., Hu, K.D., Hu, L.Y., Li, Y.H., Han, Y., Wang, H.L., et al., 2013. Hydrogen sulfide delays postharvest senescence and plays an antioxidative role in fresh-cut kiwifruit. HortScience 48, 1385–1392.

Gao, H., Zhang, Z.K., Chai, H.K., Cheng, N., Yang, Y., Wang, D.N., 2016. Melatonin treatment delays postharvest senescence and regulates reactive oxygen species metabolism in peach fruit. Postharvest Biol. Technol. 118, 103–110.

Gao, H., Lu, Z., Yang, Y., Wang, D., Yang, T., Cao, M., 2018. Melatonin treatment reduces chilling injury in peach fruit through its regulation of membrane fatty acid contents and phenolic metabolism. Food Chem. 245, 659–666.

Ge, Y., Hu, K.D., Wang, S.S., Hu, L.Y., Chen, X.Y., Li, Y.H., et al., 2017. Hydrogen sulfide alleviates postharvest ripening and senescence of banana by antagonizing the effect of ethylene. PLoS One 12 (6), e0180113. Available from: https://doi.org/10.1371/journal.pone.0180113.

Geng, B., Huang, D., Zhu, S., 2019. Regulation of hydrogen sulfide metabolism by nitric oxide inhibitors and the quality of peaches during cold storage. Antioxidants 8, 401. Available from: https://doi.org/10.3390/antiox8090401.

Gong, T., Li, C., Bian, B., Wu, Y., Dawuda, M.M., Liao, W., 2018. Advances in application of small molecule compounds for extending the shelf life of perishable horticultural products: a review. Sci. Hortic. 230, 25–34.

Gray, J., Picton, S., Shabbeer, J., Schuch, W., Grierson, D., 1992. Molecular biology of fruit ripening and its manipulation with antisense genes. 10 Years Plant Molecular Biology. Springer, Dordrecht, pp. 69–87.

Grozeff, G.E.G., Alegre, M.L., Senn, M.E., Chaves, A.R., Simontacchi, M., Bartoli, C.G., 2017. Combination of nitric oxide and 1-MCP on postharvest life of the blueberry (*Vaccinium* spp.) fruit. Postharvest Biol. Technol. 133, 72–80.

Han, S., Cai, H., An, X., Huan, C., Wu, X., Jiang, L., 2018. Effect of nitric oxide on sugar metabolism in peach fruit (cv. Xiahui 6) during cold storage. Postharvest Biol. Technol. 142, 72–80.

Hasan, M.U., Rehman, R.N.U., Malik, A.U., Haider, M.W., Ahmed, Z., Khan, A.S., 2019. Pre-storage application of L-arginine alleviates chilling injury and maintains postharvest quality of cucumber (*Cucumis sativus*). J. Hortic. Sci. Technol. 2 (4), 102–108.

Hong, J.K., Yun, B.W., Kang, J.G., Raja, M.U., Kwon, E., Sorhagen, K., et al., 2008. Nitric oxide function and signalling in plant disease resistance. J. Exp. Bot. 59 (2), 147–154.

Hong, K., Gong, D., Xu, H., Wang, S., Jia, Z., Chen, J., et al., 2014. Effects of salicylic acid and nitric oxide pretreatment on the expression of genes involved in the ethylene signalling pathway and the quality of postharvest mango fruit. N Z J. Crop. Hortic. Sci. 42 (3), 205–216.

Hu, L.Y., Hu, S.L., Wu, J., Li, Y.H., Zheng, J.L., Wei, Z.J., et al., 2012. Hydrogen sulfide prolongs postharvest shelf life of strawberry and plays an antioxidative role in fruits. J. Agric. Food Chem. 60, 8684–8693.

Hu, H., Shen, W., Li, P., 2014a. Effects of hydrogen sulphide on quality and antioxidant capacity of mulberry fruit. Int. J. Food Sci. Technol. 49, 399–409.

Hu, K.D., Wang, Q., Hu, L.Y., Gao, S.P., Wu, J., Li, Y.H., et al., 2014b. Hydrogen sulfide prolongs postharvest storage of fresh-cut pears (*Pyrus pyrifolia*) by alleviation of oxidative damage and inhibition of fungal growth. PLoS One 9 (1), e85524.

Hu, M., Yang, D., Huber, D.J., Jiang, Y., Li, M., Gao, Z., et al., 2014c. Reduction of postharvest anthracnose and enhancement of disease resistance in ripening mango fruit by nitric oxide treatment. Postharvest Biol. Technol. 97, 115–122.

Hu, H.L., Liu, D., Li, P.X., Shen, W.B., 2015. Hydrogen sulfide delays leaf yellowing of stored water spinach (*Ipomoea aquatica*) during dark-induced senescence by delaying chlorophyll breakdown, maintaining energy status and increasing antioxidative capacity. Postharvest Biol. Technol. 108, 8–20.

Hu, W., Yang, H., Tie, W., Yan, Y., Ding, Z., Liu, Y., et al., 2017. Natural variation in banana varieties highlights the role of melatonin in postharvest ripening and quality. J. Agric. Food Chem. 65 (46), 9987–9994.

Hu, M., Zhu, Y., Liu, G., Gao, Z., Li, M., Su, Z., 2019. Inhibition on anthracnose and induction of defense response by nitric oxide in pitaya fruit. Sci. Hortic. 245, 224–230.

Jannatizadeh, A., 2019. Exogenous melatonin applying confers chilling tolerance in pomegranate fruit during cold storage. Sci. Hortic. 246, 544–549.

Lai, T., Wang, Y., Li, B., Qin, G., Tian, S., 2011. Defense responses of tomato fruit to exogenous nitric oxide during postharvest storage. Postharvest Biol. Technol. 62 (2), 127–132.

Lei, Q., Wang, L., Tan, D.X., Zhao, Y., Zheng, X.D., Chen, H., et al., 2013. Identification of genes for melatonin synthetic enzymes in 'Red Fuji' apple (*Malus domestica Borkh. cv. Red*) and their expression and melatonin production during fruit development. J. Pineal Res. 55 (4), 443–451.

Leshem, Y.Y., Wills, R.B., Ku, V.V.V., 1998. Evidence for the function of the free radical gas—nitric oxide ($NO^•$) as an endogenous maturation and senescence regulating factor in higher plants. Plant Physiol. Biochem. 36 (11), 825–833.

Li, L., Wang, Y., Shen, W., 2012. Roles of hydrogen sulfide and nitric oxide in the alleviation of cadmium-induced oxidative damage in alfalfa seedling roots. Biometals. 25 (3), 617–631.

Li, T.T., Yun, Z., Zhang, D.D., Yang, C.W., Zhu, H., Jiang, Y.M., et al., 2015. Proteomic analysis of differentially expressed proteins involved in ethylene-induced chilling tolerance in harvested banana fruit. Front. Plant Sci. 6. Available from: https://doi.org/10.3389/fpls.2015.00845.

Li, D., Limwachiranon, J., Li, L., Du, R., Luo, Z., 2016. Involvement of energy metabolism to chilling tolerance induced by hydrogen sulfide in cold-stored banana fruit. Food Chem. 208, 272–278.

Li, T.T., Li, Z.R., Hu, K.D., Hu, L.Y., Chen, X.Y., Yang, Y., et al., 2017. Hydrogen sulfide alleviates kiwifruit ripening and senescence by antagonizing effect of ethylene. HortScience 52, 1556–1562.

Li, T., Wu, Q., Zhu, H., Zhou, Y., Jiang, Y., Gao, H., et al., 2019. Comparative transcriptomic and metabolic analysis reveals the effect of melatonin on delaying anthracnose incidence upon postharvest banana fruit peel. BMC Plant Biol. 19 (289), 1–15.

Lin, Y., Fan, L., Xia, X., Wang, Z., Yin, Y., Cheng, Y., 2019. Melatonin decreases resistance to postharvest green mold on citrus fruit by scavenging defense-related reactive oxygen species. Postharvest Biol. Technol. 153, 21–30.

Lindermayr, C., Saalbach, G., Bahnweg, G., Durner, J., 2006. Differential inhibition of *Arabidopsis* methionine adenosyltransferases by protein S-nitrosylation. J. Biol. Chem. 281 (7), 4285–4291.

Lisjak, M., Srivastava, N., Teklic, T., Civale, L., Lewandowski, K., Wilson, I., et al., 2010. A novel hydrogen sulfide donor causes stomatal opening and reduces nitric oxide accumulation. Plant Physiol. Biochem. 48 (12), 931–935.

Liu, J., Hou, L., Liu, G., Liu, X., Wang, X., 2011. Hydrogen sulfide induced by nitric oxide mediates ethylene-induced stomatal closure of *Arabidopsis thaliana*. Chin. Sci. Bull 56 (33), 3547–3553.

Liu, C., Zheng, H., Sheng, K., Liu, W., Zheng, L., 2018. Effects of melatonin treatment on the postharvest quality of strawberry fruit. Postharvest Biol. Technol. 139, 47–55.

Liu, J., Yang, J., Zhang, H., Cong, L., Zhai, R., Yang, C., 2019a. Melatonin inhibits ethylene synthesis via nitric oxide regulation to delay postharvest senescence in pears. J. Agric. Food Chem. 67 (8), 2279–2288.

Liu, C., Chen, L., Zhao, R., Li, R., Zhang, S., Yu, W., 2019b. Melatonin induces disease resistance to *Botrytis cinerea* in tomato fruit by activating jasmonic acid signaling pathway. J. Agric. Food Chem. 67 (22), 6116–6124.

Lizada, C., 1993. Mango. In: Seymour, G.B., Taylor, J.E., Tucker, G.A. (Eds.), Biochemistry of Fruit Ripening. Chapman and Hall, London, pp. 255–271.

Luo, Z., Li, D., Du, R., Mou, W., 2015. Hydrogen sulfide alleviates chilling injury of banana fruit by enhanced antioxidant system and proline content. Sci. Hortic. 183, 144–151.

Ma, Y., Huang, D., Chen, C., Zhu, S., Gao, J., 2019. Regulation of ascorbate-glutathione cycle in peaches via nitric oxide treatment during cold storage. Sci. Hortic. 247, 400–406.

Manjunatha, G., Lokesh, V., Neelwarne, B., 2010. Nitric oxide in fruit ripening: trends and opportunities. Biotechnol. Adv. 28 (4), 489–499.

Meng, J.F., Xu, T.F., Song, C.Z., Yu, Y., Hu, F., Zhang, L., et al., 2015. Melatonin treatment of pre-veraison grape berries to increase size and synchronicity of berries and modify wine aroma components. Food Chem. 185, 127–134.

Mittler, R., 2002. Oxidative stress, antioxidants and stress tolerance. Trends Plant Sci. 7 (9), 405–410.

Mukherjee, S., 2018. Novel perspectives on the molecular crosstalk mechanisms of serotonin and melatonin in plants. Plant Physiol. Biochem. 132, 33–45.

Mukherjee, S., 2019a. Recent advancements in the mechanism of nitric oxide signaling associated with hydrogen sulfide and melatonin crosstalk during ethylene-induced fruit ripening in plants. Nitric Oxide 82, 25–34.

Mukherjee, S., 2019b. Insights to nitric oxide-melatonin crosstalk and N-nitrosome-latonin functioning in plants: where do we stand? J. Exp. Bot. erz375.

Nasef, I.N., 2018. Short hot water as safe treatment induces chilling tolerance and antioxidant enzymes, prevents decay and maintains quality of cold-stored cucumbers. Postharvest Biol. Technol. 138, 1–10.

Ni, Z.-J., Hu, K.-D., Song, C.-B., Ma, R.-H., Li, Z.-R., Zheng, J.-L., et al., 2016. Hydrogen sulfide alleviates postharvest senescence of grape by modulating the antioxidant defenses. Oxid. Med. Cell Longev. 1–14.

Pech, J.C., Bouzayen, M., Latche, A., 2008. Climacteric fruit ripening: ethylene-dependent and independent regulation of ripening pathways in melon fruit. Plant Sci. 175, 114–120.

Prasanna, V., Prabha, T.N., Tharanathan, R.N., 2007. Fruit ripening phenomena–an overview. Crit. Rev. Food Sci. Nutr. 47 (1), 1–19.

Razzaq, K., Khan, A.S., Malik, A.U., Shahid, M., 2013. Ripening period influences fruit softening and antioxidative system of 'Samar Bahisht Chaunsa' mango. Sci. Hortic. 160, 108–114.

Riga, P., Medina, S., García-Flores, L.A., Gil-Izquierdo, Á., 2014. Melatonin content of pepper and tomato fruits: effects of cultivar and solar radiation. Food Chem. 156, 347–352.

Rodríguez-Ruiz, M., Mateos, R.M., Codesido, V., Corpas, F.J., Palma, J.M., 2017. Characterization of the galactono-1, 4-lactone dehydrogenase from pepper fruits and its modulation in the ascorbate biosynthesis. Role of nitric oxide. Redox Biol. 12, 171–181.

Ruan, J., Li, M., Jin, H., Sun, L., Zhu, Y., Xu, M., et al., 2015. UV-B irradiation alleviates the deterioration of cold-stored mangoes by enhancing endogenous nitric oxide levels. Food Chem. 169, 417–423.

Rui, H., Cao, S., Shang, H., Jin, P., Wang, K., Zheng, Y., 2010. Effects of heat treatment on internal browning and membrane fatty acid in loquat fruit in response to chilling stress. J. Sci. Food Agric. 90 (9), 1557–1561.

Selvaraj, Y., Kumar, R., Pal, D.K., 1989. Changes in sugars, organic acids, amino acids, lipid constituents and aroma characteristics of ripening mango (*Mangifera indica* L.) fruit. J. Food Sci. Technol. 26, 308–313.

Sharafi, Y., Aghdam, M.S., Luo, Z., Jannatizadeh, A., Razavi, F., Fard, J.R., et al., 2019. Melatonin treatment promotes endogenous melatonin accumulation and triggers GABA shunt pathway activity in tomato fruits during cold storage. Sci. Hortic. 254, 222–227.

Shahkoomahally, S., Ramezanian, A., Farahnaky, A., 2015. Postharvest nitric oxide treatment of persimmon (*Diospyros kaki* l.) improves fruit quality during storage. Fruits 70 (2), 63–68.

Shi, K., Liu, Z., Wang, J., Zhu, S., Huang, D., 2019. Nitric oxide modulates sugar metabolism and maintains the quality of red raspberry during storage. Sci. Hortic. 256, 108611.

Singh, S.P., Singh, Z., Swinny, E.E., 2009. Postharvest nitric oxide fumigation delays fruit ripening and alleviates chilling injury during cold storage of Japanese plums (*Prunus salicina* Lindell). Postharvest Biol. Technol. 53 (3), 101–108.

Singh, Z., Khan, A.S., Zhu, S., Payne, A.D., 2013. Nitric oxide in the regulation of fruit ripening: challenges and thrusts. Stewart Postharvest Rev. 9 (4), 1–11.

Singh, N., Kaur, H., Yadav, S., Bhatla, S.C., 2016. Does N-Nitrosomelatonin compete with S-Nitrosothiols as a long distance nitric oxide carrier in plants? Biochem. Anal. Biochem. 5, 262.

Sun, Q., Zhang, N., Wang, J., Zhang, H., Li, D., Shi, J., et al., 2015. Melatonin promotes ripening and improves quality of tomato fruit during postharvest life. J. Exp. Bot. 66 (3), 657–668.

Sun, J.W., Janisiewicz, J., Nichols, B., Jurick II, W.M., Chen, P., 2017. Composition of phenolic compounds in wild apple with multiple resistance mechanisms against postharvest blue mold decay. Postharvest Biol. Technol. 127, 68–75.

Tang, J., Hu, K.D., Hu, L.Y., Li, Y.H., Liu, Y.S., Zhang, H., 2014. Hydrogen sulfide acts as a fungicide to alleviate senescence and decay in fresh-cut sweet potato. HortScience 49, 938–943.

Tijero, V., Muñoz, P., Munné-Bosch, S., 2019. Melatonin as an inhibitor of sweet cherries ripening in orchard trees. Plant Physiol. Biochem. 140, 88–95.

Tucker, G.A., Grierson, D., 1987. Fruit ripening. In: Davies, D. (Ed.), The Biochemistry of Plants., Vol. 12. Academic Press Inc, New York, pp. 265–319.

Tucker, G., Yin, X., Zhang, A., Wang, M., Zhu, Q., Liu, X., et al., 2017. Ethylene and fruit softening. Food Qual. Saf. 1, 253–267.

Vicente, A.R., Saladié, M., Rose, J., Labavitch, K.C., John, M., 2007. The linkage between cell wall metabolism and fruit softening: looking to the future. J. Sci. Food Agric. 87, 1435–1448.

Vitalini, S., Gardana, C., Zanzotto, A., Fico, G., Faoro, F., Simonetti, P., et al., 2011. From vineyard to glass: agrochemicals enhance the melatonin and total polyphenol contents and antiradical activity of red wines. J. Pineal Res. 51 (3), 278–285.

Wang, Y., Li, L., Cui, W., Xu, S., Shen, W., Wang, R., 2012. Hydrogen sulfide enhances alfalfa (*Medicago sativa*) tolerance against salinity during seed germination by nitric oxide pathway. Plant Soil 351 (1–2), 107–119.

Wang, Y., Luo, Z., Khan, Z.U., Mao, L., Ying, T., 2015. Effect of nitric oxide on energy metabolism in postharvest banana fruit in response to chilling stress. Postharvest Biol. Technol. 108, 21–27.

Wang, F., Zhang, X., Yang, Q., Zhao, Q., 2019. Exogenous melatonin delays postharvest fruit senescence and maintains the quality of sweet cherries. Food Chem. 301, 125311.

Wang, S.Y., Shi, X.C., Wang, R., Wang, H.L., Liu, F., Laborda, P., 2020. Melatonin in fruit production and postharvest preservation: a review. Food Chem. 126642.

Wu, B., Guo, Q., Li, Q., Ha, Y., Li, X., Chen, W., 2014. Impact of postharvest nitric oxide treatment on antioxidant enzymes and related genes in banana fruit in response to chilling tolerance. Postharvest Biol. Technol. 92, 157–163.

Xu, T., Chen, Y., Kang, H., 2019. Melatonin is a potential target for improving post-harvest preservation of fruits and vegetables. Front. Plant Sci. 10. Available from: https://doi.org/10.3389/fpls.2019.01388.

Yan, B., Zhang, Z., Zhang, P., Zhu, X., Jing, Y., Wei, J., 2019. Nitric oxide enhances resistance against black spot disease in muskmelon and the possible mechanisms involved. Sci. Hortic. 256, 108650.

Yang, S.F., Hoffman, N.E., 1984. Ethylene biosynthesis and its regulation in higher plants. Annu. Rev. Plant Physiol. 35 (1), 155–189.

Yang, H., Wu, F., Cheng, J., 2011. Reduced chilling injury in cucumber by nitric oxide and the antioxidant response. Food Chem. 127, 1237–1242.

Yonggen, S., Wei, W., Weil, Z., Liqin, Z., 2015. Hydrogen sulfide facilitating enhancement of antioxidant ability and maintenance of fruit quality of kiwifruits during low-temperature storage. Transac. Chin. Soc. Agric. Eng. 31, 367–372.

Zaharah, S.S., Singh, Z., 2011a. Postharvest nitric oxide fumigation alleviates chilling injury, delays fruit ripening and maintains quality in cold-stored 'Kensington Pride' mango. Postharvest Biol. Technol. 60 (3), 202–210.

Zaharah, S.S., Singh, Z., 2011b. Mode of action of nitric oxide in inhibiting ethylene biosynthesis and fruit softening during ripening and cool storage of 'Kensington Pride' mango. Postharvest Biol. Technol. 62 (3), 258–266.

Zhai, R., Liu, J., Liu, F., Zhao, Y., Liu, L., Fang, C., 2018. Melatonin limited ethylene production, softening and reduced physiology disorder in pear (*Pyrus communis* L.) fruit during senescence. Postharvest Biol. Technol. 139, 38–46.

Zhang, W., Jiang, W., 2019. UV treatment improved the quality of postharvest fruits and vegetables by inducing resistance. Trends Food Sci. Technol. 92, 71–80.

Zhang, X., Shen, L., Li, F., Meng, D., Sheng, J., 2011. Methyl salicylate-induced arginine catabolism is associated with up-regulation of polyamine and nitric oxide levels and improves chilling tolerance in cherry tomato fruit. J. Agric. Food Chem. 59 (17), 9351–9357.

Zhang, T., Che, F., Zhang, H., Pan, Y., Xu, M., Ban, Q., 2017. Effect of nitric oxide treatment on chilling injury, antioxidant enzymes and expression of the CmCBF1 and CmCBF3 genes in cold-stored Hami melon (*Cucumis melo* L.) fruit. Postharvest Biol. Technol. 127, 88–98.

Zhang, W., Cao, J., Fan, X., Jiang, W., 2020. Applications of nitric oxide and melatonin in improving postharvest fruit quality and the separate and crosstalk biochemical mechanisms. Trend. Food Sci. Technol. 99, 531–541.

Zhang, Y., Huber, D.J., Hu, M., Jiang, G., Gao, Z., Xu, X., et al., 2018. Delay of postharvest browning in litchi fruit by melatonin via the enhancing of antioxidative processes and oxidation repair. J. Agric. Food Chem. 66 (28), 7475–7484.

Zhang, Z., Xu, J., Chen, Y., Wei, J., Wu, B., 2019. Nitric oxide treatment maintains postharvest quality of table grapes by mitigation of oxidative damage. Postharvest Biol. Technol. 152, 9–18.

Zhao, R., Sheng, J., Lv, S., Zheng, Y., Zhang, J., Yu, M., 2011. Nitric oxide participates in the regulation of LeCBF1 gene expression and improves cold tolerance in harvested tomato fruit. Postharvest Biol. Technol. 62 (2), 121–126.

Zhao, Y., Tan, D.X., Lei, Q., Chen, H., Wang, L., Li, Q.T., et al., 2013. Melatonin and its potential biological functions in the fruits of sweet cherry. J. Pineal Res. 55 (1), 79–88.

Zhao, Y., Zhu, X., Hou, Y., Wang, X., Li, X., 2019. Effects of nitric oxide fumigation treatment on retarding cell wall degradation and delaying softening of winter jujube (*Ziziphus jujuba* Mill. cv. Dongzao) fruit during storage. Postharvest Biol. Technol. 156, 110954.

Zheng, J.L., Hu, L.Y., Hu, K.D., Wu, J., Yang, F., Zhang, H., 2016. Hydrogen sulfide alleviates senescence of fresh-cut apple by regulating antioxidant defense system and senescence-related gene expression. HortScience 51, 152–158.

Zheng, X., Hu, B., Song, L., Pan, J., Liu, M., 2017. Changes in quality and defense resistance of kiwifruit in response to nitric oxide treatment during storage at room temperature. Sci. Hortic. 222, 187–192.

Zhi, H., Dong, Y., 2018. Effect of hydrogen sulfide on surface pitting and related cell wall metabolism in sweet cherry during cold storage. J. Appl. Bot. Food Qual. 91, 109–113.

Zhou, Y., Lia, S., Zenga, K., 2015. Exogenous nitric oxide-induced postharvest disease resistance in citrus fruit to *Colletotrichum gloeosporioides*. J. Sci. Food Agric. 96, 505–512.

Zhu, S., Liu, M., Zhou, J., 2006. Inhibition by nitric oxide of ethylene biosynthesis and lipoxygenase activity in peach fruit during storage. Postharvest Biol. Technol. 42 (1), 41–48.

Zhu, L., Wang, W., Shi, J., Zhang, W., Shen, Y., Du, H., et al., 2014. Hydrogen sulfide extends the postharvest life and enhances antioxidant activity of kiwifruit during storage. J. Sci. Food Agric. 94, 2699–2704.

Zhu, L., Du, H., Wang, W., Zhang, W., Shen, Y., Wan, C., et al., 2019. Synergistic effect of nitric oxide with hydrogen sulfide on inhibition of ripening and softening of peach fruits during storage. Sci. Hortic. 256, 108591.

Ziogas, V., Molassiotis, A., Fotopoulos, V., Tanou, G., 2018. Hydrogen sulfide: a potent tool in postharvest fruit biology and possible mechanism of action. Front. Plant Sci. 9, 1375.

Chapter 3

Role of hydrogen sulfide in alleviating oxidative stress in plants through induction of antioxidative defense mechanism, and modulations of physiological and biochemical components

Monika Patel[1,2] and Asish Kumar Parida[1,2,*]
[1]Plant Omics Division, CSIR-Central Salt and Marine Chemicals Research Institute (CSIR-CSMCRI), Gijubhai Badheka Marg, Bhavnagar, India, [2]Academy of Scientific and Innovative Research, Ghaziabad, India
**Corresponding author.*

3.1 Introduction

Plants often encounter multiple abiotic stresses such as extreme temperature, salinity, cold, water deficit, flooding, UV radiation, and heavy metal (HM) toxicity. Abiotic stresses mainly cause a water deficit and create osmotic stress which reduces water accessibility for crucial cellular processes and turgor pressure management. This leads to the disruption of cellular and ionic homeostasis and free radical production in the mitochondria, chloroplasts, and peroxisomes (Kaur et al., 2019). Abiotic stresses are mainly associated with overproduction of ROS, RNS, and RSS, which include molecules like superoxide ($O_2^{\bullet -}$), hydrogen peroxide (H_2O_2), hydroxyl radicals (OH^-), nitrogen oxide (NO), peroxynitrite ($ONOO^-$), disulfide-S-oxides [RS(O)2SR], thiyl radicals ($RS^\bullet$), and sulfenic acid (RSOH) (Kaur et al., 2019). These reactive species causes oxidative injury and degradation of biological molecules, such as lipid, protein, nucleic acid, pigments, and carbohydrates, causing membrane disruption and cell death. Plants protect themselves from oxidative damage

Hydrogen Sulfide in Plant Biology. DOI: https://doi.org/10.1016/B978-0-323-85862-5.00006-3

mainly by enzymatic and nonenzymatic antioxidative defense mechanisms. Contrary to previous acceptance, free radicals possess a double-sided role, on one hand they are toxic metabolite by-products, whilse on the other hand they serve as signaling molecules against various abiotic stresses like salinity (Ding et al., 2009), UV-B exposure (Mackerness et al., 2001), drought (Xiang et al., 2008), and HMs (Zhang et al., 2010b) and play a role in defense mechanisms against pathogens (Hancock et al., 2002). Furthermore, free radicals such as reactive oxygen and nitrogen species act as signaling molecules in cell expansion, growth, and development. Being sessile, plants have developed smart and complex protective mechanisms to quickly respond and adapt to sudden changes in the environment (Yao et al., 2019). In plants, specific stress tolerance generally depends on the activation and upregulation of gene encoding antioxidant enzymes. Besides, plants produce various secondary metabolites, hormones, and signaling molecules to cope with environmental stresses where certain enzymes along with specific genes are activated by various gasotransmitters to improve stress tolerance (Corpas, 2019). Gasotransmitters are small gaseous molecules produced in living organisms including plants (Bhuyan et al., 2020). The various gasotransmitters include hydrogen gas (H_2), nitric oxide (NO), hydrogen sulfide (H_2S), methane (CH_4), and carbon monoxide (CO) which each have unique effects and control specific functions. These signaling molecules function in plant growth, development, and senescence regulation and also provide resistance to adverse environmental stresses (Aroca et al., 2018; Corpas, 2019). Recently, the study of gasotransmitters in plants has gained much attention, particularly for abiotic stresses. These signaling molecules can enhance antioxidant defense systems, relieve ROS toxicity, and play important roles in synergistic interactions with phytohormones and other protective molecules to improve plant stress tolerance (Yao et al., 2019).

The novel gaseous messenger H_2S is lipophilic and capable of mediating effective response to various environmental stresses. H_2S has recently evolved as a multifunctional cell signaling molecule in animals as well as in plants (Santisree et al., 2019). H_2S has been regarded as a third gaseous transmitter along with NO and CO (Wang et al., 2012). These small gaseous molecules are highly permeable, so they easily cross the membrane for signal transduction. In recent years, significant attention has been paid for the exploration of the function and cellular interaction of H_2S with other molecules. Several studies have documented the protective function of H_2S against oxidative stress, eliciting plant resistance to various abiotic stresses ranging from water deficit (Jin et al., 2011) to HMs (Xiang et al., 2008), salinity (Christou et al., 2013), hypoxia (Cheng et al., 2013), heat (Li et al., 2013b), chilling (Liu et al., 2019), and osmotic stress (Jia et al., 2018). Recently, it has been reported that H_2S may alter the structure and activities of target proteins by persulfidation of reactive cysteine (Aroca et al., 2017). H_2S plays a regulatory role in the growth and development of plants, e.g., in seed germination, senescence in leaves, flowers and fruits, root organogenesis, stomatal movement, fruit ripening, and enhancing photosynthesis (Corpas and

Palma, 2020; Liu et al., 2020a). Moreover, H_2S also maintains postharvest quality and rhizobium–legumes symbiosis (Liu et al., 2020a). In plants, the role of H_2S in sulfur assimilation has been recognized for a long time (Calderwood and Kopriva, 2014). In *Arabidopsis* H_2S is known to inhibit autophagy through transcriptional modulation (Laureano-marín et al., 2016). Moreover, exogenous H_2S appears to enhance different component of antioxidative defense such as isozymes of SOD, catalase (CAT), and the AsA-GSH cycle which enables the reduction of H_2O_2 content and lipid peroxidation (Corpas and Palma, 2020). A low concentration of H_2S promotes seed germination and growth in some plant species, whereas high doses have a phytotoxic effect, hence regulation of H_2S homeostasis is important to protect normal cellular functions against stress conditions. Several evidences have proposed that hydrogen sulfide and reactive oxygen species (ROS) act both antagonistically and synergistically in order to regulate abiotic stress responses like protein lipid peroxidation and oxidation in plants (Corpas, 2019). These lead to the synchronized activation of antioxidative defense mechanisms. Antioxidative enzymes, such as superoxide dismutase (SOD), CAT, ascorbate peroxidase (APX), and glutathione reductase (GR), and nonenzymatic antioxidants, like ascorbic acid and reduced glutathione (GSH), participate in protecting plants from environmental stresses. These findings clearly indicate the relationship between ROS and H_2S. Root nodules of legumes are highly rich in strong reducing molecules, nitrogenase, and polyunsaturated fatty acids which promote the generation of ROS. To scavenge ROS produced due to oxidative damage, antioxidant defenses are crucial to regulate nitrogen fixation in the root nodules. Zou et al. (2020) reported that exogenously applied H_2S may function as a positive regulator of N fixation in soybean. H_2S treatment alleviates the ROS burst and cell damage by modifying the metabolism of CAT, GR, POD, and GSH antioxidant pool and redox state in Bermuda grass under salt, chilling, and osmotic stress conditions (Shi et al., 2013). The biosynthesis of various organic solutes (proline, soluble sugars, and sucrose) for osmotic adjustment is an important protective mechanism in plants against abiotic stress conditions. These osmolytes stabilize quaternary protein structure, maintain membrane stability, and allow osmotic adjustment. H_2S treatment leads to a significant accumulation of osmolytes under drought, salinity, and freezing conditions in *Cynodon dactylon*, suggesting the positive role of H_2S in the accumulation of organic solutes contributing to stress adaptation (Shi et al., 2013). Ascorbic acid and GSH are two important antioxidants which protect the plant against oxidative stress. Plants can regulate the GSH and ascorbate levels by modulating the biosynthesis and regeneration of GSH and ascorbate (Shan et al., 2011). Moreover, GSH is abundantly present in the cell and maintains cellular redox status. H_2S improves oxidative injury by regulating the GSH and ascorbate metabolism in plants against abiotic stress. Exogenous H_2S have been found to increase ascorbate and GSH metabolism and maintain redox states of GSH and ascorbate enhancing salinity tolerance in the maize (Shan et al., 2013).

Recently, hydrogen sulfide gained greater attention due to its involvement in several cellular processes. Therefore it is necessary to study its various attributes including H_2S biosynthesis and its possible involvement in various signaling pathways to alleviate various abiotic stresses. The present chapter discusses an outline of the biosynthesis of H_2S and its mechanism of action as a signaling molecule to regulate numerous physiological processes in plants. Furthermore, this chapter will highlight the efficient role of H_2S in the alleviation of various abiotic stresses by modulating antioxidant capability to manage ROS toxicity and cellular homeostasis under adverse environmental stress conditions. Moreover, this chapter will also add the beneficial effects of H_2S application in plants against environmental stress conditions.

3.2 Biosynthesis of hydrogen sulfide in plant cells under adverse environmental conditions

Significant evidence suggests that a low level of H_2S can possess signaling characteristics during the adaptation of plants to various unfavorable conditions, and for growth and development and certain physiological function by interacting with hormones, ROS, and some other signaling compounds like nitric oxide (Ziogas et al., 2018). Sulfur is a crucial macromolecule required by all plants for growth and development. Plants uptake sulfur from the soil predominantly in the form of SO_4^{2-} (Davidian and Kopriva, 2010). Sulfur and its derivatives like hydrogen sulfide, GSH, methionine, cystine, phytochelatin, protein thiol, and adenosine triphosphate (ATP) sulfurylase have been reported to enhance the antioxidant defense mechanism to scavenge excess ROS in response to various abiotic stresses. Sulfur improves the activities of APX, GR, CAT, GSH, ascorbate, nonprotein thiols, and phytochelatin content (Davidian and Kopriva, 2010). In plants, the biosynthesis of H_2S is mainly contributed by different enzymatic pathways. Five enzymatic systems have been recorded for H_2S metabolism in plant cells: LCD (EC 4.4.1.1), DCD (EC 4.4.1.15), CAS (EC 4.4.1.9), SiR (EC 1.8.7.1), and cysteine synthase [(CS) (EC 2.5.1.47)]. Among these, L-cysteine desulfhydrase (L-DES) plays a predominant function in H_2S synthesis, specifically under stress conditions in plants. (Cristiane and Modolo, 2017). In plant cells, H_2S is mainly found in various subcellular organelles where sulfur metabolism associated enzymes produce H_2S. In chloroplasts sulfite reductase catalyzes the reduction of sulfite to sulfide in the presence of ferredoxin in the sulfate assimilation pathway. DCD and LCD, which are found in cytosol, convert cystine to pyruvate and release H_2S and ammonia using pyridoxal phosphate as a cofactor. It has been documented that high NaCl concentration enhances the activity of LCD in tobacco and alfalfa (Da Silva et al., 2017; Lai et al., 2014). In mitochondria H_2S is synthesized by the action of β-cyanoalanine synthase (CAS) by converting cyanide to β-cyanoalanine at the cost of cysteine (Gotor et al., 2019). This activity helps the plants to maintain the cellular level of CN^- during the production of ethylene, as CN^- is a strong inhibitor of the mitochondrial respiratory chain (Cristiane and Modolo, 2017). Cysteine synthase (CS), which is present in chloroplasts, mitochondria, and cytoplasm, is mainly responsible for the production of O-acetyl L-serine and H_2S from L-cysteine and acetate. Moreover, the highest concentration of sulfide was reported in chloroplasts (Krueger et al., 2009). Da Silva et al. (2017) have documented the stimulated activity of CAS, CS, and the accumulation of H_2S under extreme NaCl in tobacco leaves.

3.3 Role of hydrogen sulfide as an emergent signaling molecule in plants

H_2S has long been known as an environmental toxin because of its unpleasant odor. The toxic nature of H_2S hinders the activity of cytochrome c oxidase (Santisree et al., 2019). The understanding of the role of H_2S in plants signaling has been progressing recently. Being a small gaseous molecule H_2S can simply cross the cell membrane and translocate to different tissues and organs. Additionally, the highly rich chemistry and lipophilic characteristic of H_2S allow it to control several physiological responses in plants. However, the significant role of H_2S as a gasotransmitter and the actual mode of action for the various functions of H_2S in plants is still awaiting explanation (Li et al., 2011). H_2S is a reactive molecule and associates with other signaling molecules in order to modulate their signal. The interaction of H_2S with phytohormones like abscisic acid (Scuff et al., 2014), gibberellin (Xie et al., 2014), and ethylene (Jing et al., 2012) has been reported. Moreover, H_2S also plays a role in hydrogen peroxide and NO pathways (Lisjak et al., 2011; Zhang et al., 2010a). Although several findings have revealed the role of H_2S as a signal transmitter, current knowledge proposes that the posttranslation modification of protein cysteine residues to persulfides is the important mechanism of H_2S-based cell signaling (Filipovic et al., 2018). Persulfidation can change the structure and function of protein and this modification controls the function of several proteins in plants. For example, a recent study has documented that increased levels of persulfidation of bHLH transcription factors may improve their binding capacity to the promoter of the main gene involved in the biosynthesis of cucurbitacin C, the vital secondary metabolite in cucumber for protection against abiotic stress condition (Liu et al., 2019). According to Filipovic et al. (2018), H_2S does not directly react with protein cysteine residues to form persulfides. Therefore oxidative stress at a suitable level may allow H_2S-induced signaling cascades via persulfidation under environmental stress conditions. Moreover, the complexity between H_2S and ROS can be explained by the control of plant guard cell function. The guard cells control the opening and closure of stomata during adverse environmental stress conditions like drought. Several studies have documented that H_2S and two isoforms of NADPH oxidase, which generate H_2O_2 in apoplast, are both important for ABA-dependent regulation of stomatal closure in *A. thaliana* (Scuff et al., 2014; Scuffi et al., 2018; Zhang et al., 2020). Other proteins which regulate stomatal activity, such as calcium-dependent protein kinase (CPK3 and CPK6) (Mori et al., 2006; Munemasa et al., 2018) and MAPKs (MPK3, MPK4, and MPK6) (Du et al., 2019; Gudesblat et al., 2005; Zhang et al., 2017), are targets for persulfidation and sulfenylation (Aroca et al., 2017; Huang et al., 2019). They are probable targets for understanding the mechanism of H_2O_2-mediated H_2S-dependent stomatal closure. Besides, some evidence has suggested the involvement of H_2O_2 with H_2S in polyamine-mediated adaptation from UV-B radiation in hulless barley (Li et al., 2016a,b,c). Álvarez et al. (2010) documented that OASTL (homolog of L-DES) can reversely synthesize cysteine which leads to H_2S production, thus managing H_2S homeostasis in *Arabidopsis* cell. Moreover, exogenous application of H_2S has been stated to improve the adaption capacity of plants to adverse environmental stress conditions like heat, drought, salt, flooding, cold, HM fungal infection, and UV-B by increasing H_2S signaling and altering different metabolic pathways (Filipovic and Jovanović, 2017; Hancock and Whiteman, 2014; Hossain et al., 2018; Li, 2013; Li et al., 2016a,b,c). These researches suggest the function of H_2S as a signal molecule, and its involvement in cross-adaption of stress tolerance in plants (Li et al., 2016a,b,c). In general, H_2S-induced stress tolerance crosstalks with signaling molecules like Ca^{2+}, NO, CO, and H_2O_2 and their interactions elevate plant stress tolerance by encouraging an adaptive alteration in gene expression, physiochemical reactions, growth and development, anatomical and morphological feature via diminishing oxidative stress, oxidative damage, membrane injury, protein denaturation, and maintaining ion homeostasis as well as nutrient balance (Li, 2019). Calcium is an efficient secondary messenger in the plant cell and is also involved in the signal transduction of H_2S. According to Fang et al. (2014), exogenous calcium enhances chromium-induced H_2S synthesis and reduces the toxic effects of chromium in *Setaria italic*. The crosstalk between H_2S and Ca^{2+} signaling plays a vital role in plant growth and development, reproduction, and adaption to various environmental stress conditions (Li, 2019). Another author reported that increases in endogenous ROS, NO, and Ca^{2+} levels can act as enhancement for the biosynthesis of H_2S (Jin and Pei, 2015). Therefore the function of H_2S as a signaling molecule is closely related with ROS, NO, and Ca^{2+}. In contrast, H_2S allows the influx of calcium into cytosol by opening calcium channels. Ca^{2+} along with calmodulin can function as an intermediate in the signal transduction of H_2S in genetic apparatus. Besides, NO and ROS can also act as a mediator in the recognition of H_2S effects. In general, signal transduction of H_2S perhaps depends on the concentration and presence of other signal molecules.

3.3.1 Interaction of H_2S with other signaling molecule to enhance antioxidative defense mechanism

Abiotic stress conditions induce oxidative stress which in turn increases ROS accumulation in plants. This suggests that abiotic stresses generally favor shifting between ROS generation and elimination. H_2S is the simplest thiol present in the cell which maintains ROS homeostasis in the signaling pathways. The activity of ROS scavenging enzymes such as APX, GR, DHAR, SOD, and CAT are increased by the treatment of H_2S. In vitro assays in *Arabidopsis* have revealed that in the presence of H_2S, the cytosolic APX-1 have been persulfidated at Cys32, positively altering its catalytic activity (Aroca et al., 2015). Several studies have demonstrated that NO functions in the regulation of antioxidative enzymes and stress responses. Like NO, H_2S can induce various antioxidant components in plants. Pretreatment of H_2S promotes SOD and CAT activity in soybean against water deficit-induced oxidative damage (Zhang et al., 2010a). According to Takahashi et al. (2011), an increase in GSH level signifies its role in alleviating action of H_2S on the oxidative stress generated by many abiotic stresses. GSH can eliminate excess ROS like H_2O_2 via the ascorbate–GSH pathway and $O_2^{\bullet-}$ by direct scavenging effects (Noctor et al., 2018). GSH is synthesized from glutamate, cysteine, and glycine by glutamate–cysteine ligase (GCL) and glutathione synthase (GS). It has been documented that in *Arabidopsis* both GCL and GS are the targets of persulfidation (Aroca et al., 2017). Moreover, the expression of GCL and GS have been noted to be enhanced by pretreatment with H_2S in *Medicago sativa* during cadmium stress (Cui et al., 2014).

3.4 Role of H_2S in acceleration of antioxidative mechanism in plants during abiotic stress condition

In plants, H_2S extensively interacts with other signaling molecules like NO, CO, H_2O_2, ROS, IAA, JA, GA, and ABA, subsequently forming a complex signaling network. Although a definite function of H_2S in cell signaling pathways and its relations with other signaling compounds are still not completely understood. As a new signaling molecule, H_2S can induce abiotic stress tolerance by enhancing an efficient antioxidative defense mechanism. It has been reported that in response to various abiotic stress conditions, the biosynthesis of H_2S and endogenous H_2S content increased by two to three folds to improve the stress-induced oxidative damage (Bhuyan et al., 2020). Likewise, treatment with exogenous H_2S also enhanced the endogenous level of H_2S and consequently activated stress-responsive gene expression (Bhuyan et al., 2020). Numerous evidence has proposed that the application of NaHS (H_2S donor) improves antioxidative defense and photosynthesis reducing H_2O_2 and MDA level in *Triticum aestivum* under salinity stress (Ding et al., 2019). Moreover, additional H_2S donors have been reported to enhance the survival rate in *Cynodon dactylon* under salinity, osmotic, and cold stress by reducing ion leakage, improving antioxidant response, and the accumulation of osmolytes (Shi et al., 2013).

3.4.1 Function of hydrogen sulfide in mitigation of drought stress in plants

Water deficit or drought is one of the most important environmental stresses that affects plant development and yield, thus limiting agriculture productivity. The shortage of available water or salt accumulation leads to cellular dehydration because of a decline in soil water potential causing drought stress (known as physiological drought). H_2S enhances the resistance of water stress in plants mainly in two ways. Firstly, it reduces the loss of water by inducing stomatal closure and secondly it modulates antioxidant enzymes activity (Li et al., 2017). In plants water deficit strictly affects all physiological function. Transpiration is an important pathway of water loss from plants regulated by stomata. Stomatal movement is regulated by various signaling components like Ca^{2+}, K^+, NO, CO, H_2O_2, and H_2S (Li, 2019). Numerous studies demonstrated the ameliorative role of NO and H_2S involved in plant responses after the application of H_2S during drought stress (Paul and Roychoudhury, 2020). Exogenous application of H_2S as an sodium hydrosulfide (NaSH) donor induces stomatal closure by regulating ATP-binding cassette transporters, suggesting the defensive role of H_2S in plants under water deficit (García-Mata and Lamattina, 2010). The antioxidative defense mechanism and osmotic adjustment maintains the water balance and ROS homeostasis in plants. The exogenous application of H_2S lowers ROS accumulation by modulating antioxidant enzyme activities under water deficit in wheat seedlings (Ma et al., 2016). Zhang et al. (2009) demonstrated that the spraying of NaSH promotes SOD, CAT, and APX activities while lowering the concentration of H_2O_2 in sweet potato seedlings alleviates oxidative damage against oxidative stress. Additionally, exogenous application of NaSH improves APX, dehydroascorbate reductase, GR activities, synthesis of gamma-glutamylcysteine, and also AsA, GSH, total ascorbate, and total GSH in *T. aestivum* seedlings, thus lowering electrolyte leakage and lipid peroxidation under drought condition (Shan et al., 2011). In *C. dactylon* exogenously applied NaSH has decreased the accumulation of H_2O_2 and $O_2^{\bullet -}$ by promoting the activities of CAT and APX and as a result alleviating oxidative injury (Santisree et al., 2019).

3.4.2 Hydrogen sulfide improves salinity tolerance in plants

Soil salinity is a global environmental concern because it negatively affects various physiological functions in plant. Excessive salt triggers osmotic stress, oxidative stress, ion imbalance, and toxicity in plants which directly affect various physiological and biological processes of the plant cell. To combat salt injury, plants develop various protective mechanisms such as synthesis of organic osmolytes (proline, glycine betaine, trehalose, and total soluble sugars) for osmotic adjustment; maintain translocation of ion and nutrient equilibrium; and the most important mechanism is enhancement of the antioxidant defense system (Li et al., 2016a,b,c). Salt stress generally influences osmotic stress which activates rapid production of secondary messengers like hydrogen sulfide. Salinity treatment of strawberry plants has greatly enhanced the accumulation of H_2S in leaves (Christou et al., 2013). Indeed, exogenously applied H_2S donor NaSH alleviates salt tolerance in plants. In plants salt stress-induced oxidative stress was correlated with the high production of ROS. Numerous antioxidant genes *APX-1, APX-2*, and *Cu/Zn SOD* were downregulated under saline condition. The application of NaSH could considerably inhibit salinity-mediated reduction of the transcript and antioxidant enzyme activity in alfalfa seed (Wang et al., 2012). NaSH treatment alleviates salinity-mediated oxidative damage by increasing the CAT and SOD activity in tomato (Da-Silva et al., 2018). Exogenously applied H_2S alleviated salinity-induced oxidative damage by promoting antioxidant enzymes activities including SOD, APX, CAT, GR, and DHAR in *Medicago sativa* (Lai et al., 2014), *Cucumis sativus* (Sun and Luo, 2014), rice (Mostofa et al., 2015b), *Nicotiana tabacum* (Da Silva et al., 2017), and maize (Shan et al., 2013).

3.4.3 Protective functions of H_2S in response to heavy metal and metalloid toxicity

With the development of advanced agriculture, industrial development, and urbanization, HM contamination has become a more serious abiotic stress (Jalmi et al., 2018). The use of phosphoric fertilizer, brackish water, application of sewage sludge, dust from the smelters, and disposal of industrial waste are the major human activities that contribute to HM pollution of soil (Yadav, 2010). HM toxicity negatively affects the plants in many ways, including at morphological, anatomical, biochemical, molecular, cellular, and physiological levels, that is, photosynthesis, transpiration, respiration, and reduction of plant growth and productivity (Li, 2019). The preliminary reaction of plants in response to HM toxicity is the generation of oxidative stress and generation of ROS. However, plants trigger antioxidant defense mechanisms for HM detoxification to scavenge excessive ROS and control homeostasis (Foyer and Noctor, 2011, 2009). In addition, plants also accumulate metal-binding nonprotein thiols such as GSH and phytochelatin (Cobbett, 2000; Yadav, 2010). A study conducted by Tiwari and Lata (2018) suggested that the biosynthesis of H_2S is a common reaction to metal toxicity. In plants, the function of H_2S under Cd, Cr, Al, Cu, and Pb stress have been reported (Hancock and Whiteman, 2015). Numerous studies have documented that pretreatment with NaSH enhances the metal toxicity mainly in the following ways: (1) exogenously applied H_2S alleviates oxidative damage by promoting the activity of antioxidant enzymes; (2) exogenous H_2S limits the entry of HM into the cell by H_2O_2-activated channels on the cell membrane; (3) by enhancing translocation of HM in vacuoles, thus reducing the accumulation of HM to nontoxic levels (Sun et al., 2013; Zhang et al., 2010c). Arsenic contamination increases the generation of ROS which disrupt membranes and macromolecules, limit photosynthesis, and reduce the nitrogen level. These adverse effects of arsenic toxicity have been improved by pr-treatment with NaSH. Arsenic induces a reduction in the reduced/oxidized ratio of AsA and GSH by increasing MDHAR, DHAR, and GR activities which are mainly involved in ascorbate–GSH cycle in pea plants. Furthermore, arsenic treatment disturbs the redox status of ascorbate and GSH, as observed by a decrease in their reduced/oxidized ratios. However, pretreatment of NaSH reestablished the redox status of ascorbate and GSH pools in response to arsenic treatment. Further, Singh et al. (2015) suggested that H_2S might be involved in the mitigation of ROS-induced macromolecule damage through the downregulation of As accumulation and enhance the regulation of the AsA-GSH cycle, thereby alleviating As stress in pea seedlings (Singh et al., 2015). Pretreatment of NaSH improves the activities of antioxidant enzymes, including SOD, CAT, APX, and guaiacol peroxidase by reducing lipoxygenase activity and the accumulation of H_2O_2 in *Triticum aestivum* seed under chromium toxicity (Zhang et al., 2010c). Moreover, H_2S alleviates Cr stress in cauliflower (Ahmad et al., 2019) and maize (Kharbech et al., 2017) by enhancing the antioxidant mechanism. H_2S modulates Cd-induced oxidative stress by enhancing AsA and GSH level, redox status, and upregulating various antioxidative enzymes' activity in *Oryza sativa* (Mostofa et al., 2015a,b). Exposure of H_2S ameliorates Cd stress by enhancing SOD, CAT, APX, and GR activity in *Brassica napus* (Ali et al., 2014). NaSH induces the mitigation of Cd toxicity by promoting APX, POD, and Cu/Zn-SOD activity in *M. sativus* seedlings (Li et al., 2012). Zhang et al. (2008) have documented that excess Cu triggers the accumulation of ROS which subsequently creates oxidative injury in wheat seeds and inhibits germination. However, exogenous application of NaSH reduces MDA content and electrolyte leakage, whereas it upregulates gene expression and enhances GR, DHAR activity, AsA, GSH, AsA + DHA, and GSH + GSSG content under Cu stress in *T. asetivum* seedlings (Shan et al., 2012). A similar protective role of H_2S has also been reported in *Coriandrum sativum* seedlings under Cu stress condition. Further, pretreatment of H_2S improves oxidative injury by maintaining the AsA and GSH metabolism under Cu toxicity (Karam and Keramat, 2017). Al is known to cause phytotoxic effects at the macromolecular level and severely affect plant growth in acidic soil (Horst et al., 2010). The primary symptoms of Al toxicity include root growth inhibition and iron scarcity. Al toxicity accumulates ROS in barley seedlings, causing oxidative injury. However, exposure of NaSH upregulates SOD and CAT activity to manage the oxidative stress under Al toxicity (Dawood et al., 2012). Further, the application of NaSH significantly alleviates oxidative damage by improving the antioxidant mechanism in *H. vulgare* (Chen et al., 2013) and *T. aestivum* seedlings (Zhang et al., 2010b). Valivand et al. (2019) demonstrated that seed priming with NaSH decreases Ni toxicity by enhancing antioxidant defense in zucchini seedlings. H_2S alleviates Hg toxicity by greatly activating SOD, POD, and APX in rice seedlings to scavenge Hg-induced $O_2^{\bullet-}$ and H_2O_2 (Chen et al., 2017).

3.4.4 Bioprotective nature of hydrogen sulfide on plant thermotolerance

High temperature (heat stress) is one of the major environmental stress factors that negatively affect crop development and productivity. High temperature induces oxidative burst, denaturation, and aggregation of proteins, increases membrane lipids fluidity, deactivates enzymes in mitochondria and chloroplasts, and downregulates protein synthesis, which creates secondary oxidative stress by the accumulation of ROS, subsequently leading to severe cellular injury and cell death (Paul and Roychoudhury, 2020). Consequently, plants evolved various protective strategies to survive under higher temperatures. In plants, major thermotolerance mechanisms include stimulation of the antioxidant system, biosynthesis of heat stress pretreated (HSP) and its accumulation, osmoprotectants, and secondary metabolite regulation for maintaining cellular homeostasis and restoring disrupted membranes and proteins (Santisree et al., 2019). Heat stress-induced endogenous production of H_2S has been documented in a few plant species. However, the mechanism of endogenous production of H_2S is poorly understood, and only limited studies have reported LCD-dependent biosynthesis of H_2S in tobacco. Chen et al. (2016) suggested that heat stress increased the endogenous production of H_2S by enhancing LCD activity in tobacco seedlings. However, NaSH treatment in wheat increased the generation of $O_2^{\bullet-}$ and H_2O_2 in turn NaSH stimulates the activity of SOD, CAT, and GPX to alleviate heat tolerance (Kolupaev et al., 2017). Besides some studies have found that Ca^{2+} signaling and H_2S signaling improves heat tolerance by the activation of various antioxidant components including SOD, APX, GPX, CAT, GR, ascorbate, and GSH to maintain redox homeostasis (Li, 2019). Under HSP, strawberry plants with NaSH upregulate the gene encoding AsA and GSH and maintain higher AsA/GSH homeostasis. In addition, exogenous NaSH also elevated the transcription level of enzymatic antioxidants like APX, CAT, SOD, and GR and protected the strawberry from the high-temperature damage (Christou et al., 2014). Foliar application of NaSH elevates the antioxidant enzyme activities, reduces MDA content, and alleviates thermotolerance in

Triticum aestivum (Yang et al., 2016). In *Zea mays*, NaSH treatment has elevated the activity of CAT, SOD, guaiacol peroxidase, GR, ascorbate, GSH content, AsA/(AsA + DHA) ratio, and GSH/(GSH + GSSG) ratio under heat stress condition (Li et al., 2014). Comparable effects have also been documented in *Nicotiana tabacum* cells (Li et al., 2016a,b,c).

3.4.5 Protective function of hydrogen sulfide under cold stress

Cold stress occurs when plants are exposed to chilling temperature ($>0°C$) and freezing temperature ($<0°C$). Cold stress often results in oxidative stress, osmotic stress, and dysfunction of cell membranes. Plants can combat cold stress damage by promoting the antioxidant mechanism and osmotic adjustment (Foyer and Noctor, 2011, 2009). Various studies have demonstrated that exogenously applied methyl jasmonate and salicylic acid alleviate cold stress damage in plants. However, some evidence has revealed the role of H_2S in the alleviation of cold stress damage (Cao et al., 2010; Wang et al., 2009). Fu et al. (2013) found that low temperature rapidly enhances the content of H_2S in *Vitis vinifera*. The major production of H_2S has increased SOD activity and reduces $O_2^{\bullet -}$ and MDA levels and membrane porosity by the application of exogenous NaSH in *Vitis vinifera* in response to cold stress. Meanwhile, another study on chilling stress demonstrated that exogenously applied NaSH induces cold tolerance and increases survival against freezing condition by relieving cold stress-induced oxidative stress. NaSH treatment regulates GPX, CAT, and GR activity and GSH pool to mitigate electrolyte leakage and cellular injury and to maintain redox status in *C. dactylon* (Shi et al., 2013). Further, the function of H_2S under chilling stress has been reported in diverse plants species. Luo et al. (2015) demonstrated that hydrogen sulfide elevates the activity of SOD, CAT, APX, guaiacol peroxidase, and GR, reducing the accumulation of $O_2^{\bullet -}$ and H_2O_2 and improving the antioxidant potential of banana against chilling stress. Moreover, it has been described that the activity of enzymes and proteins involved in H_2S production significantly improved at higher altitudes, thus maintaining the greater level of H_2S in the alpine plants grown on the Northern Tibetan Plateau. Pretreatment with H_2S has elevated the activity of antioxidant enzymes and GSNOR (*S*-nitrosoglutathione reductase) and improved oxidative damage, protein degradation, and accumulation of osmolytes (Ma et al., 2015). The biosynthesis of H_2S and its abiotic stress tolerance mechanisms via ionic and osmotic homeostasis, cell signaling, gene regulation, and induction of antioxidative defense systems is demonstrated Fig. 3.1.

3.5 Role of H_2S in ionic and osmotic homeostasis in plants under various abiotic stresses

It has been well established that high salinity leads to several negative influences on plant cells like ion toxicity, osmotic stress, and ultimately nutrient

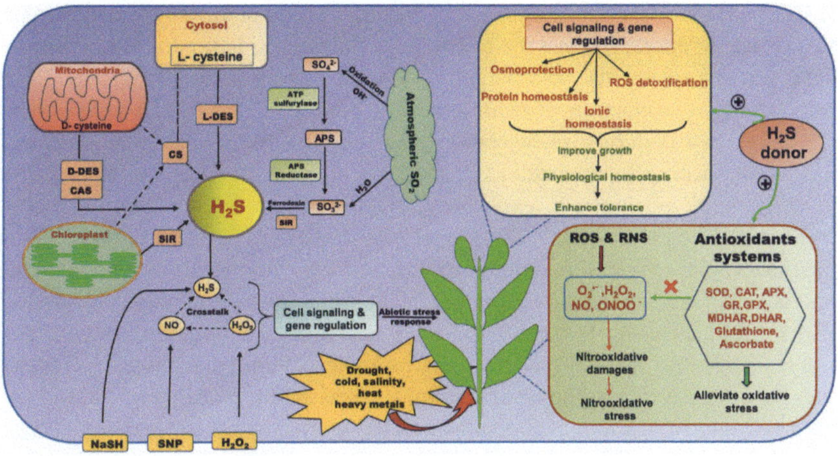

FIGURE 3.1 Schematic diagram showing the biosynthesis of H_2S and role of hydrogen sulfide in abiotic stress tolerance through ionic and osmotic homeostasis, cell signaling, gene regulation, and ROS detoxification. APS, adenosine 5′-phosphosulfate; APX, ascorbate peroxidase; ATP, adenosine triphosphate; CAT, catalase; CAS, β- cyanoalanine synthase; CS, cysteine synthase; D-DES, D-cysteine desulfhydrase; DHAR, dehydroascorbate reductase; GPX, guaiacol peroxidase; GR, glutathione reductase; H_2S, hydrogen sulfide; H_2O_2, hydrogen peroxide; L-DES, L-cysteine desulfhydrase; MDHAR, monodehydroascorbate reductase; NO, nitric oxide; NaSH, Sodium hydrosulfide; $O_2^{\bullet-}$, superoxide anion radical; $ONOO^-$, peroxynitrite; ROS, reactive oxygen species; RNS, reactive nitrogen species; SiR, sulfite reductase; SNP, sodium nitroprusside; SOD, superoxide dismutase; SO_3^{2-}, sulfite; SO_4^{2-}, sulfate.

disorder. The regulation of ion flux is important for cells to minimize the toxic ion concentration and accumulate essential ions. The plant cell maintains the K^+/Na^+ ratio in the cytosol via various transporters, as intracellular K^+ and Na^+ homeostasis is crucial for activities of cytosolic enzymes, to maintain membrane potential, and to exert osmotic adjustment under salt stress (Zhu, 2003). The efficient removal of excess Na^+ from the vacuoles and cytoplasm is crucial for plants to tolerate salt stress. According to Wang et al. (2012) pre-application of H_2S improves the salt stress-induced increase of Na^+/K^+ and enhances the K^+/Na^+ ratio in *Medicago sativa*. Moreover, the exogenous application of NaSH has been reported to repress K^+ loss from plant cells by downregulating SKOR (shaker-like K^+ outward-rectifying K channel) expression and thus maintaining K^+/Na^+ homeostasis in alfalfa seeds (Wang et al., 2012). The regulatory function of H_2S on ion homeostasis has been further reported in *Arabidopsis thaliana* root, where H_2S-mitigated salt stress-induced the increase of the Na^+/K^+ ratio and improved growth (Li et al., 2014). Christou et al. (2013) reported the contribution of H_2S in regulating *SOS4* expression and the uptake of K^+ under salt stress. Pretreatment of H_2S also enhances gene expression and phosphorylation of H^+ ATPase and Na^+/K^+ antiporter protein. Numerous studies have reported the exogenous H_2S-mediated induction of osmoregulators in *Oryza sativa* (Mostofa et al., 2015a,b),

C. dactylon (Sun and Luo, 2014), and *C. sativus* (Shi et al., 2013) under salinity. All these reports suggest the role of H_2S in the modulation of redox homeostasis by enhancing the antioxidative system, maintenance of high K^+/Na^+ ratio, and overproduction of osmolytes (Da Silva et al., 2017; Li et al., 2014; Shi et al., 2013) during plant response to salinity. Moreover, pretreatment of H_2S has been documented to increase the K^+/Na^+ homeostasis in *H. vulgare*, *T. aestivum*, *P. euphratica*, and *P. popularis* under high salt concentration (Chen et al., 2015; Deng et al., 2016; Zhao et al., 2018). Under stress conditions, the production and accumulation of osmolytes (proline, glycine betaine, sucrose, and soluble sugars) is an important mechanism to protect plants through balancing osmotic pressure and maintaining cell membrane stability. Exogenous application of NaSH promotes overproduction of the osmolytes to enhance water deficit and salinity stress in Bermuda grass (Shi et al., 2013). Similarly, additional H_2S enhances the production of proline and glycine betaine in wheat seedlings, thereby protecting the plants against drought stress (Khan et al., 2017). It has been also recorded that exogenous NaSH promoted phenylalanine ammonia lyase activity, proline, and total phenolic content in *Musa acuminate* to improve cold stress-induced damage (Luo et al., 2015). Moreover, pretreatment of NaSH enhances proline biosynthesis and lowers proline degradation under heat stress, thereby improving heat tolerance in maize seedlings (Li, 2013).

3.6 Interaction of H_2S and physiological responses in plants imposed to various abiotic stresses

H_2S has been reported to play an important role in numerous physiological functions from seed germination to the maturation and senescence of flowers in plants. Several studies have addressed the role of H_2S in guard cell signaling in *Arabidopsis* (García-Mata and Lamattina, 2010; Lisjak et al., 2010), promoting root organogenesis *I. batatas*, *S. matsudana*, and *G. max* (Zhang et al., 2009) and adventitious root formation in *Cucumis sativus* (Li et al., 2012). Pretreatment with H_2S improves the rate of germination by enhancing the activity of endosperm β-amylase and maintaining low MDA and H_2O_2 in germinating seeds of wheat (Zhang et al., 2010a). Exogenous H_2S has been documented for enhancement of the rate of germination, vigor index, germination index, and seedling growth under salt stress (Li, 2013). Moreover, preapplication of H_2S contributes to redox homeostasis and eventually reduces salt-mediated hampering of seed germination and growth of *Medicago sativa* (Wang et al., 2012). Exogenous NaSH enhances the plant biomass and reduces drought-mediated chlorophyll loss in *Glycine max* under drought stress (Zhang et al., 2010b). Likewise, H_2S maintains the RWC and stomatal conductance in *F. ananassa* against drought stress (Christou et al., 2013). Furthermore, exogenous H_2S improves drought-mediated inhibition of seed germination, radical and coleoptiles growth by enhancing the amylase and esterase activity in *T. aestivum* seedlings under drought stress (Zhang et al., 2010c). Numerous studies have demonstrated H_2S production under HM stress condition in plants (Santisree et al., 2019). NaSH treatment promotes seed germination and seedling growth under Cr, Al, and Cu toxicity in *T. aestivum* (Zhang et al., 2010d). Moreover, the exogenous application of H_2S has been reported to alleviate B toxicity in *Cucumis sativus* (Wang et al., 2010) and Cd tolerance in *Medicago sativa* (Li et al., 2012). Furthermore, NaSH application in *T. aestivum* enhances amylase and esterase activity in radical tips and lowers MDA and H_2O_2 content in germinating seeds, thereby promoting the seed germination rate under Cu and Cd stress (Zhang et al., 2008, 2010c). H_2S plays an important role in abscisic acid signaling-mediated stomatal closure (García-Mata and Lamattina, 2010). According to Jin et al. (2011), the pretreatment of NaSH regulates stomatal closure by predominantly interacting with abscisic acid to enhance drought tolerance in *Arabidopsis thaliana.* H_2S-mediated stomatal closure under drought condition has also been recorded in other species, such as *Vicia faba* and *Impatiens walleriana* (Zhang et al., 2010b). NaSH pretreatment enhances seed germination and seedling survival rate in *Z. mays* seedlings against heat stress (Li, 2013). It is well-known that the photosynthesis rate can be increased by improving the activity of RuBISCO, which plays a fundamental role in controlling the process of photosynthesis. Chen et al. (2011) have addressed the enhancement of photosynthesis in *Spinacia oleracea* through promoting the expression and activity of RuBISCO and proteins involved in thiol redox modification. Prior treatment of NaSH to the plant root have increased the Fv/Fm ratio by alleviating heat stress damage to PSII activity. Moreover, exogenous H_2S has effectively improved reduction in leaf chlorophyll fluorescence, RWC, and stomatal conductance in strawberry plants in heat stress condition (Santisree et al., 2019). Pretreatment of NaSH minimizes the drought-induced chlorophyll loss in sweet potato during osmotic stress (Zhang et al., 2010b). Exogenous H_2S has been reported to alleviate Al toxicity by enhancing the chlorophyll content, net photosynthetic rate, and maximal photochemical efficiency in *Hordeum vulgare* (Dawood et al., 2012). Similarly, the exogenous addition of NaSH has ameliorated As toxicity on photosynthesis and enhanced biomass production in pea plant in response to As stress (Singh et al., 2015). Mostofa et al. (2015a,b) have demonstrated the involvement of exogenously applied H_2S in the improvement of Cd-induced reduction in chlorophylls, carotenoids, and proteins contents, thereby enhancing overall growth and biomass in *Oryza sativa.* Furthermore, H_2S pretreatment has alleviated Ni-toxicity by enhancing plant growth and biomass, photosynthetic rate, internal CO_2 concentration, stomatal conductance, transpiration rate, photosynthesis, and pigment content in rice seedlings in response to Ni stress (Rizwan et al., 2019). Pretreatment of NaSH ameliorated heat stress-induced phenotypic damage and enhanced maximum photochemical PSII quantum yields in *F. ananassa* by activating heat stress defense-related pathways at transcriptional level in response to heat stress condition (Christou et al., 2014).

3.7 Conclusions and future prospective

As a signaling molecule hydrogen sulfide has a necessary role in various developmental responses and environmental stresses. The exogenous application of H_2S involves a signaling mechanism leading to an increase in various components of the antioxidant system at gene and protein levels. The presence of H_2S can promote plant growth and improve photosynthesis, activate antioxidant enzymes to maintain ROS level, and protect the structure of the plant cell. Moreover, it also enhances the gene responsible for phytochelatin synthesis and reduced oxidative damage by regulating GSH-based antioxidative defense. H_2S executes its antioxidative properties by enhancing ascorbate and GSH content and rebuilding the redox homeostasis. Furthermore, H_2S promotes SOD and CAT activity and balances ROS production and its detoxification, thus protecting the cell from oxidative injury. Hence, the pretreatment of H_2S has positive effects on diverse plant species under adverse environmental stress conditions. The use of H_2S donors like NaSH with appropriate concentration greatly enhance the plant's abiotic stress tolerance. Although the above stated evidence pinpoint the key function of H_2S as a signaling component and its role in regulation of antioxidant defense system under stress conditions, numerous questions remain unanswered for future research. In spite of the growth-promoting effects of H_2S, Lisjak et al. (2010) reported a negative response to H_2S for plant growth, nutrient uptake, and defoliation, thus it is necessary to understand the induction of H_2S' positive effects based on its cellular location and concentration. Besides this, new directions are needed for defining H_2S generation and the mechanism of downward signaling cascades prioritizing specific receptors and targets in plants under abiotic stress conditions. Most importantly, a more conclusive idea of H_2S-mediated stress amelioration needs the redesign of stress treatments to be nearer to physiological relevance (Tables 3.1 and 3.2).

TABLE 3.1 Involvement of hydrogen sulfide in the enhancement of enzymatic and nonenzymatic antioxidative systems to defend against oxidative damage in plants under various abiotic stresses. Upward (↑) and downward arrows (↓) respectively indicate up- and downregulations of different antioxidants' components and oxidative stress indicators.

Environmental stress	Plant species	Regulation of antioxidant components	References
Drought	*Trifolium repens*	SOD↑, CAT↑, POX↑, APX↑, DHAR↑, MDHAR↑, GR↑	Li et al. (2018)
	Triticum aestivum	APX↑, GR↑, POX↑, SOD↑, CAT↑, $O_2^{\bullet -}$↓, H_2O_2↓	Khan et al. (2017)

(Continued)

TABLE 3.1 (Continued)

Environmental stress	Plant species	Regulation of antioxidant components	References
	Fragaria ananassa	AsA and GSH redox state↑, MDA↓	Christou et al. (2013)
	Triticum aestivum	APX↑, GR↑, DHAR↑, γ-ECS↑, AsA↑, GSH↑, MDA↓, EC↓	Shan et al. (2011)
	Glycine max	SOD↑, CAT↑, LOX↓, MDA↓, $O_2^{•-}$↓, H_2O_2↓	Zhang et al. (2010a)
	Triticum aestivum	CAT↑, APX↑, LOX↓, MDA↓	Zhang et al. (2010c)
	Ipomea batatus	SOD↑, CAT↑, APX↑, LOX↓, MDA↓, H_2O_2↓	Zhang et al. (2009)
	Spinach oleracea	GSH↑	Kok et al. (1985)
Salinity	*Solanum lycopersicum*	CAT↑, SOD↑	Da-Silva et al. (2018)
	Nicotiana tabacum	SOD↑, CAT↑, APX↑	Da Silva et al. (2017)
	Oryza sativa	SOD↑, CAT↑, AsA↑, GSH↑, GR↑, GPX↑	Mostofa et al. (2015a,b)
	Cucumis sativus	SOD↑, POD↑, MDA↓	Sun and Luo (2014)
	Cynodon dactylon	CAT↑, POD↑, GR↑, AsA↑ and GSH pool↑, EC↓	Shi et al. (2013)
	Fragaria ananassa	AsA and GSH redox state↑, MDA↓	Christou et al. (2013)
	Zea mays	AsA↑, GSH↑, APX↑, GR↑, DHAR↑, γ-ECS↑, MDA↓, EC↓	Shan et al. (2013)
	Medicago sativa	SOD↑, CAT↑, POD↑, APX↑	Wang et al. (2012)
Heavy metal (HM)/metalloids	*Solanum lycopersicum* (B toxicity)	SOD↑, CAT↑, POD↑, EC↓, MDA↓, H_2O_2↓	Kaya et al. (2020)
	Brassica oleracea (Cr toxicity)	SOD↑, CAT↑, POD↑, APX↑, MDA↓, EC↓, H_2O_2↓	Ahmad et al. (2019)
	Cucurbita pepo (Ni toxicity)	AsA↑, total thiol↑, APX↓, GR↑, EC↓, H_2O_2↓	Valivand et al. (2019)
	Zea mays (Pb toxicity)	GSH↑	Zanganeh et al. (2019)
	Brassica oleracea (Pb toxicity)	SOD↓, CAT↓, POD↓, APX↓, GR↓, MDA↓, $O_2^{•-}$↓, H_2O_2↓	Chen et al. (2018)

(Continued)

TABLE 3.1 (Continued)

Environmental stress	Plant species	Regulation of antioxidant components	References
	Capsicum annuum (Zn toxicity)	SOD↑, CAT↑, POD↑, EC↓, MDA↓, H_2O_2↓	Kaya et al. (2018)
	Coriandrum sativum (Cu toxicity)	GR↑, AsA↑, DHA↑, GSH↑, MDA↓, EC↓	Karam and Keramat (2017)
	Oryza sativa (Hg toxicity)	SOD↑, CAT↑, nonprotein thiols↑, metallothioneins↑, MDA↓, H_2O_2↓	Chen et al. (2017)
	Zea mays (Cr toxicity)	SOD isoenzymes↑, CAT↑, MDA↓,	Kharbech et al. (2017)
	Oryza sativa (Cd toxicity)	SOD↑, CAT↑, GPX, DHAR↑, GR↑, AsA↑, GSH↑	Mostofa et al. (2015a,b)
	Pisum sativum (As toxicity)	APX↑, DHAR↑, MDHAR↑, GR↑, MDA↑, $O_2^{\bullet-}$↓, H_2O_2↓, •OH↓	Singh et al. (2015)
	Brassica napus (Cd toxicity)	SOD↑, POD↑, CAT↑, APX↑, GR↑, MDA↓, $O_2^{\bullet-}$↓, H_2O_2↓	Ali et al. (2014)
	Brassica napus (Al toxicity)	CAT↑, SOD↑, GR↑, APX↑, POD↑, MDA↓, H_2O_2↓	Qian et al. (2014)
	Populus euphratica (Cd toxicity)	SOD↑, CAT↑, GR↑, MDA↓, H_2O_2↓	Sun et al. (2013)
	Triticum aestivum (Al toxicity)	POD↓, CAT↓, APX↓, SOD↓, GSH↓	Chen et al. (2013)
	Hordeum vulgare (Al toxicity)	CAT↑, POD↑, MDA↓	Dawood et al. (2012)
	Medicago sativa (Cd toxicity)	SOD↑, POD↑, APX↓, MDA↓	Li et al. (2012)
	Triticum aestivum (Cu toxicity)	GR↑, DHAR↑, γ-ECS↑, AsA↑, GSH↑, MDA↓, EC↓	Shan et al. (2012)
	Triticum aestivum (Al toxicity)	GPX↑, APX↑, SOD↑, CAT↑, LOX↓, MDA↓, H_2O_2↓	Zhang et al. (2010b)
	Triticum aestivum (Cr toxicity)	SOD↑, CAT↑, APX↑, GPX↑, LOX↓, MDA↓, H_2O_2↓	Zhang et al. (2010c)

(Continued)

TABLE 3.1 (Continued)

Environmental stress	Plant species	Regulation of antioxidant components	References
Heat	*Populus trichocarpa*	MDHAR↑, DHAR↑, APX↑, GR↑, MDA↓	Cheng et al. (2018)
	Triticum aestivum	SOD↑, CAT↑, GPX↑	Kolupaev et al. (2017)
	Nicotiana tabacum	SOD↑, GR↑, APX↑, CAT↑	Li et al. (2016a, b,c)
	Fragaria ananassa	CAT↑, cAPX↑, MnSOD↑, GR↑, MDA↓, H_2O_2↓	Christou et al. (2014)
	Zea mays	CAT↑, GPX↑, SOD↑, GR↑, APX↑, AsA↑, GSH↑, MDA↓, H_2O_2↓	Li et al. (2014)
Cold	*Cucumis sativus*	GSH↑, GSH/GSSG↑, APX↑, GR↑, MDHAR↑	Liu et al. (2020b)
	Lamiophlomis rotata	APX↑, SOD↑, CAT↑, DHAR↑, GSH↑, MDA↓, H_2O_2↓	Ma et al. (2015)
	Musa acuminata	SOD↑, CAT↑,GPX↑, APX↑, GR↑, MDA↓, $O_2^{\bullet-}$↓, H_2O_2↓	Luo et al. (2015)
	Cynodon dactylon	CAT↑, POD↑, GR↑, EC↓, MDA↓, $O_2^{\bullet-}$↓, H_2O_2↓	Shi et al. (2013)
	Vitis vinifera	SOD↑, $O_2^{\bullet-}$↓, MDA↓	Fu et al. (2013)

TABLE 3.2 Role of hydrogen sulfide in modulating various physiological and biochemical processes in plants under different abiotic stress conditions. Upward (↑) and downward arrows (↓), respectively, indicate the increase or decrease of various physiological and biochemical components.

Environmental stress	Plant species	Plant responses	Reference
Drought	*Triticum aestivum*	Glycine betaine↑, proline↑, RWC↑	Khan et al. (2017)
	Fragaria ananassa	Chlorophyll fluorescence ↑, stomatal conductance↑, RWC↑	Christou et al. (2013)
	Arabidopsis thaliana	Drought associated genes↑, stomatal aperture↓	Jin et al. (2011)
	Arabidopsis thaliana	H_2S synthesis↑, stomatal closure↑	García-Mata and Lamattina (2010)

(Continued)

TABLE 3.2 (Continued)

Environmental stress	Plant species	Plant responses	Reference
	Glycine max	Chlorophyll content↑	Zhang et al. (2010a)
	Impatiens walleriana	Water loss↓, stomatal closure ↑	García-Mata and Lamattina (2010)
	Vicia faba	Stomatal closure↑, RWC↑	García-Mata and Lamattina (2010)
	Triticum aestivum	Germination↑, amylase↑, esterase↑	Zhang et al. (2010c)
	Ipomea batatus	Chlorophyll content↑	Zhang et al. (2009)
Salinity	*Triticum aestivum*	Amylase↑, esterase↑	Li et al. (2016a,b,c)
	Hordeum vulgare	K^+/ Na^+ ratio↑	Chen et al. (2015)
	Oryza sativa	Chlorophyll↑, carotenoids↑, soluble protein↑, ROS↓	Mostofa et al. (2015a,b)
	Arabidopsis thaliana	Na^+/K^+ ratio↓	Li et al. (2014)
	Cucumis sativus	Hypocotyl and radical length↑, soluble sugar↑, α-amylase↑,β-amylase↑	Sun and Luo (2014)
	Cynodon dactylon	Proline↑, sucrose↑, total soluble sugars↑	Shi et al. (2013)
	Fragaria ananassa	Chlorophyll fluorescence ↑, stomatal conductance↑, RWC↑	Christou et al. (2013)
	Medicago sativa	Seed germination↑	Wang et al. (2012)
HM/metalloids	*Solanum nigrum* (Zn toxicity)	ROS content↓, Zn uptake↓	Luo et al. (2020)
	Brassica oleracea (Cr toxicity)	Plant growth↑, chlorophyll content↑, gas exchange parameter↑	Ahmad et al. (2019)
	Cucurbita pepo (Ni toxicity)	Photosynthetic pigments↑, seedling biomass↑,	Valivand et al. (2019)
	Oryza sativa (Ni toxicity)	Plant growth and biomass↑, P_N↑, g_s↑, E↑, C_i↑, photosynthetic pigments↑	Rizwan et al. (2019)

(*Continued*)

TABLE 3.2 (Continued)

Environmental stress	Plant species	Plant responses	Reference
	Zea mays (Pb toxicity)	Plant growth↑, Pb accumulation↓, protein↑, nitrate reductase↑, free amino acids↑	Zanganeh et al. (2019)
	Brassica oleracea (Pb toxicity)	Germination↑, growth↑	Chen et al. (2018)
	Capsicum annuum (Zn toxicity)	Growth↑, fruit yield↑, water content↑, proline↑	Kaya et al. (2018)
	Oryza sativa (Hg toxicity)	Seedling growth↑	Chen et al. (2017)
	Oryza sativa (Cd toxicity)	Plant biomass↑, chlorophyll content↑, photosynthesis↑, RWC↑, stomatal conductance↑, transpiration rate↑	Mostofa et al. (2015a,b)
	Pisum sativum (As toxicity)	Plant growth↑, total nitrogen content↑, chlorophyll fluorescence↑	Singh et al. (2015)
	Brassica napus (Cd toxicity)	Plant growth↑, root morphology↑, chlorophyll content↑, photosynthetic activity↑	Ali et al. (2014)
	Brassica napus (Al toxicity)	Plant growth↑, total soluble protein content↑	Qian et al. (2014)
	Triticum aestivum (Al toxicity)	Citrate secretion and citrate transporter gene expression↑, expression of PM H^+-ATPase↑	Chen et al. (2013)
	Hordeum vulgare (Al toxicity)	Chlorophyll content↑, net photosynthetic rate (Pn)↑, maximal photochemical efficiency(Fv/Fm)↑	Dawood et al. (2012)
	Medicago sativa (Cd toxicity)	Seedling growth↑, Cd accumulation↓	Li et al. (2012)
	Cucumis sativus (B toxicity)	Root elongation↑, pectin methylesterase↓	Wang et al. (2010)
	Triticum aestivum (Cr toxicity)	Seed germination↑, esterase↑, amylase↑,	Zhang et al. (2010c)
	Triticum aestivum (Al toxicity)	Seed germination↑, seedling growth↑, amylase↑, esterase↑	Zhang et al. (2010b)
Heat	*Populus trichocarpa*	*S*-Nitrosoglutathione reductase↑, ROS and RNS↓	Cheng et al. (2018)
	Fragaria ananassa	Photochemical PSII quantum yields↑, heat shock protein↑, aquaporin↑	Christou et al. (2014)
	Zea mays	Seedling survival percentage↑	Li et al. (2014)

(*Continued*)

TABLE 3.2 (Continued)

Environmental stress	Plant species	Plant responses	Reference
	Zea mays	Proline degradation↓, proline accumulation↑, MDA↓, survival percentage↑	Li et al. (2013a)
	Arabidopsis thaliana	NO accumulation↓, stomata opening in the light	Lisjak et al. (2010)
Cold	*Cucumis sativus*	Activity and gene expression of photosynthetic enzyme↑, photoprotection of PS I and PS II ↑, protection of the photosynthetic apparatus↑, growth and carbon metabolism↑	Liu et al. (2020a)
	Cucumis sativus	Net photosynthetic rate↑, light-saturated CO_2 assimilation rate↑, Fv/Fm↑, φPSII↑, photosynthetic enzymes↑	Liu et al. (2020b)
	Vaccinium corymbosum	Photosynthetic pigments↑, photoinhibition of PS I and PS II ↓, photosynthetic capacity↑, proline↑	Tang et al. (2020)
	Arabidopsis thaliana	Cold stress-responsive gene expression↑, stomatal movement↑	Du et al. (2017)
	Musa acuminata	PAL activity↑, total phenolic content↑, proline↑	Luo et al. (2015)
	Lamiophlomis rotata	Proline↑, sugar accumulation↑, GSNOR↑, HSPs↑	Ma et al. (2015)
	Cynodon dactylon	Survival rate↑, proline↑, soluble sugar↑, sucrose↑	Shi et al. (2013)

Acknowledgments

MP is thankful to the Department of Science and Technology, New Delhi, for providing financial support in the form of INSPIRE Fellowship. This manuscript was assigned CSIR-CSMCRI Communication No. 097/2020.

References

Ahmad, R., Ali, S., Rizwan, M., Dawood, M., Farid, M., 2019. Hydrogen sulfide alleviates chromium stress on cauliflower by restricting its uptake and enhancing antioxidative system. Physiol. Plant 168, 289–300. Available from: https://doi.org/10.1111/ppl.1300.

Ali, B., Gill, R.A., Yang, S., Gill, M.B., Ali, S., Ra, M.T., et al., 2014. Hydrogen sulfide alleviates cadmium-induced morpho-physiological and ultrastructural changes in *Brassica napus*. Ecotoxicol. Environ. Saf. 110, 197–207. Available from: https://doi.org/10.1016/j.ecoenv.2014.08.027.

Álvarez, C., Calo, L., Romero, L.C., García, I., Gotor, C., 2010. An O-Acetylserine (thiol) lyase homolog with L-cysteine desulfhydrase activity regulates cysteine homeostasis in arabidopsis. Plant Physiol. 152, 656–669. Available from: https://doi.org/10.1104/pp.109.147975.

Aroca, Á., Serna, A., Gotor, C., Romero, L.C., 2015. S-Sulfhydration : a cysteine posttranslational modification in plant systems. Plant Physiol. 168, 334–342. Available from: https://doi.org/10.1104/pp.15.00009.

Aroca, A., Benito, J.M., Gotor, C., Romero, L.C., 2017. Persulfidation proteome reveals the regulation of protein function by hydrogen sulfide in diverse biological processes in *Arabidopsis* Available from: https://doi.org/10.1093/jxb/erx294.

Aroca, A., Gotor, C., Romero, L.C., 2018. Hydrogen sulfide signaling in plants: emerging roles of protein persulfidation. Front. Plant Sci. 9, 1–8. Available from: https://doi.org/10.3389/fpls.2018.01369.

Bhuyan, M.H.M.B., Hasanuzzaman, M., Parvin, K., Mohsin, S.M., 2020. Nitric oxide and hydrogen sulfide: two intimate collaborators regulating plant defense against abiotic stress. Plant Growth Regul. 90, 409–424. Available from: https://doi.org/10.1007/s10725-020-00594-4.

Calderwood, A., Kopriva, S., 2014. Hydrogen sulfide in plants: from dissipation of excess sulfur to signaling molecule. Nitric Oxide 41, 72–78. Available from: https://doi.org/10.1016/j.niox.2014.02.005.

Cao, S., Zheng, Y., Wang, K., Rui, H., Tang, S., 2010. Effect of methyl jasmonate on cell wall modification of loquat fruit in relation to chilling injury after harvest. Food Chem. 118, 641–647. Available from: https://doi.org/10.1016/j.foodchem.2009.05.047.

Chen, J., Wang, W., Wu, F., 2013. Hydrogen sulfide alleviates aluminum toxicity in barley seedlings. Plant Soil 362, 301–318. Available from: https://doi.org/10.1007/s11104-012-1275-7.

Chen, J., Wang, W.H., Wu, F.H., He, E.M., Liu, X., Shangguan, Z.P., Zheng, H.L., 2015. Hydrogen sulfide enhances salt tolerance through nitric oxide-mediated maintenance of ion homeostasis in barley seedling roots. Sci. Rep. 5, 12516. Available from: https://doi.org/10.1038/srep12516.

Chen, J., Wu, F.H., Wang, W.H., Zheng, C.J., Lin, G.H., Dong, X.J., et al., 2011. Hydrogen sulphide enhances photosynthesis through promoting chloroplast biogenesis, photosynthetic enzyme expression, and thiol redox modification in *Spinacia oleracea* seedlings. J. Exp. Bot. 62, 4481–4493. Available from: https://doi.org/10.1093/jxb/err145.

Chen, Z., Chen, M., Jiang, M., 2017. Hydrogen sulfide alleviate mercury toxicity by sequestering it in roots or regulating reactive oxygen species productions in rice seedlings. Plant Physiol. Biochem. 111, 179–192. Available from: https://doi.org/10.1016/j.plaphy.2016.11.027.

Chen, Z., Yang, B., Hao, Z., Zhu, J., Zhang, Y., Xu, T., 2018. Exogenous hydrogen sulfide ameliorates seed germination and seedling growth of cauliflower under lead stress and its antioxidant role. J. Plant Growth Regul. 37, 5–15. Available from: https://doi.org/10.1007/s00344-017-9704-8.

Cheng, W., Zhang, L., Jiao, C., Su, M., Yang, T., Zhou, L., et al., 2013. Hydrogen sulfide alleviates hypoxia-induced root tip death in *Pisum sativum*. Plant Physiol. Biochem. 70, 278–286. Available from: https://doi.org/10.1016/j.plaphy.2013.05.042.

Cheng, T., Shi, J., Dong, Y., Ma, Y., Peng, Y., Hu, X., 2018. Hydrogen sulfide enhances poplar tolerance to high-temperature stress by increasing S -nitrosoglutathione reductase (GSNOR) activity and reducing reactive oxygen / nitrogen damage. Plant Growth Regul. 84, 11–23. Available from: https://doi.org/10.1007/s10725-017-0316-x.

Christou, A., Manganaris, G.A., Papadopoulos, I., Fotopoulos, V., 2013. Hydrogen sulfide induces systemic tolerance to salinity and non-ionic osmotic stress in strawberry plants through modification of reactive species biosynthesis and transcriptional regulation of multiple defence pathways. J. Exp. Bot. 64, 1953–1966. Available from: https://doi.org/10.1093/jxb/ert055.

Christou, A., Filippou, P., Manganaris, G.A., Fotopoulos, V., 2014. Sodium hydrosulfide induces systemic thermotolerance to strawberry plants through transcriptional regulation of heat shock proteins and aquaporin. BMC Plant Biol. 14, 1–11.

Cobbett, C.S., 2000. Phytochelatins and their roles in heavy metal detoxification. Plant Physiol. 123, 825–832. Available from: https://doi.org/10.1104/pp.123.3.825.

Corpas, F.J., 2019. Hydrogen sulfide: a new warrior against abiotic stress trends in plant science abiotic stresses antioxidant systems. Trends Plant Sci. 24, 983–988. Available from: https://doi.org/10.1016/j.tplants.2019.08.003.

Corpas, F.J., Palma, J.M., 2020. H_2S signaling in plants and applications in agriculture. J. Adv. Res. 24, 131–137. Available from: https://doi.org/10.1016/j.jare.2020.03.011.

Cristiane, J., Modolo, L.V., 2017. Hydrogen sulfide: a new endogenous player in an old mechanism of plant tolerance to high salinity. Acta Bot. Brasilica 32, 150–160. Available from: https://doi.org/10.1590/0102-33062017abb0229.

Cui, W., Chen, H., Zhu, K., Jin, Q., Xie, Y., Cui, J., et al., 2014. Cadmium-induced hydrogen sulfide synthesis is involved in cadmium tolerance in *Medicago sativa* by reestablishment of reduced (Homo) glutathione and reactive oxygen species homeostases. PLoS One 9, 1–12. Available from: https://doi.org/10.1371/journal.pone.0109669.

Da Silva, C.J., Batista, E.P., Luzia, F., Modolo, V., 2017. Salinity-induced accumulation of endogenous H_2S and NO is associated with modulation of the antioxidant and redox defense systems in *Nicotiana tabacum* L. cv. Havana. Plant Sci. 256, 148–159. Available from: https://doi.org/10.1016/j.plantsci.2016.12.011.

Da-Silva, C.J., Mollica, D.C.F., Vicente, M.H., Peres, L.E.P., Modolo, L.V., 2018. NO, hydrogen sulfide does not come first during tomato response to high salinity. Nitric Oxide 76, 164–173. Available from: https://doi.org/10.1016/j.niox.2017.09.008.

Davidian, J., Kopriva, S., 2010. Regulation of sulfate uptake and assimilation- the same or not the same? Mol. Plant 3, 314–325. Available from: https://doi.org/10.1093/mp/ssq001.

Dawood, M., Cao, F., Jahangir, M.M., Zhang, G., Wu, F., 2012. Alleviation of aluminum toxicity by hydrogen sulfide is related to elevated ATPase, and suppressed aluminum uptake and oxidative stress in barley. J. Hazard Mater. 209–210, 121–128. Available from: https://doi.org/10.1016/j.jhazmat.2011.12.076.

Deng, Y.Q., Bao, J., Yuan, F., Liang, X., Feng, Z.T., Wang, B.S., 2016. Exogenous hydrogen sulfide alleviates salt stress in wheat seedlings by decreasing Na^+ content. Plant Growth Regul. 79, 391–399. Available from: https://doi.org/10.1007/s10725-015-0143-x.

Ding, Z., Li, S., An, X., Liu, X., Qin, H., Wang, D., 2009. Transgenic expression of MYB15 confers enhanced sensitivity to abscisic acid and improved drought tolerance in *Arabidopsis thaliana*. J. Genet. Genomics 36, 17–29. Available from: https://doi.org/10.1016/S1673-8527(09)60003-5.

Ding, H., Ma, D., Huang, X., Hou, J., Wang, C., Xie, Y., et al., 2019. Exogenous hydrogen sulfide alleviates salt stress by improving antioxidant defenses and the salt overly sensitive pathway in wheat seedlings. Acta Physiol. Plant 41, 1–11. Available from: https://doi.org/10.1007/s11738-019-2918-6.

Du, X., Jin, Z., Liu, D., Yang, G., Pei, Y., 2017. Hydrogen sulfide alleviates the cold stress through MPK4 in *Arabidopsis thaliana*. Plant Physiol. Biochem. 120, 112–119. Available from: https://doi.org/10.1016/j.plaphy.2017.09.028.

Du, X., Jin, Z., Zhang, L., Liu, X., 2019. H_2S is involved in ABA-mediated stomatal movement through MPK4 to alleviate drought stress in *Arabidopsis thaliana*. Plant Soil 435, 295–307. Available from: https://doi.org/10.1007/s11104-018-3894-0.

Fang, H., Jing, T., Liu, Z., Zhang, L., Jin, Z., Pei, Y., 2014. Hydrogen sulfide interacts with calcium signaling to enhance the chromium tolerance in *Setaria italica*. Cell Calcium 56, 472–481. Available from: https://doi.org/10.1016/j.ceca.2014.10.004.

Filipovic, M.R., Jovanović, V.M., 2017. More than just an intermediate: hydrogen sulfide signalling in plants. J. Exp. Bot. 68, 4733–4736. Available from: https://doi.org/10.1093/jxb/erx352.

Filipovic, M.R., Zivanovic, J., Alvarez, B., Banerjee, R., 2018. Chemical biology of H_2S signaling through persulfidation. Chem. Rev. 118, 1253–1337. Available from: https://doi.org/10.1021/acs.chemrev.7b00205.

Foyer, C.H., Noctor, G., 2009. Redox regulation in photosynthetic organisms: signaling, acclimation, and practical implications. Antioxid. Redox Signal. 11, 861–905. Available from: https://doi.org/10.1089/ars.2008.2177.

Foyer, C.H., Noctor, G., 2011. Ascorbate and glutathione: the heart of the redox hub. Plant Physiol. 155, 2–18. Available from: https://doi.org/10.1104/pp.110.167569.

Fu, P., Wang, W., Lixia Hou, X.L., 2013. Hydrogen sulfide is involved in the chilling stress response in *Vitis vinifera* L. Acta Soc. Bot. Pol. 82, 295–302. Available from: https://doi.org/10.5586/asbp.2013.031.

García-Mata, C., Lamattina, L., 2010. Hydrogen sulphide, a novel gasotransmitter involved in guard cell signalling. N. Phytol. 188, 977–984. Available from: https://doi.org/10.1111/j.1469-8137.2010.03465.x.

Gotor, C., García, I., Aroca, Á., Laureano-marín, A.M., Jurado-flores, A., Moreno, I., et al., 2019. Signaling by hydrogen sulfide and cyanide through post-translational modification. J. Exp. Bot. 16, 4251–4265. Available from: https://doi.org/10.1093/jxb/erz225.

Gudesblat, G.E., Iusem, N.D., Morris, P.C., Morris, P., 2005. Guard cell-specific inhibition of *Arabidopsis* MPK3 expression causes abnormal stomatal responses to abscisic acid and hydrogen peroxide. N. Physiol. 173, 713–721. Available from: https://doi.org/10.1111/j.1469-8137.2006.01953.x.

Hancock, J.T., Whiteman, M., 2014. Hydrogen sulfide and cell signaling: team player or referee? Plant Physiol. Biochem. 78, 37–42. Available from: https://doi.org/10.1016/j.plaphy.2014.02.012.

Hancock, J.T., Whiteman, M., 2015. Hydrogen sulfide and reactive friends: the interplay with reactive oxygen species and nitric oxide signalling pathways. In: De Kok, L.J., Hawkesford, M.J., Rennenberg, H., Saito, K., Schnug, E. (Eds.), Molecular Physiology and Ecophysiology of Sulfur. pp. 153–168. Available from: https://doi.org/10.1007/978-3-319-20137-5.

Hancock, J.T., Desikan, R., Clarke, A., Hurst, R.D., Neill, S.J., 2002. Cell signalling following plant/pathogen interactions involves the generation of reactive oxygen and reactive nitrogen species. Plant Physiol. Biochem. 40, 611–617. Available from: https://doi.org/10.1016/S0981-9428(02)01403-1.

Horst, W.J., Wang, Y., Eticha, D., 2010. The role of the root apoplast in aluminium-induced inhibition of root elongation and in aluminium resistance of plants: a review. Ann. Bot. 106, 185–197. Available from: https://doi.org/10.1093/aob/mcq053.

Hossain, M.A., Li, Z.G., Hoque, T.S., Burritt, D.J., Fujita, M., Munné-Bosch, S., 2018. Heat or cold priming-induced cross-tolerance to abiotic stresses in plants: key regulators and possible mechanisms. Protoplasma 255, 399–412. Available from: https://doi.org/10.1007/s00709-017-1150-8.

Huang, J., Willems, P., Wei, B., Tian, C., Ferreira, R.B., Bodra, N., et al., 2019. Mining for protein S-sulfenylation in *Arabidopsis* uncovers redox-sensitive sites. Proc. Natl. Acad. Sci. 116, 20256–20261. Available from: https://doi.org/10.1073/pnas.1906768116.

Jalmi, S.K., Bhagat, P.K., Verma, D., Noryang, S., Tayyeba, S., 2018. Traversing the links between heavy metal stress and plant signaling. Front. Plant Sci. 9, 1–21. Available from: https://doi.org/10.3389/fpls.2018.00012.

Jia, H., Chen, S., Liu, D., Liesche, J., Shi, C., Wang, J., et al., 2018. Ethylene-induced hydrogen sulfide negatively regulates ethylene biosynthesis by persulfidation of ACO in tomato under osmotic stress. Front. Plant Sci. 9, 1–11. Available from: https://doi.org/10.3389/fpls.2018.01517.

Jin, Z., Pei, Y., 2015. Physiological implications of hydrogen sulfide in plants: pleasant exploration behind its unpleasant odour. Oxid. Med. Cell. Longev. 1–6. Available from: https://doi.org/10.1155/2015/397502.

Jin, Z., Shen, J., Qiao, Z., Yang, G., Wang, R., Pei, Y., 2011. Biochemical and biophysical research communications hydrogen sulfide improves drought resistance in *Arabidopsis thaliana*. Biochem. Biophys. Res. Commun. 414, 481–486. Available from: https://doi.org/10.1016/j.bbrc.2011.09.090.

Jing, L.I.U., Zhi-Hui, H.O.U., Guo-Hua, L.I.U., Li-Xia, H.O.U., Xin, L.I.U., 2012. Hydrogen sulfide may function downstream of nitric oxide in ethylene- induced stomatal closure in *Vicia faba* L. J. Integr. Agric. 11, 1644–1653. Available from: https://doi.org/10.1016/S2095-3119(12)60167-1.

Karam, E.A., Keramat, B., 2017. Hydrogen sulfide protects coriander seedlings against copper stress by regulating the ascorbate-glutathione cycle in leaves. J. Plant Process. Funct. 5, 59–64.

Kaur, P., Handa, N., Verma, V., Bakshi, P., Kalia, R., Sareen, S., et al., Cross talk among reactive oxygen, nitrogen and sulfur during abiotic stress in plants. In: Hasanuzzaman, M., Fotopoulos, V., Nahar, K., Fujita, M. (Eds.), 2019. *Reactive Oxygen, Nitrogen and Sulfur Species in Plants: Production, Metabolism, Signaling and Defense Mechanisms* 2, John Wiley & Sons, Inc., US, pp. 857–871. Available from: http://doi.org/10.1002/9781119468677.ch38.

Kaya, C., Ashraf, M., Akram, N.A., 2018. Hydrogen sulfide regulates the levels of key metabolites and antioxidant defense system to counteract oxidative stress in pepper (*Capsicum annuum* L.) plants exposed to high zinc regime. Environ. Sci. Pollut. Res. 25, 12612–12618. Available from: 10.1007/s11356-018-1510-8.

Kaya, C., Sarıoğlu, A., Ashraf, M., Alyemeni, M.N., 2020. Gibberellic acid-induced generation of hydrogen sulfide alleviates boron toxicity in tomato (*Solanum lycopersicum* L.) plants. Plant Physiol. Biochem. 153, 53–63. Available from: 10.1016/j.plaphy.2020.04.038.

Khan, M.N., Mobin, M., Abbas, Z.K., Siddiqui, M.H., 2017. Nitric oxide-induced synthesis of hydrogen sulfide alleviates osmotic stress in wheat seedlings through sustaining antioxidant enzymes, osmolyte accumulation and cysteine homeostasis. Nitric Oxide 68, 91–102. Available from: 10.1016/j.niox.2017.01.001.

Kharbech, O., Houmani, H., Chaoui, A., Corpas, F.J., 2017. Alleviation of Cr (VI) -induced oxidative stress in maize (*Zea mays* L.) seedlings by NO and H_2S donors through differential organ-dependent regulation of ROS and NADPH-recycling metabolisms. J. Plant Physiol. 219, 71–80. Available from: 10.1016/j.jplph.2017.09.010.

Kok, L.J.D.E., Bosma, W., Maas, F.M., 1985. The effect of short-term H_2S fumigation on water-soluble sulphydryl and glutathione levels in spinach. Plant Cell Environ. 25, 189–194. Available from: 10.1111/1365-3040.ep11604605.

Kolupaev, Y.E., Firsova, E.N., Yastreb, T.O., Lugovaya, A.A., 2017. The participation of calcium ions and reactive oxygen species in the induction of antioxidant enzymes and heat resistance in plant cells by hydrogen sulfide donor. Appl. Biochem. Microbiol. 53, 573–579. Available from: 10.1134/S0003683817050088.

Krueger, S., Niehl, A., Martin, M.C.L., Steinhauser, D., Donath, A., Hildebrandt, T., et al., 2009. Analysis of cytosolic and plastidic serine acetyltransferase mutants and subcellular metabolite distributions suggests interplay of the cellular compartments for cysteine biosynthesis in *Arabidopsis*. Plant Cell Environ. 49, 349–367. Available from: 10.1111/j.1365-3040.2009.01928.x.

Lai, D., Mao, Y., Zhou, H., Li, F., Wu, M., Zhang, J., et al., 2014. Endogenous hydrogen sulfide enhances salt tolerance by coupling the reestablishment of redox homeostasis and preventing salt-induced K^+ loss in seedlings of *Medicago sativa*. Plant Sci. 225, 117–129. Available from: https://doi.org/10.1016/j.plantsci.2014.06.006.

Laureano-Marín, A.M., Moreno, I., Romero, L.C., Gotor, C., 2016. Negative regulation of autophagy by sulfide is independent of reactive oxygen species. Plant Physiol. 171, 1378–1391. Available from: https://doi.org/10.1104/pp.16.00110.

Li, Z.G., 2013. Hydrogen sulfide: a multifunctional gaseous molecule in plants. Russ. J. Plant Physiol. 60, 733–740. Available from: https://doi.org/10.1134/S1021443713060058.

Li, Z.G., 2019. Interplay between hydrogen sulfide and calcium signaling in plant abiotic stress response and adaptation. In: De Kok, L.J., Hawkesford, M.J., Rennenberg. H., Saito, K., Schnug, E. (Eds.), Molecular Physiology and Ecophysiology of Sulfur, John Wiley & Sons, Inc., USA, pp. 669-683.

Li, L., Rose, P., Moore, P.K., 2011. Hydrogen sulfide and cell signaling. Annu. Rev. Pharmacol. Toxicol. 51, 169–187. Available from: https://doi.org/10.1146/annurev-pharmtox-010510-100505.

Li, L., Wang, Y., Shen, W., 2012. Roles of hydrogen sulfide and nitric oxide in the alleviation of cadmium-induced oxidative damage in alfalfa seedling roots. Biometals 25, 617–631. Available from: https://doi.org/10.1007/s10534-012-9551-9.

Li, Z., Ding, X., Du, P., 2013a. Hydrogen sulfide donor sodium hydrosulfide-improved heat tolerance in maize and involvement of proline. J. Plant Physiol. 170, 741–747. Available from: https://doi.org/10.1016/j.jplph.2012.12.018.

Li, Z., Yang, S., Long, W., Yang, G., Shen, Z., 2013b. Hydrogen sulphide may be a novel downstream signal molecule in nitric oxide-induced heat tolerance of maize (*Zea mays* L.) seedlings. Plant Cell Environ. 36, 1564–1572. Available from: https://doi.org/10.1111/pce.12092.

Li, Z., Yi, X., Li, Y., 2014. Effect of pretreatment with hydrogen sulfide donor sodium hydrosulfide on heat tolerance in relation to antioxidant system in maize (*Zea mays*) seedlings. Biologia 69, 1001–1009. Available from: https://doi.org/10.2478/s11756-014-0396-2.

Li, Q., Wang, Z., Zhao, Y., Zhang, X., Zhang, S., Bo, L., et al., 2016a. Putrescine protects hulless barley from damage due to UV-B stress via H_2S- and H_2O_2-mediated signaling pathways. Plant Cell Rep. 35, 1155–1168. Available from: https://doi.org/10.1007/s00299-016-1952-8.

Li, Z., Long, W., Yang, S., Wang, Y., 2016b. Endogenous hydrogen sulfide regulated by calcium is involved in thermotolerance in tobacco *Nicotiana tabacum* L. suspension cell cultures. Acta Physiol. Plant 37, 1–11. Available from: https://doi.org/10.1007/s11738-015-1971-z.

Li, Z.G., Min, X., Zhou, Z.H., 2016c. Hydrogen sulfide: a signal molecule in plant cross-adaptation. Front. Plant Sci. 7, 1–12. Available from: https://doi.org/10.3389/fpls.2016.01621.

Li, H., Li, M., Wei, X., Zhang, X., Xue, R., Zhao, Y., 2017. Transcriptome analysis of drought-responsive genes regulated by hydrogen sulfide in wheat (*Triticum aestivum* L.) leaves. Mol. Genet. Genomics 29, 1091–1110. Available from: https://doi.org/10.1007/s00438-017-1330-4.

Li, Z., Zhu, Y., He, X., Yong, B., Peng, Y., Zhang, X., et al., 2018. The hydrogen sulfide, a downstream signaling molecule of hydrogen peroxide and nitric oxide, involves spermidine-regulated transcription factors and antioxidant defense in white clover in response to dehydration. Environ. Exp. Bot. 161, 255–264. Available from: https://doi.org/10.1016/j.envexpbot.2018.06.036.

Lisjak, M., Srivastava, N., Teklic, T., Civale, L., Lewandowski, K., Wilson, I., et al., 2010. A novel hydrogen sulfide donor causes stomatal opening and reduces nitric oxide accumulation. Plant Physiol. Biochem. 48, 931–935. Available from: https://doi.org/10.1016/j.plaphy.2010.09.016.

Lisjak, M., Teklić, T., Wilson, I.D., Wood, M., Whiteman, M., Hancock, J.T., et al., 2011. Hydrogen sulfide effects on stomatal apertures. Plant Signal. Behav. 6, 1444–1446. Available from: https://doi.org/10.4161/psb.6.10.17104.

Liu, Z., Li, Y., Cao, C., Liang, S., Ma, Y., Liu, X., et al., 2019. The role of H_2S in low temperature-induced cucurbitacin C increases in cucumber. Plant Mol. Biol. 99, 535–544. Available from: https://doi.org/10.1007/s11103-019-00834-w.

Liu, F., Fu, X., Wu, G., Feng, Y., Li, F., Bi, H., et al., 2020a. Hydrogen peroxide is involved in hydrogen sulfide-induced carbon assimilation and photoprotection in cucumber seedlings. Environ. Exp. Bot. 175, 104052. Available from: https://doi.org/10.1016/j.envexpbot.2020.104052.

Liu, F., Zhang, X., Cai, B., Pan, D., Fu, X., Bi, H., et al., 2020b. Physiological response and transcription profiling analysis reveal the role of glutathione in H_2S-induced chilling stress tolerance of cucumber seedlings. Plant Sci. 291, 110363. Available from: https://doi.org/10.1016/j.plantsci.2019.110363.

Luo, Z., Li, D., Du, R., Mou, W., 2015. Hydrogen sulfide alleviates chilling injury of banana fruit by enhanced antioxidant system and proline content. Sci. Hortic. 183, 144–151. Available from: https://doi.org/10.1016/j.scienta.2014.12.021.

Luo, S., Calderón-urrea, A., Yu, J., 2020. The role of hydrogen sulfide in plant alleviates heavy metal stress. Plant Soil. 449, 1–10. Available from: https://doi.org/10.1007/s11104-020-04471-x.

Ma, L., Yang, L., Zhao, J., Wei, J., Hu, Y.Y.Á.X., 2015. Comparative proteomic analysis reveals the role of hydrogen sulfide in the adaptation of the alpine plant *Lamiophlomis rotata* to altitude gradient in the Northern Tibetan Plateau. Planta 241, 887–906. Available from: https://doi.org/10.1007/s00425-014-2209-9.

Ma, D., Ding, H., Wang, C., Qin, H., Han, Q., Hou, J., et al., 2016. Alleviation of drought stress by hydrogen sulfide is partially related to the abscisic acid signaling pathway in Wheat. PLoS One 11, 1–16. Available from: https://doi.org/10.1371/journal.pone.0163082.

Mackerness, S.A.-H., John, C.F., Jordan, B., Thomas, B., 2001. Early signaling components in ultraviolet-B responses: distinct roles for different reactive oxygen species and nitric oxide. FEBS Lett. 489, 237–242. Available from: https://doi.org/10.1016/S0014-5793(01)02103-2.

Mori, I.C., Murata, Y., Yang, Y., Munemasa, S., Wang, Y., Andreoli, S., et al., 2006. CDPKs CPK6 and CPK3 function in ABA regulation of guard cell S-Type anion- and Ca^{2+}- permeable channels and stomatal closure. PLoS Biol 4, e327. Available from: https://doi.org/10.1371/journal.pbio.0040327.

Mostofa, M.G., Rahman, A., Ansary, M.U., 2015a. Hydrogen sulfide modulates cadmium-induced physiological and biochemical responses to alleviate cadmium toxicity in rice. Sci. Rep. 5, 14078, 1–17. Available from: https://doi.org/10.1038/srep14078.

Mostofa, M.G., Saegusa, D., Fujita, M., Tran, L.P., 2015b. Hydrogen sulfide regulates salt tolerance in rice by maintaining Na^+/K^+ balance, mineral homeostasis and oxidative metabolism under excessive salt stress. Front. Plant Sci. 6, 1–14. Available from: https://doi.org/10.3389/fpls.2015.01055.

Munemasa, S., Nakamura, Y., Murata, Y., Science, L., 2018. Guard cell salicylic acid signaling is integrated into abscisic acid signaling via the Ca^{2+}/CPK-dependent pathway. Plant Physiol. 178, 441–450. Available from: https://doi.org/10.1104/pp.18.00321.

Noctor, G., Reichheld, J.-P., Foyer, C.H., 2018. ROS-related redox regulation and signaling in plants. Semin. Cell Dev. Biol. 80, 3–12. Available from: https://doi.org/10.1016/j.semcdb.2017.07.013.

Paul, S., Roychoudhury, A., 2020. Regulation of physiological aspects in plants by hydrogen sulfide and nitric oxide under challenging environment. Physiol. Plant 168, 374–393. Available from: https://doi.org/10.1111/ppl.13021.

Qian, P., Sun, R., Ali, B., 2014. Effects of hydrogen sulfide on growth, antioxidative capacity, and ultrastructural changes in oilseed rape seedlings under aluminum toxicity. J. Plant Growth Regul. 33, 526–538. Available from: https://doi.org/10.1007/s00344-013-9402-0.

Rizwan, M., Golam, M., Zul, M., Zhou, Y., Adeel, M., Mehmood, S., et al., 2019. Hydrogen sulfide enhances rice tolerance to nickel through the prevention of chloroplast damage and the improvement of nitrogen metabolism under excessive nickel. Plant Physiol. Biochem. 138, 100–111. Available from: https://doi.org/10.1016/j.plaphy.2019.02.023.

Santisree, P., Adimulam, S.S., Bommineni, P., Sharma, K.K., 2019. Hydrogen sulfide in plant abiotic stress tolerance. In: De Kok, L.J., Hawkesford, M.J., Rennenberg. H., Saito, K., Schnug, E. (Eds.), Molecular Physiology and Ecophysiology of Sulfur, John Wiley & Sons, Inc., USA, pp. 743-775.

Scuff, D., Álvarez, C., Gotor, Lamattina, L., García-Mata, C., 2014. Hydrogen sulfide generated by L-cysteine desulfhydrase acts upstream of nitric oxide to modulate abscisic acid-dependent stomatal closure. Plant Physiol. 166, 2065–2076. Available from: https://doi.org/10.1104/pp.114.245373.

Scuffi, D., Nietzel, T., di Fino, L.M., Meyer, A.J., Lamattina, L., Schwarzlände, M., et al., 2018. Hydrogen sulfide increases production of NADPH oxidase-dependent hydrogen peroxide and phospholipase D-derived phosphatidic acid in guard cell signaling. Plant Physiol. 176, 2532–2542. Available from: https://doi.org/10.1104/pp.17.01636.

Shan, C., Zhang, S., Li, D., Zhao, Y., Tian, X., Zhao, X., et al., 2011. Effects of exogenous hydrogen sulfide on the ascorbate and glutathione metabolism in wheat seedlings leaves under water stress. Acta Physiol. Plant 33, 2533–2540. Available from: https://doi.org/10.1007/s11738-011-0746-4.

Shan, C., Dai, H., Sun, Y., 2012. Hydrogen sulfide protects wheat seedlings against copper stress by regulating the ascorbate and glutathione metabolism in leaves. Aust. J. Crop. Sci. 6, 248–254.

Shan, C., Liu, H., Zhao, L., Wang, X., 2013. Effects of exogenous hydrogen sulfide on the redox states of ascorbate and glutathione in maize leaves under salt stress. Biol. Plant. 58, 169–173. Available from: https://doi.org/10.1007/s10535-013-0366-5.

Shi, H., Ye, T., Chan, Z., 2013. Exogenous application of hydrogen sulfide donor sodium hydrosulfide enhanced multiple abiotic stress tolerance in bermudagrass (*Cynodon dactylon* (L). Pers.). Plant Physiol. Biochem. 71, 226–234. Available from: https://doi.org/10.1016/j.plaphy.2013.07.021.

Singh, V.P., Singh, S., Kumar, J., Prasad, S.M., 2015. Hydrogen sulfide alleviates toxic effects of arsenate in pea seedlings through up-regulation of the ascorbate-glutathione cycle: Possible involvement of nitric oxide. J. Plant Physiol. 181, 20–29. Available from: https://doi.org/10.1016/j.jplph.2015.03.015.

Sun, Y.D., Luo, W.R., 2014. Effects of exogenous hydrogen sulphide on seed germination and seedling growth of cucumber (*Cucumis sativus*) under sodium bicarbonate stress. Seed Sci. Technol. 42, 126–131. Available from: https://doi.org/10.15258/sst.2014.42.2.02.

Sun, J., Wang, R., Zhang, X., Yu, Y., Zhao, R., Li, Z., et al., 2013. Plant physiology and biochemistry hydrogen sulfide alleviates cadmium toxicity through regulations of cadmium transport across the plasma and vacuolar membranes in *Populus euphratica* cells. Plant Physiol. Biochem. 65, 67–74. Available from: https://doi.org/10.1016/j.plaphy.2013.01.003.

Takahashi, H., Kopriva, S., Giordano, M., Saito, K., 2011. Sulfur assimilation in photosynthetic organisms: molecular functions and regulations of transporters and assimilatory enzymes. Annu. Rev. Plant Biol. 62, 157–186. Available from: https://doi.org/10.1146/annurev-arplant-042110-103921.

Tang, X., An, B., Cao, D., Xu, R., Wang, S., Zhang, Z., et al., 2020. Improving photosynthetic capacity, alleviating photosynthetic inhibition and oxidative stress under low temperature stress with exogenous hydrogen sulfide in blueberry seedlings. Front. Plant Sci. 11, 1–13. Available from: https://doi.org/10.3389/fpls.2020.00108.

Tiwari, S., Lata, C., 2018. Heavymetal stress, signaling, and tolerance due to plant-associated microbes: an overview. Front. Plant Sci. 9, 1–12. Available from: https://doi.org/10.3389/fpls.2018.00452.

Valivand, M., Amooaghaie, R., Ahadi, A., 2019. Plant physiology and biochemistry seed priming with H_2S and Ca^{2+} trigger signal memory that induces cross- adaptation against nickel stress in zucchini seedlings. Plant Physiol. Biochem. 143, 286–298. Available from: https://doi.org/10.1016/j.plaphy.2019.09.016.

Wang, P., Zhang, B., Li, X., Xu, C., Yin, X., Shan, L., et al., 2009. Ethylene signal transduction elements involved in chilling injury in non-climacteric loquat fruit. J. Exp. Bot. 61, 179–190. Available from: https://doi.org/10.1093/jxb/erp302.

Wang, B.L., Shi, L., Li, Y.X., Zhang, W.H., 2010. Boron toxicity is alleviated by hydrogen sul W de in cucumber (*Cucumis sativus* L.) seedlings. Planta 231, 1301–1309. Available from: https://doi.org/10.1007/s00425-010-1134-9.

Wang, Y., Li, L., Cui, W., Xu, S., 2012. Hydrogen sulfide enhances alfalfa (*Medicago sativa*) tolerance against salinity during seed germination by nitric oxide pathway. Plant Soil 351, 107–119. Available from: https://doi.org/10.1007/s11104-011-0936-2.

Xiang, Y., Tang, N., Du, H., Ye, H., Xiong, L., 2008. Characterization of OsbZIP23 as a key player of the basic leucine zipper transcription factor family for conferring abscisic acid sensitivity and salinity. Plant Physiol. 148, 1938–1952. Available from: https://doi.org/10.1104/pp.108.128199.

Xie, Y., Zhang, C., Lai, D., Sun, Y., Samma, M.K., Zhang, J., 2014. Hydrogen sulfide delays GA-triggered programmed cell death in wheat aleurone layers by the modulation of glutathione homeostasis and heme oxygenase-1 expression. J. Plant Physiol. 171, 53–62. Available from: https://doi.org/10.1016/j.jplph.2013.09.018.

Yadav, S.K., 2010. Heavy metals toxicity in plants : an overview on the role of glutathione and phytochelatins in heavy metal stress tolerance of plants. South. Afr. J. Bot. 76, 167–179. Available from: https://doi.org/10.1016/j.sajb.2009.10.007.

Yang, M., Bao-ping, Q., Xue-li, M., Ping, W., Mei-ling, L., Lu-lu, C., et al., 2016. Foliar application of sodium hydrosulfide (NaHS), a hydrogen sulfide (H2S) donor, can protect seedlings against heat stress in wheat (*Triticum aestivum* L.). J. Integr. Agric. 15, 2745–2758. Available from: https://doi.org/10.1016/S2095-3119(16)61358-8.

Yao, Y., Yang, Y., Li, C., Huang, D., Zhang, J., Wang, C., et al., 2019. Research progress on the functions of gasotransmitters in plant responses to abiotic stresses. Plants 8, 605. Available from: https://doi.org/10.3390/plants8120605.

Zanganeh, R., Jamei, R., Rahmani, F., 2019. Role of salicylic acid and hydrogen sulfide in promoting lead stress tolerance and regulating free amino acid composition in *Zea mays* L. Acta Physiol. Plant 41, 1–9. Available from: https://doi.org/10.1007/s11738-019-2892-z.

Zhang, H., Hu, L., Hu, K., He, Y., Wang, S., Luo, J., 2008. Hydrogen sulfide promotes wheat seed germination and alleviates oxidative damage against copper stress. J. Integr. Plant Biol. 50, 1518–1529. Available from: https://doi.org/10.1111/j.1744-7909.2008.00769.x.

Zhang, H., Ye, Y.K., Wang, S.H., Luo, J.P., Tang, J., Ma, D.F., 2009. Hydrogen sulfide counteracts chlorophyll loss in sweetpotato seedling leaves and alleviates oxidative damage against osmotic stress. Plant Growth Regul. 58, 243–250. Available from: https://doi.org/10.1007/s10725-009-9372-1.

Zhang, H., Jiao, H., Jiang, C.X., Wang, S.H., Wei, Z.J., Luo, J.P., et al., 2010a. Hydrogen sulfide protects soybean seedlings against drought-induced oxidative stress. Acta Physiol. Plant 32, 849–857. Available from: https://doi.org/10.1007/s11738-010-0469-y.

Zhang, Hua, Tan, Z., Hu, L., Wang, S., Luo, J., Jones, R.L., 2010b. Hydrogen sulfide alleviates aluminium toxicity in germinating wheat seedlings. J. Integr. Plant Biol. 52, 556–567. Available from: https://doi.org/10.1111/j.1744-7909.2010.00946.x.

Zhang, H., Hu, L.Y., Li, P., Hu, K.D., Jiang, C.X., Luo, J.P., 2010c. Hydrogen sulfide alleviated chromium toxicity in wheat. Biol. Plant 54, 743–747. Available from: https://doi.org/10.1007/s10535-010-0133-9.

Zhang, H., Wang, M.J., Hu, L.Y., Wang, S.H., Hu, K.D., Bao, L.J., et al., 2010d. Hydrogen sulfide promotes wheat seed germination under osmotic stress. Russ. J. Plant Physiol. 57, 532–539. Available from: https://doi.org/10.1134/S1021443710040114.

Zhang, T., Li, F., Fan, C., Li, X., Zhang, F., He, J., 2017. Role and interrelationship of MEK1-MPK6 cascade, hydrogen peroxide and nitric oxide in darkness-induced stomatal closure. Plant Sci. 262, 190–199. Available from: https://doi.org/10.1016/j.plantsci.2017.06.010.

Zhang, J., Zhou, M., Ge, Z., Shen, J., Zhou, C., Gotor, C., et al., 2020. Abscisic acid-triggered guard cell L-cysteine desulfhydrase function and in situ hydrogen sulfide production contributes to heme oxygenase-modulated stomatal closure. Plant Cell Environ. 43, 624–636. Available from: https://doi.org/10.1111/pce.13685.

Zhao, N., Zhu, H., Zhang, H., Sun, J., Zhou, J., Deng, C., Zhang, Y., Zhao, R., Zhou, X., Lu, C., Lin, S., 2018. Hydrogen sulfide mediates K^+ and Na^+ homeostasis in the roots of salt-resistant and salt-sensitive poplar species subjected to NaCl stress. Front. Plant Sci. 9, 1366. Available from: https://doi.org/10.3389/fpls.2018.01366.

Ziogas, V., Molassiotis, A., Fotopoulos, V., Tanou, G., 2018. Hydrogen sulfide: a potent tool in postharvest fruit biology and possible mechanism of action. Front. Plant Sci. 9, 1–9. Available from: https://doi.org/10.3389/fpls.2018.01375.

Zhu, J.K., 2003. Regulation of ion homeostasis under salt stress. Curr. Opin. Plant Biol. 6, 441–445. Available from: https://doi.org/10.1016/S1369-5266(03)00085-2.

Zou, H., Zhang, N., Lin, X., 2020. Hydrogen sulfide is a crucial element of the antioxidant defense system in *Glycine max—Sinorhizobium fredii* symbiotic root nodules. Plant Soil 449, 209–231. Available from: https://doi.org/10.1007/s11104-020-04465-9.

Chapter 4

Regulation of metal stress toxicity in plants by the hydrogen sulfide

Nimisha Amist* and N.B. Singh
Plant Physiology Laboratory, Department of Botany, University of Allahabad, Allahabad, India
**Corresponding author.*

4.1 Introduction

Unfavorable environmental conditions like high temperature, cold, drought, salinity, heavy metals, or other pollutant accumulation, ozone, and excess ultraviolet radiation are included under the term abiotic stress. They impose negative impacts on crop productivity and result in major restrictions on global food security (Wang and Frei, 2011).

The heavy metals are one of the major factors responsible for soil pollution which is now a serious threat to agriculture. The effects of metals on plant growth have attracted great attention. The main sources of metals in soil are activities like mining, exhaust emissions, sewage water irrigation, and utilization of metal products. The anthropogenic activities have elevated the level of metals in the environment outside the normal range leading to declining environmental quality and ecosystem health. The metals copper (Cu), zinc (Zn), cadmium (Cd), aluminum (Al), arsenic (As), chromium (Cr), and lead (Pb) are of great concern. All stages of plant life are affected by metals. Inhibition of seed germination and seedling growth, damaged antioxidant enzymes systems and membranes, induction of chromosomal aberrations, and eventual plant death are the impacts induced by metal exposure. Metals also hinder plant metabolic processes by inducing nutrient imbalance, a great amount of free radicals, and oxidative stress. The most identifying feature of metal stress is the accumulation of toxic metals and root damage. The solubility, mobility, and bioavailability of metals depend on soil characteristics like pH, clay, organic matter, and redox potential. However, plants have developed parallel mechanisms to alleviate the metal stress-induced damage during the course of long-term evolution. The mechanisms of metal

Hydrogen Sulfide in Plant Biology. DOI: https://doi.org/10.1016/B978-0-323-85862-5.00013-0

tolerance developed by plants include control over metal absorption and transport in the roots, chelation with metallothioneins (MTs), phytochelatins (PCs), or organic acids, sequestration in vacuoles, and finally restoration of damaged proteins by heat shock proteins.

Recently hydrogen sulfide (H_2S) has been documented as an important secondary messenger in various developmental and stress responses in plants, stimulating signal transduction cascades in a pattern similar to nitric oxide (NO) (Zhang et al., 2008, Garcia-Mata and Lamattina, 2010, 2013; Shi et al., 2012; Simontacchi et al., 2013). H_2S is a gaseous signal molecule that is present in bacteria, invertebrates, vertebrates, plants, and mammals (Wang, 2012). H_2S has been reported to play significant roles in plant responses to environmental stress and also growth and development (Jin and Pei, 2015). The uptake, transport, and accumulation of metals in plants are inhibited by H_2S and in fact tolerance of plants against metal stress is elevated due to H_2S. H_2S with its reducing properties exhibits strong scavenging activity against reactive oxygen species (ROS), such as superoxide (Mitsuhashi et al., 2006), peroxynitrite, hypochlorite (Whiteman et al., 2004), and hydrogen peroxide (H_2O_2) (Geng et al., 2004).

In this chapter, we aim to present the potential role of H_2S in alleviating the detrimental effects of exposure of crops to metal toxicity and also understand the physiological and molecular mechanisms of H_2S that are involved in elevating metal tolerance in plants. Thus emphasis will be given to highlighting the possible role of H_2S as an emerging signaling molecule involved in the regulation of several physiological processes in plants under metal stress.

4.2 H_2S in plants: metabolism and role in plant development

H_2S is a colorless, extremely soluble, and flammable gas that exhibits toxic property (Fotopoulos et al., 2013). The relative molecular weight of H_2S is 34 and it produces a rotten egg odor. H_2S moves freely between cells (Mathai et al., 2009) but the concentration of H_2S in plants is maintained through a severe regulatory enzyme systems (Papenbrock et al., 2007; Wang, 2012). There have been relatively few studies on the biosynthesis of H_2S in plants.

H_2S metabolism includes synthesis and decomposition of H_2S in plants (Fig. 4.1). There have been six pathways identified through which H_2S is generated in higher plants. These pathways can be summarized as follows. The first pathway identified involved the degradation of L-cysteine to release the products H_2S, pyruvate, and NH_4^+ in tobacco cells (Harrington and Smith, 1980). Following this discovery several researches concluded that L-cysteine desulfhydrase (L-CDes) is the enzyme which catalyzes the degradation of L-cysteine to form H_2S in higher plants (Sekiya et al., 1982; Rennenberg, 1983; Xie et al., 2013) (Fig. 4.1, green arrows pathway). In the second pathway, L-CDes catalyzes the synthesis of L-alanine and elemental sulfur by breaking down L-cysteine and after that sulfur is usually reduced

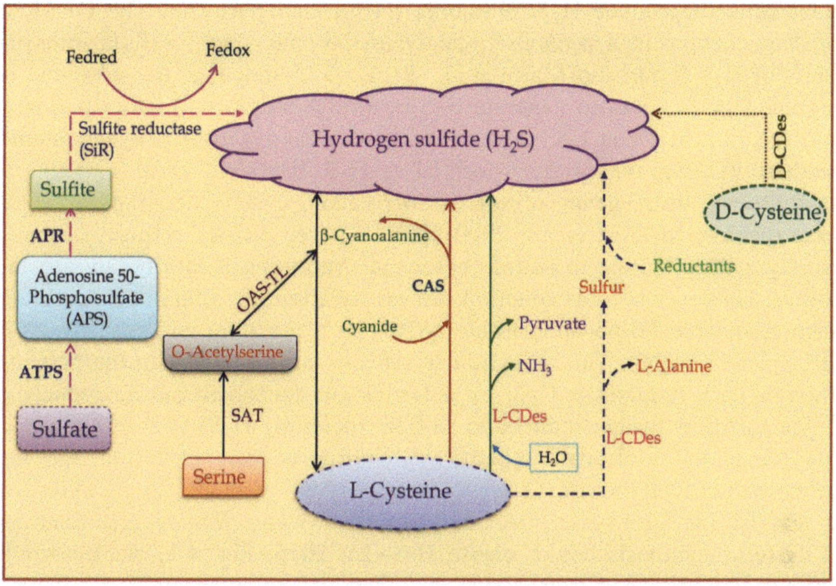

FIGURE 4.1 Hydrogen sulfide biosynthesis in plants. ATPS, ATP sulfurylase; APR, APS reductase; CAS, β-cyanoalanine synthase; D-CDes, D-cysteine desulfhydrase; Fedred, ferredoxin reduced; Fedox, ferredoxin oxidized; L-CDes, L-cysteine desulfhydrase; OAS-TL, O-acetylserine thiol lyase; SAT, serine acetyltransferase. *Modified after Huo, J., Huang, D., Zhang, J., Fang H., Wang, B., Wang, C., et al., 2018. Hydrogen sulfide: a gaseous molecule in postharvest freshness. Front. Plant Sci. 9, 1172.*

to H_2S but the reaction is dependent on the presence of the reductant (Papenbrock et al., 2007; Fig. 4.1, blue dotted arrows pathway). However, Nagasawa et al. (1985, 1988) reported the presence of another pathway in *Escherichia coli* and *Pseudomonas putida* where D-cysteine was the substrate and its decomposition in the presence of enzyme D-cysteine desulfhydrase (D-CDes) released H_2S as a product (Fig. 4.1, brown arrows pathway). Several specific cysteine desulfydrases have been identified to be involved in the biosynthesis of H_2S in plants. The identification and detection of the first desulfydrases, L- and D-cysteine amino acid-specific L- and D-cysteine desulfydrases that were involved in catalyzing the production of H_2S in chloroplasts and mitochondria or in the cytoplasm of plants, was reported by Rennenberg et al. (1987). Sulfur assimilation and H_2S biosynthesis pathways have been identified in plants due to the discovery of numerous cysteine desulfydrases (Bloem et al., 2004; Rausch and Wachter, 2005; Riemenschneider et al., 2005). The biosynthesis of H_2S is controlled by an enzyme D-CDes, in a reaction where it causes the degradation of D-cysteine resulting in the generation of pyruvate, H_2S, and NH_3 in *Arabidopsis thaliana* plants (Riemenschneider et al., 2005). L-CDes degrades L-cysteine present in the cytoplasm, nucleus, and mitochondria of plant cells to produce H_2S, ammonia (NH_3), and pyruvate. The DES1, a L-CDes enzyme in *Arabidopsis*, catalyzes the endogenous cellular biosynthesis of H_2S (Laureano-Marín et al., 2014). D-Cysteine, as the substrate, is degraded by D-CDes to generate H_2S in mitochondria. In another pathway in a reaction between L-cysteine and cyanide catalyzed by β-cyanoalanine synthase (CAS), the product formed is H_2S (Hatzfeld et al., 2000). In *Arabidopsis*, three genes (*cysc1, cys-d1*, and *cys-d2*) encoding for CAS were discovered (Jost et al., 2000) (Fig. 4.1, red arrows pathway). Plants usually reduce sulfate to sulfide before incorporating it into organic metabolites. However before reduction, the adenylation of sulfate takes place to form adenosine 50-phosphosulfate (APS), in a reaction catalyzed by ATP sulfurylase (ATPS). The reduction of APS to sulfite occurs in the plastids through APS reductase. Then by a ferredoxin-dependent sulfite reductase (SiR), sulfite is further reduced to sulfide (including H_2S) (Takahashi et al., 2011; Fig. 4.1, pink arrows pathway). Sulfide is also integrated into the amino acid skeleton of O-acetylserine (OAS) through the activity of O-acetylserine thiol lyase (OAS-TL) to form cysteine. The reverse reaction of cysteine synthesis could release H_2S (Li, 2015; Fig. 4.1, black arrow). The OAS is produced from serine in a reaction governed by serine acetyltransferase (SAT). H_2S synthesized within the plant cell acts as a signaling molecule and moves freely from one cell to another. Usually, plants manufacture H_2S through L-CDes or D-CDes, and it is in response to environmental stress, thus leading to elevation in stress tolerance. In addition, excess H_2S can be released to the air (Li, 2013; Calderwood and Kopriva, 2014; Hancock and Whiteman, 2014).

H_2S plays crucial roles in regulating plant growth and development. Numerous studies have shown that H_2S is directly or indirectly concerned with several physiological processes including seed germination (Zhang et al., 2008; Li and Liu 2012), root organogenesis (Zhang et al., 2009; Mei et al., 2019), photosynthesis (Chen et al., 2011), stomatal movement (Garcia-Mata and Lamattina, 2010; Zhang et al., 2020; Chen et al., 2020; Shen et al., 2020), and fruit ripening (Ge et al., 2017; Muñoz-Vargas et al., 2018), as well as senescence in leaves, flowers and fruits (Zheng et al., 2016). H_2S is also implicated in alleviating stresses like drought, heating, salt, and heavy metals (Deng et al., 2016; Du et al., 2019; Li et al., 2012). H_2S also is concerned with the regulation of ion channels. H_2S synchronizes stomatal closure by controlling the expression of genes coding for the protein involved in Ca^{2+} and K^+ channels in guard cells (Jin et al., 2013). H_2S helps to maintain high K^+/Na^+ ratios in cells as it regulates ion channels of the plasma membrane and represses K^+ efflux (Lai et al., 2014).

In order to contribute toward plant growth, development, and the regulation of stress, H_2S interacts with other signaling molecules and plant hormones, such as carbon monoxide (Lin et al., 2012), nitric oxide (Peng et al., 2016), Ca^{2+} (Li et al., 2012), H_2O_2 (Lv et al., 2017), abscisic acid (ABA) (Jin et al., 2013), auxin (IAA) (Jia et al., 2015), salicylic acid (Li et al., 2015), gibberellin (GA) (Xie et al., 2014), and ethylene (Eth) (Hou et al., 2013). Several physiological pathways and plant growth, development, and stress responses are modulated by microRNA (Sunkar et al., 2012). In *Arabidopsis* it was observed that H_2S played an important role in maintaining the transcription level of mir167a, mir167c, mir167d, mir393a, and mir396a, and eventually helped in elevating tolerance against drought stress (Shen et al., 2013).

It has been recognized that various interacting factors are responsible for organogenesis. Many studies have established proof of the involvement of H_2S in plant organogenesis. Sodium hydrosulfide (NaHS), being a H_2S donor, has been widely used as a source of H_2S in experimental studies. An upregulation in the expression of H_2S-anabolic genes even under normal growth conditions during exogenous treatment with NaHS was observed (Christou et al., 2013; Zhang et al., 2010a). The exogenous treatment with H_2S increased seed germination and improved the germination percentage of wheat seeds equally under normal conditions and under heavy metal stress (Zhang et al., 2010b). NaHS treatment has been found to be responsible for increasing the expression of genes involved in heme oxygenase-1/carbon monoxide (HO-1/CO)-induced adventitious root formation (Lin et al., 2012). The pretreatment with NaHS enhanced root organogenesis in *Ipomoea batatas, Salix matsudana*, and *Glycine max* (Zhang et al., 2009) and seedling growth in *Spinacia oleracea* (Chen et al., 2011). On the other hand, in *Pisum sativum* even low concentrations of H_2S increased the embryo root length, along with improvements in the root tip soluble proteins content, antioxidant enzymes activity as well as the survival rate of root border cells (Li et al., 2010). H_2S treatment to cut explants of flowers also showed delayed flower opening and senescence along with stimulated activity of antioxidant enzymes and decreased level of H_2O_2 (Zhang et al., 2011). Thus it is evident from studies that H_2S might be playing an essential role in the physiological process of plants such as organogenesis and growth. Exogenous treatment of H_2S in plants also activates expression of genes involved in the synthesis of secondary metabolites, along with a variety of compounds involved in plant defense that enhance plant development and abiotic stress resistance (Ma and Yang, 2018). However, the exact biochemistry of endogenous H_2S in plant cells and its metabolic interactions with other molecules are still unclear.

4.3 H_2S-induced protein persulfidation during oxidative stress

H_2S arbitrates various signaling reactions in plants. It has an ability to induce posttranslational modifications of cysteine (Cys) residues in macromolecular protein complexes. In fact the study of H_2S as a signaling molecule has been focused on its competence to interact with thiol ($-$SH) groups present in protein cysteine residues through the posttranslational modification (PTM) persulfidation (Corpas et al., 2019; Aroca et al., 2018). H_2S reacts with free Cys residue ($-SH-$) or different states of Cys residue (-S-S-, -S-OH, S-NO) to form $-SSH$, a process called S-sulfhydration or persulfidation (Paul and Snyder, 2012; Mustafa et al., 2009). The posttranslational modifications of Cys residues include S-nitrosation, S-sulfenylation, and S-gluthionylation. The synthesis of persulfides is catalyzed by sulfurtransferases and cysteine desulfurases. It appears that posttranslational alteration in protein through persulfidation is involved in mediating several activities within the cell (Filipovic, 2015). A detailed study of the *Arabidopsis* proteome revealed that at least 5% of proteins were persulfidated and just about 2015 endogenously persulfidated proteins were identified (Aroca et al., 2017). It was also revealed that proteins involved in vital biochemical pathways like the glycolytic cycle, tricarboxylic acid cycle, Calvin–Benson cycle, and starch biosynthesis also undergo persulfidation.

AUTOPHAGY-RELATED UBIQUITIN-LIKE 18a (Atg18a) has been identified as a target for persulfidation (Aroca et al., 2017). It seems that the posttranslational modification provides some defense to this enzyme as it shows great sensitivity to salinity, drought, and oxidative stresses (Xiong et al., 2005). The redox sulfidation induced by H_2S might be a mechanism for protecting the Cys residues of protein from unwanted oxidation (Cuevasanata et al., 2015). Cys residues are oxidized to sulfenic acids (RSOH) and further to sulfinic (RSO_2H) and sulfonic acids (RSO_3H) under long-lasting oxidative stress. However, usually these conversions of Cys residues are permanent and cannot be easily reverted back to thiols because oxidized Cys residues are situated deep within protein quaternary complexes (Cuevasanata et al., 2015). Hence, persulfidation can be treated as a defense mechanism adapted by the cell during H_2O_2-induced oxidative stress (Wedmann et al., 2016). The ROS such as hydrogen peroxide (H_2O_2), hydroxyl, and superoxide radicals are efficiently scavenged by persulfides. Persulfides form sulfonate ($RSSO_3H$) on reacting with ROS and sulfonates can be easily reduced to thiols through the action of thioredoxins (Filipovic, 2015; Inupakutika et al., 2016). The persulfides have a strong nucleophilic character due to its continued existence in a deprotonated (RSS) state under physiological conditions, and this nucleophilic character has imparted these moieties with an extensive scavenging property that hinders ROS and even reactive nitrogen species (RNS) (Filipovic and Jovanovic, 2017). NaHS treatment reduced the activity of glutamine synthetase (GS) and improved the activity of ascorbate peroxidise (APX) and glyceraldehyde-3-phosphate dehydrogenase (GAPDH) due to S-sulfhydration in *Arabidopsis* (Aroca et al., 2015). Some antioxidant enzymes like APX and catalase (CAT) are positively regulated by persulfidation and S-nitrosylation.

4.4 H_2S signaling triggered by heavy metal stress

H_2S exhibits a dual role in plant cells, that is, in a high concentration it behaves as a cytotoxin but in a low concentration it operates as a cell signaling molecule. Thus homeostasis of the amount of H_2S in plant cells is very vital for exerting its physiological functions including adaptation against abiotic stress. A balance between the generation and removal of endogenous H_2S is crucial for the adaptation of plants to metal stress. The metal stress triggers endogenous H_2S production. Metal tolerance in plants has been correlated with the enhanced level of endogenous H_2S content. Endogenous H_2S production was induced under Cd treatment in *Medicago sativa* (Cui et al., 2014). The level of endogenous H_2S increased in Chinese cabbage under Cd stress as the expression of genes (*DCD1* and *DES1*; of cysteine desulfhydrase, *OAS-TL* homogenous family) accountable for H_2S synthesis was upregulated (Zhang et al., 2015). A H_2S scavenger, hypotaurine decreased H_2S level in plants under Cd stress compared to nontreated plants indicating that elevated H_2S level is associated with Cd tolerance (Mostofa et al., 2015). In *Arabidopsis* under Cd stress, ABA-related transcription factor WRKY is activated which adjusts the transcripts of *LCD* and *DES* to produce H_2S (Liu et al., 2015a). The expression of the genes *LCD, DCD2*, and *DES* increased when seedlings were exposed to Cr^{6+} (Fang et al., 2014a). The alteration in H_2S content in plants under metal stress also depends upon various factors like species, genotypes, tissues of plants, different metals, metal concentrations, and the duration of treatment. The concentration and location of H_2S work as deciding factors as to whether it will exert toxic or protective effects in plants.

4.5 H_2S-induced metal stress tolerance

H_2S being a signaling molecule plays an essential role in plant tolerance to toxic metal stress. Multiple pathways regarding H_2S-induced metal tolerance have been identified. The metal in plants is generally reduced as H_2S modulates and controls metal transport proteins. H_2S improves the antioxidant capacity and helps in scavenging metal-induced ROS in plants. Ultimately, H_2S may also participate in the communication with other signaling pathways, protein S-sulfhydration, and microRNAs (miRNAs) in response to toxic metal stress (Fig. 4.2).

Endogenous and exogenous H_2S induces metal stress tolerance in plants. The decline in metal accumulation alleviates metal toxicity in plants. The pretreatment with the H_2S donor NaHS decreased Al-induced stress by escalating antioxidant capacity in wheat seeds (Zhang et al., 2010b). The H_2S hindered the uptake of Al and increased ATPase and photosynthetic machinery activity, thus contributing toward alleviating Al toxicity (Dawood et al., 2012). H_2S caused an enhancement in antioxidant capacity, increased citrate secretion, and improved plasma membrane H^+-ATPase activity in order to ameliorate Al toxicity (Chen et al., 2013). H_2S stimulated an increase in secretion of citrate along with upregulation in expression of an Al activated citrate transporter (*HvAACT1*) (Chen et al., 2013). The expression of *MT3A*

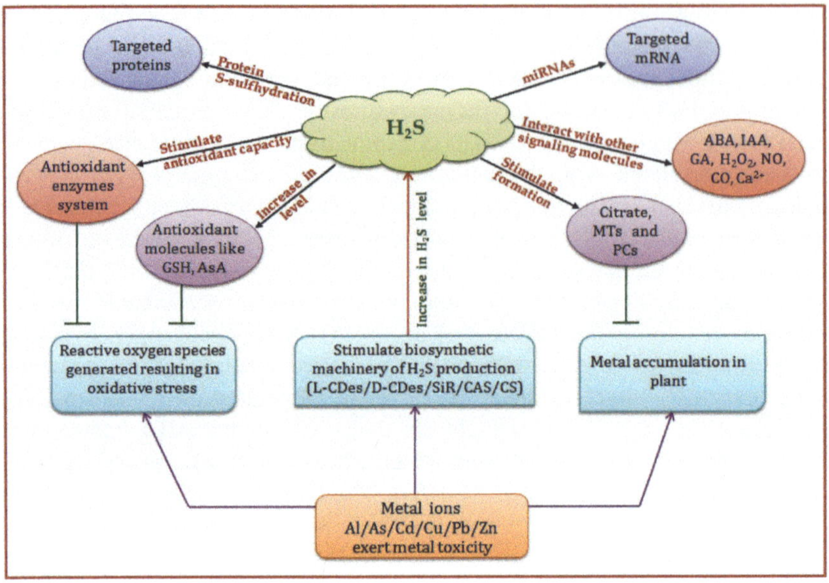

FIGURE 4.2 Diagrammatic representation on the role of H_2S in alleviating metal toxicity in plants. L-CDes, L-cysteine desulfhydrase; D-CDes, D-cysteine desulfhydrase; PCs, phytochelatins; MTs, metallothioneins.

and *PCs* was triggered by H_2S (Fang et al., 2014a) and it also interacted with Ca^{2+} to regulate metal transporters in response to Cr-induced metal stress (Fang et al., 2014b). H_2S also enhanced the expression of *PCS1, PCS2*, and *MT2A* in *Arabidopsis* to increase the accretion of PCs and MTs in response to Cr stress (Fang et al., 2016). H_2S helped the *Solanum nigrum* seedling to overcome Zn toxicity by decreasing Zn accumulation in the seedlings. The mitigating effect of H_2S can be related to the suppressed Zn uptake and decreased expression of related genes like zinc-regulated transporter(*ZRT*), iron-regulated transporter (*IRT*)-like protein (*ZIP*), and natural resistance associated macrophage protein (*NRAMP*) which are all connected with Zn homeostasis (Liu et al., 2016). H_2S also caused a high accumulation of MTs which chelated and removed the excess Zn^{2+} (Liu et al., 2016). The Cd-induced stress was alleviated through H_2S regulated methyl jasmonate (MeJA) by hindering the transcription of *NRAMP1/NRAMP6* and also escalating the expression of *MTP1/12* (metal tolerance protein), *CAX2* (Ca^{2+}/H^+ exchanger antiporter), and ZIP4 (an IRT-like protein) (Tian et al., 2017).

H_2S improved heavy metal-induced oxidative stress by stimulating the antioxidant system in plants. Plants have developed defense systems including both enzymatic and nonenzymatic antioxidants. Superoxide dismutase (SOD), CAT, ascorbate peroxidase (APX), and guaiacol peroxidase (GPX) are enzymatic antioxidants regulating ROS levels in cells. H_2S amplified

SOD and CAT activities in response to Cu stress but the activity of lipoxygenase (LOX) was reduced, consequently lowering malondialdehyde (MDA) content and H_2O_2 in wheat (Zhang et al., 2008). Exogenous H_2S upregulated the AsA-GSH cycle, which eventually decreased As-induced ROS and lipid peroxidation in pea seedlings (Singh et al., 2015). H_2S stimulated the expression of stress-related genes which repressed the accretion of ROS (Shi et al., 2015). Priming wheat seeds with NaHS helped the seedling to adapt against Cr stress because of enhanced enzymes activity, particularly of amylase and esterase, and antioxidative enzymes like APX, SOD, CAT, and GPX. However a significant decline in the LOX activity along with H_2O_2 and MDA content was observed (Zhang et al., 2010c).

An increase in ascorbic acid (AsA) and glutathione (GSH) content was recorded in response to H_2S. It appears that H_2S elevated the expression of gene glutamylcysteine synthetase (*GSH1*) and activity of glutathione reductase (GR) enzyme. Furthermore, the ratio of oxidized/reduced AsA was also affected as H_2S increased the proportion of reduced AsA. Thus H_2S activated antioxidant enzymes and antioxidant molecules such as AsA and GSH to increase tolerance to Cr stress (Fang et al., 2014a). Apart from increasing the accumulation of ROS, proteins, and membrane damage, a higher level of cysteine was also reported from seedlings under As-stress but NaHS treatment alleviated the damage (Singh et al., 2015). As- treatment suppressed the activity of the enzymes implicated in the AsA-GSH cycle. The redox ratio of AsA and GSH was disturbed, which was reflected by a sharp decline in the reduced/oxidized ratios. However, exogenously applied NaHS enhanced activities of enzymes, reinstating the redox status of the AsA and GSH pools (Singh et al., 2015). In *Arabidopsis* Cr stress was alleviated through the accumulation of PCs and MTs by increasing the expression of *PC1*, *PC2*, and *MT2A* in a process mediated by H_2S (Fang et al., 2016). It can be concluded that H_2S regulates metal stress tolerance by inhibiting membrane peroxidation, ROS accumulation, NADPH recycling metabolism, and by amplifying antioxidant enzymes activity (Kharbech et al., 2017).

H_2S signaling might be linked with plant response to metal stress. H_2S interacts with ABA, IAA, GA, JA, H_2O_2, NO, CO, and Ca^{2+} in signaling pathways (Jin and Pei, 2015). H_2S and NO are actively involved in metal stress tolerance mechanisms. H_2S and NO are involved in the alleviation of metal toxicity (Shi et al., 2014; Singh et al., 2015). An H_2S-mediated pathway contributes to the Ca^{2+}-activating antioxidant system. The interaction of H_2S with Ca^{2+} signaling mediated Cr tolerance in millet (Fang et al., 2014a). Calcium-dependent protein kinases improved Cd tolerance in *Arabidopsis* through H_2S-induced decline in oxidized glutathione (GSSH) content (Qiao et al., 2016). However detailed studies of the interaction of H_2S with ABA, CO, and H_2O_2 in plant response to metal stress are limited. High endogenous cysteine (Cys) levels are exhibited in response to NaHS treatment in seedlings under As-stress (Singh et al., 2015). This generates an opportunity for the utilization of excess Cys moieties in the formation of PCs which helps in the detoxification of the toxic metals within the cells (Banerjee and Roychoudhury, 2018). S-Sulfhydration converts cysteinyl thiolates to persulfides. The posttranslational modification, protein S-sulfhydration amends the activities of numerous proteins including actin, tubulin, and GAPDH (Mustafa et al., 2009). The H_2S-mediated sulfhydration of reactive Cys residues elevated the catalytic activity of targeted proteins (Paul and Snyder, 2012). The S-sulfhydration altered the activities of GS, APX, and GAPDH (Aroca et al., 2015). H_2S-mediated protein S- sulfhydration was indirectly implicated with increasing plant tolerance toward abiotic stresses (Liu et al., 2015b). The intensity of protein S-sulfhydration could be estimated in terms of GSSH content (Qiao et al., 2016). However, H_2S-mediated S-sulfhydration of protein in plants in response to metal stress has not been reported.

miRNAs are distributed throughout the plant kingdom and play an significant role in plant growth and development, and in stress responses through the degradation of mRNA or inhibition of its translation (Sunkar et al., 2012). Most of the miRNAs-regulated target genes were concerned in antioxidation, sulfate allocation and assimilation, miRNA biogenesis, and phytohormone signaling in plants in response to metal toxicity (Yang and Chen, 2013). However, the correlation between H_2S and miRNAs in plant reactions to metal stress has not been reported so far.

4.6 Conclusions and future perspectives

The research on H_2S is moving toward a new era. This chapter presents a better understanding of the importance of H_2S in metal tolerance by plants. Elevated H_2S levels in response to metal toxicity might be responsible for the alleviation of stress. The role of H_2S in improving plants' tolerance to metal stress can be summarized as H_2S promotes plant growth and development, reduces uptake of the metal ion, and activates the antioxidant enzymes system to scavenge the ROSs, thus consequently protecting plant cell structures. H_2S also actively enhances the expression of gene-related PCs and MTs synthesis. H_2S interacts with other signaling molecules and contributes to plant hormone signaling pathways. There are studies related to H_2S interactions with some signal molecules and hormones but a detailed pathway is not clear. Although H_2S functions as a signal molecule in plants' cross-adaptation, the signal pathways and its components have not been identified. There are no specific receptors or targets of H_2S identified but due to high lipophilicity it effortlessly penetrates the cell membrane, thus the absence of receptors can be acknowledged. However, H_2S induces the modification of some proteins through sulfhydration, so there is a need to evaluate the significance of these proteins as receptors. Protein S-sulfhydration mediated by H_2S is actively involved in the response of plants to abiotic stress. On the other hand, there is no definite information connecting

H_2S-mediated protein S-sulfhydration and heavy metal stress tolerance. There is a need to improve our understanding of the role of H_2S in alleviating heavy metal stress. It is important to reinforce the research on H_2S target molecules so that a complete signal transduction pathway is identified. In addition, the molecular mechanism of H_2S-induced tolerance against metal stress can be revealed through new studies involving the transcriptome, proteome, and metabolome. This will create a new approach for the genetic improvement of plants against metal stress.

References

Aroca, A., Benito, J.M., Gotor, C., Romero, L.C., 2017. Persulfidation proteome reveals the regulation of protein function by hydrogen sulfide in diverse biological processes in *Arabidopsis*. J. Exp. Bot. 68, 4915–4927.

Aroca, Á., Serna, A., Gotor, C., Romero, L.C., 2015. S-sulfhydration: a cysteine posttranslational modification in plant systems. Plant Physiol. 168, 334–342.

Aroca, A., Gotor, C., Romero, L.C., 2018. Hydrogen sulfide signaling in plants: emerging roles of protein persulfidation. Front. Plant Sci. 9, 1369.

Banerjee, A., Roychoudhury, A., 2018. Strigolactones: multi-level regulation of biosynthesis and diverse responses in plant abiotic stresses. Acta Physiol. Plant 40, 86.

Bloem, E., Riemenschneider, A., Volker, J., Jutta, P., Schmidt, A., Salac, I., et al., 2004. Sulphur supply and infection with *Pyrenopeziza brassicae* influence L cysteine desulphydrase activity in *Brassica napus* L. J. Exp. Bot 55, 2305–2312.

Calderwood, A., Kopriva, S., 2014. Hydrogen sulfide in plants: from dissipation of excess sulphur to signalling molecule. Nitric Oxide 41, 72–78.

Chen, J., Wang, W.H., Wu, F.H., You, C.Y., Liu, T.W., Dong, X.J., et al., 2013. Hydrogen sulfide alleviates aluminum toxicity in barley seedlings. Plant Soil. 362 (1–2), 301–318.

Chen, J., Wu, F.H., Wang, W.H., Zheng, C.J., Lin, G.H., Dong, X.J., et al., 2011. Hydrogen sulphide enhances photosynthesis through promoting chloroplast biogenesis, photosynthetic enzyme expression, and thiol redox modification in *Spinacia oleracea* seedlings. J. Exp. Bot. 62, 4481–4493.

Chen, S., Jia, H., Wang, X., Shi, C., Wang, X., Ma, P., et al., 2020. Hydrogen sulfide positively regulates abscisic acid signaling through persulfidation of SnRK2.6 in guard cells. Mol. Plant S1674–2052 (20), 30004–30006.

Christou, A., Manganaris, G.A., Papadopoulos, I., Fotopouls, V., 2013. Hydrogen sulphide induces systemic tolerance to salinity and non-ionic osmotic stress in strawberry plants through modification of reactive species biosynthesis and transcriptional regulation of multiple defence pathways. J. Exp. Bot. 64, 1953–1966.

Corpas, F.J., González-Gordo, S., Cañas, A., Palma, J.M., 2019. Nitric oxide and hydrogen sulfide in plants: which comes first? J. Exp. Bot. 70, 4391–4404.

Cuevasanata, E., Lange, M., Bonanata, J., Coitino, E.L., Ferrer-Sueta, G., et al., 2015. Reaction of hydrogen sulphide with disulfide and sulfenic acid to form the strongly nucleophilic persulfide. J. Biol. Chem. 290, 26866–26880.

Cui, W., Chen, H., Zhu, K., Jin, Q., Cui, J., Xia, Y., Zhang, J., 2014. Cadmium-induced hydrogen sulfide synthesis is involved in cadmium tolerance in *Medicago sativa* by reestablishment of reduced (homo) glutathione and reactive oxygen species homeostases. PLoS One 9 (10), e109669.

Dawood, M., Cao, F., Jahangir, M.M., Zhang, G., Wu, F., 2012. Alleviation of aluminium toxicity by hydrogen sulfide is related to elevated ATPase, and uppressed aluminium uptake and oxidative stress in barley. J. Hazard. Mater. 209–210, 121–128.

Deng, Y.Q., Bao, J., Yuan, F., Liang, X., Feng, Z.T., Wang, B.S., 2016. Exogenous hydrogen sulfide alleviates salt stress in wheat seedlings by decreasing Na^+ content. Plant Growth Regul. 79, 391–399.

Du, X., Jin, Z., Zhang, L., Liu, X., Yang, G., Pei, Y., 2019. H_2S is involved in ABA-mediated stomatal movement through MPK4 to alleviate drought stress in *Arabidopsis thaliana*. Plant Soil 435, 295–307.

Fang, H.H., Jing, T., Liu, Z.Q., Zhang, L., Jin, Z., Pei, Y., 2014a. Hydrogen sulfide interacts with calcium signaling to enhance the chromium tolerance in *Setaria italica*. Cell Calcium 56 (6), 472–481.

Fang, H.H., Liu, Z.Q., Jin, Z.P., Zhang, L.P., Liu, D.M., Pei, Y.X., 2016. An emphasis of hydrogen sulfide-cysteine cycle on enhancing the tolerance to chromium stress in *Arabidopsis*. Environ. Pollut. 213, 870–877.

Fang, H., Pei, Y., Tian, B., Zhang, L., Qiao, Z., Liu, Z., 2014b. Ca^{2+} participates in H2S induced Cr^{6+} tolerance in *Setaria italica*. Chin. J. Cell Biol. 36 (6), 758–765.

Filipovic, M.R., 2015. Persulphidation (S-sulphhydration) and H_2S. Handb. Exp. Pharmacol. 230, 29–59.

Filipovic, M.R., Jovanovic, V.M., 2017. More than just an intermediate: hydrogen sulphide signalling in plants. J Exp Bot. 68, 4733–4736.

Fotopoulos, V., Christou, A., Manganaris, G., 2013. Hydrogen sulfide as a potent regulator of plant responses to abiotic stress factors. In: Gaur, R.K., Sharma, P. (Eds.), Molecular Approaches in Plant Abiotic Stress. CRC Press, Boca Raton, FL, pp. 353–373.

Garcia-Mata, C., Lamattina, L., 2010. Hydrogen sulphide, a novel gasotransmitter involved in guard cell signalling. N. Phytol. 188, 977–984.

Garcia-Mata, C., Lamattina, L., 2013. Gasotransmitters are emerging as new guard cell signaling molecules and regulators of leaf gas exchange. Plant Sci. 201, 66–73.

Ge, Y., Hu, K.D., Wang, S.S., Hu, L.Y., Chen, X.Y., Li, Y.H., et al., 2017. Hydrogen sulfide alleviates postharvest ripening and senescence of banana by antagonizing the effect of ethylene. PLoS One 12 (6), e0180113.

Geng, B., Chang, L., Pan, C., Qi, Y., Zhao, J., Pang, Y., et al., 2004. Endogenous hydrogen sulfide regulation of myocardial injury induced by isoprotenol. Biochem. Biophys. Res. Commun. 318, 756–763.

Hancock, J.T., Whiteman, M., 2014. Hydrogen sulfide and cell signaling: team player or referee? Plant Physiol. Biochem. 78, 37–42.

Harrington, H.M., Smith, I.K., 1980. Cysteine metabolism in cultured tobacco cells. Plant Physiol. 65, 151–155.

Hatzfeld, Y., Maruyama, A., Schmidt, A., Noji, M., Ishizawa, K., Saito, K., 2000. b-cyanoalanine synthase is a mitochondrial cysteine synthase-like protein in spinach and *Arabidopsis*. Plant Physiol. 123, 1163–1172.

Hou, Z., Wang, L., Liu, J., Hou, L., Liu, X., 2013. Hydrogen sulfide regulates ethylene-induced stomatal closure in *Arabidopsis thaliana*. J. Integr. Plant Biol. 55, 277–289.

Huo, J., Huang, D., Zhang, J., Fang, H., Wang, B., Wang, C., et al., 2018. Hydrogen sulfide: a gaseous molecule in postharvest freshness. Front. Plant Sci. 9, 1172.

Inupakutika, M.A., Sengupta, S., Devireddy, A.R., Azad, R.K., Mittler, R., 2016. The evolution of reactive oxygen species metabolism. J. Exp. Bot. 67, 5933–5943.

Jia, H., Hu, Y., Fan, T., Li, J., 2015. Hydrogen sulfide modulates actin dependent auxin transport via regulating ABPs results in changing of root development in *Arabidopsis*. Sci. Rep. 5, 8251.

Jin, Z.P., Pei, Y.X., 2015. Physiological implications of hydrogen sulfide in plants: pleasant exploration behind its unpleasant odour. Oxid. Med. Cell Longev. 2015, 397502.

Jin, Z., Xue, S., Luo, Y., Tian, B., Fang, H., Li, H., et al., 2013. Hydrogen sulfide interacting with abscisic acid in stomatal regulation responses to drought stress in *Arabidopsis*. Plant Physiol. Biochem. 62, 41–46.

Jost, R., Berkowitz, O., Wirtz, M., Hopkins, L., Hawkesford, M.J., Hell, R., 2000. Genomic and functional characterization of the oas gene family encoding O-acetylserine (thiol) lyases, enzymes catalyzing the final step in cysteine biosynthesis in *Arabidopsis thaliana*. Gene 253, 237–247.

Kharbech, O., Houmani, H., Chaoui, A., Corpas, F.J., 2017. Alleviation of Cr (VI)-induced oxidative stress in maize (*Zea mays* L.) seedlings by NO and H_2S donors through differential organ dependent regulation of ROS and NADPH-recycling metabolisms. J. Plant Physiol. 219, 71–80.

Lai, D., Mao, Y., Zhou, H., Li, F., Wu, M., Zhang, J., et al., 2014. Endogenous hydrogen sulfide enhances salt tolerance by coupling the reestablishment of redox homeostasis and preventing salt-induced K^+ loss in seedlings of *Medicago sativa*. Plant Sci. 225, 117–129.

Laureano-Marín, A.M., García, I., Romero, L.C., Gotor, C., 2014. Assessing the transcriptional regulation of L-cysteine desulfhydrase 1 in *Arabidopsis thaliana*. Front. Plant Sci. 5, 683.

Li, Z.G., 2013. Hydrogen sulfide: a multifunctional gaseous molecule in plants. Russ. J. Plant Physiol. 60, 733–740.

Li, Z.G., 2015. Analysis of some enzymes activities of hydrogen sulfide metabolism in plants. Methods Enzymol. 555, 253–269.

Li, Z.G., Liu, P., 2012. Hydrogen sulfide is a mediator in HO induced seed germination in *Jatropha Curcas*. Acta Physiol. Plant 34, 2207–2213.

Li, D.B., Xiao, Z.X., Liu, L.X., Wang, J.C., Song, G.L., Bi, Y.R., 2010. Eff. exogenous hydrogen sulfide (H_2S) root tip and root border cells of *Pisum sativum*. Chin. Bull. Bot. 45, 354–362.

Li, Z.G., Gong, M., Xie, H., Yang, L., Li, J., 2012. Hydrogen sulfide donor sodium hydrosulfide-induced heat tolerance in tobacco (*Nicotiana tabacum* L) suspension cultured cells and involvement of Ca^{2+} and calmodulin. Plant Sci. 185-186, 185–189.

Li, Z.G., Xie, L.R., Li, X.J., 2015. Hydrogen sulfide acts as a downstream signal molecule in salicylic acid-induced heat tolerance in maize (*Zea mays* L.) seedlings. J. Plant Physiol. 177, 121–127.

Lin, Y.T., Li, M.Y., Cui, W.T., Lu, W., Shen, W.B., 2012. Haemoxygenase-1 is involved in hydrogen sulfide-induced cucumber adventitious root formation. J. Plant. Growth Regul. 31, 519–528.

Liu, Z.Q., Fang, H.H., Pei, Y.X., Jin, Z.P., Zhang, L.P., Liu, D.M., 2015a. WRKY transcription factors down-regulate the expression of H_2S generating genes, LCD and DES in *Arabidopsis thaliana*. Sci. Bull. 60 (11), 995–1001.

Liu, Z.Q., Pei, Y.X., Fang, H.H., Tian, B.H., 2015b. H_2S Regulates foxtail millet responsing to stress by protein S-sulfhydration. Chin. J. Biochem. Mol. Biol. 31 (10), 1085–1091.

Liu, X., Chen, J., Wang, G.H., Wang, W.H., Shen, Z.J., Luo, M.R., et al., 2016. Hydrogen sulfide alleviates zinc toxicity by reducing zinc uptake and regulating genes expression of antioxidative enzymes and metallothioneins in roots of the cadmium/zinc hyperaccumulator *Solanum nigrum* L. Plant Soil 400, 177–192.

Lv, W., Yang, L., Xu, C., Shi, Z., Shao, J., Xian, M., et al., 2017. Cadmium disrupts the balance between hydrogen peroxide and superoxide radical by regulating endogenous hydrogen sulfide in the root tip of *Brassica rapa*. Front. Plant Sci. 8, 232.

Ma, Q., Yang, J., 2018. Transcriptome profiling and identification of functional genes involved in H2S response in grapevine tissue cultured plantlets. Genes Genom. 40 (12), 1287–1300.

Mathai, J.C., Andreas, M., Philipp, K., Saparov, S.M., Zeidel, M.L., Lee, J.K., et al., 2009. No facilitator required for membrane transport of hydrogen sulfide. Proc. Natl. Acad. Sci. USA 106, 16633–16638.

Mei, Y., Zhao, Y., Jin, X., Wang, R., Xu, N., Hu, J., et al., 2019. L-Cysteine desulfhydrasedependent hydrogen sulfide is required for methane-induced lateral root formation. Plant Mol. Biol. 99 (3), 283–298.

Mitsuhashi, H., Yamashita, S., Ikeuchi, H., Kuroiwa, T., Kaneko, Y., Kiromura, K., et al., 2006. Oxidative stress-dependent conversion of hydrogen sulfide to sulfite by activated neurophils. Shock 24, 529.

Mostofa, M.G., Rahman, A., Ansary, M.M.U., Watanabe, A., Fujita, M., Tran, L.P., 2015. Hydrogen sulfide modulates cadmium-induced physiological and biochemical responses to alleviate cadmium toxicity in rice. Sci. Rep. 5, 14078.

Muñoz-Vargas, M.A., González-Gordo, S., Cañas, A., López-Jaramillo, J., Palma, J.M., Corpas, F.J., 2018. Endogenous hydrogen sulfide (H_2S) is up-regulated during sweet pepper (*Capsicum annuum* L.) fruit ripening. In vitro analysis shows that NADP-dependent isocitrate dehydrogenase (ICDH) activity is inhibited by H_2S and NO. Nitric Oxide 81, 36–45.

Mustafa, A.K., Gadalla, M.M., Sen, N., Kim, S., Mu, W., Gazi, S.K., et al., 2009. H_2S signals through protein S-sulfhydration. Sci. Signal. 2 (96), ra72.

Nagasawa, T., Ishii, T., Kumagai, H., Yamada, H., 1985. D-cysteine desulfhydrase of *Escherichia coli*: purification and characterization. FEBS J. 153, 541–551.

Nagasawa, T., Ishii, T., Yamada, H., 1988. Physiological comparison of D-cysteine desulfhydrase of *Escherichia coli* with 3-chloro-D-alanine dehydrochlorinase of *Pseudomonas putida* CR 1-1. Arch. Microbiol. 149, 413–416.

Papenbrock, J., Riemenschneider, A., Kamp, A., Schulz-Vogt, H.N., Schmidt, A., 2007. Characterization of cysteine-degrading and H_2S-releasing enzymes of higher plants-from the field to the test tube and back. Plant Biol. 9, 582–588.

Paul, B.D., Snyder, S.H., 2012. H_2S signalling through protein sulfhydration and beyond. Nat. Rev. Mol. Cell Biol. 13, 499–507.

Peng, R., Bian, Z., Zhou, L., Cheng, W., Hai, N., Yang, C., et al., 2016. Hydrogen sulfide enhances nitric oxide induced tolerance of hypoxia in maize (*Zea mays* L.). Plant Cell Rep. 35, 2325–2340.

Qiao, Z.J., Jing, T., Jin, Z., Pei, Y., 2016. CDPKs enhance Cd tolerance through intensifying H_2S signal in *Arabidopsis thaliana*. Plant Soil 398 (1), 99–110.

Rausch, T., Wachter, A., 2005. Sulfur metabolism: a versatile platform for launching defence operations. Trends Plant Sci. 10, 503–509.

Rennenberg, H., 1983. Cysteine desulfhydrase activity in cucurbit plants: stimulation by preincubation with L-or D-cysteine. Phytochemistry 22, 1557–1560.

Rennenberg, H., Arabatzis, N., Grundel, I., 1987. Cysteine desulphydrase activity in higher plants: evidence for the action of L- and D-cysteine specific enzymes. Phytochemistry 26, 1583–1589.

Riemenschneider, A., Wegele, R., Schmidt, A., Papenbrock, J., 2005. Isolation and characterization of a D-cysteine desulfhydrase protein from *Arabidopsis thaliana*. FEBS J. 272, 1291–1304.

Sekiya, J., Schmidt, A., Wilson, L.G., Filner, P., 1982. Emission of hydrogen sulfide by leaf tissue in response to L-cysteine. Plant Physiol. 70, 430–436.

Shen, J., Xing, T., Yuan, H., Liu, Z., Jin, Z., Zhang, L., et al., 2013. Hydrogen sulfide improves drought tolerance in *Arabidopsis thaliana* by microRNA expressions. PLoS One 8, e77047.

Shen, J., Zhang, J., Zhou, M., Zhou, H., Cui, B., Gotor, C., et al., 2020. Persulfidation-based modification of cysteine desulfhydrase and the NADPH oxidase RBOHD controls guard cell abscisic acid signaling. Plant Cell 32 (4), 1000–1017.

Shi, H.T., Li, R.J., Cai, W., Liu, W., Fu, Z.W., Lu, Y.T., 2012. In vivo role of nitric oxide in plant response to abiotic and biotic stress. Plant Signal. Behav. 7, 437–439.

Shi, H., Ye, T., Chan, Z., 2014. Nitric oxide-activated hydrogen sulfide is essential for cadmium stress response in bermudagrass (*Cynodon dactylon* (L). Pers.). Plant Physiol. Biochem. 74, 99–107.

Shi, H., Ye, T., Han, N., Bian, H., Liu, X., Chan, Z., 2015. Hydrogen sulfide regulates abiotic stress tolerance and biotic stress resistance in *Arabidopsis*. J. Integr. Plant Biol. 57 (7), 628–640.

Simontacchi, M., Garcia-Mata, C., Bartoli, C.G., Santa-Maria, G.E., Lamattina, L., 2013. Nitric oxide as a key component in hormone-regulated processes. Plant Cell Rep. 32, 853–866.

Singh, V.P., Singh, S., Kumar, J., Prasad, S.M., 2015. Hydrogen sulfide alleviates toxic effects of arsenate in pea seedlings through up-regulation of the ascorbate-glutathione cycle: possible involvement of nitric oxide. J. Plant Physiol. 181, 20–29.

Sunkar, R., Li, Y.F., Jagadeeswaran, G., 2012. Functions of microRNAs in plant stress responses. Trends Plant Sci. 17, 196–203.

Takahashi, H., Kopriva, S., Giordano, M., Saito, K., Hell, R., 2011. Sulfur assimilation in photosynthetic organisms: molecular functions and regulations of transporters and assimilatory enzymes. Annu. Rev. Plant Biol 62, 157–184.

Tian, B.H., Zhang, Y.J., Jin, Z.P., Liu, Z.Q., Pei, Y.X., 2017. Role of hydrogen sulfide in the methyl jasmonate response to cadmium stress in foxtail millet. Front. Biosci. 22, 530–539.

Wang, R., 2012. Physiological implications of hydrogen sulfide: a whiff exploration that blossomed. Physiol. Rev. 92, 791–896.

Wang, Y., Frei, M., 2011. Stressed food—the impact of abiotic environmental stresses on crop quality. Agri. Ecosyst. Environ. 141, 271–286.

Wedmann, R., Onderka, C., Wei, S., 2016. Improved tag-switch method reveals that thioredoxin acts as depersulfidase and controls the intracellular levels of protein persulfidation. Chem. Sci. 7, 3414–3426.

Whiteman, M., Armstrong, J.S., Chu, S.H., Jia-Ling, S., Wong, B.S., Cheung, N.S., et al., 2004. The novel neuromodulater hydrogen sulfide: an endogenous peroxynitrite 'scavenger'? J. Neurochem. 90, 765–768.

Xie, Y., Lai, D., Mao, Y., Zhang, W., Shen, W., Guan, R., 2013. Molecular cloning, characterization, and expression analysis of a novel gene encoding L-cysteine desulfhydrase from *Brassica napus*. Mol. Biotechnol. 54, 737–746.

Xie, Y., Zhang, C., Lai, D., Sun, Y., Samma, M.K., Zhang, J., et al., 2014. Hydrogen sulfide delays GA-triggered programmed cell death in wheat aleurone layers by the modulation of glutathione homeostasis and heme oxygenase-1 expression. J. Plant Physiol. 171, 53–62.

Xiong, Y., Contento, A.L., Bassham, D.C., 2005. AtATG18a is required for the formation of autophagosomes during nutrient stress and senescence in *Arabidopsis thaliana*. Plant J. 42, 535–546.

Yang, Z.M., Chen, J., 2013. A potential role of microRNAs in plant response to metal toxicity. Metallomics 5, 1184–1190.

Zhang, H., Hu, L.Y., Hu, K.D., He, Y.D., Wang, S.H., Luo, J.P., 2008. Hydrogen sulfide promotes wheat seed germination and alleviates oxidative damage against copper stress. J. Integr. Plant Biol. 50, 1518–1529.

Zhang, H., Tang, J., Liu, X.P., Wang, Y., Yu, W., Peng, W.Y., et al., 2009. Hydrogen sulfide promotes root organogenesis in *Ipomoea batatas, Salix matsudana* and *Glycine max.* J. Integr. Plant Biol. 51, 1084–1092.

Zhang, H., Dou, W., Jiang, C.X., Wei, Z.J., Liu, J., Jones, R.L., 2010a. Hydrogen sulfide stimulates b-amylase activity during early stages of wheat grain germination. Plant Signal. Behav. 5, 1031–1033.

Zhang, H., Tan, Z.Q., Hu, L.Y., Wang, S.H., Luo, J.P., Jones, R.L., 2010b. Hydrogen sulfide alleviates aluminium toxicity in germinating wheat seedlings. J. Integr. Plant Biol. 52, 556–567.

Zhang, H., Hu, L.Y., Li, P., Hu, K.D., Jiang, C.X., Luo, J.P., 2010c. Hydrogen sulfide alleviated chromium toxicity in wheat. Biol. Plant (Prague) 54, 743–747.

Zhang, H., Hu, S.L., Zhang, Z.J., Hua, L.Y., Jiang, C.X., Wei, Z.J., et al., 2011. Hydrogen sulfide acts as a regulator of flower senescence in plants. Postharvest Biol. Technol. 60, 251–257.

Zhang, L., Pei, Y., Wang, H., Jin, Z., Liu, Z., Qiao, Z., et al., 2015. Hydrogen sulfide alleviates cadmium-induced cell death through restraining ROS accumulation in roots of *Brassica rapa* L. ssp. pekinensis. Oxid. Med. Cell Longev. 2015, 1–11.

Zheng, J.L., Hu, L.Y., Hu, K.D., Wu, J., Yang, F., Zhang, H., 2016. Hydrogen sulfide alleviates senescence of fresh-cut apple by regulating antioxidant defense system and senescence-related gene expression. HortScience 51, 152–158.

Zhang, J., Zhou, M., Ge, Z., Shen, J., Zhou, C., Gotor, C., et al., 2020. ABA-triggered guard cell L-cysteine desulfhydrase function and in situ H2S production contributes to heme oxygenase-modulated stomatal closure. Plant Cell Environ. 43, 624–636.

Chapter 5

Hydrogen sulfide and lateral root development in plants under stress

Muhammad Fasih Khalid[1,2], Sajjad Hussain[1,*], Anas Fadli[2], Faisal Shahzad[2], Muhammad Akbar Anjum[1], Iqra Zakir[3] and Shakeel Ahmad[3]

[1]Department of Horticulture, Faculty of Agricultural Sciences and Technology, Bahauddin Zakariya University, Multan, Pakistan, [2]Horticultural Sciences Department, Citrus Research and Education Centre, Institute of Food and Agricultural Sciences, University of Florida, Lake Alfred, FL, United States, [3]Department of Agronomy, Faculty of Agricultural Sciences and Technology, Bahauddin Zakariya University, Multan, Pakistan

**Corresponding author.*

5.1 Introduction

Hydrogen sulfide (H_2S) is a naturally occurring, colorless, water-soluble, and flammable gas belonging to a family of related molecules called sulfur species (Corpas, 2019). Although H_2S has been historically considered to be a highly toxic and pollutant gas, there is increasing evidence supporting its role as a gaseous transmitter in animal and plant systems along with nitric oxide (NO) and carbon monoxide (CO) (García-Mata and Lamattina, 2010; Tang et al., 2006). Many researchers showed that H_2S mainly contributes to seed germination, root organogenesis, physiological mechanisms, and postharvest physiology (Fig. 5.1).

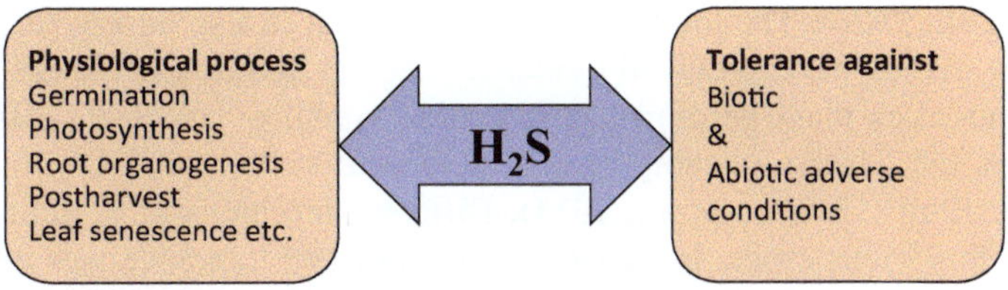

FIGURE 5.1 The main contribution of H_2S in plants.

Hydrogen Sulfide in Plant Biology. DOI: https://doi.org/10.1016/B978-0-323-85862-5.00007-5

5.2 Hydrogen sulfide and environmental stress

As sessile organisms, plants are continuously exposed to multiple environmental stresses, such as drought, salinity, heavy metals, and extreme temperatures (Khalid et al., 2019). These conditions result generally in an increase in reactive oxygen species, like hydroxyl radicals (OH^-), superoxide ions (O_2^-), and hydrogen peroxide (H_2O_2), which leads to the inactivation of enzyme activities, protein degradation, alterations in gene expression, and subsequently cellular damage (Suzuki et al., 2012).

In recent years, a number of reports have highlighted the influence of exogenous H_2S on plant response to environmental stresses, particularly its involvement in the regulation of cell signaling pathways, gene expression levels and the synthesis of antioxidant enzymes (Fotopoulos et al., 2015; Singh et al., 2019). Many works have also reported a rapid accumulation of endogenous H_2S under various abiotic stresses to levels that trigger physiological responses (Fang et al., 2014a,b; Fu et al., 2013; Lai et al., 2014; Zhang et al., 2008, 2010a). However, the molecular mechanisms and signal transduction pathways of H_2S in plant biology are still unclear. The knowledge of these mechanisms and the changes that occur in various conditions is important to fully understand how plants can respond to their environment.

Cold stress affects the growth and development of plants directly by inhibiting enzyme activity and metabolic reactions and, indirectly, through cold-induced osmotic stress (inhibition of water uptake and cellular dehydration) and oxidative stress (Chinnusamy et al., 2007). Hydrogen sulfide was reported to mitigate both osmotic and oxidative stresses induced by low temperatures by modulating the activities of antioxidant enzymes, enhancing the accumulation of osmolytes (proline, amino acids, and total soluble sugars), and reducing cell membrane permeability (Fu et al., 2013; Shi et al., 2013; Stuiver et al., 1992). More recently, the mitogen activated protein kinase (MAPK) pathway was shown to be involved in the H_2S-mediated response of *Arabidopsis* seedlings to cold stress, with a specific effect on stomatal opening (Du et al., 2017).

Heat, drought, and salinity have attained a serious concern in the context of climate change. Increasing temperature adversely affects plant growth and development, resulting in massive yield losses (Akter and Islam, 2017; Khalid et al., 2020). The tolerance of plants to heat is also improved by H_2S. Previous studies have provided evidence that the addition of NaHS, a H_2S donor, enhances plant germination and tissue viability (Li et al., 2013) and attenuates the adverse phenotypic effects of heat stress such as wilting and/or leaf curling (Christou et al., 2014). Different mechanisms of action were reported in various crops. In maize the H_2S-induced tolerance was mediated by an increase in salicylic acid (SA) and proline (Li et al., 2013, 2015). In tobacco, it was suggested that H_2S regulates the entry of extracellular calcium ions across the plasma membrane through a direct effect on calmodulin, a ubiquitous calcium-binding protein (Li et al., 2012). In strawberry H_2S alleviated oxidative stress and helped to preserve root tissues against heat-induced damage by inducing gene expression of several antioxidants and heat shock proteins, including catalase, superoxide dismutase, HSP70, HSP80, and HSP90 (Christou et al., 2014).

The involvement of H_2S in plant responses to drought has been also widely studied. Depending on exposure time and dose, H_2S may induce the closure or opening of stomata. Both effects are mediated by 8-mercapto-cGMP as a signaling molecule and involve the activity of Ca^{2+}, cADP ribose, and slow anion channel 1 (Honda et al., 2015). By knocking down the L-cysteine desulfhydrase (LCD) in *Arabidopsis* plants, Jin et al. (2013) studied the function of H_2S in stomatal movement and the relationship between H_2S and abscisic acid (ABA) pathways in signaling transduction. They concluded that H_2S is involved in the expression regulation of ABA receptor candidates and K^+ and Ca^{2+} ion-channels in guard cells. Their conclusion was further supported by the phenotype of the *lcd* mutant which showed an enlarged stomatal aperture and increased drought sensitivity. Other studies have further suggested the involvement of ethylene signaling in H_2S-induced stomatal closure (Liu et al., 2011) and/or a crosstalk between H_2S, NO, and ABA with NO being downstream of H_2S in the signaling pathway (Scuffi et al., 2014a). In addition to stomatal movement, numerous works have reported that H_2S helps in providing tolerance to drought through the accumulation of osmoregulators like proline and trehalose and the involvement of the calcium messenger system (Li et al., 2014).

Heavy metals such as Cd, Cr, Cu, and Zn are highly toxic for plants. Their accumulation in the intracellular compartments can cause DNA damage, enzyme inactivation, protein oxidation, and lipid peroxidation (Jaishankar et al., 2014; Shahid et al., 2014). The role of H_2S in plant response to heavy metal stress has been recently reviewed by Hancock and Whiteman (2015) and He et al. (2018). In cucumber seedlings, pretreatment with NaHS altered the expression of cell wall-associated proteins, such as PME (*CsPME*) and expansin (*CsExp*), leading to the alleviation of boron-induced toxicity (Wang et al., 2010). Similarly, NaHS alleviated the inhibitory effects of copper and reduced the visible symptoms in germinated seeds and radical tips.

5.3 Signaling of hydrogen sulfide

Hydrogen sulfide has been considered as a toxic gas to living organisms; however, it has emerged as a new signaling molecule in plants and animal cells, considered to be as equally important as hydrogen peroxide (H_2O_2), carbon monoxide (CO), and nitric oxide (NO) (Garcia-Mata and Lamattina, 2010; Vandiver and Snyder, 2012; Kimura, 2014). Hydrogen sulfide is a central component in plant stress responses. Many stresses may lead to the accumulation of H_2S in plants and this may lead to the alleviation of cell stress by different cellular mechanisms (Fig. 5.2). A number of reports have been published on this gasotransmitter related to growth of vegetative plants and affecting the disease resistance (Rodriguez-Kabana et al., 1965; Thompson and Kats, 1978), protective roles against stresses such as oxidative stresses (Fang et al., 2016), drought and heat tolerance (Li et al., 2012; Shen et al., 2013), osmotic and saline stresses (Shi et al., 2013), stomatal closure/opening (Papanatsiou et al., 2015), modulation of photosynthetic machinery (Chen et al., 2011), and autophagy regulation (Laureano-Marin et al., 2016). H_2S production occurs in chloroplasts via the photosynthetic sulfate-assimilation pathway and the reaction is catalyzed by sulfide reductase (Garcia et al., 2015). The enzyme cyanoalanine synthase cI also produces H_2S in mitochondria during the synthesis of β-cyanoalanine (Yamaguchi et al, 2000). In addition, cytosolic H_2S is produced mainly with the action of the enzyme L-cysteine desulhydrase 1 and it is of interest that the sulfide concentration in chloroplasts is always higher than in cytosol, due to the basic pH inside the organelles making it difficult for the sulfide to pass through the cytosol (Kabil and Banerjee, 2010). Moreover, H_2S is a very reactive molecule to modify signals in relation to the plant hormones, including gibberellic acid (GA) (Xie et al., 2014), ABA (Scuffi et al., 2014b), ethylene (Liu et al., 2012), nitric oxide (NO) (Lisjak et al., 2011), and hydrogen peroxide (H_2O_2) (Zhang et al., 2010b). With respect to the primary mechanism of action, a posttranslational modification of proteins (persulfidation) is formed when the conversion of thiol group (−SH) into persulfide group (−SSH) occurs at the reactive cysteine residues on the target proteins. Persulfidation showed an increased nucleophilicity compared with the thiol group; thus modified cysteines demonstrate greater challenge activity leading to a higher percentage of proteins compared to the reactive nitrogen and oxygen species. The modified targets of H_2S are unclear;

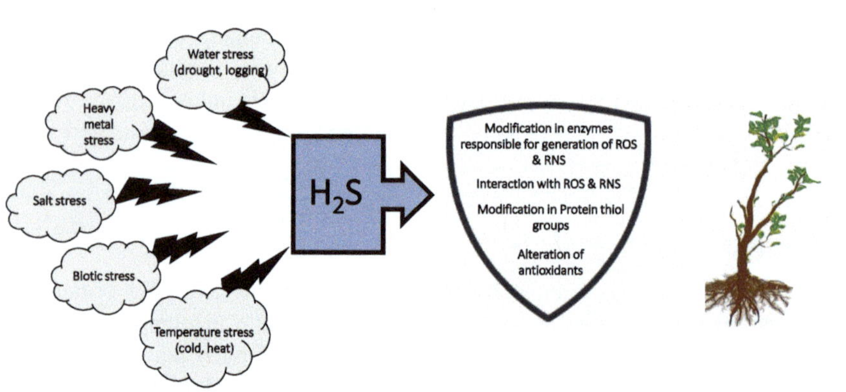

FIGURE 5.2 Influence of H_2S in cellular reactions under stress.

however, the sulfane's sulfur atom has the unique ability to bind reversibly to other sulfur atoms to generate hydrosulfides and polysulfides. Polysulfides seem to be effective in the persulfidation since these are more nucleophilic than H_2S (Toohey, 1989). These low-molecular-weight compound persulfides have newly emerged as the potential mediators in sulfide signaling. In this aspect cysteine-persulfide (Cys-SSH), glutathione persulfide, and its persulfurated species have been recognized as the redox regulators (Kimura et al., 2017). In conclusion, the extent of the interaction between signaling molecules and the same residue (cysteine) needs more investigation in order to expand our knowledge on the biochemical cascade triggered in the plant cells to different stresses.

5.4 Hydrogen sulfide and production of lateral roots

The mechanism involved in H_2S signaling related to plant growth is very complex and is intertwined between auxin, H_2S, and actin-binding proteins (ABPs) that are responsible for controlling the root growth. Auxin has been identified as an important phytohormone which plays a role in developmental and physiological processes in the plants including embryogenesis, root development, lateral root and adventitious root initiation, tropism, and vascular differentiation, suggesting the establishment of auxin concentration gradients in relation to auxin biosynthesis and polar auxin transport. Auxin is produced in the young aerial tissues and roots, especially in the meristematic part of the root apices and transported to basal parts (e.g., roots) via a specialized PAT3 delivery system (Rashotte et al., 2000; Dubrovsky et al., 2011). That creates an auxin gradient and it is maximum at the root apex that is important to root growth and for the lateral root development (Benková et al., 2003; Rahman et al., 2007). Lateral root development is initiated through the asymmetric division in a pair of the founder cells within the xylem pole pericycle cells. After getting it to the auxin threshold level, xylem pole pericycle cells sense and transduce it leading to activation of the cell cycle machinery and progression of the S phase. In addition, auxin has been reported to affect the pattering and organization of actin cytoskeleton during the cell growth. Instead, the directional transport of auxin affects the actin cytoskeleton partially through modulating the cycling of auxin efflux carriers (Mudaya and Murphy, 2002; Sun et al., 2004). Literature supports the adventitious root formation in batatas and lateral roots in the tomatoes (Fang et al., 2014a,b). The root organogenesis mechanism is regulated by auxins (Sabatini et al., 1999), and its interaction with H_2S modulated the root growth system (Zhang et al., 2009; Fang et al., 2014a,b). Auxin distribution was maximum in the root apex and it is very necessary for the meristem maintenance, being observed in QC, columella initial cells, and mature columella cells present on the root apices but reduced under the influence of a higher degree of H_2S. In addition, when the internal production of H_2S is increased by NaHS or GYY4137, the indole acetic acid (IAA) movement is also reduced in the roots. In a study on *Arabidopsis* roots, higher amounts of H_2S inhibited the distribution and transportation of auxin, implying alterations in the primary roots, lateral roots, and root hairs. This inhibition is caused by an alteration in the assembly and depolymerization of F-actin, ABPs act as downstream effectors of H_2S signal transduction that regulates the depolymerization of F-actin in root cells; ultimately, alterations in auxin transport occurs. H_2S also inhibited the primary root growth and root hair growth but also promoted the lateral root formation in *Arabidopsis*. At the end, H_2S actually triggered changes in the auxin transport and its distribution in roots of *Arabidopsis*.

5.5 Conclusions

Hydrogen sulfide is a new emerging signaling molecule in plants. H_2S is involved in many plant metabolisms, especially in tolerance machinery under unfavorable conditions (biotic and abiotic). H_2S is a central component in plant stress responses. Many stresses may lead to the accumulation of H_2S in plants and this may lead to the alleviation of cell stress by different cellular mechanism. H_2S signaling related to plant growth is very complex and is intertwined between auxin, H_2S, and ABPs that are responsible for controlling the root growth by making changes in the transportation and distribution of auxins which are responsible for root formation.

References

Akter, N., Islam, M.R., 2017. Heat stress effects and management in wheat. A review. Agron. Sustain. Dev. 37, 37.

Benková, E., Michniewicz, M., Sauer, M., Teichmann, T., Seifertová, D., Jürgens, G., et al., 2003. Local, efflux-dependent auxin gradients as a common module for plant organ formation. Cell 115, 591–602.

Chen, J., Wu, F.H., Wang, W.H., Zheng, C.J., Lin, G.H., Dong, X.J., et al., 2011. Hydrogen sulphide enhances photosynthesis through promoting chloroplast biogenesis, photosynthetic enzyme expression, and thiol redox modification in *Spinacia oleracea* seedlings. J. Exp. Bot. 62, 4481–4493.

Chinnusamy, V., Zhu, J., Zhu, J.-K., 2007. Cold stress regulation of gene expression in plants. Trends Plant Sci. 12, 444–451.

Christou, A., Filippou, P., Manganaris, G.A., Fotopoulos, V., 2014. Sodium hydrosulfide induces systemic thermotolerance to strawberry plants through transcriptional regulation of heat shock proteins and aquaporin. BMC Plant Biol. 14, 42.

Corpas, F.J., 2019. Nitric oxide and hydrogen sulfide in higher plants under physiological and stress conditions. Multidiscip. Digital Publ. Inst. 8, 457.

Du, X., Jin, Z., Liu, D., Yang, G., Pei, Y., 2017. Hydrogen sulfide alleviates the cold stress through MPK4 in *Arabidopsis thaliana*. Plant Physiol. Biochem. 120, 112–119.

Dubrovsky, J.G., Napsucialy-Mendivil, S., Duclercq, J., Cheng, Y., Shishkova, S., Ivanchenko, M.G., et al., 2011. Auxin minimum defines a developmental window for lateral root initiation. N. Phytol. 191, 970–983.

Fang, H., Jing, T., Liu, Z., Zhang, L., Jin, Z., Pei, Y., 2014a. Hydrogen sulfide interacts with calcium signaling to enhance the chromium tolerance in *Setaria italica*. Cell Calcium 56, 472–481.

Fang, T., Cao, Z., Lia, Z., Shen, W., Huang, L., 2014b. Auxin-induced hydrogen sulfide generation is involved in lateral root formation in tomato. Plant Physiol. Biochem. 76, 44–51.

Fang, H., Liu, Z., Jin, Z., Zhang, L., Liu, D., Pei, Y., 2016. An emphasis of hydrogen sulfide-cysteine cycle on enhancing the tolerance to chromium stress in *Arabidopsis*. Environ. Pollut. 213, 870–877.

Fotopoulos, V., Christou, A., Antoniou, C., Manganaris, G.A., 2015. Review article hydrogen sulphide: a versatile tool for the regulation of growth and defence responses in horticultural crops. J. Hortic. Sci. Biotechnol. 90, 227–234.

Fu, P., Wang, W., Hou, L., Liu, X., 2013. Hydrogen sulfide is involved in the chilling stress response in Vitis vinifera L. Acta Soc. Bot. Pol. 82.

Garcia, I., Gotor, C., Romero, L.C., 2015. Cysteine homeostasis. In: D'Mello, J.P.F. (Ed.), Amino Acids in Higher Plants. CABI Publishing, Wallingford, pp. 219–233.

Garcia-Mata, C., Lamattina, L., 2010. Hydrogen sulphide, a novel gasotransmitter involved in guard cell signalling. N. Phytol. 188, 977–984.

Hancock, J.T., Whiteman, M., 2015. Hydrogen sulfide and reactive friends: the interplay with reactive oxygen species and nitric oxide signalling pathways. In: de Kok, L., Hawkesford, M., Rennenberg, H., Saito, K. (eds.), Molecular Physiology and Ecophysiology of Sulfur. Springer, pp. 153–168.

He, H., Li, Y., He, L.F., 2018. The central role of hydrogen sulfide in plant responses to toxic metal stress. Ecotoxicol. Environ. Saf. 157, 403–408.

Honda, K., Yamada, N., Yoshida, R., Ihara, H., Sawa, T., Akaike, T., et al., 2015. 8-Mercapto-cyclic GMP mediates hydrogen sulfide-induced stomatal closure in *Arabidopsis*. Plant Cell Physiol. 56, 1481–1489.

Jaishankar, M., Tseten, T., Anbalagan, N., Mathew, B.B., Beeregowda, K.N., 2014. Toxicity mechanism and health effects of some heavy metals. Interdiscip Toxicol. 7, 60–72.

Jin, Z., Xue, S., Luo, Y., Tian, B., Fang, H., Li, H., et al., 2013. Hydrogen sulfide interacting with abscisic acid in stomatal regulation responses to drought stress in *Arabidopsis*. Plant Physiol. Biochem. 62, 41–46.

Kabil, O., Banerjee, R., 2010. Redox biochemistry of hydrogen sulfide. J. Biol. Chem. 285, 21903–21907.

Khalid, M.F., Hussain, S., Ahmad, S., Ejaz, S., Zakir, I., Ali, M.A., et al., 2019. Impacts of abiotic stresses on growth and development of plants. Plant Tolerance to Environmental Stress. CRC Press, pp. 1–8.

Khalid, M.F., Hussain, S., Anjum, M.A., Ahmad, S., Ali, M.A., Ejaz, S., et al., 2020. Better salinity tolerance in tetraploid vs diploid volkamer lemon seedlings is associated with robust antioxidant and osmotic adjustment mechanisms. J. Plant Physiol. 244, 153071.

Kimura, H., 2014. The physiological role of hydrogen sulfide and beyond. Nitric Oxide 41, 4–10.

Kimura, Y., Koike, S., Shibuya, N., Lefer, D., Ogasawara, Y., Kimura, H., 2017. 3-Mercaptopyruvate sulfurtransferase produces potential redox regulators cysteine- and glutathione-persulfide (Cys-SSH and GSSH) together with signaling molecules H_2S_2, H_2S_3 and H_2S. Sci. Rep. 7, 10459.

Lai, D., Mao, Y., Zhou, H., Li, F., Wu, M., Zhang, J., et al., 2014. Endogenous hydrogen sulfide enhances salt tolerance by coupling the reestablishment of redox homeostasis and preventing salt-induced K^+ loss in seedlings of *Medicago sativa*. Plant Sci. 225, 117–129.

Laureano-Marin, A.M., Moreno, I., Romero, L.C., Gotor, C., 2016. Negative regulation of autophagy by sulfide is independent of reactive oxygen species. Plant Physiol. 171, 1378–1391.

Li, Z.-G., Gong, M., Xie, H., Yang, L., Li, J., 2012. Hydrogen sulfide donor sodium hydrosulfide-induced heat tolerance in tobacco (*Nicotiana tabacum* L) suspension cultured cells and involvement of Ca^{2+} and calmodulin. Plant Sci. 185, 185–189.

Li, Z.-G., Ding, X.-J., Du, P.-F., 2013. Hydrogen sulfide donor sodium hydrosulfide-improved heat tolerance in maize and involvement of proline. J. Plant Physiol. 170, 741–747.

Li, S.-P., Hu, K.-D., Hu, L.-Y., Li, Y.-H., Jiang, A.-M., Xiao, F., et al., 2014. Hydrogen sulfide alleviates postharvest senescence of broccoli by modulating antioxidant defense and senescence-related gene expression. J. Agric. Food Chem. 62, 1119–1129.

Li, Z.-G., Xie, L.-R., Li, X.-J., 2015. Hydrogen sulfide acts as a downstream signal molecule in salicylic acid-induced heat tolerance in maize (*Zea mays* L.) seedlings. J. Plant Physiol. 177, 121–127.

Lisjak, M., Teklić, T., Wilson, I.D., Wood, M., Whiteman, M., Hancock, J.T., 2011. Hydrogen sulfide effects on stomatal apertures. Plant Signal. Behav. 6, 1444–1446.

Liu, J., Hou, L., Liu, G., Liu, X., Wang, X., 2011. Hydrogen sulfide induced by nitric oxide mediates ethylene-induced stomatal closure of *Arabidopsis thaliana*. Chin. Sci. Bull. 56, 3547–3553.

Liu, J., Hou, Z.-H., Liu, G.-H., Hou, L.-X., Liu, X., 2012. Hydrogen sulfide may function downstream of nitric oxide in ethylene-induced stomatal closure in *Vicia faba* L. J. Integr. Agric. 11, 1644–1653.

Mudaya, G.K., Murphy, A.S., 2002. An emerging model of auxin transport regulation. Plant Cell 14, 293–299.

Papanatsiou, M., Scuffi, D., Blatt, M.R., García-Mata, C., 2015. Hydrogen sulfide regulates inward-rectifying KC channels in conjunction with stomatal closure. Plant Physiol. 168, 29–35.

Rahman, A., Bannigan, A., Sulaman, W., Pechter, P., Blancaflor, E.B., Baskin, T.I., 2007. Auxin, actin and growth of the *Arabidopsis thaliana* primary root. Plant. J. 50 (2007), 514–528.

Rashotte, A.M., Brady, S.R., Reed, R.C., Ante, S.J., Muday, G.K., 2000. Basipetal auxin transport is required for gravitropism in roots of *Arabidopsis*. Plant Physiol. 122, 481–490.

Rodriguez-Kabana, R., Jordan, J.W., Hollis, J.P., 1965. Nematodes: biological control in rice fields: role of hydrogen sulfide. Science 148, 524–526.

Sabatini, S., Beis, D., Wolkenfelt, H., Murfett, J., Guilfoyle, T., Malamy, J., et al., 1999. An auxin-dependent distal organizer of pattern and polarity in the *Arabidopsis* root. Cell 99, 463–472.

Scuffi, D., Álvarez, C., Laspina, N., Gotor, C., Lamattina, L., García-Mata, C., 2014a. Hydrogen sulfide generated by L-cysteine desulfhydrase acts upstream of nitric oxide to modulate abscisic acid-dependent stomatal closure. Plant Physiol. 166, 2065–2076.

Scuffi, D., Nunez, A., Laspina, N., Gotor, C., Lamattina, L., Garcia-Mata, C., 2014b. Hydrogen sulfide generated by L-cysteine desulfhydrase acts upstream of nitric oxide to modulate ABA-dependent stomatal closure. Plant Physiol. 166, 2065–2076.

Shahid, M., Pourrut, B., Dumat, C., Nadeem, M., Aslam, M., Pinelli, E., 2014. Heavy-metal-induced reactive oxygen species: phytotoxicity and physicochemical changes in plants. In Reviews of Environmental Contamination and Toxicology. Springer Cham. 232, 1–44.

Shen, J., Xing, T., Yuan, H., Liu, Z., Jin, Z., Zhang, L., et al., 2013. Hydrogen sulfide improves drought tolerance in *Arabidopsis thaliana* by microRNA expressions. PLoS One 8, e77047.

Shi, H., Ye, T., Chan, Z., 2013. Exogenous application of hydrogen sulfide donor sodium hydrosulfide enhanced multiple abiotic stress tolerance in bermudagrass (*Cynodon dactylon* (L). Pers.). Plant Physiol. Biochem. 71, 226–234.

Singh, S., Kumar, V., Kapoor, D., Kumar, S., Singh, S., Dhanjal, D.S., et al., 2019. Revealing on hydrogen sulfide and nitric oxide signals co-ordination for plant growth under stress conditions. Physiol. Plant. 168, 301–317.

Stuiver, C.E., De Kok, L.J., Kuiper, P.C., 1992. Freezing tolerance and biochemical changes in wheat shoots as affected by H_2S fumigation. Plant Physiol. Biochem. Paris. 30, 47–55.

Sun, H., Basu, S., Brady, S.R., Luciano, R.L., Muday, G.K., 2004. Interactions between auxin transport and the actin cytoskeleton in developmental polarity of *Fucus distichus* embryos in response to light and gravity. Plant Physiol. 135, 266–278.

Suzuki, N., Koussevitzky, S., Mittler, R.O.N., Miller, G.A.D., 2012. ROS and redox signalling in the response of plants to abiotic stress. Plant Cell Environ. 35, 259–270.

Tang, C., Li, X., Du, J., 2006. Hydrogen sulfide as a new endogenous gaseous transmitter in the cardiovascular system. Curr. Vasc. Pharmacol. 4, 17–22.

Thompson, C.R., Kats, G., 1978. Effects of continuous hydrogen sulfide fumigation on crop and forest plants. Environ. Sci. Technol. 12, 550–553.

Toohey, J.I., 1989. Sulphane sulphur in biological systems: a possible regulatory role. Biochem. J. 264, 625–632.

Vandiver, M., Snyder, S., 2012. Hydrogen sulfide: a gasotransmitter of clinical relevance. J. Mol. Med. 90, 255–263.

Wang, B.-L., Shi, L., Li, Y.-X., Zhang, W.-H., 2010. Boron toxicity is alleviated by hydrogen sulfide in cucumber (Cucumis sativus L.) seedlings. Planta 231, 1301–1309.

Xie, Y., Zhang, C., Lai, D., Sun, Y., Samma, M.K., Zhang, J., et al., 2014. Hydrogen sulfide delays GA-triggered programmed cell death in wheat aleurone layers by the modulation of lutathione homeostasis and heme oxygenase-1 expression. J. Plant Physiol. 171, 53–62.

Yamaguchi, Y., Nakamura, T., Kusano, T., Sano, H., 2000. Three *Arabidopsis* genes encoding proteins with differential activities for cysteine synthase and beta-cyanoalanine synthase. Plant Cell Physiol. 41, 465–476.

Zhang, H., Hu, L.-Y., Hu, K.-D., He, Y.-D., Wang, S.-H., Luo, J.-P., 2008. Hydrogen sulfide promotes wheat seed germination and alleviates oxidative damage against copper stress. J. Integr. Plant Biol. 50, 1518–1529.

Zhang, H., Tang, J., Liu, X.P., Wang, Y., Yu, W., Peng, W.Y., et al., 2009. Hydrogen sulfide promotes root organogenesis in *Ipomoea batatas*, *Salix matsudana* and *Glycine max*. J. Integr. Plant Biol. 51, 1086–1094.

Zhang, H., Dou, W., Jiang, C.-X., Wei, Z.-J., Liu, J., Jones, R.L., 2010a. Hydrogen sulfide stimulates β-amylase activity during early stages of wheat grain germination. Plant Signal. Behav. 5, 1031–1033.

Zhang, H., Jiao, H., Jiang, C.X., Wang, S.H., Wei, Z.J., Luo, J.P., et al., 2010b. Hydrogen sulfide protects soybean seedlings against drought-induced oxidative stress. Acta Physiol. Plant 32, 849–857.

Chapter 6

General view on H_2S and abiotic stress tolerance in plants

Savita Bhardwaj and Dhriti Kapoor*
Department of Botany, School of Bioengineering and Biosciences, Lovely Professional University, Phagwara, India
**Corresponding author.*

6.1 Introduction

Plants live in constantly fluctuating environmental conditions that are generally disadvantageous or stressful for them. Abiotic stresses are the foremost constraints for plant growth and development (Ashraf et al., 2018). The abiotic stresses include drought stress, salinity stress, high and low temperature stress, and heavy metal toxicity (Vishwakarma et al., 2017). These stresses affect plants' geographical distribution in nature, limit plant productivity in agriculture, and threaten food security. As sessile organisms, diverse type of mechanisms have been adopted by plants to encounter environmental changes and plants are able to respond to and cope with various different abiotic stress or combinations of stresses. Plants escape from these factors either by completing their life cycle before the onset of stress or via avoidance and tolerance through morphological alterations and by changes in their physiological processes (Rafique et al., 2020). Abiotic stresses reduce plant development by disturbing different biochemical and physiological processes like photosynthesis, the antioxidant defense system, and phytohormones signaling (Sharma et al., 2016). Plants modify different cellular and molecular mechanisms in response to these abiotic stresses that ultimately escalate their growth and development (Sharma et al., 2019). Reactive oxygen species (ROS) such as hydrogen peroxide, superoxide, singlet oxygen, and hydroxyl radicals are generated by these abiotic stresses (Anjum et al., 2017). Increased generation of these ROS during stressful circumstances results in oxidative damage to plants (Ashraf et al., 2019). Many ROS are eliminated by the plant antioxidant system to achieve redox homeostasis in plants. Recently, hydrogen sulfide (H_2S) and its role in cell signaling has become a major focus area in plant research in order to gain a better understanding of plants protective mechanisms.

Hydrogen Sulfide in Plant Biology. DOI: https://doi.org/10.1016/B978-0-323-85862-5.00010-5

Hydrogen sulfide (known as H_2S, dihydrogen sulfide, or sulfane) has three main roles: as a substrate, as a toxin, and as a signal molecule in plants. For a long time, H_2S was known as a toxic gas, but it is now well-known as a crucial signaling molecule in various organisms such as bacteria, fungi, worms, human, and plants (Kabil et al., 2014). Hydrogen sulfide at much lower concentrations is considered to be as a signaling molecule in alliance with other reactive molecules, especially with carbon monoxide (CO), nitric oxide (NO), and ROS, while at high concentrations it is known to be harmful (Hancock and Whiteman, 2014). The H_2S, exclusively endogenous H_2S, can move freely from one plant cell to another. The research into the effects of H_2S on plants can date back to 1960s, when H_2S was reported to stimulate the comprehensive physiology of vegetative plants and to affect disease resistance (Joshi and Hollis, 1977; Rodriguez-Kabana et al., 1965). H_2S has the ability to move to various portions of the plant cells and balance the antioxidant pools by delivering sulfur to cells because of its gaseous nature and it has also exhibited resistance against plenty of detrimental environmental conditions like drought, salt, high temperature, chilling, heavy metals, and flood through a change in the levels of osmolytes, malondialdehyde (MDA), sodium/potassium ions uptake, the mechanism of H_2S biosynthesis, and the activities of antioxidative enzymes (Pandey and Gautam, 2020). H_2S boost plants capability of adapting to the various environmental conditions by mitigating injuries and toxicity caused by these adverse stressful circumstances. It also regulates plant growth, development, stomatal movement, drought tolerance, senescence, maturation, and lateral root formation (Guo et al., 2016). Being a reductive molecule, hydrogen sulfide can directly scavenge ROS such as superoxide, peroxynitrite, hypochlorite, and hydrogen peroxide, although the reducing power of H_2S is lower than that of glutathione (GSH) and cysteine (He et al., 2018). Exogenous treatment of H_2S, which is released by its donors such as sodium hydro sulfide (NaHS) and morpholin-4-ium 4-methoxypheny 1 (morpholino) phosphinodithioate(GYY4137), has displayed noteworthy enhancing effects on the germination of seed (Wojtyla et al., 2016), organogenesis and growth (Fang et al., 2014), senescence regulation (Zhang et al., 2011), and it also has showed tolerance to various stresses such as salt stress and drought stress (Christou et al., 2013), heavy metal toxicity (Chen et al., 2013), high temperature (Li, 2015), and low temperature (Fu et al., 2013).

6.2 Insight into H_2S metabolism

The synthesis and breakdown of H_2S in plants is known as H_2S metabolism. Plants reduce and add inorganic sulfur into cysteine by the process of assimilatory sulfate reduction (Filipovic and Jovanović, 2017). Inorganic sulfur almost wholly exists as oxidized sulfate. Sulfate from the soil is transported to the root epidermal cells by a H^+/SO_4^{2-} cotransport which is carried out by sulfate transporters, then carried out to the xylem vessels, and then allocated it to the whole plant parts (Leustek et al., 2000). To start the assimilatory sulfate reduction process, sulfate is reserved in vacuoles or carried out to the chloroplasts (Gotor Martínez et al., 2015) (Fig. 6.1). Before sulfate reduction, sulfate is activated to adenosine 5′-phosphosulfate (APS) which is catalyzed by enzyme ATP sulfurylase (Hatzfeld et al., 2000). Reduction of APS to H_2S is further carried out in two enzymatic steps which are located solely in chloroplasts or plastids. In the first step APS is reduced to sulfite by APS reductase (APR) using GSH as the reducing molecule and in the second step reduction of sulfite to hydrogen sulfide occurs, mediated by sulfite reductase (SiR) enzyme in a six-electron reaction using reduced ferredoxin as reductant (Fu et al., 2018). Beside hydrogen sulfide synthesis through sulfur assimilation, H_2S can also be synthesized by L-cysteine desulfhydrase (LCD), D-cysteine desulfhydrase (DCD), cyanoalanine synthase (CAS), and cysteine synthase (CS).

In the cytoplasm, nuclei, and mitochondria of plant cells, H_2S, ammonia (NH_3), and pyruvate are produced when LCD degrades L-cysteine (He et al., 2018). In this, pyridoxal phosphate acts as a cofactor to accelerate this reaction. In mitochondria, H_2S is produced by the degradation of D-cysteine by

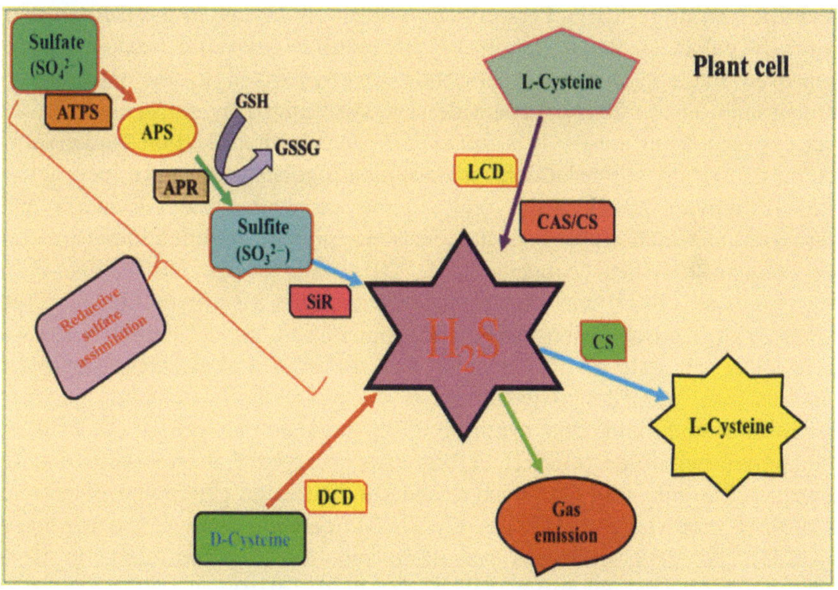

FIGURE 6.1 H_2S metabolism in plants. DCD, D-cysteine desulfhydrase; LCD, L-cysteine desulfhydrase; CAS, cyanoalanine synthase; CS, cysteine synthase, ATPS, ATP sulfurylase, APS-5′- adenosine 5′-phosphosulfate; APR, adenosine 5′-phosphosulfate reductase; SiR, sulfite reductase. *Modified after Gotor Martínez, C., Laureano Marín, A.M., Moreno González, M.I., Aroca Aguilar, Á., García Fernández, I., Romero González, L.C., 2015. Signaling in the plant cytosol: cysteine or sulfide? Amino Acids 47, 2155-2164.*

DCD. CAS formed cyanuric acid and H_2S from cyanide and L-cysteine in the cytoplasm and mitochondria of plant cells. These enzymes are responsive to abiotic stresses and hence are the products of H_2S (Hancock and Whiteman, 2014). Known as O-acetylserine (thiol) lyase, cysteine synthase forms cysteine in the cytoplasm, chloroplasts, and mitochondria from O-acetylserine (OAS) and H_2S. This reversible reaction can also yield H_2S. The enzymes which degrade cysteine and also release H_2S have been characterized in higher plants (Papenbrock et al., 2007). H_2S homeostasis is important for plants to tolerate metal toxicity, which depends on the balance between the synthesis and elimination of endogenous H_2S. Besides the importance of LCD/DCD, the influx of inorganic sulfur to cysteine is also important for H_2S synthesis. Adenosine 5′-phosphosulfate reductase (APR) controls the activity of the sulfur assimilation pathway (Fu et al., 2018).

6.3 Protein persulfidation induced by H_2S due to oxidative stress

H_2S facilitates diverse signaling responses in plants. The primary signaling mechanism of H_2S occurs through the persulfidation of reactive cysteine residues on target proteins via conversion of the thiol group (−SH) into a persulfide group (−SSH). Persulfidation, formerly known as S-sulfhydration, typically enhances the reactivity of the modified cysteine because of the improved nucleophilicity of persulfide in comparison to the thiol group (Cuevasanta et al., 2015). Persulfides are synthesized by sulfur transferases and cysteine desulfurases. It is found from protein modification via persulfidation that this posttranslational modification regulates several cellular activities (Filipovic, 2015). Sulfhydration interacts with phosphorylation and nitrosylation which are also posttranslational protein modifications to mediate cellular activities (Altaany et al., 2014; Sen et al., 2012) (Fig. 6.2). Mustafa et al. (2009) was first to describe proteins S-sulfhydration in 2009. Along with S-nitrosylation and S-glutathionylation, persulfidation is known to regulate the state of proteins and also performs a mechanism to protect proteins from oxidation (Cuevasanta et al., 2015). H_2S-producing enzymes (CSE) or cystathionine beta synthase (CBS) regulate the endogenous level of glutathione persulfide (GSSH); it was also observed that overexpression of CSE or CBS enhances GSSH level and stimulates the generation of protein S-sulfhydration (Lu et al., 2013). CysSSCys metabolism, which is mediated by CSE and CBS, generates persulfide and hence is considered to be a source of biological persulfides (Wedmann et al., 2016).

Aroca et al. (2015) found that the persulfidation of target proteins results in functional alterations in enzymatic structures and activities, which include *Arabidopsis* ascorbate peroxidase (APX), glyceraldehyde-3-phosphate dehydrogenase, and glutamine synthetase, and in subcellular localizations (Kimura, 2015; Mustafa et al., 2009; Paul and Snyder, 2015). Aroca et al.

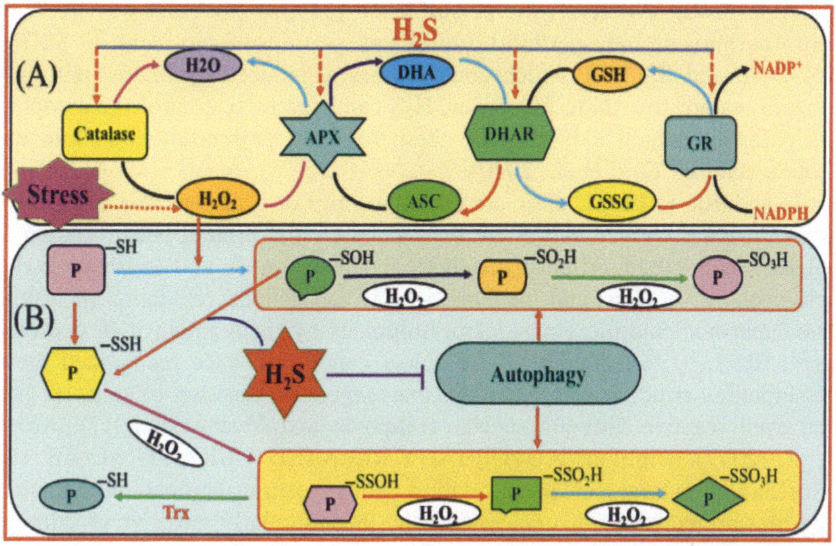

FIGURE 6.2 Proposed model for the complicated interactions of hydrogen sulfide (H_2S) with oxidative stress through its signaling function. (A) Reactive oxygen species (ROS) such as hydrogen peroxide (H_2O_2), hydroxyl (OH·), superoxide ($O_2^{-\cdot}$) radicals as well as H_2S are generated by various abiotic stress conditions. These ROS are capable of regulating antioxidant enzymes activities of catalase, ascorbate peroxidase (APX), dehydroascorbate reductase (DHAR) and glutathione reductase (GR) utilizing associated reducing substrates, i.e., ascorbic acid (Assc), oxidized glutathione (GSSG) and Nicotinamide adenine dinucleotide hydrogen phosphate (NADPH) to minimize ROS levels via directly metabolizing H_2O_2. (B) Protein cysteine thiols (P-SH) are reversibly oxidized by ROS (e.g., H_2O_2) to sulfenic acids (−SOH) upon encountering oxidative stress. Sulfenic acids are then reduced to sulfinic (SO_2H) or sulfonic acid (SO_3H) persulfidate proteins are then formed by the reaction of H_2S with −SOH. Thioredoxin (Trx) like enzyme reduce persulfidate proteins back to thiol. Moreover, persulfidated proteins interact with ROS to form perthiosulfenic (SSOH), perthiosulfinic (SSO_2H) and perthiosulfonic acid ($-SSO_3H$) which could be readily reduced by Trx enzyme to resustain free thiol. Autophagic pathway is negatively regulated by H_2S to control protein stability. *Modified after Altaany, Z., Ju, Y., Yang, G., Wang, R., 2014. The coordination of S-sulfhydration, S-nitrosylation, and phosphorylation of endothelial nitric oxide synthase by hydrogen sulfide. Sci. Signal. 7(342), ra87.*

(2017) performed a large-scale proteomics study of endogenously persulfidated proteins in wild-type *Arabidopsis* and *LCD1* defective mutant plants using the tag-switch method and found a significantly increased number of identified proteins that may be susceptible to persulfidation. The fact that a high number of proteins are persulfidated in both wild-type and des1 mutant lines indicates that the basal level of persulfidation is high under normal growth conditions.

An excess of ROS levels further caused irreversible hyperoxidation from sulfenic acid to sulfinic ($-SO_2H$) or sulfonic acids ($-SO_3H$). H_2S can react with sulfinic acid residues to form persulfidated proteins. They can be reduced back to thiol by enzymes such as thioredoxin (Trx) that can reduce disulfide bonds. Alternatively, persulfidated proteins can react with ROS to form perthiosulfenic, perthiosulfinic, and perthiosulfonic acid (−SSOH, $-SSO_2H$, and SSO_3H), which could be easily reduced by thioredoxin systems to restore free thiol. Moreover, H_2S can negatively control the autophagic pathway, which may degrade these oxidized proteins, to maintain the protein stability (Tao et al., 2020).

In addition to DES1, there are other enzymes or processes that can generate sulfurating species and it can be concluded that protein persulfidation is chiefly involved in primary metabolic pathways such as the tricarboxylic acid cycle, glycolysis, and the Calvin cycle and cellular regulation in plants. The super-nucleophilic nature of persulphides by maintaining it in deprotonated (RSS −) form under physiological conditions, have recognized these persulphides structures as extensive scavengers of reactive oxygen species and even reactive nitrogen species (Filipovic and Jovanović, 2017). Aroca et al. (2017) identified AUTOPHAGY-RELATED UBIQUITIN-LIKE 18a (Atg18a) as a target of persulfidation. Such posttranslational modification provides protection to this enzyme, which is particularly sensitive to salinity, drought, and oxidative stresses (Xiong et al., 2005). According to Cuevasanta et al. (2015), H_2S-induced redox sulfidation is a conserved mechanism of preventing protein cysteine residues from unwanted oxidation. In response to prolonged oxidative stress, Cys residues are oxidized to sulfenic acids (RSOH) and further to sulfinic (RSO_2H) and sulfonic acids (RSO_3H). These conversions are irreversible and cannot be easily reduced back to thiols if the oxidized cysteine residues are deeply buried within protein quaternary complexes (Cuevasanta et al., 2015). Hence, persulfidation can be described as a protective mechanism during H_2O_2-induced oxidative stress (Wedmann et al., 2016). The dual action of H_2S can assist plants not only to efficiently obtain the proteins that are functionally active against oxidative stress, but also to cut down energy consumption (e.g., de novo protein synthesis for growth and defense).

6.4 H_2S in abiotic stress tolerance

In recent years, there have been a number of research studies indicating that exogenous treatment with H_2S alleviates plants' tolerances to various abiotic stress conditions such as drought, salinity, heavy metals, and temperature extremes. H_2S is recognized as a key player in providing abiotic stress tolerance by upregulation and downregulation of stress-related genes and thereby modulates the associated signaling pathways (Shi et al., 2015). The complex regulatory pathway includes perceiving of stress signals from the external environment followed by initiation of the signaling pathway which consists of protein−protein interactions, various transcription factors and promoters, and modulated protein generation incorporating antioxidative defense mechanisms in the plant (Jalmi et al., 2018) (Fig. 6.3).

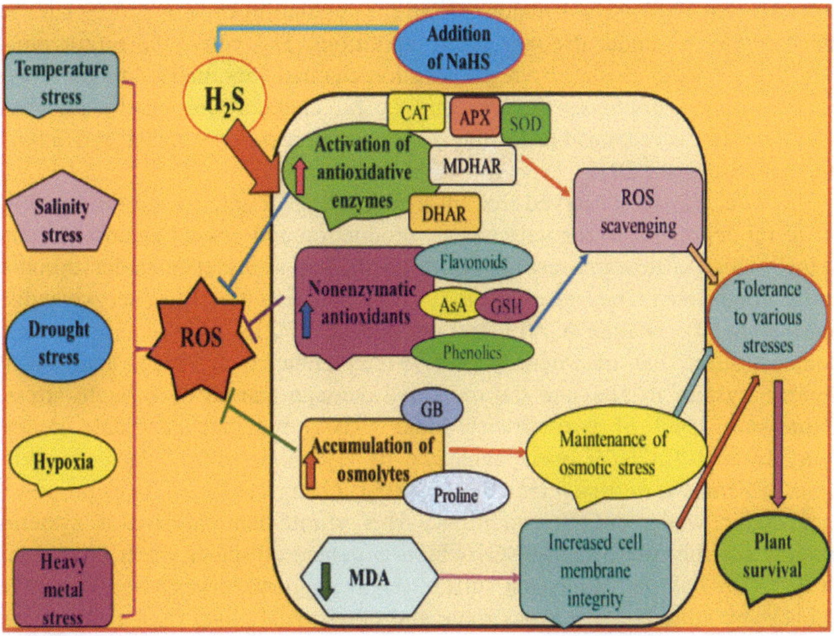

FIGURE 6.3 H_2S-mediated tolerance and defense mechanism in response to abiotic stresses in plants. Abiotic stress induced overaccumulation of reactive oxygen species (ROS) are inhibited by the application of sodium hydrosulfide (NaHS) which is a hydrogen sulfide (H_2S) donor. H_2S application increased the activities of antioxidative enzymes i.e., catalase (CAT), superoxide dismutase (SOD), ascorbate peroxidase (APX), dehydroascorbate reductase (DHAR), monodehydroascorbate reductase (MDHAR), and nonenzymatic antioxidants i.e., ascorbate (AsA), glutathione (GSH), flavonoids. It also improved the level of osmolytes and membrane integrity by declining the level of malondialdehyde (MDA) to increase plant tolerance to various abiotic stresses. *Modified after Pandey, A.K., Gautam, A., (2020). Stress responsive gene regulation in relation to hydrogen sulfide in plants under abiotic stress. Physiol Plant 168, 511–525.*

6.4.1 Ameliorative roles of H_2S during drought stress

Growth and development of plants is altered by drought conditions because water absorption by plant roots from the soil is not sufficient to satisfy the plants' transpiration demands. Due to water deficiency, there is a reduction in leaf cell turgor, which results in restricted cell expansion, enhanced leaf area and photosynthesis, thus inhibiting the biomass accumulation (Chaves et al., 2003). During water deficit and other osmotic stresses, the accumulation of compatible osmoprotectants such as soluble sugars (including glucose, fructose, and sucrose), proline, sugar alcohols, special amino acids, and glycinebetaine may provide tolerance to plants under stress conditions (Rivero et al., 2014). NaHS (H_2S donor) treatment increased the transpiration rate and stomatal conductance and higher leaf relative water content (RWC) in *Spinacia oleracea* seedlings (Chen et al., 2016a,b). Jin et al. (2013)

reported that H_2S plays a major role in the regulation of stomatal closure in the *Arabidopsis* under drought stress conditions. H_2S alters the turgor pressure and stomatal closure by regulating ion channel flux under drought stress conditions in *Arabidopsis thaliana* and the K^+ channel seems to be the chief osmolyte that is responsible for the regulatory process of stomatal movement by H_2S (Jin et al., 2017).

Jin et al. (2018) observed that in *A. thaliana* H_2S determined the capability of mitochondria to regulate energy production and protect against cellular aging, which afterward resulted in delayed leaf senescence under drought stress conditions. They also reported enhanced activities of superoxide dismutase (SOD), catalase (CAT), and peroxidases (POD) antioxidant enzymes, demonstrating that treatment with H_2S can enhance the plant antioxidant defense system to alleviate the oxidative damage caused by drought stress. Expression level of two abscisic acid (ABA) receptor candidate genes, *TaRCAR* and *TaCHLH*, were enhanced by H_2S under drought stress conditions in *Triticum aestivum* (Ma et al., 2016). Zhang et al. (2009a,b) proposed that under water deficit conditions, H_2S stimulates antioxidant systems quickly for the removal of H_2O_2 which is generated under stress conditions in *Ipomoea batatas*. Hydrogen sulfide (H_2S) application improved oxidative damage by stimulating the ascorbate and GSH metabolism by increasing the activities of antioxidative enzymes such as APX, glutathione reductase (GR), DHAR, gamma-glutamylcysteine synthetase, and the contents of reduced ascorbic acid (AsA), reduced GSH, total ascorbate, and total GSH in wheat seedlings under drought stress (Shan et al., 2011). H_2S improves the *Arabidopsis* tolerance to drought stress conditions by regulating the expression of drought-associated miRNAs genes, such as miR167, miR393, miR396, and miR398 (Shen et al., 2013). Treatment with NaHS stimulated the antioxidative enzyme activities of CAT, SOD, and Lipoxygenase and significantly increased chlorophyll content, decreased the accumulation of MDA level and decreased the production of H_2O_2 and $O_2^{\cdot -}$ in *Glycine max* seedlings, indicating that H_2S can act as an antioxidant signal during oxidative stress to increase drought tolerance (Zhang et al., 2010). Application of NaHS in *Citrus aurantium* plants increased the level of post-translational modifications (PTMs) and lessened leaf protein carbonylation, ultimately indicating that NaHS attenuated protein oxidation under drought conditions (Ziogas et al., 2015).

6.4.2 Ameliorative roles of H_2S during salinity stress

Salinity stress causes both ionic toxicity and osmotic stress in plants, which are responsible for nutritional disorders, oxidative damage, and reductions in plant growth and crop yield (Krasensky and Jonak, 2012). Oxidative burst, an uncontrolled generation of ROS, is one of the foremost acts prompted in plant tissues under salinity stress, leading to the escalation of electrolytes leakage, lipid peroxidation, and protein oxidation. Application of NaHS (H_2S donor) resulted in enhanced root length, plant height, biomass, and RWC and also enhanced the K^+ content and the K^+/Na^+ ratio in barley seedlings under salt stress (Chen et al., 2015). Lai et al. (2014) found that H_2S improved the salt tolerance of *Medicago sativa* seedlings by coupling the rectifying redox homeostasis and the enhanced K^+/Na^+ ratio. H_2S mitigated salinity-induced growth inhibition in wheat seedlings by keeping a lower Na^+ concentration through increased Na^+ extrusion and decreased Na^+ uptake by coordinating the membrane-bound translocation proteins, such as nonselective cation channels and salt overly sensitive 1 pathways (Deng et al., 2016). Christou et al. (2013) reported that H_2S provides systemic resistance to salinity by regulating the generation of reactive species through improved activity of antioxidant mechanisms in strawberry plants. Treatment with H_2S alleviates the toxicity caused by NaCl salt which results in reduced ROS accumulation and membrane lipid peroxidation in *Cucumis sativus* via enhancing the growth, photosynthetic parameters, and increasing the accumulation of antioxidants (AsA and GSH) and the activities of antioxidant enzymes (SOD, CAT, POD, and APX) (Jiang et al., 2019). Mostofa et al. (2015) observed that H_2S decreased the salt-induced increase in the levels of $O_2^{\cdot -}$, H_2O_2, and MDA in *Oryza sativa*. H_2S improved the total amylase, α-amylase, and β-amylase activities in cotyledons of cucumber (*C. sativus*) which helped in seed germination and also reduced salt-induced accumulation of proline (Yu et al., 2013). Exogenous application of H_2S sustained the redox forms of ascorbate and GSH by upregulating the metabolism of ascorbate and GSH, which in turn improved the antioxidant activity and prevented the seedlings from ROS which are generated by the salt stress and thus have an important role in enhancing salt stress tolerance in *Zea mays* (Shan et al., 2014). Application with NaHS increased the activities of α-amylase and β-amylase and also enhanced the accumulation of soluble sugar in seedlings of *C. sativus* under salinity stress conditions (Sun and Luo, 2014). H_2S improved plant responses to salinity stress by enhancing the germination rate of seeds, seedling growth, and reducing oxidative damage due to the stimulation of the antioxidant defense system as well as the reestablishment of ion homeostasis in *M. sativa* (Wang et al., 2012).

6.4.3 Ameliorative roles of H_2S during temperature stress

Temperature extremes are one of the major limiting factors in reducing plant growth and productivity. Heat stress enhances the fluidity of membrane lipids and the denaturation and aggregation of proteins, which results in the inactivation of enzymes in chloroplasts and mitochondria, disrupts synthesis of proteins, degrades and ultimately leads to cellular injury, and may even cause cell death. Regrowth capacity, survival percentage, cell vitality, MDA content, and electrolyte leakage (EL) are important heat tolerance parameters in plants (Li et al., 2012). H_2S also is responsible for nicotine biosynthesis under high-temperature conditions (Chen et al., 2016a,b). H_2S caused an enhancement in activities of major enzymatic antioxidants (cytosolic ascorbate peroxidase (cAPX), CAT, Manganese superoxide dismutase (MnSOD), GR), heat shock proteins, and aquaporin gene expression by alleviating heat shock-derived oxidative damage, lipid peroxidation, and H_2O_2 in *Fragaria ananassa* plants when exposed to high temperature (Christou et al., 2014). In plants, the accumulation of betaine, which is synthesized by betaine aldehyde dehydrogenase (BADH) enzyme, is an effective adaptive mechanism in plants to various environmental stresses including heat stress. Li et al. (2015a,b) reported that the application of the H_2S donor, sodium hydrosulfide (NaHS), significantly enhanced the activity of BADH which resulted in the accumulation of endogenous betaine, ultimately improving the survival percentage of *Z. mays* seedlings under heat stress. Treatment with H_2S donors (NaHS and GYY4137) decreased heat stress-induced H_2O_2 and $O2^-$ accumulation by improving the activity of antioxidant enzymes, including monodehydroascorbate reductase, DHAR, APX, and GR in *Populus trichocarpa* (Cheng et al., 2018). H_2S could enhance high-temperature tolerance of maize seedlings by increasing trehalose content by upregulating trehalose-6-phosphate phosphatase activity in *Z. mays* (Li et al., 2014).

H_2S acts as a novel downstream signaling molecule of salicylic acid (SA) in the procurement of heat tolerance induced by SA in *Z. mays* (Li et al., 2015a,b). Exogenous H_2S enhanced the resistance of *Nicotiana tabacum* cells to heat stress by activating Ca^{2+} signaling, which was decoded by Ca^{2+} binding protein calmodulin, and then activated the L-cysteine desulfhydrase (L-DES) enzyme activity for endogenous H_2S generation (Li et al., 2015a,b). H_2S improved the tolerance of maize seedlings to high temperature by enhancing endogenous methylglyoxal (MG) level (Li et al., 2018). H_2S showed higher antioxidant enzymes activities of SOD, CAT, and APX and their gene expression levels, and content of soluble sugar, and reduced heat stress-induced H_2O_2 and MDA content, ultimately resulting in the alleviation of oxidative damage caused by heat stress in *T. aestivum* (Min et al., 2016).

Low temperature is one of the most common factors in cold regions which adversely affects agricultural production. Perception and transduction of low-temperature signals are carried out by signal molecules and mediate chilling tolerance responses through physiological processes and transcription factors. H_2S upregulated mitogen-activated protein kinase expression levels and then both of them regulate the expression levels of the cold response genes, that is, inducer of C-repeat binding factor (CBF) expression 1 (ICE1), C repeat-binding factors (CBF3), cold responsive 15A (COR15A), and cold responsive 15B (COR15B) to mitigate cold stress in *A. thaliana* (Du et al., 2017). H_2S was involved in cold stress response in *Vitis vinifera* by reducing superoxide anion radical content, MDA content, the relative permeability of cell membrane, and enhancing SOD activity, as well as by enhancing expression of *VvICE1* and *VvCBF3I* genes (Fu et al., 2013). H_2S increases the expression of Cucurbitacin C (CuC) synthetase encoding genes, which in turn increases the content of CuC (triterpenoid secondary metabolite) which helps in enhancing stress tolerance in *C. sativus* under low-temperature conditions (Liu et al., 2019). NaHS, an exogenous H_2S donor, significantly improved the tolerance to low-temperature stress in blueberry plants by enhancing the synthesis of chlorophyll and carotenoids, the activities of PSII and PSI, and by promoting stomatal opening and photosynthetic carbon assimilation ability (Tang et al., 2020).

6.4.4 Ameliorative roles of H_2S during heavy metal stress

Toxic metals are emitted into the environment by various anthropogenic and industrial activities, which ultimately cause the contamination of soil Excess exposure to metals causes the inhibition of seed germination and seedling growth, is harmful to antioxidant enzymes and membrane system, induces chromosome aberration and even plant death. H_2S protected Chinese cabbage (*Brassica rapa* L. sp. *pekinensis*) from cell death (genomic DNA degradation) caused by oxidative stress and so alleviated Cd-mediated growth inhibition (Zhang et al., 2015). Heavy metal ions also affect plant metabolism by causing secondary stress such as nutrient imbalance, and oxidative stress. Exogenously applied H_2S showed an ameliorating role in significantly improving the plant growth, biomass, and concentration of essential nutrients in *Brassica napus* under Al stress (Ali et al., 2015). H_2S in crosstalk with nitric oxide increased lead (Pb) tolerance in *Sesamum indicum* by improving the content of essential nutrients (Ca, Mg, Fe, Zn and Mn) and decreasing the uptake and translocation of Pb in different parts of the plant (Amooaghaie and Enteshari, 2017). Liu et al. (2016) reported that H_2S prevented the uptake of Zn in *Solanum nigrum*. Dawood et al. (2012) also observed that the application of NaHS enhanced the amounts of P, S, Ca, Mg, and Fe, reduced Al accumulation, and improved the activities of both H^+-ATPase and Na^+/K^+-ATPase in *Hordeum vulgare*. Thus enhanced accumulation of minerals can be one of the mitigatory mechanisms of external NaHS. NaHS enhanced seed germination and seedling growth and increased nonprotein thiols and total GSH levels to scavenge ROS directly, thereby restricting oxidative damage in cauliflower under lead stress conditions in *Brassica oleracea* L. var *botrytis* (Chen et al., 2018). Chloroplasts are extremely susceptible to oxidative stress, which is generated by various stress conditions. Chloroplasts have an established antioxidant defense system which includes antioxidative enzymes and nonenzymatic compounds to guard against the damaging effects of oxidative damage. Dai et al. (2016) proposed that H_2S stimulates the antioxidant capacity of a chloroplast by improving the activities of APX, GR, SOD, and GPX, which enhances the antioxidant ability and prevents the activities of photosynthetic machinery of wheat seedlings against oxidative stress induced by copper stress. H_2S enhanced the synthesis of phytochelatins (PCs) by stimulating phytochelatins synthase 1 (*PCS1*) and phytochelatins synthase 2 (*PCS2*) expression levels and also activates metallothioneins accumulation by significantly improving the *MT2A* gene expression to improve the Cr^{6+} tolerance in *A. thaliana* (Fang et al., 2016). H_2S mitigated Cd toxicity in *H. vulgare* by decreasing Cd uptake and translocation and also by reducing MDA, H_2O_2, and $O_2^{\cdot -}$ levels by enhancing reduced GSH contents and ascorbate contents and increasing chlorophyll amount (Fu et al., 2019).

Application with NaHS elevated seed germination by maintaining higher amylase activity and upregulating the antioxidant defense system by decreasing ROS accumulation in wheat seedlings under cadmium stress (Huang et al., 2016). Sodium hydrosulfide (NaHS) was observed to enhance both GR activity, ascorbate (AsA), GSH, and dehydroascorbate contents and decreased the MDA content and EL in order to mitigate oxidative damage caused by copper stress conditions in *Coriandrum sativum* seedlings (Karam and Keramat, 2017). NaHS, along with thiamine, elevated cadmium tolerance in strawberry (*F. ananassa*) (Kaya and Aslan, 2020). NaHS positively overturned the harmful effects of Ni, as reflected in plant growth and biomass, photosynthetic parameters including photosynthetic rates, increased hydrogen peroxide (H_2O_2) content, and EL in *Cucurbita pepo* (Valivand et al., 2019). Application of NaHS enabled the seedlings to combat the toxic effects of Ni by improving photosynthesis and by maintaining the nitrogen metabolism both at activity as well as genetic level, which eventually resulted in an enhancement in rice growth performance under Ni stress (Rizwan et al., 2019).

6.5 H_2S and plant growth regulators

Hormones are the most important plant signaling molecules and other potential signaling molecules ultimately interact with them to generate any downstream response. The amount of endogenous H_2S is influenced by plant growth regulators (PGRs) and instead, biosynthesis, transport, and signal transduction of different PGRs are influenced by H_2S. Singh et al. (2015) have shown that the application of SA and NaHS enhanced nitrate reductase (NR) activity and protein content and helped in alleviating arsenic stress in pea plants. Ethylene in combination with H_2S alleviated oxidative stress by increased activities of the antioxidative enzymes CAT and APX in kiwifruit (Li et al., 2017). Tian et al. (2017) reported that the combination of H_2S and methyl jasmonate (MeJA) improved seedling growth in foxtail millet under cadmium stress. NaHS stimulated the biosynthesis and accumulation of ABA and also enhanced the expression of ABA responsive genes in the leaves of wheat plants exposed to drought stress (Ma et al., 2016).

H_2S alone or in conjunction with ABA stimulates K^+ transport within guard cells, which ultimately triggered stomatal closure during water stress conditions (Papanatsiou et al., 2015). H_2S was known to stimulate indole-3-acetic acid and nitric oxide (NO) signaling pathways during the formation of adventitious root in *Salix matsudana* and *G. max* (Zhang et al., 2009a,b). H_2S caused alterations in root development in *Arabidopsis* by altering auxin distribution and transport by an actin-dependent method (Jia et al., 2015). Breaking of seed dormancy and seed germination are the common physiological functions of gibberellic acid (GA) and H_2S in plants (Zhang et al., 2008). H_2S and ethylene influence the osmotic stress response via the regulation of stomata closure in *Solanum lycopersicum* (Jia et al., 2018). Crosstalk between SA and H_2S elevates heat tolerance of maize seedlings and also found that the antioxidant system and osmolytes exert a very important role in SA and H_2S interaction-induced heat tolerance in maize seedlings (Li, 2015). H_2S improved SA-induced Cd tolerance of *Arabidopsis* which diminished oxidative damage and escalated photosynthesis as well as enhanced LCD activity and also increased the amoounto of endogenous H_2S (Qiao et al., 2015). Hydrogen sulfide (H_2S) suppressed the ethylene-induced abscission of the tomato petiole in a dose-dependent manner (Liu et al., 2020). H_2S improved antioxidant activity and thus plays an important role as a fruit ripening regulator by antagonizing the effects of ethylene (Yao et al., 2018). SA and NaHS alleviated Pb-induced damage by regulating amino acids content in *Z. mays* and also decreased lead accumulation (Zanganeh et al., 2019).

6.6 Conclusion

Abiotic stresses are recognized as the foremost risk to world agricultural systems because they affect the growth and productivity of plants in almost all areas. Therefore molecules that may alleviate the toxic effects of various environmental stresses are of prime importance from both the theoretical and application points of view. To mitigate the toxicity caused by various abiotic stresses, hydrogen sulfide is recognized as a new emerging signaling molecule in higher plants with a well-known occurrence in the regulation of physiological processes of various plants. Almost all abiotic stresses have been known to generate H_2S accumulation. The emerging data have transformed the perception of H_2S from a harmful molecule to a useful signaling molecule comparable to hydrogen peroxide (H_2O_2) and nitric oxide (NO). Hydrogen sulfide has achieved global recognition as an influential regulator for plant developmental processes, such as seed germination, root growth, stomatal movement, photosynthesis, and plant growth. The interactions of different signal molecules, including H_2S, confers tolerance to different abiotic stresses, such as drought, high temperature, low temperature, salinity, and heavy metal stress. H_2S aids in plant tolerance to various abiotic stresses by stimulating the activity of various antioxidative enzymes, the expression of stress-related genes, ROS metabolism, and metabolic homeostasis. The use of H_2S donors, such as NaHS and less commonly GYY4137, greatly enables researchers to treat plants with appropriate concentrations of H_2S for the enhancement of plant tolerance to abiotic stimulus. An increasing number of studies are showing that by the exogenous application of H_2S, using donors such as NaHS, plant damage at different levels can be mitigated by stimulating enzymatic antioxidant and nonenzymatic systems. H_2S could ameliorate abiotic stresses by improving the antioxidant systems (SOD, CAT, APX, GR, GSH) to prevent oxidative stress.

References

Ali, B., Qian, P., Sun, R., Farooq, M.A., Gill, R.A., Wang, J., et al., 2015. Hydrogen sulfide alleviates the aluminum-induced changes in *Brassica napus* as revealed by physiochemical and ultrastructural study of plant. Environ. Sci. Pollut. Res. 22 (4), 3068–3081.

Altaany, Z., Ju, Y., Yang, G., Wang, R., 2014. The coordination of S-sulfhydration, S-nitrosylation, and phosphorylation of endothelial nitric oxide synthase by hydrogen sulfide. Sci. Signal. 7 (342), ra87.

Amooaghaie, R., Enteshari, S., 2017. Role of two-sided crosstalk between NO and H2S on improvement of mineral homeostasis and antioxidative defense in *Sesamum indicum* under lead stress. Ecotoxicol. Environ. Saf. 139, 210–218.

Anjum, S.A., Ashraf, U., Tanveer, M., Khan, I., Hussain, S., Shahzad, B., et al., 2017. Drought induced changes in growth, osmolyte accumulation and antioxidant metabolism of three maize hybrids. Front. Plant Sci. 8.

Aroca, Á., Serna, A., Gotor, C., Romero, L.C., 2015. S-sulfhydration: a cysteine posttranslational modification in plant systems. Plant Physiol. 168 (1), 334–342.

Aroca, A., Benito, J.M., Gotor, C., Romero, L.C., 2017. Persulfidation proteome reveals the regulation of protein function by hydrogen sulfide in diverse biological processes in *Arabidopsis*. J. Exp. Bot 68 (17), 4915–4927.

Ashraf, M.A., Akbar, A., Askari, S.H., Iqbal, M., Rasheed, R., Hussain, I., 2018. Recent advances in abiotic stress tolerance of plants through chemical priming: an overview. Advances in Seed Priming. Springer, Singapore, pp. 51–79.

Ashraf, M.A., Riaz, M., Arif, M.S., Rasheed, R., Iqbal, M., Hussain, I., et al., 2019. The role of non-enzymatic antioxidants in improving abiotic stress tolerance in plants. Plant Tolerance to Environmental Stress. Role Phytoprotectants.

Chaves, M.M., Maroco, J.P., Pereira, J.S., 2003. Understanding plant responses to drought—from genes to the whole plant. Funct. Plant Biol. 30 (3), 239–264.

Chen, J., Wang, W.H., Wu, F.H., You, C.Y., Liu, T.W., Dong, X.J., et al., 2013. Hydrogen sulfide alleviates aluminum toxicity in barley seedlings. Plant Soil 362 (1-2), 301–318.

Chen, J., Wang, W.H., Wu, F.H., He, E.M., Liu, X., Shangguan, Z.P., et al., 2015. Hydrogen sulfide enhances salt tolerance through nitric oxide-mediated maintenance of ion homeostasis in barley seedling roots. Sci. Rep. 5, 12516.

Chen, J., Shang, Y.T., Wang, W.H., Chen, X.Y., He, E.M., Zheng, H.L., et al., 2016a. Hydrogen sulfide-mediated polyamines and sugar changes are involved in hydrogen sulfide-induced drought tolerance in *Spinacia oleracea* seedlings. Front. Plant Sci. 7, 1173.

Chen, X., Chen, Q., Zhang, X., Li, R., Jia, Y., Ef, A., et al., 2016b. Hydrogen sulfide mediates nicotine biosynthesis in tobacco (*Nicotiana tabacum*) under high temperature conditions. Plant Physiol. Biochem. 104, 174–179.

Chen, Z., Yang, B., Hao, Z., Zhu, J., Zhang, Y., Xu, T., 2018. Exogenous hydrogen sulfide ameliorates seed germination and seedling growth of cauliflower under lead stress and its antioxidant role. J. Plant Growth Regul. 37 (1), 5–15.

Cheng, T., Shi, J., Dong, Y., Ma, Y., Peng, Y., Hu, X., et al., 2018. Hydrogen sulfide enhances poplar tolerance to high-temperature stress by increasing S-nitrosoglutathione reductase (GSNOR) activity and reducing reactive oxygen/nitrogen damage. Plant Growth Regul. 84 (1), 11–23.

Christou, A., Manganaris, G.A., Papadopoulos, I., Fotopoulos, V., 2013. Hydrogen sulfide induces systemic tolerance to salinity and non-ionic osmotic stress in strawberry plants through modification of reactive species biosynthesis and transcriptional regulation of multiple defence pathways. J. Exp. Bot 64 (7), 1953–1966.

Christou, A., Filippou, P., Manganaris, G.A., Fotopoulos, V., 2014. Sodium hydrosulfide induces systemic thermotolerance to strawberry plants through transcriptional regulation of heat shock proteins and aquaporin. BMC Plant Biol. 14 (1), 42.

Cuevasanta, E., Lange, M., Bonanata, J., Coitiño, E.L., Ferrer-Sueta, G., Filipovic, M.R., et al., 2015. Reaction of hydrogen sulfide with disulfide and sulfenic acid to form the strongly nucleophilic persulfide. J. Biol. Chem. 290 (45), 26866–26880.

Dai, H., Xu, Y., Zhao, L., Shan, C., 2016. Alleviation of copper toxicity on chloroplast antioxidant capacity and photosystem II photochemistry of wheat by hydrogen sulfide. Braz. J. Bot 39 (2), 787–793.

Dawood, M., Cao, F., Jahangir, M.M., Zhang, G., Wu, F., 2012. Alleviation of aluminum toxicity by hydrogen sulfide is related to elevated ATPase, and suppressed aluminum uptake and oxidative stress in barley. J. Hazard. Mater. 209, 121–128.

Deng, Y.Q., Bao, J., Yuan, F., Liang, X., Feng, Z.T., Wang, B.S., 2016. Exogenous hydrogen sulfide alleviates salt stress in wheat seedlings by decreasing Na + content. Plant Growth Regul. 79 (3), 391–399.

Du, X., Jin, Z., Liu, D., Yang, G., Pei, Y., 2017. Hydrogen sulfide alleviates the cold stress through MPK4 in *Arabidopsis thaliana*. Plant Physiol. Biochem. 120, 112–119.

Fang, T., Cao, Z., Li, J., Shen, W., Huang, L., 2014. Auxin-induced hydrogen sulfide generation is involved in lateral root formation in tomato. Plant Physiol. Biochem. 76, 44–51.

Fang, H., Liu, Z., Jin, Z., Zhang, L., Liu, D., Pei, Y., 2016. An emphasis of hydrogen sulfide-cysteine cycle on enhancing the tolerance to chromium stress in *Arabidopsis*. Environ. Pollut. 213, 870–877.

Filipovic, M.R., 2015. Persulfidation (S-sulfhydration) and H 2 S. Chemistry, Biochemistry and Pharmacology of Hydrogen Sulfide. Springer, Cham, pp. 29–59.

Filipovic, M.R., Jovanović, V.M., 2017. More than just an intermediate: hydrogen sulfide signalling in plants. J. Exp. Bot 68 (17), 4733–4736.

Fu, P., Wang, W., Hou, L., Liu, X., 2013. Hydrogen sulfide is involved in the chilling stress response in *Vitis vinifera* L. Acta Soc Bot Poloniae 82 (4).

Fu, Y., Tang, J., Yao, G.F., Huang, Z.Q., Li, Y.H., Han, Z., et al., 2018. Central role of adenosine 5′-phosphosulfate reductase in the control of plant hydrogen sulfide metabolism. Front. Plant Sci. 9, 1404.

Fu, M.M., Dawood, M., Wang, N.H., Wu, F., 2019. Exogenous hydrogen sulfide reduces cadmium uptake and alleviates cadmium toxicity in barley. Plant Growth Regul. 89 (2), 227–237.

Gotor Martínez, C., Laureano Marín, A.M., Moreno González, M.I., Aroca Aguilar, Á., García Fernández, I., Romero González, L.C., 2015. Signaling in the plant cytosol: cysteine or sulfide? Amino Acids 47, 2155–2164.

Guo, H., Xiao, T., Zhou, H., Xie, Y., Shen, W., 2016. Hydrogen sulfide: a versatile regulator of environmental stress in plants. Acta Physiol Plant 38 (1), 16.

Hancock, J.T., Whiteman, M., 2014. Hydrogen sulfide and cell signaling: team player or referee? Plant Physiol. Biochem. 78, 37–42.

Hatzfeld, Y., Lee, S., Lee, M., Leustek, T., Saito, K., 2000. Functional characterization of a gene encoding a fourth ATP sulfurylase isoform from *Arabidopsis thaliana*. Gene 248 (1-2), 51–58.

He, H., Li, Y., He, L.F., 2018. The central role of hydrogen sulfide in plant responses to toxic metal stress. Ecotoxicol. Environ. Saf. 157, 403–408.

Huang, Z.Q., Ye, S.C., Hu, L.Y., Hu, K.D., Yan, H., Li, W.J., et al., 2016. Hydrogen sulfide promotes wheat grain germination under cadmium stress. Proc. Natl Acad. Sci., India Sect. B: Biol. Sci 86 (4), 887–895.

Jalmi, S.K., Bhagat, P.K., Verma, D., Noryang, S., Tayyeba, S., Singh, K., et al., 2018. Traversing the links between heavy metal stress and plant signaling. Front. Plant Sci. 9, 12.

Jia, H., Hu, Y., Fan, T., Li, J., 2015. Hydrogen sulfide modulates actin-dependent auxin transport via regulating ABPs results in changing of root development in *Arabidopsis*. Sci. Rep. 5, 8251.

Jia, H., Chen, S., Liu, D., Liesche, J., Shi, C., Wang, J., et al., 2018. Ethylene-induced hydrogen sulfide negatively regulates ethylene biosynthesis by persulfidation of ACO in tomato under osmotic stress. Front. Plant Sci. 9, 1517.

Jiang, J., Li, L., Yu, M., Hou, R., Ren, X., 2019. H2S alleviates salinity stress in cucumber by maintaining the Na + /K + balance and regulating H2S metabolism and oxidative stress response. Front. Plant Sci. 10, 678.

Jin, Z., Xue, S., Luo, Y., Tian, B., Fang, H., Li, H., et al., 2013. Hydrogen sulfide interacting with abscisic acid in stomatal regulation responses to drought stress in *Arabidopsis*. Plant Physiol. Biochem. 62, 41–46.

Jin, Z., Wang, Z., Ma, Q., Sun, L., Zhang, L., Liu, Z., et al., 2017. Hydrogen sulfide mediates ion fluxes inducing stomatal closure in response to drought stress in *Arabidopsis thaliana*. Plant Soil 419 (1-2), 141–152.

Jin, Z., Sun, L., Yang, G., Pei, Y., 2018. Hydrogen sulfide regulates energy production to delay leaf senescence induced by drought stress in *Arabidopsis*. Front. Plant Sci. 9, 1722.

Joshi, M.M., Hollis, J.P., 1977. Interaction of Beggiatoa and rice plant: detoxification of hydrogen sulfide in the rice rhizosphere. Science 195 (4274), 179–180.

Kabil, O., Vitvitsky, V., Banerjee, R., 2014. Sulfur as a signaling nutrient through hydrogen sulfide. Annu. Rev. Nutr. 34, 171–205.

Karam, E.A., Keramat, B., 2017. Hydrogen sulfide protects coriander seedlings against copper stress by regulating the ascorbate-glutathione cycle in leaves. J. Plant Process. Funct. 5 (18).

Kaya, C., Aslan, M., 2020. Hydrogen sulphide partly involves in thiamine-induced tolerance to cadmium toxicity in strawberry (*Fragaria x ananassa* Duch) plants. Environ. Sci. Pollut. Res. 27 (1), 941–953.

Kimura, H., 2015. Signaling of hydrogen sulfide and polysulfides. Antioxid. Redox Signal. 22, 347–349.

Krasensky, J., Jonak, C., 2012. Drought, salt, and temperature stress-induced metabolic rearrangements and regulatory networks. J. Exp. Bot 63 (4), 1593–1608.

Lai, D., Mao, Y., Zhou, H., Li, F., Wu, M., Zhang, J., et al., 2014. Endogenous hydrogen sulfide enhances salt tolerance by coupling the reestablishment of redox homeostasis and preventing salt-induced K + loss in seedlings of *Medicago sativa*. Plant Sci. 225, 117–129.

Leustek, T., Martin, M.N., Bick, J.A., Davies, J.P., 2000. Pathways and regulation of sulfur metabolism revealed through molecular and genetic studies. Annu. Rev. Plant Biol. 51 (1), 141–165.

Li, Z.G., 2015. Synergistic effect of antioxidant system and osmolyte in hydrogen sulfide and salicylic acid crosstalk-induced heat tolerance in maize (*Zea mays* L.) seedlings. Plant Signal. Behav. 10 (9), e1051278.

Li, Z.G., Gong, M., Xie, H., Yang, L., Li, J., 2012. Hydrogen sulfide donor sodium hydrosulfide-induced heat tolerance in tobacco (*Nicotiana tabacum* L) suspension cultured cells and involvement of Ca2 + and calmodulin. Plant Sci. 185, 185–189.

Li, Z.G., Luo, L.J., Zhu, L.P., 2014. Involvement of trehalose in hydrogen sulfide donor sodium hydrosulfide-induced the acquisition of heat tolerance in maize (*Zea mays* L.) seedlings. Bot Stud. 55 (1), 20.

Li, Z.G., Long, W.B., Yang, S.Z., Wang, Y.C., Tang, J.H., Wen, L., et al., 2015a. Endogenous hydrogen sulfide regulated by calcium is involved in thermotolerance in tobacco *Nicotiana tabacum* L. suspension cell cultures. Acta Physiol Plant 37 (10), 219.

Li, Z.G., Xie, L.R., Li, X.J., 2015b. Hydrogen sulfide acts as a downstream signal molecule in salicylic acid-induced heat tolerance in maize (*Zea mays* L.) seedlings. J. Plant Physiol. 177, 121–127.

Li, T.T., Li, Z.R., Hu, K.D., Hu, L.Y., Chen, X.Y., Li, Y.H., et al., 2017. Hydrogen sulfide alleviates kiwifruit ripening and senescence by antagonizing effect of ethylene. HortScience 52 (11), 1556–1562.

Li, Z.G., Long, W.B., Yang, S.Z., Wang, Y.C., Tang, J.H., 2018. Signaling molecule methylglyoxal-induced thermotolerance is partly mediated by hydrogen sulfide in maize (*Zea mays* L.) seedlings. Acta Physiol Plant 40 (4), 76.

Liu, X., Chen, J., Wang, G.H., Wang, W.H., Shen, Z.J., Luo, M.R., et al., 2016. Hydrogen sulfide alleviates zinc toxicity by reducing zinc uptake and regulating genes expression of antioxidative enzymes and metallothioneins in roots of the cadmium/zinc hyperaccumulator *Solanum nigrum* L. Plant Soil 400 (1-2), 177–192.

Liu, Z., Li, Y., Cao, C., Liang, S., Ma, Y., Liu, X., et al., 2019. The role of H 2 S in low temperature-induced cucurbitacin C increases in cucumber. Plant Mol. Biol. 99 (6), 535–544.

Liu, D., Li, J., Li, Z., Pei, Y., 2020. Hydrogen sulfide inhibits ethylene-induced petiole abscission in tomato (*Solanum lycopersicum* L.). Horticulture Res. 7 (1), 1–11.

Lu, C., Kavalier, A., Lukyanov, E., Gross, S.S., 2013. S-sulfhydration/desulfhydration and S-nitrosylation/denitrosylation: a common paradigm for gasotransmitter signaling by H2S and NO. Methods 62 (2), 177–181.

Ma, D., Ding, H., Wang, C., Qin, H., Han, Q., Hou, J., et al., 2016. Alleviation of drought stress by hydrogen sulfide is partially related to the abscisic acid signaling pathway in wheat. PLoS One 11 (9).

Min, Y.A.N.G., Qin, B.P., Ping, W.A.N.G., Li, M.L., Chen, L.L., Chen, L.T., et al., 2016. Foliar application of sodium hydrosulfide (NaHS), a hydrogen sulfide (H2S) donor, can protect seedlings against heat stress in wheat (*Triticum aestivum* L.). J. Integr. Agric 15 (12), 2745–2758.

Mostofa, M.G., Saegusa, D., Fujita, M., Tran, L.S.P., 2015. Hydrogen sulfide regulates salt tolerance in rice by maintaining Na + /K + balance, mineral homeostasis and oxidative metabolism under excessive salt stress. Front. Plant Sci. 6, 1055.

Mustafa, A.K., Gadalla, M.M., Sen, N., Kim, S., Mu, W., Gazi, S.K., et al., 2009. H2S signals through protein S-sulfhydration. Sci. Signal. 2 (96), ra72.

Pandey, A.K., Gautam, A., 2020. Stress responsive gene regulation in relation to hydrogen sulfide in plants under abiotic stress. Physiol Plant 168, 511–525.

Papanatsiou, M., Scuffi, D., Blatt, M.R., García-Mata, C., 2015. Hydrogen sulfide regulates inward-rectifying K + channels in conjunction with stomatal closure. Plant Physiol. 168 (1), 29–35.

Papenbrock, J., Riemenschneider, A., Kamp, A., Schulz-Vogt, H.N., Schmidt, A., 2007. Characterization of cysteine-degrading and H2S-releasing enzymes of higher plants-from the field to the test tube and back. Plant Biol. 9 (05), 582–588.

Paul, B.D., Snyder, S.H., 2015. Protein sulfhydration, Methods in Enzymology, Vol. 555. Academic Press, pp. 79–90.

Qiao, Z., Jing, T., Liu, Z., Zhang, L., Jin, Z., Liu, D., et al., 2015. H 2 S acting as a downstream signaling molecule of SA regulates Cd tolerance in *Arabidopsis*. Plant Soil 393 (1-2), 137–146.

Rafique, S., Abdin, M.Z., Alam, W., 2020. Response of combined abiotic stresses on maize (*Zea mays* L.) inbred lines and interaction among various stresses. Maydica 64 (3), 8.

Rivero, R.M., Mestre, T.C., Mittler, R.O.N., Rubio, F., Garcia-Sanchez, F., Martinez, V., 2014. The combined effect of salinity and heat reveals a specific physiological, biochemical and molecular response in tomato plants. Plant, Cell Environ. 37 (5), 1059–1073.

Rizwan, M., Mostofa, M.G., Ahmad, M.Z., Zhou, Y., Adeel, M., Mehmood, S., et al., 2019. Hydrogen sulfide enhances rice tolerance to nickel through the prevention of chloroplast damage and the improvement of nitrogen metabolism under excessive nickel. Plant Physiol. Biochem. 138, 100–111.

Rodriguez-Kabana, R., Jordan, J.W., Hollis, J.P., 1965. Nematodes: biological control in rice fields: role of hydrogen sulfide. Science 148 (3669), 524–526.

Sen, N., Paul, B.D., Gadalla, M.M., Mustafa, A.K., Sen, T., Xu, R., et al., 2012. Hydrogen sulfide-linked sulfhydration of NF-κB mediates its antiapoptotic actions. Mol. Cell 45 (1), 13–24.

Shan, C.J., Zhang, S.L., Li, D.F., Zhao, Y.Z., Tian, X.L., Zhao, X.L., et al., 2011. Effects of exogenous hydrogen sulfide on the ascorbate and glutathione metabolism in wheat seedlings leaves under water stress. Acta Physiol Plant 33 (6), 2533.

Shan, C., Liu, H., Zhao, L., Wang, X., 2014. Effects of exogenous hydrogen sulfide on the redox states of ascorbate and glutathione in maize leaves under salt stress. Biol Plant 58 (1), 169–173.

Sharma, A., Thakur, S., Kumar, V., Kanwar, M.K., Kesavan, A.K., Thukral, A.K., et al., 2016. Pre-sowing seed treatment with 24-epibrassinolide ameliorates pesticide stress in *Brassica juncea* L. through the modulation of stress markers. Front. Plant Sci. 7, 1569.

Sharma, A., Shahzad, B., Rehman, A., Bhardwaj, R., Landi, M., Zheng, B., 2019. Response of phenylpropanoid pathway and the role of polyphenols in plants under abiotic stress. Molecules 24 (13), 2452.

Shen, J., Xing, T., Yuan, H., Liu, Z., Jin, Z., Zhang, L., et al., 2013. Hydrogen sulfide improves drought tolerance in *Arabidopsis thaliana* by microRNA expressions. PLoS One 8 (10).

Shi, H., Ye, T., Han, N., Bian, H., Liu, X., Chan, Z., 2015. Hydrogen sulfide regulates abiotic stress tolerance and biotic stress resistance in *Arabidopsis*. J. Integr. Plant Biol. 57 (7), 628–640.

Singh, V.P., Singh, S., Kumar, J., Prasad, S.M., 2015. Hydrogen sulfide alleviates toxic effects of arsenate in pea seedlings through up-regulation of the ascorbate–glutathione cycle: possible involvement of nitric oxide. J. Plant Physiol. 181, 20–29.

Sun, Y.D., Luo, W.R., 2014. Effects of exogenous hydrogen sulphide on seed germination and seedling growth of cucumber (*Cucumis sativus*) under sodium bicarbonate stress. Seed Sci. Technol. 42 (2), 126–131.

Tang, X., An, B., Cao, D., Xu, R., Wang, S., Zhang, Z., et al., 2020. Improving photosynthetic capacity, alleviating photosynthetic inhibition and oxidative stress under low temperature stress with exogenous hydrogen sulfide in blueberry seedlings. Front. Plant Sci. 11.

Tao, C., Tian, M., Han, Y., 2020. Hydrogen sulfide: a multi-tasking signal molecule in the regulation of oxidative stress responses. J. Exp. Bot 71, 2862–2869.

Tian, B., Zhang, Y., Jin, Z., Liu, Z., Pei, Y., 2017. Role of hydrogen sulfide in the methyl jasmonate response to cadmium stress in foxtail millet. Front. Biosci. 22, 530–538.

Valivand, M., Amooaghaie, R., Ahadi, A., 2019. Seed priming with H2S and Ca2 + trigger signal memory that induces cross-adaptation against nickel stress in zucchini seedlings. Plant Physiol. Biochem. 143, 286–298.

Vishwakarma, K., Upadhyay, N., Kumar, N., Yadav, G., Singh, J., Mishra, R.K., et al., 2017. Abscisic acid signaling and abiotic stress tolerance in plants: a review on current knowledge and future prospects. Front. Plant Sci. 8, 161.

Wang, Y., Li, L., Cui, W., Xu, S., Shen, W., Wang, R., 2012. Hydrogen sulfide enhances alfalfa (*Medicago sativa*) tolerance against salinity during seed germination by nitric oxide pathway. Plant Soil 351 (1-2), 107–119.

Wedmann, R., Onderka, C., Wei, S., Szijártó, I.A., Miljkovic, J.L., Mitrovic, A., et al., 2016. Improved tag-switch method reveals that thioredoxin acts as depersulfidase and controls the intracellular levels of protein persulfidation. Chem. Sci. 7 (5), 3414–3426.

Wojtyla, Ł., Lechowska, K., Kubala, S., Garnczarska, M., 2016. Different modes of hydrogen peroxide action during seed germination. Front. Plant Sci. 7, 66.

Xiong, Y., Contento, A.L., Bassham, D.C., 2005. AtATG18a is required for the formation of autophagosomes during nutrient stress and senescence in *Arabidopsis thaliana*. Plant J. 42 (4), 535–546.

Yao, G.F., Wei, Z.Z., Li, T.T., Tang, J., Huang, Z.Q., Yang, F., et al., 2018. Modulation of enhanced antioxidant activity by hydrogen sulfide antagonization of ethylene in tomato fruit ripening. J. Agric. Food Chem. 66 (40), 10380–10387.

Yu, L.X., Zhang, C.J., Shang, H.Q., Wang, X.F., Min, W.E.I., Yang, F.J., et al., 2013. Exogenous hydrogen sulfide enhanced antioxidant capacity, amylase activities and salt tolerance of cucumber hypocotyls and radicles. J. Integr. Agric 12 (3), 445–456.

Zanganeh, R., Jamei, R., Rahmani, F., 2019. Role of salicylic acid and hydrogen sulfide in promoting lead stress tolerance and regulating free amino acid composition in *Zea mays* L. Acta Physiol Plant 41 (6), 94.

Zhang, H., Hu, L.Y., Hu, K.D., He, Y.D., Wang, S.H., Luo, J.P., 2008. Hydrogen sulfide promotes wheat seed germination and alleviates oxidative damage against copper stress. J. Integr. Plant Biol. 50 (12), 1518–1529.

Zhang, H., Tang, J., Liu, X.P., Wang, Y., Yu, W., Peng, W.Y., et al., 2009a. Hydrogen sulfide promotes root organogenesis in *Ipomoea batatas*, *Salix matsudana* and *Glycine max*. J. Integr. Plant Biol. 51 (12), 1086–1094.

Zhang, H., Ye, Y.K., Wang, S.H., Luo, J.P., Tang, J., Ma, D.F., 2009b. Hydrogen sulfide counteracts chlorophyll loss in sweet potato seedling leaves and alleviates oxidative damage against osmotic stress. Plant Growth Regul. 58 (3), 243–250.

Zhang, H., Jiao, H., Jiang, C.X., Wang, S.H., Wei, Z.J., Luo, J.P., et al., 2010. Hydrogen sulfide protects soybean seedlings against drought-induced oxidative stress. Acta Physiol Plant 32 (5), 849–857.

Zhang, H., Hu, S.L., Zhang, Z.J., Hu, L.Y., Jiang, C.X., Wei, Z.J., et al., 2011. Hydrogen sulfide acts as a regulator of flower senescence in plants. Postharvest Biol. Technol. 60 (3), 251–257.

Zhang, L., Pei, Y., Wang, H., Jin, Z., Liu, Z., Qiao, Z., et al., 2015. Hydrogen sulfide alleviates cadmium-induced cell death through restraining ROS accumulation in roots of *Brassica rapa* L. ssp. pekinensis. Oxid. Med. Cell. Longev. 2015.

Ziogas, V., Tanou, G., Belghazi, M., Filippou, P., Fotopoulos, V., Grigorios, D., et al., 2015. Roles of sodium hydrosulfide and sodium nitroprusside as priming molecules during drought acclimation in citrus plants. Plant Mol. Biol. 89 (4-5), 433–450.

Chapter 7

Role of H_2S in plants: a current update

Shristy Gautam[1,#], Savita Bhardwaj[2,#] and Dhriti Kapoor[2,*]
[1]*Department of Molecular Biology and Genetic Engineering, School of Bioengineering and Biosciences, Lovely Professional University, Phagwara, India,* [2]*Department of Botany, School of Bioengineering and Biosciences, Lovely Professional University, Phagwara, India*
**Corresponding author.*

7.1 Introduction

Hydrogen sulfide (H_2S) is an important signaling molecule along with CO and NO, that has been subject to a large number of fundamental investigations. Hydrogen sulfide (H_2S) has been recognized earlier as an environmental threat and harmful gas but now H_2S has been found to be a vital gas signaling molecule in both animal and plant systems, which showed effects on several physiological progressions. Hydrogen sulfide is a gasotransmitter which is endogenously synthesized in plant tissues. It is a colorless, gaseous molecule at room temperature, a member of a family of similar molecules known as reactive sulfur species, and it is an ignitable gas with the distinctive odor of rotten eggs. Because of its unpleasant odor, it was usually recognized as a harmful gas for around 300 years. The transport of H_2S in cells and tissues is very fast and convenient because of its water-soluble and lipophilic nature (Li et al., 2009). H_2S effects in plants were discovered in the 1960s, when H_2S was observed to alter the developmental processes of vegetative plants and to disturb disease tolerance (Rodriguez-Kabana et al., 1965; Thompson and Kats, 1978). However, the endogenous synthesis of H_2S was also detected in plants. Hydrogen sulfide is associated with various responses of plants to environmental stresses, that is, both biotic and abiotic at low levels, hence it is biologically very active (Chen et al., 2013). Sodium hydrosulfide (NaHS) and GYY4137 (morpholin-4-ium 4-methoxyphenyl (morpholino) phosphinodithioate) are generally applied as H_2S donors because they can generate hydrogen sulfide when dissolved in water but sodium

[#] This chapter was equally contributed by the authors Shristy Gautam and Savita Bhardwaj.

Hydrogen Sulfide in Plant Biology. DOI: https://doi.org/10.1016/B978-0-323-85862-5.00004-X

hydrosulfide gives a small burst of hydrogen sulfide in comparison to GYY4137, which gives a longer burst of H_2S (Lisjak et al., 2013; Wang, 2012). Generally, H_2S occurs in two forms, that is, H_2S ($\approx$20%) and HS^- ($\approx$80%) in water solution but the precise concentration of H_2S in plant tissues or subcellular organelles is not clear (Li et al., 2016).

Atmospheric natural sources of hydrogen sulfide are volcanic and geothermal emissions from coal gasification plants and wastewater treatment systems. Crude oil processing industries are one of the major sources of H_2S, but it is also found naturally in drinking water (Habeeb et al., 2018). Depending on environmental conditions, H_2S has a half-life period of 12–37 hours in the air. Atmospheric H_2S uptake occurs through plant leaf's surface via stomata. At lower concentrations, hydrogen sulfide is known to be beneficial for plant growth and thus has been recognized as a signaling molecule, while it is toxic at high concentrations (Hancock and Whiteman, 2014). H_2S enhances seed germination rate, stimulates root growth, helps in increasing photosynthesis, prolonging florescence, and delaying senescence, and also increases plant tolerance to abiotic stresses (Mostofa et al., 2015). It is also observed that H_2S plays an important role in regulating stomata opening and closing (García-Mata and Lamattina, 2010). H_2S, by protecting the photosynthetic machinery, increases seed germination and plant growth, and also prevents the transport of heavy metals to the aboveground portions of the plant and the abscisic acid (ABA) signal transduction pathway that induces stomata opening and decreases leaf litter under cadmium stress in *Phlox paniculata* (Wang et al., 2020).

Hydrogen sulfide synthesize α- and β-amylase in the endosperm which helps in enhancing the seed germination rate of salt-stressed cucumber seeds by starch breakdown which finally leads to enhanced hypocotyl growth (Sun and Luo, 2014). The treatment with hydrogen sulfide improved the suppressing actions of heavy metals on root growth, ultimately reducing the harmful effects of heavy metals on plant growth and also improving plant tolerance to heavy metal stress. Due to this, it was observed that H_2S is a major signaling molecule for increasing heavy metal resistance (Walsh et al., 2018). H_2S enhanced the plant biomass and helped in the accumulation of nutrients such as nitrogen and Fe under situations of iron deficiency in soybean seedlings (Chen et al., 2018). The treatment with the H_2S donor NaHS enhanced the energy level and antioxidant defense system ability, suppressed the respiration rate, and ultimately improved the yellowing of leaves and senescence in *Ipomoea aquatica* plants (Hu et al., 2015). H_2S is known to affect energy production regulated by mitochondria and inhibit cellular senescence which results in delayed leaf senescence in *Arabidopsis* under drought stress (Jin et al., 2018). H_2S synergistically with rhizobia facilitates nitrogen assimilation and remobilization by stimulating the action of nitrogen metabolizing enzymes, their corresponding protein and gene expression, and also enhances the expression of senescence-associated genes during senescence in soybeans (Zhang et al., 2020).

H_2S increased the chlorophyll biosynthesis, photosynthesis, quantum efficiency of photosystem II, the membrane lipid stability, protein synthesis, and energy metabolism to support the *Kandelia obovata* plants in the tolerance of high-salinity conditions (Liu et al., 2020). H_2S improved the expression of photosynthetic enzymes, chloroplast biosynthesis, and thiol redox modification, hence resulting in an enhanced rate of photosynthesis of *Spinacia oleracea* seedlings (Chen et al., 2011). Hydrogen sulfide contributes to plant adaptive responses to various abiotic stresses by the regulation of antioxidative enzymes activities and nonenzymatic antioxidants. Several studies have shown that after the application of sodium hydrosulfide plants had enhanced catalase (CAT), superoxide dismutase (SOD), and peroxidase (POD) activities and decreased malondialdehyde (MDA) content and reactive oxygen species (ROS) accumulation. Hydrogen sulfide diminished the amount of chromium in various parts of cauliflower (*Brassica oleracea botrytis* L.), and also decreased the electrolyte leakage, hydrogen peroxide, and MDA amount, in addition to increasing the amount of chlorophyll, gas exchange attributes, and the activities of antioxidant enzymes in the roots and leaves of cauliflower under chromium stress conditions (Ahmad et al., 2020). H_2S aided plants by decreasing hydrogen peroxide accumulation and by regulating membrane stability and antioxidant defense system in mitochondria under salt stress conditions (Wei et al., 2019). The H_2S donor sodium hydrosulfide (NaHS) altered the oxidative status as inferred by changes in ROS accumulation and lipid peroxidation which is associated with the regulation of the activities of SOD, POD, CAT, and ascorbate peroxidase (APX) enzymes, and also enhanced the nitric oxide (NO) and total S-nitrosothiols (SNOs) amount, the activity of nitrate reductase, and decreased the S-nitrosoglutathione reductase activities in nitrate-stressed plants of *Lycopersicon esculentum* (Guo et al., 2018).

7.2 Sulfide: an emerging signal molecule

Hydrogen sulfide is a reactive molecule, and several studies have shown that H_2S interacts with other signaling molecules to modify their signaling. An increase in the number of plant studies in recent decades has led H_2S to be considered to have the same relevance as the signaling molecules NO and H_2O_2. It was observed that plant hormones, hydrogen peroxide, nitric oxide, calcium, and carbon monoxide signaling are involved in H_2S signal transduction, resulting in a complex signaling network. H_2S delays programmed cell death by the modulation of glutathione homeostasis and heme oxyenase-1 expression (Xie et al., 2014). Moreover, the progression of autophagy is negatively regulated by H_2S in a way unrelated to sulfur nutrition and by a mechanism of action independent of redox conditions (Laureano-Marín et al., 2016). H_2S can interact with the metal center of metalloproteins (Filipovic et al., 2018) and bound covalently to heme porphyrins, acting as a potent inhibitor of mitochondrial cytochrome c oxidase and inhibiting respiration in mitochondria where sulfide is detoxified (Birke et al., 2015). Cross-talk between H_2S and SNOs indicates that sulfide species play an important role in modulating the profile of these molecules through the reaction of H_2S with small or protein S-nitrosothiol molecules to form nitropersulfides ($SSNO^-$), polysulfides (HSn^-), and dinitrososulfites ($ONN(OH)SO_3^-$) (Cortese-Krott et al., 2015). Recently, polysulfides appeared as potential mediators of H_2S signaling (Fig. 7.1). H_2S modifies the proteins by the oxidation of cysteine residues to form the corresponding persulfides, resulting in persulfidation of proteins (Gotor et al., 2019). Persulfidation of *Arabidopsis* can affect either cytosolic/nuclear partitioning or its activity, as described for the cytosolic $GapC_1$ and $GapC_2$ isoforms, which are persulfidated at Cys160 as analyzed by parallel reaction monitoring (Aroca et al., 2015). H_2S-stimulated persulfide generation could not be the consequence of its reaction with protein disulfides because the incubation of the cells with diamide, an inducer of disulfide bond formation, attenuates H_2S-induced S-sulfhydration (Wedmann et al., 2016).

Hydrogen sulfide directly regulates persulfide groups formation by the posttranslational modification of cysteine residues (conversion of cysteine-SH groups to cysteine-SSH groups) (Paul and Snyder, 2012). This procedure is

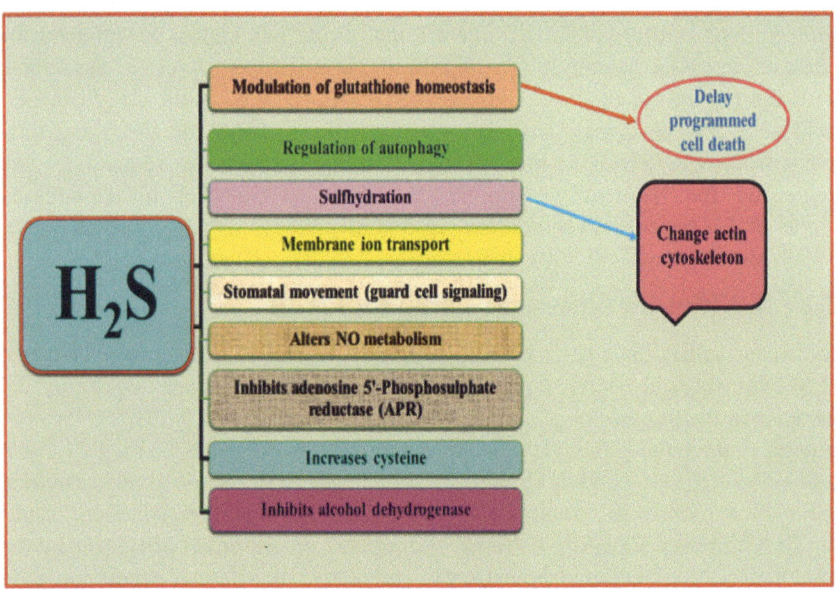

FIGURE 7.1 Schematic representation of H_2S as a signaling molecule. *Modified after Aroca, Á., Serna, A., Gotor, C., Romero, L.C. 2015. S-sulfhydration: a cysteine posttranslational modification in plant systems. Plant Physiol. 168(1), 334–342.*

known as *S*-sulfhydration and is similar to *S*-nitrosylation. Endogenous H_2S stimulates the functions of certain proteins, such as APX1 and glyceraldehyde 3-phosphate dehydrogenase (Aroca et al., 2015). H_2S can directly modify specific targets through protein sulfhydration, that is, addition of an − SH group to the thiol moiety of proteins (Mustafa et al., 2009) or the reaction with metal centers (Li and Lancaster, 2013).

Indirectly, hydrogen sulfide can also react with nitric oxide to form nitrosothiols (Whiteman et al., 2006). H_2S alters actin polymerization through the *S*-sulfhydration of actin and it was found that *S*-sulfhydration inhibited actin polymerization (Li et al., 2018). H_2S is an important key regulator of membrane ion transport as it regulates adenosine triphosphate-dependent K^+ channels (Yang et al., 2005), Ca^{2+} activated K^+ channels, T- and L-type Ca^{2+} channels, and transient receptor potential channels (Peers et al., 2012). H_2S is also involved in guard cell signaling, through the ATP - binding cassette transporter multidrug resistant protein 5, which is an anion and calcium channel regulator (García-Mata and Lamattina, 2013). In tomato plants under osmotic stress, ethylene regulates stomatal closure and stimulates the production of H_2S in guard cells, but H_2S feedback also regulates ethylene biosynthesis through inhibiting the enzymatic activity of 1-aminocyclopropane-1-carboxylic acid oxidase (ACO_1) by persulfidation at Cys60 (Jia et al., 2018).

7.3 H_2S synthesis in plants

Synthesis of endogenous H_2S in plants is totally different from that in animals. Hydrogen sulfide emission in plants was detected by De Cormis in the late 1960s and it was demonstrated that exogenous application of sulfite (SO_3^{2-}), sulfate (SO_4^{2-}), sulfur dioxide (SO_2), or L-cysteine resulted in H_2S emissions (Rennenberg, 1984). Wilson et al. (1978) reported the emission of H_2S from plants, when they quantified hydrogen sulfide from leaves of cucumber, pumpkin, cantaloupe, corn, soybean, and cotton. It was also observed that the amount of H_2S emission was significantly enhanced when leaves were fed with sulfate via their petioles or when the roots of the plant were mechanically damaged (Rennenberg, 1984). Sekiya et al. (1982) also demonstrated that the emission of H_2S is higher in younger leaves in comparison to mature leaves using cucumber as a model plant Plants contain enzymes which reduce inorganic sulfite to compounds that release H_2S upon acidification and other enzymes which reduce a carrier-bound sulfo group to a form which can be exchanged with free hydrosulfide (HS^-) and then liberated as H_2S upon acidification. Synthesis of hydrogen sulfide in plants is chiefly regulated by enzymatic and nonenzymatic pathways, although the latter only accounts for a small portion of H_2S generation. H_2S biosynthesis generally occurs in the chloroplast, to a smaller extent in the cytosol, and in the mitochondria (Aroca et al., 2018) (Fig. 7.2). H_2S is synthesized in plants via cysteine degradation which is catalyzed by DCD/LCD or through reducing SO_4^{2-} by sulfite reductase (Papenbrock et al., 2007). Two

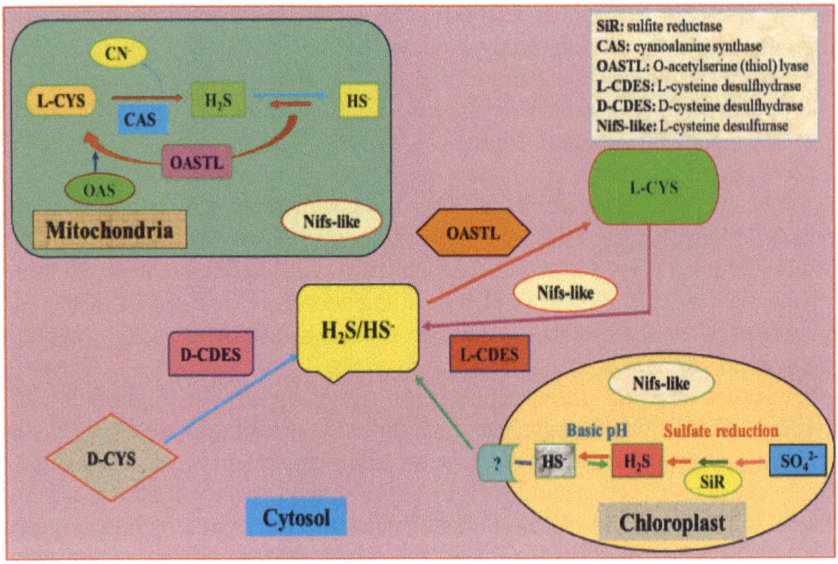

FIGURE 7.2 Hydrogen sulfide (H_2S) synthesis in chloroplast, mitochondria and cytosol of plant cells. In mitochondria, H_2S is formed via the detoxification of cyanide (CN^-) by β-cyanoalanine synthase (CAS) which utilized cysteine (L-CYS) generated by mitochondrial O-acetylserine (thiol) lyase (OASTL). Ionic form (HS^-) is also formed in the mitochondria at basic pH. In chloroplast, sulfate (SO_4^{2-}) is reduced to sulfide (H_2S) by the action of sulfite reductase (SiR) and, then H_2S is broken down to its ionic form (HS^-) that requires an unknown active transporter (?) to exit from the membrane. Moreover, in the cytosol, cysteine is primarily formed by OASTL which ultimately form H_2S by the use of cysteine-degrading enzymes, i.e. the l-cysteine (l-CDES) and d-cysteine (d-CDES) desulfhydrases, and the l-cysteine desulfurases (NifS-like). *Modified after Gotor, C., García, I., Aroca, Á., Laureano-Marín, A.M., Arenas-Alfonseca, L., Jurado-Flores, A., et al. 2019. Signaling by hydrogen sulfide and cyanide through post-translational modification. J. Exp. Bot 70(16), 4251–4265.*

specific desulfhydrases, (LCD, EC4.4.1.1; also called L-CDes or L-DES) and (DCD, EC 4.4.1.15; also called D-CDes or D-DES), have been isolated from *Arabidopsis thaliana* (Álvarez et al., 2010).

The LCD requires pyridoxal 5-phosphate as a cofactor and is regarded as the most significant enzyme for H_2S synthesis in plants which degrades cysteine into H_2S, pyruvate, and ammonia. LCD could be an important aspect in liberating H_2S during a plant defense response and its activity and expression can be upregulated when plants are attacked by pathogena. Besides different substrates, enzymatic inhibitors of DCD/LCD are also different. The activity of LCD could be repressed by amino-oxy-acetic acid. On the other hand, it had no effects on DCD activity. Subcellular localization of both DCD/LCDs is also different. LCD activity was found in chloroplasts and mitochondria whereas D-CD was mainly confined to cytoplasm (Rennenberg et al., 1987). Hydrogen sulfide is dissociated into its ionized forms due to the basic pH inside organelles and therefore it is unable to pass through the membranes to the cytosol, thus it requires specific transporters. The transcript abundance of LCD or its total enzyme activity was enhanced by drought stress, salicylic acid (SA), ABA, etc. (Xie et al., 2013). H_2S is also synthesized by sulfite reduction which is carried out by sulfite reductase enzyme (SiR, EC 1.8.7.1) using ferredoxin as an electron donor in chloroplasts and also from cysteine via cyanoalanine synthetase (CAS, EC 4.4.1.9) or cysteine synthase in the presence of HCN (Rausch and Wachter, 2005). CAS can convert L-cysteine and cyanide into H_2S and cyanuric acid in the mitochondria and cytoplasm of plant cells. In the chloroplast, mitochondria and cytoplasm, in the presence of O-acetylserine (thiol) lyase (OASTL), CS catalyzes O-acetylserine (OAS) and H_2S to form cysteine. This reversible reaction can produce H_2S (He et al., 2019). Nifs/NFS (cysteine desulphurases) is also potentially involved in H_2S generation in *Arabidopsis* for its LCD-like activity (Léon et al., 2002).

7.4 Biological significance of H_2S

H_2S played a significant role in various plant biological processes as they are involved in signaling in plants. Although highly phytotoxic, low doses of H_2S have been demonstrated to have an important role in a variety of plant life processes including plant growth and development, as well as biotic and abiotic stress responses (Jin et al., 2013; Xie et al., 2014; Zhang et al., 2008, 2009a, 2010b). H_2S stimulates the germination of the seeds and yields of many crops including rice, maize, wheat, and pea (Dooley et al., 2013). H_2S reduces various abiotic stresses, such as heavy metal stress, drought, and osmotic stress, by boosting plant antioxidants (Shi et al., 2014; Sun et al., 2013; Zhang et al., 2008). The release of H_2S from cucumber, maize, and soybean leaves and the leaves of other older plants was found to be higher than that of younger plants. In plant growth and development, H_2S' multifunctional position emphasizes the importance of regulating the endogenous generation of H_2S for adjustment to growth stages and in response to abiotic and biotic stresses. It is observed that ABA did not close the stomata in the des1 mutant which encodes LCD, which suggests that endogenous H_2S is critical for ABA signaling in the guard cells (Scuffi et al., 2014). Endogenous H_2S increases in *A. thaliana* when plants are infected with *Pseudomonas syringae* (Shi et al., 2015). Therefore DES1 activity in pathogen-infected plants is increased (Bloem et al., 2004). In cut flowers, H_2S can also delay the opening of flowers and senescence. Later H_2S would have the senescence-alleviation effect on various fruit and vegetables, including strawberries, kiwi, grapes, tomatoes, banana, and broccoli, potentially through improved antioxidant efficiency, and the antagonizing production of ethylene (Ge et al., 2018; Hu et al., 2012; Ni et al., 2016). In addition, H_2S produced in cytosol regulates autophagy and modulates *A. thaliana*'s transcriptional profile through the analysis of des1 mutant while the negative autophagic control of sulfide does not involve ROS (Álvarez et al., 2012; Laureano-Marín et al., 2016).

In response to biotic stress, H_2S also plays an important role for plants. SA contributes to the development of endogenous H_2S by increasing the LCD activities to withstand Cd stress. The positive effect of SA in LCD-knockout is decreased in *A. thaliana*, which indicates that H_2S functions downstream to control Cd tolerance (Qiao et al., 2015). Stomatal movement is essential for plant responses and H_2S has engaged with ABA in stomatal control that induces drought stress in *A. thaliana* (Jin et al., 2013). There was evidence of changes in seed germination due to exogenous H_2S treatments. Because of other sulfur-containing components from an exogenous H_2S source, NaHS, H_2S, or HS- has helped facilitate the germination of seed (Zhang et al., 2008). NaHS maintains lower MDA and hydrogen peroxide (H_2O_2) levels in germinating grain, which better affect the activity of endosperm β-amylase (Zhang et al., 2010a). Furthermore, NaHS has promoted the number and length of adventitious roots in seedling cuttings of sweet potato (Zhang et al., 2009a). HS and thiol redox modification to regulate the photosynthesis (Chen et al., 2011) are simultaneously applied. This mechanism has also been thought to include a rise in the density of the stomata (Duan et al., 2015). Through distributing the NaHS solution (Zhang et al., 2009b) osmotic reduction in the chlorophyll concentration may be alleviated.

Recent research on H_2S emissions from plants have been linked to sulfur-induced pathogens response by plants (Bloem et al., 2004). For the first time, H_2S was established as being a significant cellular signal highlighting the copper stress defense effect of H_2S. Various beneficial effects of H_2S and H_2S signaling were subsequently observed in plants. H_2S has been shown to reduce the symptoms of drought, osmotic stress, and iron, cadmium, chromium, hypoxia, and other stressors (Cheng et al., 2013; Jin et al., 2011; Li et al., 2012a, b,c, 2013). Some of the studies examined how H_2S signaling is essential to the defense of plants against stress with analogies to animal systems. In response to the environmental stimuli, stomatal movement is very critical and the advanced guard cell is a key target of H_2S signaling in plants. Recent studies have shown that H_2S causes the alleviation of drought stress in *A. thaliana* (Jin et al., 2013). In addition, during ethylene-induced stomatal closure H_2S was reported to be a novel downstream nitric oxide (NO) indicator (Liu et al., 2011). Nonetheless, H_2S has a controversial influence on stomatal movement. It is also documented that exogenous H_2S caused reduction in stomatal opening and also declined the level of NO accumulation in *Arabidopsis thaliana* cells and in *Capsicum annuum* (Lisjak et al., 2010, 2011). The goal of the stomatal closure is to reduce moisture loss due to drought stress, to increase photosynthesis, and to reduce photorespiration by induction of a stomach opening.

7.5 Crosstalk of H_2S with other signals

Plants perceive H_2S and respond to it, but there is very limited analysis of the mechanisms of H_2S functioning in plant responses to stress. H_2S is especially involved and may communicate with many other signals and change them. Multiple H_2S-perceptive and signaling routes may therefore be unwound. It is observed that H_2S and plant hormones are related in plant defenses. There was a rapid rise in indole acetic acid by the exogenous treatment with H_2S in *Arabidopsis thaliana* plants and also requires NO to induce ethylene to increase endogenous H_2S production. (Liu et al., 2011). In plants under various abiotic stresses, ABA is produced in large quantities. Drought stress led to a substantial increase in *CDes* expression and also an increase in production rate of H_2S from these plants (Jin et al., 2011). The relationship between H_2S and ABA was subsequently stated on the basis of an H_2S deficiency in the *lcd* mutant which had a favored ABA induction of stomatal closure which suggested a partial dependence on H_2S for the induction of stomatal closure by ABA. Since H_2S also participated in the ion-channel gene expression regulation, H_2S can become a critical component of stomatal closure induced by ABA over ion channels. H_2S has also affected the expression of ABA receptors and H_2S may have started to exert its effect on the ABA signaling pathway. H_2S has therefore been observed to communicate with ABA for the stomatal regulation of *A. thaliana* that is responsible for the drought stress (Jin et al., 2013). Moreover, the H_2S signal transduction cycle was performed on both gibberellic acid (GA) and jasmonic acid (JA). The GA-caused programmed cellular death in wheat aleurone cells can be reduced by H_2S (Xie et al., 2014) and the JA-induced stomatal closure of *Vicia faba* can act downstream of H_2O_2 (Hou et al., 2011). In plants, particularly in guard cells, H_2O_2 is another signaling molecule. Abiotic stress allows both H_2S and H_2O_2 to be synthesized, but the function of both molecules together in the physiological process remains uncertain. During JA-induced stomatal movement in *V. faba*, H_2S may be the new downstream element of an H_2O_2 signal cascade (Hou et al., 2011). Pretreatment of H_2O_2 can enhance the germination of *Jatropha curcas* seeds, which is the mediation of H_2S (Li et al., 2012a,b,c). The findings show that H_2O_2 is upstream of H_2S. Cadmium influx was inhibited by H_2S through H_2O_2 channels triggered in the plasma membrane calcium channel (Sun et al., 2013). The production of antioxidant systems, such as heavy metal stress, osmotic stress, thermal stress, and hypoxia stress will enhance plant resistance to abiotic stress (Li et al., 2012a,b,c; Wang et al., 2010; Cheng et al., 2013). It is shown that NaHS as well as GYY4137 have greatly reduced the NO accumulation in *A. thaliana* epidemic cells (Lisjak et al., 2010). A rapid increase in endogenous H_2S and NO was observed in sweet potato seedlings in NaHS-treated shoot tips. Recent evidence shows that H_2S also takes part in the signaling pathway of NO and carbon monoxide (CO). Sodium nitroprusside (SNP) and NaHS combined treatments have shown that H_2S-inhibitors and scavengers can prevent NO signaling, suggesting that NO-activated H_2S was necessary for the response to cadmium stress (Shi et al., 2014). However, growing evidence indicates that H_2S interacts with the signaling pathways for calcium.

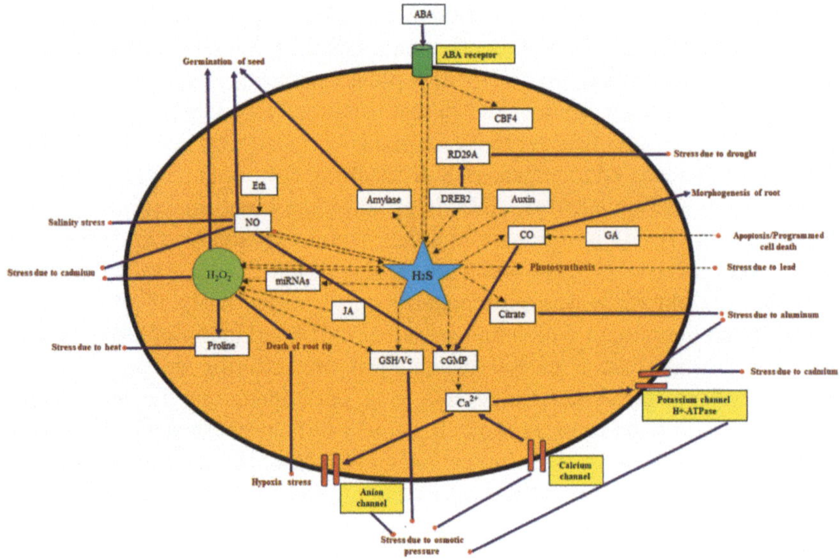

FIGURE 7.3 Cross-talk of hydrogen sulfide (H_2S) with abiotic stress. Solid line represents the stimulatory effect and dashed line represents the putative interaction. Carbon monoxide, CO; abscisic acid, ABA; nitric oxide, NO; dehydration responsive element, DREB; responsive to desiccation, RD; C-repeat binding factor, CBF; ethylene, Eth; cyclic guanosine monophosphate, cGMP; gibberellic acid, GA; hydrogen peroxide, H_2O_2; hydrogen sulfide, H_2S; jasmonic acid, JA; and glutathione, GSH. *Modified after Jin, Z., Pei, Y., 2015. Physiological implications of hydrogen sulfide in plants: pleasant exploration behind its unpleasant odour. Oxid. Med. Cell. Longev., 2015.*

Ca^{2+} confers cell wall stability and stiffness and by calmodulin regulates the plant's processes. NaHS pretreatment can help to increase the heat tolerance of extracellular Ca^{2+} into tobacco-cultivated cells, which can be regulated by intracellular calmodulin (Li et al., 2013). These results show that Ca signaling plays a major role in the H_2S mechanism (Fig. 7.3).

7.6 Conclusion

Several studies have reported the potential action of H_2S in numerous cellular events as a signaling molecule. This gasotransmitter has been widely studied and the emerging data have changed the perception of H_2S from a toxic molecule to a beneficial signaling molecule comparable to H_2O_2 and NO. The advantageous effects of H_2S are found only when applied at optimal concentrations and on a species-specific basis. H_2S has acquired worldwide acceptance as a vital regulator for plant growth and development, regulating root growth, seed germination, stomatal movement, and photosynthesis. Plants can use the H_2S gas as a source of sulfur for their growth. Oxidative stress accelerates the production of ROS which oxidize membrane systems, nucleic acids, and even target cysteine residues in proteins. H_2S is an important factor in plant tolerance, by aiding in the reduction of the oxidative stress that can be induced by a range of stressors, including heavy metal toxicity, drought stress, osmotic stress, heat stress, pathogen attack, hypoxia, and other stressors. This mechanism probably includes the stimulation of the antioxidant defense system, the induction of stomatal closure and the enhanced expression of genes encoding resistance-associated enzymes. The antioxidative enzymes are efficiently activated to scavenge ROS. Even though H_2S is a short-lived molecule, it is an extremely active one. Enhancing the generation of endogenous hydrogen sulfide in plants is an alternative technique for removing heavy metals toxicity in soil H_2S induces persulfidation of cysteine residues, which are highly susceptible to cellular oxidation. H_2S can modify many proteins through persulfidation, hence the proteins remain protected during the prevailing oxidative stress. Overall, it was concluded that H_2S is an important molecule which may help to protect plants.

References

Ahmad, R., Ali, S., Rizwan, M., Dawood, M., Farid, M., Hussain, A., et al., 2020. Hydrogen sulfide alleviates chromium stress on cauliflower by restricting its uptake and enhancing antioxidative system. Physiol. Plant 168 (2), 289–300.

Álvarez, C., Calo, L., Romero, L.C., García, I., Gotor, C., 2010. An O-acetylserine (thiol) lyase homolog with L-cysteine desulfhydrase activity regulates cysteine homeostasis in *Arabidopsis*. Plant Physiol. 152 (2), 656–669.

Álvarez, C., García, I., Moreno, I., Pérez-Pérez, M.E., Crespo, J.L., Romero, L.C., et al., 2012. Cysteine-generated sulfide in the cytosol negatively regulates autophagy and modulates the transcriptional profile in *Arabidopsis*. Plant Cell 24 (11), 4621–4634.

Aroca, Á., Serna, A., Gotor, C., Romero, L.C., 2015. S-sulfhydration: a cysteine posttranslational modification in plant systems. Plant Physiol. 168 (1), 334–342.

Aroca, A., Gotor, C., Romero, L.C., 2018. Hydrogen sulfide signaling in plants: emerging roles of protein persulfidation. Front. Plant Sci. 9, 1369.

Birke, H., Hildebrandt, T.M., Wirtz, M., Hell, R., 2015. Sulfide detoxification in plant mitochondria, Methods in Enzymology, vol. 555. Academic Press, pp. 271–286.

Bloem, E., Riemenschneider, A., Volker, J., Papenbrock, J., Schmidt, A., Salac, I., et al., 2004. Sulphur supply and infection with *Pyrenopeziza brassicae* influence L-cysteine desulphydrase activity in *Brassica napus* L. J. Exp. Bot. 55 (406), 2305–2312.

Chen, J., Shang, Y.T., Zhang, N.N., Zhong, Y., Wang, W.H., Zhang, J.H., et al., 2018. Sodium hydrosulfide modifies the nutrient ratios of soybean (*Glycine max*) under iron deficiency. J. Plant Nutr. Soil Sci. 181 (2), 305–315.

Chen, J., Wu, F.H., Wang, W.H., Zheng, C.J., Lin, G.H., Dong, X.J., et al., 2011. Hydrogen sulphide enhances photosynthesis through promoting chloroplast biogenesis, photosynthetic enzyme expression, and thiol redox modification in *Spinacia oleracea* seedlings. J. Exp. Bot. 62 (13), 4481–4493.

Chen, J., Wang, W.H., Wu, F.H., You, C.Y., Liu, T.W., Dong, X.J., et al., 2013. Hydrogen sulfide alleviates aluminum toxicity in barley seedlings. Plant Soil 362 (1-2), 301–318.

Cheng, W., Zhang, L., Jiao, C., Su, M., Yang, T., Zhou, L., et al., 2013. Hydrogen sulfide alleviates hypoxia-induced root tip death in *Pisum sativum*. Plant Physiol. Biochem. 70, 278–286.

Cortese-Krott, M.M., Kuhnle, G.G., Dyson, A., Fernandez, B.O., Grman, M., DuMond, J.F., et al., 2015. Key bioactive reaction products of the NO/H_2S interaction are S/N-hybrid species, polysulfides, and nitroxyl. Proc. Natl. Acad. Sci. 112 (34), E4651–E4660.

Dooley, F.D., Nair, S.P., Ward, P.D., 2013. Increased growth and germination success in plants following hydrogen sulfide administration. PLoS One 8 (4).

Duan, B., Ma, Y., Jiang, M., Yang, F., Ni, L., Lu, W., 2015. Improvement of photosynthesis in rice (*Oryza sativa* L.) as a result of an increase in stomatal aperture and density by exogenous hydrogen sulfide treatment. Plant Growth Regul. 75 (1), 33–44.

Filipovic, M.R., Zivanovic, J., Alvarez, B., Banerjee, R., 2018. Chemical biology of H_2S signaling through persulfidation. Chem. Rev. 118 (3), 1253–1337.

García-Mata, C., Lamattina, L., 2010. Hydrogen sulphide, a novel gasotransmitter involved in guard cell signalling. N. Phytol 188 (4), 977–984.

García-Mata, C., Lamattina, L., 2013. Gasotransmitters are emerging as new guard cell signaling molecules and regulators of leaf gas exchange. Plant Sci. 201, 66–73.

Ge, Y., Hu, K.-D., Wang, S.-S., Hu, L.-Y., Chen, X.-Y., Li, Y.-H., et al., 2018. Correction: hydrogen sulfide alleviates postharvest ripening and senescence of banana by antagonizing the effect of ethylene. PLoS One 13 (1), e0191351.

Gotor, C., García, I., Aroca, Á., Laureano-Marín, A.M., Arenas-Alfonseca, L., Jurado-Flores, A., et al., 2019. Signaling by hydrogen sulfide and cyanide through post-translational modification. J. Exp. Bot 70 (16), 4251–4265.

Guo, Z., Liang, Y., Yan, J., Yang, E., Li, K., Xu, H., 2018. Physiological response and transcription profiling analysis reveals the role of H_2S in alleviating excess nitrate stress tolerance in tomato roots. Plant Physiol. Biochem. 124, 59–69.

Habeeb, O.A., Kanthasamy, R., Ali, G.A., Sethupathi, S., Yunus, R.B.M., 2018. Hydrogen sulfide emission sources, regulations, and removal techniques: a review. Rev. Chem. Eng. 34 (6), 837–854.

Hancock, J.T., Whiteman, M., 2014. Hydrogen sulfide and cell signaling: team player or referee? Plant Physiol. Biochem. 78, 37–42.

He, H., Li, Y., He, L.F., 2019. Role of nitric oxide and hydrogen sulfide in plant aluminum tolerance. Biometals 32 (1), 1–9.

Hou, Z.H., Liu, J., Hou, L.X., Li, X.D., Liu, X., 2011. H2S may function downstream of H_2O_2 in jasmonic acid-induced stomatal closure in *Vicia faba*. Chin. Bull. Bot 46 (4), 396–406.

Hu, L.-Y., Hu, S.-L., Wu, J., Li, Y.-H., Zheng, J.-L., Wei, Z.-J., et al., 2012. Hydrogen sulfide prolongs postharvest shelf life of strawberry and plays an antioxidative role in fruits. J. Agric. Food Chem. 60 (35), 8684–8693.

Hu, H., Liu, D., Li, P., Shen, W., 2015. Hydrogen sulfide delays leaf yellowing of stored water spinach (*Ipomoea aquatica*) during dark-induced senescence by delaying chlorophyll breakdown, maintaining energy status and increasing antioxidative capacity. Postharvest Biol. Technol. 108, 8–20.

Jia, H., Chen, S., Liu, D., Liesche, J., Shi, C., Wang, J., et al., 2018. Ethylene-induced hydrogen sulfide negatively regulates ethylene biosynthesis by persulfidation of ACO in tomato under osmotic stress. Front. Plant Sci. 9, 1517.

Jin, Z., Pei, Y., 2015. Physiological implications of hydrogen sulfide in plants: pleasant exploration behind its unpleasant odour. Oxid. Med. Cell. Longev. 2015.

Jin, Z., Shen, J., Qiao, Z., Yang, G., Wang, R., Pei, Y., 2011. Hydrogen sulfide improves drought resistance in *Arabidopsis thaliana*. Biochem Biophys Res. Commun. 414 (3), 481–486.

Jin, Z., Sun, L., Yang, G., Pei, Y., 2018. Hydrogen sulfide regulates energy production to delay leaf senescence induced by drought stress in *Arabidopsis*. Front. Plant Sci. 9, 1722.

Jin, Z., Xue, S., Luo, Y., Tian, B., Fang, H., Li, H., et al., 2013. Hydrogen sulfide interacting with abscisic acid in stomatal regulation responses to drought stress in *Arabidopsis*. Plant Physiol. Biochem. 62, 41–46.

Laureano-Marín, A.M., Moreno, I., Romero, L.C., Gotor, C., 2016. Negative regulation of autophagy by sulfide is independent of reactive oxygen species. Plant Physiol. 171 (2), 1378–1391.

Léon, S., Touraine, B., Briat, J.F., Lobréaux, S., 2002. The AtNFS2 gene from *Arabidopsis thaliana* encodes a NifS-like plastidial cysteine desulphurase. Biochem J. 366 (2), 557–564.

Li, J., Chen, S., Wang, X., Shi, C., Liu, H., Yang, J., et al., 2018. Hydrogen sulfide disturbs actin polymerization via S-sulfhydration resulting in stunted root hair growth. Plant Physiol. 178 (2), 936–949.

Li, L., Hsu, A., Moore, P.K., 2009. Actions and interactions of nitric oxide, carbon monoxide and hydrogen sulphide in the cardiovascular system and in inflammation—a tale of three gases!. Pharmacol Ther 123 (3), 386–400.

Li, Q., Lancaster Jr, J.R., 2013. Chemical foundations of hydrogen sulfide biology. Nitric Oxide 35, 21–34.

Li, Y.-W., Gong, Z.-H., Mu, Y., Zhang, Y.-X., Qiao, Z.-J., Zhang, L.-P., et al., 2012a. An *Arabidopsis* mutant *atcsr-2* exhibits high cadmium stress sensitivity involved in the restriction of H_2S emission. J. Zhejiang Univ. Sci. B 13 (12), 1006–1014.

Li, Z.G., Gong, M., Liu, P., 2012b. Hydrogen sulfide is a mediator in H_2O_2-induced seed germination in *Jatropha Curcas*. Acta Physiol Plant 34 (6), 2207–2213.

Li, Z.-G., Gong, M., Xie, H., Yang, L., Li, J., 2012c. Hydrogen sulfide donor sodium hydrosulfide-induced heat tolerance in tobacco (*Nicotiana tabacum* L) suspension cultured cells and involvement of Ca^{2+} and calmodulin. Plant Sci. 185, 185–189.

Li, Z.G., Ding, X.J., Du, P.F., 2013. Hydrogen sulfide donor sodium hydrosulfide-improved heat tolerance in maize and involvement of proline. J. Plant Physiol. 170 (8), 741–747.

Li, Z.G., Min, X., Zhou, Z.H., 2016. Hydrogen sulfide: a signal molecule in plant cross-adaptation. Front. Plant Sci. 7, 1621.

Lisjak, M., Srivastava, N., Teklic, T., Civale, L., Lewandowski, K., Wilson, I., et al., 2010. A novel hydrogen sulfide donor causes stomatal opening and reduces nitric oxide accumulation. Plant Physiol. Biochem. 48 (12), 931–935.

Lisjak, M., Teklić, T., Wilson, I.D., Wood, M., Whiteman, M., Hancock, J.T., 2011. Hydrogen sulfide effects on stomatal apertures. Plant Signal. Behav. 6 (10), 1444–1446.

Lisjak, M., Teklic, T., Wilson, I.D., Whiteman, M., Hancock, J.T., 2013. Hydrogen sulfide: environmental factor or signalling molecule? Plant, Cell Environ. 36 (9), 1607–1616.

Liu, J., Hou, L., Liu, G., Liu, X., Wang, X., 2011. Hydrogen sulfide induced by nitric oxide mediates ethylene-induced stomatal closure of *Arabidopsis thaliana*. Chin. Sci. Bull. 56 (33), 3547–3553.

Liu, Y.L., Shen, Z.J., Simon, M., Li, H., Ma, D.N., Zhu, X.Y., et al., 2020. Comparative proteomic analysis reveals the regulatory effects of H_2S on salt tolerance of mangrove plant *Kandelia obovata*. Int. J. Mol. Sci. 21 (1), 118.

Mostofa, M.G., Saegusa, D., Fujita, M., Tran, L.S.P., 2015. Hydrogen sulfide regulates salt tolerance in rice by maintaining Na + /K + balance, mineral homeostasis and oxidative metabolism under excessive salt stress. Front. Plant Sci. 6, 1055.

Mustafa, A.K., Gadalla, M.M., Sen, N., Kim, S., Mu, W., Gazi, S.K., et al., 2009. H_2S signals through protein S-sulfhydration. Sci. Signal. 2 (96), ra72.

Ni, Z., Hu, K., Song, C., Ma, R., Li, Z., Zheng, J., et al., 2016. Hydrogen sulfide alleviates post-harvest senescence of grape by modulating the antioxidant defenses. Oxid. Med. Cell. Longev. 2016, 4715651.

Papenbrock, J., Riemenschneider, A., Kamp, A., Schulz-Vogt, H.N., Schmidt, A., 2007. Characterization of cysteine-degrading and H_2S-releasing enzymes of higher plants-from the field to the test tube and back. Plant Biol. 9 (05), 582–588.

Paul, B.D., Snyder, S.H., 2012. H_2S signalling through protein sulfhydration and beyond. Nat. Rev. Mol. Cell Biol. 13 (8), 499–507.

Peers, C., Bauer, C.C., Boyle, J.P., Scragg, J.L., Dallas, M.L., 2012. Modulation of ion channels by hydrogen sulfide. Antioxid. Redox Signal. 17 (1), 95–105.

Qiao, Z., Jing, T., Liu, Z., Zhang, L., Jin, Z., Liu, D., et al., 2015. H_2S acting as a downstream signaling molecule of SA regulates Cd tolerance in *Arabidopsis*. Plant Soil 393 (1-2), 137–146.

Rausch, T., Wachter, A., 2005. Sulfur metabolism: a versatile platform for launching defence operations. Trends Plant Sci. 10 (10), 503–509.

Rennenberg, H., 1984. The fate of excess sulfur in higher plants. Annu. Rev. Plant Physiol. 35 (1), 121–153.

Rennenberg, H., Arabatzis, N., Grundel, I., 1987. Cysteine desulphydrase activity in higher plants: evidence for the action of L-and D-cysteine specific enzymes. Phytochemistry 26 (6), 1583–1589.

Rodriguez-Kabana, R., Jordan, J.W., Hollis, J.P., 1965. Nematodes: biological control in rice fields: role of hydrogen sulfide. Science 148 (3669), 524–526.

Scuffi, D., Álvarez, C., Laspina, N., Gotor, C., Lamattina, L., García-Mata, C., 2014. Hydrogen sulfide generated by L-cysteine desulfhydrase acts upstream of nitric oxide to modulate abscisic acid-dependent stomatal closure. Plant Physiol. 166 (4), 2065–2076.

Sekiya, J., Schmidt, A., Wilson, L.G., Filner, P., 1982. Emission of hydrogen sulfide by leaf tissue in response to L-cysteine. Plant Physiol. 70 (2), 430–436.

Shi, H., Ye, T., Chan, Z., 2014. Nitric oxide-activated hydrogen sulfide is essential for cadmium stress response in bermudagrass (*Cynodon dactylon* (L). Pers. Plant Physiol. Biochem. 74, 99–107.

Shi, H., Ye, T., Han, N., Bian, H., Liu, X., Chan, Z., 2015. Hydrogen sulfide regulates abiotic stress tolerance and biotic stress resistance in *Arabidopsis*. J. Integr. Plant Biol. 57 (7), 628–640.

Sun, Y.D., Luo, W.R., 2014. Effects of exogenous hydrogen sulphide on seed germination and seedling growth of cucumber (*Cucumis sativus*) under sodium bicarbonate stress. Seed Sci. Technol. 42 (2), 126–131.

Sun, J., Wang, R., Zhang, X., Yu, Y., Zhao, R., Li, Z., et al., 2013. Hydrogen sulfide alleviates cadmium toxicity through regulations of cadmium transport across the plasma and vacuolar membranes in *Populus euphratica* cells. Plant Physiol. Biochem. 65, 67–74.

Thompson, C.R., Kats, G., 1978. Effects of continuous hydrogen sulfide fumigation on crop and forest plants. Environ. Sci. Technol. 12 (5), 550–553.

Walsh, T.K., Joussen, N., Tian, K., McGaughran, A., Anderson, C.J., Qiu, X., et al., 2018. Multiple recombination events between two cytochrome P450 loci contribute to global pyrethroid resistance in *Helicoverpa armigera*. PLoS One 13 (11).

Wang, B.L., Shi, L., Li, Y.X., Zhang, W.H., 2010. Boron toxicity is alleviated by hydrogen sulfide in cucumber (*Cucumis sativus* L.) seedlings. Planta 231 (6), 1301–1309.

Wang, H.R., Che, Y.H., Huang, D., Ao, H., 2020. Hydrogen sulfide mediated alleviation of cadmium toxicity in *Phlox paniculata* L. and establishment of a comprehensive evaluation model for corresponding strategy. Int. J. Phytoremediat. 1–11.

Wang, R., 2012. Physiological implications of hydrogen sulfide: a whiff exploration that blossomed. Physiol Rev. 92 (2), 791–896.

Wedmann, R., Onderka, C., Wei, S., Szijártó, I.A., Miljkovic, J.L., Mitrovic, A., et al., 2016. Improved tag-switch method reveals that thioredoxin acts as depersulfidase and controls the intracellular levels of protein persulfidation. Chem. Sci. 7 (5), 3414–3426.

Wei, G.Q., Zhang, W.W., Cao, H., Yue, S.S., Li, P., Yang, H.Q., 2019. Effects hydrogen sulfide on the antioxidant system and membrane stability in mitochondria of *Malus hupehensis* under NaCl stress. Biol Plant 63 (1), 228–236.

Whiteman, M., Li, L., Kostetski, I., Chu, S.H., Siau, J.L., Bhatia, M., et al., 2006. Evidence for the formation of a novel nitrosothiol from the gaseous mediators nitric oxide and hydrogen sulphide. Biochem Biophys Res. Commun. 343 (1), 303–310.

Wilson, L.G., Bressan, R.A., Filner, P., 1978. Light-dependent emission of hydrogen sulfide from plants. Plant Physiol. 61 (2), 184–189.

Xie, Y., Lai, D., Mao, Y., Zhang, W., Shen, W., Guan, R., 2013. Molecular cloning, characterization, and expression analysis of a novel gene encoding L-cysteine desulfhydrase from *Brassica napus*. Mol. Biotechnol. 54 (3), 737–746.

Xie, Y., Zhang, C., Lai, D., Sun, Y., Samma, M.K., Zhang, J., et al., 2014. Hydrogen sulfide delays GA-triggered programmed cell death in wheat aleurone layers by the modulation of glutathione homeostasis and heme oxygenase-1 expression. J. Plant Physiol. 171 (2), 53–62.

Yang, W., Yang, G., Jia, X., Wu, L., Wang, R., 2005. Activation of KATP channels by H_2S in rat insulin-secreting cells and the underlying mechanisms. J. Physiol. 569 (2), 519–531.

Zhang, H., Dou, W., Jiang, C.-X., Wei, Z.-J., Liu, J., Jones, R.L., 2010a. Hydrogen sulfide stimulates β-amylase activity during early stages of wheat grain germination. Plant Signal. Behav. 5 (8), 1031–1033.

Zhang, H., Hu, L.Y., Hu, K.D., He, Y.D., Wang, S.H., Luo, J.P., 2008. Hydrogen sulfide promotes wheat seed germination and alleviates oxidative damage against copper stress. J. Integr. Plant Biol. 50 (12), 1518–1529.

Zhang, H., Tan, Z.Q., Hu, L.Y., Wang, S.H., Luo, J.P., Jones, R.L., 2010b. Hydrogen sulfide alleviates aluminum toxicity in germinating wheat seedlings. J. Integr. Plant Biol. 52 (6), 556–567.

Zhang, H., Tang, J., Liu, X.P., Wang, Y., Yu, W., Peng, W.Y., et al., 2009a. Hydrogen sulfide promotes root organogenesis in *Ipomoea batatas*, *Salix matsudana* and *Glycine max*. J. Integr. Plant Biol. 51 (12), 1086–1094.

Zhang, H., Ye, Y.-K., Wang, S.-H., Luo, J.-P., Tang, J., Ma, D.-F., 2009b. Hydrogen sulfide counteracts chlorophyll loss in sweetpotato seedling leaves and alleviates oxidative damage against osmotic stress. Plant Growth Regul. 58 (3), 243–250.

Zhang, N.N., Zou, H., Lin, X.Y., Pan, Q., Zhang, W.Q., Zhang, J.H., et al., 2020. Hydrogen sulfide and rhizobia synergistically regulate nitrogen (N) assimilation and remobilization during N deficiency-induced senescence in soybean. Plant Cell Environ.

Chapter 8

Hydrogen sulfide and nitric oxide crosstalk in plants under stress

Réka Szőllősi*, Árpád Molnár, Gábor Feigl, Dóra Oláh and Zsuzsanna Kolbert
Department of Plant Biology, University of Szeged, Szeged, Hungary
**Corresponding author.*

8.1 Introduction

Nowadays, a growing number of publications have proved that not only nitric oxide (NO) but also hydrogen sulfide (H_2S) can be regarded as signaling molecules. Both seem to be effective in giving fast responses to external or internal challenges; they give specific reactions, and their concentration can increase or decrease (Erdei and Kolbert, 2008). The other common feature of them is that they are free radicals: NO belongs to reactive nitrogen species (RNS), while H_2S is a member of reactive sulfur species (RSS). In addition, both appear to have multiple physiological functions during plants' life cycles, from seed germination through to stomatal movements to senescence (Corpas, 2019). It was also discovered that the presence of these gasotransmitters is needed not only in "normal" conditions, but also for stress tolerance (reviewed by Guo et al., 2016; Corpas et al., 2019b). Although in the last few years numerous studies have been published on H_2S and its roles in physiological processes and stress tolerance, we have indisputably more knowledge about NO (Corpas, 2019). Nevertheless, several researches have attempted to determine the potential relationship between them and there are still many questions to be answered, for example, which is up- or downstream of the other one, namely who is the "real boss." In this chapter we try to summarize how NO and H_2S can work together in the very well organized and complicated organisms that are plants.

8.2 Similarities and differences between NO and H_2S

8.2.1 Physicochemical aspects

Gasotransmitters have gained a newfound importance in biology since nitric oxide (Palmer et al., 1987) and carbon monoxide were proven to play an

Hydrogen Sulfide in Plant Biology. DOI: https://doi.org/10.1016/B978-0-323-85862-5.00011-7

$$H_2S_{(gas)} \rightleftarrows H_2S_{(aq.\ solution)} \rightleftarrows HS^- + H^+ \rightleftarrows S^{2-} + 2H^+$$

FIGURE 8.1 The dynamics of H_2S in aqueous solution.

important role in the physiological processes of living organisms. Hydrogen sulfide is a novel signal transmitter molecule, which is considered to be the third gasotransmitter, as proposed first in 1996 (Abe and Kimura, 1996), and is a regulator molecule, as was proven in CBS (cystathionine β-synthase) knockout mice in 2008 (Kimura, 2011).

Hydrogen sulfide (H_2S) or sulfane (International Union of Pure and Applied Chemistry IUPAC, 2012) is a toxic gaseous molecule and exposure to 500 ppm can be lethal (Beauchamp et al., 1984). The gas itself is highly flammable, colorless, with an odor similar to rotten eggs. Its solubility in water is dependent on temperature, with 122 mM at 20°C (Li and Moore, 2008), 100 mM at 25°C (Calhoun et al., 1988), and 80 mM at 37°C (Kabil and Banerjee, 2010). The differences can be due to other experimental conditions, especially pressure, because H_2S is volatile and will equilibrate between the solution and gaseous phase (Li and Lancaster, 2013). In aqueous solutions H_2S has the chemical aspects of a weak acid, transforming and equilibrating between two anion forms, HS^- and S^{2-} (Fig. 8.1).

H_2S is lipophilic (Wagner et al., 2009, Li and Moore, 2008) with a diffusion rate of 0.5 ± 0.4 cm s^{-1} in lipid bilayers (Mathai et al., 2009). H_2S bond dissociation energy is 90 kcal mol^{-1} (Fukuto et al., 2012); similar to the thiol S—H bond of 92 ± 1 kcal mol^{-1} (Benson, 1978). Sulfur has more formal oxidation states, ranging from -2 to $+6$. The oxidation state of H_2S is -2, thus it is a reductant, with a standard reduction potential of 0.23 V (Voet and Voet, 2004).

Nitric oxide (NO) is a diatomic gaseous molecule with relatively small molecular mass of 30.006 g mol^{-1} (Kolbert et al., 2019a). As a small uncharged radical, it is highly lipophilic, resulting in high diffusion rates across plasma membranes (Goretski and Hollocher, 1988). The half-life of NO is 2–30 s (Habib and Ali, 2011), making it an ideal signal transmitter. The double bond between atoms results in an unpaired electron on the 2p-π antibonding orbital (Stamler et al., 1992). To stabilize this electron another electron is needed, which can be gained by reactions with other radicals. This explains the high reactivity of NO with other radicals, compared to other molecules (Lancaster, 2015). Further biochemical aspects of NO derivates will be discussed in the signaling pathways section.

8.2.2 Metabolic processes

H_2S is produced by various enzymes (Fig. 8.2) in chloroplasts, mitochondria, and the cytoplasm. One of the main ways to produce it is via the sulfur assimilation pathway. Plants take up sulfate from the soil which is converted

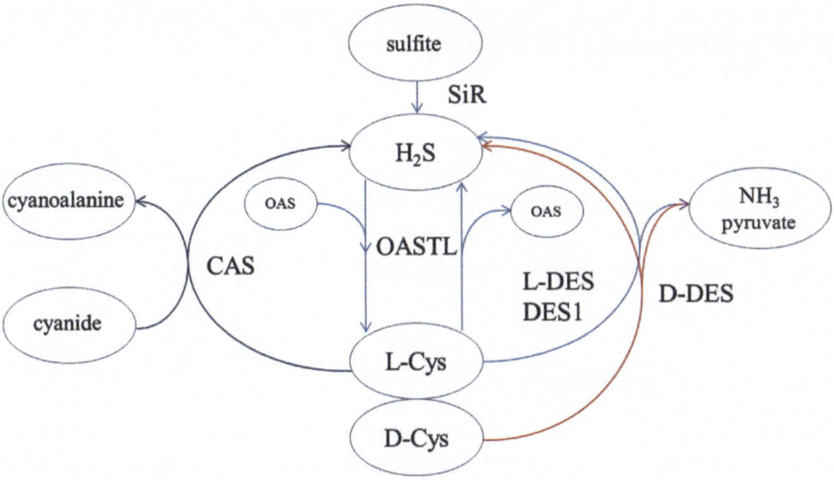

FIGURE 8.2 Metabolic pathways for sulfide production. SiR, sulfite reductase; OAS, O-acetylserine; OASTL, O-acetylserine thiol lyase; CAS, Cyanoalanine synthase; L-DES, L-cysteine desulfhydrase; DES1, Cys synthase-like L-cysteine desulfhydrase; D-DES, D-cysteine desulfhydrase.

into sulfite, which is then transformed into H_2S. The reaction is catalyzed by sulfite reductase (SiR, EC 1.8.7.1) and localized in the chloroplast (Li and Lancaster, 2013). H_2S could be made via three other reactions that all decompose cysteine (Cys). O-acetylserine thiol lyase enzymes (OASTL; EC:2.5.1.47) have a number of isoforms in different plant organelles, as described by Wirtz et al. (2004) and catalyze L-cysteine (L-Cys) synthesis from O-acetylserine (OAS) and H_2S. The same enzyme can catalyze the reverse reaction, as well, resulting in H_2S production. In mitochondria cyanoalanine synthase (CAS, EC 4.4.1.9) can breakdown L-Cys producing H_2S. D-cysteine desulfhydrase (D-DES, EC 4.4.1.15) is also localized in the mitochondria, and catalyzes the breakdown of D-Cys to ammonia, pyruvate, and H_2S. A similar reaction occurs in the cytoplasm using L-Cys, catalyzed by L-cysteine desulfhydrase (L-DES, EC:4.4.1.28, Corpas et al., 2019a) and an OSATL homolog protein Cys synthase-like L-DES (CS-LIKE, DES1; EC 4.4.1.1, Álvarez et al., 2010).

The production of nitric oxide in plants can be divided into two groups: nonenzymatic and enzymatic reactions. Nonenzymatic reactions require special conditions, for example, it has been proposed that carotenoids can reduce NO_2 to NO under light (Cooney et al., 1994). As a by-product, N_2O oxidation during nitrification and denitrification cycles could also produce NO (Wojtaszek, 2000). In an acidic environment, nitrite is spontaneously reduced into NO and nitrate (Stöhr and Ullrich, 2002). Chemical reduction of nitrite is also possible, for example in the aleurone layer of barley, where the acidic pH and presence of ascorbate favor the reaction (Bethke et al., 2004).

Polyamines, hydroxylamine, and L-arginine can be oxidized to produce NO. It is worth mentioning, that L-arginine-dependent nitric oxide synthesis was detected (Corpas et al., 2009a), however L−arginine-dependent nitric oxide synthase was not found in plants (Hancock and Neill, 2019). Nitrate reductase (NR) can directly catalyze the reduction of nitrite to NO (Rockel et al., 2002). NR has also a secondary role in NO production; it can transfer electrons from NAD(P)H to NO-forming nitrite reductase, which can also reduce nitrite to NO (Chamizo-Ampudia et al., 2016). The same reaction could be catalyzed by nitrite:NO reductase which is specific to roots (Stöhr et al., 2001).

8.2.3 Signaling pathways

H_2S and NO have similar reaction partners during signalization, which are divided into three groups: metals, oxidants, and thiols. To evaluate the cross-talk between molecules these reactions and their effects should be addressed individually.

8.2.3.1 Reaction with metals

Sulfur has two types of main reaction with metals: reduction/oxidation and ligation. In biological systems H_2S reacts with hemoproteins, the most studied example is cytochrome c oxidase (CcO) (Cooper and Brown, 2008). H_2S toxicity is most likely based on the inactivation of this enzyme in the mitochondria (Olson, 2012). H_2S interacts with the protein in the oxidized copper/iron heme oxygen binding site, reduces, and binds to it (Nicholls and Kim, 1982). Small molecular sensor hemoproteins interact with H_2S, as well. This interaction has a role in the symbiosis between mollusks and bacteria (Kraus and Wittenberg, 1990; Kraus et al., 1990; Pietri et al., 2009). Hemoglobin and myoglobin can form a covalent bond with H_2S, which used in the diagnostics of H_2S poisoning (Park and Nagel, 1984).

Metal nitrosylation is the reaction of NO and metalloproteins containing transition metals and this reaction results in metal-nitrosyl complexes. The only available data comes from animal hemoproteins, where guanylate cyclase and Cco are the targets (Henry and Guissani, 1999; Zhang et al., 2016). In plants, under hypoxia nonsymbiotic hemoglobin could be nitrosylated, especially in the root system (Dorcas et al., 2003; Gupta et al., 2011; Hebelstrup et al., 2012).

8.2.3.2 Reactions with oxidants

H_2S is prone to reacting with molecular oxygen (autoxidation), which results in various products such as sulfates and polysulfanes. The mechanism itself is not fully understood, however it has been deemed to be biologically irrelevant due to its slow kinetics (Chen, 1970). Metal ions (Tapley et al., 1999;

Zhang and Millero, 1993), aldehydes, and phenols (Chen and Morris, 1972) can speed up the reaction, proposing a role in living organisms. Reactive oxygen species (ROS), for example, superoxide radical (Searcy et al., 1995; Carballal et al., 2011) and hydrogen peroxide (Classen and Bauer, 1883; Hoffmann, 1977) chemically interact with H_2S. The exact reaction with hydrogen peroxide (H_2O_2) is yet to be found, however, in the reaction solution the pH changes between acid and base (Orban and Epstein, 1985; Rabai et al., 1992). There have been studies of RNS and H_2S. Both peroxynitrite (Carballal et al., 2011) and NO (Whiteman and Moore, 2009; Yong et al., 2011) were suggested to react with H_2S directly, however, the mechanism is still not clear. In the study of Yong et al. (2011), the crosstalk between molecules was explained with a new biological mediator. Compared to lipid nitration, H_2S can only interact with lipid hydroperoxides formed during lipid peroxidation (Jeney et al., 2009; Muellner et al., 2009), alleviating the effect of lipid peroxidation.

The reaction of NO with superoxide radical is the most important from a biological perspective. It will produce peroxynitrite, which causes macromolecule nitration, an addition of a $-NO_2$ group to amino acids (Corpas and Palma, 2018a). Protein tyrosine nitration is widely examined, however other amino acids, such as tryptophane, methionine, and cysteine are also targets for nitration (Corpas et al., 2009b; Kolbert et al., 2017). Several studies have determined nitrated proteins in higher plants (Chaki et al., 2009; Cecconi et al., 2009; Lozano-Juste et al., 2011; Begara-Morales et al., 2013; Szuba et al., 2015) and the most prevalent effect on the protein is loss of function, for example, in ascorbate peroxidase (APX) and monodehydroascorbate reductase (MDHAR). There is also a report of a neutral effect on glutathione reductase (GR) (Begara-Morales et al., 2014; Begara-Morales et al., 2015). Lipid nitration can also occur in plant cells, where the $-NO_2$ group is added to fatty acids. It has been proven, that *Arabidopsis thaliana* contains nitro-linoleic acid that is of importance under several abiotic stresses (Mata-Pérez et al., 2016a), while *Brassica napus* seeds contain free nitro-oleic acid (Vollár et al., 2020). Both of them can release NO (Mata-Pérez et al., 2016b).

8.2.3.3 Reaction with thiols

Both H_2S and thiol residues are nucleophiles and reductants, therefore a reaction would be impossible between them. For the reaction to occur, an oxidation step is needed. Protein Cys residues act as sensors to plant redox status (Spadaro et al., 2010; Klomsiri et al., 2011) and change their oxidation status up to +4 (sulfonic acid). The reaction of sulfenic acid (oxidation state 0) is favorable and will occur under biological conditions (Cuevasanta et al., 2015). Protein persulfidation is considered the main signaling pathway of H_2S to exert its effect on living organisms (Filipovic, 2015). In their study, Aroca et al. (2017) found that 5% of *Arabidopsis* proteins were persulfidated.

The effect of persulfidation on plant proteins is still unclear, however in animals a number of persulfidated proteins are identified, for example, glyceraldehyde 3-phosphate dehydrogenase (GAPDH), actin, β-tubulin, leucine aminopeptidase, catalase (CAT), and adenosine triphosphate synthase. Persulfidation activated mouse (Mustafa et al., 2009) and *Arabidopsis* (Aroca et al., 2015) GAPDH. It has also been reported that persulfidation can protect proteins from oxidation (Filipovic, 2015).

S-nitrosylation is a reaction, where NO modifies protein thiol residues with the addition of a −NO group. In higher plants it is reversible and widespread, suggesting that this posttranslational modification is in the background of most of the NO signaling in plant cells. The number of identified S-nitrosylated proteins is increasing, both under physiological and stress conditions (Lindermayr et al., 2005; Fares et al., 2011; Puyaubert et al., 2014; Sehrawat and Deswal, 2014; Gietler et al., 2016). It is important to note, that both S-nitrosoglutathion and protein S-nitrosothiol (SNO) proteins are capable of releasing NO (Begara-Morales et al., 2018).

As described above, both NO and H_2S modify thiol groups in proteins, thus the thiol derivatives can be the direct site of crosstalk between the two molecules. Aroca et al. (2018) analyzed persulfidated and nitrosylated proteins in *Arabidopsis* and found that 2330 proteins were persulfidated, while 929 nitrosylated and 612 proteins were capable of both modifications. The comprehensive analysis of the crosstalk has been reported in the case of two proteins. APX is nitrosylated (Begara-Morales et al., 2014) and persulfidated (Aroca et al., 2015), and both modifications enhance the activity of the protein. In contrast, CAT is reported to be inhibited by both modifications (Corpas et al., 2019b). Both enzymes decompose hydrogen peroxide, fine-tuning the level of ROS. H_2S can also react directly with S-nitrosothiols (Seel et al., 1985; Seel and Wagner, 1988), however the exact products are still unclear.

8.3 NO and H_2S and their role in plants at physiological conditions

Depending on their concentration NO and H_2S might act as signal molecules or in higher amounts are able to damage plant functions, either by acting synergistically or antagonistically. Their effect also depends on the plant species, organ, and circumstances during the experiments (Corpas et al., 2019b). Both molecules are able to cross membranes by diffusion and react with thiol groups of proteins, resulting in S-nitrosation (NO) or persulfidation (H_2S). CAT and APX are two known targets of these posttranslational modifications. APX is positively regulated by both NO (Begara-Morales et al., 2014; Yang et al., 2015) and H_2S (Aroca et al., 2015), while CAT activity is inhibited by NO and H_2S (Corpas et al., 2019a). The exogenous applications of

NO and H_2S derived from donors have different actions on some physiological processes, however these effects might differ from the ones caused by endogenous NO and H_2S.

8.3.1 Seed dormancy/germination

NO is able to break seed dormancy by the modulation of abscisic acid (ABA)-initiated cascades through protein tyrosine nitration, as was exhibited in numerous studies (Albertos et al., 2015; Belda-Palazon et al., 2018; Castillo et al., 2015; Signorelli and Considine, 2018; Prakash et al., 2019; Wang et al., 2015), but no such mechanism is known about H_2S (or persulfidation). H_2S metabolism however might be stimulated in seed imbibition of *Arabidopsis*, but under control conditions it does not seem to have a significant effect on its seed germination (Baudouin et al., 2016).

8.3.2 Root organogenesis

There is a lot of information available about the effect of NO in root growth (reviewed by Corpas and Barroso, 2015). Using different NO donors, they increased root elongation and also lateral root and adventitious root development (Pagnussàt et al., 2002; Correa-Aragunde et al., 2004). In sunflower, a connection was found between endogenous NO content and lignin composition, influencing root branching (Monzón et al., 2014). The effect of H_2S on lateral root formation is not exactly clear, it heavily depends on the investigated plant species or the type of donor and the time of exposition. In tomato 1 mM sodium hydrosulfide (NaHS) as H_2S donor was able to induce lateral root formation (Mei et al., 2017), while in *Arabidopsis*, the same donor at 200 and 800 μM concentration significantly decreased root growth (Zhang et al., 2017).

A possible overlap between the mechanism of action of the two molecules can be the actin cytoskeleton. Li et al. (2018) found that *Arabidopsis* actin2 is a subject to persulfidation, leading to F-actin depolymerization and resulting in inhibited root hair growth, while in corn roots treated with NO donors F-actin reorganization was also observed (Kasprowicz et al., 2009; Plohovska et al., 2019).

8.3.3 Stomatal closure

Movement of the stomata is a highly controlled process, in which (among many other factors) NO and H_2S play signaling roles. According to Scuffi et al. (2014), their interplay has a role in the ABA-induced stomatal closure. They found that H_2S acts upstream of NO, since H_2S promotes NO production, moreover, L-cysteine desulfhydrase (DES1, responsible for enzymatic H_2S production) is essential in ABA-dependent NO production. Besides that,

Honda et al. (2015) found that H_2S also induces the synthesis of the NO-mediated 8-nitro-cGMP and 8-mercapto-cGMP, which can evoke stomatal closure.

8.3.4 Leaf senescence

Until now, no evidence was found about the crosstalk of NO and H_2S in the regulation of leaf senescence, however they seem to act in an analogous manner by suppressing chlorophyll loss and delaying senescence (Mishina et al., 2007; Wei et al., 2017; Al Ubeed et al., 2019).

8.3.5 Fruit ripening

Similar to senescence processes, the exact effects of NO and H_2S on fruit ripening and postharvest fruit life is insufficient and mostly based on empirical observations, but the exogenously applied NO (Corpas and Palma, 2018b) and H_2S (Huo et al., 2018) are proved to have a positive effect on postharvest fruit life. While a little more is known about the behavior of NO and related molecules during fruit ripening (Chaki et al., 2015; Rodríguez-Ruiz et al., 2017; Corpas et al., 2018), endogenous H_2S content was only measured in pepper, where Muñoz-Vargas et al. (2018) found that its level increased during ripening, while NO level decreased.

8.4 Crosstalk between H_2S and NO in plants under stress conditions

8.4.1 Abiotic stress

It is well-known that both NO and H_2S are colorless gases with a small molecular size and have been regarded as toxic for animals and humans for a long time (Corpas et al., 2019b). Over the last 40 years thousands of reports have confirmed the pivotal role of NO in plants' vegetative and reproductive growth and nitrosative stress (reviewed by Kolbert et al., 2019b). Similarly, at the end of the last century (i.e., in the 1990s) it was discovered that H_2S also has a remarkable role in plant physiological processes such as seed germination, morphogenesis of root, photosynthesis, senescence, stomatal closure, and stress tolerance (summarized by Hancock and Whiteman, 2014; Guo et al., 2016; Corpas et al., 2019a,b; Hancock, 2019).

Moreover, in the case of both gasotransmitters the application of donors at lower concentrations (depending on plant species, exposure time, etc.), might have beneficial influences on plant growth and biotic or abiotic stress tolerance (reviewed by Li, 2013; Nabi et al., 2019).

8.4.1.1 Salt and drought stress

Nowadays, one of the most harmful abiotic stress factors is salinity accompanied by osmotic stress (i.e., drought), since these reduce plant growth and crop yield (Szepesi et al., 2009). Recent studies have demonstrated that both NO and H_2S alone or their interaction have beneficial effects in salinity and drought, namely reducing oxidative damage and enhancing plant growth (Table 8.1; review by Nabi et al., 2019). Though we have little data about the NO and H_2S crosstalk under salt stress, it seems that in the case of barley, strawberry, and alfalfa exogenously applied NaHS elevated NO production, while in tomato, inversely, the use of NO donor stimulated H_2S generation (Table 8.1).

Nevertheless, in plants exposed to osmotic stress the NO–H_2S crosstalk appears to be more univocal: NO derived from several sources can increase the synthesis of H_2S through enhancing the activity of L-DES, D-DES, or OASTL enzymes (Table 8.1; summarized by Khan et al., 2017).

8.4.1.2 Heavy metal stress

Many researchers have ascertained that the increased emission of heavy metals (HMs) to the environment originating from different sources like geological or anthropogenic activities (e.g., mining, industrial, excess application of fertilizers and pesticides, or contaminated irrigation water) may have several negative effects on plants. The main symptoms of HM stress and/or toxicity in plants are generally disturbances in photosynthesis, reduced germination rate, and abnormal development of vegetative parts (root and/or shoot) (Emamverdian et al., 2015). Not only can the toxic metals like Cd, Pb, and Hg cause nitrooxidative stress in the plant cells, but the essential HMs (e.g., Fe, Ni, Zn, and Cu) at supraoptimal concentrations can also cause ROS/RNS/RSS overproduction and consequently affect the activity of the antioxidant defense system (review by Bhuyan et al., 2020). There are numerous results demonstrating that HM stress irrespective of being essential or nonessential can alter positively or negatively H_2S production in the plant organ investigated (Chen et al., 2018; Yu et al., 2019).

Researchers also have shown that the application of either H_2S (e.g., NaHS) or NO donors alone [sodium nitroprusside (SNP) or GSNO] at a certain concentration may have significantly positive effects on plants exposed to HM stress (reviewed by Nabi et al., 2019; Shivaraj et al., 2020). At the same time, there is not much data about the interaction between NO and H_2S, but all are in accordance: H_2S seems to be the downstream signaling molecule of NO (Table 8.1).

8.4.1.3 Heat stress and chilling

Both NO and H_2S have been shown to participate in the regulation of temperature stress responses in plants (recently reviewed by Parankusam et al., 2017

TABLE 8.1 Crosstalk between H_2S and NO in plants exposed to abiotic stresses.

Stress factor	Plant name	Concentration or degree of the stress factor	Time of HM exposure	Plant organ investigated	The application form of H_2S	The application form of NO	Crosstalk between H_2S and NO[a]	References
Salinity								
	Hordeum vulgare	100–600 mM NaCl (hydroponic)	24–96 h	Roots	NaHS: 50–500 μM (cotreatment)	SNP: 100 μM (cotreatment)	H_2S increased NO production and salt tolerance	Chen et al. (2015)
	Fragaria x ananassa	100 mM NaCl (hydroponic)	7 days	Leaves	NaHS: 100 μM (pretreatment)		NO content ↓	Christou et al. (2013)
	Medicago sativa	100 mM NaCl (Petri dishes)	2 and 4 days	Seedlings	NaHS: 100 μM (cotreatment)	SNP: 100 μM (cotreatment)	NaHS triggered NO production	Wang et al. (2012)
	Solanum lycopersicon	50–300 mM NaCl (hydroponic)	48 h	Roots and leaves	NaHS: 25–200 μM (cotreatment)	GSNO: 100–500 μM (cotreatment)	NO stimulated H_2S accumulation in roots and leaves, but H_2S did not increase NO content (H_2S is downstream of NO in signaling pathway)	da-Silva et al. (2018)

				Drought				
	Fragaria x ananassa	10% (w/v) PEG-6000	7 days	Leaves	NaHS: 100 μM (pretreatment)		NO content ↓	Christou et al. (2013)
	Capsicum annuum	15% (w v^{-1}) PEG-8000 (pot experiment)	12 h	Seedlings		GSNO: 0.2 mM (cotreatment)	NO enhanced synthesis of H_2S resulting in osmotic stress tolerance	Khan et al. (2017)
	Trifolium repens	15% (w v^{-1}) PEG-6000 (pot experiment)	24 h	Leaves	NaHS: 20 μM (cotreatment)	SNP: 25 μM (cotreatment)	H_2S might be the downstream of NO signaling	Li et al. (2019)
				Heavy metal				
Co	*Triticum aestivum*	150–300 μM (hydroponic)	72 h	Leaves	NaHS: 600 μM (cotreatment)	SNP: 100 μM (cotreatment)	H_2S synthesis ↑ due to NO donor (SNP)	Ozfidan-Konakci et al. (2020)
Cd	*Arabidopsis thaliana ecotype Col-0*	40 μM (hydroponic)	8 days	Leaves	NaHS: 300 μM (cotreatment, foliar spraying)		Foliarly applied NaHS repressed Cd-induced NO accumulation	Guan et al. (2018)
	Cynodon dactylon	100–5000 μM (hydroponic)	12–48 h	Leaves	NaHS: 500 μM (cotreatment)	SNP: 250 μM (cotreatment)	NO-activated H_2S resulted in Cd tolerance	Shi et al. (2014)
	Medicago sativa	100 μM (hydroponic)	12 h	Roots	NaHS: 100 μM (cotreatment)	SNP: 100 μM (cotreatment)	NO is the upstream component of H_2S signaling	Li et al. (2012)
	Triticum aestivum	100 μM (pot experiment)	4 weeks	Leaves	NaHS: 100 μM (cotreatment)	SNP: 200 μM (cotreatment)	NO-triggered H_2S improved Cd tolerance	Kaya et al. (2020)

[a]↑ *Indicates significant increase, while* ↓ *refers to significant decrease.*

and Singh et al., 2020a, respectively) suggesting that their signal pathways may be linked in plants subjected to extreme temperatures. So far, only a few studies have examined NO—H_2S crosstalk in heat or cold stress responses of plants.

In heat-stressed (48°C, 18 hours) maize seedlings, NO donor sodium nitroprusside (SNP) pretreatment at concentrations of 100, 150, and 200 μM improved survival by lowering the heat-induced accumulation of malondialdehyde and electrolyte leakage. Additionally, 2-hour-long SNP pretreatment enhanced the activity of L-cysteine desulfhydrase (L-DES), which may be responsible for the consequent accumulation of endogenous H_2S after 4 hours of treatment. The positive effect of H_2S on improving heat tolerance was evidenced by the application of H_2S donor chemicals (NaHS and GYY4137) which effectively enhanced endogenous H_2S levels in the seedlings. The coapplication of H_2S donors, inhibitors, scavengers, and SNP revealed that H_2S may be a downstream signal molecule in NO-induced heat tolerance of maize seedlings since the inhibition or scavenging of H_2S notably reduced SNP-induced heat tolerance (Li et al., 2013).

Using similar pharmacological approaches, the same research group revealed that exogenous application of H_2O_2 intensifies thermotolerance of maize seedlings by increasing endogenous H_2O_2 levels within 1 hour post-treatment (Li et al., 2015). Additionally, NO levels were increased by 2-hour-long H_2O_2 treatment and H_2S levels were enhanced in the 4th hour following H_2O_2 supply. Among H_2S metabolic enzymes, L-DES activity was increased by H_2O_2, while the activity of SR and CS enzymes remained unchanged, indicating the primary role of L-DES activity in H_2O_2-triggered H_2S synthesis in maize seedlings. The H_2O_2-induced H_2S production could be reduced by cPTIO (2-4-carboxyphenyl-4,4,5,5-tetramethylimidazoline-1-oxyl-3-oxide) application, suggesting that NO is an upstream element of H_2S in signal transduction. In further experiments, the changes in thermotolerance of maize due to pretreatments with H_2S donors (NaHS and GYY4137) and inhibitors (propargylglycine and hypotaurine) were studied. It was found that H_2S donors further increased H_2O_2-induced tolerance and produced a higher survival percentage. The results of Li et al. (2015) collectively show the involvement of both NO and H_2S in H_2O_2-induced heat tolerance of maize and NO seems to be upstream of H_2S in the signal cascade.

In a further study, heat treatment (35°C, 7 days) of rooted poplar cuttings led to photosynthetic and membrane damage which was accompanied by the fast (6 hours of heat) induction of H_2S biosynthesis—related enzyme activities (O-acetyl-(thiol)-serinelyase (OAS-TL), β-cyanoalanine synthase, D-DES, and L-DES) and the increase of H_2S levels. Moreover, 12-hour-long heat treatment caused an increase in NO and SNO levels as well as in S-nitrosoglutathione reductase (GSNOR) expression and activity. Interestingly, H_2S donor treatments increased while H_2S suppressor decreased GSNOR activity and expression suggesting that H_2S acts upstream of GSNOR which in turn limits RNS and ROS signaling and thus enhances tolerance to heat stress in poplar (Cheng et al., 2018).

Storage at low temperatures is a common method for prolonging the post-harvest life of fruits like peach. Therefore it is important to reveal the effect of cold on fruit quality as well as the molecular responses within the fruits due to low storage temperatures. Pharmacological treatment with NO improved peach quality while treatment with cPTIO reduced peach quality, as determined by parameters like respiration rate, weight loss, browning, and ethylene production of the fruits (Geng et al., 2019). Exogenous NO application reduced H_2S levels, decreased activities of L-/D-DES, O-acetylserine (thiol)lyase (OAS-TL), and SiR, and increased activity of β-cyanoalanine synthase in peaches. These results indicate that H_2S may play a negative regulatory role in NO-induced quality improvement of peach fruits during cold storage; however, these preliminary data should be supported and supplemented by further research.

Collectively, the limited amount of research data available so far suggests that both NO and H_2S contribute to heat tolerance of plants and they influence each other's metabolism in heat-exposed plant tissues.

8.4.2 Biotic stress

Nitric oxide (NO) and hydrogen sulfide (H_2S) are both important gaseous molecules in plant biotic stress responses. Both reactive molecules take part in stomatal movements, as has been discussed above. Plants have several strategies to protect themselves against pathogens, one of the most useful is stomatal closure, because it is a physical barrier against the attack of microorganisms (Qu et al., 2019). García-Mata and Lamattina (2010) showed that exogenous application of H_2S donors such as NaHS (sodium hydrosulfide) and GYY4137 provoked stomatal closure in *Vicia Fabia*, *A. thaliana*, and *Impatiens walleriana* resulting in a physical barrier against potential pathogens. NO and H_2S can modulate the expression of multiple biotic stress-related genes; furthermore H_2S regulates carbon metabolites, for example, amino acides, sugars, and aromatic amines (Shi et al., 2015). An NO derivative, peroxynitrite can react with H_2S and activate new signaling pathways. Besides, peroxinitrite plays an essential role in plant immune defense responses against pathogens (Singh et al., 2020b).

8.5 Conclusions and future perspectives

We may conclude that both NO and H_2S are multifunctional gaseous molecules with vital roles in the physiological processes of plants either at unstressed or stressed conditions. Exogenous application of these gases (alone or together) at lower concentrations can provide protection for plants against biotic and abiotic stresses, but it is still not obvious which is the upstream or downstream player of the signaling pathway. Another question is if this signaling cascade is stress-specific or organ-specific. It would be also useful to clarify the crosstalk among H_2S, NO, and polyamines or other secondary metabolites (Szepesi, 2020) or a possible cue of nitrosothiol, the product of the interaction between NO and H_2S (He et al., 2019).

Acknowledgment

Á. Molnár was supported by UNKP-19-3-SZTE-201 New National Excellence Program of the Ministry for Innovation and Technology. G. Feigl was funded by the National Research, Development and Innovation Office (Grant no. NKFIH 131589). D. Oláh was supported by the UNKP-19-3-SZTE-216 New National Excellence Program of the Ministry for Innovation and Technology.

References

Abe, K., Kimura, H., 1996. The possible role of hydrogen sulfide as an endogenous neuromodulator. J. Neurosci. 16, 1066–1071.

Al Ubeed, H.M.S., Wills, R.B.H., Bowyer, M.C., Golding, J.B., 2019. Interaction of the hydrogen sulphide inhibitor, propargylglycine (PAG), with hydrogen sulphide on postharvest changes of the green leafy vegetable, pak choy. Postharvest. Biol. Technol. 147, 54–58.

Albertos, P., Romero-Puertas, M.C., Tatematsu, K., Mateos, I., Sánchez-Vicente, I., Nambara, E., et al., 2015. S-nitrosylation triggers ABI5 degradation to promote seed germination and seedling growth. Nat. Commun. 6, 1–10.

Álvarez, C., Calo, L., Romero, L.C., García, I., Gotor, C., 2010. An O acetylserine(thiol)lyase homolog with L-cysteine desulfhydrase activity regulates cysteine homeostasis in *Arabidopsis*. Plant Phys. 152, 656–669.

Aroca, Á., Serna, A., Gotor, C., Romero, L.C., 2015. S-sulfhydration: a cysteine posttranslational modification in plant systems. Plant Phys. 168, 334–342.

Aroca, A., Benito, J.M., Gotor, C., Romero, L.C., 2017. Persulfidation proteome reveals the regulation of protein function by hydrogen sulfide in diverse biochemical processes in *Arabidopsis*. J. Exp. Bot. 68, 4915.

Aroca, A., Gotor, C., Romero, L.C., 2018. Hydrogen sulfide signaling in plants: emerging roles of protein persulfidation. Front. Plant Sci. 9, 1369.

Baudouin, E., Poilevey, A., Hewage, N.I., Cochet, F., Puyaubert, J., Bailly, C., 2016. The significance of hydrogen sulfide for *Arabidopsis* seed germination. Front. Plant Sci. 7, 930.

Beauchamp, R.O., Bus, J.S., Popp, J.A., Boreiko, C.J., Andjelkovich, D.A., Leber, P., 1984. A critical review of the literature on hydrogen sulfide toxicity. CRC Crit. Rev. Toxicol. 13, 25–97.

Begara-Morales, J.C., Chaki, M., Sánchez-Calvo, B., Mata-Pérez, C., Leterrier, M., Palma, J.M., et al., 2013. Protein tyrosine nitration in pea roots during development and senescence. J. Exp. Bot. 64, 1121–1134.

Begara-Morales, J.C., Sánchez-Calvo, B., Chaki, M., Valderrama, R., Mata-Pérez, C., López-Jaramillo, J., et al., 2014. Dual regulation of cytosolic ascorbate peroxidase (APX) by tyrosine nitration and S-nitrosylation. J. Exp. Bot. 65, 527–538.

Begara-Morales, J.C., Sánchez-Calvo, B., Chaki, M., Mata-Pérez, C., Valderrama, R., Padilla, M.N., et al., 2015. Differential molecular response of monodehydroascorbate reductase and glutathione reductase by nitration and S-nitrosylation. J. Exp. Bot. 66, 5983–5996.

Begara-Morales, J.C., Chaki, M., Valderrama, R., Sánchez-Calvo, B., Mata-Pérez, C., Padilla, M.N., et al., 2018. NO buffering and conditional NO release in stress response. J. Exp. Bot. 69, 3425–3438.

Belda-Palazon, B., Gonzalez-Garcia, M.P., Lozano-Juste, J., Coego, A., Antoni, R., Julian, J., et al., 2018. PYL8 mediates ABA perception in the root through non-cell-autonomous and ligand-stabilization–based mechanisms. Proc. Natl. Acad. Sci. U. S. A. 115, E11857–E11863.

Benson, S.W., 1978. Thermochemistry and kinetics of sulfur-containing molecules and radicals. Chem. Rev. 78, 23–35.

Bethke, P.C., Badger, M.R., Jones, R.L., 2004. Apoplastic synthesis of nitric oxide by plant tissues. Plant Cell 16, 332–341.

Bhuyan, M.B., Hasanuzzaman, M., Parvin, K., Mohsin, S.M., Al Mahmud, J., Nahar, K., et al., 2020. Nitric oxide and hydrogen sulfide: two intimate collaborators regulating plant defense against abiotic stress. Plant Growth Reg. 90, 409–424.

Calhoun, D.B., Englander, S.W., Wright, W.W., Vanderkooi, J.M., 1988. Quenching of room temperature protein phosphorescence by added small molecules. Biochemistry 27, 8466–8474.

Carballal, S., Trujillo, M., Cuevasanta, E., Bartesaghi, S., Moller, M.N., Folkes, L.K., et al., 2011. Reactivity of hydrogen sulfide with peroxynitrite and other oxidants of biological interest. Free Radic. Biol. Med. 50, 196–205.

Castillo, M.C., Lozano-Juste, J., González-Guzmán, M., Rodriguez, L., Rodriguez, P.L., León, J., 2015. Inactivation of PYR/PYL/RCAR ABA receptors by tyrosine nitration may enable rapid inhibition of ABA signaling by nitric oxide in plants. Sci. Signal. 8, ra89-ra89.

Cecconi, D., Orzetti, S., Vandelle, E., Rinalducci, S., Zolla, L., Delledonne, M., 2009. Protein nitration during defense response in *Arabidopsis thaliana*. Electrophoresis 30, 2460–2468.

Chaki, M., Valderrama, R., Fernández-Ocaña, A.M., Carreras, A., López-Jaramillo, J., Luque, F., et al., 2009. Protein targets of tyrosine nitration in sunflower (*Helianthus annuus* L.) hypocotyls. J. Exp. Bot. 60, 4221–4234.

Chaki, M., Álvarez de Morales, P., Ruiz, C., Begara-Morales, J.C., Barroso, J.B., Corpas, F.J., et al., 2015. Ripening of pepper (*Capsicum annuum*) fruit is characterized by an enhancement of protein tyrosine nitration. Ann. Bot. 116, 637–647.

Chamizo-Ampudia, A., Sanz-Luque, E., Llamas, Á., Ocaña-Calahorro, F., Mariscal, V., Carreras, A., et al., 2016. A dual system formed by the ARC and NR molybdoenzymes mediates nitrite-dependent NO production in Chlamydomonas. Plant Cell Environ. 39, 2097–2107.

Chen, K.Y., 1970. Oxidation of Aqueous Sulfide by O_2 (PhD thesis/dissertation). Harvard University, Division of Engineering and Applied Physics. Cambridge, Massachusetts, USA.

Chen, K.Y., Morris, J.C., 1972. Oxidation of sulfide by O2: catalysis and inhibition. J. Sanit. Eng. Div. 98, 215–227.

Chen, J., Wang, W.H., Wu, F.H., He, E.M., Liu, X., Shangguan, Z.P., et al., 2015. Hydrogen sulfide enhances salt tolerance through nitric oxide-mediated maintenance of ion homeostasis in barley seedling roots. Sci. Rep. 5, 12516.

Chen, Z., Yang, B., Hao, Z., Zhu, J., Zhang, Y., Xu, T., 2018. Exogenous hydrogen sulfide ameliorates seed germination and seedling growth of cauliflower under lead stress and its antioxidant role. J. Plant Growth Reg. 37, 5–15.

Cheng, T., Shi, J., Dong, Y., Ma, Y., Peng, Y., Hu, X., et al., 2018. Hydrogen sulfide enhances poplar tolerance to high-temperature stress by increasing S-nitrosoglutathione reductase (GSNOR) activity and reducing reactive oxygen/nitrogen damage. Plant Growth Regul. 84, 11–23.

Christou, A., Manganaris, G.A., Papadopoulos, I., Fotopoulos, V., 2013. Hydrogen sulfide induces systemic tolerance to salinity and non-ionic osmotic stress in strawberry plants through modification of reactive species biosynthesis and transcriptional regulation of multiple defence pathways. J. Exp. Bot. 64, 1953–1966.

Classen, A., Bauer, O., 1883. Use of hydrogen peroxide in analytical chemistry. J. Chem. Soc. Abstr. 44, 934.

Cooney, R.V., Harwood, P.J., Custer, L.J., Franke, A.A., 1994. Light-mediated conversion of nitrogen dioxide to nitric oxide by carotenoids. Environ. Health Persp 102, 460–462.

Cooper, C.E., Brown, G.C., 2008. The inhibition of mitochondrial cytochrome oxidase by the gases carbon monoxide, nitric oxide, hydrogen cyanide and hydrogen sulfide: chemical mechanism and physiological significance. J. Bioenerg. Biomembr. 40, 533–539.

Corpas, F.J., 2019. Nitric oxide and hydrogen sulfide in higher plants under physiological and stress conditions. Antioxidants 8, 457. Available from: https://doi.org/10.3390/antiox8100457.

Corpas, F.J., Barroso, J.B., 2015. Functions of nitric oxide (NO) in roots during development and under adverse stress conditions. Plants 4, 240–252.

Corpas, F.J., Palma, J.M., 2018a. Assessing nitric oxide (NO) in higher plants: an outline. Nitrogen 1, 12–20.

Corpas, F.J., Palma, J.M., 2018b. Nitric oxide on/off in fruit ripening. Plant Biol. 20, 805–807.

Corpas, F.J., Palma, J.M., DelRío, L.A., Barroso, J.B., 2009a. Evidence supporting the existence of L-arginine-dependent nitric oxide synthase activity in plants. N. Phytol. 184, 9–14.

Corpas, F.J., Chaki, M., Leterrier, M., Barroso, J.B., 2009b. Protein tyrosine nitration: a new challenge in plants. Plant Signal. Behav. 4, 920–923.

Corpas, F.J., Freschi, L., Rodríguez-Ruiz, M., Mioto, P.T., González-Gordo, S., Palma, J.M., 2018. Nitro-oxidative metabolism during fruit ripening. J. Exp. Bot. 69, 3449–3463.

Corpas, F.J., Barroso, J.B., González-Gordo, S., Muñoz-Vargas, M.A., Palma, J.M., 2019a. Hydrogen sulfide: a novel component in *Arabidopsis* peroxisomes which triggers catalase inhibition. J. Integr. Plant Biol. 61, 871–883.

Corpas, F.J., González-Gordo, S., Cañas, A., Palma, J.M., 2019b. Nitric oxide and hydrogen sulfide in plants: which comes first? J. Exp. Bot. 70, 4391–4404. Available from: https://doi.org/10.1093/jxb/erz031.

Correa-Aragunde, N., Graziano, M., Lamattina, L., 2004. Nitric oxide plays a central role in determining lateral root development in tomato. Planta 218, 900–905.

Cuevasanta, E., Lange, M., Bonanata, J., Coitiño, E.L., Ferrer-Sueta, G., Filipovic, M.R., et al., 2015. Reaction of hydrogen sulfide with disulfide and sulfenic acid to form the strongly nucleophilic persulfide. J. Biol. Chem. 290, 26866–26880.

Dordas, C., Rivoal, J., Hill, R.D., 2003. Plant haemoglobins, nitric oxide and hypoxic stress. Ann. Bot. 91, 173–178.

Emamverdian, A., Ding, Y., Mokhberdoran, F., Xie, Y., 2015. Heavy metal stress and some mechanisms of plant defense response. Sci. World J. 756120. Available from: https://doi.org/10.1155/2015/756120.

Erdei, L., Kolbert, Z., 2008. Nitric oxide as a potent signalling molecule in plants. Acta Biol. Szeged. 52, 1–5.

Fares, A., Rossignol, M., Peltier, J.B., 2011. Proteomics investigation of endogenous S-nitrosylation in *Arabidopsis*. Biochem. Biophys. Res. Commun. 416, 331–336.

Filipovic, M.R., 2015. Persulfidation (S-sulfhydration) and H2S. Handb. Exp. Pharmacol. 230, 29–59.

Fukuto, J.M., Carrington, S.J., Tantillo, D.J., Harrison, J.G., Ignarro, L.J., Freeman, B.A., et al., 2012. Small molecule signaling agents: the integrated chemistry and biochemistry of nitrogen oxides, oxides of carbon, dioxygen, hydrogen sulfide, and their derived species. Chem. Res. Toxicol. 25, 769–793.

García-Mata, C., Lamattina, L., 2010. Hydrogen sulphide, a novel gasotransmitter involved in guard cell signalling. N. Phytol. 188, 977–984.

Geng, B., Huang, D., Zhu, S., 2019. Regulation of hydrogen sulfide metabolism by nitric oxide inhibitors and the quality of peaches during cold storage. Antioxidants 8, 401.

Gietler, M., Nykiel, M., Orzechowski, S., Fettke, J., Zagdańska, B., 2016. Proteomic analysis of S-nitrosylated and S-glutathionylated proteins in wheat seedlings with different dehydration tolerances. Plant Physiol. Biochem. 108, 507–518.

Goretski, J., Hollocher, T.C., 1988. Trapping of nitric oxide produced during denitrification by extracellular hemoglobin. J. Biol. Chem. 263, 2316–2323.

Guan, M.Y., Zhang, H.H., Pan, W., Jin, C.W., Lin, X.Y., 2018. Sulfide alleviates cadmium toxicity in *Arabidopsis* plants by altering the chemical form and the subcellular distribution of cadmium. Sci. Total. Environ. 627, 663–670.

Guo, H., Xiao, T., Zhou, H., Xie, Y., Shen, W., 2016. Hydrogen sulfide: a versatile regulator of environmental stress in plants. Acta Phys. Plant 38, 16. Available from: https://doi.org/10.1007/s11738-015-2038-x.

Gupta, K.J., Hebelstrup, K.H., Mur, L.A., Igamberdiev, A.U., 2011. Plant hemoglobins: important players at the crossroads between oxygen and nitric oxide. FEBS Lett. 585, 3843–3849.

Hancock, J.T., 2019. Hydrogen sulfide and environmental stresses. Environ. Exp. Bot. 161, 50–56.

Hancock, J.T., Whiteman, M., 2014. Hydrogen sulfide and cell signaling: team player or referee? Plant Phys. Biochem. 78, 37–42.

Hancock, J.T., Neill, S.J., 2019. Nitric oxide: its generation and interactions with other reactive signaling compounds. Plants 8, 41.

Habib, S., Ali, A., 2011. Biochemistry of nitric oxide. Ind. J. Clin. Biochem. 26, 3–17.

Hebelstrup, K.H., van Zanten, M., Mandon, J., Voesenek, L.A., Harren, F.J., Cristescu, S.M., et al., 2012. Haemoglobin modulates NO emission and hyponasty under hypoxia-related stress in *Arabidopsis thaliana*. J. Exp. Bot. 63, 5581–5591.

Henry, Y., Guissani, A., 1999. Interactions of nitric oxide with hemoproteins: roles of nitric oxide in mitochondria. Cell. Mol. Life Sci. 55, 1003–1014.

He, H., Li, Y., He, L.F., 2019. Role of nitric oxide and hydrogen sulfide in plant aluminum tolerance. Biometals 32, 1–9.

Hoffmann, M.R., 1977. Kinetics and mechanism of oxidation of hydrogen sulfide by hydrogen peroxide in acidic solution. Environ. Sci. Technol. 11, 61–66.

Honda, K., Yamada, N., Yoshida, R., Ihara, H., Sawa, T., Akaike, T., et al., 2015. 8-Mercapto-cyclic GMP mediates hydrogen sulfide-induced stomatal closure in *Arabidopsis*. Plant Cell Phys. 56, 1481–1489.

Huo, J., Huang, D., Zhang, J., Fang, H., Wang, B., Wang, C., et al., 2018. Hydrogen sulfide: a gaseous molecule in postharvest freshness. Front. Plant Sci. 9, 1172.

International Union of Pure and Applied Chemistry (IUPAC), 2012. Compendium of Chemical Terminology Gold Book, Version 2.3.2. Blackwell Scientific, Oxford.

Jeney, V., Komódi, E., Nagy, E., Zarjou, A., Vercellotti, G.M., Eaton, J.W., et al., 2009. Supression of hemin-mediated oxidation of low-density lipoprotein and subsequent endothelial reactions by hydrogen sulfide (H2S). Free Radical Biol. Med. 46, 616–623.

Kabil, O., Banerjee, R., 2010. Redox biochemistry of hydrogen sulfide. J. Biol. Chem. 285, 21903–21907.

Kasprowicz, A., Szuba, A., Volkmann, D., Baluška, F., Wojtaszek, P., 2009. Nitric oxide modulates dynamic actin cytoskeleton and vesicle trafficking in a cell type-specific manner in root apices. J. Exp. Bot. 60, 1605–1617.

Kaya, C., Ashraf, M., Alyemeni, M.N., Ahmad, P., 2020. Responses of nitric oxide and hydrogen sulfide in regulating oxidative defence system in wheat plants grown under cadmium stress. Phys. Plant 168, 345–360.

Khan, M.N., Mobin, M., Abbas, Z.K., Siddiqui, M.H., 2017. Nitric oxide-induced synthesis of hydrogen sulfide alleviates osmotic stress in wheat seedlings through sustaining antioxidant enzymes, osmolyte accumulation and cysteine homeostasis. Nitric Oxide 68, 91–102.

Kimura, H., 2011. Hydrogen sulfide: its production, release and functions. Amino Acids 41, 113–121.

Klomsiri, C., Karplus, P.A., Poole, L.B., 2011. Cysteine-based redox switches in enzymes. Antioxid. Redox Sign 14, 1065–1077.

Kolbert, Z., Feigl, G., Bordé, Á., Molnár, Á., Erdei, L., 2017. Protein tyrosine nitration in plants: present knowledge, computational prediction and future perspectives. Plant Physiol. Biochem. 113, 56–63.

Kolbert, Z., Barroso, J.B., Brouquisse, R., Corpas, F.J., Gupta, K.J., Lindermayr, C., et al., 2019a. A forty year journey: the generation and roles of NO in plants. Nitric Oxide 93, 53–70.

Kolbert, Z., Feigl, G., Freschi, L., Poór, P., 2019b. Gasotransmitters in action: nitric oxide-ethylene crosstalk during plant growth and abiotic stress responses. Antioxid. (Basel) 8, 167.

Kraus, D.W., Wittenberg, J.B., 1990. Hemoglobins of the *Lucina pectinata*/bacteria symbiosis. I. Molecular properties, kinetics and equilibria of reactions with ligands. J. Biol. Chem. 265, 16043–16053.

Kraus, D.W., Wittenberg, J.B., Lu, J.F., Peisach, J., 1990. Hemoglobins of the *Lucina pectinata*/ bacteria symbiosis. II. An electron paramagnetic resonance and optical spectral study of the ferric proteins. J. Biol. Chem. 265, 16054–16059.

Lancaster, J.R., 2015. Nitric oxide: a brief overview of chemical and physical properties relevant to therapeutic applications. Future Sci. OA 1.

Li, Z.G., 2013. Hydrogen sulfide: a multifunctional gaseous molecule in plants. Russ. J. Plant Physiol. 60, 733–740.

Li, L., Moore, P.K., 2008. Putative biological roles of hydrogen sulfide in health and disease: a breath of not so fresh air? Trends Pharmacol. Sci. 29, 84–90.

Li, L., Wang, Y., Shen, W., 2012. Roles of hydrogen sulfide and nitric oxide in the alleviation of cadmium-induced oxidative damage in alfalfa seedling roots. Biometals 25, 617–631.

Li, Q., Lancaster Jr, J.R., 2013. Chemical foundations of hydrogen sulfide biology. Nitric Oxide 35, 21–34.

Li, Z.G., Yang, S.Z., Long, W.B., Yang, G.X., Shen, Z.Z., 2013. Hydrogen sulphide may be a novel downstream signal molecule in nitric oxide-induced heat tolerance of maize (*Zea mays* L.) seedlings. Plant Cell Environ. 36, 1564–1572.

Li, Z.G., Luo, L.J., Sun, Y.F., 2015. Signal crosstalk between nitric oxide and hydrogen sulfide may be involved in hydrogen peroxide-induced thermotolerance in maize seedlings. Russ. J. Plant Physiol. 62, 507–514.

Li, J., Chen, S., Wang, X., Shi, C., Liu, H., Yang, J., et al., 2018. Hydrogen sulfide disturbs actin polymerization via S-sulfhydration resulting in stunted root hair growth. Plant Phys. 178, 936–949.

Li, Z., Zhu, Y., He, X., Yong, B., Peng, Y., Zhang, X., et al., 2019. The hydrogen sulfide, a downstream signaling molecule of hydrogen peroxide and nitric oxide, involves spermidine-regulated transcription factors and antioxidant defense in white clover in response to dehydration. Environ. Exp. Bot. 161, 255–264.

Lindermayr, C., Saalbach, G., Durner, J., 2005. Proteomic identification of S-nitrosylated proteins in *Arabidopsis*. Plant Physiol. 137, 921–930.

Lozano-Juste, J., Colom-Moreno, R., León, J., 2011. In vivo protein tyrosine nitration in *Arabidopsis thaliana*. J. Exp. Bot. 62, 3501–3517.

Mata-Pérez, C., Sánchez-Calvo, B., Padilla, M.N., Begara-Morales, J.C., Luque, F., Melguizo, M., et al., 2016a. Nitro-fatty acids in plant signaling: nitro-linolenic acid induces the molecular chaperone network in *Arabidopsis*. Plant Physiol. 170, 686–701.

Mata-Pérez, C., Sánchez-Calvo, B., Begara-Morales, J.C., Carreras, A., Padilla, M.N., Melguizo, M., et al., 2016b. Nitro-linolenic acid is a nitric oxide donor. Nitric Oxide 5, 57–63.

Mathai, J.C., Missner, A., Kügler, P., Saparov, S.M., Zeidel, M.L., Lee, J.K., et al., 2009. No facilitator required for membrane transport of hydrogen sulfide. Proc. Natl. Acad. Sci. 106, 16633–16638.

Mei, Y., Chen, H., Shen, W., Shen, W., Huang, L., 2017. Hydrogen peroxide is involved in hydrogen sulfide-induced lateral root formation in tomato seedlings. BMC Plant Biol. 17, 162.

Mishina, T.E., Lamb, C., Zeier, J., 2007. Expression of a nitric oxide degrading enzyme induces a senescence programme in *Arabidopsis*. Plant Cell Environ 30, 39–52.

Monzón, G.C., Pinedo, M., Di Rienzo, J., Novo-Uzal, E., Pomar, F., Lamattina, L., et al., 2014. Nitric oxide is required for determining root architecture and lignin composition in sunflower. Supporting evidence from microarray analyses. Nitric Oxide 39, 20–28.

Muellner, M.K., Schreier, S.M., Laggner, H., Hermann, M., Esterbauer, H., Exner, M., et al., 2009. Hydrogen sulfide destroys lipid hydroperoxides in oxidized LDL. Biochem. J. 420, 277–281.

Muñoz-Vargas, M.A., González-Gordo, S., Cañas, A., López-Jaramillo, J., Palma, J.M., Corpas, F.J., 2018. Endogenous hydrogen sulfide (H2S) is up-regulated during sweet pepper (*Capsicum annuum* L.) fruit ripening. In vitro analysis shows that NADP-dependent isocitrate dehydrogenase (ICDH) activity is inhibited by H2S and NO. Nitric Oxide 81, 36–45.

Mustafa, A.K., Gadalla, M.M., Sen, N., Kim, S., Mu, W., Gazi, S.K., et al., 2009. H2S signals through protein S-sulfhydration. Sci. Signal. 2, ra72.

Nabi, R.B.S., Tayade, R., Hussain, A., Kulkarni, K.P., Imran, Q.M., Mun, B.G., et al., 2019. Nitric oxide regulates plant responses to drought, salinity, and heavy metal stress. Environ. Exp. Bot. 161, 120–133.

Nicholls, P., Kim, J.K., 1982. Sulphide as an inhibitor and electron donor for the cytochrome c oxidase system. Can. J. Biochem. 60, 613–623.

Olson, K.R., 2012. A practical look at the chemistry and biology of hydrogen sulfide. Antioxid. Redox Signal. 17, 32–44.

Orban, M., Epstein, I.R., 1985. A new halogen-free chemical oscillator: the reaction between sulfide ion and hydrogen peroxide in a CSTR. J. Am. Chem. Soc. 107, 2302–2305.

Ozfidan-Konakci, C., Yildiztugay, E., Elbasan, F., Kucukoduk, M., Turkan, I., 2020. Hydrogen sulfide (H2S) and nitric oxide (NO) alleviate cobalt toxicity in wheat (*Triticum aestivum* L.) by modulating photosynthesis, chloroplastic redox and antioxidant capacity. J. Haz. Mat. 122061.

Pagnussat, G.C., Simontacchi, M., Puntarulo, S., Lamattina, L., 2002. Nitric oxide is required for root organogenesis. Plant Phys. 129, 954–956.

Palmer, R.M., Ferrige, A.G., Moncada, S., 1987. Nitric oxide release accounts for the biological activity of endothelium-derived relaxing factor. Nature 327, 524–526.

Parankusam, S., Adimulam, S.S., Bhatnagar-Mathur, P., Sharma, K.K., 2017. Nitric oxide (NO) in plant heat stress tolerance: current knowledge and perspectives. Front. Plant Sci. 8, 1582.

Park, C.M., Nagel, R.L., 1984. Sulfhemoglobinemia. Clinical and molecular aspects. N. Engl. J. Med. 310, 1579–1584.

Pietri, R., Lewis, A., Leon, R.G., Casabona, G., Kiger, L., Yeh, S.R., et al., 2009. Factors controlling the reactivity of hydrogen sulfide with hemeproteins. Biochemistry 48, 4881–4894.

Plohovska, S.H., Krasylenko, Y.A., Yemets, A.I., 2019. Nitric oxide modulates actin filament organization in *Arabidopsis thaliana* primary root cells at low temperatures. Cell Biol. Int. 43, 1020–1030.

Prakash, V., Singh, V.P., Tripathi, D.K., Sharma, S., Corpas, F.J., 2019. Crosstalk between nitric oxide (NO) and abscisic acid (ABA) signalling molecules in higher plants. Environ. Exp. Bot. 161, 41–49.

Puyaubert, J., Fares, A., Rézé, N., Peltier, J.B., Baudouin, E., 2014. Identification of endogenously S-nitrosylated proteins in *Arabidopsis* plantlets: effect of cold stress on cysteine nitrosylation level. Plant Sci. 215–216, 150–156.

Qu, X., Cao, B., Kang, J., Wang, X., Han, X., Jiang, W., et al., 2019. Fine-tuning stomatal movement through small signaling peptides. Front. Plant Sci. 10.

Rabai, G., Orban, M., Epstein, I.R., 1992. A model for the pH-regulated oscillatory reaction between hydrogen peroxide and sulfide ion. J. Phys. Chem. 96, 5414–5419.

Rockel, P., Strube, F., Rockel, A., Wildt, J., Kaiser, W.M., 2002. Regulation of nitric oxide (NO) production by plant nitrate reductase in vivo and in vitro. J. Exp. Bot. 53, 103–110.

Rodríguez-Ruiz, M., Mioto, P., Palma, J.M., Corpas, F.J., 2017. S-nitrosoglutathione reductase (GSNOR) activity is down-regulated during pepper (*Capsicum annuum* L.) fruit ripening. Nitric Oxide 68, 51–55.

Scuffi, D., Álvarez, C., Laspina, N., Gotor, C., Lamattina, L., García-Mata, C., 2014. Hydrogen sulfide generated by L-cysteine desulfhydrase acts upstream of nitric oxide to modulate abscisic acid-dependent stomatal closure. Plant Phys. 166, 2065–2076.

Searcy, D.G., Whitehead, J.P., Maroney, M.J., 1995. Interaction of Cu, Zn superoxide dismutase with hydrogen sulfide. Arch. Biochem. Biophys. 318, 251–263.

Seel, F., Wagner, M., 1988. Über die umsetzung von sulfiden mit stickstoffmonoxid in wäßrigen lösungen. Z. Anorg. Allg. Chem. 558, 189–192.

Seel, F., Kuhn, R., Simon, G., Wagner, M., 1985. Naturforsch. Teil. B 40, 1607.

Sehrawat, A., Deswal, R., 2014. S-nitrosylation analysis in *Brassica juncea* apoplast highlights the importance of nitric oxide in cold-stress signaling. J. Proteome Res. 13, 2599–2619.

Shivaraj, S.M., Vats, S., Bhat, J.A., Dhakte, P., Goyal, V., Khatri, P., et al., 2020. Nitric oxide and hydrogen sulfide crosstalk during heavy metal stress in plants. Phys. Plant 168, 437–455.

Shi, H., Ye, T., Chan, Z., 2014. Nitric oxide-activated hydrogen sulfide is essential for cadmium stress response in bermudagrass (*Cynodon dactylon* (L). Pers.). Plant Phys. Biochem. 74, 99–107.

Shi, H., Ye, T., Han, N., Bian, H., Liu, X., Chan, Z., 2015. Hydrogen sulfide regulates abiotic stress tolerance and biotic stress resistance in *Arabidopsis*. J. Integr. Plant Biol. 57, 628–640.

Signorelli, S., Considine, M.J., 2018. Nitric oxide enables germination by a four-pronged attack on ABA-induced seed dormancy. Front. Plant Sci. 9, 296.

Singh, S., Kumar, V., Kapoor, D., Kumar, S., Singh, S., Dhanjal, D.S., et al., 2020a. Revealing on hydrogen sulfide and nitric oxide signals co-ordination for plant growth under stress conditions. Phys. Plant 168, 301–317.

Singh, V.P., Tripathi, D.K., Fotopoulos, V., 2020b. Hydrogen sulfide and nitric oxide signal integration and plant development under stressed/non-stressed condition. Phys. Plant 168, 239–240.

Spadaro, D., Yun, B.W., Spoel, S.H., Chu, C., Wang, Y.Q., Loake, G.J., 2010. The redox switch: dynamic regulation of protein function by cysteine modifications. Phys. Plant 138, 360–371.

Stamler, J.S., Singel, D.J., Loscalzo, J., 1992. Biochemistry of nitric oxide and its redox-activated forms. Science 258, 1898–902.

Stöhr, C., Ullrich, W.R., 2002. Generation and possible roles of NO in plant roots and their apoplastic space. J. Exp. Bot. 53, 2293–2303.

Stöhr, C., Strube, F., Marx, G., Ullrich, W.R., Rockel, P., 2001. A plasma membrane-bound enzyme of tobacco roots catalyses the formation of nitric oxide from nitrite. Planta 212, 835–841.

Szepesi, Á., 2020. Role of metabolites in abiotic stress tolerance. In: Tripathi, D.K., et al., (Eds.), Plant Life under Changing Environment. Academic Press, UK, pp. 755–774.

Szepesi, Á., Csiszár, J., Gémes, K., Horváth, E., Horváth, F., Simon, M.L., et al., 2009. Salicylic acid improves acclimation to salt stress by stimulating abscisic aldehyde oxidase activity and abscisic acid accumulation, and increases Na+ content in leaves without toxicity symptoms in *Solanum lycopersicum* L. J. Plant Phys. 166, 914–925.

Szuba, A., Kasprowicz-Maluśki, A., Wojtaszek, P., 2015. Nitration of plant apoplastic proteins from cell suspension cultures. J. Proteom. 120, 158–168.

Tapley, D., Buettner, G.R., Shick, J.M., 1999. Free radicals and chemiluminescence as products of the spontaneous oxidation of sulfide in seawater, and their biological implications. Biol. Bull. 196, 52–56.

Voet, D., Voet, J.G., 2004. Introduction to metabolism. In: Harris, D., Fitzgerald, P. (Eds.), Biochemistry. John Wiley & Sons, Inc, Hoboken, NJ, p. 573.

Vollár, M., Feigl, G., Oláh, D., Horváth, A., Molnár, Á., Kúsz, N., et al., 2020. Nitro-oleic acid in seeds and differently developed seedlings of *Brassica napus* L. Plants 9, 406. Available from: https://doi.org/10.3390/plants9030406.

Wagner, F., Asfar, P., Calzia, E., Radermacher, P., Szabó, C., 2009. Bench-to-bedside review: hydrogen sulfide–the third gaseous transmitter: applications for critical care. Crit. Care 13, 213.

Wang, Y., Li, L., Cui, W., Xu, S., Shen, W., Wang, R., 2012. Hydrogen sulfide enhances alfalfa (*Medicago sativa*) tolerance against salinity during seed germination by nitric oxide pathway. Plant Soil 351, 107–119.

Wang, P., Zhu, J.K., Lang, Z., 2015. Nitric oxide suppresses the inhibitory effect of abscisic acid on seed germination by S-nitrosylation of SnRK2 proteins. Plant Signal. Behav. 10, e1031939.

Wei, B., Zhang, W., Chao, J., Zhang, T., Zhao, T., Noctor, G., et al., 2017. Functional analysis of the role of hydrogen sulfide in the regulation of dark-induced leaf senescence in *Arabidopsis*. Sci. Rep. 7, 1–16.

Whiteman, M., Moore, P.K., 2009. Hydrogen sulfide and the vasculature: a novel vasculoprotective entity and regulator of nitric oxide bioavailability? J. Cell Mol. Med. 13, 488–507.

Wirtz, M., Droux, M., Hell, R., 2004. O-acetylserine (thiol) lyase: an enigmatic enzyme of plant cysteine biosynthesis revisited in *Arabidopsis thaliana*. J. Exp. Bot. 55, 1785–1798.

Wojtaszek, P., 2000. Nitric oxide in plants: to NO or not to NO. Phytochemistry 54, 1–4.

Yang, H., Mu, J., Chen, L., Feng, J., Hu, J., Li, L., et al., 2015. S-nitrosylation positively regulates ascorbate peroxidase activity during plant stress responses. Plant Phys. 167, 1604–1615.

Yong, Q.C., Cheong, J.L., Hua, F., Deng, L.W., Khoo, Y.M., Lee, H.S., et al., 2011. Regulation of heart function by endogenous gaseous mediators—crosstalk between nitric oxide and hydrogen sulfide. Antioxid. Redox Sign 14, 2081–2091.

Yu, Y., Zhou, X., Zhu, Z., Zhou, K., 2019. Sodium hydrosulfide mitigates cadmium toxicity by promoting cadmium retention and inhibiting its translocation from roots to shoots in *Brassica napus*. J. Agric. Food Chem. 67, 433–440.

Zhang, J.Z., Millero, F.J., 1993. The products from the oxidation of H2S in seawater. Geochim. Cosmochim. Acta 57, 1705–1718.

Zhang, S., Melzer, M.M., Sen, S.N., Çelebi-Ölçüm, N., Warren, T.H., 2016. A motif for reversible nitric oxide interactions in metalloenzymes. Nat. Chem. 8, 663–669.

Zhang, P., Luo, Q., Wang, R., Xu, J., 2017. Hydrogen sulfide toxicity inhibits primary root growth through the ROS-NO pathway. Sci. Rep. 7, 1–11.

Chapter 9

Gene regulation by H_2S in plants

Alireza Iranbakhsh[1], Zahra Oraghi Ardebili[2,]* and Narges Oraghi Ardebili[1]

[1]*Department of Biology, Science and Research Branch, Islamic Azad University, Tehran, Iran,*
[2]*Department of Biology, Garmsar Branch, Islamic Azad University, Garmsar, Iran*
**Corresponding author.*

9.1 Introduction

Plant cells govern their reactions to the recognized physicochemical internal and external signals through dynamic complex signal recognition and transduction events (Ghasempour et al., 2019; Moghanloo et al., 2019). Putative receptors, kinases, Ca^{2+}, ion channels, phytohormones, reactive oxygen/nitrogen species (ROS/RNS), protein kinases/phosphatases, calcium-dependent protein kinases (CDPKs), mitogen-activated protein kinases (MAPKs), and transcription factors are major components contributing to signal recognition and transduction (Khan et al., 2018; Babajani et al., 2019a,b; Iranbakhsh et al., 2020a,b,c). Sulfur and its derivatives, especially hydrogen sulfide (H_2S), cysteine (Cys), methionine (Met), and glutathione (GSH), function in various signaling routes through which they modulate the defense machinery. H_2S is a bioreactive and gasotransmitter molecule that interacts with other signaling agents, especially hydrogen peroxide (H_2O_2) and nitric oxide (NO), as well as phytohormones. The close crosslink of H_2S with NO, reactive oxygen species (ROS), Ca^{+2}, phytohormones, and other signaling molecules provides a great opportunity for this signaling molecule to directly or indirectly influence various aspects of plant growth, development, metabolism, and defense. Among these interactions, the interplay of H_2S with plant hormones is of great significance. Like NO, H_2S with its gaseous nature acts as a signaling molecule at optimal doses, while at high concentrations it can exhibit cytotoxicity. Therefore its homeostasis is of great importance. In this regard, the H_2S homeostasis is highly mediated through the catalytic functions of several enzymes, including L-cysteine

Hydrogen Sulfide in Plant Biology. DOI: https://doi.org/10.1016/B978-0-323-85862-5.00014-2

desulfhydrase (LCD), cyanoalanine synthase (CAS), sulfite reductase (SiR), and cysteine synthase, and D-cysteine desulfhydrase (DCD) in cytosol, mitochondria, and chloroplasts (Banerjee et al., 2018).

Like NO, H_2S, which may diffuse between cells and intracellular domains, plays a secondary messenger role, thereby activating signal transduction cascades (Shivaraj et al., 2019; Iranbakhsh et al., 2020a,b). Owing to its small size and an uncharged lipophilic nature, it can easily diffuse through the lipid bilayer part of plasma membranes or organelle intracellular membrane in a transporter-independent manner (Shivaraj et al., 2019). However, H_2S transportation through aquaporins and the regulatory role of H_2S toward aquaporins are controversial (Shivaraj et al., 2019). H_2S acts as an active component of the complicated signaling network with the contributions of Ca^{+2}, H_2O_2, NO, and phytohormones (Fotopoulos et al., 2015; Aroca et al., 2018; Shivaraj et al., 2019). It is important to mention that the accurate equilibrium between H_2S, H_2O_2, and NO is considered to be a determining factor in regulating biological processes in response to internal and/or external cues (Shivaraj et al., 2019).

Taking H_2S reactivity into account, H_2S may associate with the following reactions (Filipovic et al., 2017):

1. Reaction with ROS/RNS and generating other signaling species.
2. Binding and/or redox reaction with metal in metalloproteins.
3. Persulfidation; modification of protein at cysteine residues.

H_2S has a high affinity to target Fe-containing proteins like cytochrome oxidase (Pandey and Gautam, 2020) and can therefore influence respiration, photosynthesis, and metabolism. As is well-known, H_2S signaling is associated with protein persulfidation. Persulfidation (wrong term: S-sulfhydration) is a posttranslational modification of proteins in which a Cys residue in target proteins mainly undergoes a conversion of the thiol group (-SH) into a persulfide group (-SSH). This modification may influence activities or subcellular localization of the modified proteins. Taking the biological function of persulfidation into account, this change may associate with the functional modifications in protein structures, enzyme activities, and subcellular localization (Mustafa et al., 2009; Paul and Snyder, 2015; Aroca et al., 2017b). Considering comparative data on modified proteomes, the number of proteins undergoing persulfidation is greater than those undergoing S-nitrosylation change (Shivaraj et al., 2019). However, some proteins can be subjected to a combination of both modifications (Aroca et al., 2017a; Shivaraj et al., 2019). In *Arabidopsis*, persulfidation of proteins contributes to the regulation of plenty of vital biological events, like growth, development, carbon metabolism, stress responses, and RNA translation (Aroca et al., 2017a). The persulfidation reaction is also involved in the regulation of the cytosolic/nuclear switches in localization of glyceraldehyde-3-phosphate dehydrogenase (Aroca et al., 2017b) which plays important roles in the regulation of the transcriptional program. Besides, a posttranslational reversible modification of persulfidation plays a critical protective role toward Cys residues of the target protein against irreversible oxidation conversion (Wedmann et al., 2016; Aroca et al., 2018; Banerjee et al., 2018).

Plenty of convincing reports confirm the regulatory roles of H_2S toward diverse important plant cellular organelles/biological processes, including seed germination (Chen et al., 2019), cell cycle (Mei et al., 2017), cytoskeleton (Li et al., 2018a), senescence (Jin et al., 2018), photomorphogenesis (Liu et al., 2019a), nutrition (especially Fe, Zn, Mg, and Ca) (Kaya and Ashraf, 2019), stomatal conductance (Ma et al., 2019), transport (Kabała et al., 2019), hormones (Banerjee et al., 2018; Zhang et al., 2020b), photosynthesis (Kabała et al., 2019; Liu et al., 2019b; Liu et al., 2020b; Ozfidan-Konakci et al., 2020), chloroplast (Liu et al., 2019a), mitochondria (Jin et al., 2018; Meng et al., 2018; Li et al., 2020), respiration (Liu et al., 2020b), stress responses (Kaya et al., 2020; Santisree et al., 2019; Zhang et al., 2020b; Liu et al., 2020b), fruit ripening (Mukherjee, 2019; Lin et al., 2020), symbiosis (Zhang et al., 2020a), and the transcriptome (Li et al., 2017b; Li et al., 2019; Lin et al., 2020). In this chapter, there will be a particular focus on the functions of H_2S in gene regulation at different epigenetic, transcriptional, posttranscriptional, translational, and posttranslational levels and its crosslink with phytohormones, NO, ROS, and some other signaling cascades.

9.2 Multilayered transcriptional regulation

In plants, outstanding dynamic crosstalk among various regulatory elements, including RNA polymerases, numerous transcription factors, the DNA sequence, the transcribed RNA, and the translated protein, modulates the nuclear gene expression patterns in responses to different internal, developmental, and environmental cues. It is worth mentioning that several methods, like microarrays, real-time PCR, semiquantitative PCR, nuclease protection, northern analysis, and next-generation sequencing (NGS), have been employed to explore the intracellular gene expression profiles or quantify the abundance of a specific transcript. In plants, the transcriptional landscape of cells is regulated at various stages, including pretranscriptional, transcriptional, posttranscriptional, translational, and posttranslational levels. The presence of DNA sequence regulatory elements in the promoter architecture plays a crucial determining role in the regulation of cellular transcription profile in responses to various environmental cues. Dependent on the differentiation process, internal and external factors, a plethora of gene families differentially express owing to their varied promoters at the upstream domain of the gene start coding region. The H_2S treatment can lead to remodeling in the transcriptional program. Guo et al. (2018) reported H_2S application in tomato plants exposed to excess nitrate stress resulted in differential expressions of genes, among which 213 genes were upregulated whereas 271 genes were downregulated. The function of these differentially expressed genes in response to H_2S was mainly related to signal transduction, transcription factors, hormone response, protein biosynthesis and modifications, defense response, metabolism, cell wall, and transporters (Guo et al., 2018).

9.3 H_2S and epigenetics

Epigenetic controlling machinery ascertains chromatin conformation, gene accessibility, and transcription program. This chromatin remodeling system mediates the following modifications and interacts with various signaling cascades.

9.3.1 Histone modification (histone methylation/demethylation or acetylation/deacetylation) by catalytic functions of histone acetyltransferases, histone deacetylases, methyltransferases, and demethylases

Histone acetylation catalyzed by histone acetyltransferases (transferring acetyl groups from Acetyl-CoA on histone Lys residues) is known to be a critical vital epigenetic control mechanism. By neutralizing the positive charge of the histones upon actions of histone acetyltransferases resulting in a loose chromatin structure, the expression of genes can be enhanced. Removing the acetyl group of the histone tails through the catalytic function of histone deacetylases, in contrast, condenses chromatin structure and consequently suppresses gene transcription. Histone deacetylases (HDACs; an acetyl agent removing lysine residue from histone tails) play key regulatory roles as a transcriptional corepressor involved in epigenetic modulation of gene expression by chromatin remodeling, change in gene accessibility, and subsequent gene repression. Furthermore, histone tails or DNA methylation by the enzymatic function of methyltransferases and demethylases are other factors implicated in remodeling chromatin conformation and gene transcription events. As can be noted in Table 9.1, the H_2S mediates changes in epigenetic chromatin remodeling through histone and/or DNA modifications. Recent reports highlighted the H_2S-associated changes in histone deacetylation (Behera et al., 2018), histone acetylation (Rios et al., 2016), and histone methyltransferase/demethylase (Rios et al., 2016; Fawcett et al., 2019). According to the proteomic analysis, H_2S can epigenetically affect nuclear proteins of the chromatin-remodeling system such as methyltransferases, acetyltransferases, and histones by the posttranslational modification of persulfidation (Aroca et al., 2017a), thereby modulating transcription through influencing their subcellular localization/binding affinities (Aroca et al., 2018). In this regard, there are however knowledge gaps, especially in plants. A schematic design on the potential H_2S-associated epigenetic modifications is shown in Fig. 9.1.

TABLE 9.1 The H_2S-mediated changes in epigenetic chromatin remodeling through the histone and/or DNA modifications.

Species/ organelle	Epigenetic target	Change type	References
Mitochondria	Demethylation of TFAM promoter	DNA modification	Li et al. (2015)
Mice	DNA methylation	DNA modification	Ide et al. (2019)
Mice	DNA hyper-methylation in the OPG gene HDACs	Histone and DNA modification	Behera et al. (2018)
Worm	Histone methyltransferase and demethylase H3K4me2 (at specific promoters)	Histone modification	Fawcett et al. (2019)
THP1 monocytes	Histone acetylation (the activity of HDACs) and Histone methylation	Histone modification	Rios et al. (2016)

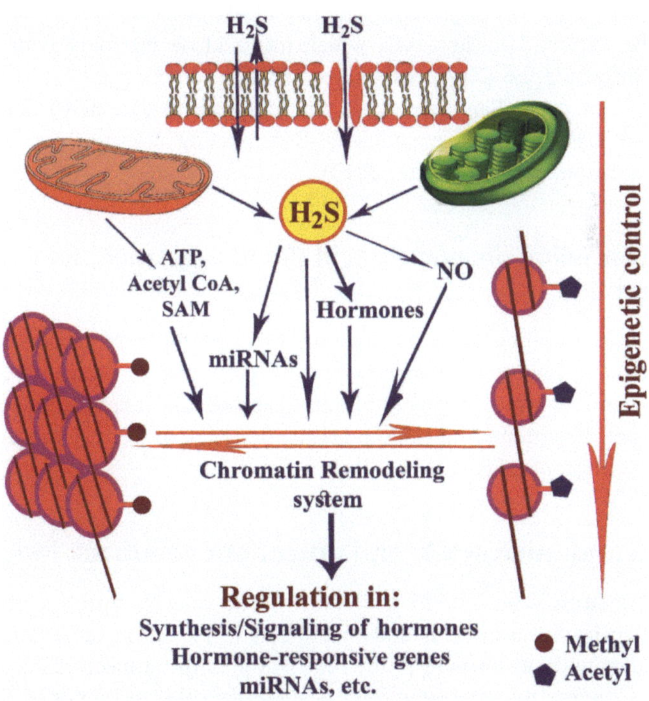

FIGURE 9.1 Potential H_2S-associated epigenetic control of gene regulation.

9.3.2 DNA cytosine methylation

In higher eukaryotes, the DNA methylation profile as a constituent of the epigenetic modification program may be influenced or perturbed as a response to particular internal and/or external factors (Rajaee Behbahani et al., 2020; Sotoodehnia-Korani et al., 2020). Plants exhibit the highest rates (up to 50%) of DNA methylation relative to other kinds of eukaryotic living organisms. Furthermore, DNA methylation as a critical epigenetic marker plays vital roles in coordinating the nuclear transcriptional profile and maintaining genome stability (Li et al., 2017b). Considering DNA methylation, the non-CpG methylation (with interplay with transposable elements/small interfering RNAs (siRNAs)) present in plants is the most important difference to animals (Suzuki and Bird, 2008). The implications of small RNAs (siRNAs and miRNAs) in DNA methylation and their interplay also become evident (Murr, 2010). The findings of numerous experiments underline this hypothesis that cytosine DNA methylation keeps the suppressed chromatin status, leading to the silenced promoter (Rajaee Behbahani et al., 2020; Sotoodehnia-Korani et al., 2020). It is worth mentioning that the RdDM pathway which targets euchromatic regions represses target genes and transposons at the transcriptional level, thereby regulating plant responses to environmental factors, especially stress conditions. Several reports underline this hypothesis that H_2S can associate with changes in DNA demethylation (Li et al., 2015; Behera et al., 2018; Ide et al., 2019), but it needs to be explored further. NO, which has a close interplay with H_2S, can indirectly influence the DNA methylation through affecting the SAM (S-adenosyl methionine) pool (Santisree et al., 2019; Abedi et al., 2020). Besides, the transcription rates of genes encoding H_2S-generating enzymes correlated with DNA methylation rates (Ide et al., 2019).

9.3.3 Noncoding ribonucleic acid (RNA) regulation; fully discussed in the next title

Therefore these epigenetic modifications through the remodeling chromatin conformation influence the gene accessibility for the transcriptional system. It is important to note that this chromatin-remodeling machinery is regulated in a manner dependent on redox status which can be controlled by the endogenous content of H_2S.

9.4 H_2S and microRNA, and epigenetic feedback loop

MicroRNAs (miRNAs; 18–25 nucleotides; encoded by miRNA genes) play vital roles in both the nucleus and cytoplasm where these molecular biomarkers can efficiently contribute to the regulation of the transcriptional program of genes. Convincing scientific reports underline the tremendous versatility in miRNA function at both cytoplasmic and nuclear levels (Catalanotto et al., 2016), indicating that changes in miRNAs can associate with enormous modifications in both the transcriptome and proteome. Moreover, miRNAs at the transcriptional level can govern the activation/repression of expression of a target gene. At the posttranscriptional level, they also control mRNA stability and the alternative splicing process (Catalanotto et al., 2016). Like protein-encoding genes, the transcription of miRNAs is modulated by both genetic and epigenetic strategies. These small noncoding RNAs contribute to the RNA-mediated interference (RNAi) system involved in degrading mRNA, suppressing translation, and leading to gene silencing. It is noteworthy to mention that a plethora of mRNAs at the 3' UTR region can be targeted by each miRNA, exhibiting the high influence of miRNAs on both transcriptomes and proteomes (Yao et al., 2019). They do not only control their own biogenesis but also efficiently participate in signal transduction, protein degradation and regulation of plant development, and modulation of plant responses to environmental cues (Shen et al., 2013).

There is a close crosslink between miRNAs and epigenetics. The interconnection between miRNAs and epigenetics can be considered at the following levels.

9.4.1 MicroRNAs can influence both histone and DNA modifications through targeting the responsible enzymes

It has become evident that miRNAs can directly target enzymes involved in chromatin-modifying machinery, thereby modulating chromatin conformation (Humphries et al., 2019). DNA and/or histone modifications can epigenetically control miRNAs. On the other hand, the epigenetic chromatin-remodeling system can be influenced by miRNAs at the following levels (Moutinho and Esteller, 2017);

1. Controlling enzymes implicated in the epigenetic chromatin-remodeling system.
2. Binding to a specific sequence in the promoter region.
3. Recruiting specific protein complexes modulating chromatin conformation and gene transcription.

9.4.2 The epigenetic machinery controls gene transcriptions of microRNAs

The occurrence of the methylation of CpG islands within the promoter region of miRNAs prevents binding of transcription factors and RNA polymerase to the promoter, leading to the suppressed gene expression (Yao et al., 2019). Nearly half of miRNAs contain CpG islands in their promoters, reflecting the potential involvements of miRNA on DNA methylation status (Yao et al., 2019). Apart from DNA methylation, histone modification can associate with the repression or activation of transcription of miRNAs (Yao et al., 2019; Zhao et al., 2017).

Taking H_2S and miRNA interplay into account, the H_2S fluctuations can associate with alterations in the expression of miRNAs. In this regard, H_2S can improve plant tolerance to stress conditions by modulating the expression of stress-responsive miRNAs. The reports on the H_2S/miRNA crosslink are highlighted in Table 9.2. The interplay between H_2S and miRNA can be monitored in the following two ways.

9.4.3 MicroRNAs control H_2S biosynthesis

Endogenous H_2S synthesis is partly regulated by the involvement of miRNAs (Zhai et al., 2017). The antioxidative and cytoprotective functions of H_2S have been attributed to the crosstalk of H_2S and miRNAs (Zhai et al., 2017). Moreover, there is evidence indicating several miRNAs, including miR21, miR22, and miR30 are implicated in the regulation of the expression of the CSE gene encoding the H_2S-generating enzyme (Zhai et al., 2017; Shen et al., 2015; Wang et al., 2015). Furthermore, there are potential miRNA binding sites in *OsH2S2* and *OsH2S14* genes which are responsible for regulating H_2S homeostasis in rice (Alagarasan and Ramalingam, 2018).

TABLE 9.2 H_2S/miRNA interplay recorded in different plant species and animals.

Species	Target	References
Tomato	miR390a and miR160 and their target genes	Mei et al. (2017)
Arabidopsis	miR167, miR393, miR396, and miR398	Shen et al. (2013)
Arabidopsis	miR393	Shi et al. (2015)
Rice	miRNA binding site in genes of H_2S homeostasis	Alagarasan and Ramalingam (2018)
Mice	miR133	Liu et al. (2011)
Mice	miR21	Toldo et al. (2014)
Rat	miR21, miR22, and miR30	Zhai et al. (2017), Shen et al. (2015), Wang et al. (2015)
Broiler trachea	lncRNA3037/miR-15a/BCL2-A20	Li et al. (2020)

9.4.4 H_2S modulates microRNAs

H_2S mediates reprogramming of transcriptions of downstream target genes through its effects on miRNAs. Shen et al. (2013) provided convincing evidence in *Arabidopsis*. Drought-associated H_2S accumulation led to variations in the expression of miRNAs (in particular miR167, miR393, miR396, and miR398) and their target genes, thereby conferring an improved *Arabidopsis* resistance against stress (Shen et al., 2013). Moreover, the exogenously applied H_2S led to upregulations in miR133 (Liu et al., 2011) and miR21 (Toldo et al., 2014). In *Arabidopsis*, exogenous H_2S triggered modifications in several miRNAs (*miR167a*, *miR167c*, *miR167d*, *mIR393a*, *miR396a*, *miR398a*, and *miR398c*) which are responsible for plant tolerance to drought stress (Shen et al., 2013), manifesting the close interplay between miRNAs and H_2S. The H_2S-supplemented seedlings or the cysteine desulfhydrases overexpressing *Arabidopsis* also displayed high potency to cope with biotic stress through increases in endogenous H_2S level as well as activation of the miR393-mediated auxin signaling pathway. The application of NaH as an H_2S donor stimulated the transcription of *miR393*a/b and their target genes, including *AFB1*, *AFB2*, *AFB3*, and *TIR1* as an auxin signaling pathway in tomato (Shi et al., 2015). Exogenous NaHS also upregulated the expression of several miRNAs, including *miR390a* and *miR160* in tomato seedlings (Mei et al., 2017). In broiler trachea, it has become evident that H_2S triggers alteration in the miRNA and lncRNA (IncRNA3037/miR-15a/BCL2-A20) profiles, and in turn the signaling caused apoptosis and necroptosis (Li et al., 2020).

9.5 H_2S, hormones, epigenetics, and transcriptional landscape

Plants manage their program of growth, development, metabolism, and protection highly through hormone signaling as the most significant signaling molecules. Plant responses to environmental fluctuations are governed by an organized interplay of hormones, epigenetic chromatin-remodeling machinery, and transcriptional program. H_2S and phytohormones have synergistic (like H_2S and salicylic acid) or antagonistic (like H_2S and ethylene) interplay. As it is depicted in Fig. 9.2, H_2S can modulate gene regulation through diverse mechanisms, among which its interplay with hormones is of critical importance. H_2S can function upstream and/or downstream of hormones. We can classify H_2S and hormones crosstalk into the two following levels:

9.5.1 Phytohormones involves in H_2S production and homeostasis

Plant hormones, like salicylic acid (Li et al., 2015), spermidine (Li et al., 2019), auxin (Li et al., 2014; Fang et al., 2014a,b), abscisic acid (Scuffi et al., 2014), gibberellic acid (Xie et al., 2014), jasmonic acid (Deng et al., 2020), and

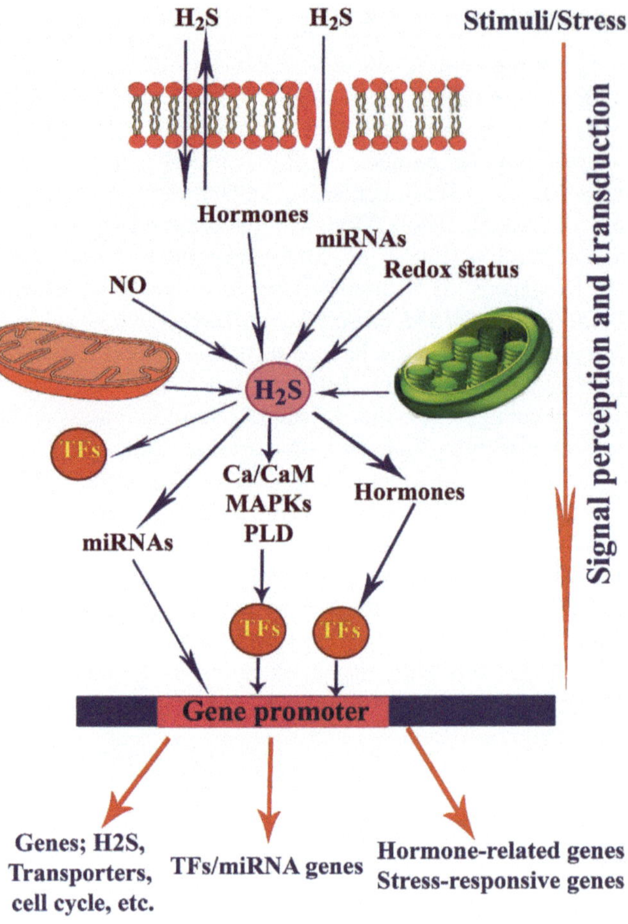

FIGURE 9.2 Various mechanisms by which H_2S can be involved in gene regulation.

ethylene (Hou et al., 2013; Ge et al., 2017) control the H_2S production and hemostasis. It has been reported that abscisic acid modulates the expression of the DES1 gene and Cys homeostasis in *Arabidopsis* (Álvarez et al., 2012). There is a correlation between DES activity and resistance against stress conditions (Da-Silva and Modolo, 2018).

9.5.2 H_2S contributes to the modulation of biosynthesis, signaling, and/or functions of phytohormones

H_2S participates in regulation of biosynthesis, signaling, and/or functions of plant hormones, like auxin (Zhang et al., 2020b), ABA (Ma et al., 2019; Aroca et al., 2017a), ethylene (Ge et al., 2017; Lin et al., 2020; Liu et al., 2020a), gibberellic acid (Xie et al., 2014), and jasmonic acid (Tian et al., 2017). H_2S might also alter the distribution and signaling of phytohormones by sulfhydration of reactive Cys residues as a protective posttranslational modification of targeted proteins (He et al., 2019). Transcriptome analysis in drought-treated wheat seedlings revealed that H_2S transcriptionally modified signaling of several key hormones, including ethylene, brassinosteroid, cytokinin, auxin, jasmonic acid, and salicylic acid, abscisic acid, and gibberellin (Li et al., 2017b). In response to drought stress, *AUX/IAA*, *ARF*, *SAUR*, *BARR*, *A-ARR*, *GID1*, *PP2C*, *EBF1/2*, *EIN3*, *BRI1*, *SnRK2*, *ABF*, *JAR1*, *AUX1*, *GH3*, *BSK*, *JAZ*, *MYC2*, *NPR1*, *AHP*, *PYR/PYL*, *EIN2*, *ETR*, *COI1*, and *PR-1* exhibited the differential expression patterns attributed to the H_2S signaling (Li et al., 2017b).

Within this framework, it is believed that there is crosstalk between hormones and epigenetic chromatin-modifying machinery (Maury et al., 2019). In plants, hormonal balances and chromatin-remodeling machinery appears to be epigenetically interconnected (Maury et al., 2019). This interplay may occur at the following multiple levels:

1. Constituents of signaling pathways of various hormones regulate the activity of chromatin-modifying enzymes.
2. Modifications in chromatin conformation modulate phytohormone synthesis, signaling, and reactions.
3. Direct control of hormone-responsive genes by chromatin remodeling system and gene accessibility.

Several reports indicate that there is a correlation between histone (Li et al., 2018b) and DNA modifications, transcription factors, as well as variations in plant hormones (Lafon-Placette et al., 2018).

9.6 H_2S, mitochondria, chloroplast, epigenetics, and gene regulation

Mitochondria and chloroplasts are intracellular organelles with a semiautonomous nature that contribute to modulating metabolism, ROS, and H_2S levels. H_2S plays a critical role in maintaining the integrity and functions of mitochondria and chloroplasts. In *Arabidopsis* exposed to drought stress, H_2S protected cells and delayed senescence via maintaining mitochondrial ultrastructure and function, improving energy generation in mitochondria through the modified ATPase activity rate, and controlling senescence-associated genes (Jin et al., 2018). Based on the current knowledge, H_2S is effectively involved in maintaining and improving both the mitochondrial structure, biogenesis, transcription of genes (especially TFAM; a major activator of mitochondrial transcription and genome replication), and functions (Li et al., 2015). Exogenously applied H_2S suppressed oxidative stress via improvement in the function of mitochondria, modification in mitochondrial ultrastructure, upregulation in *SIRT3*, *OPA1*, *MFN1*, and *MFN2* genes, and downregulation in *DRP1* and *FIS1* genes (Meng et al., 2018). H_2S can improve photosynthesis performance and protect chloroplast ultrastructure through inductions in the transcription of genes encoding functional photosynthesis proteins (especially RUBISCO), and thiol redox modification (Chen et al., 2011). H_2S-mediated modifications in carboxylation efficiency by RUBISCO, stomatal conductance, photosynthetic pigments, PSII activity, and antioxidant system resulted in the maintenance of a high net photosynthetic rate under low light stress condition in tall fescue seedlings (Liu et al., 2019a).

Both mitochondria and chloroplasts are involved in ROS and H_2S homeostasis. ROS and redox status can epigenetically affect the chromatin structure. Besides, the mitochondria are involved in the generation of epigenetically significant metabolites, like acetyl-CoA, S-adenosyl methionine (SAM), and nicotinamide adenine dinucleotide (NAD), that serve as coenzymes (Shrishrimal et al., 2019). As shown in Fig. 9.2, the mitochondria and chloroplast organelles, as well as their interactions with H_2S, play therefore critical roles in epigenetic modifications and, consequently, cellular transcription profile.

9.7 H_2S and transcription factors

Transcription factors are proteins recognizing DNA at particular sequences in the promoter of genes to modulate the nuclear transcription profile (Iranbakhsh et al., 2020a,b,c). It is important to note that some genes are constitutively expressed, whereas several are stimuli-responsive. In response to diverse signals/stimuli, the cytosolic/nuclear translocations and functions of the majority of these regulatory proteins are controlled. Transcription factors that function in the control of photosynthesis, abiotic stress responses, and immunity are vital components of the transcription regulatory system through which gene expression patterns are modulated to control growth, development, metabolism, and protection. A plethora of transcription factors can be targeted by H_2S. Moreover, it has been reported that some gene families of transcription factors, like *WRKY*s, *MYB*s, *bZIP*s, *HSF*s, and *ERF*s, are H_2S-responsive. As can be seen in Table 9.3, transcription factors and their genes are among the important targets of H_2S signaling. In tomato, the H_2S-improved tolerance to excess nitrate has been attributed to regulation of RNS/ROS signaling and downstream transcriptional profile, especially genes related to signal transduction, transcription factors, and defense system (Guo et al., 2018). *MADS-box*, *WRKY1*, *WRKY3*, *NAC1*, *C2H2-type zinc finger protein*, *THM1*, *IIIA*, *JERF1*, *TCP13*, *TCP19*, *TCP5*, *HSFA3*, *TAGL12*, *bHLH*, *bZIP*, *MTSHP*, *HSP*, *histidine decarboxylase*, *MYB12*, *NAM1*, and *CaM-binding transcription factors* were the most upregulated transcription factor genes following H_2S application in tomato (Guo et al., 2018).

TABLE 9.3 The H_2S-responsive transcription factors or signal transduction components.

Species	Target	Function	References
White clover	*bZIP37, bZIP107, DREB2, DREB4, WRKY108715*	Transcription factor	Li et al. (2019)
Kiwifruit	Downregulation in ethylene receptor 2 (*ETR2*), and *ERF016, ERF003*, and *ERF5* Upregulation in the *ERF4* and *ERF113*	Transcription factor	Liu et al. (2020a)
Tomato	*MADS-box, WRKY1, WRKY3, NAC1, C2H2-type zinc finger protein, THM1, IIIA, JERF1, TCP13, TCP19, TCP5, HSFA3, TAGL12, bHLH, bZIP, MTSHP, HSP, histidine decarboxylase, MYB12, NAM1*, and *calmodulin-binding*	Transcription factor	Guo et al. (2018)
Wheat	*MYB, NAC4, bHLH, bZIP, WRKY17, WRKY19, Orphans, MYB*, and *ARF*	Transcription factor	Li et al. (2017b)
Arabidopsis	*TGA3* (a bZIP transcription factor)	Transcription factor	Fang et al. (2017)
Arabidopsis	*DREB2A, DREB2B*, and *CBF4*	Transcription factor	Jin et al. (2011)
Strawberry	*DREB transcription factor*	Transcription factor	Christou et al. (2014)
Arabidopsis	*ARF8, GRF1, GRF2*, and *GRF3*	Transcription factor	Shen et al. (2013)
Rice	*bZIP, NAC, WRKY, MYB*, and *AP2/EREBP*	Transcription factor	Alagarasan and Ramalingam (2018)
Tomato	*MPK3, MAPK4, MAPK11, MAPK14, MAPK7, PK, CDPK, CaM, cdkb1, phospholipase*, and *CaM-binding*	Signal transduction	Guo et al. (2018)
Wheat	Calcium-dependent protein kinase (*CDPK*) and phospholipase D (*PLD*)	Signal transduction	Li et al. (2017b)
Arabidopsis	Ca^{2+}/*CaM2*	Signal transduction	Fang et al. (2017)

(Continued)

TABLE 9.3 (Continued)

Species	Target	Function	References
Wheat	*AUX/IAA, ARF, SAUR, BARR, A-ARR, GID1, PP2C, EBF1/2, EIN3, BRI1, SnRK2, ABF, JAR1, AUX1, GH3, BSK, JAZ, MYC2, NPR1, AHP, PYR/PYL, EIN2, ETR, COI1*, and *PR-1*	Signaling of hormones	Li et al. (2017b)
Arabidopsis	*TIR1, AFB2, AFB3, TIR1/AFB F-box proteins*	Auxin receptors	Shen et al. (2013)
Mice	Nuclear factor κB (*NF-κB*; an antiapoptotic)	Transcription factor	Sen et al. (2012)
Mice	Stimulation in *SIRT3* promoter activity	Transcription factor	Xie et al. (2016)
Staphylococcus aureus	*CstR* (CsoR-like sulfurtransferase repressor)	Transcription factor	Luebke et al. (2014)
Cupriavidus pinatubonensis	*FisR* as a RSS-responsive transcriptional activator	Transcription factor	Li et al. (2017a)
E. coli	*YcjW*	Transcription factor	Luhachack et al. (2019)
Rhodobacter capsulatus	transcriptional repressor *SqrR*	Transcription factor	Shimizu et al. (2017)
Fish	Upregulated transcription factor *MafG* and *MafK*	Transcription factor	Kelley et al. (2016)

Moreover, H_2S conferred the improved tolerance through induction in DES (H_2S production), DREB2A and DREB2B (two transcription factors responsive to drought in an ABA-independent manner), CBF4 (an ABA-responsive transcription factor), and Rd29A (a drought-responsive gene) in *Arabidopsis* (Jin et al., 2011). Under drought stress condition, H_2S modified expressions of transcription factors (499 distributed in 22 gene families), among which *MYB*, *NAC4*, *bHLH*, *bZIP*, *WRKY17*, *WRKY19* were the top differentially expressed (Li et al., 2017b). In *Arabidopsis*, the Ca^{2+}/CaM2 mechanism enhanced the binding of the TGA3 transcription factor to the LCD promoter, thereby increasing H_2S generation and improving chromium tolerance (Fang et al., 2017). Transcriptome analysis in H_2S-treated kiwifruit revealed that H_2S transcriptionally downregulated ethylene receptor 2 (*ETR2*), and *ERF016*, *ERF003*, and *ERF5*, whereas it upregulated the *ERF4* and *ERF113* transcription factors (Liu et al., 2020a). In white clover, dehydration/exogenously applied spermidine-mediated H_2S accumulation and its crosstalk with Ca^{+2}, NO, and

H_2O_2 led to increases in transcript levels of several responsive transcription factors, including *DREB2*, *DREB4*, *bZIP107*, *bZIP37*, and *WRKY108715* (Li et al., 2019). It has been addressed that H_2S mediated modifications in the expression patterns of different transcription factors, like *DREB* transcription factor in strawberry (Christou et al., 2014), *ARF8*, *GRF1*, *GRF2*, and *GRF3* in *Arabidopsis* (Shen et al., 2013), as well as *bZIP*, *NAC*, *WRKY*, *MYB*, and *AP2/EREBP* in rice (Alagarasan and Ramalingam, 2018). Likewise, some transcription factors, like *NF-Kb* (Sen et al., 2012), *YcjW* (Luhachack et al., 2019), *MafG*, and *MafK* (Kelley et al., 2016) have been reported to be H_2S-responsive in other living organisms. The H_2S functions in gene regulation via the contribution of transcription factors are highlighted in Fig. 9.2.

9.8 Ca^{2+}, Ca^{2+}/CaM, H_2S, and gene regulation

Ca^{2+} as a core secondary messenger is a vital signal transducer and participates in the modulation of cellular reactions to environmental cues. Following stress/stimulus perception, signal transduction may be mediated through the transient fluctuation in cytosolic Ca^{2+}, its interference with calmodulin (CaM; intracellular Ca^{2+} receptor), and Ca^{2+}/CaM downstream components (in particular target transcription factors), thereby activating/suppressing transcriptions of target stimuli-responsive genes (Fang et al., 2017). There is a complicated interplay between Ca^{2+} signaling, H_2S homeostasis, and H_2S signaling during plant adaptation/acclimation to diverse environmental cues. In response to transient fluctuation in cytosolic Ca^{2+} concentration, its subsequent sophisticated interplay with CaM, and target proteins (especially transcription factors), is a mechanism through which the expression of target transcription factors, transcription of genes responsible for H_2S and Cys production, H_2S-Cys signaling, and downstream-related molecular/physiological responses can be regulated. For instance, exposure to Cr^{6+} through a Ca^{2+}/CaM-mediated pathway led to subsequent transcriptional induction in the cysteine synthesis-related genes (*SAT1*, *SAT5*, and *OASTLa*; involved in H_2S−cysteine signaling) and downstream events conferred the tolerance in *Arabidopsis* (Fang et al., 2016). The H_2S/Ca^{2+}-CaM crosslink can be presented in the two following states:

9.8.1 Ca^{2+}/CaM signaling regulates H_2S production and homeostasis

In *Arabidopsis* counteracted with heavy metal stress, a Ca^{2+}/CaM-mediated pathway caused upregulation in the transcription of L-cysteine desulfhydrase (*LCD*; H_2S production) through the binding of TGA3 (a member of bZIP transcription factor; "TGACG"-binding factor family) to the promoter of *LCD* which in turn promoted H_2S production and improved plant resistance (Fang et al., 2017). In *Arabidopsis*, the reinforced tolerance to Cd has been attributed to induction in Ca^{2+}-dependent protein kinase 3 (*CDPK3*), *LCD*, and an increase in endogenous H_2S level (Qiao et al., 2016). Likewise, Ca^{2+}/CaM-triggered H_2S accumulation through induction in H_2S-generating responsible gene (CSE) was also supported by the findings of Yang et al. (2008) in mice.

9.8.2 H_2S can effectively participate in the modulation of Ca^{2+} signaling

In wheat, the H_2S-mediated upregulation in the transcription of Ca^{2+}-dependent protein kinase (*CDPK*) and Ca^{2+}/CaM-dependent protein kinases have been reported by Li et al. (2017b), manifesting interplay between H_2S and Ca^{2+} signaling under drought stress. Fang et al. (2014a) also reported that H_2S interacted with Ca^{2+} signaling and upregulated *MT3A* and *PCS* genes (involved in the synthesis of chelators) in *Setaria italic* to cope with chromium-associated stress. Furthermore, microarray-based analysis in H_2S-treated tomato seedlings revealed the H_2S-mediated changes in Ca^{2+}/CaM signaling (Guo et al., 2018). In zucchini seedlings, Valivand et al. (2019) also reported that the Ca^{2+}/CaM mechanism functions both upstream and downstream of H_2S through which systemic acquired acclimation to heavy metal stress is mediated.

The convergence of the Ca^{2+} and H_2S signaling pathways can occur at several checkpoints through regulating ion channels or acting as a secondary messenger. Moreover, the mutual crosslink of H_2S and phytohormones signaling can occur by regulations of ion channels or acting as a secondary messenger in the response to diverse plant hormones (He et al., 2019). In maize seedlings, the NO treatment led to an increase in H_2S, Ca^{2+} concentration, and activity of alcohol dehydrogenase (Ca^{2+}-dependent), thereby inducing the defense system and improving hypoxia tolerance (Peng et al., 2016). Observations of Li et al. (2019) also provided evidence on the cross-talk among Ca^{2+}, and H_2S, H_2O_2, and NO, thereby conferring dehydration tolerance. It has been hypothesized that the function of the CSE/H_2S route and the intracellular Ca^{2+} concentration play crucial roles protecting the Golgi system (Zhang et al., 2020c).

9.9 H_2S, mitogen-activated protein kinase, signal transduction, and transcriptional control

As a critical component of posttranslational regulatory machinery, mitogen-activated protein kinases (MAPKs; usually localized in the cytosol and/or nucleus) mediate the phosphorylation of a plethora of proteins by which the modifications in the subcellular localization and activity of proteins are managed. By perceptions of environmental stimuli, MAPKs as a downstream component of a complex network of signaling pathways can participate in the signal transduction event to the nucleus, thereby reprogramming transcription processes. Tyrosine, threonine, and serine residues can be targeted by protein kinases (Bigeard and Hirt, 2018). It is important to note that following perceptions of some developmental or environmental signal/stimuli, the nucleocytoplasmic shuttling of MAPKs can be performed. It is believed that the important targets of MAPKs are transcription factors that are involved in the transcriptional reprogramming. In response to diverse internal or external signals/stimuli, the interplay between several members of MAPKs and transcription factors, like MPK4/WRKY (Andreasson et al., 2005), MPK3,6/WRKY (Li et al., 2012; Sheikh et al., 2016), MPK3/6/HSFA4A (Iranbakhsh et al., 2018), MPK3,6/ERF6 (Meng et al., 2013), MPK3/bZIP (Djamei et al., 2007), MPK3/MYB (Pitzschke et al., 2009) have become evident. As is well-known, MYBs, WRKYs, HSFs, and bZIPs are implicated in the transcription reprogramming upon the stress-associated oxidative burst (Iranbakhsh et al., 2020a). Moreover, the key modulatory role of MAPKs (especially MPK3 and MPK6) toward the transcriptional landscape has been well illustrated (Iranbakhsh et al., 2018).

Taking H_2S-triggered signal transduction into account, H_2S upregulated the MAPKs, phospholipase C, and phospholipase D (having functional roles in signal transduction) in tomato seedlings exposed to excess nitrate (Guo et al., 2018). *MAPK14*, *MAPK7*, *MAPK4*, *MAPK11*, *MPK3*, *protein kinases*, *CaM-binding protein*, calcium-dependent protein kinase (*CDPK*), *CaM*, *cdkb1* (B1-type cyclin-dependent kinase), and phospholipase were upregulated in response to exogenous H_2S (Guo et al., 2018). H_2S-modulated root development is mediated through the signal transduction route with the interplay among ROS, MPK6, and NO (Zhang et al., 2017). As further supporting evidence, H_2S stimulated expression of MAPKs, especially *MPK4*, thereby relieving the sign of cold stress in *Arabidopsis* seedlings (Du et al., 2017). Qi et al., (2019) provided convincing data on the interplay among H_2S, MAPKs, and NO in cucumber exposed to nitrate stress. They found that H_2S improved plant protection against the risk of excess nitrate through the MAPKs/NO signaling.

9.10 Crosstalk between NO and H_2S

Both NO and H_2S as the secondary messengers contribute to the regulation of the plant responses to the diverse physicochemical and biotic stress factors (Nazerieh et al., 2018). Hence, their potential interactions are of great importance. As is well-known, H_2S signaling is mediated through the posttranslational modification of thiol (-SH) of Cys residue in proteins into a persulfide (-SSH) through a process called persulfidation (S-sulfhydration) (Aroca et al., 2018). In plants, both NO and H_2S (a signaling molecule with a wide range of functions similar to those of NO) have efficient potencies to act synergistically or antagonistically as signals. The current knowledge underlines the biological connections between these two bioactive signaling agents in dose-, cell-, tissue-, organ-, and species-dependent manners. Taking Cys-based redox switches into account, proteins may undergo two reversible modifications, including H_2S-associated persulfidation and NO-mediated S-nitrosation, both of which alter activity, stability, and functionality of the posttranslationally modified proteins (Abedi et al., 2020). Target cysteine residue in proteins may undergo S-nitrosylation or S-sulfhydration (Aroca et al., 2018; Abedi et al., 2020) dependent on stimuli, signal transduction, cell type, and responses.

H_2S may function upstream (Scuffi et al., 2014) or downstream of NO (He et al., 2019). The exposure to Cd improved the endogenous NO and H_2S concentrations in Bermuda grass (Shi et al., 2013). The exogenously-applied NO or H_2S mitigated the Cd-associated toxicity through regulating enzymatic and nonenzymatic antioxidant machinery. The utilization of NO and H_2S inhibitors revealed that the NO-mediated activation of H_2S signaling (H_2S as the downstream signaling of NO) toward the Cd tolerance in *Cynodon dactylon* (Shi et al., 2013). Scuffi et al. (2014) investigated the crosstalk among NO, ABA, and H_2S toward the stomatal closure using the *Arabidopsis* mutant (*des1*; defect in *AT3G62130* gene coding l-cysteine desulfhydrase which is involved in H_2S production). They found that H_2S stimulated NO accumulation and subsequently the NO molecule performs a critical role downstream of H_2S in the ABA-mediated stomatal closure. Likewise, the exogenously applied H_2S induced the stomatal closure via regulating ATP-binding cassette (ABC) transporters, whereas the ABA-mediated stomatal closure was prevented by an H_2S scavenger (Garcia-Mata and Lamattina, 2010). In barley, Chen et al. (2015) provided evidence of the effects of NO and H_2S on the transcriptional regulations of several pumps and transporters (HvSOS1 and H^+-ATPase (HvHA1)) through which plant resistance against salinity was improved. Moreover, several reports underline this hypothesis that H_2S contributes to the signaling routes of NO (Lisjak et al., 2011) and H_2O_2 (Zhang et al., 2013). An increase in NO following the application of exogenous H_2S has been considered as a mechanism by which H_2S can confer the improved resistance of plants under abiotic stress conditions (Shi et al., 2013; Li et al., 2012; Li, 2013; Chen et al., 2015). Taking mutual interplay into account, exogenous NO utilization is also associated with stimulation in the H_2S level, reflecting NO as an upstream signaling component for H_2S generation (Li et al., 2012; Shi et al., 2013). H_2S at toxic doses along with NO led to suppression in the auxin distribution and reduction in the cell division in the root meristem (Zhang et al., 2017). H_2S and NO have close interplay by which they can govern the regulation of multiple processes through the cyclic guanosine monophosphate (cGMP) and mitogen-activated protein kinase (MAPK) signal route (He et al., 2019). This evidence shows that the strong interplay between NO and H_2S offers plant tolerance to environmental stress.

9.11 Crosstalk between H_2O_2 and H_2S

Plant defense responses against environmental cues are partly managed through the critical crosslink among the redox-based signaling network, ROS, and H_2S. Moreover, crosstalk between H_2S and the ROS-associated signaling network can occur at multilayer transcriptional or posttranslational levels (Tao et al., 2020). H_2S and ROS interplay can be considered in the following two contrary ways:

1. H_2S govern preventing reactions of ROS overaccumulation
2. Synergistic crosstalk of H_2S and ROS signals

Findings of Mei et al. (2017) indicated that H_2O_2 acted as the downstream messenger of H_2S signaling, thereby modulating the development of lateral roots in *Solanum lycopersicum*. H_2S/H_2O_2 interplay was found to control the development of lateral roots through regulation of transcriptions of cell cycle-related genes, including *SlCYCA2;1*, *SlCYCA3;1*, *SlCDKA1*, and *SlKRP2* (Mei et al., 2017). In cucumber, Kabała et al. (2019) also reported the protecting roles of H_2S in alleviating Cd toxicity through its interplay with H_2O_2. Guo et al. (2018) also reported H_2S-improved tolerance to excess nitrate via crosstalk between H_2S and RNS/ROS signaling. Based on the H_2S-mediated changes in transcriptome, H_2S is involved in regulation of stress-triggered ROS homeostasis through the activation of heat shock proteins (HSPs) (Guo et al., 2018), which are molecular chaperones that contribute to control protein conformation and protection of proteins (Iranbakhsh et al., 2018; Safari et al., 2018). During the response of white clover to dehydration, H_2S acts downstream of H_2O_2 and NO (Li et al., 2019).

9.12 H_2S and redox-based gene regulation

H_2S is a redox bioactive multitask signaling substance that can act as a biomessenger in both plants and animals. The redox-triggered modifications in the transcriptome, proteome, and metabolome are considered to be critical mechanisms through which plant growth, development, metabolism, and protection are modulated. Signaling networks triggered by the temporal alterations in redox state reprogram cellular metabolism through which growth and development is adjusted. The balance between oxidation and reduction reactions confers the cellular redox homeostasis, which is one of the most important checkpoints toward the cellular transcription program. It is worth mentioning that redox-based regulation of the nuclear transcriptional pattern is performed through the close interplay of RSS along with RNS and ROS (Hasanuzzaman et al., 2018). Cao et al. (2019) concluded that communication between miRNAs, their targets, and redox status is a part of the regulatory machinery toward transcription program/translation suppression. Moreover, the various kinds of transcription factors (especially bZIPs and

MYBs) have been found to be regulated in a redox-based way. Furthermore, the implication of redox-based regulation in binding activities of several transcription factors (bZIP16, bZIP68, and GBF1) has been manifested (Li and Loake, 2016). TCP transcription factors act as a redox sensor and exhibit a redox-dependent modified activity (Viola et al., 2013). TCP transcription factors (H_2S-responsive) are implicated in the regulation of plant cell proliferation, growth, and development (Manassero et al., 2013).

9.13 H_2S, cytoskeleton, and gene regulation

As is well-known, the plant cytoskeleton (complicated machinery of microtubules, actin filaments, and associated proteins functioning as the effectors of signaling pathways) also plays significant roles in morphogenesis, organogenesis, and plant responses to internal and external stimuli (Li et al., 2018a). In this regard, H_2S can also target the cytoskeleton system. It is believed that actin is a major component of the cellular skeleton that is involved in responding to the signaling of bioactive agents (Li et al., 2018a). Actin can be also involved in gene regulation at different levels, including genomic integrity, chromatin remodeling, and gene activation (Viita and Vartiainen, 2016). In *Arabidopsis*, it has been reported that the ultrastructure of the actin component of cytoskeleton and actin dynamics can be modulated in response to the H_2S signal (Jia et al., 2015; Li et al., 2018a). However, more studies should be performed to clarify the potential H_2S-mediated changes in the cytoskeleton and its subsequent regulatory role in the nuclear transcription program.

9.14 H_2S-mediated mechanisms of transcription reprogramming, concluding remarks, and future perspective

In this chapter, the mechanisms by which H_2S can be involved in the transcriptional regulation of genes are discussed. The most important mechanisms are as follows:

1. H_2S and epigenetic landscape;
   - **a.** H_2S can modify histone methylation/demethylation or acetylation/deacetylation.
   - **b.** H_2S can modify DNA cytosine methylation/demethylation.
   - **c.** miRNAs and epigenetics:

i. miRNAs can influence both histone and DNA modifications.

ii. the epigenetic machinery controls gene transcriptions of miRNAs.

   - **d.** H_2S, Hormones, and epigenetics:

i. Constituents of signaling pathways of various hormones regulate the activity of chromatin-modifying enzymes.

   - ii. Modifications in chromatin conformation modulate phytohormone synthesis, signaling, and reactions.
   - iii. Direct control of hormone-responsive genes by chromatin remodeling system and gene accessibility.
 - e. Crosslink among mitochondria, chloroplast, and epigenetics
   - i. H_2S production and homeostasis.
   - ii. Providing epigenetically significant metabolites and coenzymes, like SAM, Acetyl-CoA, and NAD.

2. H_2S, signal transduction, and transcriptional landscape:
   - a. miRNAs:
     - i. miRNAs control H2S biosynthesis.
     - ii. H_2S targets miRNAs.
   - b. H_2S crosstalk with phytohormones:
     - i. Phyto-hormones involves in H_2S production and homeostasis.
     - ii. H_2S contributes to the modulation of biosynthesis, signaling, and/or functions of phytohormones.
   - c. H_2S, mitochondria, and chloroplast.
   - d. H_2S interferes with transcription factors.
   - e. Ca^{2+}, Ca^{2+}/CaM, H_2S, and gene regulation:
     - i. Ca^{2+}/CaM signaling regulates H_2S production and homeostasis.
     - ii. H_2S can effectively participate in the modulation of Ca^{2+} signaling.
   - f. The interplay of H_2S and MAPKs.
   - g. Crosstalk between NO and H_2S.
   - h. Crosstalk between ROS (especially H_2O_2), and H_2S.
   - i. H_2S mediates gene regulation through the redox-based gene regulation.
   - j. H_2S can target cytoskeleton and gene regulation.

As highlighted in this chapter, H_2S can epigenetically affect chromatin conformation and gene accessibility through modifying DNA/histone acetylation and/or methylation, as well as miRNAs. The current knowledge underlines this hypothesis that the regulatory roles of chloroplasts and mitochondria in the H_2S production and homeostasis, as well as the supplying cofactors (acetyl-CoA, SAM, and ATP) necessary for the epigenetic modifications, are of critical importance. H_2S crosstalk with other signaling molecules, especially hormones, ROS, RNS, and Ca^{+2}/CaM, regulates gene expression through posttranslational modification of proteins (especially transcription factors and MAPKs). It appears therefore that in response to diverse internal and/or external cues, H_2S fluctuations and its choreographed downstream signaling along with its interplay with other signaling networks epigenetically and/or transcriptionally regulates/remodels the cellular transcriptional program of genes, thereby modulating plant growth, development, and metabolism and improving plant protection. Taking knowledge gaps into account, further convincing studies are required to gain a better insight into the way in which H_2S can regulate gene expression at epigenetic and/or pre-/posttranscriptional levels in plant cells compared to animal counterparts.

References

Abedi, S., Iranbakhsh, A., Oraghi Ardebili, Z., Ebadi, M., 2020. Nitric oxide and selenium nanoparticles confer changes in growth, metabolism, antioxidant machinery, gene expression, and flowering in chicory (Cichorium intybus L.): potential benefits and risk assessment. Environ. Sci. Pollut. Res. Available from: https://doi.org/10.1007/s11356-020-10706-2.

Alagarasan, G., Ramalingam, J., 2018. Reaction of hydrogen sulfide homeostasis genes under biotic and abiotic stress condition in rice—computational approach. bioRxiv 269639. Available from: https://doi.org/10.1101/269639.

Álvarez, C., García, I., Moreno, I., Pérez-Pérez, M.E., Crespo, J.L., Romero, L.C., Gotor, C., 2012. Cysteine-generated sulfide in the cytosol negatively regulates autophagy and modulates the transcriptional profile in Arabidopsis. Plant Cell 24 (11), 4621–4634.

Andreasson, E., Jenkins, T., Brodersen, P., Thorgrimsen, S., Petersen, N.H., Zhu, S., et al., 2005. The MAP kinase substrate MKS1 is a regulator of plant defense responses. EMBO J 24 (14), 2579–2589.

Aroca, A., Benito, J.M., Gotor, C., Romero, L.C., 2017a. Persulfidation proteome reveals the regulation of protein function by hydrogen sulfide in diverse biological processes in *Arabidopsis*. J. Exp. Bot. 68 (17), 4915–4927.

Aroca, A., Schneider, M., Scheibe, R., Gotor, C., Romero, L.C., 2017b. Hydrogen sulfide regulates the cytosolic/nuclear partitioning of glyceraldehyde-3-phosphate dehydrogenase by enhancing its nuclear localization. Plant Cell Physiol 58, 983–992. Available from: https://doi.org/10.1093/pcp/pcx056.

Aroca, Á., Serna, A., Gotor, C., Romero, L.C., 2015. S-sulfhydration: a cysteine posttranslational modification in plant systems. Plant Physiology 168 (1), 334–342.

Aroca, A., Gotor, C., Romero, L.C., 2018. Hydrogen sulfide signaling in plants: emerging roles of protein persulfidation. Front. Plant Sci. 9, 1369.

Babajani, A., Iranbakhsh, A., Ardebili, Z.O., Eslami, B., 2019a. Differential growth, nutrition, physiology, and gene expression in *Melissa officinalis* mediated by zinc oxide and elemental selenium nanoparticles. Environ. Sci. Pollut. Res. 26, 24430–24444. Available from: 10.1007/s11356-019-05676-z.

Babajani, A., Iranbakhsh, A., Ardebili, Z.O., Eslami, B., 2019b. Seed priming with non-thermal plasma modified plant reactions to selenium or zinc oxide nanoparticles: cold plasma as a novel emerging tool for plant science. Plasma Chem. Plasma Process. 39 (1), 21–34.

Banerjee, A., Tripathi, D.K., Roychoudhury, A., 2018. Hydrogen sulphide trapeze: environmental stress amelioration and phytohormone crosstalk. Plant Physiol. Biochem. 132, 46–53.

Behera, J., George, A.K., Voor, M.J., Tyagi, S.C., Tyagi, N., 2018. Hydrogen sulfide epigenetically mitigates bone loss through OPG/RANKL regulation during hyperhomocysteinemia in mice. Bone 114, 90–108.

Bigeard, J., Hirt, H., 2018. Nuclear signaling of plant MAPKs. Front. Plant Sci. 9, 469.

Cao, J., Gulyás, Z., Kalapos, B., Boldizsár, Á., Liu, X., Pál, M., et al., 2019. Identification of a redox-dependent regulatory network of miRNAs and their targets in wheat. J. Exp. Bot 70 (1), 85–99.

Catalanotto, C., Cogoni, C., Zardo, G., 2016. MicroRNA in control of gene expression: an overview of nuclear functions. Int. J. Mol. Sci. 17 (10), 1712.

Chen, J., Wu, F.H., Wang, W.H., Zheng, C.J., Lin, G.H., Dong, X.J., et al., 2011. Hydrogen sulphide enhances photosynthesis through promoting chloroplast biogenesis, photosynthetic enzyme expression, and thiol redox modification in *Spinacia oleracea* seedlings. J. Exp. Bot 62 (13), 4481–4493.

Chen, J., Wang, W.H., Wu, F.H., He, E.M., Liu, X., Shangguan, Z.P., et al., 2015. Hydrogen sulfide enhances salt tolerance through nitric oxide-mediated maintenance of ion homeostasis in barley seedling roots. Sci. Rep. 5, 12516.

Chen, Z., Huang, Y., Yang, W., Chang, G., Li, P., Wei, J., et al., 2019. The hydrogen sulfide signal enhances seed germination tolerance to high temperatures by retaining nuclear COP1 for HY5 degradation. Plant Sci. 285, 34–43.

Christou, A., Filippou, P., Manganaris, G.A., Fotopoulos, V., 2014. Sodium hydrosulfide induces systemic thermotolerance to strawberry plants through transcriptional regulation of heat shock proteins and aquaporin. BMC Plant Biol. 14 (1), 42.

Da-Silva, C.J., Modolo, L.V., 2018. Hydrogen sulfide: a new endogenous player in an old mechanism of plant tolerance to high salinity. Acta Bot Brasilica 32 (1), 150–160.

Deng, G., Zhou, L., Wang, Y., Zhang, G., Chen, X., 2020. Hydrogen sulfide acts downstream of jasmonic acid to inhibit stomatal development in *Arabidopsis*. Planta 251 (2), 42.

Djamei, A., Pitzschke, A., Nakagami, H., Rajh, I., Hirt, H., 2007. Trojan horse strategy in *Agrobacterium* transformation: abusing MAPK defense signaling. Science 318 (5849), 453–456.

Du, X., Jin, Z., Liu, D., Yang, G., Pei, Y., 2017. Hydrogen sulfide alleviates the cold stress through MPK4 in *Arabidopsis thaliana*. Plant Physiol. Biochem. 120, 112–119.

Fang, H., Jing, T., Liu, Z., Zhang, L., Jin, Z., Pei, Y., 2014a. Hydrogen sulfide interacts with calcium signaling to enhance the chromium tolerance in *Setaria italica*. Cell Calcium 56 (6), 472–481.

Fang, T., Cao, Z., Li, J., Shen, W., Huang, L., 2014b. Auxin-induced hydrogen sulfide generation is involved in lateral root formation in tomato. Plant Physiol. Biochem. 76, 44–51.

Fang, H., Liu, Z., Jin, Z., Zhang, L., Liu, D., Pei, Y., 2016. An emphasis of hydrogen sulfide-cysteine cycle on enhancing the tolerance to chromium stress in *Arabidopsis*. Environ. Pollut. 213, 870–877.

Fang, H., Liu, Z., Long, Y., Liang, Y., Jin, Z., Zhang, L., et al., 2017. The Ca2 + /calmodulin2-binding transcription factor TGA 3 elevates LCD expression and H2S production to bolster Cr6 + tolerance in *Arabidopsis*. Plant J. 91 (6), 1038–1050.

Fawcett, E.M., Johnson, J.K., Braden, C., Garcia, E., Miller, D.L., 2019. Epigenetic bookmarking of H2S exposure in *Caenorhabditis elegans*. bioRxiv 619734.

Filipovic, M.R., Zivanovic, J., Alvarez, B., Banerjee, R., 2017. Chemical biology of H2S signaling through persulfidation. Chem. Rev. 118 (3), 1253–1337.

Fotopoulos, V., Christou, A., Antoniou, C., Manganaris, G.A., 2015. Review article hydrogen sulphide: a versatile tool for the regulation of growth and defence responses in horticultural crops. J. Hortic Sci. Biotechnol. 90 (3), 227–234.

García-Mata, C., Lamattina, L., 2010. Hydrogen sulphide, a novel gasotransmitter involved in guard cell signalling. New Phytologist 188 (4), 977–984.

Ge, Y., Hu, K.D., Wang, S.S., Hu, L.Y., Chen, X.Y., Li, Y.H., et al., 2017. Hydrogen sulfide alleviates postharvest ripening and senescence of banana by antagonizing the effect of ethylene. PLoS One 12 (6), e0180113.

Ghasempour, M., Iranbakhsh, A., Ebadi, M., Ardebili, Z.O., 2019. Multi-walled carbon nanotubes improved growth, anatomy, physiology, secondary metabolism, and callus performance in *Catharanthus roseus*: an in vitro study. 3 Biotech. 9 (11), 404.

Guo, Z., Liang, Y., Yan, J., Yang, E., Li, K., Xu, H., 2018. Physiological response and transcription profiling analysis reveals the role of H2S in alleviating excess nitrate stress tolerance in tomato roots. Plant Physiol. Biochem. 124, 59–69.

Hasanuzzaman, M., Bhuyan, M.H.M.B., Mahmud, J.A., Nahar, K., Mohsin, S.M., Parvin, K., et al., 2018. Interaction of sulfur with phytohormones and signaling molecules in conferring abiotic stress tolerance to plants. Plant Signal. Behav. 13 (5), e1477905.

He, H., Garcia-Mata, C., He, L.F., 2019. Interaction between hydrogen sulfide and hormones in plant physiological responses. Plant Growth Regul. 87 (1), 175–186.

Hou, Z., Wang, L., Liu, J., Hou, L., Liu, X., 2013. Hydrogen sulfide regulates ethylene-induced stomatal closure in *Arabidopsis thaliana*. J. Integr. Plant Biol. 55 (3), 277–289.

Humphries, B., Wang, Z., Yang, C., 2019. MicroRNA regulation of epigenetic modifiers in breast cancer. Cancers 11 (7), 897.

Ide, M., Ohnishi, T., Toyoshima, M., Balan, S., Maekawa, M., Shimamoto-Mitsuyama, C., et al., 2019. Excess hydrogen sulfide and polysulfides production underlies a schizophrenia pathophysiology. EMBO Molecular Medicine, **11**(12), p.e10695.

Iranbakhsh, A., Ardebili, N.O., Ardebili, Z.O., Shafaati, M., Ghoranneviss, M., 2018. Non-thermal plasma induced expression of heat shock factor A4A and improved wheat (*Triticum aestivum* L.) growth and resistance against salt stress. Plasma Chem. Plasma Process. 38 (1), 29–44.

Iranbakhsh, A., Oraghi Ardebili, Z., Molaei, H., Oraghi Ardebili, N., Amini, M., 2020a. Cold plasma up-regulated expressions of WRKY1 transcription factor and genes involved in biosynthesis of cannabinoids in hemp (*Cannabis sativa* L.). Plasma chemistry and plasma processing **40** (2), 527–537. Available from: https://doi.org/10.1007/s11090-020-10058-2.

Iranbakhsh, A., Oraghi Ardebili, Z., Oraghi Ardebili, N., 2020b. Gene Regulation by NO in Plants, in Nitric Oxide in Plant Biology. Elsevier.

Iranbakhsh, A., Oraghi Ardebili, Z., Oraghi Ardebili, N., 2020c. Synthesis and Characterization of Zinc Oxide Nanoparticles and their Impact on Plants; in Plant Responses to Nanomaterials. Springer.

Jia, H., Hu, Y., Fan, T., Li, J., 2015. Hydrogen sulfide modulates actin-dependent auxin transport via regulating ABPs results in changing of root development in *Arabidopsis*. Sci. Rep. 5, 8251.

Jin, Z., Shen, J., Qiao, Z., Yang, G., Wang, R., Pei, Y., 2011. Hydrogen sulfide improves drought resistance in *Arabidopsis thaliana*. Biochem Biophys Res. Commun. 414 (3), 481–486.

Jin, Z., Sun, L., Yang, G., Pei, Y., 2018. Hydrogen sulfide regulates energy production to delay leaf senescence induced by drought stress in *Arabidopsis*. Front. Plant Sci. 9, 1722.

Kabała, K., Zboińska, M., Głowiak, D., Reda, M., Jakubowska, D., Janicka, M., 2019. Interaction between the signaling molecules hydrogen sulfide and hydrogen peroxide and their role in vacuolar H + -ATPase regulation in cadmium-stressed cucumber roots. Physiol Plant 166 (2), 688–704.

Kaya, C., Ashraf, M., 2019. The mechanism of hydrogen sulfide mitigation of iron deficiency-induced chlorosis in strawberry (Fragaria × ananassa) plants. Protoplasma 256 (2), 371–382.

Kaya, C., Ashraf, M., Alyemeni, M.N., Ahmad, P., 2020. Responses of nitric oxide and hydrogen sulfide in regulating oxidative defence system in wheat plants grown under cadmium stress. Physiol Plant. 168 (2), 345–360.

Kelley, J.L., Arias-Rodriguez, L., Patacsil Martin, D., Yee, M.C., Bustamante, C.D., Tobler, M., 2016. Mechanisms underlying adaptation to life in hydrogen sulfide–rich environments. Mol. Biol. Evol 33 (6), 1419–1434.

Khan, S.A., Li, M.Z., Wang, S.M., Yin, H.J., 2018. Revisiting the role of plant transcription factors in the battle against abiotic stress. Int. J. Mol. Sci. 19 (6), 1634.

Lafon-Placette, C., Gac, Le, Chauveau, A.L., Segura, D., Delaunay, V., Lesage-Descauses, A., et al., 2018. Changes in the epigenome and transcriptome of the poplar shoot apical meristem in response to water availability affect preferentially hormone pathways. J. Exp. Bot 69 (3), 537–551.

Li, Z.G., 2013. Hydrogen sulfide: a multifunctional gaseous molecule in plants. Russian J. Plant Physiol. 60 (6), 733–740.

Li, Y., Loake, G.J., 2016. Redox-regulated plant transcription factors. Plant Transcription Factors. Academic Press, pp. 373–384.

Li, Z.G., Gong, M., Xie, H., Yang, L., Li, J., 2012. Hydrogen sulfide donor sodium hydrosulfide-induced heat tolerance in tobacco (*Nicotiana tabacum* L) suspension cultured cells and involvement of Ca2 + and calmodulin. Plant Sci. 185, 185–189.

Li, Y.J., Chen, J., Xian, M., Zhou, L.G., Han, F.X., Gan, L.J., et al., 2014. In site bioimaging of hydrogen sulfide uncovers its pivotal role in regulating nitric oxide-induced lateral root formation. PLoS One 9 (2), e90340.

Li, Z.G., Xie, L.R., Li, X.J., 2015. Hydrogen sulfide acts as a downstream signal molecule in salicylic acid-induced heat tolerance in maize (*Zea mays* L.) seedlings. J. Plant Physiol. 177, 121–127.

Li, H., Li, J., Lü, C., Xia, Y., Xin, Y., Liu, H., et al., 2017a. F is R activates σ54-dependent transcription of sulfide-oxidizing genes in *Cupriavidus pinatubonensis* JMP 134. Mol. Microbiol 105 (3), 373–384.

Li, H., Li, M., Wei, X., Zhang, X., Xue, R., Zhao, Y., et al., 2017b. Transcriptome analysis of drought-responsive genes regulated by hydrogen sulfide in wheat (*Triticum aestivum* L.) leaves. Mol. Genet. Genomics 292 (5), 1091–1110.

Li, J., Chen, S., Wang, X., Shi, C., Liu, H., Yang, J., et al., 2018a. Hydrogen sulfide disturbs actin polymerization via S-sulfhydration resulting in stunted root hair growth. Plant Physiol. 178 (2), 936–949.

Li, Q.F., Lu, J., Yu, J.W., Zhang, C.Q., He, J.X., Liu, Q.Q., 2018b. The brassinosteroid-regulated transcription factors BZR1/BES1 function as a coordinator in multisignal-regulated plant growth. Biochim. Biophys Acta (BBA)-Gene Regulatory Mechanisms 1861 (6), 561–571.

Li, Z., Zhu, Y., He, X., Yong, B., Peng, Y., Zhang, X., et al., 2019. The hydrogen sulfide, a downstream signaling molecule of hydrogen peroxide and nitric oxide, involves spermidine-regulated transcription factors and antioxidant defense in white clover in response to dehydration. Environ. Exp. Bot 161, 255–264.

Li, X., Chen, M., Shi, Q., Zhang, H., Xu, S., 2020. Hydrogen sulfide exposure induces apoptosis and necroptosis through lncRNA3037/miR-15a/BCL2-A20 signaling in broiler trachea. Sci. Total. Environ. 699, 134296.

Lin, X., Yang, R., Dou, Y., Zhang, W., Du, H., Zhu, L., et al., 2020. Transcriptome analysis reveals delaying of the ripening and cell wall degradation of kiwifruit by hydrogen sulfide. J. Sci. Food Agric. 100 (5), 2280–2287.

Lisjak, M., Teklić, T., Wilson, I.D., Wood, M., Whiteman, M., Hancock, J.T., 2011. Hydrogen sulfide effects on stomatal apertures. Plant Signal. Behav. 6 (10), 1444–1446.

Liu, J., Hao, D.D., Zhang, J.S., Zhu, Y.C., 2011. Hydrogen sulphide inhibits cardiomyocyte hypertrophy by up-regulating miR-133a. Biochem Biophys Res. Commun. 413 (2), 342–347.

Liu, Z., Cao, C., Li, Y., Yang, G., Pei, Y., 2019a. Light regulates hydrogen sulfide signalling during skoto-and photo-morphogenesis in foxtail millet. Funct. Plant Biol. 46 (10), 916–924.

Liu, Y.H., Zhang, X.H., Liu, B.W., Ao, B., Liu, Q., Wen, S.Y., et al., 2019b. Hydrogen sulfide regulates photosynthesis of tall fescue under low-light stress. Photosynthetica 57 (3), 714–723.

Liu, D., Li, J., Li, Z., Pei, Y., 2020a. Hydrogen sulfide inhibits ethylene-induced petiole abscission in tomato (*Solanum lycopersicum* L.). Hortic Res 7 (1), 1–11.

Liu, Y.L., Shen, Z.J., Simon, M., Li, H., Ma, D.N., Zhu, X.Y., et al., 2020b. Comparative proteomic analysis reveals the regulatory effects of H2S on salt tolerance of mangrove plant *Kandelia obovata*. Int. J. Mol. Sci. 21 (1), 118.

Luebke, J.L., Shen, J., Bruce, K.E., Kehl-Fie, T.E., Peng, H., Skaar, E.P., et al., 2014. The CsoR-like sulfurtransferase repressor (CstR) is a persulfide sensor in *Staphylococcus aureus*. Mol. Microbiol. 94, 1343–1360.

Luhachack, L., Rasouly, A., Shamovsky, I., Nudler, E., 2019. Transcription factor YcjW controls the emergency H2S production in E. coli. Nat. Commun. 10 (1), 1–8.

Ma, Y., Zhang, W., Niu, J., Ren, Y., Zhang, F., 2019. Hydrogen sulfide may function downstream of hydrogen peroxide in salt stress-induced stomatal closure in *Vicia faba*. Funct. Plant Biol. 46 (2), 136–145.

Manassero, N.G.U., Viola, I.L., Welchen, E., Gonzalez, D.H., 2013. TCP transcription factors: architectures of plant form. Biomol Concepts 4 (2), 111–127.

Maury, S., Sow, M.D., Le Gac, A.L., Genitoni, J., Lafon-Placette, C., Mozgova, I., 2019. Phytohormone and chromatin crosstalk: the missing link for developmental plasticity? Front. Plant Sci. 10, 395.

Mei, Y., Chen, H., Shen, W., Shen, W., Huang, L., 2017. Hydrogen peroxide is involved in hydrogen sulfide-induced lateral root formation in tomato seedlings. BMC Plant Biol. 17 (1), 162.

Meng, X., Xu, J., He, Y., Yang, K.Y., Mordorski, B., Liu, Y., Zhang, S., 2013. Phosphorylation of an ERF transcription factor by Arabidopsis MPK3/MPK6 regulates plant defense gene induction and fungal resistance. Plant Cell 25 (3), 1126–1142.

Meng, G., Liu, J., Liu, S., Song, Q., Liu, L., Xie, L., et al., 2018. Hydrogen sulfide pretreatment improves mitochondrial function in myocardial hypertrophy via a SIRT3-dependent manner. Br. J. Pharmacol 175 (8), 1126–1145.

Moghanloo, M., Iranbakhsh, A., Ebadi, M., Ardebili, Z.O., 2019. Differential physiology and expression of phenylalanine ammonia lyase (PAL) and universal stress protein (USP) in the endangered species *Astragalus fridae* following seed priming with cold plasma and manipulation of culture medium with silica nanoparticles. 3 Biotech. 9 (7), 288.

Moutinho, C., Esteller, M., 2017. MicroRNAs and epigenetics, Advances in Cancer Research, vol. 135. Academic Press, pp. 189–220.

Mukherjee, S., 2019. Recent advancements in the mechanism of nitric oxide signaling associated with hydrogen sulfide and melatonin crosstalk during ethylene-induced fruit ripening in plants. Nitric Oxide 82, 25–34.

Murr, R., 2010. Interplay between different epigenetic modifications and mechanisms, Advances in Genetics, 70. Academic Press, pp. 101–141.

Mustafa, A.K., Gadalla, M.M., Sen, N., Kim, S., Mu, W., Gazi, S.K., et al., 2009. H_2S signals through protein S-sulfhydration. Sci. Signal. 2 (96), ra72-ra72.

Nazerieh, H., Ardebili, Z.O., Iranbakhsh, A., 2018. Potential benefits and toxicity of nanoselenium and nitric oxide in peppermint. Acta Agric. Slovenica 111 (2), 357–368.

Ozfidan-Konakci, C., Yildiztugay, E., Elbasan, F., Kucukoduk, M., Turkan, I., 2020. Hydrogen sulfide (H2S) and nitric oxide (NO) alleviate cobalt toxicity in wheat (*Triticum aestivum* L.) by modulating photosynthesis, chloroplastic redox and antioxidant capacity. J. Hazard. Mater. 122061.

Pandey, A.K., Gautam, A., 2020. Stress responsive gene regulation in relation to hydrogen sulfide in plants under abiotic stress. Physiol Plant 168 (2), 511–525.

Paul, B.D., Snyder, S.H., 2015. Protein sulfhydration, Methods in Enzymology, 555. Academic Press, pp. 79–90.

Peng, R., Bian, Z., Zhou, L., Cheng, W., Hai, N., Yang, C., et al., 2016. Hydrogen sulfide enhances nitric oxide-induced tolerance of hypoxia in maize (*Zea mays* L.). Plant Cell Rep. 35 (11), 2325–2340.

Pitzschke, A., Schikora, A., Hirt, H., 2009. MAPK cascade signalling networks in plant defence. Curr. Opin. Plant Biol. 12 (4), 421–426.

Qi, Q., Guo, Z., Liang, Y., Li, K., Xu, H., 2019. Hydrogen sulfide alleviates oxidative damage under excess nitrate stress through MAPK/NO signaling in cucumber. Plant Physiol. Biochem. 135, 1–8.

Qiao, Z., Jing, T., Jin, Z., Liang, Y., Zhang, L., Liu, Z., et al., 2016. CDPKs enhance Cd tolerance through intensifying H2S signal in *Arabidopsis thaliana*. Plant Soil 398 (1-2), 99–110.

Rajaee Behbahani, S., Iranbakhsh, A., Ebadi, M., Majd, A., Ardebili, Z.O., 2020. Red elemental selenium nanoparticles mediated substantial variations in growth, tissue differentiation, metabolism, gene transcription, epigenetic cytosine DNA methylation, and callogenesis in bittermelon (*Momordica charantia*); an *in vitro* experiment. PLoS ONE 15 (7), e0235556. Available from: https://doi.org/10.1371/journal.pone.0235556.

Rios, E.C., Soriano, F.G., Olah, G., Gerö, D., Szczesny, B., Szabo, C., 2016. Hydrogen sulfide modulates chromatin remodeling and inflammatory mediator production in response to endotoxin, but does not play a role in the development of endotoxin tolerance. J. Inflamm. 13 (1), 10.

Safari, M., Ardebili, Z.O., Iranbakhsh, A., 2018. Selenium nano-particle induced alterations in expression patterns of heat shock factor A4A (HSFA4A), and high molecular weight glutenin subunit 1Bx (Glu-1Bx) and enhanced nitrate reductase activity in wheat (*Triticum aestivum* L.). Acta Physiol Plant 40 (6), 117.

Santisree, P., Adimulam, S.S., Bommineni, P., Bhatnagar-Mathur, P., Sharma, K.K., 2019. Hydrogen sulfide in plant abiotic stress tolerance: progress and perspectives. Reactive Oxygen, Nitrogen and Sulfur Species in Plants: Production, Metabolism, Signaling and Defense Mechanisms, Wiley, pp. 743–775. Available from: https://doi.org/10.1002/9781119468677.ch33.

Scuffi, D., Álvarez, C., Laspina, N., Gotor, C., Lamattina, L., García-Mata, C., 2014. Hydrogen sulfide generated by L-cysteine desulfhydrase acts upstream of nitric oxide to modulate abscisic acid-dependent stomatal closure. Plant Physiol. 166 (4), 2065–2076.

Sen, N., Paul, B.D., Gadalla, M.M., Mustafa, A.K., Sen, T., Xu, R., et al., 2012. Hydrogen sulfide-linked sulfhydration of NF-κB mediates its antiapoptotic actions. Mol. Cell 45 (1), 13–24.

Sheikh, A.H., Eschen-Lippold, L., Pecher, P., Hoehenwarter, W., Sinha, A.K., Scheel, D., et al., 2016. Regulation of WRKY46 transcription factor function by mitogen-activated protein kinases in *Arabidopsis thaliana*. Front. Plant Sci. 7, 61.

Shen, J., Xing, T., Yuan, H., Liu, Z., Jin, Z., Zhang, L., et al., 2013. Hydrogen sulfide improves drought tolerance in *Arabidopsis thaliana* by microRNA expressions. PLoS One 8 (10), e77047.

Shen, Y., Shen, Z., Miao, L., Xin, X., Lin, S., Zhu, Y., et al., 2015. miRNA-30 family inhibition protects against cardiac ischemic injury by regulating cystathionine-γ-lyase expression. Antioxid. Redox Signal. 22 (3), 224–240.

Shi, H., Ye, T., Chan, Z., 2013. Comparative proteomic and physiological analyses reveal the protective effect of exogenous polyamines in the bermudagrass (*Cynodon dactylon*) response to salt and drought stresses. J. Proteome Res. 12 (11), 4951–4964.

Shi, H., Ye, T., Han, N., Bian, H., Liu, X., Chan, Z., 2015. Hydrogen sulfide regulates abiotic stress tolerance and biotic stress resistance in *Arabidopsis*. J. Integr. Plant Biol. 57 (7), 628–640.

Shimizu, T., Shen, J., Fang, M., Zhang, Y., Hori, K., Trinidad, J.C., et al., 2017. Sulfide-responsive transcriptional repressor SqrR functions as a master regulator of sulfide-dependent photosynthesis. Proc Natl Acad Sci 114 (9), 2355–2360.

Shivaraj, S.M., Vats, S., Bhat, J.A., Dhakte, P., Goyal, V., Khatri, P., et al., 2019. Nitric oxide and hydrogen sulfide crosstalk during heavy metal stress in plants. Physiol Plant .

Shrishrimal, S., Kosmacek, E.A., Oberley-Deegan, R.E., 2019. Reactive oxygen species drive epigenetic changes in radiation-induced fibrosis. Oxid. Med. Cell. Longev. 2019.

Sotoodehnia-Korani, S., Iranbakhsh, A., Ebadi, M., Majd, A., Ardebili, Z.O., 2020. Selenium nanoparticles induced variations in growth, morphology, anatomy, biochemistry, gene expression, and epigenetic DNA methylation in *Capsicum annuum*; an in vitro study. Environ. Pollut. 114727. Available from: https://doi.org/10.1016/j.envpol.2020.114727.

Suzuki, M.M., Bird, A., 2008. DNA methylation landscapes: provocative insights from epigenomics. Nat. Rev. Genet. 9 (6), 465.

Tao, C., Tian, M., Han, Y., 2020. Hydrogen sulfide: a multi-tasking signal molecule in the regulation of oxidative stress responses. J Exp Bot. 71 (10), 2862–2869.

Tian, B., Zhang, Y., Jin, Z., Liu, Z., Pei, Y., 2017. Role of hydrogen sulfide in the methyl jasmonate response to cadmium stress in foxtail millet. Front. Biosci. 22, 530–538.

Toldo, S., Das, A., Mezzaroma, E., Chau, V.Q., Marchetti, C., Durrant, D., et al., 2014. Induction of microRNA-21 with exogenous hydrogen sulfide attenuates myocardial ischemic and inflammatory injury in mice. Circ Cardiovasc Genet. 7 (3), 311–320.

Valivand, M., Amooaghaie, R., Ahadi, A., 2019. Interplay between hydrogen sulfide and calcium/calmodulin enhances systemic acquired acclimation and antioxidative defense against nickel toxicity in zucchini. Environ. Exp. Bot 158, 40–50.

Viita, T., Vartiainen, M.K., 2016. From cytoskeleton to gene expression: actin in the nucleus. The Actin Cytoskeleton. Springer, Cham, pp. 311–329.

Viola, I.L., Güttlein, L.N., Gonzalez, D.H., 2013. Redox modulation of plant developmental regulators from the class I TCP transcription factor family. Plant Physiol. 162 (3), 1434–1447.

Wang, L., Tang, Z.P., Zhao, W., Cong, B.H., Lu, J.Q., Tang, X.L., et al., 2015. MiR-22/Sp-1 links estrogens with the up-regulation of cystathionine γ-lyase in myocardium, which contributes to estrogenic cardioprotection against oxidative stress. Endocrinology 156 (6), 2124–2137.

Wedmann, R., Onderka, C., Wei, S., Szijártó, I.A., Miljkovic, J.L., Mitrovic, A., et al., 2016. Improved tag-switch method reveals that thioredoxin acts as depersulfidase and controls the intracellular levels of protein persulfidation. Chem. Sci. 7 (5), 3414–3426.

Xie, Y., Zhang, C., Lai, D., Sun, Y., Samma, M.K., Zhang, J., et al., 2014. Hydrogen sulfide delays GA-triggered programmed cell death in wheat aleurone layers by the modulation of glutathione homeostasis and heme oxygenase-1 expression. J. Plant Physiol. 171 (2), 53–62.

Xie, L., Feng, H., Li, S., Meng, G., Liu, S., Tang, X., et al., 2016. SIRT3 mediates the antioxidant effect of hydrogen sulfide in endothelial cells. Antioxid. Redox Signal. 24 (6), 329–343.

Yang, G., Wu, L., Jiang, B., Yang, W., Qi, J., Cao, K., et al., 2008. H2S as a physiologic vasorelaxant: hypertension in mice with deletion of cystathionine γ-lyase. Science 322 (5901), 587–590.

Yao, Q., Chen, Y., Zhou, X., 2019. The roles of microRNAs in epigenetic regulation. Curr. Opin. Chem. Biol. 51, 11–17.

Zhai, Y., Tyagi, S.C., Tyagi, N., 2017. Cross-talk of MicroRNA and hydrogen sulfide: a novel therapeutic approach for bone diseases. Biomed Pharmacother 92, 1073–1084.

Zhang, J., Chen, S., Liu, H., Zhang, B., Zhao, Y., Ma, K., Zhao, D., Wang, Q., Ma, H., Zhang, Z., 2013. Hydrogen sulfide prevents hydrogen peroxide-induced activation of epithelial sodium channel through a PTEN/PI (3, 4, 5) P 3 dependent pathway. PloS one 8, p.e64304.

Zhang, P., Luo, Q., Wang, R., Xu, J., 2017. Hydrogen sulfide toxicity inhibits primary root growth through the ROS-NO pathway. Sci. Rep. 7 (1), 868.

Zhang, N.N., Zou, H., Lin, X.Y., Pan, Q., Zhang, W.Q., Zhang, J.H., et al., 2020a. Hydrogen sulfide and rhizobia synergistically regulate nitrogen (N) assimilation and remobilization during N deficiency-induced senescence in soybean. Plant, Cell Environ 43 (5), 1130–1147.

Zhang, X.W., Liu, F.J., Zhai, J., Bi, H.G., Ai, X.Z., 2020b. Auxin acts as a downstream signaling molecule involved in hydrogen sulfide-induced chilling tolerance in cucumber. Planta 251 (3), 1–19.

Zhang, Y., Wang, Y., Read, E., Fu, M., Pei, Y., Wu, L., et al., 2020c. Golgi stress response, hydrogen sulfide metabolism, and intracellular calcium homeostasis. Antioxid. Redox Signal. 32 (9), 583–601.

Zhao, Q., Li, S., Li, N., Yang, X., Ma, S., Yang, A., et al., 2017. miR-34a targets HDAC1-regulated H3K9 acetylation on lipid accumulation induced by homocysteine in foam cells. J. Cell Biochem. 118, 4617–4627.

Chapter 10

Hydrogen sulfide and reactive oxygen species crosstalk and acquisition of abiotic stress tolerance

Krishna Kumar Choudhary[1] **and Nivedita Chaudhary**[2,*]
[1]*Botany Section, Mahila Mahavidyalaya, Banaras Hindu University, Varanasi, India,*
[2]*Department of Environmental Science, School of Earth Sciences, Central University of Rajasthan, Kishangarh, Ajmer, India*
**Corresponding author.*

10.1 Introduction

Hydrogen sulfide (H_2S), otherwise known as dihydrogen sulfide or sulfane, is considered to be a water-soluble, inorganic, and flammable gas with an evident rotten egg odor (Corpas et al., 2019; Zhang et al., 2017). As a gaseous molecule H_2S is small, lipophilic, and later appeared as an adapting signaling cellular component in plants and animals (Santisree et al., 2019). Previously, H_2S gas was only thought of as a toxic and pollutant; however, now it has been recognized as a gasotransmitter, that is, a signaling molecule, existing both in animal and plant cells that causes various physiological effects (Kimura, 2014). Investigations into hydrogen sulfide over the past four decades has found that its endogenous generation in plant cells stimulates growth and development, and resistance against disease and stress conditions (Aroca et al., 2018). Fotopoulos et al. (2015) suggested that H_2S is an advanced signaling molecule that regulates plants' metabolism, particularly under stress conditions (Fotopoulos et al., 2015). The level of H_2S is important for proper functioning, therefore at low levels, H_2S is considered to be beneficial for plants during germination, growth, and development, although at higher levels it has phytotoxic effects causing negative impacts on plant processes, and thus plants contain enzymes to produce and eliminate excessive levels of H_2S to regulate homeostasis in plant cells (Santisree et al., 2019). Plants are stimulated to synthesized H_2S (particularly when exposed

Hydrogen Sulfide in Plant Biology. DOI: https://doi.org/10.1016/B978-0-323-85862-5.00016-6

to sulfate, sulfite, SO_2, cysteine, and most importantly under various stress conditions), therefore, H_2S is crucial for normal plant functioning and also is a signaling molecule against various stresses (Zhang et al., 2008; Jin et al., 2011). Corpas and Palma (2020) suggested that H_2S is engaged directly or indirectly in various plant physiological processes, for instance, seed germination, photosynthesis, stomatal control, fruit ripening, maturity of the plant, and in addition H_2S is also involved in the response mechanism of the plant under different environmental stress conditions. Therefore in plants at low concentration, H_2S is observed to perform as a gaseous signaling molecule analogous to other secondary messengers, for instance, hydrogen peroxide, abscisic acid, nitric oxide, and Ca^{2+}. The prompt production of endogenous H_2S takes place in plants under various stress conditions (Li et al., 2016a).

In this chapter, we will discuss the production and metabolism of H_2S in plants, its importance as a signaling molecule, and the crosstalk between H_2S and reactive oxygen species (ROS) during stress condition in plants.

10.2 Role of hydrogen sulfide in plants

Hydrogen sulfide (H_2S) is generally considered to be a naturally reduced form of sulfur, however, its atmospheric presence is due to biogenic emissions and anthropogenic activities (Bacci et al., 2000). Natural processes constitute the emission from plants of H_2S, which is the most important volatile S-containing compound released by higher plants (Lakkineni et al., 2003). Studies suggested that plants are efficient at consuming atmospheric H_2S for their sulfur requirements (De Kok et al., 2000), and in addition the endogenous production of H_2S allows its participation in the growth and development of plants, for instance, germination of seeds, stomatal control, root development, maturity and senescence regulation, enriching photosynthesis, and as a signaling molecule for gaining tolerance against stress in plants (Santisree et al., 2019).

10.2.1 Generation of H_2S in plants

Wilson et al. (1978) initially reported the significant and varied rate of emission of volatile sulfur compounds, mainly H_2S, in squash, cucumber, soyabean, and cotton. Plants use H_2S as a sulfur nutrient for various plant processes, however, a surplus amount of H_2S causes negative impacts (De Kok et al., 2002). Optimum concentrations of H_2S help to develop resistance against pathogens and abiotic stresses (Zhang et al., 2008). Therefore, Jin and Pie (2015) suggested that H_2S participated not only in the management of plant physiological processes, such as germination, root morphogenesis, photosynthesis, and senescence of flowers but also acted as a signaling molecule required to stimulate the plant's defense mechanisms against both abiotic and biotic stresses. The generation of H_2S in plants occurred via both enzymatic and nonenzymatic pathways, although primarily with the former process (Jin and Pie, 2015).

Corpas et al. (2019) described the production of H_2S in the plant from different sources and associated enzymes; for instance, in the cytosol, chloroplasts, and mitochondria where the subcellular compartments accompanied with Cys metabolism and the enzymes participates are L- and D-cysteine desulfhydrase (DES/D-DES), cysteine synthase (CS), sulfite reductase (SiR), and cyano alanine synthase (CAS). Desulfuration occurs of L-Cys to sulfide plus ammonia and pyruvate, catalyzed by enzymes such as Cys synthase-like and At5g28030. It is considered that in plants, the production of H_2S takes place primarily in chloroplasts through the photosynthetic sulfate-assimilation pathway due to a reaction initiated through enzyme SiR, as reported by Garcia et al. (2015). The major production of H_2S occurs by the enzyme L-cysteine desulfhydrase 1 (DES1) in the cytosol, although other enzymes such as D-DES and Nifs-like proteins are also reported to participate in H_2S production. The concentration of sulfide in chloroplasts is greater than in the cytosol, due to limitations in passage of H_2S to cytosol because ionized forms of sulfide are formed due to the low pH inside organelles (Aroca et al., 2018). Bloem et al. (2019) proposed that the concentration of H_2S produced in plants varied but did not depend completely on stress conditions. The magnitude of H_2S released/emitted under natural conditions may also depend on GSH and glucosinolates and other secondary S compounds. A plant's nutritional status also plays a part, which depends on sulfur dioxide, sulfate, and L-cysteine. Giovanelli (1990) reported the release of H_2S subsequently, where primarily two enzymes participate: L-cysteine desulfydrase (LCD) and O-acetyl-L-serine(thiol)lyase (OAS-TL). The former stimulates cysteine to pyruvate, H_2S decomposition reaction, and ammonia, whereas the latter accounts for the merging of S (inorganic) into cysteine, which consequently convert into S-containing compounds.

10.2.1.1 Endogenous production of H_2S under environmental stresses

H_2S has a significant role in plants' responses to environmental stress. Several studies have suggested that the rapid triggering of endogenous H_2S in plants is similar to H_2O_2, abscisic acid, and nitric oxide. Plants stresses such as drought, heat, metal stress, cold, salt, fungal, radiation, and flooding initiate H_2S signaling (Li et al., 2016b). Corpas and Palma (2020) reviewed studies conducted for the exogenous treatment of H_2S where it helped to alleviate environmental stress particularly heavy metals, metalloids, drought, salinity, and extreme temperatures, thus suggesting that H_2S signaling elevates antioxidative enzymes, such as superoxide dismutase and catalase (CAT), mainly to regulate the lipid peroxidation and increase H_2O_2.

10.2.2 H_2S metabolism in plants

Lisjak et al. (2013) suggested that H_2S concentration depends on the sulfur demand in plants and sulfur in the atmosphere, therefore the release of H_2S from plants and its takeup varied in different conditions. The importance of H_2S in plants depends on the sulfur metabolism along with signaling. In plants, sulfur is a significant macronutrient that is required for growth, development, and crop yield. Stress (biotic/abiotic) induces sulfur-induced resistance (SiR) (Papenbrock et al., 2007) for protection by inducing endogenous sulfur-comprising defense compounds (SDCs), such as glutathione (GSH), glucosinolates, phytoalexins, phytochelatins, sulfur-rich proteins, elemental sulfur, and secondary metabolites. This process stimulates sulfate uptake and assimilation therefore, H_2S plays a role which may also belong to SDCs; therefore its concentration may also vary (Rausch and Wachter, 2005). In plants, under stress conditions, H_2S is also expressed by the reduction of sulfate, and this sulfate is transferred into the plastids where it binds to ATP (adenosine triphosphate) to form APS (adenosine-5'-phosphosulfate). Further, enzymes APR (APS-reductase) and SiR triggered the reduction of APS into sulfide. This sulfur mechanism uphold the redox equilibrium and significantly affects the regulation of oxidative stress by increasing the level of GSH, cysteine, and phytochelatin for protection Wang (2012). Therefore, the production of endogenous H_2S in the sub-cellular compartments of plant cell required many enzymes. For instance, in the cytosol DES1 and L-cysteine desulfhydrase (L-DES), in mitochondria, cyano alanine synthase (CAS) and D-cysteine desulfhydrase (D-DES); and in chloroplast sulfite reductase (SiR) enzymes takes part in the production of H_2S (Corpas and Palma 2020).

10.2.3 Importance of H_2S in plants as a signaling molecule

The properties of H_2S influence its significance in biological systems as H_2S is a weak acid and is disassociated to sulfide anions (S_2^-) and hydrosulfide (HS^-) in an aqueous solution, making it a lipophilic molecule that disperses through cell membranes and reacts with thiol ($-SH$) groups. This further facilitates protein posttranslational modifications (PTMs), persulfidation, and S-nitrosation, and the process stimulates the thiol group of Cys residues (Corpas et al., 2019). Therefore H_2S is a predominantly reactive molecule, released due to the increase in ROS because of the S-oxidation states in proteins in plants. This process causes an alteration from thiol (-2) to sulfonic acid ($+4$) in specific Cys, however, the process is adjustable depending on the concentration of ROS (Hancock et al., 2011). Zhang et al. (2008) described that in plants under biotic/abiotic stress H_2S is obtained due to significant metal stress, such as copper, and later various studies suggested that H_2S ameliorates various metal, heat, drought, and hypoxia stresses (Jin and Pei, 2015).

Certain enzymes in plants regulate the synthesis of H_2S, for instance, reduction of L-cysteine to H_2S by using enzyme L-cysteine desulfhydrase (DES), D-cysteine under go the process of decomposition to form H_2S by utilizing enzyme D-cysteine desulfhydrase (D-DES), sulfite reductase (SiR) used to reduce sulfite to release H_2S, cysteine form H_2S by the usage of enzyme cyanoalanine synthase (CAS), and cysteine synthase (CS) used to form cysteine and revere reaction induces the formation of H_2S (Li et al., 2016a).

Aroca et al. (2018) reviewed H_2S' significant role in the nitric oxide and hydrogen peroxide signaling pathways. H_2S can act together with thiol (−SH) groups existing in protein cysteine residues via the PTM of persulfidation. In this persulfidation process reactive cysteine residues on proteins are altered via the conversion of the thiol (−SH) into a persulfide (−SSH) group. Persulfide adducts represent an amplified nucleophilicity associated with the thiol group and, consequently, altered cysteines exhibit a better reactivity, and thus persulfidation affects proteins more than ROS and NO. Therefore, H_2S plays an adjusting role by involving protein thiol groups with multiple interactions, helping to regulate target proteins. In addition, there is the importance of H_2S in competing with GSH, nitric oxide (NO), fatty acids, and cyanide to produce the PTMs of S-nitrosation, S-cyanylation, S-glutathionylation, and S-acylation (Corpas and Palma 2020).

Postharvest and during fruit ripening, H_2S endogenous levels are also significant for the fruit maturation and protecting against environmental stress-induced ROS. However, more importantly, studies have been conducted on the exogenous application of H_2S to elevate the shelf life of vegetables and fruits, due to its characteristic of increasing the antioxidant systems, thus inhibiting excess ROS production and the subsequent oxidative damage Liu et al. (2017). In *Rhizobium*−legume symbiosis, H_2S and rhizobia maintain the nitrogen content in leguminous plants, such as soybean, by altering the enzyme levels for the protein and genes (Zhang et al., 2020).

10.3 Hydrogen sulfide and reactive oxygen species crosstalk during abiotic stress in plants

H_2S is produced naturally inside the plant cells (Corpas et al., 2019) and it is now well established as an active gas that is involved in several signaling mechanisms to acclimatize or to develop tolerance capacity against numerous abiotic stresses like heat, cold, salinity, drought, heavy metals, etc. (Banerjee et al., 2018; Tao et al., 2020). H_2S and ROS play the combative or collective role and both are responsible for the adjustment of plant responses, and their interactions are well explained during oxidative stress under numerous abiotic stress factors (Scuffi et al., 2018; Aroca et al., 2018; Li et al., 2019; Gotor et al., 2019; Corpas, 2019; Hancock, 2019). Still, our knowledge regarding the interaction of H_2S with ROS during abiotic stresses in plants is limited and needs to explore. Therefore, in this section, we mainly emphasize the crosstalk of H_2S and ROS during oxidative stress in plants under abiotic stress conditions.

10.3.1 H_2S and antioxidative defense mechanism

During stressful conditions, plants observe higher accumulation of ROS like H_2O_2, singlet oxygen, and superoxide (Choudhary et al., 2017). Upholding the balance between ROS production and its scavenging, along with the survival of plants, depends on the capability of plants' antioxidative defense mechanisms (Choudhary et al., 2017). The role of H_2S in direct scavenging of ROS is very little known and needs to explore much more; however, the indirect impact of H_2S on balancing ROS through antioxidative enzymes or signaling pathways involve is being well explored (Tao et al., 2020). Activities of several enzymes like APX, CAT, glutathione reductase, superoxide dismutase, and DHAR are increased after H_2S treatment and have a considerable role in ROS scavenging (Tao et al., 2020). In *Arabidopsis*, persulfidation of cytosolic APX1 is responsible for the positive regulation of its catalytic activity under the presence of H_2S (Aroca et al., 2015). Aroca et al. (2017) suggested that about 57 antioxidants, including CAT1, CAT2, CAT3, thylakoid APX, stromal APX, GR1, GR2, DHAR1, DHAR2, and DHAR3, undergo persulfidation. The proteins undergoing persulfidation have functional variations in their enzymatic activities, structural modifications, and changes in their subcellular localizations (Aroca et al., 2015). Persulfidation is a key monitoring process for carbon metabolism, growth, and development, RNA translation, and reactions to abiotic stresses in plants (Aroca et al., 2017).

Further, it is also been suggested that during abiotic stress or under adverse environmental conditions H_2S and NO have mutual roles in the regulation of oxidative stress (Corpas et al., 2019). Increased APX activity has also been documented for *Arabidopsis* due to S-nitrosation of APX1 under the influence of NO (Yang et al., 2015). Therefore a decrease in oxidative stress in plants under various abiotic stresses due to H_2S is well correlated with the thiol modifications mediated via NO, along with the increased GSH (Takahashi et al., 2011). GSH, as a principle constituent of the ascorbate--GSH cycle, is mainly helpful in the scavenging of H_2O_2 and superoxide, which is produced in the plants through cysteine, glutamate, and glycine via reactions catalyzed by glutathione synthetase (GS) and glutamate–cysteine ligase (GCL) (Noctor et al., 2012). H_2S helps to increase GSH content, as it is involved in the production of cysteine (a key component for the production of GSH) inside the plant cell via O-acetylserine (OAS). Aroca et al. (2017) revealed that in *Arabidopsis* GS and GCL are persulfidation targets, and their enhanced activity was reported under Cd stress (Cui et al., 2014) after pretreatment with H_2S.

10.3.2 H_2S and reactive oxygen species

Plants encounter numerous abiotic stresses in the environment, which lead to oxidative damages through the accumulation of several ROS and RNS (Choudhary et al., 2017; Schieber and Chandel, 2014). Under stressful conditions in plants, cysteine thiols undertake several modifications like S-sulfenylation (−SHO), S-glutathionylation (−SSG), and S-nitrosylation (−SNO). ROS is responsible for the oxidation of cysteine to sulfenic acid (−SOH) through the sulfenylation process, later these proteins are reversibly regulated via thioredoxin systems (Ren et al., 2017). Further, higher generation or accumulation of ROS is responsible for the conversion of sulfenic to sulfinic ($-SO_2H$) and sulfonic ($-SO_3H$) acids via oxidation leading to a loss in functionality of proteins (Paulsen and Carroll, 2013). Also, the cysteine residues present in CAT, APX, and DHAR are vulnerable to oxidation by ROS (Kneeshaw et al., 2017; Calabrese et al., 2019). In response to these modifications, H_2S forms persulfides after interaction with sulfenic/sulfinic or sulfonic acids, to activate the antioxidative defense mechanism of plants (Filipovic et al., 2018). Later on, these persulfides reduce back to thiols and are involved in the normal functioning inside the cells, thus preventing the hyperoxidation of antioxidative enzymes to thiol modification.

In *Arabidopsis* 643 proteins have been reported for their sensitivity against both protein modifications, that is, sulfenylation and persulfidation (Aroca et al., 2017; Huang et al., 2019). Therefore these proteins related to H_2S signaling have a significant role in the sensitivity/tolerance of plants against various abiotic stresses. The role of H_2S has also been explored in the regulation of photosynthesis in spinach (Chen et al., 2011) and nitrogen metabolism in rice (Rizwan et al., 2019). Since persulfides have higher reactivity to ROS than thiols, they are hence are oxidized by ROS to form perthiosulfenic (−SSOH), perthiosulfinic ($-SSO_2H$), and perthiosulfonic ($-SSO_3H$) acids. This suggests an alternative mechanism of protection from oxidative stress leading to protein modifications, regulated via H_2S in plants (Tao et al., 2020).

Moreover, the interactions of H_2S, ROS, and autophagy have also been explored during protein modifications in plants. Autophagy has been observed during abiotic stress (Xiong et al., 2007) and carbon or nitrogen starvation (Phillips et al., 2008) has been observed for the degradation of cytoplasmic components as well as the degradation of oxidized proteins in *Arabidopsis* (Xiong et al., 2007). During stressful conditions, metabolic reactions of plants need to be rapidly adjusted for their survival. This situation is well supported by H_2S-dependent persulfidation, as it is a more competent and balanced mechanism for the better growth and defense of plants under stress (Tao et al., 2020). In *Arabidopsis*, Alvarez et al. (2012) suggested negative characteristics of H_2S for controlling autophagy via preventing the accumulation of ATG8 (autophagy-related ubiquitin-like proteins), and exogenous H_2S supply in *Arabidopsis* alleviated nitrogen deficiency and inhibited the autophagy mechanism under nitrogen deprivation. Later, Laureano-Marín et al. (2016) described that elevated accumulation of ROS is well connected with these H_2S modifications under nitrogen-deprived conditions in *Arabidopsis*. The susceptibility of autophagy-related proteins like ATG3, ATG5, ATG7, and ATG18a has been demonstrated in Arabidopsis for persulfidation (Aroca et al., 2017). These results suggested that sulfides might be regulating the autophagy in plants cells through persulfidation (Aroca et al., 2018).

10.3.3 Cooperative interactions of H_2S and reactive oxygen species in plants

Besides the antagonistic behavior of H_2S and ROS, a synergistic role has also been explored under stressful conditions in plants. H_2S is unable to interact directly with a cysteine residue, and the reaction of H_2S with sulfenic acid is responsible for the persulfide formation in plants (Filipovic et al., 2018). Therefore, it reflects that the appropriate intensity of oxidative stress or ROS accumulation is necessary for the activation of signaling cascades regulated by H_2S, that is, persulfidation under stressful conditions in plants. Since guard cells are responsible for the regulation of stomatal opening/closure under various environmental stresses. In *Arabidopsis* it has been found that two isoforms respiratory burst oxidase homolog protein D (RBOHD) and respiratory burst oxidase homolog protein F (RBOHF) of nicotinamide adenine dinucleotide phosphate (NADPH) oxidase along with H_2S are responsible for the accumulation of H_2O_2 in the apoplasts, leading to stomatal closure via ABA-dependent regulation (Scuffi et al., 2018; Zhang et al., 2019). Recent reports also suggest that various proteins related to stomatal regulation are targets for sulfenylation and persulfidation reactions (Aroca et al., 2017; Huang et al., 2019), like calcium-dependent protein kinases (CPK3 and CPK6) and mitogen-activated protein kinases (MPK3, MPK4, and MPK6) (Mori et al., 2006; Zhang et al., 2017; Prodhan et al., 2018; Du et al., 2019).

Cooperation of H_2S with H_2O_2 is well reflected by increased tolerance capacity in barley (Li et al., 2016b) under UV-B radiation and dehydration in white clover (Li et al., 2019). Mei et al. (2017) had explored the participation of H_2O_2 in the tomato plant for inducing root development via H_2S. Sodium hydrosulfide (NaHS) treatment led to the accumulation of H_2O_2 through stimulating NADPH oxidase and RBOH1, and it was responsible for the promotion of lateral roots in tomato plants. Ion homeostasis in the roots of *Arabidopsis* is well mediated by H_2S, which was dependent on H_2O_2 under saline environment (Li et al., 2014). H_2S-mediated improved germination in *Jatropha curcus* seeds has been observed after pretreatment with H_2O_2 (Li et al., 2012).

10.4 Conclusion and future perspectives

H_2S is produced naturally inside the plant cells and besides its consideration as a toxic gas, it is also involved in several signaling mechanisms in plants to acclimatize or to increase tolerance capacity against numerous abiotic stresses like heat, cold, salinity, drought, heavy metals, etc. H_2S and ROS play an antagonistic as well as a synergistic role and both are responsible for the regulation of plant responses, during oxidative stress. Still, our knowledge about the interaction of H_2S with ROS during abiotic stresses in plants is limited and needs to be explored. Cysteine residues of various antioxidative enzymes are vulnerable to oxidation by ROS and lots of proteins are sensitive to protein modifications like sulfenylation and persulfidation. Therefore, the elucidation or establishment of the connection between sulfenylation, persulfidation, and nitrosation with key enzymes will open the door for significant research in the field of protein modifications and H_2S signaling in plants. Further, the role of autophagy, ROS, and H_2S must be elucidated further in order to enhance our understanding of redox regulation and various abiotic stresses. Results from these studies will be helpful for crop improvement under climate changing scenarios globally.

Acknowledgments

KKC is thankful to UGC-BSR startup grant, New Delhi, India and to Seed grant, Institute of Eminence (IoE), Banaras Hindu University, Varanasi, India. NC is thankful to UGC-BSR startup grant, New Delhi, India.

References

Alvarez, C., Garcia, I., Moreno, I., Perez-Perez, M.E., Crespo, J.L., Romero, L.C., et al., 2012. Cysteine-generatedsulfide in the cytosol negatively regulates autophagy and modulates the transcriptional profile in *Arabidopsis*. Plant Cell 24, 4621–4634.

Aroca, A., Serna, A., Gotor, C., Romero, L.C., 2015. S-sulfhydration: a cysteine posttranslational modification in plant systems. Plant Physiol. 168, 334–342.

Aroca, A., Benito, J.M., Gotor, C., Romero, L.C., 2017. Persulfidation proteome reveals the regulation of protein function by hydrogen sulfide in diverse biological processes in *Arabidopsis*. J. Exp. Bot. 68, 4915–4927.

Aroca, A., Gotor, C., Romero, L.C., 2018. Hydrogen sulfide signaling in plants: emerging roles of protein persulfidation. Front. Plant Sci. 9, 1369.

Bacci, E., Gaggi, C., Lanzillotti, E., Ferrozzi, S., Valli, L., 2000. Geothermal power plants at Mt. Amiata (Tuscany-Italy): mercury and hydrogen sulphide deposition revealed by vegetation. Chemosphere 40, 907–911.

Banerjee, A., Tripathi, D.K., Roychoudhury, A., 2018. Hydrogen sulphide trapeze: environmental stress amelioration and phytohormone crosstalk. Plant Physiol. Biochem. 132, 46–53.

Bloem, E., Riemenschneider, A., Volker, J., Papenbrock, J., Schmidt, A., Salac, I., et al., 2019. Sulphur supply and infection with pyrenopezizabrassicae influence L-cysteine desulphydrase activity in *Brassica napus* L. J. Exp. Bot. 70 (17), 4391–4404.

Calabrese, G., Peker, E., Amponsah, P.S., Hoehne, M.N., Riemer, T., Mai, M., et al., 2019. Hyperoxidation of mitochondrial peroxiredoxin limits H_2O_2-induced cell death in yeast. EMBO J. 38 (18), e101552.

Chen, J., Wu, F.H., Wang, W.H., Zheng, C.J., Lin, G.H., Dong, X.J., et al., 2011. Hydrogen sulphide enhances photosynthesis through promoting chloroplast biogenesis, photosynthetic enzyme expression, and thiol redox modification in *Spinaciaoleracea*seedlings. J. Exp. Bot. 62, 4481–4493.

Choudhary, K.K., Chaudhary, N., Agrawal, S.B., Agrawal, M., 2017. Reactive oxygen species: generation, damage and quenching in plants during stress. In: Singh, V.P., Singh, S., Tripathi, D.K., Prasad, S.M., Chauhan, D.K. (Eds.), Revisiting the Role of Reactive Oxygen Species (ROS) in Plants: ROS Boon or Bane for Plants? John Wiley & Sons Limited, Chichester, West Sussex, pp. 85–115.

Corpas, F.J., 2019. Hydrogen sulfide: a new warrior against abiotic stress. Trends Plant Sci. 11, 983–988.

Corpas, F.J., Palma, J.M., 2020. H_2S signaling in plants and applications in agriculture. J. Adv. Res. 24, 131–137.

Corpas, F.J., Barroso, J.B., González-Gordo, S., Muñoz-Vargas, M.A., Palma, J.M., 2019. Hydrogen sulfide (H_2S): a novel component in *Arabidopsis* peroxisomes which triggers catalase inhibition. J. Integr. Plant Biol. 61 (7), 871–883.

Cui, W., Chen, H., Zhu, K., Jin, Q., Xie, Y., Cui, J., et al., 2014. Cadmium-induced hydrogen sulfide synthesis is involved in cadmium tolerance in *Medicago sativa* by reestablishment of reduced (homo)glutathione and reactive oxygen species homeostases. PLoS One 9, e109669.

De Kok, L.J., Stuiver, C.E.E., Westerman, S., Stulen, I., 2000. Atmospheric H_2S pollution- deposition and impact on sulphur metabolism in plants. In: Yunus, M., Singh Nand De Kok, L.J. (Eds.), Environmental Stress: Indication, Mitigation, and Eco-Conservation. Kluwer Academic Publishers, Dordrecht, pp. 135–141.

De Kok, L.J., Castro, A., Durenkamp, M., Stuiver, C.E.E., Westerman, S., Yang L., et al., 2002. Sulphur in plant physiology. In: Proceedings No. 500. The International Fertiliser Society, New York, pp. 1–26.

Du, X.Z., Jin, Z.P., Zhang, L.P., Liu, X., Yang, G.D., Pei, Y.X., 2019. H_2S is involved in ABA-mediated stomatal movement through MPK4 to alleviate drought stress in *Arabidopsis thaliana*. Plant Soil 435, 295–307.

Filipovic, M.R., Zivanovic, J., Alvarez, B., Banerjee, R., 2018. Chemical biology of H_2S signaling through persulfidation. Chem. Rev. 118, 1253–1337.

Fotopoulos, V., Christou, A., Antoniou, C., Manganaris, G.A., 2015. Hydrogen sulphide: a versatile tool for the regulation of growth and defense responses in horticultural crops. J. Hortic. Sci. Biotech. 90, 227–234.

Garcia, I., Gotor, C., Romero, L.C., 2015. Cysteine homeostasis. In: D'Mello, J.P.F. (Ed.), Amino Acids in Higher Plants. CABI Publishing, Wallingford, pp. 219–233.

Giovanelli, J., 1990. Regulatory aspects of cysteine and methionine biosynthesis. In: Rennenberg, H., Brunold, C., De Kok, L.J., Stulen, I. (Eds.), Sulfur Nutrition and Sulfur Assimilation in Higher Plants; Fundamental, Environmental and Agricultural Aspects. SPB Academic Publishing bv, The Hague, pp. 33–48.

Gotor, C., Garcia, I., Aroca, A., Laureano-Marin, A.M., Arenas-Alfonseca, L., Jurado-Flores, A., et al., 2019. Signaling by hydrogen sulfide and cyanide through post-translational modification. J. Exp. Bot. 70, 4251–4265.

Hancock, J.T., 2019. Hydrogen sulfide and environmental stresses. Environ. Exp. Bot. 161, 50–56.

Huang, J., Willems, P., Wei, B., Tian, C., Ferreira, R.B., Bodra, N., et al., 2019. Mining for protein S-sulfenylation in *Arabidopsis* uncovers redox-sensitive sites. Proc. Natl. Acad. Sci. USA 116, 21256–21261.

Jin, Z., Pei, Y., 2015. Physiological implications of hydrogen sulfide in plants: pleasant exploration behind its unpleasant odour. Oxid. Med. Cell. Longev. 397502, 6.

Jin, Z., Shen, J., Qiao, Z., Yang, G., Wang, R., Pei, Y., 2011. Hydrogen sulfide improves drought resistance in Arabidopsis thaliana. Biochem. Bioph. Res. Co. 414 (3), 481–486.

Kimura, H., 2014. The physiological role of hydrogen sulfide and beyond. Nitric Oxide 41, 4–10.

Kneeshaw, S., Keyani, R., Delorme-Hinoux, V., Imrie, L., Loake, G.J., Le, et al., 2017. Nucleoredoxin guards against oxidative stress by protecting antioxidant enzymes. Proc. Natl. Acad. Sci. USA 114, 8414–8419.

Lakkineni, K.C., Ahmad, A., Abrol, Y.P., 2003. Hydrogen sulphide: emission and utilization by plants. In: Abrol, Y.P., Ahmad, A. (Eds.), Sulphur in Plants. Kluwer Academic Publisher, pp. 265–278.

Laureano-Marín, A.M., Moreno, I., Romero, L.C., Gotor, C., 2016. Negative regulation of autophagy by sulfide Is independent of reactive oxygen species. Plant Physiol. 171, 1378–1391.

Li, Z.G., Gong, M., Liu, P., 2012. Hydrogen sulfide is a mediator in H_2O_2-induced seed germination in *Jatropha Curcas*. Acta Physiol. Plant 34 (6), 2207–2213.

Li, Y.J., Shi, Z.Q., Gan, L.J., Chen, J., 2014. Hydrogen sulfide is a novel gasotransmitter with pivotal role in regulating lateral root formation in plants. Plant Signal. Behav. 9, e29127.

Li, Z.-G., Min, X., Zhou, Z.-H., 2016a. Hydrogen sulfide: a signal molecule in plant cross-adaptation. Front. Plant Sci. 7, 1621.

Li, Q., Wang, Z., Zhao, Y., Zhang, X., Zhang, S., Bo, L., et al., 2016b. Putrescine protects hulless barley from damage due to UV-B stress via H_2S-and H_2O_2-mediated signaling pathways. Plant Cell Rep. 5, 1155–1168.

Li, Z., Zhu, Y., He, X., Yong, B., Peng, Y., Zhang, X., et al., 2019. The hydrogen sulfide, a downstream signaling molecule of hydrogen peroxide and nitric oxide, involves spermidine-regulated transcription factors and antioxidant defense in white clover in response to dehydration. Environ. Exp. Bot. 161, 255–264.

Lisjak, M., Teklic, T., Wilson, I.D., Whiteman, M., Hancock, J.T., 2013. Hydrogen sulfide: environmental factor or signalling molecule? Plant Cell Env. 36, 1607–1616.

Liu, D., Xu, S., Hu, H., Pan, J., Li, P., Shen, W., 2017. Endogenous hydrogen sulfide homeostasis is responsible for the alleviation of senescence of postharvest daylily flower via increasing antioxidant capacity and maintained energy status. J. Agr. Food Chem. 65 (4), 718–1726.

Mei, Y., Chen, H., Shen, W., Huang, L., 2017. Hydrogen peroxide is involved in hydrogen sulfide-induced lateral root formation in tomato seedlings. BMC Plant Biol 17, 162.

Mori, I.C., Murata, Y., Yang, Y., Munemasa, S., Wang, Y.F., Andreoli, S., et al., 2006. CDPKs CPK6 and CPK3 function in ABA regulation of guard cell S-type anion-and Ca2 + -permeable channels and stomatal closure. PLoS Biol. 4, e327.

Noctor, G., Mhamdi, A., Chaouch, S., Han, Y., Neukermans, J., Marquez-Garcia, B., et al., 2012. Glutathione in plants: an integrated overview. Plant Cell Environ. 35, 454–484.

Papenbrock, J., Riemenschneider, A., Kamp, A., Schulz-Vogt, H.N., Schmidt, A., 2007. Characterization of cysteine-degrading and H_2S-releasing enzymes of higher plants: from the field to the test tube and back. Plant Biol. 9, 582–588.

Paulsen, C.E., Carroll, K.S., 2013. Cysteine-mediated redox signaling: chemistry, biology, and tools for discovery. Chem. Rev. 113, 4633–4679.

Phillips, A.R., Suttangkakul, A., Vierstra, R.D., 2008. The ATG12-conjugating enzyme ATG10 is essential for autophagic vesicle formation in *Arabidopsis thaliana*. Genetics 178, 1339–1353.

Prodhan, M.Y., Munemasa, S., Nahar, M.N., Nakamura, Y., Murata, Y., 2018. Guard cell salicylic acid signaling is integrated into abscisic acid signaling via the Ca2 + /CPK-dependent pathway. Plant Physiol. 178, 441–450.

Rausch, T., Wachter, A., 2005. Sulfur metabolism: a versatile platform for launching defence operations. Trends Plant Sci. 10, 503–509.

Ren, X., Zou, L., Zhang, X., Branco, V., Wang, J., Carvalho, C., et al., 2017. Redox signaling mediated by thioredoxin and glutathione systems in the central nervous system. Antioxid. Redox Sig 27, 989–1010.

Rizwan, M., Mostofa, M.G., Ahmad, M.Z., Zhou, Y., Adeel, M., Mehmood, S., et al., 2019. Hydrogen sulfide enhances rice tolerance to nickel through the prevention of chloroplast damage and the improvement of nitrogen metabolism under excessive nickel. Plant Physiol. Biochem. 138, 100–111.

Santisree, P.S., Adimulam, S.S., Bommineni, P., Bhatnagar-Mathur, P., Sharma, K.K., 2019. Hydrogen sulfide in plant abiotic stress tolerance: progress and perspectives. In: Hasanuzzaman, M., Fotopoulos, V., Nahar, K., Fujita, M. (Eds.), Reactive oxygen, Nitrogen and Sulfur Species in Plants: Production, Metabolism, Signaling and Defense Mechanisms. John Wiley & Sons Ltd.

Schieber, M., Chandel, N.S., 2014. ROS function in redox signaling and oxidative stress. Curr. Biol. 24, 453–462. Available from: 10.1016/j.cub.2014.03.034.

Scuffi, D., Nietzel, T., Di Fino, L.M., Meyer, A.J., Lamattina, L., Schwarzlander, M., et al., 2018. Hydrogen sulfide increases production of NADPH oxidase-dependent hydrogen peroxide and phospholipase D-Derived phosphatidic acid in guard cell signaling. Plant Physiol. 176, 2532–2542.

Takahashi, H., Kopriva, S., Giordano, M., Saito, K., Hell, R., 2011. Sulfur assimilation in photosynthetic organisms: molecular functions and regulations of transporters and assimilatory enzymes. Annu. Rev. Plant Biol. 62, 157–184.

Tao, C., Tian, M., Han, Y., 2020. Hydrogen sulfide: a multi-tasking signal molecule in the regulation of oxidative stress responses. J. Exp. Bot. 71 (10), 2862–2869.

Wang, R., 2012. Physiological implications of hydrogen sulfide: a whiff exploration that blossomed. Physiol. Rev. 92, 791–896.

Wilson, L.G., Bressan, R.A., Filner, P., 1978. Light-dependent emission of hydrogen sulphide from plants. Plant Physiol. 61, 184–189.

Xiong, Y., Contento, A.L., Nguyen, P.Q., Bassham, D.C., 2007. Degradation of oxidized proteins by autophagy during oxidative stress in *Arabidopsis*. Plant Physiol. 143, 291–299.

Yang, H., Mu, J., Chen, L., Feng, J., Hu, J., Li, L., et al., 2015. *S*-nitrosylation positively regulates ascorbate peroxidase activity during plant stress responses. Plant Physiol. 167, 1604–1615.

Zhang, H., Hu, L.-Y., Hu, K.-D., He, Y.-D., Wang, S.-H., Luo, J.-P., 2008. Hydrogen sulfide promotes wheat seed germination and alleviates oxidative damage against copper stress. J. Integr. Plant Biol. 50 (12), 1518–1529.

Zhang, T.Y., Li, F.C., Fan, C.M., Li, X., Zhang, F.F., He, J.M., 2017. Role and interrelationship of MEK1-MPK6 cascade, hydrogen peroxide and nitric oxide in darkness-induced stomatal closure. Plant Sci. 262, 190–199.

Zhang, J., Zhou, M., Ge, Z., Shen, J., Zhou, C., Gotor, C., et al., 2019. Abscisic acid-triggered guard cell L-cysteine desulfhydrase function and in situ hydrogen sulfide production contributes to heme oxygenase-modulated stomatal closure. Plant Cell Environ. 43 (3), 624–636.

Zhang, N.N., Zou, H., Lin, X.Y., Pan, Q., Zhang, W.Q., Zhang, J.H., et al., 2020. Hydrogen sulfide and rhizobia synergistically regulate nitrogen (N) assimilation and remobilization during N deficiency-induced senescence in soybean. Plant Cell Environ. 43 (5), 1130–1147.

Chapter 11

Regulation of salinity stress by hydrogen sulfide in plants

Owias Iqbal Dar[1], Kirpal Singh[1], Junaid Aslam[2], Sunil Sharma[1], Arvinder Kaur[1], Renu Bhardwaj[2] and Anket Sharma[3,]*

[1]Department of Zoology, Guru Nanak Dev University, Amritsar, India, [2]Department of Botanical and Environmental Sciences, Guru Nanak Dev University, Amritsar, India, [3]State Key Laboratory of Subtropical Silviculture, Zhejiang A&F University, Hangzhou, P.R. China
**Corresponding author.*

11.1 Introduction

Salt stress is one of the major abiotic stress globally that mainly influence 20% of irrigated land (Pitman and Läuchli, 2002; Qadir et al., 2014). Salinity stress is the accumulation of excessive salt content in soil. It occurs either naturally and/or by anthropogenic activities (Hafiz Che-Othman et al., 2017). In plants high salinity results in osmotic tension, oxidative stress, cytotoxicity, and ultimately nutritional disorders (Munns and Tester, 2008), which greatly influence plant growth, development, and photosynthesis and cause significant reductions in crop yield (Husen et al., 2016). High salinity also adversely affects the mechanism of germination (Ibrar et al., 2003; Ulfat et al., 2007; Carpıcı et al., 2009; Akbarimoghaddam et al., 2011; Xu et al., 2011; Khodarahmpour et al., 2012; Fernández-Torquemada and Sánchez-Lizaso, 2013).

To modify the adaptability of plants against the adverse effects resulting from various abiotic stresses, small biological molecules may be considered to play a role as powerful tools (Mostofa et al., 2015a). In numerous signaling events, biologically active gases have appeared as key regulator molecules and effectors. Gasotransmitters are gaseous signaling biologically active molecules (Moore et al., 2003; Mustafa et al., 2009). Hydrogen sulfide (H_2S) has been recognized as the third bioactive gasotransmitter signaling molecule after nitric oxide (NO) and carbon monoxide (CO), implicated in various health benefits in animal systems (Mancardi et al., 1787; Toit, 2015), but there have been relatively fewer studies on the role of H_2S in plant systems. In the past few years, H_2S is receiving considerable attention in research carried out with plants, while it was thought to be only an environmental hazards and body toxic agent for a long

Hydrogen Sulfide in Plant Biology. DOI: https://doi.org/10.1016/B978-0-323-85862-5.00001-4

time (Fotopoulos et al., 2013). From an environmental point of view H_2S can have negative as well as positive effects in plants. High level exposure of exogenous H_2S gas in plants can show phytotoxic effects like reduced oxygen production, nutrient uptake, and increased leaf damage (Thompson et al., 1979). It was Wilson et al. (1978) who observed that the plants could synthesize and release H_2S, this may be the earliest report of the study of H_2S in higher plants. H_2S has recently emanated in plant biology because of its adaptability toward harsh environmental conditions. In recent years, researchers have evolved the impact of H_2S on plant physiology by using its water-soluble donors, which showed significant tolerance to a number of abiotic stresses (Christou et al., 2013; Shi et al., 2013; Ali et al., 2015; Mostofa et al., 2015b). H_2S has the ability to enhance plant growth and vigor, it also reduces stress caused by biotic and abiotic factors and increase crop yields (Dooley et al., 2013). Moreover, plants exposed to H_2S have shown increased rates of photosynthesis (Chen et al., 2011), prolonged vase-life of cut flowers, guard cell signaling, formation of adventitious roots, germination, seedling growth, and shelf life of fruit (Zhang et al., 2009, 2011; García-Mata and Lamattina, 2010; Liu et al., 2011; Hu et al., 2012; Li et al., 2012, 2014a,b; Lin et al., 2012; Dooley et al., 2013; Gao et al., 2013; Hou et al., 2013; Fang et al., 2014a, b;,Zhu et al., 2014). Keeping in mind the salinity protective roles of H_2S, in the current review we will discuss how this gaseous molecule ameliorates salt stress in plants by regulating various plant processes.

11.2 Impact of salinity stress on plants

One of the hardest environmental conditions influencing the growth, development, and productivity of plants is salinity stress (Parihar et al., 2015). It adversely reduces growth, photosynthetic rate, and increases oxidative stress, thereby causing a decline in productivity.

11.2.1 Impact of salinity on growth

The growth in plants is greatly influenced by saline soil (Rakhmankulova et al., 2016; Fall et al., 2017). Salinity stress reduces area and thickness of leaves, causes abscissions in leavef, shoot necrosis, and reduces internode length (Parida and Das, 2005). Husen et al. (2016) observed that the length of root and shoot, length, area and number of leaves and basal diameter of stem in pea plants were significantly declined on exposure to salt stress. The percentage reductions in relative shoot height, shoot fresh mass, and dry matter mass were 20%, 17%, and 22%, respectively, and reductions in relative root elongation, root fresh, and dry masses were 39%, 32%, and 9% respectively in *Amaranthus cruentus* under salt stress (Gandonou et al., 2018). Zeeshan et al. (2020) noticed that exposure of salt stress in barley wheat plants caused significant wilting, necrosis, chlorosis, and inhibited growth, resulting in biomass reduction.

11.2.2 Impact of salinity on photosynthesis

The decline in photosynthetic pigments with salinity stress is directly correlated with photosynthetic efficiency (Kirschbaum, 2011). Salinity stress increases the production of reactive oxygen species (ROS) and causes oxidative damage, which decreases pigment production by damaging chloroplast membranes, photosystem II, and affects CO_2 fixation (Woodward and Bennett, 2005). The reduced photosynthetic rate was observed in wheat plants under salinity stress due to reduced chlorophyll a and carotenoid synthesis (Ahanger et al., 2019). Husen et al. (2016) determined that salt treatment to pea plants caused a significant reduction in photosynthetic pigments. Nxele et al. (2017) exposed sorghum plants to salinity stress and observed a significant decline in chlorophyll content. The photosynthetic rate, chlorophyll and carotenoid content, and stomatal conductance declined in tomato plants under salt stress (Ahanger et al., 2020). Yilmaz et al. (2020) observed that chlorophyll a, chlorophyll b, total chlorophyll, and carotenoid pigments were increased in sweet sorghum due to salinity.

11.2.3 Impact of salinity on oxidative stress

The salinity stress results in the increased generation of ROS, i.e., superoxide (O_2^-), hydroxyl radical (OH^-), hydrogen peroxide (H_2O_2), singlet oxygen, etc. (Pan et al., 2006). To combat against oxidative stress, plants produce osmolytes, nonenzymatic and enzymatic antioxidants to avoid cellular damage (Mittova et al., 2003). Nxele et al. (2017) observed a significant increase in H_2O_2 content and malonaldehyde (MDA) level in leaves of *Sorghum* after salinity treatment. Increased H_2O_2 content and MDA levels induced oxidative damage in soya beans under salt stress conditions (Sadak et al., 2020). Ahanger and Agarwal (2017) noted that salt treatment enhanced antioxidative enzyme activities such as superoxide dismutase (SOD), catalase (CAT), GR, guaiacol peroxidase (GPX), and ascorbate peroxidase (APX) at the preflowering stage in *Triticum aestivum*. Yilmaz et al. (2020) noted that salinity stress caused a higher accumulation of MDA in the leaves of sweet sorghum plants. They also noted that salt stress elevated the activity of CAT, APx, GR, glutathione-S-transferase (GST), and SOD. Table 11.1 provides an overview of plants' antioxidative defense responses under salt stress.

11.3 Role of H_2S in plant biology

Hydrogen sulfide (H_2S) is lipophilic and biological active gas that acts as an important signaling molecule in plants (Gotor et al., 2019). H_2S is endogenously produced in different subcellular compartments, that is, chloroplast, mitochondria, cytosol, etc., where enzymes associated with sulfur metabolism are present. It plays an important role in different physiological

TABLE 11.1 Effect of salt stress on antioxidative system.

Plant name	Impact on antioxidative system	References
Gossypium hirsutum L.	High levels of antioxidants (increased SOD, GPX and GR) and an active ascorbate–glutathione cycle	Gosset et al. (1994)
Morus alba L.	CAT, GPX, SOD, and GR were affected	Sudhakar et al. (2001)
Sorghum bicolor	SOD and CAT increased in leaves and roots	Costa et al. (2005)
Zea mays	SOD, APX, GPX, and GR increased	Azevedo Neto et al. (2006)
Triticum aestivum (KRL-19 and WH-542)	CAT, GR, SOD, APX, and POD increased	Mandhania et al. (2006)
Triticum aestivum	SOD, CAT, and POD increased	Athar et al. (2007)
Solanum tuberosum	APX and GR declined	Aghaei et al. (2009)
Zea mays L.	SOD and GPX declined	Carrasco-Ríos and Pinto (2014)
Triticum aestivum (HD 2329 and Kharchia-65)	SOD declined and POD, APX, CAT, and GR increased	Singh et al. (2015)
Triticum monococcum	SOD and POD increased	Lv et al. (2016)
Triticum aestivum (Jimai 22)	SOD, POD, CAT, GR, APX, and DHAR increased	Zou et al. (2016)

functions related to the growth and development of plants, i.e., seed germination, root development, organogenesis, movement of stomata, photosynthesis, etc. (Lisjak et al., 2013; Zhang et al., 2015). Chen et al. (2011) observed that H_2S elevates chloroplast synthesis and expression of photosynthetic enzymes which in turn increases photosynthetic efficiency. In *Dendrobium officinale* H_2S was found to enhance photosynthesis by stimulating the activity of antioxidant enzymes (Fan et al., 2014). Abscission is a critical physiological process involving shedding of unwanted organs from the main body of the plants. Liu et al. (2020) found that abscission of the tomato petiole is inhibited by the H_2S treatment in a dose-dependent manner. H_2S increases the absorption of nutrients and reduces metal uptake in plants and also elevates the antioxidant enzyme activities to prevent the damage caused by oxidative stress (Ali et al., 2013; Chen et al., 2013; Rizwan et al., 2019; Ahmad et al., 2020). H_2S inhibits autophagy, induces adventitious roots, stimulates chlorophyll biogenesis, and also plays a role in ethylene-induced fruit ripening (Zhang et al., 2009; Álvarez et al., 2012; Lin et al., 2012; Ziogas et al., 2018; Rizwan et al., 2019).

11.4 Role of H_2S in plants under salinity stress

H_2S has been recognized as a defender of different abiotic stresses like chilling, drought, heat, and salinity (Arnao and Hernández-Ruiz, 2015) by modulating their signaling pathways (Shi et al., 2015). It plays a significant role in the regulation of homeostasis in plant cells (Papanatsiou et al., 2015). H_2S protects the plants from different abiotic stresses by crosstalk with other signaling molecules and cross tolerance (Niu and Liao, 2016). The accumulation of osmolytes, nutrient balance, and induction of antioxidant enzymes are common H_2S-induced functions during stress conditions (Mostofa et al., 2015 a). H_2S interactions with other signaling molecules (CO, NO, Ca^{2+}, H_2O_2) and plant hormones (auxin, gibberellins, ethylene, abscisic acid) play key roles in regulating stress, growth, and development in plants (Pandey and Gautam, 2020). Kaya et al. (2018) observed that during salt stress H_2S treatment results in decreased H_2O_2 and MDA levels showing crop protection from oxidative damage. H_2S elevated expression of stress-related genes and improved stress tolerance in *Arabidopsis* (Shen et al., 2013). The inhibitory effects of NaCl stress, such as reduced seed germination and improper plant growth and development in *Medicago sativa*, are significantly alleviated by NaHS treatment (Wang et al., 2012). The electrolyte leakage and ionic imbalance caused by NaCl stress are properly maintained with NaHS treatment by elevating the expression and phosphorylation of different transporters of the plasma membrane (Li et al., 2014).

11.4.1 Impact on photosynthesis

Photosynthesis, one of the key metabolic processes in plants, is greatly affected by environmental stress (Mehta et al., 2011). During salt stress, NaCl accumulates in chloroplasts and chlorophyll, which reduces the photosynthetic electron transport chain (Boyer, 1976; Zhang et al., 2005) and reduces the level of photosynthesis (Downton, 1977; Hasegawa et al., 2000; Munns, 2002; Ashraf and Shahbaz, 2003; Kao et al., 2003; Sayed, 2003). Moreover, salt stress lowers chlorophyll content in various plants (Lapina and Popov, 1970; Seemann and Critchley, 1985; Abdullah and Ahmed, 1990; Hamada and El-Enany, 1994; Lòpez-Climent et al., 2008; Saha et al., 2010; Amirjani, 2011). Therefore a reduction in the content of chlorophyll has been used as an early indicator of the cellular metabolic state under salt stress (Chutipaijit et al., 2011).

H_2S enhanced photosynthesis by elevating the biogenesis of chloroplasts, thus increasing the photosynthetic rate by modulating the expression of genes involved in photosynthesis (Chen et al., 2011). Chen et al. (2015) observed that H_2S improves photosynthesis by enhancing the expression of ribulose-1,5-bisphosphate carboxylase (RuBISCO) and phosphoenolpyruvate carboxylase (PEPC) in maize seedlings. In *Oryza sativa* L. the photosynthetic rate, stomatal aperture, and the leaf photosynthetic area increased upon treatment of H_2S (Duan et al., 2014). The content of chlorophyll a, chlorophyll b, and total chlorophyll showed the significant increase in *Oryza sativa* L. when treated with NaHS (Mostofa et al., 2015a). In addition, the exposure of H_2S to cucumber showed improvement in photosynthetic parameters in one of the studies of Jiang et al. (2019). They observed that H_2S inhibits the depletion of photosynthetic pigments under saline stress.

11.4.2 Impact on oxidative stress

Oxidative stress is one of the major causes of plant cell damage during environmental stress. ROS mainly comprise superoxide anions ($O_2^{\bullet -}$), hydroxyl radicals (•OH), perhydroxyl radicals (•O_2H), and hydrogen peroxide (H_2O_2) (Apel and Hirt, 2004; Foyer and Noctor, 2005). Excessive salinity in plant cells induces different kinds of harmful effects, such as ionic stress, osmotic stress, and oxidative stress (Munns, 2005; Gupta and Huang, 2014). Increased levels of ROS in plant cells cause immense damage to proteins, DNA, and lipids, disrupting redox homeostasis in cells (Ceron-Garcia et al., 2012). To measure the damage caused due to oxidative stress, levels of H_2O_2 and MDA (lipid peroxidation product) are recognized as early biomarkers (Miller et al., 2010). Salt stress to plants results in the accumulation of ROS and MDA associated with oxidative stress and lipid peroxidation (Hossain et al., 2005).

Under salinity stress plants produce increased levels of a highly toxic compound called methylglyoxal, an α-ketoaldehyde that reacts with DNA, RNA, and proteins, and as such increases oxidative stress (Yadav et al., 2005; Saito et al., 2011; Hossain et al., 2012). The exposure of NaHS to rice reduced methylglyoxylate levels by increasing glyoxalase I and glyoxalase II activities (Mostofa et al., 2015a). They also noticed that H_2S regulates ROS homeostasis, thereby protecting plants from salt-induced oxidative stress. Ding et al. (2019) noted that exposure of H_2S to wheat plants improved antioxidant defense and declined MDA and H_2O_2 content under salinity. Exposure of exogenous H_2S reduced the effect of NaCl-induced oxidative stress in alfalfa, Bermuda grass, strawberry, and cucumber (Wang et al., 2012; Christou et al., 2013; Shi et al., 2013; Lai et al., 2014; Sun and Luo, 2014). H_2S also declined the salt-induced oxidative damage by decreasing the contents of O_2^-, H_2O_2, MDA, and electrolyte leakage in cucumber (Jiang et al., 2019), rice (Mostofa et al., 2015a), and spinach (Chen et al., 2016).

11.4.3 Impact on antioxidative system

Redox homeostasis during stress is maintained by antioxidant machinery to employ an efficient ROS scavenging system such as antioxidative/

detoxification enzymes, i.e., SOD, GPX, APX, GST, and CAT, and non-enzymatic low-molecular compounds like ascorbic acid, reduced glutathione (GSH), α-tocopherol, carotenoids, phenolics, flavonoids, and proline (Gill and Tuteja, 2010; Miller et al., 2010; Gill et al., 2011). H_2S regulates stress tolerance by activating these antioxidant defense mechanisms (Corpas et al., 2019). H_2S has been observed to be helpful in salt stress reduction by gearing up antioxidant activities and enhancing the ROS detoxification system (Chen et al., 2011; Mostofa et al., 2015b). Christou et al. (2013) noted that strawberry plants treated with NaHS under salt stress showed greater redox state and enhanced antioxidant level. Wang et al. (2012) observed that NaHS declined the oxidative damage induced by NaCl by activating antioxidant enzymes such as SOD, CAT, POD, and APX in *Medicago sativa* L. Jiang et al. (2019) found that H_2S counteracted the decline in CAT, GR, and POD activities and the content of ASA and GSH under salt stress in cucumber. Studies of Wang et al. (2012), Shi et al. (2013), Shan et al. (2014), and Mostofa et al. (2015a) observed that H_2S reduces the content of ROS and cell damage by adjusting antioxidant enzymes and regulating the redox states of AsA and GSH under salinity stress.

11.5 Role of H_2S in plant productivity under salinity stress

In recent years, detrimental abiotic stresses are becoming more common and have adverse impacts on crop productivity that poses major threats to global food security (Wang and Frei, 2011). These adverse environmental conditions suggests the occurrence of a greater loss of yield globally (Krasensky and Jonak, 2012). Salt stress can significantly influence plant growth by reducing the Pn and Tr of the plant canopy, thereby decreasing biomass production (Chen et al., 2016). Exposure of seeds or roots to traces of dissolved H_2S increased adventitious root formation, seed germination, seedling growth, biomass accumulation, and yields (Zhang et al., 2009; Dooley et al., 2013). In another study H_2S increased the seedling biomass, soluble protein, and chlorophyll content in spinach plants (Chen et al., 2011). Also, when treated with H_2S salt-stressed rice plants showed improved overall growth and biomass of the plant (Mostofa et al., 2015a).

11.6 Conclusion

Hydrogen sulfide is an efficient way to protect crop plants from salt stress. This gaseous molecule stimulates the amelioration of oxidative stress by triggering the antioxidative system (Fig. 11.1). However, more in-depth molecular studies are needed to open the way for the introduction of new salt-resistant crop varieties with H_2S-responsive genes.

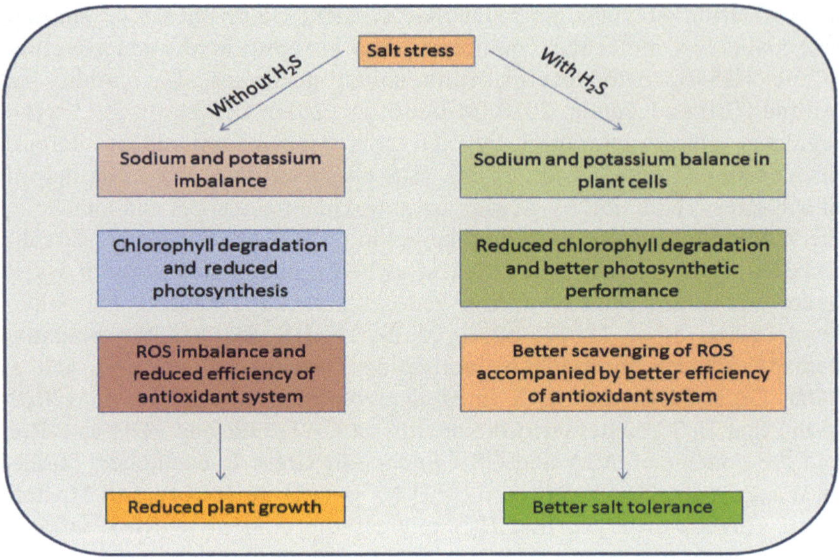

FIGURE 11.1 Overview explaining role of H_2S in ameliorating salt stress in plants.

References

Abdullah, Z., Ahmed, R., 1990. Effect of pre and post kinetin treatment on salt tolerance of different potato cultivars growing on saline soils. J. Agron. Crop. Sci. 165, 94–102.

Aghaei, K., Ehsanpour, A.A., Komatsu, S., 2009. Potato responds to salt stress by increased activity of antioxidant enzymes. J. Integr. Plant Biol. 51, 1095–1103.

Ahanger, M.A., Agarwal, R.M., 2017. Salinity stress induced alterations in antioxidant metabolism and nitrogen assimilation in wheat (*Triticum aestivum* L) as influenced by potassium supplementation. Plant Physiol. Biochem. 115, 449–460.

Ahanger, M.A., Qin, C., Begum, N., Maodong, Q., Dong, X.X., El-Esawi, M., et al., 2019. Nitrogen availability prevents oxidative effects of salinity on wheat growth and photosynthesis by up-regulating the antioxidants and osmolytes metabolism, and secondary metabolite accumulation. BMC Plant Biol. 19, 1–12.

Ahanger, M.A., Mir, R.A., Alyemeni, M.N., Ahmad, P., 2020. Combined effects of brassinosteroid and kinetin mitigates salinity stress in tomato through the modulation of antioxidant and osmolyte metabolism. Plant Physiol. Biochem. 147, 31–42.

Ahmad, R., Ali, S., Rizwan, M., Dawood, M., Farid, M., Hussain, A., et al., 2020. Hydrogen sulfide alleviates chromium stress on cauliflower by restricting its uptake and enhancing antioxidative system. Physiol. Plant 168, 289–300.

Akbarimoghaddam, H., Galavi, M., Ghanbari, A., Panjehkeh, N., 2011. Salinity effects on seed germination and seedling growth of bread wheat cultivars. Trakia J. Sci. 9, 43–50.

Ali, B., Qian, P., Sun, R., Farooq, M.A., Gill, R.A., Wang, J., et al., 2015. Hydrogen sulfide alleviates the aluminum-induced changes in *Brassica napus* as revealed by physiochemical and ultrastructural study of plant. Environ. Sci. Pollut. Res. 22, 3068–3081.

Ali, S., Farooq, M.A., Hussain, S., Yasmeen, T., Abbasi, G.H., Zhang, G., 2013. Alleviation of chromium toxicity by hydrogen sulfide in barley. Environ. Toxicol. Chem. 32, 2234–2239.

Álvarez, C., Ángeles Bermúdez, M., Romero, L.C., Gotor, C., García, I., 2012. Cysteine homeostasis plays an essential role in plant immunity. N. Phytol. 193, 165–177.

Amirjani, M.R., 2011. Effect of salinity stress on growth, sugar content, pigments and enzyme activity of rice. Int. J. Bot. 7, 73–81.

Apel, K., Hirt, H., 2004. Reactive oxygen species: metabolism, oxidative stress, and signal transduction. Annu. Rev. Plant Biol. 55, 373–399.

Arnao, M.B., Hernández-Ruiz, J., 2015. Functions of melatonin in plants: a review. J. Pineal Res. 59, 133–150.

Ashraf, M., Shahbaz, M., 2003. Assessment of genotypic variation in salt tolerance of early CIMMYT hexaploid wheat germplasm using photosynthetic capacity and water relations as selection criteria. Photosynthetica 41, 273–280.

Athar, H.U.R., Khan, A., Ashraf, M., 2007. Exogenously applied ascorbic acid alleviates salt-induced oxidative stress in wheat. Environ. Exp. Bot. 63, 224–231.

Azevedo Neto, A.D., Prisco, J.T., Enéas-Filho, J., Abreu, C.E.B., Gomes-Filho, E., 2006. Effect of salt stress on antioxidative enzymes and lipid peroxidation in leaves and roots of salt-tolerant and salt-sensitive maize genotypes. Environ. Exp. Bot. 56, 87–94.

Boyer, J.S., 1976. Water deficits and photosynthesis. In: Kozlowsky, T.T. (Ed.), Water Deficit and Plant Growth, vol. 4. Academic Press, New York, San Francisco, London, pp. 153–190.

Carpıcı, E.B., Celık, N., Bayram, G., 2009. Effects of salt stress on germination of some maize (*Zea mays* L.) cultivars. Afr. J. Biotechnol. 8, 4918–4922.

Carrasco-Ríos, L., Pinto, M., 2014. Effect of salt stress on antioxidant enzymes and lipid peroxidation in leaves in two contrasting corn, 'Lluteno' and 'Jubilee'. Chilean J. Agric. Res. 74, 89–95.

Ceron-Garcia, A., Vargas-Arispuro, I., Aispuro-Hernandez, E., Angel, M., Oligoglucan elicitor effects during plant oxidative stress. Cell Metabolism—Cell Homeostasis and Stress Response, 2012, IntechOpen Limited, London, UK, 1–12. https://doi.org/10.5772/26057.

Chen, J., Wu, F.-H., Wang, W.-H., Zheng, C.-J., Lin, G.-H., Dong, X.-J., et al., 2011. Hydrogen sulphide enhances photosynthesis through promoting chloroplast biogenesis, photosynthetic enzyme expression, and thiol redox modification in *Spinacia oleracea* seedlings. J. Exp. Bot. 62, 4481–4493.

Chen, J., Shang, Y.T., Wang, W.H., Chen, X.Y., He, E.M., Zheng, H.L., et al., 2016. Hydrogen sulfide-mediated polyamines and sugar changes are involved in hydrogen sulfide-induced drought tolerance in *Spinacia oleracea* seedlings. Front. Plant Sci. 7, 1173.

Chen, J., Wang, W.H., Wu, F.H., You, C.Y., Liu, T.W., Dong, X.J., et al., 2013. Hydrogen sulfide alleviates aluminum toxicity in barley seedlings. Plant Soil 362, 301–318.

Chen, J., Wu, F.H., Shang, Y.T., Wang, W.H., Hu, W.J., Simon, M., et al., 2015. Hydrogen sulphide improves adaptation of *Zea mays* seedlings to iron deficiency. J. Exp. Bot. 66, 6605–6622.

Christou, A., Manganaris, G.A., Papadopoulos, I., Fotopoulos, V., 2013. Hydrogen sulfide induces systemic tolerance to salinity and non-ionic osmotic stress in strawberry plants through modification of reactive species biosynthesis and transcriptional regulation of multiple defence pathways. J. Exp. Bot. 64, 1953–1966.

Chutipaijit, S., Cha-um, S., Sompornpailin, K., 2011. High contents of proline and anthocyanin increase protective response to salinity in *Oryza sativa* L. spp. indica. Aust. J. Crop. Sci. 5, 1191–1198.

Corpas, F.J., González-Gordo, S., Cañas, A., Palma, J.M., 2019. Nitric oxide and hydrogen sulfide in plants: which comes first? J. Exp. Bot. 70, 4391–4404.

Costa, P.H.A., Azevedo, Neto, A.D., Bezerra, M.A., Prisco, J.T., Gomes-Filho, E., 2005. Antioxidant-enzymatic system of two sorghum genotypes differing in salt tolerance. Braz. J. Plant Physiol. 17, 353–361.

Ding, H., Ma, D., Huang, X., Hou, J., Wang, C., Xie, Y., et al., 2019. Exogenous hydrogen sulfde alleviates salt stress by improving antioxidant defenses and the salt overly sensitive pathway in wheat seedlings. Acta. Physiol. Plant 41, 123.

Dooley, F.D., Nair, S.P., Ward, P.D., 2013. Increased growth and germination success in plants following hydrogen sulfide administration. PLoS One 8, e62048.

Downton, W.J.S., 1977. Photosynthesis in salt-stressed grapevines. Aust. J. Plant Physiol. 4, 183–192.

Duan, B., Ma, Y., Jiang, M., Yang, F., Ni, L., Lu, W., 2014. Improvement of photosynthesis in rice (*Oryza sativa* L.) as a result of an increase in stomatal aperture and density by exogenous hydrogen sulfide treatment. Plant Growth Regul. 75, 33–44.

Fall, F., Diouf, D., Fall, D., Bakhoum, N., Thioye, B., Kane, A., et al., 2017. Growth and physiological responses of *Sporobolus robustus kunth* seedlings to salt stress. Arid Land Res. Manag. 31, 46–56.

Fan, H., Guan, L., Li, T., Wu, Q., Wu, M., Cai, Y., et al., 2014. Hydrogen sulphide alleviates oxidative damage and enhances light energy transformation under high light for *Dendrobium officinale*. Sci. Hortic. 177, 47–52.

Fang, T., Cao, Z., Li, J., Shen, W., Huang, L., 2014a. Auxininduced hydrogen sulfide generation is involved in lateral root formation in tomato. Plant Physiol. Biochem. 76, 44–51.

Fang, T., Li, J., Cao, Z., Chen, M., Shen, W., Huang, L., 2014b. Heme oxygenase-1 is involved in sodium hydrosulfide-induced lateral root formation in tomato seedlings. Plant Cell Rep. 33, 969–978.

Fernández-Torquemada, Y., Sánchez-Lizaso, J.L., 2013. Effects of salinity on seed germination and early seedling growth of the Mediterranean seagrass *Posidonia oceanica* (L.) Delile. Estuarine, Coast Shelf Sci. 119, 64–70.

Fotopoulos, V., Christou, A., Manganaris, G., 2013. Hydrogen sulfide as a potent regulator of plant responses to abiotic stress factors. In: Gaur, R.K., Sharma, P. (Eds.), Molecular Approaches in Plant Abiotic Stress. CRC Press, Boca Raton, FL, pp. 353–373.

Foyer, C.H., Noctor, G., 2005. Redox homeostasis and antioxidant signaling: a metabolic interface between stress perception and physiological responses. Plant Cell 17, 1866–1875.

Gandonou, C.B., Prodjinoto, H., Zanklan, S.A., Wouyou, A.D., Lutts, S., Montcho, D.H., et al., 2018. Effects of salinity stress on growth in relation to gas exchanges parameters and water status in amaranth (*Amaranthus cruentus*). Int. J. Plant Physiol. Biochem. 10, 19–27.

Gao, S.P., Hu, K.D., Hu, L.Y., Li, Y.H., Han, Y., Wang, H.L., et al., 2013. Hydrogen sulfide delays postharvest senescence and plays an antioxidative role in freshcut kiwifruit. HortScience 48, 1385–1392.

García-Mata, C., Lamattina, L., 2010. Hydrogen sulphide, a novel gasotransmitter involved in guard cell signalling. N. Phytol. 188, 977–984.

Gill, S.S., Khan, N.A., Anjum, N.A., Tuteja, N., 2011. Amelioration of cadmium stress in crop plants by nutrients management: morphological, physiological and biochemical aspects. In: Anjum, N.A., Lopez-Lauri, F. (Eds.), Plant Nutrition and Abiotic Stress Tolerance III, Plant Stress 5 (Special Issue 1). Global Science Books Ltd., UK, Ikenobe, pp. 1–23.

Gill, S.S., Tuteja, N., 2010. Reactive oxygen species and antioxidant machinery in abiotic stress tolerance in crop plants. Plant Physiol. Biochem. 48, 909–930.

Gosset, D.R., Millhollon, E.P., Lucas, M.C., 1994. Antioxidant response to NaCl stress in salt-tolerant and salt-sensitive cultivars of cotton. Crop. Sci. 34, 706–714.

Gotor, C., García, I., Aroca, Á., Laureano-Marín, A.M., Arenas-Alfonseca, L., Jurado-Flores, A., et al., 2019. Signaling by hydrogen sulfide and cyanide through post-translational modification. J. Exp. Bot. 70, 4251–4265.

Gupta, B., Huang, B., 2014. Mechanism of salinity tolerance in plants: physiological, biochemical, and molecular characterization. Int. J. Genomics 2014, 1–18.

Hafiz Che-Othman, M., Harvey, Millar, A., Taylor, N.L., 2017. Connecting salt stress signaling pathways with salinity induced changes in mitochondrial metabolic processes in C3 plants. Plant, Cell Environ. 40, 2875–2905.

Hamada, A.M., El-Enany, A.E., 1994. Effect of NaCl salinity on growth, pigment and mineral element contents, and gas exchange of broad bean and pea plants. Biol. Plant 36, 75–81.

Hasegawa, P.M., Bressan, R.A., Zhu, J.K., Bohnert, H.J., 2000. Plant cellular and molecular responses to high salinity. Annu. Rev. Plant Biol. 51, 463–499.

Hossain, M., Naher, F., Shahabuddin, Q., 2005. Food security and nutrition in Bangladesh: progress and determinants. Electron. J. Agric. Dev. Econ. 2, 103–132.

Hossain, M.A., Piyatida, P., Teixeira da Silva, J.A., Fujita, M., 2012. Molecular mechanism of heavy metal toxicity and tolerance in plants: central role of glutathione in detoxification of reactive oxygen species and methylglyoxal and in heavy metal chelation. J. Bot. 2012, 1–37.

Hou, Z., Wang, L., Liu, J., Hou, L., Liu, X., 2013. Hydrogen sulfide regulates ethylene-induced stomatal closure in *Arabidopsis thaliana*. J. Integr. Plant Biol. 55, 277–289.

Hu, L.Y., Hu, S.L., Wu, J., Li, Y.H., Zheng, J.L., Wei, Z.J., et al., 2012. Hydrogen sulfide prolongs postharvest shelf-life of strawberry and plays an antioxidative role in fruits. J. Agric. Food Chem. 60, 8684–8693.

Husen, A., Iqbal, M., Aref, I.M., 2016. IAA-induced alteration in growth and photosynthesis of pea (L.) plants grown under salt stress *Pisum sativum*. J. Environ. Biol. 37, 421–429.

Ibrar, M., Jabeen, M., Tabassum, J., Hussain, F., Ilahi, I., 2003. Salt tolerance potential of *Brassica juncea* Linn. J. Sci. Tech. Univ. Peshawar 27, 79–84.

Jiang, J.L., Tian, Y., Li, L., Yu, M., Hou, R.P., Ren, X.M., 2019. H_2S alleviates salinity stress in cucumber by maintaining the Na^+/K^+ balance and regulating H_2S metabolism and oxidative stress response. Front. Plant Sci. 10, 678.

Kao, W.Y., Tsai, T.T., Shih, C.N., 2003. Photosynthetic gas exchange and chlorophyll a fluorescence of three wild soybean species in response to NaCl treatments. Photosynthetica 41, 415–419.

Kaya, C., Ashraf, M., Akram, N.A., 2018. Hydrogen sulfide regulates the levels of key metabolites and antioxidant defense system to counteract oxidative stress in pepper (*Capsicum annuum* L.) plants exposed to high zinc regime. Environ. Sci. Pollut. Res. 25, 12612–12618.

Khodarahmpour, Z., Ifar, M., Motamedi, M., 2012. Effects of NaCl salinity on maize (*Zea mays* L.) at germination and early seedling stage. Afr. J. Biotechnol. 11, 298–304.

Kirschbaum, M.U., 2011. Does enhanced photosynthesis enhance growth? Lessons learned from CO_2 enrichment studies. Plant Physiol. 155, 117–124.

Krasensky, J., Jonak, C., 2012. Drought, salt, and temperature stress-induced metabolic rearrangements and regulatory networks. J. Exp. Bot. 63, 1593–1608.

Lai, D., Mao, Y., Zhou, H., Li, F., Wu, M., Zhang, J., et al., 2014. Endogenous hydrogen sulfide enhances salt tolerance by coupling the reestablishment of redox homeostasis and preventing salt-induced K^+ loss in seedlings of *Medicago sativa*. Plant Sci. 225, 117–129.

Lapina, L.P., Popov, B.A., 1970. Effect of sodium chloride on photosynthetic apparatus of tomatoes. Fiziol. Rast. 17, 580–584.

Li, J., Jia, H., Wang, J., Cao, Q., Wen, Z., 2014. Hydrogen sulfide is involved in maintaining ion homeostasis via regulating plasma membrane Na^+/H^+ antiporter system in the hydrogen peroxide-dependent manner in salt-stress *Arabidopsis thaliana* root. Protoplasma 251, 899–912.

Li, S.P., Hu, K.D., Hu, L., Li, Y., Jiang, Y.H., Xiao, A.M., et al., 2014a. Hydrogen sulfide alleviates postharvest senescence of broccoli by modulating antioxidant defense and senescence-related gene expression. J. Agric. Food Chem. 62, 1119–1129.

Li, Y.J., Chen, J., Xian, M., Zhou, L.G., Han, F.X., Gan, L.J., et al., 2014b. Bioimaging of hydrogen sulfide uncovers its pivotal role in regulating nitric oxide-induced lateral root formation. PLoS One 9, e90340.

Li, Z.G., Gong, M., Liu, P., 2012. Hydrogen sulfide is a mediator in H_2O_2-induced seed germination in *Jatropha curcas*. Acta Physiol. Plant 34, 2207–2213.

Lin, Y.T., Li, M.Y., Cui, W.T., Lu, W., Shen, W.B., 2012. Haem oxygenase-1 is involved in hydrogen sulfide-induced cucumber adventitious root formation. J. Plant Growth Regul. 31, 519–528.

Lisjak, M., Teklic, T., Wilson, I.D., Whiteman, M., Hancock, J.T., 2013. Hydrogen sulfide: environmental factor or signalling molecule? Plant, Cell Environ. 36, 1607–1616.

Liu, D., Li, J., Li, Z., Pei, Y., 2020. Hydrogen sulfide inhibits ethylene-induced petiole abscission in tomato (*Solanum lycopersicum* L.). Hortic. Res. 7, 1–11.

Liu, J., Hou, L., Liu, G., Liu, X., Wang, X., 2011. Hydrogen sulfide induced by nitric oxide mediates ethylene-induced stomatal closure of *Arabidopsis thaliana*. Chin. Sci. Bull. 56, 3547–3553.

Lòpez-Climent, M.F., Arbona, V., Pérez-Clemente, R.M., Gòmez-Cadenas, A., 2008. Relationship between salt tolerance and photosynthetic machinery performance in citrus. Environ. Exp. Bot. 62, 176–184.

Lv, D.W., Zhu, G.R., Zhu, D., Bian, Y.W., Liang, X.N., Cheng, Z.W., et al., 2016. Proteomic and phosphoproteomic analysis reveals the response and defense mechanism in leaves of diploid wheat *Traticum monococcum* under salt stress and recovery. J. Proteom. 143, 93–105.

Mancardi, D., Penna, C., Merlino, A., Del Soldato, P., Wink, D.A., Pagliaro, P., 2009. Physiological and pharmacological features of the novel gasotransmitter: hydrogen sulfide. Biochim. Biophys. Acta 1787, 864–872.

Mandhania, S., Madan, S., Sawhney, V., 2006. Antioxidant defense mechanism under salt stress in wheat seedlings. Biol. Plant 50, 227–231.

Mehta, P., Kraslavsky, V., Bharti, S., Allakhverdiev, S.I., Jajoo, A., 2011. Analysis of salt-stress induced changes in photosystem II heterogeneity by prompt fluorescence and delayed fluorescence in wheat (*Triticum vulgare*) leaves. J. Photochem. Photobiol. B 104, 308–313.

Miller, G., Suzuki, N., Ciftci-Yilmaz, S., Mittler, R., 2010. Reactive oxygen species homeostasis and signalling during drought and salinity stresses. Plant Cell Environ. 33, 453–467.

Mittova, V., Tal, M., Volokita, M., Guy, M., 2003. Up-regulation of the leaf mitochondrial and peroxisomal antioxidative systems in response to salt-induced oxidative stress in the wild salt-tolerant tomato species *Lycopersicon pennellii*. Plant Cell Environ. 26, 845–856.

Moore, P.K., Bhatia, M., Moochhala, S., 2003. Hydrogen sulfide: from the smell of the past to the mediator of the future? Trends Pharmacol. Sci. 24, 609–611.

Mostofa, M.G., Rahman, A., Ansary, M.M.U., Watanabe, A., Fujita, M., Tran, L.S., 2015b. Hydrogen sulfide modulates cadmium-induced physiological and biochemical responses to alleviate cadmium toxicity in rice. Sci. Rep. 5, 14078.

Mostofa, M.G., Saegusa, D., Fujita, M., Tran, L.S., 2015a. Hydrogen sulfide regulates salt tolerance in rice by maintaining $Na+/K+$ balance, mineral homeostasis and oxidative metabolism under excessive salt stress. Front. Plant Sci. 6, 1055.

Munns, R., 2002. Comparative physiology of salt and water stress. Plant Cell Environ. 25, 239–250.

Munns, R., 2005. Genes and salt tolerance: bringing them together. N. Phytol. 167, 645–663.

Munns, R., Tester, M., 2008. Mechanisms of salinity tolerance. Annu. Rev. Plant Biol. 59, 651–681.

Mustafa, A.K., Gadalla, M.M., Snyder, S.H., 2009. Signaling by gasotransmitters. Sci. Signal. 2, re2.

Niu, L., Liao, W., 2016. Hydrogen peroxide signaling in plant development and abiotic responses: crosstalk with nitric oxide and calcium. Front. Plant Sci. 7, 230.

Nxele, X., Klein, A., Ndimba, B.K., 2017. Drought and salinity stress alters ROS accumulation, water retention, and osmolyte content in sorghum plants. S. Afr. J. Bot. 108, 261–266.

Pan, X., Geng, Y., Zhang, W., Li, B., Chen, J., 2006. The influence of abiotic stress and phenotypic plasticity on the distribution of invasive *Alternanthera philoxeroides* along a riparian zone. Acta Oecol. 30, 333–341.

Pandey, A.K., Gautam, A., 2020. Stress responsive gene regulation in relation to hydrogen sulfide in plants under abiotic stress. Physiol. Plant 168, 511–525.

Papanatsiou, M., Scuffi, D., Blatt, M.R., García-Mata, C., 2015. Hydrogen sulfide regulates inward-rectifying K + channels in conjunction with stomatal closure. Plant Physiol. 168, 29–35.

Parida, A.K., Das, A.B., 2005. Salt tolerance and salinity effects on plants: a review. Ecotoxicol. Environ. Saf. 60, 324–349.

Parihar, P., Singh, S., Singh, R., Singh, V.P., Prasad, S.M., 2015. Effect of salinity stress on plants and its tolerance strategies: a review. Environ. Sci. Pollut. Res. 22, 4056–4075.

Pitman, M.G., Läuchli, A., 2002. Global impact of salinity and agricultural ecosystems. *Salinity: Environment-Plants-Molecules*. Springer, Dordrecht, pp. 3–20.

Qadir, M., Quillérou, E., Nangia, V., Murtaza, G., Singh, M., Thomas, R.J., et al., 2014. Economics of salt-induced land degradation and restoration. Nat. Resour. Forum 38, 282–295.

Rakhmankulova, Z.F., Shuyskaya, E.V., Suyundukov, Y.T., Usmanov, I.Y., Voronin, P.Y., 2016. Different responses of two ecotypes of C 3–C 4 xero-halophyte *Bassia sedoides* to osmotic and ionic factors of salt stress. Russ. J. Plant Physiol. 63, 349–357.

Rizwan, M., Mostofa, M.G., Ahmad, M.Z., Zhou, Y., Adeel, M., Mehmood, S., et al., 2019. Hydrogen sulfide enhances rice tolerance to nickel through the prevention of chloroplast damage and the improvement of nitrogen metabolism under excessive nickel. Plant Physiol. Biochem. 138, 100–111.

Sadak, M.S., El-Hameid, A.R.A., Zaki, F.S., Dawood, M.G., El-Awadi, M.E., 2020. Physiological and biochemical responses of soybean (*Glycine max* L.) to cysteine application under sea salt stress. Bull. Natl. Res. Cent. 44, 1–10.

Saha, P., Chatterjee, P., Biswas, A.K., 2010. NaCl pretreatment alleviates salt stress by enhancement of antioxidant defense system and osmolyte accumulation in mungbean (*Vigna radiata* L.Wilczek). Indian J. Exp. Biol. 48, 593–600.

Saito, R., Yamamoto, H., Makino, A., Sugimoto, T., Miyake, C., 2011. Methylglyoxal functions as Hill oxidant and stimulates the photoreduction of O_2 at photosystem I: a symptom of plant diabetes. Plant Cell Environ. 34, 1454–1464.

Sayed, O.H., 2003. Chlorophyll fluorescence as a tool in cereal crop research. Photosynthetica 41, 321–330.

Seemann, J.R., Critchley, C., 1985. Effects of salt stress on the growth, ion content, stomatal behaviour and photosynthetic capacity of a salt-sensitive species, *Phaseolus vulgaris* L. Planta 164, 151–162.

Shan, C., Liu, H., Zhao, L., Wang, X., 2014. Effects of exogenous hydrogen sulfide on the redox states of ascorbate and glutathione in maize leaves under salt stress. Biol. Plant 58, 169–173.

Shen, J., Xing, T., Yuan, H., Liu, Z., Jin, Z., Zhang, L., et al., 2013. Hydrogen sulfide improves drought tolerance in *Arabidopsis thaliana* by microRNA expressions. PLoS One 8, e77047.

Shi, H., Ye, T., Chan, Z., 2013. Exogenous application of hydrogen sulfide donor sodium hydrosulfide enhanced multiple abiotic stress tolerance in bermudagrass (*Cynodon dactylon* (L). Pers. Plant Physiol. Biochem. 71, 226–234.

Shi, H., Ye, T., Han, N., Bian, H., Liu, X., Chan, Z., 2015. Hydrogen sulfide regulates abiotic stress tolerance and biotic stress resistance in *Arabidopsis*. J. Integr. Plant Biol. 57, 628–640.

Singh, A., Bhushan, B., Gaikwad, K., Yadav, O.P., Kumar, S., Rai, R.D., 2015. Induced defence responses of contrasting bread wheat genotypes under differential salt stress imposition. Indian. J. Biochem. Biophys. 52, 75–85.

Sudhakar, C., Lakshmi, A., Giridarakumar, S., 2001. Changes in the antioxidant enzyme efficacy in two high yielding genotypes of mulberry (*Morus alba* L.) under NaCl salinity. Plant Sci. 161, 613–619.

Sun, Y.D., Luo, W.R., 2014. Effects of exogenous hydrogen sulphide on seed germination and seedling growth of cucumber (*Cucumis sativus*) under sodium bicarbonate stress. Seed Sci. Technol. 42, 126–131.

Thompson, C.R., Kats, G., Lennox, R.W., 1979. Effects of fumigating crops with hydrogen sulfide or sulphur dioxide. Calif. Agr. 33, 9–10.

Toit, A.D., 2015. Metabolism: the health benefits of hydrogen sulphide. Nat. Rev. Mol. Cell Biol. 16, 68.

Ulfat, M., Athar, H., Ashraf, M., Akram, N.A., Jamil, A., 2007. Appraisal of physiological and biochemical selection criteria for evaluation of salt tolerance in canola (*Brassica napus* L.). Pak. J. Bot. 39, 1593–1608.

Wang, Y., Frei, M., 2011. Stressed food. The impact of abiotic environmental stresses on crop quality. Agric. Ecosyst. Environ. 141, 271–286.

Wang, Y., Li, L., Cui, W., Xu, S., Shen, W., Wang, R., 2012. Hydrogen sulfide enhances alfalfa (*Medicago sativa*) tolerance against salinity during seed germination by nitric oxide pathway. Plant Soil 351, 107–119.

Wilson, L.G., Bressan, R.A., Filner, P., 1978. Light-dependent emission of hydrogen sulfide from plants. Plant Physiol. 61, 184–189.

Woodward, A.J., Bennett, I.J., 2005. The effect of salt stress and abscisic acid on proline production, chlorophyll content and growth of in vitro propagated shoots of Eucalyptus camaldulensis. Plant Cell, Tissue Organ. Cult. 82, 189–200.

Xu, S., Hu, B., He, Z., Ma, F., Feng, J., Shen, W., et al., 2011. Enhancement of salinity tolerance during rice seed germination by presoaking with hemoglobin. Int. J. Mol. Sci. 12, 2488–2501.

Yadav, S.K., Singla-Pareek, S.L., Ray, M., Reddy, M.K., Sopory, S.K., 2005. Methylglyoxal levels in plants under salinity stress are dependent on glyoxalase I and GSH. Biochem. Biophys. Res. Commun. 337, 61–67.

Yilmaz, S., Temizgül, R., Yürürdurmaz, C., Kaplan, M., 2020. Oxidant and antioxidant enzyme response of redbine sweet sorghum under NaCl salinity stress. Bioagro 32, 31–38.

Zeeshan, M., Lu, M., Sehar, S., Holford, P., Wu, F., 2020. Comparison of biochemical, anatomical, morphological, and physiological responses to salinity stress in wheat and barley genotypes deferring in salinity tolerance. Agronomy 10, 127.

Zhang, H., Hu, S.L., Zhang, Z.J., Hu, L.Y., Jiang, C.X., Wei, Z.J., et al., 2011. Hydrogen sulfide acts as a regulator of flower senescence in plants. Postharvest Biol. Technol. 60, 251–257.

Zhang, H., Tang, J., Liu, X.P., Wang, Y., Yu, W., Peng, W.Y., et al., 2009. Hydrogen sulfide promotes root organogenesis in *Ipomoea batatas*, *Salix matsudana* and *Glycine max*. J. Integr. Plant Biol. 51, 1086–1094.

Zhang, L., Pei, Y., Wang, H., Jin, Z., Liu, Z., Qiao, Z., et al., 2015. Hydrogen sulfide alleviates cadmium-induced cell death through restraining ROS accumulation in roots of *Brassica rapa* L. ssp. pekinensis. Oxid. Med. Cell. Longev. 2015, 804603.

Zhu, L., Wang, W., Shi, J., Zhang, W., Shen, Y., Du, H., et al., 2014. Hydrogen sulfide extends the postharvest life and enhances antioxidant activity of kiwifruit during storage. J. Sci. Food Agric. 94, 2699–2704.

Zhang, M.H., Qin, Z.H., Liu, X., 2005. Remote sensed spectral imagery to detect late blight in field tomatoes. Precis. Agric. 6, 489–508.

Ziogas, V., Molassiotis, A., Fotopoulos, V., Tanou, G., 2018. Hydrogen sulfide: a potent tool in postharvest fruit biology and possible mechanism of action. Front. Plant Sci. 9, 1375.

Zou, P., Li, K., Liu, S., He, X., Zhang, X., Xing, R., et al., 2016. Effect of sulfated chitooligosaccharides on wheat seedlings (*Triticum aestivum* L.) under salt stress. J. Agric. Food Chem. 64, 2815–2821.

Chapter 12

Regulation of drought stress by hydrogen sulfide in plants

Owias Iqbal Dar[1], Kirpal Singh[1], Sunil Sharma[1], Junaid Aslam[2], Arvinder Kaur[1], Renu Bhardwaj[2] and Anket Sharma[3,*]

[1]Department of Zoology, Guru Nanak Dev University, Amritsar, India, [2]Department of Botanical and Environmental Sciences, Guru Nanak Dev University, Amritsar, India, [3]State Key Laboratory of Subtropical Silviculture, Zhejiang A&F University, Hangzhou, P.R. China

**Corresponding author.*

12.1 Introduction

Drought is a condition caused by the absence of water to plants and is one of the prevalent stress factors for plants globally (Salehi-Lisar and Bakhshayeshan-Agdam, 2016). Drought takes place when the humidity of soil and atmosphere is low and the temperature of air is high (Lamaoui et al., 2018). It occurs as a consequence of drastic and rapid climatic changes (Hussain et al., 2018). Stress caused by drought alters morphological, physiological, and molecular responses in plants such as germination, photosynthesis, chlorophyll synthesis, respiration, nutrient metabolism, carbohydrate metabolism, ion uptake, and translocation, thereby reducing plant growth (Jaleel et al., 2008; Farooq et al., 2009; Li et al., 2011; Hussain et al., 2018). It causes increased production of reactive oxygen species (ROS) by impairing cellular metabolic activities leading to DNA, RNA, protein, and membrane damage (Priestley, 1986).

To handle environmental stresses, plants have adapted to retaliate morphologically, physiologically, and biochemically for survival (Huber and Bauerle, 2016). Among the physiological processes, H_2S has been revealed as an important messenger molecule in plants against various abiotic stresses at physiological concentrations (Jin and Pei, 2015). It has been shown to regulate normal physiological and biochemical functions, such as seed germination, root morphogenesis, photosynthesis, fluorescence, activities of antioxidant enzyme, and content of malondialdehyde (MDA) and H_2O_2 (Ma et al., 2016). H_2S also acts as a nutrient as its treatment to plants increases amino acids (cysteine) for the synthesis of proteins (Rostami et al., 2019). It also increases the availability of reduced sulfur for the synthesis of glutathione (Calderwood and Kopriva, 2014), which acts as a key player in removing ROS and reducing oxidative stress.

Hydrogen Sulfide in Plant Biology. DOI: https://doi.org/10.1016/B978-0-323-85862-5.00008-7

Therefore H_2S plays a crucial role in plant growth, development, senescence, maturation, and as a modulator of defense responses during drought stress (Li, 2013). In this chapter we discuss recent advancement that deepen our understanding about the role of H_2S in regulating drought stress in plants.

12.2 Impact of drought stress on plants

In plants, drought stress is represented by reduced leaf water potential and turgor pressure, which causes the closure of stomata and a decline in cell growth and enlargement (Farooq et al., 2009). It reduces the growth of plants by affecting photosynthesis, chlorophyll content, respiration, metabolism of carbohydrate and other nutrients, and the translocation of ions (Hussain et al., 2018). Drought stress increases ROS production and gives rise to oxidative damage in cells, inactivation of enzymes, protein degradation, DNA damage, ionic imbalance, and cellular membrane disruption (Tuteja and Tuteja, 2001; Baier et al., 2005; Tarchoune et al., 2010). The increased production of ROS and disruption of cellular homeostasis during drought stress in plants is regulated by the action of various enzymes and nonenzymatic antioxidants (Hussain et al., 2018).

12.2.1 Impact on growth and photosynthesis

Drought stress is known to cause reduction in plant growth and development. It has been reported to inhibit plant height, leaf area, stem diameter, and plant biomass (Apel and Hirt, 2004; Farooq et al., 2009; Zheng et al., 2016). Drought stress caused an increase in the root length of sunflower and *Catharanthus roseus* (Tahir et al., 2002; Jaleel et al., 2008) but did not influence the root growth of wheat and maize (Sacks et al., 1997). Studies of Almaghrabi (2012) and Amoah et al. (2019) noticed that drought stress in wheat cultivars reduced shoot length, root length, shoot weight, root weight, root numbers, and seedlings length. Drought stress sharply decreased shoot length, root length, and dry matter in eight upland varieties of rice (Lum et al., 2014). Hussain et al. (2019) noticed reduced plant height, shoot weight, stem diameter, and leaf area in maize hybrids under drought stress. Rezayian et al. (2018) observed that drought stress significantly increases root length in RGS003 and Sarigol cultivars of Canola (*Brassica napus*) but seedlings height in RGS003 cultivar only. Gao et al. (2020) observed that the total biomass of *Adonis amurensis* and *Adonis pseudoamurensis* significantly decreased due to drought stress.

Photosynthesis in plants depends upon many components, that is, photosystems, photosynthetic pigments, the electron transport chain, and CO_2 reduction pathways (Lamaoui et al., 2018). However, the stress-induced damage on any component in these systems may lead to a decline in photosynthesis. Drought stress results in reduced photosynthetic pigments, photosynthetic efficiency, and damage of photosystem II in the seedlings of *Sophora davidii* (Wu et al., 2008). Arabzadeh. (2013) reported that drought stress on *Haloxylon aphyllum* and *Haloxylon persicum* significantly reduced chlorophyll content and photosynthetic quantum yield but elevated chlorophyll a/b ratio. Allahverdiyev (2016) determined that water stress reduced photosynthetic pigments and the ratio of chlorophyll (a + b)/carotenoids (x + c), but increased the ratio of chlorophyll (a/b) in various genotypes of durum and bread wheat. Studies of Koller et al. (2013) and Epila et al. (2018) noted a decline in net photosynthesis in *Quercus* species and *Maesopsis eminii* under drought stress.

12.2.2 Impact on oxidative stress

During normal metabolic activities, highly reactive compounds (free radicals) are generated in plants, however when plants are under stress there is overproduction of these free radicals, especially free oxygen radicals, as well as ROS. These molecules are inherently unstable and hence become highly reactive and their reactions with cellular biomolecules result in various morbid conditions like lipid peroxidation, protein degradation, and enzyme inactivation. To counteract the effect of ROS on cellular molecules, plants use both enzymatic defenses, such as superoxide dismutase (SOD), peroxidase (POD), ascorbate peroxidase (APX), polyphenol oxidase (PPO), proline oxidase (POX), glutathione-S-transferase (GST), glutathione peroxidase (GPX), and catalase (CAT), and nonenzymatic defenses, such as ascorbic acid (AsA), reduced glutathione (GSH), α-tocopherol, carotenoids, phenolics, flavonoids, and proline (Gill and Tuteja, 2010).

Fazeli et al. (2007) observed a significant increase in the content of MDA (lipid peroxidation product) and activities of SOD, POX CAT, and PPO in the leaves of Darab 14 and Yekta plants under drought stress. Wei et al. (2013) noted that drought stress in *Diosphyros* seedlings increased the content of ROS, H_2O_2, MDA, GSH, and AsA and the activities of SOD, POD, CAT, and APX. Elsayed et al. (2019) noted that exposure of drought stress to *Silybum mariamum* caused an increase in the content of H_2O_2 and proline and activities of SOD, POD CAT, APX, and GR but a decrease in the content of AsA and GSH. They also noted that drought stress caused an increase in the transcription levels of APX, SOD, and CAT. The study of Amoach et al. (2019) demonstrated that drought stress to wheat caused an increase in the contents of ROS, MDA, and activities of SOD, POD, CAT, and APX. Hussain et al. (2019) observed an increase in ROS, H_2O_2, MDA, GSH, and AsA content and the activity of SOD but a decrease in the activities of CAT, POD, and APX in maize plants after drought stress.

12.3 Role of hydrogen sulfide in plant biology

Hydrogen sulfide (H_2S), an endogenous messenger, is a colorless and lipophilic gas which can freely penetrate through cell membranes. In the past it has been considered to be a toxic gas and an environmental hazard to living organisms but in recent years H_2S has emerged as an important signaling molecule which modulates various important physiological processes in plants (Zhang et al., 2008). Its endogenous production, free permeability, physiological functions, and rapid degradation make it an important signaling molecule in plants. H_2S plays a key role in seed germination, organogenesis, stomatal movement, autophagy, senescence, and photosynthesis (Lisjak et al., 2013; Zhang et al., 2015).

H_2S has been found to induce the growth of adventitious roots in *Ipomoea batatas*, *Salix matsudana*, *Glycine max* (Zhang et al., 2009), and cucumber (Lin et al., 2012). It improved the process of photosynthesis in *Spinacia oleracea* seedlings through increased biogenesis of chloroplasts, expression of photosynthetic enzymes, and thiol redox modification (Chen et al., 2011). H_2S has been also found to improve photosynthesis in *Dendrobium officinale* under high light through the stimulation of antioxidant enzyme activities (Fan et al., 2014). H_2S as a signaling molecule may stimulate the biogenesis of chloroplasts by increasing the quantity of grana lamellae and chlorophyll biosynthesis which both by their additive effect boosted the capacity of the photosynthesis of nickel-exposed rice plants (Rizwan et al., 2019). Low concentrations of H_2S enhance stomatal apertures by decreasing the expression level of ABA receptor candidates of *Arabidopsis* (Jin et al., 2013). Garcia-Mata and Lamattina (2010) suggest that low content of H_2S represses stomatal opening through an ABA-dependent signaling network. H_2S is also involved in regulating the closure of stomata induced by abscisic acid (ABA), jasmonic acid, ethylene, salt, and drought (Hou et al., 2011; Hou et al., 2012; Che et al., 2012; Che et al., 2015; Wang et al., 2012; Jin et al., 2013; Scuffi et al., 2014; Ma et al., 2018).

H_2S reduces autophagy by preventing the accumulation of ATG8 (Álvarez et al., 2012). According to Gotor et al. (2013) the process of autophagy inhibition might be due to the persulfidation of the enzyme involved in autophagosome formation. H_2S plays a great role in inhibiting fruit ripening and delaying senescence. Li et al. (2017) reported that the activity of H_2S in delaying senescence was associated with fruit ripening in kiwifruit (*Actinidia deliciosa*). H_2S inhibits the activity of polygalacturonase, an enzyme responsible for cell wall loosening (Zhu et al., 2019). It decreased phenylalanine and tyrosine content as a substrate of phenylpropanoid pathway, but increased the level of proline and histidine as an osmolyte and antioxidant, respectively (Fakhari et al., 2019).

H_2S has been observed to maintain the external color of certain horticultural crops (Zhang et al., 2011). Chang et al. (2014) and Hu et al. (2012) exposed strawberry fruit to 0.8 mmol L^{-1} NaHS treatment and observed that it maintained a better reed color. Water spinach leaves after treatment with 2.4 mmol L^{-1} NaHS maintained their green color for 8 days by delayed chlorophyll loss and reduced chlorophyllase activity (Hu et al., 2015). The role of H_2S in the symbiotic association between soybean (*G. max*) and rhizobium (*Sinorhizobium fredii*) were demonstrated by Zou et al. (2019). They observed that NaHS treatment increases soybean growth, nodulation, and nitogenase activity while the expression of symbiotic genes nodA, nodC, and nodD was increased in *S. fredii*. They suggested that H_2S may act as an important signaling molecule in the symbiotic association between soybean and rhizobium to enhance the nitrogen fixation ability of the system.

12.4 Role of H_2S in plants under drought stress

The potential physiological functions of H_2S during the response of plants to abiotic stress include enhancement of photosynthesis, control of ROS, activation of antioxidant system, organogenesis, and growth. Some previous studies found that H_2S can effectively enhance drought resistance in a variety of plants (Zhang et al., 2010a; Shan et al., 2011; Jin et al., 2013). Thus if treated with appropriate concentration, an H_2S donor (NaHS) could prevent the negative impact of continuous drought on the biology of plants.

12.4.1 Impact on photosynthesis

Photosynthesis, one of the most important biochemical pathways, is negatively affected by low water stress. Drought stress inhibits the assimilation of carbon and damages photosynthetic frameworks by disrupting the basic organization structure in cells (Ali and Ashraf, 2011; Golldack et al., 2011). Photosynthetic reduction is attributed to the restrictions in stomata regarding the supply of CO_2, the decline in turgor pressure, limitation of gas exchange, impaired photosystem mostly PSI and PSII, and increased metabolite fluxes (Angelopoulos et al., 1996; Chaves et al., 2003; Jaleel et al., 2009). Drought stress also caused a reduction in chlorophyll content in *C. roseus* (Jaleel et al., 2008) and *Gossypium hirsutum* (Massacci et al., 2008).

H_2S enhanced photosynthesis in *S. oleracea* by elevating the biogenesis of chloroplasts and biosynthesis of chlorophyll, thus remarkably increasing the photosynthetic rate (Chen et al., 2011). H_2S enhanced the photosynthetic rate, leaf photosynthetic area, and the stomatal aperture in *Oryza sativa* L (Duan et al., 2014). H_2S induced stomatal closure in *A. thaliana, Vicia faba,* and *Impatiens walleriana* under drought stress (Garcia-Mata and Lamattina 2010; Jin et al., 2011). ABA, known for important regulatory roles in plants' life cycle, also regulates stomatal movements in response to drought stress (MacRobbie, 1998). In response to drought stress, H_2S increases the expression of ABA receptor coding genes which induced the closure of stomata by controlling ion channel expression at the transcriptional level in *Arabidopsis* (Jin et al., 2013). In addition, the protein expression of the Rubisco large subunit and the activity of genes encoding ribulose-1,5-bisphosphate carboxylase (Rubisco) were also significantly upregulated by optimal NaHS concentration, ultimately enhancing photosynthesis (Chen et al., 2011). In another study, Zhang et al. (2010a) found that H_2S alleviated drought-induced decline in chlorophyll of soybean leaves. Under drought stress the treatment of NaHS (H_2S donor) to *S. oleracea* seedlings increase the net

CO_2 assimilation rate, stomatal conductance, and leaf transpiration rate (Chen et al., 2016). They also demonstrated that NaHS treatment increased the light compensation point (L_{cp}), light saturation point (L_{sp}), and net photosynthetic rate (P_{max}), which show the overall increase in photosynthesis.

12.4.2 Impact on oxidative stress

During environmental stress, levels of ROS in plants increases sharply (Devasagayam et al., 2004) and their accumulation causes disruption of the redox homeostasis in cells, which results in damage to cellular structures such as proteins, DNA, and lipids, thereby affecting normal cellular functions (Apel and Hirt, 2004; Foyer and Noctor, 2005). When a plant experiences a stress, the known features of the oxidative burst promote the formation of the MDA end-product of lipid peroxidation (Fazlieva et al., 2012) and ROS accumulation (O_2^-, OH^-, H_2O_2) in different cell compartments, especially mitochondria (Gautam et al., 2017). Plants suffering from drought stress often exhibit symptoms resulting in the increased production of ROS, which can cause protein breakdown, photooxidative damage to chlorophyll, lipid peroxidation, and increased production of MDA (Mehta et al., 1992; Blokhina et al., 2003; Nield et al., 2004; Song et al., 2006). Similar to other gasotransmitters, such as CO and NO, H_2S acts as a signaling molecule under oxidative stress. The application of NaHS to soybean seedlings provides a way to combat drought stress by increasing survival rate and decreasing the $O_2^{\bullet-}$, H_2O_2, and MDA levels (Zhang et al., 2010a), which indicates the protective role of the H_2S donor against drought-induced oxidative damage in plant Chen et al. (2016) also noted that NaHS declined drought-induced accumulation of ROS and increased MDA levels.

12.4.3 Impact on antioxidative system

To combat the harmful oxidative damage in response to water deficit stress, plants exhibit a versatile scavenging system. Under drought stress H_2O_2 and other ROSs generated were removed by increased activities and gene expressions of antioxidant enzymes, that is, APX, POD, CAT, SOD, and GR (ElSayed et al., 2019). H_2S has proven to be involved in activating the signaling pathway of the antioxidative system. The activities of APX and CAT significantly increased under NaHS treatment and reduced lipoxygenase (LOX) activity and MDA and H_2O_2 accumulation in wheat seeds (Zhang et al., 2010b). Further experiments showed that low concentrations of H_2S increased the activities of APX, SOD, and POD in *Pisum sativum* (Li et al., 2010). In soybean seedlings, H_2S enhanced the CAT and SOD activities, lowered levels of LOX, and decreased the overproduction of, MDA, H_2O_2, and $O_2^{\bullet-}$ in response to drought stress conditions (Zhang et al., 2010a). It is evident that an H_2S donor has the ability to increase gene expressions of antioxidant enzymes and their activities. Exogenous H_2S treatment of rice

TABLE 12.1 Role of H_2S on antioxidative system of plants under drought stress.

Effect on antioxidative system	Plant	References
• Enhanced CAT and SOD activities • Reduced LOX activity, MDA, H_2O_2, and O_2^- contents	Soybean (*Glycine max*)	Zhang et al. (2010a)
• Enhanced APX, DHAR, total ascorbate, and AsA activities	Wheat (*Triticum aestivum*)	Shan et al. (2011)
• Enhanced SOD, POD, and CAT activities • Reduced MDA and H_2O_2 contents	Wheat (*Triticum aestivum*)	Ma et al. (2016)
• Enhanced SOD, POD, CAT, and APX activities • Reduced MDA content	Rice (*Oryza sativa*)	Liu et al. (2017)
• Enhanced CAT and guaiacol peroxidase activities • Reduced H_2O_2 and MDA contents	Wheat (*Triticum aestivum*)	Kolupaev et al. (2019)

seedlings effectively reduced MDA content in rice seeds and decreased the permeability of the plasma membrane; it also boosted the activities of SOD, POD, CAT, and APX of rice during germination in response to the drought stress (Liu et al., 2017). Drought simultaneously decreased the CAT activity and increased H_2O_2 and MDA contents in the leaves of wheat plants, however, when treated with the hydrogen sulfide donor, it prevented the decrease of CAT and guaiacol peroxidase activities caused by the drought stress (Kolupaev et al., 2019). Moreover, under drought stress NaHS significantly increased the SOD, POD, and CAT activities and reduced the contents of MDA and H_2O_2, implying that H_2S improved the drought resistance of wheat plants and protected them from harmful oxidative damage (Ma et al., 2016). The activities of APX, dehydroascorbate reductase (DHAR), and total ascorbate and the content of AsA also increased in wheat plants under water stress when treated with H_2S (Shan et al., 2011). Table 12.1 summarizes the regulation of the antioxidative system in plants by H_2S under drought stress.

12.5 Role of H_2S in plant productivity under drought stress

Droughts can have an adverse effect on ecology and crop production with a substantial impact on agriculture and the ecosystem. Agricultural production is affected by drought critically because of its negative impacts on plant growth which eventually reduces the crop yield (Karl et al., 2009). If drought persists for an extended period of time, it can eventually cause damage to our economy.

Drought during early reproductive and vegetative growth hinders the most sensitive stage of the plant, that is, developmental stage, hence decreasing the yield. Severe drought can cause a higher mortality rate of vegetation due to carbon starvation ultimately limiting photosynthesis (Da-Silva et al., 2013). Drought-like conditions have negative impacts on the growth and survival of plants, as reported by Golldack et al. (2014) and Xiong et al. (2006), who found that drought stress affected plant survival, distribution, growth, and productivity. Moreover, production in crops under water stress such as coffee reduced by as much as 80% in dry years (DaMatta and Ramalho, 2006).

H_2S has been known to be involved in plant organogenesis and growth. Dooley et al. (2013) demonstrated that by treating seeds or roots of beans and pea with extremely low concentration of H_2S, seed germination, seedling growth, biomass accumulation, and yields were significantly increased. So it can be proposed that H_2S can have positive effects on the productivity and improved yield of plants under drought stress. Reliable information and evidence is needed which will facilitate further insights into the effective role of H_2S on productivity, growth, and yield of various plants under drought-like conditions.

12.6 Conclusion

Hydrogen sulfide protects plants from the ill effects of drought stress by regulating key plant processes like photosynthesis and the ROS detoxification mechanism. Fig. 12.1 shows the protective roles of H_2S in plants under drought stress.

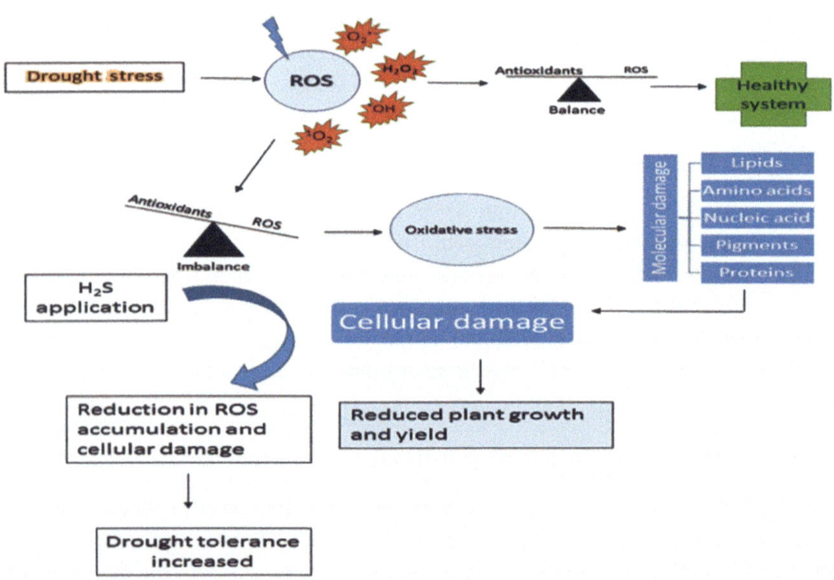

FIGURE 12.1 Generalized scheme of drought-induced oxidative stress in plants and its regulation by hydrogen sulfide.

References

Ali, Q., Ashraf, M., 2011. Induction of drought tolerance in maize (*Zea mays* L.) due to exogenous application of trehalose: growth, photosynthesis, water relations and oxidative defence mechanism. J. Agron. Crop. Sci. 197, 258–271.

Allahverdiyev, T.I., 2016. Impact of soil water deficit on some physiological parameters of durum and bread wheat genotypes. Poljoprivreda i Sumar. 62, 131.

Almaghrabi, O.A., 2012. Impact of drought stress on germination and seedling growth parameters of some wheat cultivars. Life Sci. J. 9, 590–598.

Álvarez, C., Ángeles Bermúdez, M., Romero, L.C., Gotor, C., García, I., 2012. Cysteine homeostasis plays an essential role in plant immunity. N. Phytol. 193, 165–177.

Amoah, J.N., Ko, C.S., Yoon, J.S., Weon, S.Y., 2019. Effect on drought acclimation on oxidative stress and transcript expression in wheat (*Triticum aestivum* L.). J. Plant Interact. 14, 492–505.

Angelopoulos, K., Dichio, B., Xiloyannis, C., 1996. Inhibition of photosynthesis in olive trees (*Olea europaea L.*) during water stress and rewatering. J. Exp. Bot. 47, 1093–1100.

Apel, K., Hirt, H., 2004. Reactive oxygen species: metabolism, oxidative stress, and signal transduction. Annu. Rev. Plant Biol. 55, 373–399.

Arabzadeh, N., 2013. The impact of drought stress on photosynthetic quantum yield in *Haloxylon aphyllum* and *Haloxylon persicum*. Afr. J. Plant Sci. 7, 185–189.

Baier, M., Kandlbinder, A., Golldack, D., Dietz, K., 2005. Oxidative stress and ozone: perception, signalling and response. Plant Cell Environ. 28, 1012–1020.

Blokhina, O., Virolainen, E., Fagerstedt, K.V., 2003. Antioxidants, oxidative damage and oxygen deprivation stress: a review. Ann. Bot. 91, 179–194.

Calderwood, A., Kopriva, S., 2014. Hydrogen sulfide in plants: from dissipation of excess sulfur to signaling molecule. Nitric Oxide 41, 72–78.

Chang, Z., Jingying, S., Liqin, Z., Changle, L., Qingguo, W., 2014. Cooperative effects of hydrogen sulfide and nitric oxide on delaying softening and decay of strawberry. Int. J. Agric. Biol. Eng. 7, 114–122.

Chaves, M.M., Maroco, J.P., Pereira, J.S., 2003. Understanding plant responses to drought—from genes to the whole plant. Funct. Plant Biol. 30, 239.

Che, Y.M., Zhang, D.D., Hou, L.X., Wang, L.X., Liu, X., 2015. Extracellular adenosine triphosphate functions downstream of hydrogen sulfide in ethylene-induced stomatal closure in *Arabidopsis thaliana*. Chin. Bull. Bot. 50, 22–31.

Che, Y.M., Zou, X., Wang, L.X., Zhang, D.D., Liu, X., 2012. H_2S signals salt-induced stomatal closure in *Arabidopsis thaliana* by SOS pathway. Plant Physiol. J. 48, 1098–1104.

Chen, J., Wu, F.H., Wang, W.H., Zheng, C.J., Lin, G.H., Dong, X.J., et al., 2011. Hydrogen sulphide enhances photosynthesis through promoting chloroplast biogenesis, photosynthetic enzyme expression, and thiol redox modification in *Spinacia oleracea* seedlings. J. Exp. Bot. 62, 4481–4493.

Chen, J., Shang, Y.T., Wang, W.H., Chen, X.Y., He, E.M., Zheng, H.L., et al., 2016. Hydrogen sulfide-mediated polyamines and sugar changes are involved in hydrogen sulfide-induced drought tolerance in *Spinacia oleracea* seedlings. Front. Plant Sci. 7, 1173.

DaMatta, F.M., Ramalho, J.D.C., 2006. Impacts of drought and temperature stress on coffee physiology and production: a review. Braz. J. Plant Physiol. 18, 55–58.

Da-Silva, E.C., de-Albuquerque, M.B., de-Azevedo Neto, A.D., and da-Silva Junior, C.D. 2013. Drought and its consequences to plants—from individual to ecosystem, responses of organisms to water stress. Sener Akıncı, IntechOpen.

Devasagayam, T.P., Tilak, J.C., Boloor, K.K., Sane, K.S., Ghaskadbi, S.S., Lele, R.D., 2004. Free radicals and antioxidants in human health: current status and future prospects. J. Assoc. Phys India. 52, 794–804.

Dooley, F.D., Nair, S.P., Ward, P.D., 2013. Increased growth and germination success in plants following hydrogen sulfide administration. PLoS One 8, e62048.

Duan, B., Ma, Y., Jiang, M., Yang, F., Ni, L., Lu, W., 2014. Improvement of photosynthesis in rice (*Oryza sativa L.*) as a result of an increase in stomatal aperture and density by exogenous hydrogen sulfide treatment. Plant Growth Regul. 75, 33–44.

ElSayed, A.I., El-hamahmy, M.A.M., Rafudeen, M.S., Mohamed, A.H., Omar, A.A., 2019. The impact of drought stress on antioxidant responses and accumulation of flavonolignans in milk thistle (*Silybum marianum* (L.) Gaertn). Plants 8, 611.

Epila, J., Hubeau, M., Steppe, K., 2018. Drought effects on photosynthesis and implications of photoassimilate distribution in 11c-labeled leaves in the african tropical tree species *Maesopsis eminii*. Engl. Forests 9, 109.

Fakhari, S., Sharifi, M., De Michele, R., Ghanati, F., Safaie, N., Sadeghnezhad, E., 2019. Hydrogen sulfide directs metabolic flux towards the lignan biosynthesis in *Linum album* hairy roots. Plant Physiol. Biochem. 135, 359–371.

Fan, H., Guan, L., Li, T., Wu, Q., Wu, M., Cai, Y., et al., 2014. Hydrogen sulphide alleviates oxidative damage and enhances light energy transformation under high light for *Dendrobium officinale*. Sci. Hortic. 177, 47–52.

Farooq, M., Wahid, A., Kobayashi, N., Fujita, D., Basra, S.M.A., 2009. Plant drought stress: effects, mechanisms and management. Agron. Sustain. Dev. 29, 185–212.

Fazeli, F., Ghorbanli, M., Niknam, V., 2007. Effect of drought on biomass, protein content, lipid peroxidation and antioxidant enzymes in two sesame cultivars. Biol. Plant 51, 98–103.

Fazlieva, E.R., Kiseleva, I.S., Zhuikova, T.V., 2012. Antioxidant activity in the leaves of *Melilotus albus* and *Trifolium medium* from man-made disturbed habitats in the middle urals under the influence of copper. Russ. J. Plant Physiol. 59, 333–338.

Foyer, C.H., Noctor, G., 2005. Redox homeostasis and antioxidant signaling: a metabolic interface between stress perception and physiological responses. Plant Cell 17, 1866–1875.

Gao, S., Wang, Y., Yu, S., Huang, Y., Liu, H., Chen, W., et al., 2020. Effects of drought stress on growth, physiology and secondary metabolites of two adonis species in Northeast China. Sci. Hortic. 259, 108795.

Garcia-Mata, C., Lamattina, L., 2010. Hydrogen sulphide, a novel gasotransmitter involved in guard cell signalling. N. Phytol. 188, 977–984.

Gautam, V., Kaur, R., Kohli, S.K., Verma, V., Kaur, P., Singh, R., et al., 2017. ROS compartmentalization in plant cells under abiotic stress condition. In: Khan, M.I.R., Khan, N.A. (Eds.), Reactive Oxygen Species and Antioxidant Systems in Plants: Role and Regulation under Abiotic Stress. Springer, Singapore, pp. 89–114.

Gill, S.S., Tuteja, N., 2010. Reactive oxygen species and antioxidant machinery in abiotic stress tolerance in crop plants. Plant Physiol. Biochem. 48, 909–930.

Golldack, D., Li, C., Mohan, H., Probst, N., 2014. Tolerance to drought and salt stress in plants: unraveling the signaling networks. Front. Plant Sci. 5, 151.

Golldack, D., Lüking, I., Yang, O., 2011. Plant tolerance to drought and salinity: stress regulating transcription factors and their functional significance in the cellular transcriptional network. Plant Cell Rep. 30, 1383–1391.

Gotor, C., García, I., Crespo, J.L., Romero, L.C., 2013. Sulfide as a signaling molecule in autophagy. Autophagy 9, 609–611.

Hou, Z.H., Che, Y.M., Wang, L.X., Hou, L.X., Liu, X., 2012. H_2S functions downstream of H_2O_2 in mediating ethylene-induced stomatal closure in *Arabidopsis thaliana*. Plant Physiol. J. 48, 1193–1199.

Hou, Z.H., Liu, J., Hou, L.X., Li, X.D., Liu, X., 2011. H_2S may function downstream of H_2O_2 in jasmonic acid-induced stomatal closure in *Vicia faba*. Chin. Bull. Bot. 46, 396–406.

Hu, H., Liu, D., Li, P., Shen, W., 2015. Hydrogen sulfide delays leaf yellowing of stored water spinach (*Ipomoea aquatica*) during dark-induced senescence by delaying chlorophyll breakdown, maintaining energy status and increasing antioxidative capacity. Postharvest Biol. Technol. 108, 8–20.

Hu, L.Y., Hu, S.L., Wu, J., Li, Y.H., Zheng, J.L., Wei, Z.J., et al., 2012. Hydrogen sulfide prolongs postharvest shelf life of strawberry and plays an antioxidative role in fruits. J. Agric. Food Chem. 60, 8684–8693.

Huber, A.E., Bauerle, T.L., 2016. Long-distance plant signaling pathways in response multiple stressors: the gap in knowledge. J. Exp. Bot. 67, 2063–2079.

Hussain, H.A., Hussain, S., Khaliq, A., Ashraf, U., Anjum, S.A., Men, S., et al., 2018. Chilling and drought stress in crop plants: implications, cross talk, and potential management opportunities. Front. Plant Sci. 9, 393.

Hussain, H.A., Men, S., Hussain, S., Chen, Y., Ali, S., Zhang, S., et al., 2019. Interactive effects of drought and heat stress on morpho-physiological attributes, yield, nutrient uptake and oxidative status in maize hybrids. Sci. Rep. 9, 3890.

Jaleel, C.A., Manivannan, P., Lakshmanan, G.M.A., Gomathinayagam, M., Panneerselvam, R., 2008. Alterations in morphological parameters and photosynthetic pigment responses of *Catharanthus roseus* under soil water deficits. Colloids Surf. B: Biointerfaces 61, 298–303.

Jaleel, C.A., Manivannan, P., Wahid, A., Farooq, M., Al-Juburi, H.J., Somasundaram, R., et al., 2009. Drought stress in plants: a review on morphological characteristics and pigments composition. Int. J. Agric. Biol. 11, 100–105.

Jin, Z., Pei, Y., 2015. Physiological implications of hydrogen sulfide in plants: pleasant exploration behind its unpleasant odour. Oxid. Med. Cell Longev. 2015, 397502.

Jin, Z.P., Shen, J.J., Qiao, Z.J., Yang, G.D., Wang, R., Pei, Y.X., 2011. Hydrogen sulfide improves drought resistance in *Arabidopsis thaliana*. Biochem. Biophys. Res. Commun. 414, 481e486.

Jin, Z., Xue, S., Luo, Y., Tian, B., Fang, H., Li, H., et al., 2013. Hydrogen sulfide interacting with abscisic acid in stomatal regulation responses to drought stress in *Arabidopsis*. Plant Physiol. Biochem. 62, 41–46.

Karl, T.R., Melillo, J.M., Peterson, T.C., 2009. Global Climate Change Impacts in the United States. Cambridge University Press, Cambridge.

Koller, S., Holland, V., Brüggemann, W., 2013. Effects of drought stress on the evergreen *Quercus ilex* L., the deciduous *Q. robur* L. and their hybrid *Q.* × *turneri* Willd. Photosynthetica 51, 574–582.

Kolupaev, Y.E., Firsova, E.N., Yastreb, T.O., Ryabchun, N.I., Kirichenko, V.V., 2019. Effect of hydrogen sulfide donor on antioxidant state of wheat plants and their resistance to soil drought. Russ. J. Plant Physiol. 66, 26–34.

Lamaoui, M., Jemo, M., Datla, R., Bekkaoui, F., 2018. Heat and drought in crops and approaches for their mitigation. Front. Chem. 6, 26.

Li, C., Jiang, D., Wollenweber, B., Li, Y., Dai, T., Cao, W., 2011. Waterlogging pretreatment during vegetative growth improves tolerance to waterlogging after anthesis in wheat. Plant Sci. 180, 672–678.

Li, D.B., Xiao, Z.X., Liu, L.X., Wang, J.C., Song, G.L., Bi, Y.R., 2010. Effects of exogenous hydrogen sulfide (H_2S) on the root tip and root border cells of *Pisum sativum*. Chin. Bull. Bot. 45, 354–362.

Li, T.T., Li, Z.R., Hu, K.D., Hu, L.Y., Chen, X.Y., Li, Y.H., et al., 2017. Hydrogen sulfide alleviates kiwifruit ripening and senescence by antagonizing effect of ethylene. HortScience 52, 1556–1562.

Li, Z.G., 2013. Hydrogen sulfide: a multifunctional gaseous molecule in plants. Russ J. Plant Physiol. 60, 733–740.

Lin, Y.T., Li, M.Y., Cui, W.T., Lu, W., Shen, W.B., 2012. Haem oxygenase-1 is involved in hydrogen sulfide-induced cucumber adventitious root formation. J. Plant Growth Regul. 31, 519–528.

Lisjak, M., Teklic, T., Wilson, I.D., Whiteman, M., Hancock, J.T., 2013. Hydrogen sulfide: environmental factor or signaling molecule? Plant Cell Environ. 36, 1607–1616.

Liu, J., Zhang, H., Yin, Y., Chen, H., 2017. Effects of exogenous hydrogen sulfide on antioxidant metabolism of rice seed germinated under drought stress. J. South. Agric. 48, 31–37.

Lum, M.S., Hanafi, M.M., Rafii, Y.M., Akmar, A.S.N., 2014. Effect of drought stress on growth, proline and antioxidant enzyme activities of upland rice. J. Anim. Plant Sci. 24, 1487–1493.

Ma, D., Ding, H., Wang, C., Qin, H., Hou, J., Lu, H., et al., 2016. Alleviation of drought stress by hydrogen sulfide is partially related to the abscisic acid signaling pathway in wheat. PLoS One 11, e0163082.

Ma, Y., Niu, J., Zhang, W., Wu, X., 2018. Hydrogen sulfide may function downstream of hydrogen peroxide in mediating darkness-induced stomatal closure in *Vicia faba*. Funct. Plant Biol. 45, 553–560.

MacRobbie, E.A., 1998. Signal transduction and ion channels in guard cells. Philos. Trans. R. Soc. Lond. Ser. B 353, 1475e1488.

Massacci, A., Nabiev, S.M., Pietrosanti, L., Nematov, S.K., Chernikova, T.N., Thor, K., et al., 2008. Response of the photosynthetic apparatus of cotton (*Gossypium hirsutum*) to the onset of drought stress under field conditions studied by gas-exchange analysis and chlorophyll fluorescence imaging. Plant Physiol. Biochem. 46, 189–195.

Mehta, R.A., Fawcett, T.W., Porath, D., Mattoo, A.K., 1992. Oxidative stress causes rapid membrane translocation and in vivo degradation of ribulose-1,5-biphosphate carboxylase/oxygenase. J. Biol. Chem. 267, 2810–2816.

Nield, J., Redding, K., Hippler, M., 2004. Remodeling of light-harvesting protein complexes in chlamydomonas in response to environmental changes. Eukaryot. Cell 3, 1370–1380.

Priestley, D.A., 1986. Seed Ageing: Implications for Seed Storage and Persistence in the Soil. Comstock Associates, Ithaca, NY, p. 304.

Rezayian, M., Niknam, V., Ebrahimzadeh, H., 2018. Effects of drought stress on the seedling growth, development, and metabolic activity in different cultivars of canola. Soil Sci. Plant Nutr. 64, 360–369.

Rizwan, M., Mostofa, M.G., Ahmad, M.Z., Zhou, Y., Adeel, M., Mehmood, S., et al., 2019. Hydrogen sulfide enhances rice tolerance to nickel through the prevention of chloroplast damage and the improvement of nitrogen metabolism under excessive nickel. Plant Physiol. Biochem. 138, 100–111.

Rostami, F., Nasibi, F., Kalantari, K.M., 2019. Alleviation of UV-B radiation damages by sodium hydrosulfide (H_2S donor) pre-treatment in borage seedlings. J. Plant Interact. 14, 519–524.

Sacks, M.M., Silk, W.K., Burman, P., 1997. Effect of water stress on cortical cell division rates within the apical meristem of primary roots of maize. Plant Physiol. 114, 519–527.

Salehi-Lisar, S.Y., Bakhshayeshan-Agdam, H., 2016. Drought stress in plants: causes, consequences, and tolerance. In: Hossain, M., Wani, S., Bhattacharjee, S., Burritt, D., Tran, L.S. (Eds.), Drought Stress Tolerance in Plants, vol 1. Springer, Cham.

Scuffi, D., Álvarez, C., Laspina, N., Gotor, C., Lamattina, L., García-Mata, C., 2014. Hydrogen sulfide generated by L-cysteine desulfhydrase acts upstream of nitric oxide to modulate abscisic acid-dependent stomatal closure. Plant Physiol. 166, 2065–2076.

Shan, C.J., Zhang, S.L., Li, D.F., Zhao, Y.Z., Tian, X.L., Zhao, X.L., et al., 2011. Effects of exogenous hydrogen sulfide on the ascorbate and glutathione metabolism in wheat seedlings leaves under water stress. Acta Physiol. Plant 33, 2533–2540.

Song, Y.G., Liu, B., Wang, L.F., Li, M.H., Liu, Y., 2006. Damage to the oxygen-evolving complex by superoxide anion, hydrogen peroxide, and hydroxyl radical in photoinhibition of photosystem II. Photosynth. Res. 90, 67–78.

Tahir, M.H.N., Imran, M., Hussain, M.K., 2002. Evaluation of sunflower (*Helianthus annuus* L.) inbred lines for drought tolerance. Int. J. Agric. Biol. 3, 398–400.

Tarchoune, I., Sgherric, C., Izzo, R., Lachaal, M., Ouerghi, Z., 2010. Navari-Izzo, F. Antioxidative responses of Ocimum basilicum to sodium chloride or sodium sulphate salinization. *Plant Physiology and Biochemistry*, 48, 772–777.

Tuteja, N., Tuteja, R., 2001. Unravelling DNA repair in human: molecular mechanisms and consequences of repair defect. Crit. Rev. Biochem. Mol. Biol. 36, 261–290.

Wang, L.X., Hou, Z.H., Hou, L.X., Zhao, F.G., Liu, X., 2012. H_2S induced by H_2O_2 mediates drought-induced stomatal closure in *Arabidopsis thaliana*. Chin. Bull. Bot. 47, 217–225.

Wei, P., Yang, Y., Wang, F., Chen, H., 2013. Effect of drought stress on the antioxidant system in the three species of *Diospyros* L. Hortic. Environ. Biotechnol. 56, 597–605.

Wu, F.Z., Bao, W.K., Li, F.L., Wu, N., 2008. Effects of water stress and nitrogen supply on leaf gas exchange and fluorescence parameters of *Sophora davidii* seedlings. Photosynthetica 46, 40–48.

Xiong, L., Wang, R.G., Mao, G., Koczan, J.M., 2006. Identification of drought tolerance determinants by genetic analysis of root response to drought stress and abscisic acid. Plant Physiol. 142, 1065–1074.

Zhang, H., Hu, L.Y., Hu, K.D., He, Y.D., Wang, S.H., Luo, J.P., 2008. Hydrogen sulfide promotes wheat seed germination and alleviates oxidative damage against copper stress. J. Integr. Plant Biol. 50, 1518–1529.

Zhang, H., Hu, S.L., Zhang, Z.J., Hu, L.Y., Jiang, C.X., Wei, Z.J., et al., 2011. Hydrogen sulfide acts as a regulator of flower senescence in plants. Postharvest Biol. Technol. 60, 251–257.

Zhang, H., Jiao, H., Jiang, C.X., Wang, S.H., Wei, Z.J., Luo, J.P., et al., 2010a. Hydrogen sulfide protects soybean seedlings against drought-induced oxidative stress. Acta Physiol. Plant 32, 849–857.

Zhang, H., Tang, J., Liu, X.P., Wang, Y., Yu, W., Peng, W.Y., et al., 2009. Hydrogen sulfide promotes root organogenesis in *Ipomoea batatas*, *Salix matsudana* and *Glycine max*. J. Integr. Plant Biol. 51, 1086–1094.

Zhang, H., Wang, M.J., Hu, L.Y., Wang, S.H., Hu, K.D., Bao, L.J., et al., 2010b. Hydrogen sulfide promotes wheat seed germination under osmotic stress. Russ. J. Plant Physiol. 57, 532–539.

Zhang, L., Pei, Y., Wang, H., Jin, Z., Liu, Z., Qiao, Z., et al., 2015. Hydrogen sulfide alleviates cadmium-induced cell death through restraining ROS accumulation in roots of *Brassica rapa* L. ssp. pekinensis. Oxid. Med. Cell. Longev. 2015.

Zheng, M., Tao, Y., Hussain, S., Jiang, Q., Peng, S., Huang, J., et al., 2016. Seed priming in dry direct-seeded rice: consequences for emergence, seedling growth and associated metabolic events under drought stress. Plant Growth Regul. 78, 167–178.

Zhu, L., Du, H., Wang, W., Zhang, Y.S., Wan, C., Chen, J., 2019. Synergistic effect of nitric oxide with hydrogen sulfide on inhibition of ripening and softening of peach fruits during storage. Sci. Hortic. 256, 108591.

Zou, H., Zhang, N.N., Pan, Q., Zhang, J.H., Chen, J., Wei, G.H., 2019. Hydrogen sulfide promotes nodulation and nitrogen fixation in soybean–rhizobia symbiotic system. Mol. Plant Microbe Interact. MPMI-01.

Chapter 13

Hydrogen sulfide and phytohormones crosstalk in plant defense against pathogen attack

Shivam Jasrotia[1], Raman Jasrotia[2] and Anket Sharma[3,*]
[1]Government Degree College, Kathua, Jammu and Kashmir, India, [2]Department of Zoology, University of Jammu, Jammu, India, [3]State Key Laboratory of Subtropical Silviculture, Zhejiang A&F University, Hangzhou, P.R. China
**Corresponding author.*

13.1 Introduction

Hydrogen sulfide (H_2S), a gaseous transmitter, also performs several plant growth regulatory functions at low concentrations. This molecule has also been recorded to exhibit positive effects on development, growth, and against various types of stresses in plants (Wang, 2012), thereby fulfiling many functions along with the modulation of several physiological processes in plants (Li, 2013). Like other gaseous molecules, that is, carbon monoxide (CO) and nitric oxide (NO) in plants, H_2S also plays a role in developmental processes and resistance against environmental stress and pathogens (Rausch and Wachter, 2005; Wang, 2010; Lisjak et al., 2013; Calderwood and Kopriva, 2014; Hancock and Whiteman, 2014). The notable contributions of H_2S in plant physiology includes root organogenesis (Zhang et al., 2009), germination of seed (Zhang et al., 2010a,b), acceleration of photosynthesis (Chen et al., 2011a), and formation of lateral roots (Fang et al., 2014). Further, the protective role of H_2S against abiotic stress, such as drought, heat, salt, chilling, and osmotic variations (Zhang et al., 2010b, 2011a; Wang et al., 2010; Jin et al., 2011, 2013; Shen et al., 2012, 2013; Xie et al., 2014), and the biotic stress of pathogens (Shi et al., 2015) is phenomenal. H_2S interacts with phytohormones to confer resistance against unfavorable conditions and pathogen attacks (Jin et al., 2013; Shi et al., 2015).

Moreover, recently, research has shown that disease-related genes are expressed and this expression is upregulated by using some physiological

Hydrogen Sulfide in Plant Biology. DOI: https://doi.org/10.1016/B978-0-323-85862-5.00015-4

concentrations of H_2S, thereby exhibiting the capability of biotic resistance (Zhang et al., 2019). However, antibacterial tolerance of plants was found to be affected via the auxin pathway by the expression of miR393 a/b and its associated genes of targets that include AFB2, AFB1, AFB3, and TIR1, that are mediated by the ability of H_2S. (Shi et al., 2015; Bai et al., 2017). Even the production of high levels of H_2S was noted in pathogen-infected plants (Shi et al., 2015).

13.2 Impact of pathogens on plant physiology

Plant pathogens affect plants, leading to several diseases in plants which ultimately cause a 16% loss of crops globally (Oerke, 2006). Due to pathogen infestation in plants, effector-triggered immunity (ETI) is induced that leads to programmed cell death on the site of infection, a strong and rapid resistance response. There are certain other defense responses also, like cell wall enhancement proteins or toxic metabolite accumulation and hormone alteration (Leach et al., 2014).

Around the world, many important plant diseases are caused by viruses which affect the agricultural crops badly, decreasing the crop yield on a significant rate (Rybicki, 2015). No doubt crops are under various types of stresses whether they are dealing with live organisms like insects, nematodes, viruses, bacteria, or weeds. But the severity of infections and therefore the damage caused by them in terms of crop loss also depend upon various factors like different environmental conditions and other causal organisms (Walters, 2010). Diseases in the plants are associated with different plant pathogens which are considered to be the main accountable pathogens like bacteria, fungi, viruses, and nematodes (Moustafa-Farag et al., 2020).

Nematodes pose a great threat to the plants. Plant pathogens like plant parasitic nematodes distort many functions in the plants by causing oxidative damage. This oxidative damage can be calculated in terms of disruptions in the contents of various nonenzymatic antioxidants and antioxidative enzymes (Jasrotia and Ohri, 2016). Also an increase in the H_2O_2 content can simply justify the increased oxidative damage in the tomato plants due to nematode stress. A decrease in the levels of stress indicators like photosynthetic pigments was also observed in plants under nematode stress (Jasrotia and Ohri, 2017). Plant parasitic nematodes thrive on the plant cell contents and then all the plant parts are attacked by the nematodes. Nematodes are also found to ease other pathogens' (soil-borne) entry into the root system that ultimately cause symptoms related to nutrient deficiency like wilting or stunting. Likewise a systemic damage is caused by viruses also, leading to chlorosis, stunting, and malformation in plants (Agrios, 2005).

Both negative and positive impacts of fungi were also reported on host fitness, and fungi also have symbiotic relationships with plants (Rodriguez et al., 2009). Those symbionts exhibiting positive effects on the health of plants include *Epichloe festucae* var. lolii (grass endophytes) and mycorrhizal fungal species (e.g., *Rhizophagus irregularis*). Also those showing negative effects on host plants' fitness include *Sclerotinia sclerotiorum* and *Botrytis cinerea* (Heijden et al., 2015, Johnson et al., 2013, Stukenbrock and McDonald, 2008). Moreover these fungal species also show their interactions with some other microorganisms, such as bacteria (Kobayashi and Crouch, 2009, vanOverbeek and Saikkonen, 2016), which can have a direct or indirect impact on plants by the promotion or inhibition of plants' fitness by affecting the fungal fitness (Frey-Klett et al., 2011). The impact of fungal pathogens also affects the position and morphology of chloroplasts, as both were found to be changed during disease. The function and morphology of chloroplasts of rice were found to be severely disturbed due to the fungal pathogen *Rhizoctonia solani*. When the diseased tissue was examined 3 days after infestation, membrane structures of chloroplasts were found to be disintegrated structurally. Also, photosynthetic performance was found to be reduced and the formation of reactive oxygen species (ROS) was also recorded from those infested chloroplasts. Along with this, a reduction in the rate of electron transport, PSII's maximum quantum yield, and nonphotochemical quenching was also recorded (Ghosh et al., 2017).

Due to the diversity of interactions between host and the pathogens, complicated plant–pathogen interactions occur. Pathogens can destroy the host tissue rapidly by producing cell wall-degrading enzymes (Oliver and Ipcho, 2004). Certain toxins are also produced by the pathogens that ultimately pose certain effects on photosynthetic metabolism, whether direct or indirect (Walters et al., 2008). However, complicated defense systems in plants have also been evolved after the attack of pathogens like fungi, bacteria, and oomycetes. Such defense systems include ETI and pathogen-associated molecular pattern-triggered immunity (PTI). Both ETI and PTI were contributed by the siRNAs and miRNAs by gene silencing or by refining the plant hormones (Gasciolli et al., 2005, Katiyar-Agarwal et al., 2006). So plant immunity was found to be induced by RNA silencing against the pathogen infection. RNA silencing machinery also acted upon transcripts of viriods and the genomic RNA of viruses, and the suppression of RNA accumulation was observed (Yu et al., 2008, Howell et al., 2007, Vazquez, 2006).

Every year there occurs a large economic loss to crops due to fungal pathogens. Wheat is the major crop grown throughout the world. It is also found to be infected by several fungal pathogens like *Puccinia striiformis* that causes wheat strip rust, *Puccinia graminis* that is responsible for wheat stem rust, *Blumeria graminis* that leads to wheat powdery mildew, *Bipolaris sorokiniana* that is responsible for wheat spot blotch, leaf spot disease, and root rot, and *Fusarium culmorum* that is responsible for foot and root rot and Fusarium head blight (FHB) disease (Gill et al., 2004).

13.3 Effect of H_2S on plant morphology

H_2S is an important signaling molecule that helps in performing many functions in plants (Wang, 2012). Basically in plants, H_2S is involved in performing various functions that includes stomatal closure, growth and organogenesis, senescence and regulation, germination of seeds, photosynthesis enhancement, and responses against abiotic stressors. Light-induced stomatal opening was stimulated by H_2S and this molecule also caused the closure of stomata in the dark by decreasing the NO accumulation in *Arabidopsis thaliana* (Hou et al., 2011). In the seedlings of six cultivars of rice it was found that nutrient uptake was suppressed due to a decrease in the release of oxygen induced by H_2S (Wang, 2012, Rennenberg, 1984), thereby supporting the role of H_2S as a phytotoxin. However, in some cultivars of rice, H_2S was observed to promote the uptake of nutrients. The promotion of stomatal opening by H_2S application in different plant species also showed enhanced photosynthesis (Wang, 2012).

Since the involvement of H_2S in the growth and organogenesis in plants has been recorded, it was observed by Lin et al. (2012) NaHS treatment upregulated target genes that are responsible for heme oxygenase-1 [carbon monoxide (HO-1/Co)], ultimately leading to the formation of adventitious roots (Lin et al., 2012). Also, as reported by Li et al., 2010, H_2S at low concentration showed enhancement in the root length of *Pisum sativum* embryo along with root border cells' rate of survival and the activities of peroxidases, ascorbate peroxidase, and superoxide dismutase. Even root organogenesis was also promoted in *Ipomoea batatas*, *Glycine max*, and *Salix matsudana* due to pretreatment of NaHS (Zhang et al., 2009). However, an increase in seedling growth was also recorded in NAHS-treated *Spinacia oleracea* (Chen et al., 2011a). Even flower opening and senescence was also found to be delayed in various flower plants, as observed by Zhang et al. (2011b), when treated with different concentrations of NaHS. They also noted the delayed senescence of cut flowers and branches of different plants when treated with concentrations of NaHS in a dose-dependent manner.

13.4 Hydrogen sulfide–plant hormone crosstalk under pathogen attack

Hydrogen sulfide–plant hormone crosstalk in plants under biotic stress factors regulates various key physiological processes. In the coming sections, we discuss each crosstalk of H_2S with hormones.

13.4.1 Crosstalk between hydrogen sulfide and abscisic acid

The proteins directly related to H_2S production in the small flowering plant *Arabidopsis* are D-cysteine desulfhydrase 1 (DCD1), D-cysteine desulfhydrase 2 (DCD2), L-cysteine desulfhydrase (LCD), and L-cysteine desulfhydrase 1 (DES1) (Bloem et al., 2004; Riemenschneider et al., 2005; Papenbrock et al., 2007; Alvarez et al., 2010, 2012a,b; Shen et al., 2013). It is suggested that H_2S revealed its crosstalk with phytohormone abscisic acid

(ABA) in the regulation of stomatal opening and closing in responss to harsh conditions of drought stress in *Arabidopsis* (Jin et al., 2013). Detailed analysis by real-time polymerase chain reaction disclosed an increase in the levels of *LCD* and *DCD1* transcripts along with a significant increase in the endogenous H_2S upon exposure to cold, dehydration, salt, hydrogen peroxide, and bacterial pathogen in *Arabidopsis* (Shi et al., 2015). The transcripts of various nonbiotic stress-related genes, such as C-Repeat Binding factor (CBF), *CBF1, CBF3, CBF4;* Dehydration Responsive Element Binding (DREB), *DREB2A, DREB2B;* Rat brain (RAB: encodes RAS-like G proteins; name RAB derived from the mammalian (rat) counterpart), *RAB18*; and Responsive to dehydration/dessication (RD), *RD22, RD29A*, and *RD29B*, showed elevation in *LCD* and *DCD1* overexpressing plants and sodium hydrosulfide (NaHS) (H_2S donor)-treated plants. However, a marked decrease in *LCD* and *DCD1* has been observed in knockdown plants and hypotaurine (H_2S scavenger)-treated plants displaying the large effects of H_2S on levels of stress-related gene transcripts (Shi et al., 2015).

By quantifying the number of bacteria in the infested leaves of *Arabidopsis* at 3 days after infection it was found that *DCD1* and *LCD* overexpressing plants possessed fewer bacteria than the wild-type (WT) plants, whereas the *DCD1* and *LCD* knockdown plants possessed more bacteria than WT. Also, NaHS-treated plants were found to exhibit improved defense resistance, whereas the hypotaurine-treated plants showed reduced defense resistance suggesting a direct involvement of hydrogen sulfide in protection against bacteria (Shi et al., 2015).

13.4.2 Crosstalk between hydrogen sulfide and salicylic acid

Salicylic acid (SA) is an essential phytohormone which is known to trigger the defense response in plants against biotrophic and semibiotrophic phytopathogens (Loake and Grant, 2007) and also confers protection against abiotic or environmental stress (Dong et al., 2014; Wani et al., 2017). SA induces a huge set of genes of defense including those which encode pathogenesis-related (PR) proteins. Mutations which impair the production of SA develop susceptibility to virulent and avirulent pathogens. Acetyl SA (aspirin) provides resistance against Tobacco Mosaic Virus in *Nicotiana tabacum* cv. Xanthi-nc (White, 1979). Acetyl SA and SA confer better tolerance to chilling, heat, and drought stresses in tomato plants (*Lycopersicon esculentum L.*) and beans (*Phaseolus vulgaris L.*) (Senaratna et al., 2000). SA also plays a crucial role in the potentiation of the expression of local defense genes (phenylalanine ammonia-lyase, *PAL*; and *Asparagus officinalis* pathogenesis-related, *Ao-PR1*) that do not respond directly to SA but get induced on pathogen attack or wounding as found in tobacco (Mur et al., 1996).

Phytoalexin deficient (PAD) genes (*PAD1, PAD2, PAD3*, and *PAD4*) in *Arabidopsis* encode regulatory factors which act against eukaryotic biotroph *Peronospora parasitica* and confer downy mildew resistance. Mutation in *PAD1, PAD2*, and *PAD4* genes lead to increased susceptibility to bacterial pathogen *Pseudomonas syringae* (Glazebrook et al., 1997). Enhanced disease susceptibility (EDS) genes encode proteins which function like lipases and work in resistance (*R*) gene-directed and play role in plant disease resistance. The *EDS1* is also crucial for pathogen-induced accumulation of *PAD4* mRNA (Feys et al., 2001). It has been observed in *Arabidopsis* that the *PAD4* and *EDS1* transcripts that are associated with biosynthesis of SA were found to be constitutively activated in the plants possessing elevated H_2S levels, but were observed to be reduced in plants with decreased levels of H_2S (Shi et al., 2015). The accumulation of SA is essential for the activation of SA downstream gene expression such as *PRs* (Wang et al., 2007; Kazan and Manners, 2009; Fu and Dong, 2013). Amplified expression of SA-mediated *PR* genes was found in the plants having higher H_2S levels and consequently improved immunity resistance and vice versa (Shi et al., 2015).

13.4.3 Crosstalk between hydrogen sulfide and auxins

Auxin and SA might act separately or through antagonistic crosstalk and the balance between the two is crucial for the plant–pathogen relations (Wang et al., 2007). The transcripts of the multiple miRNAs such as *MIR393, MIR398, MIR167*, and *MIR396* were considerably upregulated by H_2S. *MIR393* among them has been found to be implicated prominently in plant–pathogen communication or interaction (Navarro et al., 2006; Robert-Seilaniantz et al., 2011). Overexpression of a microRNA, *MIR393*, which targets auxin receptors (*AFB1, AFB2, AFB3*, and *TIR1*), provides more resistance to the plants against bacterial pathogens, whereas the overexpression of *AFB1* increases the susceptibility of plants to bacterial pathogens (Navarro et al., 2006; Robert-Seilaniantz et al., 2011; Chen et al., 2011b; Bian et al., 2012; Liu et al., 2013). *MIR393a/b* transcript levels and their genes of target (*AFB1, AFB2, AFB3*, and *TIR1*) were downregulated and upregulated by H_2S, respectively. As the *MIR393* transcript is induced by indole acetic acid (IAA) (Chen et al., 2011b) and H_2S (Shen et al., 2013; Shi et al., 2015), the connection between H_2S and auxin is clearly associated with bacterial resistance.

Shi et al. (2015) found that levels of *MIR393a* and *MIR393b* transcripts were upregulated by H_2S, whereas levels of the Transport Inhibitor Response1 (*TIR1*), Auxin Signaling F Box Protein1 (AFB1), AFB2, and AFB3 transcripts were downregulated by H_2S. Real-time assay showed that the GUS activities of *pMIR393a:GUS* and *pMIR393b:GUS* genetically modified plants were considerably activated by exogenous treatment of NaHS, but were largely decreased by exogenous treatment of HT. They also suggested that H_2S-mediated antibacterial protection might be primarily due to *MIR393*-related auxin signaling.

13.4.4 Crosstalk between hydrogen sulfide and jasmonic acid

Jasmonic acid also performed to its best for the control of plant growth, development, and responses to stress (Ren and Dai, 2012). In plants H_2S acts as a gaseous signaling molecule. H_2S applied exogenously leads to the induction of stress responses in plants (Wang et al., 2012a, Li et al., 2012, Chen et al., 2016). Hou et al. (2011) described the involvement of H_2S in the process of stomatal closure induced by JA.

MEK1/2, an important signal molecule, helps in the regulation of certain biochemical and physiological processes in plants. It has phosphorylated as well as nonphosphorylated forms where the phosphorylated form is the active one. So, the promotion of the phosphorylated form of MEK1/2 by the growth regulators ultimately leads to the regulation of its physiological activity in plants. Anyhow, participation of MEK1/2 in the JA-regulated stomatal closure process has also been observed (Liu et al., 2011). In the leaves of *A. thaliana*, the redox state of ascorbate is found to be regulated by the relationship between H_2S and mitogen-activated protein kinase (MEK1/2) in jasmonic acid (Shan et al., 2019). A significant enhancement in the endogenous H_2S production, MEK1/2 phosphorylation, and reduced ascorbate to dehydroascorbate ratio (AsA/DHA) was observed in *A. thaliana* (wild-type) due to jasmonic acid treatment. A reduction in jasmonic acid was found to be due to hypotaurine (HT), an H_2S scavenger, leading to the induction of the phosphorylation level of MEK1/2, endogenous H_2S production, and the AsA/DHA ratio in wild-type host. These indicators were found to be increased due to sodium hydrosulfide application which acted as a H_2S donor in the mutant-type plants of *A. thaliana*. However mutant-type plants treated with HT and JA, when given an application of NaHS, failed to reverse the effects of hypotaurine on those JA-induced indicators.

13.4.5 Crosstalk between hydrogen sulfide and ethylene

Banana is a climacteric fruit in which the process of ripening is initiated by the increased production of ethylene hormone. Extensive postharvest losses are generally observed in ripe fruits because of the ethylene-induced softening of texture, deterioration of peel, and vulnerability to a variety of diseases cause by pathogens (Liu et al., 1999). Hydrogen sulfide was found to delay the ripening of fruit and senescence in banana by producing an antagonistic effect to ethylene. It also sustained chlorophyll content, a higher level of carotenoids, the soluble proteins, and largely the antioxidant capability of the fruit. It is achieved by the blockage of ethylene signaling by H_2S (Ge et al., 2017). Stomatal closure is also found to be induced by ethylene which is observed by the generation of H_2S from D-/L-CDes, indicating a precursor of ethylene, where ACC is posing its most significant effect. In *Arabidopsis*, synthetic inhibitors of H_2S showed a significant inhibition of ethylene-induced stomatal closure. An increase in AtD-/L-CDes activity and production of H_2S were observed in *Arabidopsis* leaves due to the treatment of ethylene. Plants, where AtD-/L-CDes are overexpressed, showed stomatal closure rather than the ethylene treated wild-type, anyhow AtL-Cdes and AtD-Cdes mutants didn't exhibit that effect. So conclusively their results suggested the role of H_2S generated by D-/L-CDes in regulating stomatal closure induced by ethylene in *A. thaliana* (Hou et al., 2013).

13.4.6 Crosstalk between hydrogen sulfide and gibberellic acid

In plants H_2S is produced to act as a cellular signal. Certain studies revealed that Cd-induced programmed cell death (PCD) was significantly mitigated in cells of *Populus euphratica* due to the pretreatment of NAHS, a H_2S donor (Sun et al., 2013). In wheat aleurone layers, programmed cell death triggered by gibberelins (GA) was found to be delayed due to H_2S by modulating heme oxygenase-1 (HO-1) expression and glutathione homeostasis (Xie et al., 2014). The same was also documented in aleurone layers of barley which were isolated from embryoless imbibed grains of barley by acting as an antioxidant (Zhang et al., 2015). Study on wheat aleurone cells showed that H_2S production and activity of L-cysteine desulfydrase were found to be decreased gradually due to gibberellic acid treatment rather than the responses exhibited by ABA. After observing such a decrease, they also documented the occurrence of their programmed cell death. In wheat aleurone cells PCD triggered by GA was found to be alleviated by using exogenous H_2S that also blocks the reduction of endogenous H_2S release (Xie et al., 2014). Several defense responses can be evoked by H_2S against a wide range of stresses. Various studies have revealed that H_2S when applied exogenously led to the activation of antioxidative enzymes and nonenzymatic antioxidants, thereby leading to the mitigation of plant damage (Francisco, 2019). Moreover other sulfur-containing compounds also play a key role in stress response and plant metabolism and it is also well accepted that phytohormones mark their presence in maintaining the development and growth of plants (Jahan et al., 2019).

13.5 Conclusion

Hydrogen sulfide interacts with plant hormones resulting in the regulation of plant biology. This crosstalk modulates physiological processes in plants under biotic stress by regulation of transcription and posttranslation changes in plant cells. Fig. 13.1 summarizes the main aspects of plant biology regulated by H_2S—hormone crosstalk under pathogen attack.

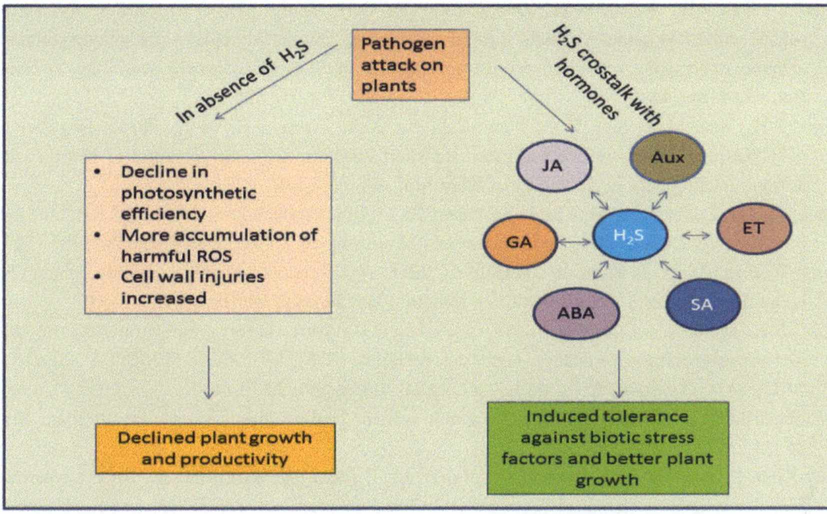

FIGURE 13.1 Overview of hydrogen sulfide and hormone crosstalk-mediated regulation of plant growth.

References

Agrios, G.N., 2005. Preface. In: Agrios, G.N. (Ed.), Plant Pathology, 5th ed. Academic Press, San Diego, CA.

Alvarez, C., Calo, L., Romero, L.C., García, I., Gotor, C., 2010. An O-acetylserine(thiol)lyase homolog with L-cysteine desulfhydrase activity regulates cysteine homeostasis in *Arabidopsis*. Plant Physiol. 152, 656–669.

Alvarez, C., Bermúdez, M.Á., Romero, L.C., Gotor, C., García, I., 2012a. Cysteine homeostasis plays an essential role in plant immunity. N. Phytol. 193, 165–177.

Alvarez, C., García, I., Moreno, I., Pérez-Pérez, M.E., Crespo, J.L., Romero, L.C., et al., 2012b. Cysteine-generated sulfide in the cytosol negatively regulates autophagy and modulates the transcriptional profile in *Arabidopsis*. Plant Cell 24, 4621–4634.

Bai, B., Bian, H., Zeng, Z., et al., 2017. miR393-Mediated auxin signaling regulation is involved in root elongation inhibition in response to toxic aluminum stress in barley. Plant Cell Physiol. 58, 426–439.

Bian, H., Xie, Y., Guo, F., Han, N., Ma, S., Zeng, Z., et al., 2012. Distinctive expression patterns and roles of the miRNA393/TIR1 homolog module in regulating flag leaf inclination and primary and crown root growth in rice (*Oryzasativa*). N. Phytol. 196, 149–161.

Bloem, E., Riemenschneider, A., Volker, J., Papenbrock, J., Schmidt, A., Salac, I., et al., 2004. Sulphur supply and infection with pyrenopezizabrassicae influence l-cysteine desulphydrase activity in *Brassica napus* L. J. Exp. Bot. 55, 2305–2312.

Calderwood, A., Kopriva, S., 2014. Hydrogen sulphide in plants: from dissipation of excess sulfur to signaling molecule. Nitric Oxide 41, 72–78.

Chen, J., Shang, Y.T., Wang, W.H., Chen, X.Y., He, E.M., Zheng, H.L., et al., 2016. Hydrogen sulfide-mediated polyamines and sugar changes are involved in hydrogen sulfide-induced drought tolerance in *Spinacia oleracea* seedlings. Front. Plant Sci. 7, 1–18. Available from: https://doi.org/10.3389/fpls.2016.01173.

Chen, J., Wu, F.H., Wang, W.H., Zheng, C.J., Lin, G.H., Dong, X.J., et al., 2011a. Hydrogen sulfide enhances photosynthesis through promoting chloroplast biogenesis, photosynthetic enzyme expression, and thiol redox modification in *Spinacia oleracea* seedlings. J. Exp. Bot. 62, 4481–4493.

Chen, Z.H., Bao, M.L., Sun, Y.Z., Yang, Y.J., Xu, X.H., Wang, J.H., et al., 2011b. Regulation of auxin response by miR393-targeted transport inhibitor response protein 1 is involved in normal development in *Arabidopsis*. Plant Mol. Biol. 77, 619–629.

Dong, C.J., Li, L., Shang, Q.M., Liu, X.Y., Zhang, Z.G., 2014. Endogenous salicylic acid accumulation is required for chilling tolerance in cucumber (*Cucumissativus* L.) seedlings. Planta 240, 687–700.

Fang, T., Cao, Z., Li, J., Shen, W., Huang, L., 2014. Auxin-induced hydrogen sulfide generation is involved in lateral root formation in tomato. Plant Physiol. Biochem. 76, 44–51.

Feys, B.J., Moisan, L.J., Newman, M.A., Parker, J.E., 2001. Direct interaction between the *Arabidopsis* disease resistance signaling proteins, EDS1 and PAD4. EMBO J. 20 (19), 5400–5411. Available from: https://doi.org/10.1093/emboj/20.19.5400.

Francisco, J.C., 2019. Hydrogen sulfide: a new warrior against abiotic stress. Trends Plant Sci. 24 (11), 983–988.

Frey-Klett, P., Burlinson, P., Deveau, A., Barret, M., Tarkka, M., Sarniguet, A., 2011. Bacterial-fungal interactions: hyphens between agricultural, clinical, environmental, and food microbiologists. Microbiol. Mol. Biol. Rev. 75, 583–609.

Fu, Z.Q., Dong, X.N., 2013. Systemic acquired resistance: turning local infection into global defense. Annu. Rev. Plant Biol. 64, 839–863.

Gasciolli, V., Mallory, A.C., Bartel, D.P., Vaucheret, H., 2005. Partially redundant functions of *Arabidopsis* DICER-like enzymes and a role for DCL4 in producing trans-acting siRNAs. Curr. Biol. 15 (16), 1494–1500.

Ge, Y., Hu, K.D., Wang, S.S., Hu, L.Y., Chen, X.Y., Li, Y.H., et al., 2017. Hydrogen sulfide alleviates postharvest ripening and senescence of banana by antagonizing the effect of ethylene. PLoS One 12, e0180113. Available from: https://doi.org/10.1371/journal.pone.0180113.

Ghosh, S., Kanwar, P., Jha, G., 2017. Alterations in rice chloroplast integrity, photosynthesis and metabolome associated with pathogenesis of *Rhizoctonia solani*. Sci. Rep. 7, 41610.

Gill, B.S., Appels, R., Botha-Oberholster, A.M., Buell, C.R., Bennetzen, J.L., Chalhoub, B., et al., 2004. A workshop report on wheat genome sequencing: International Genome Research on Wheat Consortium. Genetics 168 (2), 1087–1096.

Glazebrook, J., Zook, M., Mert, F., et al., 1997. Phytoalexin-deficient mutants of *Arabidopsis* reveal that PAD4 encodes a regulatory factor and that four PAD genes contribute to downy mildew resistance. Genetics 146 (1), 381–392.

Hancock, J.T., Whiteman, M., 2014. Hydrogen sulfide and cell signaling: team player or referee? Plant Physiol. Biochem. 78, 37–42.

Heijden, M.G.A., Martin, F.M., Selosse, M.-A., Sanders, I.R., 2015. Mycorrhizal ecology and evolution: the past, the present, and the future. N. Phytol. 205, 1406–1423.

Hou, Z.H., Liu, J., Hou, L.X., Li, X.D., Liu, X., 2011. H_2S may function downstream of H_2O_2 in jasmonic acid_induced stomatal closure in *Vicia faba*. Chin. Bull. Bot. 46, 396–406. Available from: https://doi.org/10.3724/SP.J.1259.2011.00396.

Hou, Z., Wang, L., Liu, J., Hou, L., Liu, X., 2013. Hydrogen sulfide regulates ethylene-induced stomatal closure in *Arabidopsis thaliana*. J. Integr. Plant Biol. 55 (3), 277–289.

Howell, M.D., Fahlgren, N., Chapman, E.J., Cumbie, J.S., Sullivan, C.M., Givan, S.A., et al., 2007. Wide analysis of the RNA-DEPENDENT RNA POLYMERASE6/DICERLIKE4 pathway in *Arabidopsis* reveals dependency on miRNA- and tasiRNA-directed targeting. Plant Cell 19 (3), 926–942.

Jahan, B., Sehar, Z., Masood, A., Anjum, N.A., Khan, M.I.R., Khan, N.A., 2019. Sulfur availability potentiates phytohormones-mediated action in plants. Plant Signaling Molecules. Available from: https://doi.org/10.1016/B978-0-12-816451-8.00017-4.

Jasrotia, S., Ohri, P., 2016. 24-Epibrassinolide regulated physiological and biochemical alterations in nematode infected tomato cultivar. Int. J. Pharma Biol Sci. 7, 1179–1185.

Jasrotia, S., Ohri, P., 2017. 24-Epibrassinolide reduces stress in nematode-infected tomato (*Solanum lycopersicum* L.) plants cultured *in vitro*. Vitro Cell Dev. Biol—Plant. Available from: https://doi.org/10.1007/s11627-017-9859-9.

Jin, Z., Shen, J., Qiao, Z., Yang, G., Wang, R., Pei, Y., 2011. Hydrogen sulfide improves drought resistance in *Arabidopsis thaliana*. Biochem. Biophys Res. Commun. 414, 481–486.

Jin, Z., Xue, S., Luo, Y., Tian, B., Fang, H., Li, H., et al., 2013. Hydrogen sulfide interacting with abscisic acid in stomatal regulation responses to drought stress in *Arabidopsis*. Plant Physiol. Biochem 62, 41–46.

Johnson, L., de Bonth, A.M., Briggs, L., Caradus, J., Finch, S., Fleetwood, D., et al., 2013. The exploitation of epichloae endophytes for agricultural benefit. Fungal Divers 60, 171–188.

Katiyar-Agarwal, S., Morgan, R., Dahlbeck, D., Borsani, O., Villegas Jr., A., Zhu, J.K., et al., 2006. A pathogen-inducible endogenous siRNA in plant immunity. Proc. Natl. Acad. Sci. U. S. A. 103 (47), 18002–18007.

Kazan, K., Manners, J.M., 2009. Linking development to defense: auxin in plant-pathogen interactions. Trends Plant Sci. 2009 (14), 1360–1385.

Kobayashi, D.Y., Crouch, J.A., 2009. Bacterial/fungal interactions: from pathogens to mutualistic endosymbionts. Annu. Rev. Phytopathol. 47, 63–82.

Leach, J.E., Leung, H., Tisserat, N.A., 2014. Plant disease and resistance. In: Van Alfen, N.K. (Ed.), Encyclopedia of agriculture and food systems, vol 4. Elsevier, San Diego, pp. 360–374.

Li, D.B., Xiao, Z.X., Liu, L.X., Wang, J.C., Song, G.L., Bi, Y.R., 2010. Effects of exogenous hydrogen sulfide (H_2S) on the root tip and root border cells of *Pisum sativum*. Chin. Bull. Bot. 45, 354–362.

Li, L., Wang, Y., Shen, W., 2012. Roles of hydrogen sulfide and nitric oxide in the alleviation of cadmium-induced oxidative damage in alfalfa seedling roots. Biometals 25, 617–631. Available from: https://doi.org/10.1007/s10534-012-9551-9.

Li, Z.G., 2013. Hydrogen sulfide: a multifunctional gaseous molecule in plants. Russ J. Plant Physiol. 60 (6), 733–740.

Lin, Y.T., Li, M.Y., Cui, W.T., Lu, W., Shen, W.B., 2012. Haem oxygenase_1 is involved in hydrogen sulfide_induced cucumber adventitious root formation. J. Plant Growth Regul. 31, 519–528.

Lisjak, M., Teklic, T., Wilson, I.D., Whiteman, M., Hancock, J.T., 2013. Hydrogen sulfide: environmental factor or signalling molecule? Plant Cell Environ. 36, 1607–1616.

Liu, J., L X H., Liu, G.H., Liu, X., Wang, X.C., 2011. Hydrogen sulfide induced by nitric oxide mediates ethylene-induced stomatal closure of *Arabidopsis thaliana*. Chin. Sci. Bull. 56, 2515–2522. Available from: https://doi.org/10.1007/s11434-011-4819-y.

Liu, X., Shiomi, S., Nakatsuka, A., Kubo, Y., Nakamura, R., Inaba, A., 1999. Characterization of ethylene biosynthesis associated with ripening in banana fruit. Plant Physiol. 121, 1257–1265. Available from: https://doi.org/10.1104/pp.121.4.1257.

Liu, X., Zhang, H., Zhao, Y., Feng, Z., Li, Q., Yang, H.Q., et al., 2013. Auxin controls seed dormancy through stimulation of abscisic acid signaling by inducing ARF-mediated ABI3 activation in *Arabidopsis*. Proc. Natl Acad Sci USA 110, 15485–15490.

Loake, G., Grant, M., 2007. Salicylic acid in plant defence—the players and protagonists. Curr. Opin. Plant Biol 10, 466–472.

Moustafa-Farag, M., Almoneafy, A., Mahmoud, A., Elkelish, A., Arnao, M.B., Li, L., et al., 2020. Melatonin and its protective role against biotic stress. Impacts on plants. Biomolecules 10, 54. Available from: https://doi.org/10.3390/biom10010054.

Mur, L.A.J., Naylor, G., Warner, S.A.J., Sugars, J.M., White, R.F., Draper, J., 1996. Salicylic acid potentiates defence gene expression in tissue exhibiting acquired resistance to pathogen attack. Plant J. 9 (4), 559–571.

Navarro, L., Dunoyer, P., Jay, F., Arnold, B., Dharmasiri, N., Estelle, M., et al., 2006. A plant miRNA contributes to antibacterial resistance by repressing auxin signaling. Science 312, 436–439.

Oerke, E.C., 2006. Crop losses to pests. J. Agric. Sci. 144, 31–43.

Oliver, R.P., Ipcho, S.V.S., 2004. *Arabidopsis* pathology breathes new life into the necrotrophs-vs.-biotrophs classification of fungal pathogens. Mol. Plant Pathol. 5, 347–352.

Papenbrock, J., Riemenschneider, A., Kamp, A., Schulz-Vogt, H.N., Schmidt, A., 2007. Characterization of cysteine-degrading and H_2S-releasing enzymes of higher plants-from the field to the test tube and back. Plant Biol. 9, 582–588.

Rausch, T., Wachter, A., 2005. Sulfur metabolism: a versatile platform for launching defence operations. Trends Plant Sci. 10, 1360–1385.

Ren, C., Dai, C., 2012. Jasmonic acid is involved in the signaling pathway for fungal endophyte-induced volatile oil accumulation of *Atractylodeslancea* plantlets. BMC Plant Biol. 12, 128–133. Available from: https://doi.org/10.1186/1471-2229-12-128.

Rennenberg, H., 1984. The fate excess of sulfur in higher plants. Annu. Rev. Plant Physiol. 35, 121–153.

Riemenschneider, A., Wegele, R., Schmidt, A., Papenbrock, J., 2005. Isolation and characterization of a D-cysteine desulfhydrase protein from *Arabidopsis thaliana*. FEBS J. 272, 1291–1304.

Robert-Seilaniantz, A., MacLean, D., Jikumaru, Y., Hill, L., Yamaguchi, S., Kamiya, Y., et al., 2011. The microRNA miR393 re-directs secondary metabolite biosynthesis away from camalexin and towards glucosinolates. Plant J. 67, 218–231.

Rodriguez, R.J., White Jr., J.F., Arnold, A.E., Redman, R.S., 2009. Fungal endophytes: diversity and functional roles. N. Phytol. 182, 314–330.

Rybicki, E.P., 2015. A top ten list for economically important plant viruses. Arch. Virol. 160, 17–20.

Senaratna, T., Touchell, D., Bunn, E., et al., 2000. Acetyl salicylic acid (Aspirin) and salicylic acid induce multiple stress tolerance in bean and tomato plants. Plant Growth Regul. 30, 157–161.

Shan, C., Sun, H., Zhou, Y., Wang, W., 2019. Jasmonic acid-induced hydrogen sulfide activates MEK1/2 in regulating the redox state of ascorbate in *Arabidopsis thaliana* leaves. Plant Signal Behav 14((8), e1629265.

Shen, J.J., Qiao, Z.J., Xing, T.J., Zhang, L.P., Liang, Y.L., Jin, Z.P., et al., 2012. Cadmium toxicity is alleviated by *AtLCD* and *AtDCD* in *Escherichia coli*. J. Appl Microb 113, 1130–1138.

Shen, J., Xing, T., Yuan, H., Liu, Z., Jin, Z., Zhang, L., et al., 2013. Hydrogen sulfide improves drought tolerance in *Arabidopsis thaliana* by microRNA expressions. PLoS One 8, e77047.

Shi, H., Ye, T., Han, N., Bian, H., Liu, X., Chan, Z., 2015. Hydrogen sulfide regulates abiotic stress tolerance and biotic stress resistance in *Arabidopsis*. J. Integr. Plant Biol. 57 (7), 628–640.

Stukenbrock, E.H., McDonald, B.A., 2008. The origins of plant pathogens in agro-ecosystems. Annu. Rev. Phytopathol. 46, 75–100.

Sun, J., Wang, R., Zhang, X., Yu, Y., Zhao, R., Li, Z., et al., 2013. Hydrogen sulfide alleviates cadmium toxicity through regulations of cadmium transport across the plasma and vacuolar membranes in *Populus euphratica* cells. Plant Physiol Biochem 65, 67–74.

van Overbeek, L.S., Saikkonen, K., 2016. Impact of bacterial–fungal interactions on the colonization of the endosphere. Trends Plant Sci. 21, 230–242.

Vazquez, F., 2006. *Arabidopsis* endogenous small RNAs: highways and byways. Trends Plant Sci. 11 (9), 460–468.

Walters, D.R. (Ed.), 2010. Why do plants need defenses? In: Plant Defense. Wiley-Blackwell; Edinburgh. pp. 1–14.

Walters, D.R., McRoberts, N., Fitt, B.D.L., 2008. Are green islands red herrings? Significance of green islands in plant interactions with pathogens and pests. Biol. Rev. Camb. Philos. Soc. 83, 79–102.

Wang, B.L., Shi, L., Li, Y.X., Zhang, W.H., 2010. Boron toxicity is alleviated by hydrogen sulfide in cucumber (*Cucumissativus* L.) seedlings. Planta 231, 1301–1309.

Wang, D., Pajerowska-Mukhtar, K., Culler, A.H., Dong, X., 2007. Salicylic acid inhibits pathogen growth in plants through repression the auxin signaling pathway. Curr Biol 17, 1784–1790.

Wang, L., Hou, Z., Hou, L., Zhao, F., Liu, X., 2012a. H2S induced byH2O2 mediates drought-induced stomatal closure in *Arabidopsis thaliana*. Chin. Bull. Bot. 47, 217–225. Available from: https://doi.org/10.3724/SP.J.1259.2012.00217.

Wang, R., 2010. Hydrogen sulfide: the third gasotransmitter in biology and medicine. Antioxid. Redox Signal 12, 1061–1064.

Wang, R., 2012. Physiological implications of hydrogen sulfide: a whiff exploration that blossomed. Physiol. Rev. 92, 791–896.

Wani, A.B., Chadar, H., Wani, A.H., Singh, S., Upadhyay, N., 2017. Salicylic acid to decrease plant stress. Environ. Chem. Lett. 15, 101–123.

White, R.F., 1979. Acetylsalicylic-acid (aspirin) induces resistance to tobacco mosaic-virus in tobacco. Virology 99 (2), 410–412.

Xie, Y., Zhang, C., Lai, D., Sun, Y., Samma, M.K., Zhang, J., et al., 2014. Hydrogen sulfide delays GA-triggered programmed cell death in wheat aleurone layers by the modulation of glutathione homeostasis and heme oxygenase-1 expression. J. Plant Physiol. 171, 53–62.

Yu, B., Bi, L., Zheng, B., Ji, L., Chevalier, D., Agarwal, M., et al., 2008. The FHA domain proteins DAWDLE in *Arabidopsis* and SNIP1 in humans act in small RNA biogenesis. Proc. Natl. Acad. Sci. U. S. A. 105 (29), 10073–10078.

Zhang, H., Hu, L.Y., Hu, K.D., He, Y.D., Wang, S.H., Luo, J.P., 2010a. Hydrogen sulfide promotes wheat seed germination and alleviates oxidative damage against copper stress. J. Integr. Plant Biol. 50, 1518–1529.

Zhang, H., Tan, Z.Q., Hu, L.Y., Wang, S.H., Luo, J.P., Jones, R.L., 2010b. Hydrogen sulfide alleviates aluminum toxicity in germinating wheat seedlings. J. Integr. Plant Biol. 52, 556–567.

Zhang, H., Dou, W., Jiang, C.X., Wei, Z.J., Liu, J., Jones, R.L., 2011a. Hydrogen sulfide protects soybean seedlings against drought-induced oxidative stress. Acta Physiol Plant 32, 849–857.

Zhang, H., Hu, S.L., Zhang, Z.J., Hua, L.Y., Jiang, C.X., Wei, Z.J., et al., 2011b. Hydrogen sulfide acts as a regulator of flower senescence in plants. Postharvest Biol. Technol. 60, 251–257.

Zhang, H., Tang, J., Liu, X.P., Wang, Y., Yu, W., Peng, W.Y., et al., 2009. Hydrogen sulfide promotes root organogenesis in *Ipomoea batatas*, *Salix matsudana* and *Glycine max*. J. Integr. Plant Biol. 51, 1086–1094.

Zhang, Y.X., Hu, K.D., Lv, K., Li, Y.H., Hu, L.Y., Zhang, X.Q., et al., 2015. The hydrogen sulfide donor NaHS delays programmed cell death in barley aleurone layers by acting as an antioxidant. Oxid. Med. Cell Longev. 714756. Available from: https://doi.org/10.1155/2015/714756.

Zhang, Y., Pei, Y., Yang, G., 2019. Hydrogen sulphide: a new gasotransmitter in plant defenses. In: 1st ed. Hasanuzzaman, M., Fotopoulos, V., KamrunNahar, Fujita, M. (Eds.), Reactive Oxygen, Nitrogen and Sulfur Species in Plants: Production, Metabolism, Signaling and Defense Mechanisms, vol. 2. John Wiley & Sons Ltd.

Chapter 14

Role of H_2S in plants against radiation stress

Shivam Jasrotia
Government Degree College, Kathua, Jammu and Kashmir, India

14.1 Introduction-role of H_2S

Hydrogen sulfide (H_2S) is associated with the growth and developmental attributes of many horticultural plants. Along with their growth and development of certain tolerance responses were also recorded after stress exposure and in the physiology of fruits in the postharvest period. Enhancement of tolerance has been well recorded by H_2S against many abiotic stress factors, where the effects of H_2S were highlighted in redox regulation, signaling antioxidant capacity, and transcriptional regulation of components involved in cellular defense in plants. Also H_2S has been observed to deal with the germination of seeds, enhancement in growth, organogenesis of roots, and the increase in the fruit's antioxidant capacity and postharvest performance (Fotopoulos et al., 2015)

There is knowledge of H_2S regarding its release and synthesis by plants from a long time ago, but despite this there are few similar studies about the same. H_2S is considered to be a biologically active gaseous molecule and its functioning against various stresses in plants has increased attention toward itself when it is used at low concentrations. It has been seen in *Pisum sativum* that embryonic root length is found to be promoted by H_2S at low concentrations (Li et al., 2010), thereby reducing accumulation of NO and ultimately causing stomatal opening (Lisjak et al., 2010). The role of H_2S in inducing stomata closure and its participation in guard cells for abscisic acid (ABA)-dependent signaling by ATP-binding cassette (ABC) transporters' regulation were reported by García-Mata and Lamattina (2010). Earlier studies also proved H_2S alleviated boron-induced root elongation inhibition in seedlings of cucumber (Wang et al., 2010). Even water has also been found to impart a good role in responses of plants against osmotic or drought stress (Zhang et al., 2009a,b, 2010b), and copper and chromium toxicity (Zhang et al., 2008, 2010a).

Hydrogen Sulfide in Plant Biology. DOI: https://doi.org/10.1016/B978-0-323-85862-5.00012-9

H_2S was noted to induce signaling in *Arabidopsis thaliana* in its guard cells (Lisjak et al., 2010; García-Mata and Lamattina, 2010; Liu et al., 2011). Along with this H_2S is also found to be involved in promoting organogenesis of roots in *Salix matsudana*, *Glycine max*, and *Ipomoea batatas* (Zhang et al., 2009a,b), the formation of adventious roots in cucumber (Lin et al., 2012) plants, elongating the life span of cut flowers (Zhang et al., 2011), photosynthetic enhancement through expression of photosynthetic enzymes, chloroplast biogenesis, and the promotion and modification of thiol redox in the seedlings of *Spinacia oleracea* (Chen et al., 2011).

14.2 Effect of radiation

We know that the solar UV radiation that reaches Earth is UV-A and UV-B; the spectral regions are UV-A 320–400 nm and UV-B 280–320 nm. Due to the depletion of the stratospheric ozone layer, an increase in the intensity of harmful UV-B radiation was observed (Scientific assessment of ozone depletion: 2018, https://www.esrl.noaa.gov/csd/assessments/ozone/2018/). According to Jenkins (2009), it is UV-B spectrum's large proportion that can cause damage to all the living organisms. No doubt growth and metabolism in plants are influenced by solar ultraviolet light but the deleterious effects of the same were also observed in plants due to its high doses, whereas natural doses of UV light are found to be important for morphogenesis in several plant species, including those used in horticulture. In horticultural production greenhouses and some cladding materials are widely used that can absorb UV light strongly. In some plant species there may be limited development due to lower amounts of UV light reaching them. In certain cases light supplementation is also provided by using UV tubes to overcome the UV deficiency (Robson et al., 2015).

Ultraviolet B radiation reaching the Earth's surface was found to have increased significantly due to a decrease in the ozone layer (Blumthaler and Amback, 1990). Wide intra- and interspecific differences to UV-B irradiation have been noted regarding the production and growth of dry matter and changes at the biochemical level (Kramer et al., 1991). Anthocyanins are found to be accumulated and total phenols, flavonoids, and certain UV-absorbing compounds were found to be accumulated after UV irradiation (Balakumar et al., 1993; Lindoo and Caldwell, 1978; Mazza et al., 1999; Dunning et al., 1994). They may help the plants by acting as solar screens to absorb UV before it reaches chloroplasts and other organelles (i.e., the UV sensitive targets). Even enhancements in the soluble phenols and anthocyanins were also observed after UV irradiation in pea or wheat leaves (Alexieva et al., 2001).

Certain physiological changes are also recorded in plants because of UV-B radiation, such as stature and biomass reduction, photosynthesis inhibition, and UV-absorbing compounds' accumulation (Brosche and Strid, 2003;

Bjorn, 1996; Jansen et al., 1998; Frohnmeyer and Staiger, 2003; Rozema et al., 1997; Jordan, 1996). Along with this reactive oxygen species and oxidative stress overproduction was also observed due to high UV-B radiation (Zhao et al., 2015).

The effects of UV-A were enhanced photosynthesis and activation of the photoprotective responses (Verdaguer et al., 2017), for example, photosynthetic rate can be enhanced by 8%–10% by using UV-A in *Sorghum halepense*, *Poa annua*, and *Nerium oleander*. The same was also found to be enhanced in *Pimelea ligustrina* by 12% (Turnbull et al., 2013). On the other hand, an inhibitory effect was also observed on biomass accumulation and growth of cucumber by UV-B (Krizek et al., 1997). An effect of ultraviolet radiation was also found on the chlorophyll content of the plants. A reduction in the total chlorophyll, chlorophyll a, and chlorophyll b content was observed in plants treated with acid rain or UV-B (Zhao et al., 2007). Chl content was found to be reduced due to UV-B due to the inhibition of chl composition because of destruction of chl structure as well as its precursor (Liang et al., 2004; Mirecki and Teramura, 1984; Sullivan and Teramura, 1988). Certain other studies also reported the damage in chl membrane, increasing chlorophyll decomposition and the aggravation of membrane lipid peroxidation, and all this ultimately led to the reduction of chlorophyll content (Caldwell et al., 1998; Allen et al., 2010). For photosynthesis plants require sunlight so there is direct exposure of ultraviolet radiation (UV-B; 280–315 nm). Decreased growth and productivity in plants was observed due to this UV-B radiation. It was also observed that this radiation if provided artificially to the plants, led to damage in the plants causing a reduction in the production of economically important crops like rice, and if this UV-B radiation was excluded, such damage can be prevented, and thus enhancement in the plant growth was observed (Krizek et al., 1997; Krizek et al., 1998). Elevated CO_2 levels caused damage to row crops due to UV-B as dictated by the previous studies (Singh et al., 2010; Koti et al., 2007).

Although chlorophyll content, leaf area, and photosynthesis all were found to be impaired due to enhanced ultraviolet radiation (Correia et al., 1999) and oxidative stress (Costa et al., 2002; A-H-Mackerness et al., 2001). Cellular free active oxygen species (AOS) were found to be generated by UV-B radiation and such AOS include singlet oxygen ($^1O_2$), superoxide anions ($O_2^{\bullet-}$), hydrogen peroxide (H_2O_2), and hydroxyl radicals. Biomolecules like proteins and lipids are found to be damaged by AOS, ultimately progressing toward cell death. However, due to ROS generation and DNA damage caused by elevated UV levels (UV-B; 280-315 nm), a negative effect on the assimilation of carbon (Klem et al., 2012; Guidi et al., 2011; Urban et al., 2006) and the development (Gruber et al., 2010; Wargent et al., 2009) of plants has been recorded. However, despite this it is also notable that UV-B radiation susceptibility also depends upon repair, protection, damage, and on other responses by plants, which are found to vary depending upon different environmental factors, genotypes, UV-spectral composition, dose, and acclimation time (Kataria et al., 2014; Jordan, 2002).

14.3 Plant's response against radiation

As UV-B radiation poses many adverse effects on the plants, many strategies have been developed by plants to repair or prevent such damage by accumulating those UV-absorbing compounds in their leaves (Jordan, 1996). Basically these UV-absorbing compounds help in reducing the UV-B penetration by creating a barrier on the surface of the leaf (Rozema et al., 1997; Bornman et al., 1997). Generally, accumulation of UV-absorbing compounds of a phenolic nature and increased enzymatic and nonenzymatic scavengers of ROS are observed in UV-B acclimated plants (Tilbrook et al., 2013; Jenkins, 2009).

Sturdier and more compact plants were observed when supplemented with UV-A radiation, whereas those enriched with UV-B radiation showed a lack of a sturdy stem and smaller size. Studies revealed that supplementation of UV-A light that acts as an alternative to chemical growth regulator, leads to the production of sturdy horticultural plants. Plants exposed to UV particularly UV-B (280–315 nm) exhibited thicker smaller leaves and increased number of branches (Robson et al., 2015).

14.4 H_2S helps in the amelioration of radiation stress by interacting with other compounds

No doubt interactions of H_2S with other hormones are also observed in plants associated with UV-B stress (Honda et al., 2015; Chen et al., 2015; Jin et al., 2013; Fang et al., 2014; Scuffi et al., 2014; Li et al., 2015; Shi et al., 2014), but still these studies need more clarification. In plants UV-absorbing compounds may prove themselves to be good candidates for preventing UV irradiation–induced damage (Kolb et al., 2001; Flint et al., 1985). Such UV-absorbing compounds are regarded as ROS scavangers (Agati et al., 2009). Gitz et al. (2004) observed a significant induction in leaf accumulation of UV-absorbing compounds in hulless barley seedlings. H_2S is further found to induce the contents of UV-absorbing compounds. Along with H_2S, putrescine (Put) was also found to increase the same against UV-B stress. The interaction of H_2S with Put and H_2O_2 has been observed in hulless barley plants and it was found that plants in which UV-B-absorbing compounds were produced at low levels had higher malondialdehyde (MDA) and EL content than those bringing about high levels of UV-absorbing compounds. They found that H_2S, Put, and H_2O_2 when interacting showed the accumulation of the same compounds in leaves, thus playing an important role against UV-B stress by improving the tolerance of plants to UV-B radiation. By regulating the plasma membrane - Nicotinamide adenine dinucleotide phosphate (PM-NADPH) oxidase and L-cysteine desulfhydrase (L-DES), a significant induction in the accumulation of H_2O_2 and H_2S has been recorded due to the application of putrescine. It was found that it was H_2S that mediated the regulation of H_2O_2 production. Also in creating tolerance against UV-B radiation in hulless barley, the accumulation of flavone and redox homeostasis maintenance helped a lot that were also stimulated by H_2O_2 (Li et al., 2016).

Moreover, the addition of inhibitors of H_2S, putrescine, and H_2O_2 led to the enhancement of some negative effects that were otherwise reduced by the exogenous application of H_2S, Put, and H_2O_2. Also against the damage by UV-B radiation in hulless barley, Put marked its protective role which was impaired by hydroxylamine (a H_2S scavenger), DL-propargyl glycine (a H_2S biosynthesis inhibitor), dimethyl thiourea (a ROS scavenger), and diphenylene iodonium (a PM-NADPH oxidase inhibitor). H_2S production was increased against UV-B stress due to Put, and induction in the activity of L-DES was also observed subsequently. Thus their study concluded that H_2S might play a key role as a novel downstream signaling molecule in hulless barley seedlings against UV tolerance mediated by putrescine (Li et al., 2016).

14.5 H_2S itself is triggered by UV radiation

Li et al. (2016) found that the production of H_2S was even induced by UV-B radiation itself. The production of H_2S was induced in the leaves of barley seedlings due to radiation that in turn ultimately promoted the accumulation of UV-absorbing compounds like anthocyanins and flavonoids.

14.6 Antistress role of H_2S exhibited by regulating antioxidative enzyme activities and antioxidant contents

ROS are generated in response to UV-B radiation in plants and cause oxidative damage. In borage plants, H_2S was observed to alleviate the damage due to UV-B radiation stress when a H_2S donor, that is, NaHS was used. Plants exposed to UV-B radiation showed an increase in protein oxidation, lipid peroxidation, and H_2O_2 content. Activities of various antioxidative enzymes like catalase, guaiacol peroxidase, ascorbate peroxidase, and reduced glutathione (GSH) were not found to exhibit significant changes, whereas a considerable increase in the concentration of flavonoids was observed. Borage plants when pretreated with NaHS showed a decrease in flavonoid content, as determined by HPLC analysis, and an increase in the activities of antioxidative enzymes—guaiacol peroxidase and ascorbate peroxidase. Thus these enhanced activities of antioxidative enzymes due to the use of H_2S proved the alleviation of UV-B stress in borage seedlings, which was indicated by ROS generation. This alleviation may be confirmed by a decrease in flavonoid content in H_2S-treated plants which was otherwise found to increase as a result of stress by UV radiation exposure (Fatemeh et al., 2019).

Certain studies also revealed the regulation of antioxidative enzymes and nonenzymatic antioxidants against oxidative damage caused by biotic stress in plants (Jasrotia and Ohri, 2016, 2017). The generation of free radicals leads to membrane lipid peroxidation which is the indication of cellular damage induced by stress (Jain et al., 2001). Since peroxidation leads to oxidative damage, often the MDA level is used as an indicator of oxidative damage. This lipid peroxidation was observed to change significantly in UV-B radiation–treated seedlings and this was found to be controlled by the activities of antioxidative enzymes (Shen et al., 2010). Studies discussed in this chapter have shown that H_2S exhibits a key role in regulating antioxidative enzyme activities and the antioxidant content helps in the amelioration of radiation stress in plants. This chapter provides some insights of studies on the alleviation of radiation stress in plants but still more research is required in this area (Fig. 14.1).

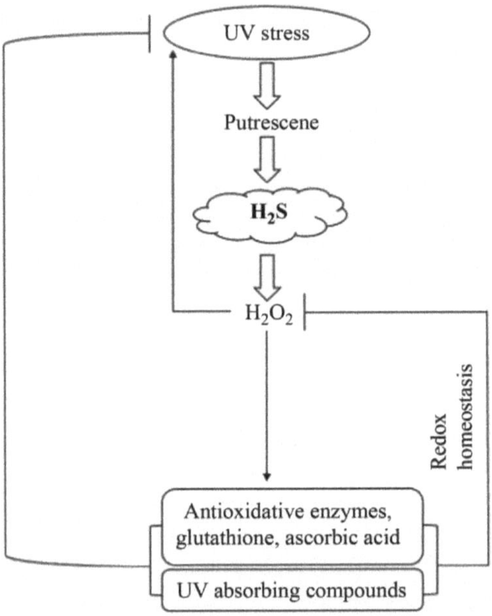

FIGURE 14.1 Model displaying the role of H_2S in relation to putrescine and H_2O_2 in intracellular redox homeostasis maintenance and the accumulation of UV-absorbing compounds in hulless barley seedlings against UV-B radiation stress. H_2S was found to be stimulated by putrescine, whose level was found to be enhanced due to UV-B radiation stress, and stimulation of H_2S ultimately led to the production of H_2O_2. Then this H_2O_2 undergoes downstream signaling that includes the accumulation of UV-absorbing compounds and cellular redox state regulation, thus finally leading to the induction of hulless barley seedling tolerance against UV-B stress. *Li Q., Wang Z., Zhao Y., Zhang X., Zhang S., Bo L., et al., 2016. Putrescine protects hulless barley from damage due to UV-B stress via H2S- and H2O2-mediated signaling pathways. Plant Cell Rep. 35, 1155–1168.*

References

Agati, G., Stefano, G., Biricolti, S., Tattini, M., 2009. Mesophyll distribution of 'antioxidant' flavonoid glycosides in *Ligustrum vulgare* leaves under contrasting sunlight irradiance. Ann. Bot. 104, 853–861.

Alexieva, V., Sergiev, I., Mapelli, S., Karanov, E., 2001. The effect of drought and ultraviolet radiation on growth and stress markers in pea and wheat. Plant Cell Environ. 24, 1337–1344.

Allen, D.J., Nogués, S., Morison, J.I.L., Greenslade, P.D., Mcleod, A.R., Baker, N.R., 2010. A thirty percent increase in UV-B has no impact on photosynthesis in well-watered and droughted pea plants in the field. Glob. Change Biol. 5, 235–244.

Balakumar, T., Hani Babu Vincent, V., Paliwal, K., 1993. On the interaction of UV-B radiation (280–315 nm) with water stress in crop plants. Physiol Plant 87, 217–222.

Bjorn, L.O., 1996. Effects of ozone depletion and increased UV-B on terrestrial ecosystems. Int. F. Environ. Stud. 51, 217–243.

Blumthaler, M., Amback, W., 1990. Indication of increasing solar UV-B radiation flux in alpine regions. Science 248, 206–208.

Bornman, J.F., Reuber, S., Cen, Y.-P., Weissenbock, G., 1997. Ultraviolet radiation as a stress factor and the role of protective pigments. In: Lumsden, P. (Ed.), Plants and UV-B: Responses to Environmental Change. Cambridge University Press, Cambridge, pp. 157–168.

Brosche, M., Strid, A., 2003. Molecular events following perception of UV-B radiation by plants. Physiol. Plant 117, 1–10.

Caldwell, M.M., Teramura, A., Tevini, M., 1998. Effects of increased solar ultraviolet radiation on terrestrial ecosystems. J. Photochem. Photobiol. B Biol. 46, 40–52.

Chen, J., Wu, F.-H., Wang, W.-H., Zheng, C.-J., Lin, G.-H., Dong, X.-J., et al., 2011. Hydrogen sulphide enhances photosynthesis through promoting chloroplast biogenesis, photosynthetic enzyme expression, and thiol redox modifi cation in *Spinacia oleracea* seedlings. J. Exp. Bot. 62, 4481–4493.

Chen, J., Wang, W.H., Wu, F.H., He, E.M., Liu, X., Shangguan, Z.P., et al., 2015. Hydrogen sulfide enhances salt tolerance through nitric oxide-mediated maintenance of ion homeostasis in barley seedling roots. Sci. Rep. 5, 12516.

Correia, C.M., Torres-Pereira, M.S., Torres-Pereira, J.M.G., 1999. Growth, photosynthesis and UV-B absorbing compounds of Portuguese Barbela wheat exposed to ultraviolet-B radiation. Environ. Pollut. 104, 383–388.

Costa, H., Gallego, S.M., Tomaro, M.L., 2002. Effect of UV-B radiation on antioxidant defense system in sunflower cotyledons. Plant Sci. 162, 939–945.

Dunning, C.A., Chalker-Scott, L., Scott, J.D., 1994. Exposure to UV-B radiation increased cold hardiness in *Rhododendron*. Physiol Plant 92, 516–520.

Fang, T., Cao, Z., Li, J., Shen, W., Huang, L., 2014. Auxin-induced hydrogen sulfide generation is involved in lateral root formation in tomato. Plant Physiol. Biochem. 76, 44–51.

Fatemeh, R., Fatemeh, N., Khosrow, M.K., 2019. Alleviation of UV-B radiation damages by sodium hydrosulfide (H_2S donor) pre-treatment in Borage seedlings. J. Plant Interact. 14 (1), 519–524. Available from: https://doi.org/10.1080/17429145.2019.1662100.

Flint, S.D., Jordan, P.W., Caldwell, M.M., 1985. Plant protective response to enhanced UV-B radiation under field conditions: leaf optical properties and photosynthesis. Photochem. Photobiol. 41, 95–99.

Fotopoulos, V., Christou, A., Antoniou, C., Manganaris, G.A., 2015. Review Article. Hydrogen sulphide: a versatile tool for the regulation of growth and defence responses in horticultural crops. J. Hortic Sci. Biotechnol. 90 (3), 227–234.

Frohnmeyer, H., Staiger, D., 2003. Ultraviolet-B radiation-mediated responses in plants. Balancing damage and protection. Plant Physiol. 133, 1420–1428.

García-Mata, C., Lamattina, L., 2010. Hydrogen sulphide, a novel gasotransmitter involved in guard cell signalling. N. Phytol 188, 977–984.

Gitz, D.C., Liu-Gitz, L., McClure, J.W., Huerta, A.J., 2004. Effects of a PAL inhibitor on phenolic accumulation and UV-B tolerance in *Spirodela intermedia* (Koch.). J. Exp. Bot. 55, 919–927.

Gruber, H., Heijde, M., Heller, W., Albert, A., Seidlitz, H.K., Ulm, R., 2010. Negative feedback regulation of UV-B induced photomorphogenesis and stress acclimation in *Arabidopsis*. PNAS 107, 20132e20137.

Guidi, L., Degl'Innocenti, E., Remorini, D., Biricolti, S., Fini, A., Ferrini, F., et al., 2011. The impact of UV-radiation on the physiology and biochemistry of *Ligustrum vulgare* exposed to different visible-light irradiance. Environ. Exp. Bot. 70, 88e95.

Honda, K., Yamada, N., Yoshida, R., Ihara, H., Sawa, T., Akike, T., et al., 2015. 8-Mercapto-cyclic GMP mediates hydrogen sulfide- induced stomatal closure in *Arabidopsis*. Plant Cell Physiol. 56, 1481–1489.

Jain, M., Mathur, G., Koul, S., Sarin, N.B., 2001. Ameliorative effects of proline on salt stress-induced lipid peroxidation in cell lines of groundnut (*Arachis hypogea* L.). Plant Cell Rep. 20, 463–468.

Jansen, M.A.K., Gaba, V., Greenberg, B.M., 1998. Higher plants and UV-B radiation: balancing damage, repair and acclimation. Trends Plant Sci. 3, 131–135.

Jasrotia, S., Ohri, P., 2016. 24-Epibrassinolide regulated physiological and biochemical alterations in nematode infected tomato cultivar. Int. J. Pharma Biol Sci. 7, 1179–1185.

Jasrotia, S., Ohri, P., 2017. 24-Epibrassinolide reduces stress in nematode-infected tomato (*Solanum lycopersicum* L.) plants cultured *in vitro*. Vitro Cell. Dev. Biol. Plant. Available from: https://doi.org/10.1007/s11627-017-9859-9.

Jenkins, G.I., 2009. Signal transduction in responses to UV-B radiation. Annu. Rev. Plant Biol. 60, 407–431.

Jin, Z., Xue, S., Luo, Y., Tian, B., Fang, H., Li, H., et al., 2013. Hydrogen sulfide interacting with abscisic acid in stomatal regulation responses to drought stress in *Arabidopsis*. Plant Physiol. Biochem. 62, 41–46.

Jordan, B.R., 1996. The effects of UV-B radiation on plants: a molecular perspective. Adv. Bot. Res. 22, 97–162.

Jordan, B.R., 2002. Review: molecular response of plant cells to UV-B stress. Funct. Plant Biol. 29, 909e916.

Kataria, S., Jajoo, A., Guruprasad, K.N., 2014. Impact of increasing Ultraviolet-B (UVB) radiation on photosynthetic processes. J. Photochem. Photobiol. B Biol. 137, 55e66.

Klem, K., Ac, A., Holub, P., Kovac, D., Spunda, V., Robson, T.M., et al., 2012. Interactive effects of PAR and UV radiation on the physiology, morphology and leaf optical properties of two barley varieties. Environ. Exp. Bot. 75, 52e64.

Kolb, C.A., Kaser, M.A., Kopecky, J., Zotz, G., Riederer, M., Pfundel, E.E., 2001. Effects of natural intensities of visible and ultraviolet radiation on epidermal ultraviolet screening and photosynthesis in grape leaves. Plant Physiol. 127, 863–875.

Koti, S., Reddy, K.R., Kakani, V.G., Zhao, D., Gao, W., 2007. Effects on carbon dioxide, temperature and ultraviolet-B radiation and their interactions on soybean (*Glycine max* L.) growth and development. Environ. Exp. Bot. 60, 1–10.

Kramer, G.F., Norman, H.A., Krizek, D.T., Mirecki, R.M., 1991. Influence of UV-B radiation on polyamines, lipid peroxidation and membrane lipids in cucumber. Phytochemistry 30, 2101–2108.

Krizek, D.T., Britz, S.J., Mirecki, R.M., 1998. Inhibitory effects of ambient levels of solar UV-A and UV-B radiation on growth of cv. New Red Fire lettuce. Physiol Plant 103, 1–7.

Krizek, D.T., Mirecki, R.M., Britz, S.J., 1997. Inhibitory effects of ambient levels of solar UV-A and UV-B radiation on growth of cucumber. Physiol. Plant 100, 886–893.

Li, D., Xiao, Z., Liu, L., Wang, J., Song, G., Bi, Y., 2010. Effects of exogenous hydrogen sulfide (H_2S) on the root tip an root border cells of *Pisum sativum*. Chin. Bull. Bot. 45, 354–362.

Li, Z.G., Xie, L.R., Li, X.J., 2015. Hydrogen sulfide acts as downstream signal molecule in salicylic acid-induced heat tolerance in maize (*Zea mays* L.) seedlings. J. Plant Physiol. 177, 121–127.

Li, Q., Wang, Z., Zhao, Y., Zhang, X., Zhang, S., Bo, L., et al., 2016. Putrescine protects hulless barley from damage due to UV-B stress via H_2S- and H_2O_2-mediated signaling pathways. Plant Cell Rep. 35, 1155–1168. Available from: 10.1007/s00299-016-1952-8.

Liang, C.J., Zhou, Q., Shen, D.X., Zeng, Q.L., Zhang, G.S., Huang, X.H., 2004. Effect of UV-B radiation and acid rain on rape seedling growth. J. Agro-Environ. Sci. 23, 231–234.

Lin, Y.-T., Li, M.-Y., Cui, W.-T., Lu, W., Shen, W.-B., 2012. Haem oxygenase-1 is Involved in hydrogen sulfide-induced cucumber adventitious root Formation. J. Plant Growth Regul. 31, 519–528.

Lindoo, S.J., Caldwell, M.M., 1978. Ultraviolet-B radiationinduced inhibition of leaf expansion and promotion of anthocyanin production. Plant Physiol. 61, 278–282.

Lisjak, M., Srivastava, N., Teklic, T., Civale, L., Lewandowski, K., Wilson, I., et al., 2010. A novel hydrogen sulfi de donor causes stomatal opening and reduces nitric oxide accumulation. Plant Physiol. Biochem. 48, 931–935.

Liu, J., Hou, L., Liu, G., Liu, X., Wang, X., 2011. Hydrogen sulfi de induced by nitric oxide mediates ethylene-induced stomatal closure of *Arabidopsis thaliana*. Chin Sci. Bull. 56, 3547–3553.

Mackerness, S.A.-H., Fred John, C., Jordan, B., Thomas, B., 2001. Early signaling components in ultraviolet-B responses: distinct roles for different reactive oxygen species and nitric oxide. FEBS Lett. 489, 237–242.

Mazza, C.A., Battista, D., Zima, A.M., Szwarcberg-Bracchitta, M., Giordano, C.V., Acevedo, A., et al., 1999. The effects of solar ultraviolet-B radiation on the growth and yield of barley are accompanied by increased DNA damage and antioxidant responses. Plant Cell Environ. 22, 61–70.

Mirecki, R.M., Teramura, A.H., 1984. Effects of ultraviolet-B irradiance on soybean: V. The dependence of plant sensitivity on the photosynthetic photon flux density during and after leaf expansion. Plant Physiol. 74, 475–480.

Robson, T.M., Klem, K., Urban, O., Jansen, M.A.K., 2015. Re-interpreting plant morphological responses to UV-B radiation. Plant Cell Environ. 38, 856–866.

Rozema, J., Van De Staaij, J., Bjorn, L.O., Caldwell, M., 1997. UV-B as an environmental factor in plant life: stress and regulation. Trends Ecol. Evol. 12, 22–28.

Scuffi, D., Alvarez, C., Laspina, N., Gotor, C., Lamattina, L., Garcia-Mata, C., 2014. Hydrogen sulfide generated by L-cysteine desulfhydrase acts upstream of nitric oxide to modulate abscisic acid-dependent stomatal closure. Plant Physiol. 166, 2065–2076.

Shen, X., Li, X., Li, Z., Li, J., Duan, L., Eneji, A.E., 2010. Growth, Physiological Attributes and Antioxidant Enzyme Activities in Soybean Seedlings Treated with or Without Silicon Under UV-B Radiation Stress, 196. Blackwell, Verlag GmbH, pp. 431–439.

Shi, H., Ye, T., Chan, Z., 2014. Nitric oxide-activated hydrogen sulfide is essential for cadmium stress response in bermudagrass (*Cynodon dactylon* (L). Pers.). Plant Physiol. Biochem. 74, 99–107.

Singh, S.K., Kakani, V.G., Surabhi, G.K., Reddy, K.R., 2010. Cowpea (*Vigna unguiculata* L. Walp.) genotypes response to multiple abiotic stresses. J. Photochem. Photobiol. B Biol. 100, 135–146.

Sullivan, J.H., Teramura, A.H., 1988. Effects of ultraviolet-B irradiation on seedling growth in the *Pinaceae*. Am. J. Bot. 75, 225–230.

Tilbrook, K., Arongaus, A.B., Binkert, M., Heijde, M., Yin, R., Ulm, R., 2013. The UVR8 UV-B photoreceptor: perception, signaling and response. Arabidopsis Book. 11, e0164. Available from: https://doi.org/10.1199/tab.0164.

Turnbull, T.L., Barlow, A.M., Adams, M.A., 2013. Photosynthetic benefits of ultraviolet-A to *Pimelea ligustrina*, a woody shrub of sub-alpine Australia. Oecologia 173, 375–385.

Urban, O., Tuma, I., Holub, P., Marek, M.V., 2006. Photosynthesis and growth response of *Calamagrostis arundinacea* and *C. villosa* to enhanced UV-B radiation. Photosynthetica 44, 215e220.

Verdaguer, D., Jansen, M.A.K., Llorens, L., Morales, L.O., Neugart, S., 2017. UV-A radiation effects on higher plants: exploring the known unknown. Plant Sci. 2017 (255), 72–81.

Wang, B.-L., Shi, L., Li, Y.-X., Zhang, W.-H., 2010. Boron toxicity is alleviated by hydrogen sulfide in cucumber (*Cucumis sativus* L.) seedlings. Planta 231, 1301–1309.

Wargent, J.J., Moore, J.P., Ennos, A.R., Paul, N.D., 2009. Ultraviolet radiation as a limiting factor in leaf expansion and development. Photochem. Photobiol. 85, 279e286.

Zhang, H., Hu, L.-Y., Hu, K.-D., He, Y.-D., Wang, S.-H., Luo, J.-P., 2008. Hydrogen sulfide promotes wheat seed germination and alleviates oxidative damage against copper stress. J. Integr. Plant Biol. 50, 1518–1529.

Zhang, H., Tang, J., Liu, X.P., Wang, Y., Yu, W., Peng, W.Y., et al., 2009a. Hydrogen sulfide promotes root organogenesis in *Ipomoea batatas, Salix matsudana Glycine max*. JIPB 51, 1086–1094.

Zhang, H., Ye, Y.-K., Wang, S.-H., Luo, J.-P., Tang, J., Ma, D.-F., 2009b. Hydrogen sulfide counteracts chlorophyll loss in sweetpotato seedling leaves and alleviates oxidative damage against osmotic stress. Plant Growth Regul. 58, 243–250.

Zhang, H., Hu, L.Y., Li, P., Hu, K.D., Jiang, C.X., Luo, J.P., 2010a. Hydrogen sulfide alleviated chromium toxicity in wheat. Biologia Plant 54, 743–747.

Zhang, H., Jiao, H., Jiang, C.-X., Wang, S.-H., Wei, Z.-J., Luo, J.-P., et al., 2010b. Hydrogen sulfide protects soybean seedlings against drought-induced oxidative stress. Acta Physiol. Plant 32, 849–857.

Zhang, H., Hu, S.-L., Zhang, Z.-J., Hu, L.-Y., Jiang, C.-X., Wei, Z.-J., et al., 2011. Hydrogen sulfi de acts as a regulator of fl ower senescence in plants. Postharvest Biol. Tec. 60, 251–257.

Zhao, X.L., Zheng-Hua, H.U., Wang, Z.M., Zheng, Y.F., Jian-Qiang, X.U., 2007. Interactive effects of enhanced UV-B radiation and acid rain on the seedling growth and physiological traits of pakchoi and lettuce. J. Henan Agric. Sci. 2, 73–77 (in Chinese).

Zhao, C., Wang, X., Wang, X., Wu, K., Li, P., Chang, N., et al., 2015. Glucose-6-phosphate dehydrogenase and alternative oxidase are involved in the cross tolerance of highland barley to salt stress and UV-B radiation. J. Plant Physiol. 181, 83–95.

Chapter 15

Hydrogen sulfide and phytohormones crosstalk in plant defense against abiotic stress

Jaspreet Kour[1], Kanika Khanna[1], Pooja Sharma[1,2], Arun Dev Singh[1], Indu Sharma[3], Priya Arora[1], Pardeep Kumar[1], Kamini Devi[1], Mohd Ibrahim[1], Puja Ohri[4], Bilal Ahmad Mir[5], Ashutosh Sharma[6] and Renu Bhardwaj[1,*]

[1]Department of Botanical and Environmental Sciences, Guru Nanak Dev University, Amritsar, India, [2]Department of Microbiology, DAV University, Jalandhar, India, [3]Sant Baba Bhag Singh University, Padhiana, Khiala, Jalandhar, India, [4]Department of Zoology, Guru Nanak Dev University, Amritsar, India, [5]Department of Botany, School of Life Sciences, Satellite Campus Kargil, University of Kashmir, Jammu and Kashmir, India, [6]Department of Agriculture Sciences, DAV University, Jalandhar, India

**Corresponding author.*

15.1 Introduction

A small gaseous transmitter and lipophilic molecule, H_2S (hydrogen sulfide) was considered a phytotoxin until recent studies reported multifaceted roles of this small molecule when applied at lower concentrations (Abe and Kimura, 1996; Li, 2013). The existence of L- and D-cysteine amino acid-specific L- and D-cysteine desulfydrases that catalyzed the production of H_2S in chloroplasts, mitochondria, or in the cytoplasm of cucurbit plants or cultured tobacco cells was first reported by Rennenberg et al. 1987. Hydrogen sulfide generation in plants is not only important for plant growth and development, but also infuences plant responses to environmental stresses (Jin and Pei, 2015). H_2S acts as a reductive substance that directly scavenges reactive oxygen species (ROS), superoxide peroxynitrite, hypochlorite, and hydrogen peroxide (H_2O_2) (Mitsuhashi et al., 2005; Whiteman et al., 2004; Geng et al., 2004). In late 2002, H_2S was reported to induce germination of seeds and metal stress amelioration (Zhang et al., 2008). It is a distinct gaseous molecule that dynamizes

Hydrogen Sulfide in Plant Biology. DOI: https://doi.org/10.1016/B978-0-323-85862-5.00009-9

the antioxidant system and effectively intervenes to ameliorate water stress (Corpas, 2019). Similarly H_2S is also reported to be helpful in ameliorating heat stress and salt stress, which are discussed in detail in this chapter. Phytohormones are important molecules in plants and interact with potential signaling molecules in order to trigger a downstream response (Cheng et al., 2013). Since the physiology of plants is highly dependent on the homeostasis between different phytohormones, it is evident that a signaling molecule like H_2S must interact with these phytohormones at multiple levels to overcome the stress in plants and enhance the growth and development of plants. In this book chapter we have discussed in detail the in-depth knowledge of the role of H_2S in ameliorating various abiotic stresses and crosstalk nodes exist between the gaseous molecule and plant hormones like abscisic acid (ABA), auxins, ethylene, and salicylic acid (SA). This discussion also aims at identifying the role of H_2S as a potential signaling agent involved in the systemic establishment of plants under variable environmental challenges.

15.2 Generation of H_2S in plants

H_2S is generated in the plant cells as an intermediary of an assimilatory sulfate reduction (Filipovic and Jovanovic, 2017) process. The most important enzymes responsible for maintaining H_2S homeostasis in plant cells are:

1. L-cysteine desulfhydrase (LCD)
2. D-cysteine desulfhydrase (DCD)
3. Sulfite reductase (SiR)
4. Cyanoalanine synthase (CAS)
5. Cysteine synthase (CS)

Prominently, the LCD/DCD equilibrium breaks down L-/D-cysteine to amine, pyruvate, and H_2S (Li, 2013). These enzymes are receptive to abiotic stresses and consequently the production of H_2S (Hancock and Whiteman, 2014). The interactions between cysteine and cyanide catalyzed by CAS yields cyanoalanine along with H_2S (Aroca et al., 2017). The ferredoxin-dependent reduction of sulfite to H_2S is catalyzed by SiR enzymes (Calderwood and Kopriva, 2014). In the biosynthesis of cysteine, O-acetylserine (OAS) thiol lyase combines H_2S with O-acetyl-L-serine. This enzymatic reaction is nonetheless reversible and therefore can also degrade cysteine to H_2S and O-acetyl-L-serine. Plants usually intercede stomatal release of the excess H_2S into the air (Li et al., 2015a,b,c).

Laureano-Marín et al. (2014) reported that the LCD enzyme in *Arabidopsis*, DES1, catalyzes the endogenous cellular biosynthesis of H_2S. Fusion of the *GFP* reporter gene with the *DES1* promoter revealed its expression to be localized in the cytosol of epidermal, mesophyll, and guard cells.

1. H_2S as an effective mechanism for dissipation of surplus sulfur

   H_2S is vital for the fate of excess sulfur. Plants are able to emit H_2S from the leaves in reaction to feeding with sulfite or sulfate, predominantly in detached leaves or through wounded roots (Wilson et al., 1978). The emissions can be diminished by feeding the cysteine precursor, that is, OAS. Thus plants tend to emit H_2S when the rate of sulfate reduction surpasses the rate of provision of carbon skeletons for cysteine biosynthesis. Nevertheless, an overload of reduced sulfur can also lead to H_2S emissions. H_2S was emitted from leaves fed with cysteine, liberated either by the reverse reaction of OAS-TL (O-acetylserine thiol lyase) or by L-cysteine desulfhydrase (LCD) (Sekiya et al., 1982). Consequently, H_2S emission as a mechanism to dissipate surplus sulfur is very crucial to prevent the accumulation of sulfite and sulfide which are potential triggers of oxidative stress and cysteine, which at elevated concentrations would affect the cellular redox balance in plants.
2. Role of H_2S in plant nutrition

   The key source of sulfur for plants is sulfate present in the soil. However, plants are also capable of utilizing other sulfur sources, such as a range of sulfur-containing gases in the atmosphere. Conventionally, these comprise sulfur dioxide and H_2S, which were considered major atmospheric pollutants before the 1980s (Lamers et al., 2013). Although other detoxification mechanisms exist, the way that plants cope up with exposure to these toxic gases is mainly by their assimilation into organic sulfur compounds (Yarmolinsky et al., 2013). Thus exposure to SO_2 leads to both increased reduction of the gas to sulfide and synthesis of cysteine and glutathione (GSH) (Brunold et al., 1983). H_2S, on the other hand, is mostly assimilated into the thiols (Durenkamp and De Kok, 2004). The rapid assimilation of the gases is vital, as they rapidly cause the accumulation of ROS leading to oxidative stress. Hence, the increased synthesis of cysteine, with sulfide as an intermediate, serves two significant purposes: (1) reducing the active concentration of the gases in leaves, and (2) making more GSH available to protect against oxidative stress in plants. Enhanced availability of cysteine in the fumigated plants reduces demand for sulfate uptake and reduction and it was observed that sulfate uptake and APS reductase are downregulated in SO_2- or H_2S-exposed plants (Koralewska et al., 2008). Interestingly, the effects of fumigation and the mechanisms of the regulation vary from plant to plant (Durenkamp et al., 2007). Usually, fumigation with H_2S mitigates symptoms of sulfate starvation. Plants grown in nutrient solution without sulfate resume growth when simultaneously exposed to H_2S. The H_2S assimilated in the leaves can therefore provide adequate sulfur for the synthesis of proteins and all other cellular components vital for growth, and can be considered as a nutrient (Lamers et al., 2013).

Analysis of S isotopes by Lamers et al. (2013) revealed that in many marsh plants and sea grasses more than 50% of sulfur originates from

H_2S. Nonetheless, this H_2S is not atmospheric, but originates in the soil through the anaerobic metabolism of bacteria. Therefore whilst the function of H_2S as a nutrient and source of sulfur for plant growth and development may be marginal for the traditional plants and crops, it is highly essential in specific soils and ecosystems.

15.3 Role of H_2S in plants under abiotic stress

15.3.1 Salt stress

Salinity is one of the abiotic stresses that lead to nutritional disorders, oxidative stress, and cause damage to the photosynthetic apparatus, which affect plant growth and crop yield (Ismail et al., 2014). Various effects produced due to salinity in plants are given in Table 15.1. H_2S has been found to play a protective role in plants under salt stress. In 2010, Zhang et al. (2010a,b,c) studied the effect of various treatments of H_2S donor (NaHS) during osmotic stress in wheat. It was reported that with the increase of osmotic stress, the rate of seed germination was decreased. NaHS treatment could improve the seed germination; improve the activities of enzymes amylase, esterase, ascorbate peroxidase, and catalase; reduce the amounts of malondialdehyde (MDA) and hydrogen peroxide (H_2O_2); and decreased the activity of lipoxygenase in wheat seeds under osmotic stress. Also, the treatments of H_2S donors during osmotic stress resulted in retaining higher endogenous H_2S concentrations in the seeds of wheat.

In another study, the effect of H_2S donors (NaHS) was investigated in the mitigation of stress-induced by 150 and 300 mM concentration of NaCl (salt), 15% and 30% PEG6000 (osmotic), and 4°C (cold) in Bermuda grass (Shi et al., 2013). It was observed that salinity, osmotic, and cold stress-induced the accumulation of endogenous levels of H_2S, whereas exogenous treatments of NaHS alleviated stress tolerances by decreasing electrolyte leakage, ROS imbalance, and cell damage, thereby enhancing survival rate under stressed conditions in Bermuda grass. The effect of the H_2S donor (NaHS) in the hypocotyls and radicles of cucumber under salt (100 mmol L^{-1} NaCl) stress was observed by Yu et al. (2013) on growth, oxidative damage, and antioxidant capacity of salt-stressed cucumber. The stress alleviating role of NaHS was observed in terms of a significantly reduced accumulation of thiobarbituric acid-reactive substances and H_2O_2, whereas both activities of antioxidant enzymes and antioxidant capacity was enhanced under salt stress.

Li et al. (2014a,b,c) reported that treatment of NaHS (H_2S donor) could maintain the ion (Na^+/K^+ ratio) homeostasis in an H_2O_2-dependent manner in salt (NaCl at100 mM concentration)-stressed *Arabidopsis* root. This was possible because H_2O_2 was an integral part of the H_2S-induced salt tolerance pathway and H_2S could induce the H_2O_2 production endogenously through

TABLE 15.1 Effect of H_2S in plants under salt stress.

Plant studied	Treatments applied	Parameters studied	Effect/role on plant during stress	References
Wheat (*Triticum aestivum* L.) seeds	NaHS: a sustained H_2S donor	• Seed germination activities of enzymes amylase, esterase, ascorbate peroxidase, superoxide dismutase, catalase • Contents of hydrogen peroxide and malondialdehyde • Lipoxygenase activity Endogenous levels of H_2S	H_2S donor, NaHS, played protective role in osmotic stress amelioration in rice seeds by enhancing activities of antioxidant enzymes, esterase and amylase. The rate of lipid peroxidation was decreased, whereas levels of endogenous H_2S were retained.	Zhang et al. (2010a,b,c)
Bermuda grass [*Cynodon dactylon* (L). Pers.]	NaHS: the H_2S donor	• Electrolyte leakage • ROS burst • Cell damage • Activities of catalase, peroxidase and glutathione reductase • Amounts of glutathione	Exogenous treatments of NaHS alleviated salt, osmotic and cold stress in Bermuda grass by maintaining ROS balance, reducing cell damage, enhancing osmolyte accumulation; and, modulating both enzymatic and nonenzymatic antioxidant defense system.	Shi et al. (2013)
Cucumber hypocotyls and radicles	NaHS: the H_2S donor	• Content of H_2O_2 • Amylase activity • Length of hypocotyls and radicles • Rate of lipid peroxidation • Activities of antioxidant enzymes • Antioxidant capacity in terms of DPPH (**2,2**-diphenyl-1-picrylhydrazyl) scavenging activity, chelating activity of ferrous ions and hydroxyl radical (−OH) scavenging activity	Application of NaHS at 400 $\mu mol\ L^{-1}$ concentration was found most effective dose in alleviating salt stress. NaHS application reduced the rate of lipid peroxidation, increased amylase activities in cotyledons, promoted length of hypocotyls and radicles, and modulated the activities of antioxidant enzymes and capacity of antioxidants under salt stress in cucumber.	Yu et al. (2013)
Arabidopsis thaliana root	NaHS (H_2S donor) and H_2O_2	• Ion (Na^+/K^+ ratio) homeostasis	H_2S application maintained the Na^+/K^+ homeostasis in the H_2O_2-dependent manner in salt (NaCl at 100 mM concentration) stressed *Arabidopsis* root.	Li et al. (2014a,b,c)

(*Continued*)

TABLE 15.1 (Continued)

Plant studied	Treatments applied	Parameters studied	Effect/role on plant during stress	References
Maize (*Zea mays* L.) leaves	NaHS: the H_2S donor	• Redox states of ascorbate, glutathione • Activities of antioxidant enzymes • Electrolyte leakage • Ascorbate and glutathione metabolism	Exogenous applications of H_2S maintained the redox states of glutathione and ascorbate through the upregulation of glutathione and ascorbate metabolism in salt stress management in maize leaves.	Shan et al. (2014)
Alfalfa (*Medicago sativa*) seedlings	Sodium hydrosulfide (NaHS; 100 μM), a H_2S donor	• *In planta* changes of H_2S homeostasis • Activity of L-cysteine desulfhydrase • Seedling growth • Lipid peroxidation	Endogenous H_2S enhanced salt (NaCl) stress tolerance in *Medicago* seedlings by maintaining K^+/Na^+ homeostasis. Also, it prevented efflux (loss) of K^+ under salt stress by impairing the expression of Shaker-like K^+ outward-rectifying channel.	Lai et al. (2014)
Rice (*Oryza sativa* L. cv. BRRI dhan52) plants	0, 25, 50, 100, and 200 μM NaHS 500 μM Hypotaurine (scavenger of H_2S)	• Mineral homeostasis • Na^+/K^+ balance • Oxidative metabolism	H_2S treatments restricted the excessive uptake of Na^+ ions and thus maintained Na^+/K^+ balance during salt stress. H_2S applications maintained mineral homeostasis, enhanced photosynthetic capacity, and reduced the salt stress-induced oxidative damage in rice plants.	Mostofa et al. (2015a,b)
Tobacco (*Nicotiana tabacum* L. cv. Havana)	NO and H_2S	• Endogenous levels of H_2S and NO • Stomatal conductance • Activities of enzymes of biosynthetic pathway of H_2S and NO • Activities of enzymes of redox and antioxidant defense system	Salinity stress stimulated the accumulation of endogenous NO and H_2S via regulating the activities of enzymes of their biosynthetic pathway, antioxidant and redox defense system, metabolic and biochemical changes in tobacco. The interaction of both NO and H_2S levels resulted in the survival of tobacco plants under high salt stress conditions.	da Silva et al. (2017)

the regulation of activities of enzymes, like plasma membrane bound NADPH oxidase and glucose-6-phosphate dehydrogenase. Shan et al. (2014) studied the effect of exogenous application of an H_2S donor (NaHS) on the redox states of ascorbate, GSH, and activities of antioxidant enzymes in the leaves of maize under salt stress (100 mM NaCl). Pretreatments of NaHS increased the activities of antioxidant enzymes such as Γ-glutamylcysteine synthetase, dehydroascorbate reductase (DHAR), glutathione reductase (GR), *L*-galactono-1,4-lactone dehydrogenase, and ascorbate peroxidase under NaCl stress. NaHS pretreatments reduced the content of MDA and leakage of electrolytes during salinity stress in maize leaves. Lai et al. (2014) reported the effect of H_2S in alfalfa seedlings under salt stress. The investigation of *in planta* changes during H_2S homeostasis revealed that endogenous H_2S increased salt stress tolerance through the reestablishment of Na^+/K^+ balance and restricting the loss of K^+ in alfalfa seedlings under salinity stress.

Mostofa et al. (2015a,b) studied the role of H_2S donor (NaHS) on endogenous H_2S levels, mineral homeostasis, Na^+/K^+ balance, accumulations of MDA and H_2O_2 in rice plants under salt (150 mM NaCl) stress. It has been reported that H_2S treatments helped to maintain the homeostasis of Na^+ and K^+ ions which is a vital factor for the survival of plants under salt stress. At the cellular level, H_2S application inhibited the excessive uptake of Na^+ ions and maintained the balance of K^+ ions during salinity stress in salt-sensitive crop (rice). Besides this, H_2S treatments increased fresh weight and plant height during stress. Further, the endogenous H_2S levels were also observed to be altered under salinity stress in rice. The treatment of a H_2S scavenger (hypotaurine), decreased the plant growth and biomass under salt stress. NaHS (a H_2S donor) treatments enhanced the levels of carotenoid, soluble proteins, chlorophylls, ascorbic acid, redox states, and GSH, whereas it reduced the accumulation of various ROS, thereby emphasizing the salt stress protective roles of H_2S in rice plants.

Further, Deng et al. (2016) reported stress protective effects of NaHS in seedlings of wheat (*Triticum aestivum* L. cv LM 15, a salt-sensitive cultivar) subjected to NaCl stress. It has been reported that pretreatments with 0.05 mM NaHS remarkably ameliorated the salt stress-induced growth, decreased the Na^+ levels, maintained Na^+/K^+ homeostasis, promoted selective absorption and transport for K^+ instead of Na^+, and enhanced selective transport for K^+ under salt stress. This study emphasized that exogenous treatments of H_2S ameliorated salt-stress induced growth inhibition through the reestablishment of lower levels of Na^+ in wheat seedlings during salinity stress.

The endogenous levels of H_2S and NO (nitric oxide) in the leaves of tobacco were studied under salinity (NaCl) stress (da Silva et al., 2017). Salt stress enhanced the levels of both H_2S and NO in tobacco leaves which was due to enhancement in levels of L-arginine and L-cysteine. This increase was correlated with the altered stomatal conductance; induction of activities of enzymes such as arginase, cyanoalanine synthase (CAS), LCD, cysteine synthase (CS), nitrate reductase, superoxide dismutase, and catalase; and the nonenzymatic antioxidant defense system (GSH) in salt-stressed tobacco leaves.

In 2018, da-Silva and Modolo (2018) reviewed the role of both NO (nitric oxide) and H_2S in mitigation of salinity stress in plants. It has been reported that plant growth and development has been affected due to salinity stress-induced osmotic stress in plants. Due to applications of H_2S and NO, cell signaling induced a cascade of subsequent biochemical events triggered during osmotic stress. The studies by Shi et al. (2013), Yu et al. (2013), Shan et al. (2014), Lai et al. (2014), Deng et al. (2016) and da-Silva et al. (2017) emphasized that H_2S can be employed in controlling ROS via regulation of the antioxidant defense system, maintaining the homeostasis of the Na^+/K^+ ratio, and accumulation of osmolytes in plant stress management and protection against salinity stress. Further, the stress protective roles of the effect of H_2S have been studied in stomatal closure (Ma et al., 2019), and the antioxidant defense system and stability of mitochondrial membrane, respectively, in *Vicia faba* and *Malus hupehensis* under salinity stress.

On the basis of present information, a model for the role of H_2S in plant salt stress amelioration has been proposed (Fig. 15.1). High concentrations of salt lead to the activation of a series of biochemical events and an imbalance results in oxidative damage leading to cell death. However, the application of the H_2S donor triggers a cascade of sequential biochemical events. Salt stress leads to the production of LCD which results in the production of endogenous H_2S (da-Silva and Modolo, 2018). Besides this, the condensation of L-cysteine to cyanide (CN^-) results in the production of endogenous H_2S, which is catalyzed by a mitochondrial enzyme, namely β-cyanoalanine synthase (Hatzfeld et al., 2000; Li et al., 2015a,b,c; da-Silva and Modolo, 2018).

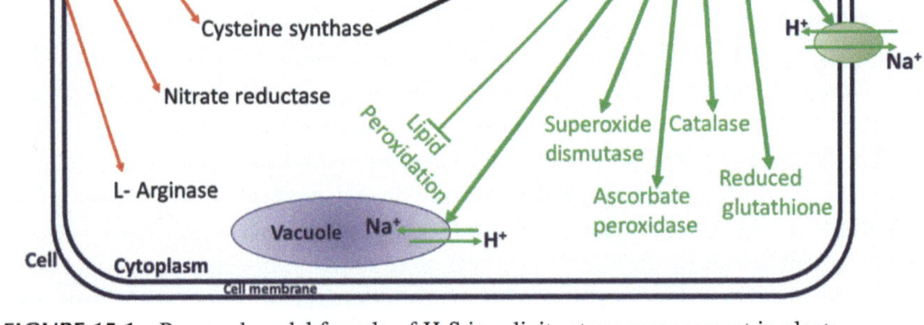

FIGURE 15.1 Proposed model for role of H_2S in salinity stress management in plants.

In cytosol, chloroplasts, and mitochondria the enzyme CS catalyzes the reaction (reversible) between L-cysteine and acetate so as to form *O*-acetyl-L-serine and produce H_2S (Wirtz and Hell, 2006; Li, 2015; da-Silva and Modolo, 2018). Furthermore, it has been reported that NaCl at higher concentrations leads to the stimulation of CS and β-cyanoalanine synthase in tobacco which subsequently leads to the accumulation of H_2S in tobacco leaves (da Silva et al., 2017; da-Silva and Modolo, 2018). Therefore high salt stress results in enhanced levels of endogenous H_2S in stressed plants through the activation of enzymes of its biosynthetic pathway. The prolonged periods of salt stress and exogenous H_2S applications result in modulating the enzymatic and nonenzymatic antioxidant defense systems, selective uptake of K^+ and Na^+, and enhanced levels of osmolytes and inhibition of lipid peroxidation. These processes thereby lead to stress management in plants. Recent reports by Singh et al. (2020), Pandey and Gautam (2020) and Liu et al. (2020) also emphasized the stress protective role of H_2S in plants subjected to salt stress. Further detailed investigations into the interplay of H_2S with other molecules may help in understanding the detailed complex cascade of events responsible for plant stress management, and such studies may provide a solid foundation for designing and developing salt-tolerant crop varieties through the management of endogenous H_2S levels.

15.3.2 Water stress

Drought/water/osmotic stress is considered to be the most important atmospheric factor that reduces the yield of crops (Boyer, 1982). Osmotic stress leads to limiting the growth of the plant by reducing water uptake which affects the biosynthesis of important proteins and enzymes. Due to this, there is the increase of hydrogen peroxide and other oxygen reactive species that cause oxidative stroke to the cell membrane (Mittler, 2002). Van Breusegem et al. (2001) reported that the high accumulation of oxygen reactive species cause photoinhibition leading to chlorosis of the leaf. To mitigate the osmotic stress, plants secrete hydrogen sulfide which further triggers the expression of antioxidant enzymes like catalase, dismutase, etc. that scavenge the ROS and improve the oxidative toxicity. During stress, it acts as a trigger which provokes the osmolytes and antioxidants (Banerjee et al., 2018) that scavenge the oxygen reactive species. Several osmoprotectants like sugars, beatinealdehyde, glycinebeatine, sugar alcohols (sorbitol, pinnitol), amino acids (proline), polyamines, etc., are elevated during drought stress (Rivero et al., 2014). To conquer water stress-induced oxidative deflation, plants adopt an intoxicant defense system. Under water stress, the ascorbate and GSH cycle also plays an ameliorating role in balancing their redox state. The cycle involves four enzymes: APX (ascorbate peroxidise), GR, MDHAR (monodehydroascorbate reductase), and DHAR (Shan et al., 2018). Hydrogen sulfide protects stress-sensitive proteins and has diverse role in H_2S-mediated signaling pathways and interactions with phytohormones (Banerjee Roychoudhury 2018). There is an increase in H_2S level in the leaves and an enhancement of the expressions of L/D_CD and H_2O_2 content in the guard cells during drought stress (Li et al., 2013). H_2S alters the expression of stress-sensitive proteins (like MAP kinases) and antioxidants. It is also reported that GSH is managed by H_2S in time of osmotic stress. Water stress leads to oxidative damage causing the amassing of ROS which further reduces the absorption of GSH, thus boosting GSH internally by H_2S, which has a decisive effect on cell signaling (Hancock, 2019).

H_2S generally provides osmotic tolerance in plants by two means: Firstly, it can diffuse out from cell membranes and organelle matrixes to discard the extra ROS (Fan et al., 2017) and, secondly, H_2S reduces transpiration rate by closing stomata (Zupin et al., 2017). It has been found that there is cross talk between plant hormones and Hydrogen sulphide under drought stress in *V. faba, Impatiens walleriana* and *Arabidopsis thaliana* which causes closing of stomata via ABC transporters that are convoluted in ABA biosynthesis (García-Mata and Lamattina, 2010). To ameliorate the drought stress, H_2S production or the application of hydrogen donors, for example, NaHS, is increased due to the enhanced expression of LCD and DCDes. This leads to the closing of stomata and alleviates drought stress in *A. thaliana* (Jin et al., 2011a,b).

Various reports have been compiled to explain the role of H_2S in plants under water stress in Table 15.2.

15.3.3 Heat stress

Unremitting increase in temperature due to global warming is of great concern for agriculture globally, because heat stress alters cellular homeostasis, hinders plant growth and development, and eventually causes death and limits the crop yield. Heat stress damages the photosynthetic machinery of plants, hinders carbon assimilation processes, and disturbs water status by increased transpiration and membrane stability (Hasanuzzaman et al., 2013; Cheng et al., 2018). Plants have evolved different mechanisms to tolerate elevated heat stress. H_2S plays an important role in combating heat stress. The role of H_2S in various plant species under heat stress is summarized in Table 15.3.

15.3.4 Metal stress

Heavy metal stress has become a widespread problem in all types of terrestrial ecosystems. In recent times extensive industrialization, exhaust emissions, sewage irrigation, and the use of metal products have caused the accumulation of heavy metals in plants which directly or indirectly interfere with plant growth and development by influencing various physiological and

TABLE 15.2 Role of H_2S under water stress.

S. no.	Plant species	Concentration of H_2S/ H_2S donor (NaHS)	Effect of H_2S	References
1	*Vicia faba,*	100 μM NaHS	Stimulates stomatal closure, upregulates the ATP-binding cassette (ABC) transporters.	García-Mata and Lamattina (2010); Aimar et al. (2011)
2	*Arabidopsis thaliana*			
3	*Impatiens walleriana*			
4	*Arabidopsis thaliana*	50 μmol L^{-1} NaHS	Upregulates the expression of *LCD, DCD1, NFS1, NFS2*, and *DES1*, induces the ABA-dependent signaling pathway.	Shen et al. (2013)
5	*Arabidopsis thaliana*	80 μM NaHS	Stimulates the expression of drought-responsive genes including *DREB2A, DREB2B, RD29A*, and *CBF4.*	Jin et al. (2013)
6	*Triticum aestivum*	0.6 mM NaHS	↑CAT↑APX↓ MDA↓H_2O_2	Zhang et al. (2010a,b,c)
7	*Arabidopsis thaliana*	80 μM NaHS	Stimulates drought associated genes (DREB2 and CBF4), reduces the aperture of stomata.	Jin et al. (2011a,b)
8	*Vicia faba*	100 μM NaHS	Inflates relative water content.	García-Mata and Lamattina (2010)
9	*Triticum aestivum*	1 mM NaHS	Increases activities of APX, GR, DHAR, gamma-glutamylcysteine synthetase, and reduces MDA and H_2O_2 accumulation.	Shan et al. (2011)
10	*Ipomoea batata*	0.05 mmol L^{-1} NaHS	Restores chlorophyll content.	Zhang et al. (2009a,b,c)

(*Continued*)

TABLE 15.2 (Continued)

S. no.	Plant species	Concentration of H_2S/ H_2S donor (NaHS)	Effect of H_2S	References
11	*Arabidopsis thaliana*	50 μmol L^{-1} NaHS 100 μM NaHS	Triggers regulation of MIR393 and also stimulates the expression of AFB1, AFB2, and AFB3.	Shen et al. (2013); Shi et al. (2015)
12	*Triticum aestivum*	Endogenous	Increases electrolyte leakage of roots, declines vitality of tissue and accumulation of MDA.	Wu et al. (2013)
13	*Ipomoea batatas*	0.02 mmol L^{-1} NaHS	Stimulates organogenesis of root.	Zhang et al. (2009a,b,c)
14	*Salix matsudana*			
15	*Glycine max*			
16	*Arabidopsis thaliana*	Endogenous	Stomata closed under short-term H_2S exposure whereas prolonged exposure inflates stomatal aperture.	Honda et al. (2015)
17	*Arabidopsis thaliana*	Endogenous	Stimulates ethylene signaling.	Liu et al. (2011a,b)
18	*Triticum aestivum*	500 μM NaHS	Alters the expression level of ABA in roots and leaves.	Ma et al. (2016a,b)
19	*Arabidopsis thaliana*	100 μM NaHS	Triggers the NO production regulate ABA-induced closing of stomata.	Scuffi et al. (2014a,b)
20	*Triticum aestivum*	0.4 mM NaHS	Increases level of functional and regulatory proteins like LEA, HSPs, SOD, MDHAR, DHAR, transcription factors (*bHLH, bZIP.*)	Li et al. (2017)

CAT, catalase; MDA, malondialdehyde; APX, ascorbate peroxidise; GR, glutathione reductase; DHAR, dehydroascorbate reductase; AFB, auxin signaling F, box protein; LCD, L-cysteine (Cys) desulfhydrase; DCD1, D-cys desulfhydrase 1; NO, nitric oxide; MDHAR, monodehydroascorbate reductase; SOD, superoxide dismutase; LEA, late embryogenesis abundant; HSPs, heat shock proteins; ABA, abscisic acid; DREB2, dehydration responsive element-binding protein 2.

TABLE 15.3 Role of H_2S under Heat Stress.

S. no	H_2S treatment	Plant	Effect	References
1	NaHS	Wheat plantlets	Increased in the content of H_2O_2 and NO in roots is reported.	Karpets et al. (2019)
2	100 μM NaHS for 24 h	*Triticum aestivum* L.	Increased production of the superoxide anion radical and elevated H_2O_2 content. Enhanced activities of antioxidant enzymes—guaiacol peroxidase, superoxide dismutase and catalase.	Kolupaev et al. (2017)
3	25, 50, 75 or 100 μM NaHS	*Nicotiana tabacum* L.	Significantly enhanced survival percentage of tobacco suspension cultured cells regrowth capability.	Li et al. (2012)
4	0.5 mM NaHS	*Zea mays* L.	Pretreatment with NaHS-induced the accretion of endogenous betaine by enhancing BADH activity.	Li and Zhu (2015)
5	NaHS; 100 μM for 48 h	*Fragaria x ananassa* cv. *Camarosa*	Ascorbate/glutathione homeostasis conserved. NaHS pretreatment of root resulted in stimulation of gene expression levels of protective molecules, such as enzymatic antioxidants, heat shock proteins and aquaporins.	Christou et al. (2014)
6	0.1, 0.5, 1.0, and 1.5 mM NaHS for 12 h	*Zea mays* L.	Significantly improved germination rate of seeds and survival rate of seedlings of maize under heat stress, and reduced electrolyte leakage of roots. Improved activity of 1-pyrroline-5-carboxylate synthetase (P5CS) and lower proline dehydrogenase (ProDH) activity, which resulted in increased concentration of endogenous proline in seedlings of maize.	Li et al. (2013)

(*Continued*)

TABLE 15.3 (Continued)

S. no	H_2S treatment	Plant	Effect	References
7	0.1, 0.3, 0.5, 0.7, 0.9, 1.5, 2.0 and 3.0 mM NaHS for 6 h	*Zea mays*	Enhanced survival rate of maize seedlings under heat stress. NaHS-treated seedlings showed higher activities of antioxidant enzymes and antioxidants levels and also maintained the ratio of reduced antioxidants to total antioxidants.	Li et al. (2014a,b,c)
8	H_2S	*Populus trichocarpa*	Elevated temperature rapidly induced the production of H_2S, Enhanced activities of enzymes involved in H_2S biosynthesis.	Cheng et al. (2018)
9	0–1.5 mmol L^{-1}	*Triticum aestivum* L.	Exogenous application of NaHS alleviated oxidative damage and improved heat tolerance by modifying the antioxidant system.	Min et al. (2016)
10	0.3, 0.6, 0.9 and 1.2 mM NaHS	*Zea mays* L.	NaHS treatment increased the activity of trehalose-6-phosphate phosphatase which resulted in the accumulation of endogenous trehalose.	Li et al. (2014a,b,c)
11	0.25 μM hematin, 50 μM NaHS; 0.25 μM hematin +50 μM NaHS	*Nicotiana tabaccum* L.	Hematin treatment increased the activity of L-cysteine desulfhydrase, an important enzyme in H_2S biosynthesis, which resulted in induced buildup of endogenous H_2S in tobacco cells. Hematin-induced heat tolerance was improved by supplementation of NaHS.	Li and Gu (2016)

molecular activities (Hassan et al., 2017). The effect of heavy metals on plant productivity, protein and lipid metabolism, and balancing the components of thylakoid membranes have attracted recent scientific attention. Singh et al. (2015) reported the rate of change of uptake, accumulation, and translocation of essential elements in plants under superoptimal concentration of heavy metals. Heavy metals cause serious consequences both in plants and animals. In plants the inhibition of photosynthesis, low biomass accumulation, altered nutrient assimilation, water balance, and senescence are some of the most detected toxic effects which lead to plant death (Ali et al., 2011). Mostly heavy metals are not essential for plant growth but a few, such as Mo, Mn, Zn, Cu, Ni, and Co, are indispensable (Shahid et al., 2015). However, these heavy metals along with Al, As, Hg, Cd, and Be prove toxic to plants when their concentration rises beyond a certain level (Pierart et al., 2015). The Agency for Toxic Substances and Disease Registry (ATSDR, 2003) classified As, Pb, Cd, and Hg as the most toxic metals based on their frequency of occurrence, toxicity, and their exposure potential to various flora and fauna. The morphological abnormalities and metabolic disorders from these toxic metals results in the production of ROS (e.g., O_2^-, H_2O_2, OH^-) leading to the disruption of redox homeostasis of cells which is considered to be the major cause of HM toxicity in plants (Ibrahim et al., 2015; Shahid et al., 2015).

Plants alleviate heavy metal stress by adopting several defense strategies such as restricting metal uptake by roots, sequestration of metals in vacuole, repair of damaged proteins by heat shock proteins, activation of antioxidants, etc., which are triggered by certain signaling molecules (Shahid et al., 2015). After NO and CO, H_2S is considered to be the third gaseous signal molecule that is produced in response to various environmental stresses like salt stress, temperature stress, drought stress, chilling stress, and heavy metal stress (Wang, 2012; Jin and Pei, 2015). Though H_2S plays a crucial role in plant growth and development, it also provide defense to metal stress by limiting or restricting the absorption, accumulation, and transport of metals, ameliorates the stress tolerance of plants (Kabil and Banerjee, 2010), and traps the ROS, for example, O_2^-, H_2O_2, OH^- (He et al., 2018). He et al. (2018) reported a link between H_2S and ROS production and the subsequent alleviation of metal toxicity in plants. In many crop plants, the involvement of NO and H_2S as signaling molecules to tolerate heavy metal stress has been reported (Hancock and Whiteman, 2016). Studies suggest that the endogenous increase of H_2S in plants triggered by metal stress is linked with metal tolerance. For instance, the emission of H_2S was induced by Cd^{2+} in *Medicago sativa* (Cui et al., 2014) and a 20% rise of H_2S was induced by 0.5 mM Cd stress in rice seedlings (Mostofa et al., 2015a,b). Exogenous application of H_2S is also reported to induce stress tolerance to toxic metals like Cd, Cr, Cu, Al, As, Pb, and Zn, in many plants. Some of the recent reports of both endogenous and exogenous applications are shown in Table 15.4.

TABLE 15.4 Role of H_2S under Metal Stress.

S. no.	Plant species	H_2S/ donor	Metal stress	Mechanism	References
1.	*Hordium vulgare* (Barley)	NaHS	Al	Expression of H^+-ATPase is upregulated, induction of citrate transporter (*HvAACT1*) activity, and purification of the antioxidant system.	Chen et al. (2013)
2.	*Oryza sativa* (Rice)	H_2S	Al	PME activity in roots is decreased by decreasing the pectin and hemicellulose content which blocks Al-binding sites in the root cell walls, expression of *OsSATR1* and *OsSTAR2* is augmented, and *OsFRDL4*'s induction of citrate secretion is increased. Increased antioxidant enzyme activity and sequestration of Al to vacuole.	Zhu et al. (2018)
3.	*Arabidopsis thaliana*	H_2S	Cd	Upregulates cysteine biosynthetic genes *SAT1* and *OASA1* that cause increase in antioxidant activity, induces MTs and PCs gene expression that counteracts Cd toxicity.	Jia et al. (2016)
4.	*Medichago sativa* (Alfalfa)	NaHS	Cd	Alleviates Cd toxicity whereas the scavengers of H_2S block NaHS-induced Cd tolerance.	Shivaraj et al. (2019)
5.	*Pisum sativum* (Pea)	Exogenous H_2S	Ar	H_2S alleviates Ar toxicity through upregulation of the AsA-GSH cycle in pea seedlings.	Singh et al. (2015)
6.	*Populus euphratica*	Exogenous H_2S	Cd	H_2S alleviates Cd toxicity by activation of antioxidant enzymes and tonoplast Cd^{2+}/H^+ antiporters.	Sun et al. (2013)
7.	*Cynodon dactylon* (Bermuda grass)	Exogenous H_2S	Cd	Together with NO through ROS and osmolytes modulation.	Shi et al. (2014)
8.	*Brassica rapa*	Exogenous H_2S	Cd	Through upregulation of antioxidant enzymes.	Zhang et al. (2015a,b)

9.	*Triticum aestivum* (Wheat)	Exogenous NaHS	Cr	Seed germination under Cr stress is alleviated through increasing the antioxidant enzyme activity.	Zhang et al. (2010a,b,c)
10.	*Seteria italica* (Foxtail millet)	Exogenous H_2S	Cr	Ca^{2+}-activated heavy metal chelators and antioxidant system.	Fang et al. (2014a,b)
11.	*Brassica napus* (Oilseed rape)	Exogenous H_2S	Pb	Through enhancing nutrient uptake and antioxidant system.	Ali et al. (2014)
12.	*Triticum* (Wheat)	Exogenous H_2S	Cu	Through regulation of AsA-GSH metabolism.	Shan et al. (2012)
13.	*Solanum nigrum*	Exogenous H_2S	Zn	Through enhancement of MTs expression and antioxidant enzymes system.	Liu et al. (2016)
14.	*Setaria italica*	H_2S	Cr	H_2S induce the expression of *MT3A* and *PCs*.	Fang et al. (2014a,b)
15.	*Brassica oleracea* (Cauliflower)	Exogenous H_2S	Cr	Enhances mineral nutrient uptake, chlorophyll content, gas exchange attribute and antioxidant system.	Ahmad et al. (2019)

15.4 Ameliorative role of phytohormones in plants under abiotic stress

Phytohormones serves as messenger molecules that are produced at particular sites in the plants and often translocated to different parts to play crucial roles in the regulation of stress responses within them (Asif et al., 2019). They enable the plants to combat a plethora of abiotic stress conditions via inducing many adaptive or resistance mechanisms varying upon stress type (Javid et al., 2011). However, plants respond to different adversities by a series of processes comprising different signaling pathways involving phytohormones that act in both antagonistic and synergistic manners in order to regulate growth, development, and defense actions via modulating genetic systems (Khan et al., 2020). Normally, genes that are overexpressed through these hormones enable the plants to acclimatize to stress environment, via the generation of stress resistance. The majority of studies in relation to stress responses have been found to be involved in ABA signaling and ABA-response genes. In addition, the role of cytokinins (CK), auxins (IAA), gibberellins (GA), brassinosteroids (BRs), jasmonic acid (JA), ABA, SA, etc. under abiotic stress conditions have also been explored (Peleg and Blumwald, 2011). Under freezing stress, the expression levels of *ARRs* are upregulated along with the downregulation of expression levels of all CK receptors (Argueso et al., 2009). Meanwhile, in drought condition, minimal levels of CK within xylem sap of plants have been recorded (Bano et al., 1994). Studies determining the crosstalk among different phytohormones mainly comprise antagonistic and synergistic relations that are vital during abiotic stress conditions (Peleg and Blumwald, 2011). Moreover, CK is antagonistic towards ABA, however, under water-deprived situations, it reduces the CK levels (Argueso et al., 2009). Together, GA and BR are substantial for regulating different physiological processes of plants. Researchers found that *OsGSR1*, a member of the GA-induced transcript gene family, plays a significant role during BR and GA signaling and mediates crosstalk among them (Wang et al., 2009). In addition, ABA suppresses the BR-mediated responses under abiotic stress conditions (Divi et al., 2010). There are many findings that consider the crosstalk of BR, CK, GA, IAA, JA, SA, and ET, that culminate in the modulation of expression levels of hormonal biosynthetic genes along with the signaling intermediate compounds (Bouquin et al., 2001).

Interestingly, a closed connection among expression profiles and diverse regulatory motifs of stress-controlled genes has been reported. This illustrates that a complex network is primitive for regulation of stress genes, and it is mainly laid on transcriptomic analysis data revealed through theoretical approaches (Ma and Bohnert, 2007). A wide spectra of transcription factors are involved in the process, for instance, the *DREB1/CBF* family constitutes *APETALA2 (AP2)*-type transcription factor recognizing *DRE/CRT* along with factors actively linked in low temperature stress responses. *DREB1A/CBF3* and *DREB1C/CBF2* expression profiles are curbed at transcriptomic level through the induction of CBF expression levels (ICE1) or Calmodulin-binding transcriptomic activator (Agarwal et al., 2006). Earlier studies also projected that the expression levels of expansin were also regulated by phytohormones. For example, *MaExp1* from *Musa paradiciaca* encoding ripening was upregulated by ET (Trivedi and Nath, 2004). In addition, IAA stimulated expansin gene expression alongside *EXP-1* and *EXP-B* (Ding et al., 2008). Likewise, Azeez et al. (2010), revealed the GA-responsive expansin gene *GgEXPA1* in many crop plants. Similarly, Yang Han et al. (2012) showed the role of expansin genes in response to stress and its regulation by cloning TaEXPB23 in wheat plants and further transforming it in tobacco. They concluded that *TaEXPB23* transcripts were stimulated via NaCl and methyl jasmonate (MeJA) but reduced under high-temperature conditions, and in the presence of ET and napthylacetic acid. They also reported that overexpressed *TaEXPB23* led to alterations of the root patterns of transgenic tobacco plants and induced tolerance to saline conditions.

Nevertheless, the association of plant hormones plays an influential role under stress conditions by the modulation of Ca^{2+} levels that act as intracellular messengers. Inositol phosphate is generated in conjunction with ROS during abiotic stresses, that is, drought, freezing, saline, and high temperature, that in turn induce ABA levels within plant cells (Leshem et al., 2007). Remarkably, plant defense hormones considerably act against stresses, while their level gets enhanced with the stress intensity (Bari and Jones, 2009). ABA is concerned with both short- and long-term responses via closing stomatal pores so as to maintain the water balance and activating stress responsive genes. Moreover, the ABA also gets triggered during osmotic stress via induced expression of ABA biosynthetic genes, namely, *AAO3* encoding aldehyde oxidase, *ZEP* encoding Zeaxanthin epoxide, *LOS6* encoding Low Expression Osmotic-Stress-Responsive gene 6/*ABA1*, *MCSU* encoding Molybdenum Cofactor Sulfurase and *NCED3* encoding 9-cis-epoxycarotenoid dioxygenase respectively (Xiong et al., 2002). Within the promoter site of ABA-response genes cis elements were also found and termed as ABREBs/ABA binding factors (ABFs). Additionally, in the case of *Arabidopsis* the association between ABA and JA was also reported as being mediated through *MYC2* and their homologs (*MYC3* and *MYC4*). The ABA/JA wounding response expression of *LAP* and *PIN2* are also curbed by *MYC2* orthodox (*JAM YC2* and *JAM YC10*) (Fernández-Calvo et al., 2011). The corelation of ABA and SA is important to maintain the water balance in plants along with regulating stomatal conductance, osmotic homeostasis, plant senescence, and translocation of photoassimilates. Furthermore, ET also coordinates with ABA by antagonistically acting via the regulation of stomatal apertures and morphological growth under drought conditions (Gazzarrini and McCourt, 2003). Usually, under excessive irrigation, ABA antagonizes ET during functioning of adventitious roots and hyponastic movements. Moreover, the JA-GA associations are synergistic and involved in different metabolic actions of plants, like trichome formation and sequitripenoid synthesis (Tanaka et al., 2006). The *WRKY23* gene regulated by IAA curbs the morphological and developmental processes of the plant.

In addition to the above, the IAA has also been found to interact with flavonoids to battle for binding of flavonoids with glutathione 5-transferase ϕ-2 (STF2), which controls the synthesis and translocation of IAA within the stress sites, to actively facilitate shielding against them. On the contrary, ET also activates ABI1 and ABI2 that negatively regulate ABA signaling (Asselbergh et al., 2008). Another phytohormonal association among JA and GA also regulates the growth and defense trade-offs. For instance, JA antagonizes GA-mediated growth actions through JAZs-DELLAs. During JA-deprived conditions, *Arabidopsis* JAZ9 also binds onto DELLA proteins RGA, in order to prevent them from suppressing TF PIF3 (Phytochrome-Interacting Factor 3). Moreover, it also prevents the GA- and JA-mediated DELLA degradation from inhibiting GA-dependent growth (Yang et al., 2012). Phytohormones also initiate the mitogen activated protein kinase (MAPK) pathway that integrates the signaling factors to control hormonal crosstalk. The transducing environmental factors are also controlled by MAPK cascades. Plants have a diversity of MAPK fragments with regard to different stressors. When plants encounter any abiotic stress, a chain of MAPK cascades, such as MEKKI/MEKK2/MPK4 and MPK6, are triggered and they work together to initiate innate immunity in plants (Rodriguez et al., 2010). A study reported on MAPK cascade mutants with defensive nature and found that MKKI and MKK2 double mutants showed instant death and constitutional gene expression of PR proteins (Gao et al., 2018). Also, Ludwig et al. (2005) revealed that CDKP and MAPK pathways stimulated during abiotic stresses were due to ET that governs the target for manipulating stress responses of plants via regulating gene expression. Furthermore, *Arabidopsis CML42*, a calmodulin-like protein 42 restricts the ABA accumulation during drought stress (Khan et al., 2020). Moreover, some of the *WRKY* transcription factors lowered ABA-inducing promoters during stress condition and antagonized ABA functioning. To elucidate, Chen et al. (2017) found that WRKY transcription factors curb crosstalk of growth of plants and water scarcity stress tolerance by regulating BR signaling. However, WRKY transcription factors *WRKY46, WRKY54, WRKY70* etc. mediate BR-controlled drought responses as they act very well with BR transcription factors BZRI/BESI (Rohila and Yang, 2007). However, a study reported that *WRKY70* induced JA but inhibited GA biosynthetic pathways and signaling in *Oryza sativa*, thereby, indicating that these responses are quite specific in nature (Foyer and Noctor, 2016).

Apart from this, SA has also been found to improve stress tolerance in plants against many abiotic stresses, such as osmotic stress, heavy metal

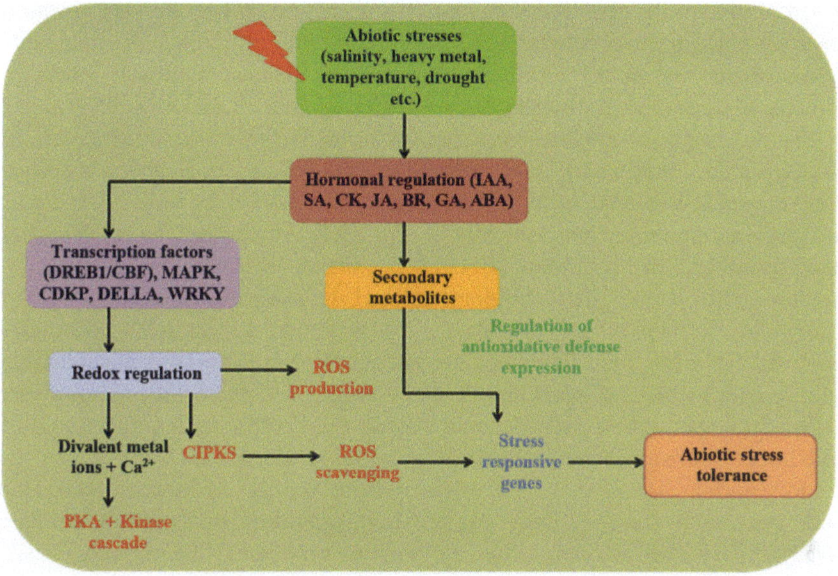

FIGURE 15.2 Regulation of abiotic stresses in plants through phytohormones.

stress, drought, and heat stress (Khan et al., 2014). It modulates the antioxidative defense expression and also amends the synthesis of different HSPs (Heat Shock Proteins) (Clarke et al., 2004). Taking all the considerations together, it can be elucidated that SA, JA, and ET defensive signaling upholds the mechanistic strategies to adapt the defense responses of plants in the best possible manner. A diagrammatic presentation of the regulation of abiotic stresses in plants through plant hormones are described in Fig. 15.2.

15.5 Crosstalk of H_2S with phytohormones under various abiotic stresses

Plant growth regulators (PGRs) or phytohormones are the most important signaling molecules instigating the developmental processes in plants. Phytohormones, like auxin, gibberellins, brassinosteroids, SA, JA, ethylene, are ABA, are crucial endogenous modulators which are known to regulate various physiological and molecular responses, associated with growth and development and other stress responses (biotic and abiotic). In recent studies H_2S has emerged as a signaling molecule which triggers the downstream response elements in plants (Cheng et al., 2013). As a potential signaling molecule H_2S has been known to interact with other signaling molecules like ROS (Hancock and Whiteman, 2014), CO (Lin et al., 2012), NO (Singh et al., 2015), and phytohormones like IAA (Zhang et al., 2009a,b,c), ABA (Jin et al., 2011a,b; Scuffi et al., 2016), GA (Xie et al., 2013), JA (Hou et al., 2011), and ethylene.

In plants, molecules like H_2S have a dual nature, either they are released as toxic cellular by-products, or as potential signaling molecules under different stress conditions (Ahmad et al., 2016). Under high salt stress in tomato, a study reveals that H_2S works downstream of NO (da-Silva et al., 2018) and helps to mitigate oxidative stress. H_2S also works downstream to SA (Li et al., 2015a,b,c). H_2S reduces the chances of postharvest ripening and senescence in banana as it reduces oxidative stress as well as acting as an ethylene signaling inhibitor (Ge et al., 2017). In further studies H_2S has been found to work upstream of NADPH oxidase (RBOH1) and H_2O_2 accumulation (Mei et al., 2017). In guard cells of *Arabidopsis*, H_2S molecules are known to regulate NADPH oxidase activity which directs ROS accumulation. At the same time they regulate phospholipase D-derived phosphatidic acid levels, to block further signaling pathways in cells. Studies shows that H_2S works upstream to MAPKs pathway as they both play a similar kind of role under cold stress conditions in *A. thaliana* (Du et al., 2017).

Researchers found that H_2S released from sodium hydrosulfide (NaHS) and morpholin-4-ium 4-methoxyphenyl (morpholino) phosphinodithiolate (*GYY4137*) promotes various developmental processes like organogenesis, seed germination, and also regulates the advent of senescence (Fang et al., 2014a,b; Wojtyla et al., 2016). Thus H_2S as a signaling molecule participates in various crosstalk networks between No, Ca^{2+}, H_2O_2 and the universal stress hormone, ABA (Guo et al., 2016) under stress conditions. Since the developmental processes in plants are highly dependent on the homeostasis and crosstalk between various phytohormones, it is evident that signaling molecules like H_2S are playing significant roles at multiple levels. These novel signaling molecules can be targeted for future genetically modified crop production and for the introduction of disease-resistant traits in the susceptible crops.

15.5.1 H_2S and auxin interaction

Auxin controls a wide range of physiological and developmental processes like apical dominance, cell wall acidification, embryogenesis, induction of lateral and main roots, tropism, and vascular differentiation in the plant kingdom (Du and Scheres, 2018). Auxin being a polar molecule forms a gradient due to polar auxin transport (PAT) across the membrane from the source tissues to the sink. At higher concentration, H_2S disrupts the PAT which leads to altered root structures. H_2S is also known to disrupt the functioning of ACTIN-BINDING PROTEINS (ABPs) which lowers the occupancy of F-acting bundles in *Arabidopsis* roots. PIN-proteins along with ABCB cassette proteins cause the PAT. PIN distribution along the membrane is purely dependent on ABPs, which can be significantly altered under the H_2S gradient (Jia et al., 2015). Thus the report indicates a tightly regulated signaling network between the auxin, H_2S, ABPs, and the PIN distribution.

In *Arabidopsis* the exogenous treatment of NaHS affects the growth in primary roots (Zhang et al., 2017). Higher H_2S cellular gradients triggers the production of NO, which leads to the activation of MITOGEN-ACTIVATED PROTEIN KINASE 6 (MPK6) and ROS generation and primary root growth inhibition. The potential interactions of H_2S with NO, CO, auxin, and indole-3 acetic acid (IAA) promotes the activity of lateral root formation.

Fang et al. (2014a) reported that the exogenous treatment of NaHS to *Solanum lycopersicum* seedlings showed a synergistic communication between H_2S and auxins during the lateral root formation. At low auxin concentration the activity of H_2S-producing enzymes like DES1 is limited to some extent. However the treatment of exogenous auxins, naphthalene acetic acid (NAA), to these cells significantly increases the DES1 expression as well as the H_2S concentration. During the lateral root growth, NAA and H_2S also triggers the upregulation of certain cell cycle genes like cyclin-dependent kinases, CDK1 and CYCA2, and downregulation of cyclin-dependent kinase inhibitors like KRP2 in tomato plant (Fang et al., 2014b). These gene regulations finally led to the cell divisions and organogenesis in plants.

15.5.2 H_2S and ethylene interaction

Ethylene a dynamic gaseous plant hormone with a wide ranging effects and well defined signaling pathway. The gaseous hormone regulates a number of physiological processes that promote senescence and induce epinasty and climacteric fruit ripening in banana, apple, and mango. This fruit ripening leads to the breakdown of starch, chloroplasts, and cell wall components as a result of which certain aromas will be produced and further deposition of carotenoids occurs. Ge et al. (2017) investigated that exogenous treatment with 1 mM NaHS solution can reverse the ethylene-induced fruit ripening and softening in banana by reduced polygalacturonase activity. Coordination and synchronization of NaHS and ethephon (synthetic ethylene donor) solutions promotes the antioxidant activities and ROS degradation which further delays the chloroplast degradation and triggers the increase in phenolic and flavonoid contents. H_2S downregulates or suppresses the ethylene biosynthetic genes like 1-aminocyclopropane -1-carboxylate synthase 1 (ACS1), ACS2, I-aminocyclopropane-1 carboxylate oxidase 1 (ACO1), pectate lyase, and upregulates the expression of ethylene receptor encoding genes, ETHYLENE RECEPTOR (ETR), ETHYLENE RESPONSE SENSOR1;2 (ERS1, ERS2). Also the exogenous treatment of H_2S in banana fruits arrests ethylene production, maintains low MDA content, and minimizes ATP degradation and cold stress injuries. Moreover H_2S treatment enhances the activities of H^+-ATPase, cytochrome c oxidase, succinate dehydrogenase, Ca^{2+}-ATPase and maintains the overall regulatory activities and senescence in banana fruits (Li et al., 2016a,b).

15.5.3 H_2S and abscisic acid interaction

ABA is well-known to regulate its stress responses against all kinds of biotic and abiotic stresses ranging from oxidative stress, salinity, temperature, heavy metal toxicity, etc. (Roychoudhury and Banerjee, 2017). H_2S is also known to have crosstalk with the ABA signalosome. NaHS pretreatment to wheat plants upregulates the synthesis of ABA and further activates the ABA responsive genes in the leaves under drought stress Ma et al. (2016a,b). H_2S accumulation by NaHS pretreatment also promotes the ABA-mediated gene regulation encoding for ABA receptors under drought. These pretreated plants also possess low MDA and H_2O_2 content due to the antioxidative defense system activated by the ABA level. Another report says that the sulfur-containing compounds are continuously required in plants' systems to produce the required amount of ABA molecules. SULFATE TRANSPORTER 3;1 (SULTR3;1) mutants in *Arabidopsis* show low aldehyde oxidase activity with low ABA level. In *A. thaliana* S-metabolism and ABA synthesis comediates the transcription regulation of ABA3 and 9-cis-epoxycarotenoid dioxygenase 3 (NCED 3) (Cao et al., 2014).

Stomatal closure in low water or stress conditions is known to be regulated by the ABA levels. H_2S is also known to coregulate the stomatal movements along with the ABA in the cell epidermal tissues of *V. faba* and *A. thaliana* (García-Mata and Lamattina, 2010). H_2S helps the guard cells to regulate the relative water content and also stimulates the ABC-transporter proteins of guard cells in a ABA-dependent manner to promote stomatal closure. DES1, a H_2S biosynthetic enzyme, mediates crosstalk between NO and H_2S in an ABA-dependent manner to promote stomatal closure; on the other side, ABA alone is unable to induce stomatal closure in des1 *Arabidopsis* mutant. However exogenous H_2S treatment promotes the stomatal closure in pyr1, pyl2, pyl4 ABA receptor mutants (Scuffi et al., 2014a,b). The functional protein phosphatase 2 C, ABA INSENSTIVE 1 is essential for H_2S-mediated stomatal closure in *Arabidopsis*.

15.5.4 H_2S and gibberellic acid interaction

Gibberellic acid (GA) is a plant growth hormone that profoundly influences the process of seed germination and internode elongation. *Hordeum* barley grains imbibed in 0.25 mM NaHS solution triggered the activation of antioxidative enzymes in the aleurone layer, such as CAT, POD, APX, and SOD, which means low levels of MDA, H_2O_2, and other superoxide radicals (Zhang et al., 2015a,b). In *T. aestivum*, GA accelerates programmed cell death (PCD) in the aleurone layer by arresting the regulatory mechanism of LCD, and H_2S synthesis (Xie et al., 2014). Exogenous NaHS-treated seeds show a lesser amount of GA-induced PCD in the aleurone layer by alleviating the endogenous GSH levels. GSH levels regulated by H_2S concentration promotes the activity of genes like HEME OXYGENASE-1 (HO-1) which helps to downregulate the aleurone layer apoptosis and further PCD. So regulatory crosstalk occurs between the GA, H_2S, HO-1, and GSH levels in the aleurone layer.

15.5.5 H_2S and jasmonic acid

Jasmonic acid (JA) is one of the signaling molecules which are downregulated by H_2S to inhibit stomatal development in *Arabidopsis* cotyledons (Deng et al., 2020). MeJA induced by H_2S is involved in reducing Cd stress by providing natural resistance associated macrophage protein (NRAMP1/NRAMP6) and expression of metal tolerance protein (MTP1/12) in *Setaria italica* (Tian et al., 2017). During heat stress H_2S synthesis is induced and JA accumulated, which promotes the synthesis of nicotine in *Nicotiana tabacum* (Chen et al., 2016). The H_2S level is increased by JA, which helps downstream of H_2O_2 in the closure of stomata in *V. faba* (Hou et al., 2011). In *A. thaliana* JA−induced H_2S activates mitogen-activated protein kinase (MEK1/2) by phosphorylation (Liu et al., 2011a,b); on the activation of MEK1/2 the AsA/DHA ratio (ascorbate (AsA) content to dehydroascorbate (DHA)) was upregulated (Shan et al., 2018). AsA is used as an antioxidant to protect plants against drought stress in wheat (Roy et al., 2017) and salt stress in Egyptian lentil cultivars (Gaafar and Seyam, 2018).

15.5.6 An outline of crosstalk between H_2S and phytohormones

Apart from the direct effects, sulfur (S) interacts with several plant biomolecules such as phytohormones, polyamines, NO, and Ca^{2+} to counteract abiotic stress. S and its derivatives like hydrogen sulfide (H_2S), methionine (Met), and GSH play a vital role in regulating many foremost plant processes. Ethylene biosynthesis significantly relies on the S bioavailability and Met metabolism. Also the S-metabolism itself requires certain phytohormones to mediate its functions. However this complex signaling network between H_2S, phytohormones, and other secondary molecules is poorly understood (Fig. 15.3).

SO_4^{2-} (sulfate) deficiency in cells leads to disruption in the biosynthesis of ABA and ET. Further it reduces the release of SAM as well as low cysteine levels. SO_4^{2-} and ABA levels regulate the stomatal closure during stress conditions. Efficient S-metabolism is also required for the biosynthesis of auxin (IAA). Reduced GSH levels cause a sharp decline of auxin gradients in root tips thus arresting lateral root formation. GSH reduction in cells also interferes in JA synthesis.

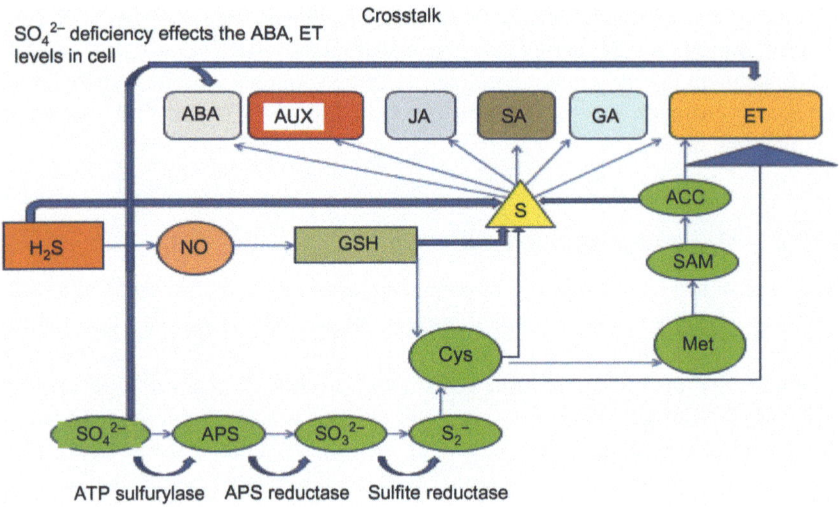

FIGURE 15.3 Outline showing the crosstalk of H_2S with other phytohormones.

15.6 Conclusion and future perspectives

H_2S is a lipophilic gaseous molecule that plays a pivotal role in plant responses to abiotic stress (Santisree et al., 2019). H_2S functions as a signaling molecule involved in the regulation of phytohormone signaling pathways (Li et al., 2016a,b). We have discussed in detail the multifaceted role of H_2S signaling in plants under abiotic stresses and its crosstalk with other phytohormones to ameliorate various stress conditions in plants. A lot of work has been done on the exogenous application of H_2S in plants which appears to be a promising strategy to palliate the abiotic stress conditions. But there appears to be a gap in basic research on understanding the endogenous H_2S to investigate its role under abiotic stresses and its interactions with other signaling molecules and its crosstalk with the phytohormones.

References

Abe, K., Kimura, H., 1996. The possible role of hydrogen sulfide as an endogenous neuromodulator. J. Neurosci. 16 (3), 1066–1071.

Agarwal, P.K., Agarwal, P., Reddy, M.K., Sopory, S.K., 2006. Role of DREB transcription factors in abiotic and biotic stress tolerance in plants. Plant Cell Rep. 25 (12), 1263–1274.

Ahmad, P., Abdel Latef, A.A., Hashem, A., Abd-Allah, E.F., Gucel, S., Tran, L.S.P., 2016. Nitric oxide mitigates salt stress by regulating levels of osmolytes and antioxidant enzymes in chickpea. Front. Plant Sci. 7, 347.

Ahmad, W., Sethupathi, S., Kanadasan, G., Lau, L.C., Kanthasamy, R., 2019. A review on the removal of hydrogen sulfide from biogas by adsorption using sorbents derived from waste. Reviews in Chemical Engineering, 1(ahead-of-print).

Aimar, D., Calafat, M., Andrade, A.M., Carassay, L., Abdala, G.I., Molas, M.L., 2011. Drought tolerance and stress hormones: from model organisms to forage crops. Plants Environ. 137–164.

Ali, S., Bai, P., Zeng, F., Cai, S., Shamsi, I.H., Qiu, B., et al., 2011. The ecotoxicological andinteractive effects of chromium and aluminum on growth, oxidative damage and antioxidantenzymes on two barley genotypes differing in Al tolerance. Environ. Exp. Bot. 70, 185–191.

Ali, B., Mwamba, T.M., Gill, R.A., Yang, C., Ali, S., Daud, M.K., et al., 2014. Improvement of element uptake and antioxidative defense in *Brassica napus* underlead stress by application of hydrogen sulfide. Plant Growth Regul. 74, 261–273.

Argueso, C.T., Ferreira, F.J., Kieber, J.J., 2009. Environmental perception avenues: the interaction of cytokinin and environmental response pathways. Plant, Cell Environ. 32 (9), 1147–1160.

Aroca, A., Benito, J.M., Gotor, C., Romero, L.C., 2017. Persulfidation proteome reveals the regulation of protein function by hydrogen sulfide in diverse biological processes in *Arabidopsis*. J. Exp. Bot. 68, 4915–4927.

Asif, M., Jamil, H.M.A., Hayat, M.T., Mahmood, Q., Ali, S., 2019. Use of phytohormones to improve abiotic stress tolerance in wheat. Wheat Production in Changing Environments. Springer, Singapore, pp. 465–479.

Asselbergh, B., De Vleesschauwer, D., Höfte, M., 2008. Global switches and fine-tuning—ABA modulates plant pathogen defense. Mol. Plant Microbe Interact. 21 (6), 709–719.

ATSDR, U., 2003. Air exposure pathway and assessment: Fallon Leukemia Cluster Investigation. US Department of Health and Human Services, Agency for Toxic Substances and Disease Registry.

Azeez, A., Sane, A.P., Tripathi, S.K., Bhatnagar, D., Nath, P., 2010. The gladiolus GgEXPA1 is a GA-responsive alpha-expansin gene expressed ubiquitously during expansion of all floral tissues and leaves but repressed during organ senescence. Postharvest Biol. Technol. 58 (1), 48–56.

Banerjee, A., Tripathi, D.K., Roychoudhury, A., 2018. Hydrogen sulphide trapeze: environmental stress amelioration and phytohormone crosstalk. Plant Physiology and Biochemistry 132, 46–53.

Banerjee, A., Roychoudhury, A., 2018. Interactions of brassinosteroids with major phytohormones: antagonistic effects. Journal of plant growth regulation 37 (4), 1025–1032.

Bano, A., Hansen, H., Dörffling, K., Hahn, H., 1994. Changes in the contents of free and conjugated abscisic acid, phaseic acid and cytokinins in xylem sap of drought stressed sunflower plants. Phytochemistry 37 (2), 345–347.

Bari, R., Jones, J.D., 2009. Role of plant hormones in plant defence responses. Plant Mol. Biol. 69 (4), 473–488.

Bouquin, T., Meier, C., Foster, R., Nielsen, M.E., Mundy, J., 2001. Control of specific gene expression by gibberellin and brassinosteroid. Plant Physiol. 127 (2), 450–458.

Boyer, J.S., 1982. Plant productivity and environment. Science 218 (4571), 443–448.

Brunold, C., Landolt, W., Lavanchy, P., 1983. SO2 and assimilatory sulfate reduction in beech leaves. Physiol. Plant 59, 313–318.

Calderwood, A., Kopriva, S., 2014. Hydrogen sulfide in plants: from dissipation of excess sulfur to signaling molecule. Nitric Oxide 41, 72–78.

Cao, M.J., Wang, Z., Zhao, Q., Mao, J.L., Speiser, A., Wirtz, M., et al., 2014. Sulfate availability affects ABA levels and germination response to ABA and salt stress in *Arabidopsis thaliana*. Plant J. 77 (4), 604–615.

Chen, J., Wang, W.H., Wu, F.H., You, C.Y., Liu, T.W., Dong, X.J., et al., 2013. Hydrogen sulphide alleviates aluminum toxicity in barley seedlings. Plant Soil 362, 301–318.

Chen, X., Chen, Q., Zhang, X., Li, R., Jia, Y., Ef, A., et al., 2016. Hydrogen sulfide mediates nicotine biosynthesis in tobacco (*Nicotiana tabacum*) under high temperature conditions. Plant Physiol. Biochem. 104, 174–179.

Chen, J., Nolan, T.M., Ye, H., Zhang, M., Tong, H., Xin, P., et al., 2017. *Arabidopsis* WRKY46, WRKY54, and WRKY70 transcription factors are involved in brassinosteroid-regulated plant growth and drought responses. Plant Cell 29 (6), 1425–1439.

Cheng, W., Zhang, L., Jiao, C., Su, M., Yang, T., Zhou, L., et al., 2013. Hydrogen sulfide alleviates hypoxia-induced root tip death in *Pisum sativum*. Plant Physiol. Biochem. 70, 278–286.

Cheng, T., Shi, J., Dong, Y., Ma, Y., Peng, Y., Hu, X., et al., 2018. Hydrogen sulfide enhances poplar tolerance to high-temperature stress by increasing S-nitrosoglutathione reductase (GSNOR) activity and reducing reactive oxygen/nitrogen damage. Plant Growth Regul. 84 (1), 11–23.

Christou, A., Filippou, P., Manganaris, G.A., Fotopoulos, V., 2014. Sodium hydrosulfide induces systemic thermotolerance to strawberry plants through transcriptional regulation of heat shock proteins and aquaporin. BMC Plant Biol. 14 (1), 42.

Clarke, S.M., Mur, L.A., Wood, J.E., Scott, I.M., 2004. Salicylic acid dependent signaling promotes basal thermotolerance but is not essential for acquired thermotolerance in *Arabidopsis thaliana*. Plant J. 38 (3), 432–447.

Corpas, F.J., 2019. Hydrogen sulfide: a new warrior against abiotic stress. Trends Plant Sci. 24 (11), 983–988.

Cui, W., Chen, H., Zhu, K., Jin, Q., Xie, Y., Cui, J., et al., 2014. Cadmium-induced hydrogen sulfide synthesis is involved in cadmium tolerance in Medicago sativa by reestablishment of reduced (Homo) glutathione and reactive oxygen species homeostases. PLoS One 9 (10).

da-Silva, C.J., Modolo, L.V., 2018. Hydrogen sulfide: a new endogenous player in an old mechanism of plant tolerance to high salinity. Acta Bot Brasilica 32 (1), 150–160.

da Silva, C.J., Fontes, E.P.B., Modolo, L.V., 2017. Salinity-induced accumulation of endogenous H_2S and NO is associated with modulation of the antioxidant and redox defense systems in *Nicotiana tabacum* L. cv. Havana. Plant Sci. 256, 148–159.

da-Silva, C.J., Mollica, D.C., Vicente, M.H., Peres, L.E., Modolo, L.V., 2018. NO, hydrogen sulfide does not come first during tomato response to high salinity. Nitric Oxide 76, 164–173.

Deng, Y.Q., Bao, J., Yuan, F., Liang, X., Feng, Z.T., Wang, B.S., 2016. Exogenous hydrogen sulfide alleviates salt stress in wheat seedlings by decreasing Na^+ content. Plant Growth Regul. 79 (3), 391–399.

Deng, G., Zhou, L., Wang, Y., Zhang, G., Chen, X., 2020. Hydrogen sulfide acts downstream of jasmonic acid to inhibit stomatal development in *Arabidopsis*. Planta 251 (2), 42.

Ding, X., Cao, Y., Huang, L., Zhao, J., Xu, C., Li, X., et al., 2008. Activation of the indole-3-acetic acid–amido synthetase GH3-8 suppresses expansin expression and promotes salicylate-and jasmonate-independent basal immunity in rice. Plant Cell 20 (1), 228–240.

Divi, U.K., Rahman, T., Krishna, P., 2010. Brassinosteroid-mediated stress tolerance in *Arabidopsis* shows interactions with abscisic acid, ethylene and salicylic acid pathways. BMC Plant Biol. 10 (1), 151.

Du, Y., Scheres, B., 2018. Lateral root formation and the multiple roles of auxin. J. Exp. Bot 69 (2), 155–167.

Du, J., Jin, H., Yang, L., 2017. Role of hydrogen sulfide in retinal diseases. Front. Pharmacol. 8, 588.

Durenkamp, M., De Kok, L.J., 2004. Impact of pedospheric and atmospheric sulphur nutrition on sulphur metabolism of *Allium cepa* L., a species with a potential sink capacity for secondary sulphur compounds. J. Exp. Bot. 55, 1821–1830.

Durenkamp, M., De Kok, L.J., Kopriva, S., 2007. Adenosine 50-phosphosulphate reductase is regulated differently in *Allium cepa* L. and *Brassica oleracea* L. upon exposure to H_2S. J. Exp. Bot. 58, 1571–1579.

Fan, H.F., Ding, L., Xu, Y.L., Du, C.X., 2017. Antioxidant system and photosynthetic characteristics responses to short-term PEG-induced drought stress in cucumber seedling leaves. Russ. J. Plant Physiol. 64 (2), 162–173.

Fang, H.H., Jing, T., Liu, Z.Q., Zhang, L., Jin, Z., Pei, Y., 2014a. Hydrogen sulfide interactswith calcium signaling to enhance the chromium tolerance in *Setaria italica*. CellCalcium 56 (6), 472–481.

Fang, T., Cao, Z., Li, J., Shen, W., Huang, L., 2014b. Auxin-induced hydrogen sulfide generation is involved in lateral root formation in tomato. Plant Physiol. Biochem. 76, 44–51.

Fernández-Calvo, P., Chini, A., Fernández-Barbero, G., Chico, J.M., Gimenez-Ibanez, S., Geerinck, J., et al., 2011. The *Arabidopsis* bHLH transcription factors MYC3 and MYC4 are targets of JAZ repressors and act additively with MYC2 in the activation of jasmonate responses. Plant Cell 23 (2), 701–715.

Filipovic, M.R., Jovanovic, V.M., 2017. More than just an intermediate: hydrogen sulphide signalling in plants. J. Exp. Bot. 68, 4733–4736.

Foyer, C.H., Noctor, G., 2016. Stress-triggered redox signalling: what's in pROSpect? Plant, Cell Environ. 39 (5), 951–964.

Gaafar, R.M., Seyam, M.M., 2018. Ascorbate–glutathione cycle confers salt tolerance in Egyptian lentil cultivars. Physiol. Mol. Biol. Plants 24 (6), 1083–1092.

Gao, Y., Wang, W., Zhang, T., Gong, Z., Zhao, H., Han, G.Z., 2018. Out of water: The origin and early diversification of plant R-genes. Plant Physiol. 177 (1), 82–89.

García-Mata, C., Lamattina, L., 2010. Hydrogen sulphide, a novel gasotransmitter involved in guard cell signalling. N. Phytol 188 (4), 977–984.

Gazzarrini, S., McCourt, P., 2003. Cross-talk in plant hormone signalling: what *Arabidopsis* mutants are telling us. Ann. Bot 91 (6), 605–612.

Ge, Y., Hu, K.D., Wang, S.S., Hu, L.Y., Chen, X.Y., Li, Y.H., et al., 2017. Hydrogen sulfide alleviates postharvest ripening and senescence of banana by antagonizing the effect of ethylene. PLoS One 12 (6).

Geng, B., Chang, L., Pan, C., Qi, Y., Zhao, J., Pang, Y., et al., 2004. Endogenous hydrogen sulfide regulation of myocardial injury induced by isoproterenol. Biochemical and biophysical research communications 318 (3), 756–763.

Guo, M., Liu, J.H., Ma, X., Luo, D.X., Gong, Z.H., Lu, M.H., 2016. The plant heat stress transcription factors (HSFs): structure, regulation, and function in response to abiotic stresses. Front. Plant Sci. 7, 114.

Hancock, J.T., Whiteman, M., 2014. Hydrogen sulphide and cell signaling: team player or referee? Plant Physiol. Biochem. 78, 37–42.

Hancock, J.T., Whiteman, M., 2016. Hydrogen sulfide signaling: interactions with nitric oxide and reactive oxygen species. Annals of the New York Academy of Sciences 1365 (1).

Hancock, J.T., 2019. Hydrogen sulfide and environmental stresses. Environmental and Experimental Botany 161, 50–56.

Hasanuzzaman, M., Nahar, K., Alam, M.M., Roy chowdhury, R., Fujita, M., 2013. Physiological, biochemical, and molecular mechanisms of heat stress tolerance in plants. Int. J. Mol. Sci. 14, 9643–9684.

Hassan, T.U., Bano, A., Naz, I., 2017. Alleviation of heavy metals toxicity by the application of plant growth promoting rhizobacteria and effects on wheat grown in saline sodic field. Int. J. Phytoremediat. 19 (6), 522–529.

Hatzfeld, Y., Maruyama, A., Schmidt, A., Noji, M., Ishizawa, K., Saito, K., 2000. β-Cyanoalanine synthase is a mitochondrial cysteine synthase-like protein in spinach and *Arabidopsis*. Plant Physiol. 123 (3), 1163–1172.

He, H., Li, Y., He, L.F., 2018. The central role of hydrogen sulfide in plant responses to toxic metal stress. Ecotoxicology and environmental safety 157, 403–408.

Honda, K., Yamada, N., Yoshida, R., Ihara, H., Sawa, T., Akaike, T., et al., 2015. 8-mercapto-cyclic GMP mediates hydrogen sulfide-induced stomatal closure in *Arabidopsis*. Plant Cell Physiol. 56, 1481–1489.

Hou, Z.H., Liu, J., Hou, L.X., Li, X.D., Liu, X., 2011. H_2S may function downstream of H_2O_2 in jasmonic acid-induced stomatal closure in Vicia faba. Chin. Bull. Bot 46 (4), 396–406.

Ibrahim, M.A., Khan, P.R., Hegazy, S.S., Hashim, E.A., Azamal, H., Ansari, M.K.A., et al., 2015. Improving the phytoextraction capacity of plants to scavenge heavy-metal infested sites. Environ. Rev. 23, 1–22.

Ismail, A., Takeda, S., Nick, P., 2014. Life and death under salt stress: same players, different timing? Journal of Experimental Botany 65 (12), 2963–2979.

Javid, M.G., Sorooshzadeh, A., Moradi, F., Modarres Sanavy, S.A.M., Allahdadi, I., 2011. The role of phytohormones in alleviating salt stress in crop plants. Aus. J. Crop. Sci. 5 (6), 726.

Jia, H., Hu, Y., Fan, T., Li, J., 2015. Hydrogen sulfide modulates actin-dependent auxin transport via regulating ABPs results in changing of root development in *Arabidopsis*. Sci. Rep. 5, 8251.

Jia, H., Wang, X., Dou, Y., Liu, D., Si, W., Fang, H., et al., 2016. Hydrogen sulfidecysteinecycle system enhances cadmium tolerance through alleviating cadmium-inducedoxidative stress and ion toxicity in *Arabidopsis* roots. Sci. Rep. 6, 39702.

Jin, Z., Pei, Y., 2015. Physiological implications of hydrogen sulfide in plants: pleasant exploration behind its unpleasant odour. Oxid. Med. Cell. Longev. 2015.

Jin, S., Kanagaraj, A., Verma, D., Lange, T., Daniell, H., 2011a. Release of hormones from conjugates: chloroplast expression of β-glucosidase results in elevated phytohormone levels associated with significant increase in biomass and protection from aphids or whiteflies conferred by sucrose esters. Plant Physiol. 155 (1), 222–235.

Jin, Z., Shen, J., Qiao, Z., Yang, G., Wang, R., Pei, Y., 2011b. Hydrogen sulfide improves drought resistancein *Arabidopsis thaliana*. Biochem. Biophys. Res. Commun. 414, 481–486.

Jin, Z., Xue, S., Luo, Y., Tian, B., Fang, H., Li, H., et al., 2013. Hydrogen sulfide interacting with abscisic acid in stomatal regulation responses to drought stress in *Arabidopsis*. Plant Physiol. Biochem. 62, 41–46.

Kabil, O., Banerjee, R., 2010. Redox biochemistry of hydrogen sulfide. J. Biol. Chem. 285 (29), 21903–21907.

Karpets, Y.V., Kolupaev, Y.E., Yastreb, T.O., Horielova, O.I., Shkliarevskyi, M.A., & Dmitriev, O.P., 2019. Role of reactive oxygen and nitrogen species in the induction of heat resistance of wheat plantlets by exogenous hydrogen sulfide. 3, 89–97.

Khan, M.I.R., Asgher, M., Khan, N.A., 2014. Alleviation of salt-induced photosynthesis and growth inhibition by salicylic acid involves glycinebetaine and ethylene in mungbean (*Vigna radiata* L.). Plant Physiol. Biochem. 80, 67–74.

Khan, N., Bano, A., Ali, S., Babar, M.A., 2020. Crosstalk amongst phytohormones from planta and PGPR under biotic and abiotic stresses. Plant Growth Regul. 1–15.

Kolupaev, Y.E., Firsova, E.N., Yastreb, T.O., Lugovaya, A.A., 2017. The participation of calcium ions and reactive oxygen species in the induction of antioxidant enzymes and heat resistance in plant cells by hydrogen sulfide donor. Appl. Biochem. Microbiol 53 (5), 573–579.

Koralewska, A., Stuiver, C.E.E., Posthumus, F.S., Kopriva, S., Hawkesford, M.J., De, L.J., 2008. Kok, Regulation of sulfate uptake, expression of the sulfate transporters Sultr1;1 and Sultr1;2, and APS reductase in Chinese cabbage (*Brassica pekinensis*) as affected by atmospheric H(2)S nutrition and sulfate deprivation. Funct. Plant Biol. 35, 318–327.

Lai, D., Mao, Y., Zhou, H., Li, F., Wu, M., Zhang, J., et al., 2014. Endogenous hydrogen sulfide enhances salt tolerance by coupling the reestablishment of redox homeostasis and preventing salt-induced K^+ loss in seedlings of *Medicago sativa*. Plant Sci. 225, 117–129.

Lamers, L.P., Govers, L.L., Janssen, I.C., Geurts, J.J., Van der Welle, M.E., Van Katwijk, M.M., et al., 2013. Sulfide as a soil phytotoxin-a review. Front. Plant Sci. 4, 268.

Laureano-Marín, A.M., García, I., Romero, L.C., Gotor, C., 2014. Assessing the transcriptional regulation of L-cysteine desulfhydrase 1 in *Arabidopsis thaliana*. Front. Plant Sci. 18 (5), 683.

Leshem, Y., Seri, L., Levine, A., 2007. Induction of phosphatidylinositol 3-kinase-mediated endocytosis by salt stress leads to intracellular production of reactive oxygen species and salt tolerance. Plant J. 51 (2), 185–197.

Li, N., Li, S., Zhang, K., Chen, W., Zhang, B., Wang, D., et al., 2017. ThMYC4E, candidate Blue aleurone 1 gene controlling the associated trait in Triticum aestivum. PloS one 12 (7), 0181116.

Li, Z.G., 2013. Hydrogen sulfide: a multifunctional gaseous molecule in plants. Russ. J. Plant. Physiol. 60, 733–740.

Li, Z.G., 2015. Analysis of some enzymes activities of hydrogen sulfide metabolism in plants, Methods in Enzymology, vol. 555. Academic Press, pp. 253–269.

Li, Z.G., Gong, M., Xie, H., Yang, L., Li, J., 2012. Hydrogen sulfide donor sodium hydrosulfide-induced heat tolerance in tobacco (Nicotiana tabacum L) suspension cultured cells and involvement of Ca^{2+} and calmodulin. Plant Science 185, 185–189.

Li, Z.G., Zhu, L.P., 2015. Hydrogen sulfide donor sodium hydrosulfide-induced accumulation of betaine is involved in the acquisition of heat tolerance in maize seedlings. Braz. J. Bot 38 (1), 31–38.

Li, Z.G., Gu, S.P., 2016. Hydrogen sulfide as a signal molecule in hematin-induced heat tolerance of tobacco cell suspension. Biol Plant 60 (3), 595–600.

Li, Z.G., Ding, X.J., Du, P.F., 2013. Hydrogen sulfide donor sodium hydrosulfide-improved heat tolerance in maize and involvement of proline. J. Plant Physiol. 170 (8), 741–747.

Li, J., Jia, H., Wang, J., Cao, Q., Wen, Z., 2014a. Hydrogen sulfide is involved in maintaining ion homeostasis via regulating plasma membrane Na^+/H^+ antiporter system in the hydrogen peroxide-dependent manner in salt-stress *Arabidopsis thaliana* root. Protoplasma 251 (4), 899–912.

Li, Z.G., Luo, L.J., Zhu, L.P., 2014b. Involvement of trehalose in hydrogen sulfide donor sodium hydrosulfide-induced the acquisition of heat tolerance in maize (*Zea mays* L.) seedlings. Bot Stud. 55 (1), 20.

Li, Z.G., Yi, X.Y., Li, Y.T., 2014c. Effect of pretreatment with hydrogen sulfide donor sodium hydrosulfide on heat tolerance in relation to antioxidant system in maize (*Zea mays*) seedlings. Biologia 69 (8), 1001–1009.

Li, H., Gao, M.Q., Xue, R.L., Wang, D., Zhao, H.J., 2015a. Effect of hydrogen sulfide on D1 protein in wheat under drought stress. Acta Physiol Plant 37 (11), 225.

Li, Z.G., Xie, L.R., Li, X.J., 2015b. Hydrogen sulfide acts as a downstream signal molecule in salicylic acid-induced heat tolerance in maize (*Zea mays* L.) seedlings. J. Plant Physiol. 177, 121–127.

Li, Z.G., Xie, L.R., Li, X.J., 2015c. Hydrogen sulfide acts as a downstream signal molecule in salicylic acid-induced heat tolerance in maize (*Zea mays* L.) seedlings. J. Plant Physiol. 177, 121–127.

Li, Z.G., Min, X., Zhou, Z.H., 2016a. Hydrogen sulfide: a signal molecule in plant cross-adaptation. Front. Plant Sci. 7, 1621.

Li, D., Limwachiranon, J., Li, L., Du, R., Luo, Z., 2016b. Involvement of energy metabolism to chilling tolerance induced by hydrogen sulfide in cold-stored banana fruit. Food Chem. 208, 272–278.

Lin, Y.T., Li, M.Y., Cui, W.T., Lu, W., Shen, W.B., 2012. Haem oxygenase-1 is involved in hydrogen sulfide-induced cucumber adventitious root formation. J. Plant Growth Regul. 31 (4), 519–528.

Liu, J., Hou, L., Liu, G., Liu, X., Wang, X., 2011a. Hydrogen sulfide induced by nitric oxide mediates ethylene-induced stomatal closure of *Arabidopsis thaliana*. Chin. Sci. Bull. 56 (33), 3547–3553.

Liu, J., Hou, L.-X., Liu, G.-H., Liu, X., Wang, X.-C., 2011b. Hydrogen sulfide induced by nitric oxide mediates ethylene-induced stomatal closure of *Arabidopsis thaliana*. Chin Sci. Bull. 56, 3547–3553.

Liu, X., Chen, J., Wang, G.H., Wang, W.H., Shen, Z.J., Luo, M.R., et al., 2016. Hydrogen sulfide alleviates zinc toxicity by reducingzinc uptake and regulating genes expression of antioxidative enzymes and metallothioneins in roots of the cadmium/zinc hyperaccumulator *Solanum nigrum*. L. Plant Soil 400, 177–192.

Liu, Y.L., Shen, Z.J., Simon, M., Li, H., Ma, D.N., Zhu, X.Y., et al., 2020. Comparative proteomic analysis reveals the regulatory effects of H_2S on salt tolerance of mangrove plant *Kandeliaobovata*. Int. J. Mol. Sci. 21 (1), 118.

Ludwig, A.A., Saitoh, H., Felix, G., Freymark, G., Miersch, O., Wasternack, C., et al., 2005. Ethylene-mediated cross-talk between calcium-dependent protein kinase and MAPK signaling controls stress responses in plants. Proc. Natl. Acad. Sci. 102 (30), 10736–10741.

Ma, S., Bohnert, H.J., 2007. Integration of *Arabidopsis thaliana* stress-related transcript profiles, promoter structures, and cell-specific expression. Genome Biol. 8 (4), R49.

Ma, D., Ding, H., Wang, C., Qin, H., Han, Q., Hou, J., et al., 2016a. Alleviation of drought stress by hydrogen sulfide is partially related to the abscisic acid signaling pathway in wheat. PLoS One 11 (9), e0163082.

Ma, D., Ding, H., Wang, C., Qin, H., Han, Q., Hou, J., et al., 2016b. Alleviation of drought stress by hydrogen sulfide is partially related to the abscisic acid signaling pathway in wheat. PLoS One 11.

Ma, Y., Zhang, W., Niu, J., Ren, Y., Zhang, F., 2019. Hydrogen sulfide may function downstream of hydrogen peroxide in salt stress-induced stomatal closure in *Viciafaba*. Funct. Plant Biol. 46 (2), 136–145.

Mei, Y., Chen, H., Shen, W., Shen, W., Huang, L., 2017. Hydrogen peroxide is involved in hydrogen sulfide-induced lateral root formation in tomato seedlings. BMC Plant Biol. 17 (1), 162.

Min, Y.A.N.G., Qin, B.P., Ping, W.A.N.G., Li, M.L., Chen, L.L., Chen, L.T., et al., 2016. Foliar application of sodium hydrosulfide (NaHS), a hydrogen sulfide (H_2S) donor, can protect seedlings against heat stress in wheat (*Triticum aestivum* L.). J. Integr. Agric. 15 (12), 2745–2758.

Mittler, R., 2002. Oxidative stress, antioxidants and stress tolerance. Trends in plant science 7 (9), 405–410.

Mitsuhashi, H., Yamashita, S., Ikeuchi, H., Kuroiwa, T., Kaneko, Y., Hiromura, K., et al., 2005. Oxidative stress-dependent conversion of hydrogen sulfide to sulfite by activated neutrophils. Shock 24 (6), 529–534.

Mostofa, M.G., Saegusa, D., Fujita, M., Tran, L.S.P., 2015a. Hydrogen sulfide regulates salt tolerance in rice by maintaining Na^+/K^+ balance, mineral homeostasis and oxidative metabolism under excessive salt stress. Front. Plant Sci. 6, 1055.

Mostofa, M.G., Rahman, A., Ansary, M.M.U., Watanabe, A., Fujita, M., Tran, L.P., 2015b. Hydrogen sulfide modulates cadmium-induced physiological and biochemical responsesto alleviate cadmium toxicity in rice. Sci. Rep. 5, e14078.

Pandey, A.K., Gautam, A., 2020. Stress responsive gene regulation in relation to hydrogen sulfide in plants under abiotic stress. Physiologia Plantarum 168, 511–525.

Peleg, Z., Blumwald, E., 2011. Hormone balance and abiotic stress tolerance in crop plants. Curr. Opin. Plant Biol. 14 (3), 290–295.

Pierart, A., Shahid, M., Séjalon-Delmas, N., Dumat, C., 2015. Antimony bioavailability: knowledge and research perspectives for sustainable agricultures. J. Hazard. Mater. 289, 219–234.

Rennenberg, H., Arabatzis, N., Grundel, I., 1987. Cysteine desulphydrase activity in higher plants: evidence for the action of L-and D-cysteine specific enzymes. Phytochemistry 26 (6), 1583–1589.

Rivero, R.M., Mestre, T.C., Mittler, R.O., Rubio, F., Garcia-Sanchez, F.R., Martinez, V., 2014. The combined effect of salinity and heat reveals a specific physiological, biochemical and molecular response in tomato plants. Plant Cell Environ. 37 (5), 1059–1073.

Rodriguez, M., Rodriguez, A., Bayer, J., Vilaseca, F., Girones, J., Mutje, P., 2010. Determination of corn stalk fibers'strength through modeling of the mechanical properties of its composites. BioResources 5 (4), 2535–2546.

Rohila, J.S., Yang, Y., 2007. Rice mitogen-activated protein kinase gene family and its role in biotic and abiotic stress response. J. Integr. Plant Biol. 49 (6), 751–759.

Roy, S., Arora, A., Chinnusamy, V., Singh, V.P., 2017. Endogenous reduced ascorbate: an indicator of plant water deficit stress in wheat. Indian. J. Plant Physiol. 22 (3), 365–368.

Roychoudhury, A., Banerjee, A., 2017. Abscisic acid signaling and involvement of mitogen activated protein kinases and calcium-dependent protein kinases during plant abiotic stress. Mechanism of plant hormone signaling under stress 1, 197–241.

Santisree, P., Adimulam, S.S., Bommineni, P., Bhatnagar-Mathur, P., Sharma, K.K., 2019. Hydrogen sulfide in plant abiotic stress tolerance: progress and perspectives. Reactive Oxygen, Nitrogen and Sulfur Species in Plants: Production, Metabolism, Signaing and Defense Mechanisms, John Wiley and Sons Ltd. pp. 743–775.

Scuffi, D., Álvarez, C., Laspina, N., Gotor, C., Lamattina, L., García-Mata, C., 2014a. Hydrogen sulfide generated by L-cysteine desulfhydrase acts upstream of nitric oxide to modulate abscisic acid-dependent stomatal closure. Plant Physiol. 166 (4), 2065–2076.

Scuffi, D., Álvarez, C., Laspina, N., Gotor, C., Lamattina, L., García-Mata, C., 2014b. Hydrogen sulfide generated by L-cysteine desulfhydrase acts upstream of nitric oxide to modulate abscisic acid-dependent stomatal closure. Plant Physiol. 166, 2065–2076.

Scuffi, D., Lamattina, L., García-Mata, C., 2016. Gasotransmitters and stomatal closure: is there redundancy, concerted action, or both? Front. Plant Sci. 7, 277.

Sekiya, J., Schmidt, A., Wilson, L.G., Filner, P., 1982. Emission of hydrogen sulfide by leaf tissue in response to L-Cysteine. Plant Physiol. 70, 430–436.

Shahid, M., Khalid, S., Abbas, G., Shahid, N., Nadeem, M., Sabir, M., et al., 2015. Heavy metal stress and crop productivity. Crop Production and Global Environmental Issues. Springer, Cham, pp. 1–25.

Shan, C.J., Zhang, S.L., Li, D.F., Zhao, Y.Z., Tian, X.L., Zhao, X.L., et al., 2011. Effects of exogenous hydrogen sulfide on the ascorbate and glutathione metabolism in wheat seedlings leaves under water stress. Acta Physiol. Plant 33, 2533–2540.

Shan, C., Dai, H., Sun, Y., 2012. Hydrogen sulfide protects wheat seedlings against copperstress by regulating the ascorbate and glutathione metabolism in leaves. Aust. J. Crop Sci 6 (2), 248–254.

Shan, C., Liu, H., Zhao, L., Wang, X., 2014. Effects of exogenous hydrogen sulfide on the redox states of ascorbate and glutathione in maize leaves under salt stress. Biol Plant 58 (1), 169–173.

Shan, C., Wang, T., Zhou, Y., Wang, W., 2018. Hydrogen sulfide is involved in the regulation of ascorbate and glutathione metabolism by jasmonic acid in *Arabidopsis thaliana*. Biol Plant 62 (1), 188–193.

Shen, J., Xing, T., Yuan, H., Liu, Z., Jin, Z., Zhang, L., et al., 2013. Hydrogen sulfide improves drought tolerance in *Arabidopsis thaliana* by microRNA expressions. PLoS One 8, e77047.

Shi, H., Ye, T., Chan, Z., 2013. Exogenous application of hydrogen sulfide donor sodium hydrosulfide enhanced multiple abiotic stress tolerance in bermudagrass (*Cynodondactylon* (L). Pers.). Plant Physiol. Biochem. 71, 226–234.

Shi, H., Ye, T., Chan, Z., 2014. Nitric oxide-activated hydrogen sulfide is essential forcadmium stress response in bermudagrass (*Cynodon dactylon* (L). Pers.). Plant Physiol. Biochem. 74, 99–107.

Shi, H., Ye, T., Han, N., Bian, H., Liu, X., Chan, Z., 2015. Hydrogen sulfide regulates abiotic stress tolerance and biotic stress resistance in *Arabidopsis*. J. Integr. Plant Biol. 57, 628–640.

Shivaraj, S.M., Vats, S., Bhat, J.A., Dhakte, P., Goyal, V., Khatri, P., et al., 2019. Nitric oxide and hydrogen sulfide crosstalk during heavy metal stress in plants. Physiologia Plantarum 168 (2), 437–455.

Singh, V.P., Singh, S., Kumar, J., Prasad, S.M., 2015. Hydrogen sulfide alleviates toxic effects of arsenate inpea seedlings through up-regulation of the ascorbate–glutathione cycle: possible involvement ofnitric oxide. J. Plant Physiol. 181, 20–29.

Singh, V.P., Tripathi, D.K., Fotopoulos, V., 2020. Hydrogen sulfide and nitric oxide signal integration and plant development under stressed/non-stressed conditions. Physiol Plant 168 (2), 239–240.

Sun, J., Wang, R., Zhang, X., Yu, Y., Zhao, R., Li, Z., et al., 2013. Hydrogen sulphide alleviates cadmium toxicity through regulations of cadmium transport across theplasma and vacuolar membranes in Populus euphratica cells. Plant Physiol. Biochem. 65, 67–74.

Tanaka, Y., Sano, T., Tamaoki, M., Nakajima, N., Kondo, N., Hasezawa, S., 2006. Cytokinin and auxin inhibit abscisic acid-induced stomatal closure by enhancing ethylene production in *Arabidopsis*. J. Exp. Bot 57 (10), 2259–2266.

Tian, B., Zhang, Y., Jin, Z., Liu, Z., Pei, Y., 2017. Role of hydrogen sulfide in the methyl jasmonate response to cadmium stress in foxtail millet. Front. Biosci. 22, 530–538.

Trivedi, P.K., Nath, P., 2004. MaExp1, an ethylene-induced expansin from ripening banana fruit. Plant Sci. 167 (6), 1351–1358.

Van Breusegem, F., Vranová, E., Dat, J.F., Inzé, D., 2001. The role of active oxygen species in plant signal transduction. Plant Science 161 (3), 405–414.

Wang, L., Wang, Z., Xu, Y., Joo, S.H., Kim, S.K., Xue, Z., et al., 2009. OsGSR1 is involved in crosstalk between gibberellins and brassinosteroids in rice. Plant J. 57 (3), 498–510.

Wang, R., 2012. Physiological implications of hydrogen sulfide: a whiff exploration that blossomed. Physiol Rev. 92 (2), 791–896.

Whiteman, M., Armstrong, J.S., Chu, S.H., Jia-Ling, S., Wong, B.S., Cheung, N.S., et al., 2004. The novel neuromodulator hydrogen sulfide: an endogenous peroxynitrite 'scavenger'? Journal of neurochemistry 90 (3), 765–768.

Wilson, L.G., Bressan, R.A., Filner, P., 1978. Light-dependent emission of hydrogen sulfide from plants. Plant Physiol. 61, 184–189.

Wirtz, M., Hell, R., 2006. Functional analysis of the cysteine synthase protein complex from plants: structural, biochemical and regulatory properties. J. Plant Physiol. 163 (3), 273–286.

Wojtyla, Ł., Lechowska, K., Kubala, S., Garnczarska, M., 2016. Different modes of hydrogen peroxide action during seed germination. Front. Plant Sci. 7, 66.

Wu, D.H., Li, Y.L., Xia, X., Pu, Z.P., Liao, J.M., Huang, K., et al., 2013. Hydrogen sulfide donor sodium hydrosulfide pretreatment improved multiple resistance abilities of wheat to high temperature and drought stress. J. Yunnan. Norm. Univ. Nat. Sci. 33, 29–35.

Xie, R.J., Deng, L., Jing, L., He, S.L., Ma, Y.T., Yi, S.L., et al., 2013. Recent advances in molecular events of fruit abscission. Biol Plant 57 (2), 201–209.

Xie, Z.Z., Shi, M.M., Xie, L., Wu, Z.Y., Li, G., Hua, F., et al., 2014. Sulfhydration of p66Shc at cysteine59 mediates the antioxidant effect of hydrogen sulfide. Antioxid. & Redox Signal. 21 (18), 2531–2542.

Xiong, L., Lee, H., Ishitani, M., Zhu, J.K., 2002. Regulation of osmotic stress-responsive gene expression by thelos6/aba1 locus in *Arabidopsis*. J. Biol. Chem. 277 (10), 8588–8596.

Yang, D.L., Yao, J., Mei, C.S., Tong, X.H., Zeng, L.J., Li, Q., et al., 2012. Plant hormone jasmonate prioritizes defense over growth by interfering with gibberellin signaling cascade. Proc Natl Acad Sci 109 (19), E1192–E1200.

Yang Han, Y., xiu Li, A., Li, F., rong Zhao, M., Wang, W., 2012. Characterization of a wheat (*Triticum aestivum* L.) expansin gene, TaEXPB23, involved in the abiotic stress response and phytohormone regulation. Plant Physiol. Biochem. 54, 49–58.

Yarmolinsky, D., Brychkova, G., Fluhr, R., Sagi, M., 2013. Sulfite reductase protects plants against sulfite toxicity. Plant Physiol. 161, 725–743.

Yu, L.X., Zhang, C.J., Shang, H.Q., Wang, X.F., Wei, M., Yang, F.J., et al. 2013, Exogenous hydrogen sulfide enhanced antioxidant capacity, amylase activities and salt tolerance of cucumber hypocotyls and radicles.

Zhang, H., Hu, L.Y., Hu, K.D., He, Y.D., Wang, S.H., Luo, J.P., 2008. Hydrogen sulfide promotes wheat seed germination and alleviates oxidative damage against copper stress. J. Integr. Plant Biol. 50 (12), 1518–1529.

Zhang, H., Tang, J., Liu, X.P., Wang, Y., Yu, W., Peng, W.P., et al., 2009a. Hydrogen sulfide promotes root organogenesis in *Ipomoea batatas*, *Salix matsudana* and *Glycine max*. J. Integr. Plant Biol. 51, 1084–1092.

Zhang, H., Ye, Y.K., Wang, S.H., Luo, J.P., Tang, J., Ma, D.F., 2009b. Hydrogen sulfide counteracts chlorophyll loss in sweet potato seedling leaves and alleviates oxidative damage against osmotic stress. Plant Growth Regul. 58, 243–250.

Zhang, S.W., Li, C.H., Cao, J., Zhang, Y.C., Zhang, S.Q., Xia, Y.F., et al., 2009c. Altered architecture and enhanced drought tolerance in rice via the down-regulation of indole-3-acetic acid by TLD1/OsGH3. 13 activation. Plant Physiol. 151 (4), 1889–1901.

Zhang, H., Wang, M.F., Hua, L.Y., Wang, S.H., Hua, K.D., Bao, L.J., et al., 2010a. Hydrogen sulfide promotes wheat seed germination under osmotic stress. Russ. J. Plant Physiol. 57, 532–539.

Zhang, H., Hu, L.Y., Li, P., Hu, K.D., Jiang, C.X., Luo, J.P., 2010b. Hydrogen sulfide alleviated-chromium toxicity in wheat. Biol. Plant 54, 743–747.

Zhang, H., Wang, M.J., Hu, L.Y., Wang, S.H., Hu, K.D., Bao, L.J., et al., 2010c. Hydrogen sulfide promotes wheat seed germination under osmotic stress. Russ J. Plant Physiol. 57 (4), 532–539.

Zhang, L., Pei, Y., Wang, H., Jin, Z., Liu, Z., Qiao, Z., et al., 2015a. Hydrogen sulfide alleviates cadmium-induced cell death through restraining ROS accumulation in roots of *Brassica rapa* L. ssp. pekinensis. Oxid. Med. Cell. Longev. 2015.

Zhang, L.P., Pei, Y.X., Wang, H.J., Jin, Z., Liu, Z., Qiao, Z., et al., 2015b. Hydrogen sulfide alleviates cadmium-induced cell death through restraining ROS accumulation in roots of *Brassica rapa* L. ssp. pekinensis. Oxid. Med. Cell Longev. 2015, 804603.

Zhang, F., Jin, H., Wu, L., Shao, J., Zhu, X., Chen, A., et al., 2017. Diallyl trisulfide suppresses oxidative stress-induced activation of hepatic stellate cells through production of hydrogen sulfide. Oxid. Med. Cell. Longev. 2017.

Zhu, C.Q., Zhang, J.H., Sun, L.M., Zhu, L.F., Abliz, B., Hu, W.J., et al., 2018. Hydrogen sulfide alleviates aluminum toxicity via decreasing apoplast and symplast Al contents in rice. Front. Plant Sci. 9, 294.

Zupin, M., Sedlar, A., Kidrič, M., Meglič, V., 2017. Drought-induced expression of aquaporin genes in leaves of two common bean cultivars differing in tolerance to drought stress. J. Plant Res. 130 (4), 735–745.

Chapter 16

Implication of H_2S and Ca signaling in abiotic stress tolerance in plants

Shalini Dhiman[1], Palak Bakshi[1], Arun Dev Singh[1], Tamanna Bhardwaj[1], Neerja Sharma[1], Nitika Kapoor[2], Dhriti Kapoor[3], Ravdeep Kaur[1], Upma Arora[4], Anket Sharma[5], Bilal Ahmad Mir[6] and Renu Bhardwaj[1,*]

[1]*Department of Botanical and Environmental Sciences, Guru Nanak Dev University, Amritsar, India,* [2]*PG Department of Botany, Hans Raj Mahila Maha Vidyalaya, Jalandhar, India,* [3]*Department of Botany, School of Bioengineering and Biosciences, Lovely Professional University, Phagwara, India,* [4]*Department of Botany, Lyallpur Khalsa College, Jalandhar, India,* [5]*State Key Laboratory of Subtropical Silviculture, Zhejiang A&F University, Hangzhou, P.R. China,* [6]*Department of Botany, School of Life Sciences, Satellite Campus Kargil, University of Kashmir, Jammu and Kashmir, India*

**Corresponding author.*

16.1 Introduction

Industrialization, urbanization, and diverse environmental stressors as well as the growing and lethal climate changes harshly affect world agriculture, which leads to a considerable decline in crop productivity as well as impairing food security (He et al., 2018). Conversely, plants have well-developed adaptive potential signaling molecules, for example, hydrogen sulfide (H_2S) and Ca^{2+}, to combat the damage caused by these environmental stressors. Moreover to cope with stresses, plants develop various adaptive mechanisms and external and internal responses. Internal responses involve the perception of stress signals and signal transduction, which is actually the involvement of various stress signals, their crosstalk, and finally the formation of signal networks responsible for plant defense, stress responses, and tolerance. These signaling molecules are also called secondary messengers, and include calcium (in the form of calcium ion: Ca^{2+}), reactive sulfur species (RSS; such as hydrogen sulfide: H_2S), reactive carboxyl species (RCS, such as methylglyoxal: MG), reactive nitrogen species (RNS, such as nitric oxide: NO), and reactive oxygen species (ROS, in the form of superoxide radical: $O_2^{\bullet 2}$; hydrogen peroxide: H_2O_2; and hydroxyl radical: $^{\bullet}OH$) (Campbell, 2017). These secondary messengers help the plant to adapt under stress conditions

Hydrogen Sulfide in Plant Biology. DOI: https://doi.org/10.1016/B978-0-323-85862-5.00002-6

(Zhu, 2016). These also play a crucial role in plant metabolism, growth, and development.

Calcium (Ca^{2+}) signaling is the essential core of signal interaction in plants' abiotic stress response and adaptation (Edel et al., 2017). There are both extracellular and intracellular Ca^{2+} pools in plants cell. Exracellular Ca^{2+} pools mainly refer to cell walls, whereas intracellular Ca^{2+} pools include various cell organelles like mitochondria, chloroplast, endoplasmic reticulum, and vacuoles. There is good documentation of huge changes in cytosolic free calcium ion concentration under most abiotic stimuli, including salinity, cold, light, low and high temperature, touch, hyper osmotic condition, drought, flooding, heavy metals, pesticides, and oxidative stresses (Sanders et al., 2002; Aldon et al., 2018). The studies reveal that external Ca^{2+} supplementation help plants to recover from stress (Robertson, 2013) by regulating the expression of genes associated with stress and induces the acquisition of stress tolerance (Reddy et al., 2011).

In plant systems, hydrogen sulfide (H_2S) is a new emerging small lipophilic third gasotransmitter molecules which is as important as nitric oxide (NO), carbon monoxide (CO), and hydrogen peroxide (H_2O_2) (Kimura, 2014) and its role is similar to Ca^{2+}, which acts as both a nutrient and signal transduction molecule in plant growth, development, and response to environmental stress (Hancock and Whiteman, 2016; Filipovic and Jovanović, 2017; Hossain et al., 2018). Further, these hydrogen sulfide (H_2S) sulfur-mediated small lipophilic molecules have high permeability which allows them to cross biological membranes and thus allow them to act as signaling molecules. H_2S is mainly involved in many physiological and pathological processes in animals as well as inside plant systems, including cell proliferation, apoptosis, inflammation, and hypoxia protection (Olas, 2015). In plants, H_2S has a protective role mainly against oxidative and metal stresses (Zhang et al., 2010a,b,c,d; Fang et al., 2016), drought (Shen et al., 2013) and heat tolerance (Li et al., 2012a,b,c,d,e), osmotic and saline stresses (Shi et al., 2013), and various other kinds of biotic and abiotic stresses. H_2S is also involved in regulating essential physiological processes in plants, such as stomatal closure (Papanatsiou et al., 2015), photosynthetic modulation (Chen et al., 2011), and autophagy regulation (Laureano-Marín et al., 2016).

Various reports have shown that H_2S enhances the toxic heavy metal tolerance in plant species by interacting with Ca^{2+} signaling by mainly activating the heavy metal chelaters and regulating the antioxidant system (Fang et al., 2014a,b). Thus plant abiotic stress tolerance is improved by hydrogen sulfide (H_2S), calcium signaling, and their crosstalk as well as through modification of the expression of genes connected with stress response and adaptation, and physiological and biochemical processes (Hancock and Whiteman, 2016; Filipovic and Jovanović, 2017; Hossain et al., 2018).

16.2 Role of different signaling molecules in plants during abiotic stresses

A signaling molecule usually is one that meets the following criteria: firstly, it is produced when and where required; secondly, it conveys particular information either within the cell or to another cell; and lastly, it is removed when no longer required (Hancock, 2017). Apart from the phytohormones, various reactive intermediates, especially ones derived from nitrogen (RNS), oxygen (ROS), and sulfur (RSS), are involved in stress-related cell signaling events (Hancock, 2017; Mostofa et al., 2018). Ideally, rather than working in isolation different signaling molecules interplay with one another influencing each other's production or detoxification (Hancock, 2017).

16.2.1 Calcium

Calcium is one of the main signaling mediators that play a crucial role in plant growth and stress responses. Diverse environmental stimuli activate Ca^{2+} channels and/or pumps that allow a transient influx of calcium ions, usually designated as calcium signatures, into the cytoplasmic matrix. These signatures differ in respect of frequency, amplitude, duration, and spatial distribution depending upon the particular stimulus (Tuteja and Sopory, 2008; Ranty et al., 2016). Ca^{2+} sensors, namely, *calmodulins (CaMs)* and *CaM-like proteins (CaMs/CMLs)*, *calcineurin B-like proteins (CBLs)*, and *Ca^{2+}-dependent protein kinases (CDPKs)*, decode and relay these stimulus-specific signals downstream of the signaling cascades via direct binding and activating various targets to produce stress-specific responses (Tuteja and Sopory, 2008; Yuan et al., 2018). Ca^{2+}-sensing proteins are the heart of stress responses as they interact and regulate protein kinases, transcription factors, phosphatases, ion channels/pumps/antiporters and enzymes, and thereby control intercellular Ca^{2+} levels, metal uptake, ROS, and nitric oxide (NO) homeostasis, etc. (Zeng et al., 2015; Virdi et al., 2015; Ranty et al., 2016; Atif et al., 2019).

16.2.2 Reactive sulfur species

Hydrogen sulfide, one of three postulated gasotransmitters, participates in signal transduction pathways as a signaling molecule just like carbon monoxide (CO) and NO (Gadalla and Snyder, 2010; Hancock and Whiteman, 2016; Guo et al., 2016). H_2S-mediated signaling pathways play a role in the regulation of vital physiological and metabolic processes in plants such as seed germination (Li et al., 2012a,b,c,d,e), photosynthesis (Chen et al., 2011), stomatal closure (Gadalla and Snyder, 2010), root organogenesis (Zhang et al., 2009), flower senescence (Zhang et al., 2011), and also environmental stress tolerance and cross-adaptation (Shi et al., 2015; Guo et al., 2016; Li et al., 2016a,b).

Biochemically, H_2S exerts its action through protein persulfidation, a process of oxidative posttranslational modification of cysteine residues (−SH) to persulfides (−SSH) in proteins (Aroca et al., 2017). However, protein persulfidation requires an oxidation step. During oxidative stress cysteine residues or thiols are oxidized to reversible sulfenic acids (RSOH) or irreversible sulfinic (RSOOH) and sulfonic acids (RSOOOH); H_2S prevents further oxidation of cysteine residues by reacting with sulfenic acids forming persulfides which could be restored to free thiols by the thioredoxin system. In this way protein persulfidation not only reduces oxidized thiols but also enhances the antioxidant capacity of thiols (Filipovic and Jovanović, 2017). Hence, H_2S acts as a novel warrior against nitrooxidative stress (a typical consequence of different abiotic stresses) by stimulating the plant's antioxidant system (Corpas, 2019).

H_2S-induced abiotic stress tolerance is a consequence of intricate signaling networks in which H_2S interplays with other signaling molecules such as Ca^{2+}, nitric oxide, ROS, CO, and phytohormones (particularly, abscisic acid—ABA; indole-3-acetic acid—IAA; gibberellic acid—GA; and jasmonic acid—JA); nonetheless the exact mechanisms of crosstalk are yet to be elucidated (Guo et al., 2016; Kolupaev et al., 2017; Banerjee et al., 2018; He et al., 2018).

16.2.3 Reactive carbonyl species

MG is a ubiquitous highly reactive dicarbonyl that is an inevitable by-product of metabolic processes such as glycolysis and lipid peroxidation. In plants, under optimal conditions, the concentration of MG typically ranges between 40 and 75 μM; however, during abiotic stress, its level may soar by up to sixfold (Hoque et al., 2012a). At higher concentrations, MG acts as a cytotoxin causing detrimental effects on various developmental processes. Although, predominantly, the glyoxalase pathway consisting of glyoxalase I (GLYI) and glyoxalase II (GLYII) enzymes detoxifies excessive MG, glyoxalase III (GLYIII) and D-lactate dehydrogenase (D-LDH) have also been reported to have a role in its detoxification (Jain et al., 2018).

On the other hand, at low cellular levels, it is reported to act as a crucial signaling molecule that regulates growth, development, and stress adaptation (Hoque et al., 2012b; Hoque et al., 2016). Its signaling role is principally attributed to the fact that at low concentrations MG modulates stomatal movements. Stomatal closure is an adaptive response to several environmental stresses (Hoque et al., 2016). MG regulates the stomatal closure independent of the endogenous concentration of both ABA and methyl jasmonate by promoting ROS and cytosolic calcium accumulation, and inhibiting light-induced stomatal opening through the restriction of potassium influx into the guard cells (Hoque et al., 2012a; Hoque et al., 2012c). Moreover, stress-induced MG regulates stress-responsive genes such as *RESPONSIVE TO DEHYDRATION (RD29B)* and *RESPONSIVE TO ABA (RAB18)* in *Arabidopsis* using endogenous ABA, an important signaling molecule in plants (Hoque et al., 2012b). The crosstalk of MG with other hormones during stress further establishes MG's role as a signaling molecule, for instance, MG interplays with the methyl jasmonate (MeJA) signaling pathway in response to biotic stress (Proietti et al., 2019). Furthermore, a study conducted by Kaur et al. (2015) to determine the molecular basis of MG using microarray analysis in rice revealed that MG responsive signal transduction pathways are possibly inextricably linked with stress-responsive ones as *MG altered stress-inducible protein kinases (mitogen-activated protein kinases (MAPKs), CDPKs, histidine kinase, OsRR2 type-A response regulator, receptor-like kinase (RLK)*, and *Ser/Thr protein kinase*), transcription factors (*AP2 domain-containing protein, basic-leucine zipper* (*bZIP*), *NAC, WRKY, dehydration-responsive element binding-type* (*DREB*), and *zinc finger proteins*), and signal transduction genes, such as hormonal and cell-to-cell signaling and even chromatin remodeling.

Innumerable recent reports have confirmed the signaling role of MG in abiotic stress. Exogenous application of MG induces abiotic stress tolerance by upregulation of the glyoxalase system and antioxidative defense system, and also the accumulation of osmolytes (Li et al., 2017; Li et al., 2018; Wang et al., 2019). MG pretreatment induces frost-hardiness in wheat via transient oxidative burst which in turn triggers enzymatic defense mechanisms, synthesis of sorbitol, and increase in omega-3 fatty acid desaturases, which help to retain membrane integrity and facilitate photosynthesis (Majláth et al., 2020).

16.2.4 Reactive oxygen species

ROS, such as hydrogen peroxide (H_2O_2), hydroxyl radical ($OH^{\cdot}$), superoxide anion ($O_2^{\cdot -}$), and singlet oxygen, are highly reactive by-products of aerobic metabolic processes which are considered not only as unavoidable toxic metabolites that cause oxidative stress but also as indispensable signaling molecules that regulate various physiological processes and stress responses during stress (Miller et al., 2010; Mittler, 2017). Depending upon the cellular concentration of ROS, it acts as a toxin or messenger (Sharma et al., 2012). Organisms possess a well-developed antioxidative defense system, comprising nonenzymatic antioxidants and ROS-scavenging enzymes, to counteract the ROS buildup and maintain ROS homeostasis (Sharma et al., 2012; Hancock, 2017). Overproduced ROS are also capable of perturbing redox homeostasis and causing cell death, ultimately culminating in reduced growth and yield (Czarnocka and Karpiński, 2018). Nevertheless, the damaging role of ROS is becoming questionable as the cell death which was previously considered as a consequence of oxidative stress is now contemplated as programmed or physiological cell death (Mittler, 2017). Moreover, some basal ROS concentration is essential for survival and adaption against unfavorable conditions, for example, systemic acquired resistance and systemic acquired acclimation (Mittler, 2017; Czarnocka and Karpiński, 2018). Hence, it is usually concluded that as long as there is efficient scavenging machinery to detoxify excessive ROS, their beneficial effects far outweigh the detrimental effects in abiotic stress (Choudhury et al., 2017).

Diverse abiotic stress produces different ROS signatures. Specific sensors deciphering these signatures can generate a stress-specific signal that alters protein redox reactions (e.g., via oxidation of sulfur-containing residues of proteins and formation of disulfide bonds) which in turn modifies transcription and translation eliciting activation of the required acclimation response (Choudhury et al., 2017). Moreover, ROS can also activate *MAPK* cascades which are implicated in plant response to stresses and act downstream of ROS signaling and RLKs to control ROS-related gene expression (Liu and He, 2017).

16.2.5 Reactive nitrogen species

NO is a tiny, transient, and bioactive gasotransmitter that is often termed as a plant growth regulator owing to its potential to mediate diverse growth and developmental processes in plants, for example, seed germination, maturation, senescence, stomatal closure, floral regulation, and hormone-mediated development (Parankusam et al., 2017; Sami et al., 2018). Moreover, it performs a vital role in redox signaling and the plant's abiotic and biotic stress responses (Parankusam et al., 2017; Chakraborty and Acharya, 2017). On encountering unfavorable circumstances, plants activate various stress-responsive signaling pathways and accumulation of signaling such as ROS and NO is the first step of these responses (Yang et al., 2015). NO regulates ROS homeostasis by regulating ROS generation and scavenging via S-nitrosylation (Yun et al., 2011; Yang et al., 2015). Just like H_2S, the main route of bioactivity of NO is S-nitrosylation, a posttranslational modification of cysteine residues (-SH) in target proteins to *S*-nitrosothiol (−SNO) (Yu et al., 2013). Another type of protein posttranslational modification, methylation—particularly protein arginine methylation catalyzed by protein *arginine methyltransferases (PRMTs)*—in addition to NO-induced S-nitrosylation is also involved in the regulation of diverse biological processes. Precisely, the NO signal is transduced to the protein methylation apparatus via S-nitrosylation of PRMT5 (Hu et al., 2017). Besides protection against oxidative stress, NO also participates as a signaling intermediate along with other players (ROS, Ca^{2+}, phospholipids, and others) in ABA (stress hormone)-induced stomatal closure by triggering cyclic guanosine monophosphate production and *MAPKs* which are necessary for ABA signaling (Neill et al., 2008; Gayatri et al., 2013).

16.3 Hydrogen sulfide (H_2S) and Ca^{2+} signaling during different types of abiotic stress tolerance in plants

16.3.1 Salt stress

Oxidative burst is the first event elicited in plant cells upon salt stress, causing the intensification of electrolytes leakage, lipid peroxidation, and protein oxidation. Mitigation of oxidative stress in salt-stressed plants is one of the most studied roles of H_2S. The antioxidant activities [like superoxide dismutase (SOD), CAT, ascorbate peroxidase (APX), glutathione reductase (GR), GPX, and DHAR] in stressed cucumber seedlings were increased by treating them with NaHS (an H_2S donor) while H_2O_2 and lipid peroxide levels decreased under the same experimental conditions (Yu et al., 2013). Suppression of endogenous H_2S was reported following the infiltration of hypotaurine into *Nicotiana tabacum* and this also negatively affected the antioxidant activities like SOD, CAT, and APX (Da Silva et al., 2017). The activity of enzymes involved in GSH (γ-glutamylcysteine synthetase) and AsA (L-galactono-1,4 lactone dehydrogenase) biosyntheses and further increase of glutathione (GSH)/oxidized GSH and AsA/DHR ratios, were stimulated by NaHS in the leaves of salt-treated maize (Shan et al., 2014). Similarly, NaHS-controlled methylglyoxylate levels in rice by increasing the activity of glyoxalase I and glyoxalase II (Mostofa et al., 2015).

16.3.1.1 H_2S signaling during salt stress

In the plant system, the enzymes cyanoalanine synthase, sulfite reductases, cysteine desulfhydrases, etc. are required for H_2S metabolism. These enzymes are responsible for the synthesis of H_2S and trigger the transformation of H_2S into cysteine (Filipovic and Jovanović, 2017; Hossain et al., 2018). Among these enzymes, *cysteine desulfhydrases (L-DES)* contributes to the synthesis of H_2S even under salinity stress (Li et al., 2016a,b; Hossain et al., 2018). Salt stress conditions may trigger H_2S signaling in plants by altering different metabolic reactions (Li et al., 2016a,b; Filipovic and Jovanović, 2017).

External priming of H_2S modulates the internal content, which develops the adaptive potential of plants against salinity stress conditions (Li et al., 2016a,b; Filipovic and Jovanović, 2017; Hossain et al., 2018). Reports indicated that H_2S also contributes as a signaling molecule in the creation of cross-adaptation in plants (Li et al., 2016a,b). Generally, H_2S-mediated salt stress tolerance is included in diverse signaling molecules like Ca^{2+}, H_2O_2, NO, H_2S, and CO and their interactions. These lead to encouraging adaptive alteration in the expression of genes, certain physiochemical activities, histochemical and morphological features, growth, and development, thus consequently raising the plant stress tolerance via forwarding oxidative damage, membrane injury, osmotic stress, and protein denaturation, along with maintaining ion and nutrient balance (Hossain et al., 2018).

The maintenance of high K^+/Na^+ ratio in plant cells under salt stress was also induced by H_2S. Wheat seedlings treated with 50 μM NaHS, followed by 100 mM NaCl exposure exhibited increased K^+/Na^+ ratio with an increase of selective transport of K^+ over Na^+ through nonselective cation channels, salt overly sensitive 1 (SOS1), and a plasma membrane Na^+/H^+ antiporter (Deng et al., 2016). Induction of plasma membrane Na^+/H^+ antiporter genes (e.g., *SOS2, SOS3*, and *SOS4*) by NaHS was also described in strawberry plants under high salinity condition, indicating the importance of H_2S in K^+ uptake (Christou et al., 2013). The K^+/Na^+ homeostasis in salt-treated alfalfa was shown to be maintained by NaHS through the prevention of K^+ efflux likely triggered by lower expression of shaker-like K^+ outward-rectifying channel genes (Lai et al., 2014). Similar results were also observed under salt-treated barley seedlings roots in the presence of NaHS (Chen et al., 2015). Moreover, H_2S maintained low Na^+ levels in cells by increasing the transcription of genes that encode for plasma membrane H^+-ATPase, H^+-ATPase subunit β, vacuolar Na^+/H^+ antiporter and augmenting Na^+ compartmentation in vacuoles (Chen et al., 2015).

16.3.1.2 Ca^{2+} signaling during salt stress

Previous reports suggested the role of Ca^{2+} signals in salinity stress conditions and strongly proved that this signaling process definitely seems to be an initial and protruding characteristic of the responses of plants to the environmental prompt (Filipovic and Jovanović, 2017). It was noted that plants show a quick enhancement in calcium amount in cytosoles within a very short time of being exposed to salinity (Feng et al., 2018). The likeness of Ca^{2+} signs persuaded by either salts or nonionic osmotic activates has up to now prohibited definite discernment of whether salts as an ion particularly boosts a Ca^{2+} response autonomously of its osmotic impact. Also, in the salinity stress signaling, recognition of functional salinity receptors is still not known which is the primary and significant step in the signaling (Köster et al., 2019). Many reports indicated that the level of Ca^{2+} modulates in the roots of the plants during salt stress (Kiegle et al., 2000; Tracy et al., 2008). Cell specificity and tissue heterogeneity of such responses were indicated by the Ca^{2+} signal. Drastic enhancement in Ca^{2+} in the cytoplasm was analyzed by Kiegle and colleagues and also in the endodermal cells throughout the root under salinity stress (Kiegle et al., 2000).

Ca^{2+} signals lead to the triggering of responses against salt stress, provided by the treatment of advanced Ca^{2+} reporter proteins. Systemic and wavy Ca^{2+} elevations are observed in the leaves of the plants due to salinity exposed roots (Xiong et al., 2014). Additionally, in the salt exposed roots of *Arabidopsis* extended signals of Ca^{2+} were analyzed in wavy manner within a few seconds (Choi et al., 2014). These reports indicated that Ca^{2+} signaling is a complicated process and it is not just limited to a single cell, but expands throughout the plant during salt stress. Besides, the previous research also indicated that Ca^{2+} signals play a fundamental role in composing the complete plant responses to soil-borne signals which are generated by salinity stress. Tonoplast-localized cation channel TPC1 and the plasma membrane-localized NADPH oxidase RBOHD significantly lead to the suitable production of salinity-triggered long-distance signals (Choi et al., 2014). Such outcomes strengthen the evolving complication of these activities as they specify the contribution of numerous cellular compartments and the involvement of different signaling pathways to the complete appearance of Ca^{2+} signaling in the presence of salinity stress (Evans et al., 2016).

Ca^{2+} signaling has capacity to interconnect salinity stress responses with certain physiological activities at the level of Ca^{2+} sensor function. In *Arabidopsis*, CBL10 interacts with the K^+ channel *Arabidopsis* K^+ Transporter 1 (AKT1) and in such conditions the accumulation of AKT1 at the plasma membrane occurs for the regulation of K^+ uptake (Ren et al., 2013; Yang and Guo, 2018). Such a process should be relevant to maintaining an optimal K^+/Na^+ ratio against salinity in plants. The *RLK FERONIA (FER)* is essential for single cell-specific Ca^{2+} signaling regulating the cell wall integrity during salinity stress. *FER*-dependent signaling provokes cell-specific Ca^{2+} signals during the process. Na^+ removal from the cytosol is obtained by the activity of the Na^+/H^+ antiporter SOS1 (Filipovic and Jovanović, 2017). During stress, phosphorylation of the transporter occurs and is stimulated by *SOS2/CIPK24 kinases*, that is, controlled by the interaction of Ca^{2+} sensor protein *SOS3/CBL4*. Apart from this, *CBL4*, another members of the *CBL protein* family, makes functional complexes with *SOS2/CIPK24*, like *CBL10* and *CBL1*. The geminivirus Rep-interacting kinases 1 and 2, play significant functions in sugar/energy signaling by phosphorylation of *SnRK1-type kinases*. Na^+ sequestration into the vacuole is a mechanism to hold cytoplasmic Na^+ doses in a tolerable range. *CBL10/CIPK24* complexes stimulate the process of Na^+ movement into the vacuole (Manishankar et al., 2018) (Fig. 16.1).

16.3.2 Hydrogen sulfide (H_2S) and Ca^{2+} signaling during temperature stress

Temperature stress in plants can be either due to exposure to extremes of temperature like low temperature (cold stress) or high temperature (heat stress). Temperature stress leads to low germination rates, growth retardation, reduced photosynthesis, and often death of plants. High temperatures impact photosynthesis, carbon assimilatory processes, membrane stability,

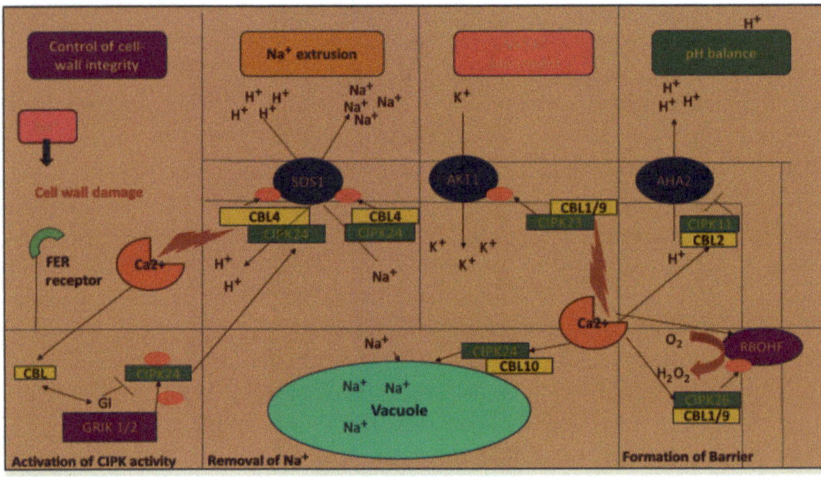

FIGURE 16.1 Ca^{2+} signaling in different facets of plant salinity stress responses. AHA2, Autoinhibited H^+-ATPase 2; AKT1, *Arabidopsis* K^+ Transporter 1; Ca^{2+}, calcium; CBL, *Calcineurin B-like protein*; CIPK, CBL-interacting protein kinase; FER receptor, FERONIA receptor; GI, GIGANTEA; GRIK, Geminivirus rep-interacting kinases; Na^+/K^+, sodium/potassium transporter; RBOHF, respiratory burst oxidase homologues F; SOS1, salt overly sensitive 1. *Modified after Manishankar, P., Wang, N., Köster, P., Alatar, A.A., Kudla, J., 2018. Calcium signaling during salt stress and in the regulation of ion homeostasis. J. Exp. Bot. 69(17), 4215–4226.*

and the plant water status (Santisree et al., 2018). Heat stress induces oxidative burst, and enhances membrane fluidity, protein denaturation/aggregation, enzyme inactivation, macromolecule damage, and severe cellular injury, which leads to plant cell death (Santisree et al., 2018). Cold stress is caused when plants are exposed to low nonfreezing chilling temperatures ($>0°C$) and freezing temperatures ($<0°C$). Cold stress often results in dysfunction of the cell membrane and osmotic and oxidative stress (Chongchatuporn et al., 2013). Cold stress greatly affects plant metabolism both by direct inhibition of metabolic enzymes by cold temperatures and reprogramming of gene expression (Chinnusamy et al., 2007). Membrane fluidity and cytoskeleton and calcium influx are found to be involved in the cold stress regulation of *MAPKs* and *COR* genes. Heat stress also activates *MAPKs*, leading to heat-induced changes in membrane fluidity and calcium signaling that further regulate *heat shock protein* (*HSP*) gene expression and thermotolerance (Sangwan et al., 2002). Heat stress induces the expression of HSPs, which function as molecular chaperones to prevent protein denaturation and maintain protein homeostasis (Scharf et al., 2012). Common features between cold- and heat-stress signaling, along with membrane fluidity changes, calcium signaling, and *MAPK* activation, include the involvement of ROS, NO, phospholipid signals, protein sumoylation, and proteosomal degradation (Zhu, 2016). H_2S is gaining increasing attention from plant researchers due to its involvement in resistance to various plant stress conditions.

16.3.2.1 Hydrogen sulfide (H_2S) signaling during temperature stress

Hydrogen sulfide (H_2S), nitric oxide (NO), carbon monoxide (CO), and hydrogen peroxide (H_2O_2), referred to as gasotransmitters, possess high permeability that allows them to cross biological membranes and to act as signaling molecules (Aroca et al., 2017). H_2S has been reported to exhibit regulatory properties with protective effects against oxidative and metal stresses (Fang et al., 2016) drought and heat stresses (Li et al., 2012b; Shen et al., 2013), and osmotic and saline stresses (Shi et al., 2013). H_2S ameliorates different stress factors by various metabolic strategies like activation of antioxidant system, increase of K^+/Na^+ ratio, stimulation of seed germination and plant growth, and induction of osmoregulators accumulation.

Endogenous H_2S generation in response to high temperature has been reported in some plant species. Chen et al. (2016) reported L−cysteine desulfhydrase (LCD)-dependent production of H_2S in *N. tabacum*. High temperature treatment led to an increase in the production of endogenous H_2S in 3-week old *N. tabacum* seedlings, and the released H_2S in turn induced the JA accumulation followed by the stimulation of nicotine synthesis. A pronounced increase in H_2S content has been observed by Christou et al. (2014) in *Fragaria ananassa* seedlings after exposure to 42°C that was also found to be regulated by NaHS pretreatment in heat-stressed *F. ananassa* seedlings. In another study, the effects of exogenous H_2S donor treatment on heat tolerance has been explored in a suspension of cultured cells of *N. tabacum* (Li et al., 2012b). Exogenous H_2S donor pretreatment markedly increased the survival percentage and regrowth ability of *N. tabacum* suspension cultured cells under heat stress by reducing electrolyte leakage and malondialdehyde (MDA) accumulation rates in pretreated cells. The enhancement of NaHS-induced heat tolerance by the application of exogenous Ca^{2+} and its ionophore A23187, while significant inhibition by ethylene diamine tetra acetic acid (EDTA), a channel blocker of plasma membrane, and *calmodulin (CaM)* antagonists, chlorpromazine and trifluoperazine, suggest the importance of transmembrane transport and the Ca^{2+} signal transducer during H_2S-mediated heat tolerance. ABA-induced heat tolerance has also been shown to be enhanced by the addition of NaHS, but reduced by H_2S biosynthesis inhibitor, propargylglycine (PAG) and H_2S scavenger, HT suggesting the significance of H_2S in ABA-mediated heat tolerance of *N. tabacum* suspension cultured cells (Li and Jin, 2016). H_2S-induced an increase in the activity of *1-pyrroline-5-carboxylate synthetase (P5CS)*, while decreasing the activity of *proline dehydrogenase (ProDH)*, resulting in an accumulation of proline, thus indicating the involvement of proline synthesis in H_2S improved heat tolerance of *Zea mays* (Li et al., 2013). Photosynthesis is the plant process most vulnerable to high-temperature stress. Many studies demonstrated the reduction of photochemical efficiency (Fv/Fm) in leaves of several plant species under heat stress conditions. Transient preexposure of *F. ananassa* plant roots to NaHS resulted in higher Fv/Fm ratio by mitigating the heat stress injuries to PSII activity. Moreover NaHS root pretreatment alleviated heat stress derived oxidative damage, lipid peroxidation, and H_2O_2 leaf content preceded by the enhanced transcript levels and activities of major enzymatic antioxidants (*cAPX, CAT, MnSOD, GR*) and also led to enhanced transcription of *HSPs (HSP70, HSP80, HSP90)* and aquaporins (Christou et al., 2014). An increase in the contents of GSH and AsA and the ratio of the reduced antioxidants to total antioxidants under heat stress culture conditions has been observed in NaHS pretreated maize seedlings and tobacco suspension cells (Li, et al., 2013; Li, 2015). H_2S can thus easily be described as a singling molecule activating a coordinated network of defense-related pathways toward induced heat tolerance in plants.

Exogenous application of H_2S has also been found to be helpful in mitigating cold stress injuries in plants. An initial study by Fu et al. (2013) demonstrated that the H_2S content in *Vitis vinifera* is enhanced by low temperature. This study investigates the major production of H_2S in response to 4°C acclimation which further increases the activities and expression of VvLCD and VvDCD genes in *V. vinifera.* In addition, enhanced activity of SOD, low levels of superoxide anion radical, low MDA content, and cell membrane permeability followed by NaHS treatment confirmed the participation of H_2S in *V. vinifera* response to chilling stress. Meanwhile, Shi et al. (2013) reported that NaHS pretreatment induced freezing tolerances in *C. dactylon* by alleviating low temperature induced oxidative stress by preserving redox homeostasis and by regulating the activities of antioxidant enzymes CAT, GPX, and GR. Ma et al. (2016) revealed that the enzyme activities of proteins involved in H_2S biosynthesis are remarkably enhanced at higher altitudes, and thus maintain a higher H_2S level as an adaptive response to altitude gradient in the alpine plant grown in the Northern Tibetan Plateau. Exogenous H_2S application stimulates antioxidant enzyme activities and *S*-nitrosoglutathione reductase activity, resulting in protein degradation, and osmolyte accumulation has been observed to be reversed by deploying exogenous H_2S biosynthesis inhibitor, PAG, and H_2S scavenger. These results reinforce the central role of H_2S in increasing alpine stress adaptation in *Lamiophlomis rotata.* Low concentration of H_2S has also been known to delay the senescence of postharvest fruit due to its antioxidative role (Hu et al., 2014). H_2S fumigation in chilling stressed banana fruits has maintained peel firmness, higher values of lightness, and reduced the accumulation of MDA mainly through increasing antioxidant enzymes capacity, resulting mainly from stimulating both enzymatic and nonenzymatic compounds biosynthesis, including major antioxidant enzymes, PPO (Phenylalanine ammonia lyase), and total phenolics content (Luo et al., 2015). Interestingly, ethylene synthesis was reduced by H_2S fumigation which shows a negative relationship between ethylene and H_2S in maintaining postharvest quality during cold storage. Further analysis also found that H_2S fumigation stimulated proline content by activating *P5CS* activity and slightly suppressing *ProDH*, which may possibly be related to cold injury tolerance. These results suggest that H_2S can improve the cold tolerance in plants through synergistic effects mainly by enhancing antioxidant capacity, proline accumulation, and also altered gene expression (Santisree et al., 2019).

Hydrogen sulfide (H_2S) is involved in plant cross-adaptation. H_2S is found to accumulate endogenously as a response of plants to environmental stress, including salt, heavy metal (HM), drought, heat, and cold stress, as well as pathogen infection, (Calderwood and Kopriva, 2014; Hancock and Whiteman, 2016). Interestingly, exogenously applied H_2S, released from donors such as NaHS and *morpholin-4-ium 4-methoxy phenyl (morpholino) phosphino dithioate (GYY4137)*, shows significant positive effects on seed germination, organogenesis and growth, the regulation of senescence, as well as the acquisition of stress tolerance, such as salt, heavy metals, drought, heat, and cold tolerance(Li, 2019).

The bioprotective behavior of H_2S on plant heat tolerance is well documented but the molecular targets and mechanisms through which it exerts its intracellular effects are only partially known. A number of proteins are covalently modified by H_2S through sulfhydration of specific cysteine residues. However, the functional role of this posttranslational modification needs to be investigated in more detail. Potassium channels, particularly *KATP*, are considered key mediators of H_2S-induced effects, and their sulfhydration has been clearly demonstrated. Since calcium signaling is involved in all cell processes, there is growing consensus about H_2S interaction with Ca^{2+} channels and its functional outcomes (Munaron et al., 2013).

16.3.2.2 Calcium signaling during temperature stress

Ca^{2+} is present at relatively high concentrations (0.1–80 mM) in cell walls and organelles, and the cytoplasmic levels of Ca^{2+} are maintained at ~100 nM. Both the deficiency as well as the high concentrations of Ca^{2+} causes localized cell death so the transport of Ca^{2+} throughout the plant needs to be tightly regulated. Thus the signal transduction in plants requires the ability to mobilize and sequester Ca^{2+} from both internal and external Ca^{2+} stores. Cells contain stores of Ca^{2+} in the apoplast and in various compartments within the cell. Cytoplasmic Ca^{2+} is kept low to prevent interference with phosphate-containing pathways. Signal transduction uses discrete Ca^{2+} fluxes to connect stimuli with adaptive responses. The Ca^{2+} signals differ in spatiotemporal properties according to the nature, strength, and duration of the stimulus. However, Ca^{2+} signals conversion into biological responses requires Ca^{2+} sensors for decoding and relaying. The occurrence of *calmodulin (CaM)* inside plant systems and also other sets of plant-specific Ca^{2+} sensors such as *calmodulin-like proteins (CMLs)*, *CDPKs*, and *CBLs* indicate that plants possess specific tools and machineries to convert Ca^{2+} signals into appropriate responses (Aldon et al., 2018). The pioneering work of Trewavas's group and his collaborators demonstrated the occurrence of Ca^{2+} variations in plant cells using the fluorescent Fluo3 Ca^{2+} indicator and caged Ca^{2+} or IP3 as triggers. Then through the application of the noninvasive, organelle-addressable, and highly dynamic Ca^{2+} probe aequorin allowed tremendous progress in the current understanding of how Ca^{2+} controls plant adaptive responses. High-temperature responsive signaling pathway in the transgenic *Arabidopsis* plants have *Calcineurin B-like protein-interacting protein kinases* (CIPKs) as important components of the Ca21-mediated *CBL-CIPK* network (Guduru et al., 2019).

Plant response to cold stress is a complex process. This response has been reported to require signal pathways like oxidative pathway, *mitogen-activated protein kinases (MAPKs)*, phytohormone, ABA, as well as *Arabidopsis response regulators (ARRs)* (Li et al., 2012a). Ca^{2+}, as a second messenger plays a critical role in cold stress response in plants. Under cold stress, signal transduction involves the activation of Ca^{2+} channels and/or Ca^{2+} pumps, which finally induces Ca^{2+} influx (Ca^{2+} signature) in plant cells. Under cold stress, Ca^{2+} signals triggered by cold stimulus are relayed by Ca^{2+} sensors, such as *CaMs*, *CMLs*, *(CPKs/CDPKs)*, and *CBLs*. These Ca^{2+} sensors, together with other components in Ca^{2+} signaling, decode Ca^{2+} signals into downstream signaling events, such as phosphorylation, transcriptional reprogramming, activation of *MAPKs* cascade, as well as the accumulation of ROS or nitric oxide (NO) (Yuan et al., 2018). Aequorin-based Ca^{2+} imaging revealed that low temperature (0 ºC) induced a rapid and transient Ca^{2+} influx in whole seedlings of *A. thaliana* (Jeon and Kim, 2013). Transient Ca^{2+} changes triggered by cold stress were also observed by using another Ca^{2+} sensor; Yellow Cameleon (YC3.6). The patch-clamp technique provided evidence of a predicted nonselective Ca^{2+}-permeable cation channel in *Arabidopsis* mesophyll cells, and this Ca^{2+} channel is activated by cold stress. Accumulation of ROS triggered by cold stress results in Ca^{2+} transients in plant cells which suggests that the crosstalk between Ca^{2+} signaling and ROS signaling plays a role in the perception of cold stimulus. COLD1 is a plasma membrane and endoplasmic reticulum localized transmembrane protein which senses cold signals and induces cold stress tolerance in plants. It has been reported that *Ca^{2+}/CaM-regulated receptor-like kinases 1 (CRLK1)*, encoding a PM-associated serine/threonine kinase regulated by Ca^{2+} signaling, plays a critical role in plant response to temperature stress (Furuya et al., 2014). It has also been suggested that *MAPK* signaling pathways play a positive role in the regulation of cold response in plants. The low temperature activates Ca^{2+} signal-mediated *CRLK1*, which subsequently phosphorylates and activates *MEKK1*, an upstream component in *MAPK* cascade of *MKK2*. Activation of *MKK2* by *MEKK1* induced phosphorylation of *MPK4* and *MPK6*. *CPKs* are reported to be involved in Ca^{2+}-mediated signal transductions by connecting cold stress-triggered Ca^{2+} transients to downstream phosphorylation events (Lv et al., 2018). Ca^{2+}/CaM-mediated transcription factors relay Ca^{2+} transients triggered by cold stress to transcriptional reprogramming.

Calcium is one of the important secondary messengers. Calcium ions regulate ion homeostasis, thereby enhancing stress tolerance. It acts as an essential signaling molecule in both abiotic and biotic stress tolerance. In plant cells, calcium ions are located in the endoplasmic reticulum, vacuoles, plastids, and mitochondria. Moreover, internal concentrations of calcium levels are maintained by various channels, pumps, and transporters (Dikilitas et al., 2019). It regulates gene expression and defense mechanisms. Exogenous Ca^{2+} application showed negative correlation with lipid peroxidation and positive correlation with total antioxidant capacity of the cell system under the elevated temperature stress (Guerret et al., 2016). Plants are often simultaneously or sequentially exposed to multiple biotic and abiotic stresses that result in substantial yield loss. Simultaneous occurrence of an abiotic stress with a biotic stress either aggravates or inhibits the effect of the latter, which leads to either enhanced or reduced susceptibility to pathogens. Overlap responses between crosstalk of signaling pathways can be a source of potential stress tolerance traits that can be engineered into crops to confer multiple stress resistance into plants (Kissoudis et al., 2016). Whenever a plant encounters any stress either biotic or abiotic, were further sensed by the potential sensors, which finally causes the generation of secondary messengers (e.g., ABA, ROS, Ca^{2+}, inositol phosphates, etc.). The secondary messengers alter the levels of intracellular Ca^{2+}. Any changes in cytoplasmic Ca^{2+} is sensed by Ca^{2+} sensors like calcium-binding proteins responsible for regulating stress-related genes. The expression of stress-regulated genes leads to the production of various kinds of regulatory molecules like the phytohormones ABA, ET, and SA and transcription factors, etc. Such signaling networks are deeply interconnected at many levels, further transferring information to finally generate the plant's adaptive response. Such kinds of signal transduction pathways demand proper coordination of all kinds of signaling molecules (Suzuki and Katano, 2018). Under stress condition signaling molecules were responsible for the activation of the defense-related genes (Li et al., 2008). For example, an increase of cytosolic calcium (Ca^{2+}) levels in response to biotic stress such as pathogen attack or any abiotic factor, like osmotic stress, water stress, cold, or wounding, triggers simultaneous pathways and activates calcium interacting proteins such as *CDPKs* and calmodulin.

16.3.2.3 Interplay between H_2S and calcium signaling during plant response to temperature stress

Calcium/calmodulin (CaM), a core component of the calcium messenger system, is a messenger that is involved in multiple stress tolerances including heat tolerance in plants. It is fast emerging that hydrogen sulfide has similar functions, but Ca^{2+} and H_2S crosstalk in the acquisition of thermotolerance. NaHS (an H_2S donor), Ca^{2+}, and CaM treatment significantly improved the thermotolerance of tobacco cells. CaM, a core component of the calcium messenger system, plays a key role (Li, 2019). The recent findings show that both H_2S and Ca^{2+} signaling interact in plant abiotic stress response and adaptation. Plant growth, development, and response and adaptation to abiotic stress commonly are achieved by many signaling molecules and their interplay. As stated above, Ca^{2+} is a universal signaling molecule in all organisms including plants, while H_2S is a novel gaseous signaling molecule after NO and CO. The cross action of Ca^{2+} signaling and H_2S signaling in some plant species under heat stress has been reported for the first time by Li et al. (2012b), who concluded that pretreatment of tobacco (*N. tabacum*) suspension cultured cells with NaHS markedly elevated the survival percentage and the vitality of cells under heat stress and regeneration capacity after heat stress, and reduced the accumulation of the membrane lipid peroxidation product, MDA. Further this improvement was reinforced by exogenously applied Ca^{2+} and A23187 (Ca^{2+} ionophore, helping Ca^{2+} enter into cytoplasm), respectively (Li et al., 2012b). H_2S may activate Ca^{2+} channels located in plasma membrane, which in turn initiate and propagate Ca^{2+} signaling, which is then decoded by Ca^{2+} sensors (especially CaM), thus inducing the formation of heat tolerance. Similarly, in wheat (*Triticum aestivum* L.) coleoptiles, treatment with 100 μM NaHS led to a transient increase in ROS, namely superoxide anion radical ($O2^{\bullet}-$) and H_2O_2. H_2S is able to activate NADPH oxidase activity, which in turn is responsible for inducing ROS generation, maintain calcium homeostasis, activating antioxidant enzymes, and thus improving plant heat tolerance in plants. In addition, further studies have reported that Ca^{2+} signaling and H_2S signaling alone or in combination causes the amelioration of temperature stress. These strategies include the activation of the antioxidant defense system composed of antioxidant enzymes (SOD, CAT, GPX, GR, and APX) and nonenzymatic antioxidants (GSH and AsA) which maintain redox homeostasis, enhance glyoxalase system (including glyoxalase I, II, and III) to scavenge excess MG, and stimulate accumulation of osmolytes [such as proline, soluble sugar, trehalose, and glycine betaine (GB)] in order to maintain cell turgor pressure. The amelioration process also involves the synthesis of *HSPs (HSP70, HSP80, and HSP90)* which maintain the spatial structures of proteins and balance the expression of transporters (aquaporin)-related genes in order to retain water, ion, and nutrient balance (Hossain et al., 2018). All of the abovementioned findings suggest the possible role of H_2S and calcium signaling inside plants system which can be interdependently induced. Such interplay helps the plant to tolerate heat stress by activating the antioxidant defense system, glyoxalase system, osmotic adjustment, synthesizing HSPs, and maintaining ion, nutrient, and water balance inside plant cells (Li et al., 2016a).

16.3.3 Hydrogen sulfide (H_2S) and Ca^{2+} signaling during heavy metal stress

The health of the environment is continuously deteriorating due to anthropogenic activities like toxic emissions from industries, mining, sewage irrigation, and the use of metal products etc. These activities are mainly responsible for soil pollution and metal stress in plants. The presence of metals like cadmium (Cd), chromium (Cr), copper (Cu), aluminum (Al), zinc (Zn), and lead (Pb) beyond the normal range deteriorate the health of plants by affecting various physiological activities including growth, yield, antioxidative defense system, membrane integration, and may induce chromosomal aberrations. In addition to this, these heavy metal ions also induce secondary stress such as nutrient imbalance, increase in production of free radicals, such as superoxide radical ($O_2^{\cdot -}$), hydrogen peroxide (H_2O_2), hydroxyl radical (OH·), and alkoxyl radical (RO·), and oxidative stress (Melegari et al., 2013; He et al., 2018).

However, plants try to overcome the damage of metal stress by mechanisms like the restriction of metal absorption and transport, vacuolar compartmentation, chelation with the help of phytochelatins (PCs), metallothioneins (MTs), or organic acids, and the repair of damaged proteins by *HSPs* (Hooda, 2010). In addition to these mechanisms, plants also adopt another way to overcome different environmental stresses, including heavy metal stress. Hydrogen sulfide (H_2S) is considered to be the third gaseous signal molecule, similar to nitric oxide (NO) and carbon monoxide (CO) (Wang, 2012). Originally the emission of H_2S was detected from cucumber, pumpkin, squash, corn, cantaloupe, cotton, and soybean (Wilson et al., 1978). Hydrogen sulfide not only regulates plant growth and development, but also affects plant responses to different environmental stresses (Jin and Pei, 2015). H_2S improves the stress tolerance of plants against heavy metal stress by inhibiting the absorption and transport of metals and their accumulation in plants. The reducing capacity of H_2S is lower than that of GSH and cysteine (Kabil and Banerjee, 2010), but it can directly scavenge ROS, such as superoxide radicals, hydrogen peroxide, hypochlorite, and peroxynitrite (Whiteman et al., 2004; Mitsuhashi et al., 2005). It has been reported that exogenous H_2S can induce stress tolerance in plants against various heavy metals, thereby alleviating metal toxicity in plants (Table 16.1).

H_2S ameliorates the metal toxicity in plants by regulating metal transport proteins, improving the antioxidant capacity of plant cells, and by scavenging metal-induced ROS in plants. H_2S might also interact with signaling pathways, protein *S-sulfhydration*, and microRNAs (miRNAs) to regulate heavy metal toxicity (Dawood et al., 2012). H_2S also reduces the

TABLE 16.1 H_2S-induced adaptation in plants against heavy metal stress.

S. no.	Plant species	Heavy metal	H_2S signaling effect and plant responses	References
1	*Solanum nigrum*	Zinc (Zn)	Alleviates Zn toxicity, reduces Zn uptake, regulates antioxidant enzymes and metallothioneins	Liu et al., 2016
2	*Brassica pekinesis*	Copper (Cu)	Influences Cu toxicity and affects sulfur metabolites and related genes expression	Shahbaz et al., 2014
3	*Triticum* (Wheat)	Copper (Cu)	Alleviates the inhibitory effect of Cu stress	Zhang et al., 2008
4	*Brassica napus*	Lead (Pb)	H_2S alleviates Pb stress, improves nutrient uptake, enhances antioxidant defense system, enhances photosynthetic activity and improve cell structure	Ali et al., 2014
5	*Arabidopsis thaliana*	Cadmium (Cd)	Enhances Cd tolerance and intensify H_2S signal	Qiao et al., 2016
6	B*rassica rape*	Cadmium (Cd)	Alleviates cell death and restricts ROS accumulation	Zhang et al., 2015
7	*Arabidopsis thaliana*	Cadmium (Cd)	Regulates Cd tolerance and acts as a downstream signal of SA	Qiao et al., 2015
8	*Oryza sativa*	Cadmium (Cd)	Alleviates Cd toxicity, reduces oxidative stress and maintains mineral homeostasis	Mostofa et al., 2015
9	*Medicago sativa*	Cadmium (Cd)	Enhances Cd tolerance, reestablishes reduced (homo) glutathione and ROS homeostatsis	Cui et al., 2014
10	*Cynodon dactylon*	Cadmium (Cd)	Improves Cd tolerance, modulates ROS and osmolytes together with NO	Shi et al., 2013
11	*Populus euphratica*	Cadmium (Cd)	Mitigates the Cd-induced programmed cell	Sun et al., 2013
12	*Brassica napus*	Aluminum (Al)	Alleviates Al toxicity, improves photosynthetic parameters, nutrient concentration, and ultrastructural changes	Ali et al., 2015

(Continued)

TABLE 16.1 (Continued)

S. no.	Plant species	Heavy metal	H_2S signaling effect and plant responses	References
13	*Hordeum vulgare* (Barley)	Aluminum (Al)	Mitigates Al-induced citrate secretion and oxidative stress	Chen et al., 2013
14	*Hordeum vulgare* (Barley)	Aluminum (Al)	Alleviates Al-induced oxidative stress	Dawood et al., 2012
15	*Pisum sativum*	Arsenic (As)	Alleviates toxic effects of As and upregulates ASA-GSH cycle	Singh et al., 2015
16	*Setaria italica*	Chromium (Cr)	Alleviates the damage from Cr stress	Fang et al., 2014a,b
17	*Triticum aestivum*	Cromium (Cr)	Alleviates seed germination and enhances the activities of antioxidant enzymes	Zhang et al., 2010a,b,c,d
18	*Cucumis sativus* (Cucumber)	Boron (B)	Alleviates the inhibition of root elongation	Wang et al., 2010

accumulation of heavy metals by increasing citrate secretion and activating citrate transporter (HvAACT1) expression. H_2S also enhanced the production of metalothiones and phytochelatins by upregulating the expression of PCS1, PCS2, and MT2A in *Arabidopsis* to cope with heavy metal toxicity (Fang et al., 2016). H_2S also escalated the expression of macrophage protein *(NRAMP1/NRAMP6)*, *metal tolerance protein (MTP1/12)*, *Ca^{2+}/ H^+exchanger antiporter (CAX2)*, and *iron-regulated transporter like protein (ZIP4)* in Cd-stressed foxtail millet (Tian et al., 2017).

H_2S overcomes the heavy metal-induced oxidative stress by activating the antioxidative defense system of plants. Enhancement of activities of various antioxidative enzymes like SOD, guaiacol peroxidase (POD), CAT, ascorbate peroxidase (APOX), GR, and the reduction in levels of H_2O_2 and MDA were observed in metal-stressed plants under the influence of H_2S (Zhang et al., 2015; Liu et al., 2016). H_2S might be involved in the Ca^{2+}-activating antioxidant system in *Setaria italica* (Fang et al., 2014a,b). H_2S, a gaseous signal molecule, interacts with various phytohormones and Ca^{2+} in signaling pathways results in activation of the Ca^{2+} ions-dependent antioxidative system (Jin and Pei, 2015). It has been observed that H_2S-mediated amelioration of heavy metal toxicity is regulated by Ca^{2+}. Ca^{2+} signaling helps H_2S to enhance Cr tolerance in millet by downregulating genes encoding proteins promoting heavy metals uptake expression (*ZIP1, ZIP3, ZIP4,* and *ZIP6*), as well as upregulating genes encoding proteins promoting heavy metals efflux expression, such as *HMA3−1, HMA3−2, MTPC1, MTPC2*. (Fang et al., 2014a). It is supported by the fact that the production of H_2S by *CSE* and *CAT/3-MST* pathway is regulated by Ca^{2+} (Guo et al., 2016).

16.3.4 Hydrogen sulfide (H_2S) and Ca^{2+} signaling during drought stress

16.3.4.1 Effect of H_2S on plants under drought conditions

Drought stress has been considered as one of the crop performance limiting factors and a threat for successful crop production. Severe drought conditions bring about accumulation of salts, and ions in the upper layer of soil around the root create osmotic stress and ion toxicity. Water scarcity in leaves causes closure of stomata, turgor loss, and a decrease in photosynthetic activity and therefore carbon assimilation (Chaves et al., 2002; Signorelli et al., 2013). This results in poor accumulation of dry matter in affected plants (Li et al., 2017;Mathobo et al., 2017). Drought stress leads to the generation of ROS such as superoxide ($O_2^{\cdot-}$), hydrogen peroxide (H_2O_2), hydroxyl radical (OH·), and alkoxyl radical (RO·), hydrogen peroxide (H_2O_2), singlet oxygen (O_2) etc. So due to high ROS accumulation, oxidative stress causes oxidation of lipids, proteins, and nucleic acids (Signorelli et al., 2013; Foyer and Noctor, 2000).

To mitigate the osmotic stress, plants synthesize compatible solutes (osmolytes) such as proline (Pro) and GB that stabilize biomolecules and membrane stability (Gadallah, 1999; Mäkelä et al., 2000; Bandurska et al., 2008; DaCosta and Huang, 2006). To counteract oxidative stress, plants possess a system of antioxidant enzymes such as SOD, peroxidase (POX), catalase (CAT), etc. that play an important role in scavenging the excess ROS caused by their overproduction.

In response to abiotic stresses, these defense systems are activated by a network of signaling molecules. Of these, H_2S is an important gasotransmitter which protect plants against osmotic and oxidative stress through enhancing antioxidant enzymes and accumulation of osmolytes (Khan et al., 2017; Li et al., 2014). The endogenous H_2S level depends on the availability of Cys in the cells. The results show that the upregulation of L-cysteine desulfhydrases and D-cysteine desulfhydrases increased the production of endogenous H_2S which led to the stomatal closure and improvement of drought stress in *A. thaliana* (Jin et al., 2011). Various H_2S signals triggered by drought stress, such as application of PEG 8000 on *Arabidopsis* seedlings (simulation of drought stress), increase the production of endogenous H_2S which alleviates the drought stress (Shen et al., 2013). At an early stage of osmotic exposure (PEG 6000 on wheat seeds), H_2S content in seeds increased from 1.5 to 3.5 μmol g^{-1} dry weight (DW). It also increased CAT and APX activities, reducing lipoxygenase activity (Zhang et al., 2010a,d) (Table 16.2).

16.3.4.2 Ca^{2+} signaling in plants under drought stress

Calcium is investigated as an important second messenger during abiotic stress signaling (Sanders et al., 1999; Knight and Knight, 2000). It is involved in all stages of plant life cycle such as growth, development, reproduction, circadian rhythms, immunity, hormone biosynthesis, and responses to biotic and abiotic stresses. Abiotic stresses and oxidative stress induce increases in cytosolic Ca^{2+} concentration. The specific calcium signatures generated are perceived by sensors, that is, *CDPKs*, *Ca^{2+}/calmodulin-dependent protein kinases*, and *CBLs* (Cheng et al., 2002; Albrecht et al., 2003; Chiasson et al., 2005; Ray et al., 2007). Calcium sensors are encoded by multiple genes and most of them expressed during stress conditions (DeFalco et al., 2010).

Plants possess two types of Ca^{2+} sensors, that is, sensor relays and sensor responders.

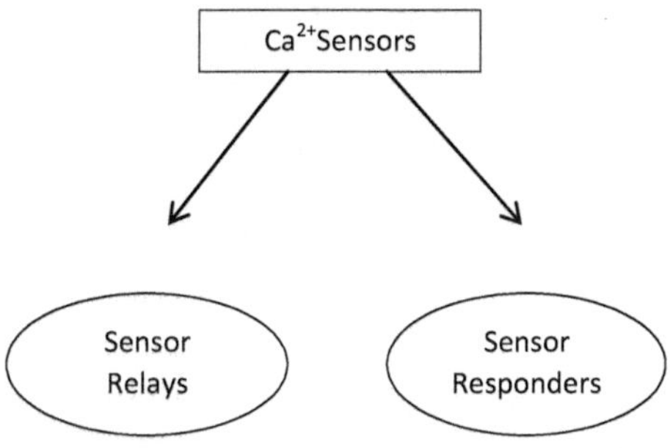

-Calmodulin (CaM) -Ca^{2+} dependent protein kinases (CDPKs)
-CaM-related proteins -Ca^{2+}/CaM dependent protein kinases
-Calcinuerin B-like protein (CCaMKs)

Calmodulin (CaM)—CDPKs
CaM-related proteins—Ca^{2+}/CaM-dependent protein kinases
Calcinuerin B-like protein (CCaMKs)

Drought stress activates signaling cascades involving protein kinases/phosphatases (e.g., *RPK1*, *SRK2C, CDPKs*, and *ABI1*) and TFs (e.g., *AREBs* and *DREBs*) and upregulates the release of chaperones and molecules participating in osmoprotectant metabolism. ABA plays an important role during drought stress and a rising level of ABA causes stomatal closure. The promoter of ABA-responsive genes contains elements, known as *ABRE*. Most of the Ca^{2+} regulated genes possess these elements, suggesting that ABA may regulate

TABLE 16.2 H_2S-induced adaptation in plants against drought stress.

S. no.	Plant species	H_2S signaling effects and plants responses	Physiological effects	References
1	*Vicia faba* (L.) var. *major* and *Impatiens walleriana*	Exogenous application of H_2S donor NaHS	Increases in relative water content of plant	Gadalla and Snyder, 2010
2	*Arabidopsis thaliana*	Pretreatment of *Arabidopsis* seedlings with H_2S donor NaHS via interacting with ABA	Stomatal closure results in protection of plant from drought stress	Jin et al., 2013; Lisjak et al., 2011
3	*Glycine max*	Exogenous application of H_2S donor NaHS	Increases the survival rate of soyabean plant, enlarged higher biomass of both leaf and root, alleviates the decreased chlorophyll content, and decreases the accumulation of MDA content and production of H_2O_2 in the plant	Zhang et al., 2010a,b,c,d
4	*Triticum aestivum*	Exogenous application of H_2S	Rise in D1 turnover of PSII which was damaged by drought stress	Li et al. (2015a,b)
5	*Spinacea oleracea*	Application of NaSH	Rise in the level of soluble sugars and polyamines (PA) due to which relative water content of leaves increases. Also there is upregulation of PA related genes and enzymes such as aldehyde dehydrogenase and choline monooxygenase	Chen et al. (2016)

ABA-responsive genes through cellular Ca^{2+} changes (Kaplan et al., 2006). *MYB2*, a *calcium-binding protein*, functions as a transcriptional factor under drought stress (Abe et al., 2003). Upregulation of *MYB2* mRNA is also detected upon treatment with ABA. The study found that exogenous ABA treatment can also induce significant CaM gene expression, and that H_2O_2 is involved in ABA-induced *CaM* gene expression in the late regulation period.

16.3.4.3 Interactions between H_2S and Ca^{2+} lead to drought tolerance by stomatal closure

H_2S can increase the Ca^{2+} concentration in guard cells by influxing these ions across the plasma membrane, which in turn closes stomata. *NADPH oxidase (NOX)* and *cell wall peroxidase (CWPOD)* are activated by H_2S and also by exogenous Ca^{2+} (Li et al., 2015a). Recently, in *A. thaliana*, it has been reported that there is an interconnection between H_2S and other signaling molecules. Briefly, H_2S stimulates *NOX* and produces H_2O_2 in guard cells, which causes the accumulation of *phosphatidic acid (PA)* by activating enzyme *phospholipase D (PLD)* and finally induces stomatal closure (Scuffi et al., 2018). ABA induces the expression of *L-DES (L-cysteine desulfhydrases)* gene, which in turn accumulates H_2S and generates NO, ultimately leading to the closing of stomatal pores. The signaling network constituted by H_2S, Ca^{2+}, ROS, NO, ABA, ETH, and JA regulates the stomatal movement and transpiration thus leading to enhancing of the tolerance of plants to drought stress (shown in Fig. 16.2).

16.3.5 Hydrogen sulfide (H_2S) and Ca^{2+} signaling during flooding stress

Flooding or waterlogging generates hypoxic and sometimes anoxic conditions (prolonged flooding) for the surrounding plant system. Hypoxic conditions result in physiological dysfunctioning that includes the generation of oxidative stress, altered membrane integrity, pH of cell cytoplasm, cytoplasmic enzymes activity, protein degradation, and limits the yield of agriculture produce. Arnell et al. (2016) in his study documented that with increased frequency of flooding annual damage will exceed €60 billion by 2050. In mitochondria, during normoxic conditions H_2S easily gets oxidized to sulfide or sulfite, whereas in hypoxic conditions the oxidation process is altered resulting in an enhanced concentration of H_2S in cytosol. Olson (2011) suggested that during hypoxic conditions in an animal system, increased concentration of H_2S acts as an oxygen sensor and may serve as a signal molecule for various physiological responses but in a plant system there is no detailed information available in response to hypoxic conditions. A study conducted by Cheng et al. (2013) on *Pisum sativum* reported that hypoxic situations may induce the H_2S biosynthetic enzymes (*DCD, LCD, CS, OAS-TL*) pointing to a signaling role for H_2S in hypoxia stress tolerance. The above hypothesis was confirmed by exogenous application of H_2S, which

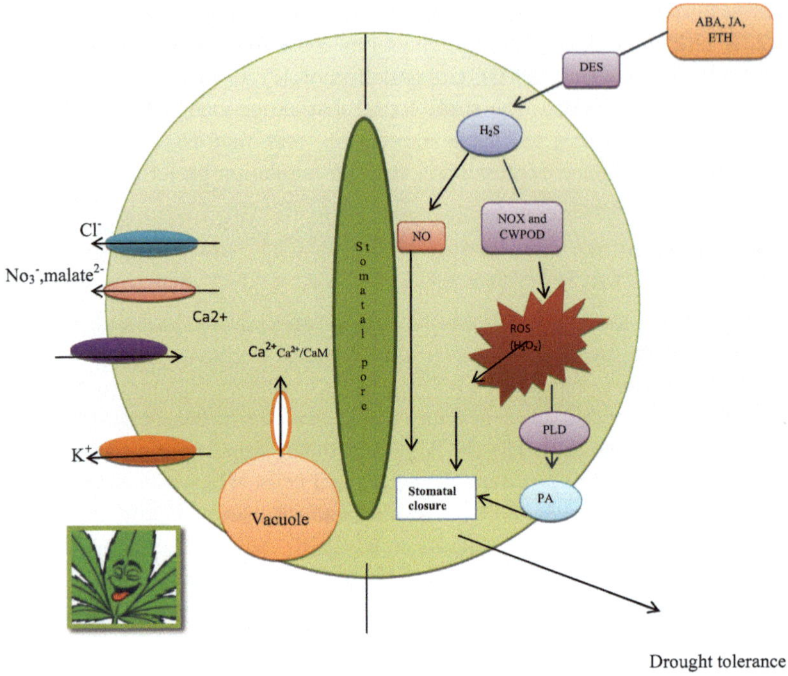

FIGURE 16.2 Hydrogen sulfide (H_2S) and calcium signaling interplay regulates stomatal movement and plant drought stress tolerance. ABA, Abscisic acid; CWPOD, cell wall peroxidase; DES, desulfhydrases; ETH, ethylene; JA, jasmonic acid; NO, nitric oxide; NOX, NADPH oxidase; PA, phosphatidic acid; PLD, phospholipase D; ROS, reactive oxygen species.

significantly reduced the root tip death induced by hypoxic conditions, and by limiting the ethylene production, thus increasing the plants' tolerance to hypoxic stress. Peng et al. (2016) documented that treatment of an NO donor sodium nitroprusside (SNP) in *Z. mays* under hypoxic conditions induced the accumulation of H_2S, thereby enhancing the survival rate of the maize seedlings. The increased endogenous level of H_2S in maize seedlings also upregulated the Ca^{2+} concentration, Ca^{2+}-dependent ADH activity, and antioxidants activity by modulating the GSH level, CAT, SOD, and APOX.

Calcium-calmodulin (CaM) proteins are involved in plant responses to various stresses. Folzer et al. (2005) studied the different isoforms of *CaM* in *Quercus petraco*. A study done by He et al. (2018) unraveled the role of Ca^{2+} in hypoxic stress in cucumber. It revealed that exogenous application of Ca^{2+} increased the productivity and improved the fruit quality of cucumber. Improved photochemical efficiency of PSII, net photosynthetic rate, intercellular CO_2, and stomatal conductance was also reported. Application of Ca^{2+} on soyabean plant under flooding stress enhanced the root length by suppressing the root tip death (Oh et al., 2014). In maize roots, the Ca^{2+}-mediated signaling process involves the formation of lysigenous aerenchyma. Rajhi et al. (2011) reported the enhanced expression of genes encoding *Calcineurin B-like protein* and *Calmodulin-like protein* during the waterlogging condition. The cortical cells showed enhanced expression of these genes compared with the stelar cells.

16.3.6 Hydrogen sulfide (H_2S) and Ca^{2+} signaling during other stresses

Plants live in a constantly changing stressful environment which is unfavorable for its growth and development. Plants are being exposed to various other stresses like pesticides, herbicides, insecticides, organic as well as inorganic pollutants, herbivore chewing, anoxia (caused by waterlogging), nutrient starvation/excess, UV radiations, pathogen attack, mechanical injury by strong wind, grazing, and many other unknown factors (Cramer et al., 2011; Zhu, 2016). In order to survive under these adverse conditions, plants have developed sophisticated signaling network, and specific changes in gene expression, metabolism, and physiology. Moreover, these plants also show cross-adaptation (Li et al., 2016a,b), which is the phenonmenon whereby exposure of plants to a moderate stress induces resistance to other stresses (Li and Gong, 2011; Foyer et al., 2016; Hossain et al., 2016). H_2S is considered an important signaling molecule for cross-adaptation to HM, salt, drought, cold, heat, and flooding stress mainly by enhancing the antioxidant system activity by reducing the level of overgenerated ROS, decreasing peroxidation of lipids (Corpas, 2019), accumulating osmolytes, synthesizing HSPs, and through maintaining ion and nutrient balance. Various investigatory reports have documented H_2S being a member of this cross-adaptaion signal network in plants (Calderwood and Kopriva, 2014; Hancock and Whiteman, 2016; Fotopoulos et al., 2015; Guo et al., 2016). Ca^{2+} signaling also plays an essential role in a plethora range of plant stresses. In various plants species stress factors like nutrient starvation were responsible for activation of Ca^{2+} signaling, such as in the case of the model plant *A. thaliana*, where nitrate-starved roots causes a rapid (within seconds) and monophasic increase in cytosolic free Ca^{2+} ($[Ca^{2+}]$cyt), followed by an increase in nuclear Ca^{2+} (Riveras et al., 2015; Liu et al., 2017). However, potassium (K^+) deficiency triggers cytosolic calcium ($[Ca^{2+}]$cyt) elevations within the *Arabidopsis* root, the first response occurring within 1–4 minutes, and the secondary response within 18–32 hours (Behera et al., 2017). One more investigatory report shows that at a high Mg^{2+} level, $[Ca^{2+}]$cyt signaling is found to play a necessary role for vacuolar magnesium (Mg^{2+}) detoxification (Tang et al., 2015). There was one investigatory report which showed that exogenous application of Ca^{2+} ameliorated acid rain stress on soybean productivity and quality by maintaining nutrient absorption (Li and Liang, 2019). The protective role of H_2S has been also reported for various plant species under various stress (Li et al., 2012a; Shen et al., 2013; Fang et al., 2016; Shi et al., 2013). Li et al. (2016a,b) reported 12 hours of exposure to UV-B radiation in leaves of barley seedlings could induce H_2S production which in turn promoted the accumulation of UV-absorbing compounds such as flavonoids and anthocyanins. Biotic stresses, especially phathogen infection such as fungal infection *Sclerotinia sclerotiorum*, in oilseed rape (*B napus* L.) seedlings lead to an increase in H_2S. Therefore suggesting that the release of H_2S seems to be part of the plant response to fungal infection (Dawood et al., 2012). H_2S and Ca^{2+} signaling is still evolving as an important signaling molecule with the further future discoveries of more and more examples of processes regulated by these signaling molecules in plants.

16.4 Crosstalk between H_2S and Ca^{2+} signaling during plant responses to abiotic stress

Calcium regulates various defense mechanisms and gene expressions (Dikilitas et al., 2019). It has been reported that the treatment of seedlings and mature plants with Ca^{2+} resulted in improved plant tolerance to heavy metals such as Cu and Cd (Ahmad et al., 2015; González et al., 2012). Various experiments were performed to study and understand molecular mechanisms and crosstalk between H_2S and Ca^{2+}. Recent findings have shown that H_2S signaling and Ca^{2+} signaling have an amicable relationship between them in terms of stress response and adaptations (Aldon et al., 2018). The application of exogenous Ca^{2+} and *CaM* stimulated *L-DES* and *OASTL* in tobacco cultured cells, which resulted in endogenous H_2S production (Li et al., 2015a,b; Min, 2017). As per Fan et al. (2016), in the mycelia of *P. linetus* H_2S elevates the content of Ca^{2+} and *CaM*, which favors the accumulation of betulin.

As per Valivand et al. (2019) seed priming with sodium hydro sulfide (NaHS) and $CaCl_2$ in Zucchini seedlings improved biomass, content of photosynthetic pigments, and enhanced GR activity in leaves. The results indicate that NaHS-induced cross-adaptation is mediated via Ca^{2+} signaling. Their findings indicate that a two-side crosstalk between Ca^{2+} and H_2S is involved in acquiring signal memory in seed embryo cells. As studied by Fang et al. (2014a,b), H_2S interacts with Ca^{2+} to ameliorate Cr^{6+} tolerance in *S. italica* by regulating the antioxidant system and the production of heavy metal chelators. The H_2S-dependent pathway is a component of the Ca^{2+} activating antioxidant (Fang et al., 2014a,b). The role of H_2S and Ca^{2+} signaling in mending heat stress was first reported by Li et al. (2012a,b,c,d,e) as they found out that pretreatment of tobacoo suspension culture with (NaHS, a H_2S donor) increased survival percentage and vitality of cells under heat stress and reduced the accumulation of MDA. As per Li et al. (2013), NaHS pretreatment could improve the entry of extracellular Ca^{2+} into tobacco suspension cultured cells mediated by intracellular calmodulin to increase the heat tolerance.

On exposure to stress, levels of Ca^{2+} and H_2S rise intracellularly. Both together play an important role in regulating plant defense machinery.

They stimulate the synthesis of antioxidative enzymes against active ROS. They are essential molecules as they play a protective role in plants. As mentioned in various reports above, together they have shown an amazingly positive synergistic relationship in order to work under stressful conditions (Fig. 16.3).

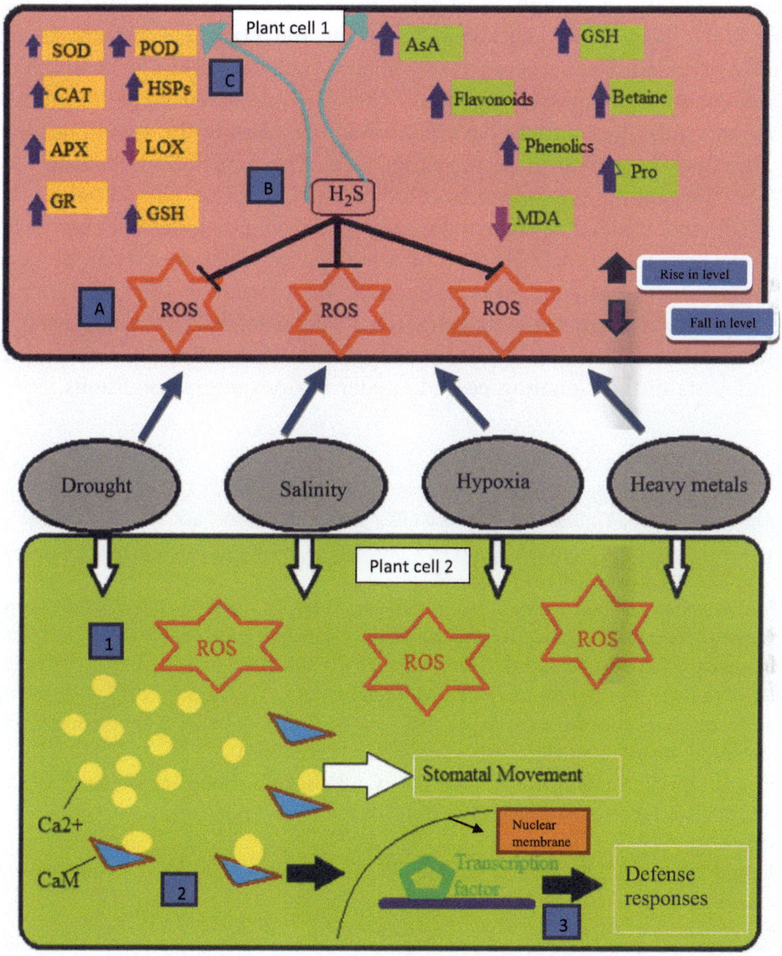

FIGURE 16.3 Diagram illustrates calcium and hydrogen sulfide signaling in plant cells on exposure of plant to various stresses. In plant cell 1, A: Indicates the production of ROS on exposure to various stresses; B: indicates the synthesis of H_2S on exposure to stress in a plant cell; C: arrows show regulation in levels of antioxidative defense enzymes which helps the plant to fight against the stress. In plant cell 2, 1: indicates the production of ROS on exposure to various stresses; 2: rise in concentration of intracellular calcium, which binds to CaM. The complex regulates stomata movement; 3: calcium regulates transcript levels and hence plays a role in the protective defense response to plant stress.

16.5 Conclusion and future perspective

In this chapter, we have construed a fundamental understanding of signaling molecules that mainly covers H_2S and Ca^{2+}. These signaling molecules have pivotal roles in growth. H_2S and Ca^{2+} maintain ion (K^+/Na^+, uptake of toxic metal ions) and nutrient balance, seed germination, growth, various developmental process, and improve photosynthetic machinery and photosynthetic activity. Although hydrogen sulfide (H_2S) and calcium (Ca^{2+}) signaling have been well investigated, so far the complete picture is only partially revealed. Future perspectives should include the scientifically generated blueprint encompassing the Ca^{2+} and H_2S-mediated signaling network and its crosstalk. It would be beneficial in understanding the signaling hierarchy amongst the global plant signaling web. The associations between H_2S and Ca^{2+} signaling are yet to be elucidated. Complete physiological, biochemical, and molecular data regarding these unknown interactions would definitely help in the identification of a molecular target which can be genetically maneuvered to yield stress tolerance. Identification of the alterations in gene expression from such investigations can also help in elucidating the H_2S-Ca^{2+} crosstalk nodes. Consequently, both of these molecules are vital integral parts of the signaling cascade under various stress conditions.

References

Abe, H., Urao, T., Ito, T., Seki, M., Shinozaki, K., Yamaguchi-Shinozaki, K., 2003. *Arabidopsis* AtMYC2 (bHLH) and AtMYB2 (MYB) function as transcriptional activators in abscisic acid signaling. Plant Cell 15 (1), 63–78.

Ali, B., Mwamba, T.M., Gill, R.A., Yang, C., Ali, S., Daud, M.K., et al., 2014. Improvement of element uptake and antioxidative defense in Brassica napus under lead stress by application of hydrogen sulfide. Plant Growth Regulation 74 (3), 261–273.

Ahmad, P., Sarwat, M., Bhat, N.A., Wani, M.R., Kazi, A.G., Tran, L.S.P., 2015. Alleviation of cadmium toxicity in *Brassica juncea* L.(Czern. & Coss.) by calcium application involves various physiological and biochemical strategies. PLoS One 10 (1).

Albrecht, V., Weinl, S., Blazevic, D., D'Angelo, C., Batistic, O., Kolukisaoglu, Ü., et al., 2003. The calcium sensor CBL1 integrates plant responses to abiotic stresses. Plant J. 36 (4), 457–470.

Aldon, D., Mbengue, M., Mazars, C., Galaud, J.P., 2018. Calcium signalling in plant biotic interactions. Int. J. Mol. Sci. 19, 665–683.

Ali, B., Qian, P., Sun, R., Farooq, M.A., Gill, R.A., Wang, J., et al., 2015. Hydrogen sulfide alleviates the aluminum-induced changes in *Brassica napus* as revealed by physiochemical and ultrastructural study of plant. Environ. Sci. Pollut. Res. 22 (4), 3068–3081.

Arnell, N.W., Brown, S., Gosling, S.N., Gottschalk, P., Hinkel, J., Huntingford, C., et al., 2016. The impacts of climate change across the globe: a multi-sectoral assessment. Climatic Change 134 (3), 457–474.

Aroca, A., Benito, J.M., Gotor, C., Romero, L.C., 2017. Persulfidation proteome reveals the regulation of protein function by hydrogen sulfide in diverse biological processes in *Arabidopsis*. J. Exp. Bot 68 (17), 4915–4927.

Atif, R.M., Shahid, L., Waqas, M., Ali, B., Rashid, M.A.R., Azeem, F., et al., 2019. Insights on calcium-dependent protein kinases (CPKs) signaling for abiotic stress tolerance in plants. Int. J. Mol. Sci. 20 (21), 5298.

Bandurska, H., Gorny, A., Zielezinska, M., 2008. Effects of water deficit on the relative water content, proline accumulation and injury of cell membranes in leaves of old and modern cultivars of winter wheat. Acta Physiol. Plant. (524).

Banerjee, A., Tripathi, D.K., Roychoudhury, A., 2018. Hydrogen sulphide trapeze: environmental stress amelioration and phytohormone crosstalk. Plant Physiol. Biochem. 132, 46–53.

Behera, S., Long, Y., Schmitz-Thom, I., Wang, X.P., Zhang, C., Li, H., et al., 2017. Two spatially and temporally distinct Ca^{2+} signals convey *Arabidopsis thaliana* responses to K^+ deficiency. N. Phytol 213 (2), 739–750.

Calderwood, A., Kopriva, S., 2014. Hydrogen sulfide in plants: from dissipation of excess sulfur to signaling molecule. Nitric Oxide 41, 72–78.

Campbell, A.K., 2017. Fundamentals of Intracellular Calcium. John Wiley & Sons, p. 89121.

Chakraborty, N., Acharya, K., 2017. NO way"! Says the plant to abiotic stress. Plant Gene 11, 99–105.

Chaves, M.M., Pereira, J.S., Maroco, J., Rodrigues, M.L., Ricardo, C.P.P., Osório, M.L., et al., 2002. How plants cope with water stress in the field? Photosynthesis and growth. Ann. Bot 89 (7), 907–916.

Chen, J., Wu, F.H., Wang, W.H., Zheng, C.J., Lin, G.H., Dong, X.J., et al., 2011. Hydrogen sulphide enhances photosynthesis through promoting chloroplast biogenesis, photosynthetic enzyme expression, and thiol redox modification in *Spinacia oleracea* seedlings. J. Exp. Bot 62 (13), 4481–4493.

Chen, J., Wang, W.H., Wu, F.H., You, C.Y., Liu, T.W., Dong, X.J., et al., 2013. Hydrogen sulfide alleviates aluminum toxicity in barley seedlings. Plant Soil 362 (1–2), 301–318.

Chen, J., Wang, W.H., Wu, F.H., He, E.M., Liu, X., Shangguan, Z.P., et al., 2015. Hydrogen sulfide enhances salt tolerance through nitric oxide-mediated maintenance of ion homeostasis in barley seedling roots. Sci. Rep. 5, 12516.

Chen, X., Chen, Q., Zhang, X., Li, R., Jia, Y., Ef, A., et al., 2016. Hydrogen sulfide mediates nicotine biosynthesis in tobacco (*Nicotiana tabacum*) under high temperature conditions. Plant Physiol. Biochem. 104, 174–179.

Cheng, S.H., Willmann, M.R., Chen, H.C., Sheen, J., 2002. Calcium signaling through protein kinases. The *Arabidopsis* calcium-dependent protein kinase gene family. Plant Physiol. 129 (2), 469–485.

Cheng, W., Zhang, L., Jiao, C., Su, M., Yang, T., Zhou, L., et al., 2013. Hydrogen sulfide alleviates hypoxia-induced root tip death in *Pisum sativum*. Plant Physiol. Biochem. 70, 278–286.

Chiasson, D., Ekengren, S.K., Martin, G.B., Dobney, S.L., Snedden, W.A., 2005. Calmodulin-like proteins from *Arabidopsis* and tomato are involved in host defense against *Pseudomonas syringae* pv. tomato. Plant Mol. Biol. 58 (6), 887–897.

Chinnusamy, V., Zhu, J., Zhu, J.K., 2007. Cold stress regulation of gene expression in plants. Trends Plant Sci. 12 (10), 444–451.

Choi, W.G., Toyota, M., Kim, S.H., Hilleary, R., Gilroy, S., 2014. Salt stress-induced Ca^{2+} waves are associated with rapid, long-distance root-to-shoot signaling in plants. Proc. Natl. Acad. Sci. 111 (17), 6497–6502.

Chongchatuporn, U., Ketsa, S., van Doorn, W.G., 2013. Chilling injury in mango (*Mangifera indica*) fruit peel: relationship with ascorbic acid concentrations and antioxidant enzyme activities. Postharvest Biol. Technol. 86, 409–417.

Choudhury, F.K., Rivero, R.M., Blumwald, E., Mittler, R., 2017. Reactive oxygen species, abiotic stress and stress combination. Plant J. 90 (5), 856–867.

Christou, A., Manganaris, G.A., Papadopoulos, I., Fotopoulos, V., 2013. Hydrogen sulfide induces systemic tolerance to salinity and non-ionic osmotic stress in strawberry plants through modification of reactive species biosynthesis and transcriptional regulation of multiple defence pathways. J. Exp. Bot 64 (7), 1953–1966.

Christou, A., Manganaris, G.A., Fotopoulos, V., 2014. Systemic mitigation of salt stress by hydrogen peroxide and sodium nitroprusside in strawberry plants via transcriptional regulation of enzymatic and non-enzymatic antioxidants. Environ. Exp. Bot 107, 46–54.

Corpas, F.J., 2019. Hydrogen sulfide: a new warrior against abiotic stress. Trends Plant Sci. 24 (11), 983–988.

Cramer, G.R., Urano, K., Delrot, S., Pezzotti, M., Shinozaki, K., 2011. Effects of abiotic stress on plants: a systems biology perspective. BMC Plant Biol. 11 (1), 163.

Cui, W., Chen, H., Zhu, K., Jin, Q., Xie, Y., Cui, J., et al., 2014. Cadmium-induced hydrogen sulfide synthesis is involved in cadmium tolerance in *Medicago sativa* by reestablishment of reduced (Homo) glutathione and reactive oxygen species homeostases. PLoS One 9 (10).

Czarnocka, W., Karpiński, S., 2018. Friend or foe? Reactive oxygen species production, scavenging and signaling in plant response to environmental stresses. Free Radic. Biol. Med. 122, 4–20.

Da Silva, C.J., Fontes, E.P.B., Modolo, L.V., 2017. Salinity-induced accumulation of endogenous H_2S and NO is associated with modulation of the antioxidant and redox defense systems in *Nicotiana tabacum* L. cv. Havana. Plant Sci. 256, 148–159.

DaCosta, M., Huang, B., 2006. Osmotic adjustment associated with variation in bentgrass tolerance to drought stress. J. Am. Soc. Horticultural Sci. 131 (3), 338–344.

Dawood, M., Cao, F., Jahangir, M.M., Zhang, G., Wu, F., 2012. Alleviation of aluminum toxicity by hydrogen sulfide is related to elevated ATPase, and suppressed aluminum uptake and oxidative stress in barley. J. Hazard. Mater. 209, 121–128.

DeFalco, T.A., Bender, K.W., Snedden, W.A., 2010. Breaking the code: Ca^{2+} sensors in plant signalling. Biochem J. 425 (1), 27–40.

Deng, Y.Q., Bao, J., Yuan, F., Liang, X., Feng, Z.T., Wang, B.S., 2016. Exogenous hydrogen sulfide alleviates salt stress in wheat seedlings by decreasing Na^+ content. Plant Growth Regul. 79 (3), 391–399.

Dikilitas, M., Simsek, E., Karakas, S., 2019. Stress responsive signaling molecules and genes under stressful environments in plants. Plant Signaling Molecules. Woodhead Publishing, pp. 19–42.

Edel, K.H., Marchadier, E., Brownlee, C., Kudla, J., Hetherington, A.M., 2017. The evolution of calcium-based signalling in plants. Curr. Biol. 27 (13), R667–R679.

Evans, M.J., Choi, W.G., Gilroy, S., Morris, R.J., 2016. A ROS-assisted calcium wave dependent on the AtRBOHD NADPH oxidase and TPC1 cation channel propagates the systemic response to salt stress. Plant Physiol. 171 (3), 1771–1784.

Fan, G., Jian, D., Sun, M., Zhan, Y., Sun, F., 2016. Endogenous and exogenous calcium involved in the betulin production from submerged culture of *Phellinus linteus* induced by hydrogen sulfide. Appl. Biochem. Biotechnol. 178 (3), 594–603.

Fang, H., Jing, T., Liu, Z., Zhang, L., Jin, Z., Pei, Y., 2014a. Hydrogen sulfide interacts with calcium signaling to enhance the chromium tolerance in *Setaria italica*. Cell Calcium 56 (6), 472–481.

Fang, T., Cao, Z., Li, J., Shen, W., Huang, L., 2014b. Auxin-induced hydrogen sulfide generation is involved in lateral root formation in tomato. Plant Physiol. Biochem. 76, 44–51.

Fang, H., Liu, Z., Jin, Z., Zhang, L., Liu, D., Pei, Y., 2016. An emphasis of hydrogen sulfide-cysteine cycle on enhancing the tolerance to chromium stress in *Arabidopsis*. Environ. Pollut. 213, 870–877.

Feng, W., Kita, D., Peaucelle, A., Cartwright, H.N., Doan, V., Duan, Q., et al., 2018. The FERONIA receptor kinase maintains cell-wall integrity during salt stress through Ca^{2+} signaling. Curr. Biol. 28 (5), 666–675.

Filipovic, M.R., Jovanović, V.M., 2017. More than just an intermediate: hydrogen sulfide signalling in plants. J. Exp. Bot 68 (17), 4733–4736.

Folzer, H., Capelli, N., Dat, J., Badot, P.M., 2005. Molecular cloning and characterization of calmodulin genes in young oak seedlings (*Quercus petraea* L.) during early flooding stress. Biochim. Biophys Acta (BBA) Gene Struct Expr. 1727 (3), 213–219.

Fotopoulos, V., Christou, A., Antoniou, C., Manganaris, G.A., 2015. Review article hydrogen sulphide: a versatile tool for the regulation of growth and defence responses in horticultural crops. J. Hortic Sci. Biotechnol. 90 (3), 227–234.

Foyer, C.H., Noctor, G., 2000. Tansley review No. 112. Oxygen processing in photosynthesis: regulation and signalling. N. Phytol 146 (3), 359–388.

Foyer, C.H., Rasool, B., Davey, J.W., Hancock, R.D., 2016. Cross-tolerance to biotic and abiotic stresses in plants: a focus on resistance to aphid infestation. J. Exp. Bot 67 (7), 2025–2037.

Fu, P., Wang, W., Hou, L., Liu, X., 2013. Hydrogen sulfide is involved in the chilling stress response in *Vitis vinifera* L. Acta Soc Bot Poloniae 82 (4).

Furuya, T., Matsuoka, D., Nanmori, T., 2014. Membrane rigidification functions upstream of the MEKK1-MKK2-MPK4 cascade during cold acclimation in *Arabidopsis thaliana*. FEBS Lett. 588, 2025–2030.

Gadalla, M.M., Snyder, S.H., 2010. Hydrogen sulfide as a gasotransmitter. J. Neurochem 113, 14–26. Available from: https://doi.org/10.1111/j.1471-4159.2010.06580.x.

Gadallah, M.A.A., 1999. Effects of proline and glycinebetaine on *Vicia faba* responses to salt stress. Biol Plant 42 (2), 249–257.

Gayatri, G., Agurla, S., Raghavendra, A.S., 2013. Nitric oxide in guard cells as an important secondary messenger during stomatal closure. Front. Plant Sci. 4, 425.

González, A., de Los Ángeles Cabrera, M., Henríquez, M.J., Contreras, R.A., Morales, B., Moenne, A., 2012. Cross talk among calcium, hydrogen peroxide, and nitric oxide and activation of gene expression involving calmodulins and calcium-dependent protein kinases in Ulva compressa exposed to copper excess. Plant Physiol. 158 (3), 1451–1462.

Guduru, K.K., Akila, C.S., Puli, C.O.R., 2019. Engineering signaling molecules to improve abiotic stress tolerance in crop plants. Plant Signaling Molecules. Woodhead Publishing, pp. 43–62.

Guerret, M.G., Barbetti, M.J., You, M.P., Jones, R.A., 2016. Effects of temperature on disease severity in plants of subterranean clover infected singly or in mixed infection with Bean yellow mosaic virus and *Kabatiella caulivora*. J. Phytopathol 164 (9), 608–619.

Guo, H., Xiao, T., Zhou, H., Xie, Y., Shen, W., 2016. Hydrogen sulfide: a versatile regulator of environmental stress in plants. Acta Physiol Plant 38 (1), 16.

Hancock, J.T., 2017. Harnessing evolutionary toxins for signaling: reactive oxygen species, nitric oxide and hydrogen sulfide in plant cell regulation. Front. Plant Sci. 8, 189.

Hancock, J.T., Whiteman, M., 2016. Hydrogen sulfide signaling: interactions with nitric oxide and reactive oxygen species. Ann. N. Y. Acad. Sci. 1365, 5–14.

He, H., Li, Y., He, L.F., 2018. The central role of hydrogen sulfide in plant responses to toxic metal stress. Ecotoxicol. Environ. Saf. 157, 403–408.

Hooda, P.S., 2010. Assessing bioavailability of soil trace elements. In: Hooda, P.S. (Ed.), Trace Elements in Soils. Wiley, pp. 229–265.

Hoque, T.S., Okuma, E., Uraji, M., Furuichi, T., Sasaki, T., Hoque, M.A., et al., 2012a. Inhibitory effects of methylglyoxal on light-induced stomatal opening and inward K^+ channel activity in *Arabidopsis*. Biosci Biotechnol Biochem. 76 (3), 617–619.

Hoque, T.S., Uraji, M., Tuya, A., Nakamura, Y., Murata, Y., 2012b. Methylglyoxal inhibits seed germination and root elongation and up-regulates transcription of stress-responsive genes in ABA-dependent pathway in *Arabidopsis*. Plant Biol. 14 (5), 854–858.

Hoque, T.S., Uraji, M., Ye, W., Hossain, M.A., Nakamura, Y., Murata, Y., 2012c. Methylglyoxal-induced stomatal closure accompanied by peroxidase-mediated ROS production in *Arabidopsis*. J. Plant Physiol. 169 (10), 979–986.

Hoque, T.S., Hossain, M.A., Mostofa, M.G., Burritt, D.J., Fujita, M., Tran, L.S.P., 2016. Methylglyoxal: an emerging signaling molecule in plant abiotic stress responses and tolerance. Front. Plant Sci. 7, 1341.

Hossain, M.A., Burritt, D.J., Fujita, M., 2016. Cross-stress tolerance in plants: molecular mechanisms and possible involvement of reactive oxygen species and methylglyoxal detoxification systems. In: Tuteja, N., Gill, S.S. (Eds.), Abiotic Stress Response in Plants. Wiley, Chennai, pp. 323–375.

Hossain, M.A., Li, Z.G., Hoque, T.S., Burritt, D.J., Fujita, M., Munné-Bosch, S., 2018. Heat or cold priming-induced cross-tolerance to abiotic stresses in plants: key regulators and possible mechanisms. Protoplasma 255 (1), 399–412.

Hu, K.D., Wang, Q., Hu, L.Y., 2014. Hydrogen sulfide prolongs postharvest storage of fresh-cut pears (*Pyrus pyrifolia*) by alleviation of oxidative damage and inhibition of fungal growth. PLoS One 9, e85524.

Hu, J., Yang, H., Mu, J., Lu, T., Peng, J., Deng, X., et al., 2017. Nitric oxide regulates protein methylation during stress responses in plants. Mol. Cell 67 (4), 702–710.

Jain, M., Nagar, P., Sharma, A., Batth, R., Aggarwal, S., Kumari, S., et al., 2018. GLYI and D-LDH play key role in methylglyoxal detoxification and abiotic stress tolerance. Sci. Rep. 8 (1), 5451.

Jeon, J., Kim, J., 2013. Cold stress signaling networks in Arabidopsis. J. Plant Biol. 56, 69–76.

Jin, Z., Pei, Y., 2015. Physiological implications of hydrogen sulfide in plants: pleasant exploration behind its unpleasant odour. Oxid. Med. Cell. Longev. 2015.

Jin, Z., Shen, J., Qiao, Z., Yang, G., Wang, R., Pei, Y., 2011. Hydrogen sulfide improves drought resistance in *Arabidopsis thaliana*. Biochem Biophys Res. Commun. 414 (3), 481–486.

Jin, Z., Xue, S., Luo, Y., Tian, B., Fang, H., Li, H., et al., 2013. Hydrogen sulfide interacting with abscisic acid in stomatal regulation responses to drought stress in *Arabidopsis*. Plant Physiol. Biochem. 62, 41–46.

Kabil, O., Banerjee, R., 2010. Redox biochemistry of hydrogen sulfide. J. Biol. Chem. 285 (29), 21903–21907.

Kaplan, B., Davydov, O., Knight, H., Galon, Y., Knight, M.R., Fluhr, R., et al., 2006. Rapid transcriptome changes induced by cytosolic Ca^{2+} transients reveal ABRE-related sequences as Ca^{2+}-responsive cis elements in *Arabidopsis*. Plant Cell 18 (10), 2733–2748.

Kaur, C., Kushwaha, H.R., Mustafiz, A., Pareek, A., Sopory, S.K., Singla-Pareek, S.L., 2015. Analysis of global gene expression profile of rice in response to methylglyoxal indicates its possible role as a stress signal molecule. Front. Plant Sci. 6, 682.

Khan, M.N., Mobin, M., Abbas, Z.K., Siddiqui, M.H., 2017. Nitric oxide-induced synthesis of hydrogen sulfide alleviates osmotic stress in wheat seedlings through sustaining antioxidant enzymes, osmolyte accumulation and cysteine homeostasis. Nitric Oxide 68, 91–102.

Kiegle, E., Moore, C.A., Haseloff, J., Tester, M.A., Knight, M.R., 2000. Cell-type-specific calcium responses to drought, salt and cold in the *Arabidopsis* root. Plant J. 23 (2), 267–278.

Kimura, H., 2014. The physiological role of hydrogen sulfide and beyond. Nitric Oxide 41, 4–10.

Kissoudis, C., Sunarti, S., Van De Wiel, C., Visser, R.G., Van Der Linden, C.G., Bai, Y., 2016. Responses to combined abiotic and biotic stress in tomato are governed by stress intensity and resistance mechanism. J. Exp. Bot 67 (17), 5119–5132.

Knight, H., Knight, M.R., 2000. Imaging spatial and cellular characteristics of low temperature calcium signature after cold acclimation in *Arabidopsis*. J. Exp. Bot 51 (351), 1679–1686.

Kolupaev, Y.E., Firsova, E.N., Yastreb, T.O., Lugovaya, A.A., 2017. The participation of calcium ions and reactive oxygen species in the induction of antioxidant enzymes and heat resistance in plant cells by hydrogen sulfide donor. Appl. Biochem. Microbiol 53 (5), 573–579.

Köster, P., Wallrad, L., Edel, K.H., Faisal, M., Alatar, A.A., Kudla, J., 2019. The battle of two ions: Ca^{2+} signalling against Na^+ stress. Plant Biol. 21, 39–48.

Lai, D., Mao, Y., Zhou, H., Li, F., Wu, M., Zhang, J., et al., 2014. Endogenous hydrogen sulfide enhances salt tolerance by coupling the reestablishment of redox homeostasis and preventing salt-induced K^+ loss in seedlings of *Medicago sativa*. Plant Sci. 225, 117–129.

Laureano-Marín, A.M., Moreno, I., Romero, L.C., Gotor, C., 2016. Negative regulation of autophagy by sulfide is independent of reactive oxygen species. Plant Physiol. 171 (2), 1378–1391.

Li, Z.G., 2015. Analysis of some enzymes activities of hydrogen sulfide metabolism in plants, Methods in Enzymology, 555. Academic Press, pp. 253–269.

Li, Z.G., 2019. Interplay between hydrogen sulfide and calcium signaling in plant abiotic stress response and adaptation. In: Hasanuzzaman, M., Fotopoulos, V., Nahar, K., Fujita, M. (Eds.), Reactive Oxygen, Nitrogen and Sulfur Species in Plants: Production, Metabolism, Signaling and Defense Mechanisms. John Wiley & Sons Ltd, pp. 83–103.

Li, Z.G., Gong, M., 2011. Mechanical stimulation-induced cross-adaptation in plants: an overview. J. Plant Biol. 54, 358–364. Available from: https://doi.org/10.1007/s12374-011-9178-3.

Li, Z.G., Jin, J.Z., 2016. Hydrogen sulfide partly mediates abscisic acid-induced heat tolerance in tobacco (*Nicotiana tabacum* L.) suspension cultured cells. Plant Cell, Tissue Organ Cult. 125, 207–214.

Li, Y., Liang, C., 2019. Exogenous application of Ca^{2+} mitigates simulated acid rain stress on soybean productivity and quality by maintaining nutrient absorption. Environ. Sci. Pollut. Res. 26 (5), 4975–4986.

Li, F., Vallabhaneni, R., Wurtzel, E.T., 2008. PSY3, a new member of the phytoene synthase gene family conserved in the Poaceae and regulator of abiotic stress-induced root carotenogenesis. Plant Physiol. 146 (3), 1333–1345.

Li, Z.G., Gong, M., Xie, H., Yang, L., Li, J., 2012a. Hydrogen sulfide donor sodium hydrosulfide-induced heat tolerance in tobacco (*Nicotiana tabacum* L) suspension cultured cells and involvement of ${Ca_2}^+$ and calmodulin. Plant Sci. 185, 185–189.

Li, Y.W., Gong, Z.H., Mu, Y., Zhang, Y.X., Qiao, Z.J., Zhang, L.P., et al., 2012b. An Arabidopsis mutant atcsr-2 exhibits high cadmium stress sensitivity involved in the restriction of H_2S emission. J. Zhejiang Univ Sci B 13 (12), 1006–1014.

Li, Z., Li, Z., Gao, X., Chinnusamy, V., Bressan, R., Wang, Z.X., et al., 2012c. ROP11 GTPase negatively regulates ABA signaling by protecting ABI1 phosphatase activity from inhibition by the ABA receptor RCAR1/PYL9 in *Arabidopsis*. J. Integrative Plant Bio. 54, 180–188.

Li, Z.G., Gong, M., Liu, P., 2012d. Hydrogen sulfide is a mediator in H_2O_2-induced seed germination in *Jatropha Curcas*. Acta Physiol. Plant 34 (6), 2207–2213.

Li, Z.G., Gong, M., Liu, P., 2012e. Hydrogen sulfide is a mediator in H_2O_2 induced seed germination in *Jatropha curcas*. Acta Physiol. Plant 34, 2207–2213.

Li, Z.G., Yang, S.Z., Long, W.B., Yang, G.X., Shen, Z.Z., 2013. Hydrogen sulphide may be a novel downstream signal molecule in nitric oxide-induced heat tolerance of maize (*Zea mays* L.) seedlings. Plant, Cell Environ. 36 (8), 1564–1572.

Li, Z.G., Luo, L.J., Zhu, L.P., 2014. Involvement of trehalose in hydrogen sulfide donor sodium hydrosulfide-induced the acquisition of heat tolerance in maize (*Zea mays* L.) seedlings. Bot Stud. 55 (1), 20.

Li, Z.G., Long, W.B., Yang, S.Z., Wang, Y.C., Tang, J.H., Wen, L., et al., 2015a. Endogenous hydrogen sulfide regulated by calcium is involved in thermotolerance in tobacco *Nicotiana tabacum* L. suspension cell cultures. Acta Physiol Plant 37 (10), 219.

Li, H., Gao, M.Q., Xue, R.L., Wang, D., Zhao, H.J., 2015b. Effect of hydrogen sulfide on D1 protein in wheat under drought stress. Acta Physiol Plant 37 (11), 225.

Li, Z., Waadt, R., Schroeder, J.I., 2016a. Release of GTP exchange factor mediated down-regulation of abscisic acid signal transduction through ABA induced rapid degradation of RopGEFs. PLoS Biol. 14, e1002461.

Li, Z.G., Min, X., Zhou, Z.H., 2016b. Hydrogen sulfide: a signal molecule in plant cross-adaptation. Front. Plant Sci. 7, 1621.

Li, Z.G., Duan, X.Q., Min, X., Zhou, Z.H., 2017. Methylglyoxal as a novel signal molecule induces the salt tolerance of wheat by regulating the glyoxalase system, the antioxidant system, and osmolytes. Protoplasma 254 (5), 1995–2006.

Li, Z.G., Nie, Q., Yang, C.L., Wang, Y., Zhou, Z.H., 2018. Signaling molecule methylglyoxal ameliorates cadmium injury in wheat (*Triticum aestivum* L) by a coordinated induction of glutathione pool and glyoxalase system. Ecotoxicol. Environ. Saf. 149, 101–107.

Lisjak, M., Teklić, T., Wilson, I.D., Wood, M., Whiteman, M., Hancock, J.T., 2011. Hydrogen sulfide effects on stomatal apertures. Plant Signal. Behav. 6 (10), 1444–1446.

Liu, Y., He, C., 2017. A review of redox signaling and the control of MAP kinase pathway in plants. Redox Biol. 11, 192–204.

Liu, X., Chen, J., Wang, G.H., Wang, W.H., Shen, Z.J., Luo, M.R., et al., 2016. Hydrogen sulfide alleviates zinc toxicity by reducing zinc uptake and regulating genes expression of antioxidative enzymes and metallothioneins in roots of the cadmium/zinc hyperaccumulator *Solanum nigrum* L. Plant Soil 400 (1-2), 177–192.

Liu, K.H., Niu, Y., Konishi, M., Wu, Y., Du, H., Chung, H.S., et al., 2017. Discovery of nitrate–CPK–NLP signalling in central nutrient–growth networks. Nature 545 (7654), 311–316.

Luo, Z., Li, D., Du, R., Mou, W., 2015. Hydrogen sulfide alleviates chilling injury of banana fruit by enhanced antioxidant system and proline content. Sci. Hortic 183, 144–151.

Lv, X., Li, H., Chen, X., Xiang, X., Guo, Z., Yu, J., et al., 2018. The role of calcium-dependent protein kinase in hydrogen peroxide, nitric oxide and ABA-dependent cold acclimation. J. Exp. Bot 69, 4127–4139.

Ma, D., Ding, H., Wang, C., Qin, H., Han, Q., Hou, J., et al., 2016. Alleviation of drought stress by hydrogen sulfide is partially related to the abscisic acid signaling pathway in wheat. PLoS One 11, 0163082.

Majláth, I., Éva, C., Tajti, J., Khalil, R., Elsayed, N., Darko, E., et al., 2020. Exogenous methylglyoxal enhances the reactive aldehyde detoxification capability and frost-hardiness of wheat. Plant Physiol. Biochem. 149, 75–85.

Mäkelä, P., Kärkkäinen, J., Somersalo, S., 2000. Effect of glycinebetaine on chloroplast ultrastructure, chlorophyll and protein content, and RuBPCO activities in tomato grown under drought or salinity. Biol Plant 43 (3), 471–475.

Manishankar, P., Wang, N., Köster, P., Alatar, A.A., Kudla, J., 2018. Calcium signaling during salt stress and in the regulation of ion homeostasis. J. Exp. Bot 69 (17), 4215–4226.

Mathobo, R., Marais, D., Steyn, J.M., 2017. The effect of drought stress on yield, leaf gaseous exchange and chlorophyll fluorescence of dry beans (*Phaseolus vulgaris* L.). Agric. Water Manag. 180, 118–125.

Melegari, S.P., Perreault, F., Costa, R.H.R., Popovic, R., Matias, W.G., 2013. Evaluation of toxicity and oxidative stress induced by copper oxide nanoparticles in the green alga *Chlamydomonas reinhardtii*. Aquat. Toxicol. 142, 431–440.

Miller, G., Suzuki, N., Ciftci-Yilmaz, S., Mittler, R., 2010. Reactive oxygen species homeostasis and signalling during drought and salinity stresses. Plant, Cell Environ. 33 (4), 453–467.

Min, X., 2017. Regulation of heat resistance of tobacco cells by interaction of H_2S signal with calcium signal. Thesis of Master, Yunnan Normal University, Kunming, China (In Chinese).

Mitsuhashi, H., Yamashita, S., Ikeuchi, H., Kuroiwa, T., Kaneko, Y., Hiromura, K., et al., 2005. Oxidative stress-dependent conversion of hydrogen sulfide to sulfite by activated neutrophils. Shock 24 (6), 529–534.

Mittler, R., 2017. ROS are good. Trends Plant Sci. 22 (1), 11–19.

Mostofa, M.G., Rahman, A., Ansary, M.M.U., Watanabe, A., Fujita, M., Tran, L.S.P., 2015. Hydrogen sulfide modulates cadmium-induced physiological and biochemical responses to alleviate cadmium toxicity in rice. Sci. Rep. 5 (1), 1–17.

Mostofa, M.G., Ghosh, A., Li, Z.G., Siddiqui, M.N., Fujita, M., Tran, L.S.P., 2018. Methylglyoxal–a signaling molecule in plant abiotic stress responses. Free. Radic. Biol. Med. 122, 96–109.

Munaron, L., Avanzato, D., Moccia, F., Mancardi, D., 2013. Hydrogen sulfide as a regulator of calcium channels. Cell Calcium 53 (2), 77–84.

Neill, S., Barros, R., Bright, J., Desikan, R., Hancock, J., Harrison, J., et al., 2008. Nitric oxide, stomatal closure, and abiotic stress. J. Exp. Bot 59 (2), 165–176.

Oh, M., Nanjo, Y., Komatsu, S., 2014. Gel-free proteomic analysis of soybean root proteins affected by calcium under flooding stress. Front. Plant Sci. 5, 559.

Olas, B., 2015. Hydrogen sulfide in signaling pathways. Clin Chim. Acta 439, 212–218.

Olson, K.R., 2011. The therapeutic potential of hydrogen sulfide: separating hype from hope. Am. J. Physiol.Regulatory, Integr. Comp. Physiol 301 (2), R297–R312.

Papanatsiou, M., Scuffi, D., Blatt, M.R., García-Mata, C., 2015. Hydrogen sulfide regulates inward-rectifying K^+ channels in conjunction with stomatal closure. Plant Physiol. 168 (1), 29–35.

Parankusam, S., Adimulam, S.S., Bhatnagar-Mathur, P., Sharma, K.K., 2017. Nitric oxide (NO) in plant heat stress tolerance: current knowledge and perspectives. Front. Plant Sci. 8, 1582.

Peng, R., Bian, Z., Zhou, L., Cheng, W., Hai, N., Yang, C., et al., 2016. Hydrogen sulfide enhances nitric oxide-induced tolerance of hypoxia in maize (*Zea mays* L.). Plant Cell Rep. 35 (11), 2325–2340.

Proietti, S., Falconieri, G.S., Bertini, L., Baccelli, I., Paccosi, E., Belardo, A., et al., 2019. GLYI4 plays a role in methylglyoxal detoxification and jasmonate-mediated stress responses in *Arabidopsis thaliana*. Biomolecules 9 (10), 635.

Qiao, Z., Jing, T., Liu, Z., Zhang, L., Jin, Z., Liu, D., et al., 2015. H_2S acting as a downstream signaling molecule of SA regulates Cd tolerance in *Arabidopsis*. Plant Soil 393 (1–2), 137–146.

Qiao, Z., Jing, T., Jin, Z., Liang, Y., Zhang, L., Liu, Z., et al., 2016. CDPKs enhance Cd tolerance through intensifying H_2S signal in *Arabidopsis thaliana*. Plant Soil 398 (1–2), 99–110.

Rajhi, I., Yamauchi, T., Takahashi, H., Nishiuchi, S., Shiono, K., Watanabe, R., et al., 2011. Identification of genes expressed in maize root cortical cells during lysigenous aerenchyma formation using laser microdissection and microarray analyses. N. Phytol 190 (2), 351–368.

Ranty, B., Aldon, D., Cotelle, V., Galaud, J.P., Thuleau, P., Mazars, C., 2016. Calcium sensors as key hubs in plant responses to biotic and abiotic stresses. Front. Plant Sci. 7, 327.

Ray, S., Agarwal, P., Arora, R., Kapoor, S., Tyagi, A.K., 2007. Expression analysis of calcium-dependent protein kinase gene family during reproductive development and abiotic stress conditions in rice (*Oryza sativa* L. ssp. indica). Mol. Genet. Genomics 278 (5), 493–505.

Reddy, A.S., Ali, G.S., Celesnik, H., Day, I.S., 2011. Coping with stresses: roles of calcium-and calcium/calmodulin-regulated gene expression. Plant Cell 23 (6), 2010–2032.

Ren, X.L., Qi, G.N., Feng, H.Q., Zhao, S., Zhao, S.S., Wang, Y., et al., 2013. Calcineurin B-like protein CBL 10 directly interacts with AKT 1 and modulates K^+ homeostasis in *Arabidopsis*. Plant J. 74 (2), 258–266.

Riveras, E., Alvarez, J.M., Vidal, E.A., Oses, C., Vega, A., Gutiérrez, R.A., 2015. The calcium ion is a second messenger in the nitrate signaling pathway of *Arabidopsis*. Plant Physiol. 169 (2), 1397–1404.

Robertson, D.N., 2013. Modulating plant calcium for better nutrition and stress tolerance. ISRN Bot 2013.

Sami, F., Faizan, M., Faraz, A., Siddiqui, H., Yusuf, M., Hayat, S., 2018. Nitric oxide-mediated integrative alterations in plant metabolism to confer abiotic stress tolerance, NO crosstalk with phytohormones and NO-mediated post translational modifications in modulating diverse plant stress. Nitric Oxide 73, 22–38.

Sanders, D., Brownlee, C., Harper, J.F., 1999. Communicating with calcium. Plant Cell 11 (4), 691–706.

Sanders, D., Pelloux, J., Brownlee, C., Harper, J.F., 2002. Calcium at the crossroads of signaling. Plant Cell 14 (suppl 1), S401–S417.

Sangwan, V., Örvar, B.L., Beyerly, J., Hirt, H., Dhindsa, R.S., 2002. Opposite changes in membrane fluidity mimic cold and heat stress activation of distinct plant MAP kinase pathways. Plant J. 31 (5), 629–638.

Santisree, P., Bhatnagar-Mathur, P., Sharma, K.K., 2018. Molecular insights into the functional role of nitric oxide (NO) as a signal for plant responses in chickpea. Funct. Plant Biol. 45, 267–283.

Santisree, P., Adimulam, S.S., Bommineni, P., Bhatnagar-Mathur, P., Sharma, K.K., 2019. Hydrogen sulfide in plant abiotic stress tolerance—progress and perspectives. In: Hasanuzzaman, M., Fotopoulos, V., Nahar, K., Fujita, M. (Eds.), Reactive Oxygen, Nitrogen and Sulfur Species in Plants: Production, Metabolism, Signaling and Defense Mechanisms. John Wiley & Sons Ltd, pp. 743–775.

Scharf, K.D., Berberich, T., Ebersberger, I., Nover, L., 2012. The plant heat stress transcription factor (Hsf) family: structure, function and evolution. Biochim. Biophys Acta 1819 (2), 104–119.

Scuffi, D., Nietzel, T., Di Fino, L.M., Meyer, A.J., Lamattina, L., Schwarzländer, M., et al., 2018. Hydrogen sulfide increases production of NADPH oxidase-dependent hydrogen peroxide and phospholipase D-derived phosphatidic acid in guard cell signaling. Plant Physiol. 176 (3), 2532–2542.

Shahbaz, M., Stuiver, C.E.E., Posthumus, F.S., Parmar, S., Hawkesford, M.J., De Kok, L.J., 2014. Copper toxicity in Chinese cabbage is not influenced by plant sulphur status, but affects sulphur metabolism-related gene expression and the suggested regulatory metabolites. Plant Biol. 16 (1), 68–78.

Shan, C., Liu, H., Zhao, L., Wang, X., 2014. Effects of exogenous hydrogen sulfide on the redox states of ascorbate and glutathione in maize leaves under salt stress. Biol Plant 58 (1), 169–173.

Sharma, P., Jha, A.B., Dubey, R.S., Pessarakli, M., 2012. Reactive oxygen species, oxidative damage, and antioxidative defense mechanism in plants under stressful conditions. J. Bot 2012.

Shen, J., Xing, T., Yuan, H., Liu, Z., Jin, Z., Zhang, L., et al., 2013. Hydrogen sulfide improves drought tolerance in *Arabidopsis thaliana* by microRNA expressions. PLoS One 8 (10).

Shi, H., Ye, T., Chan, Z., 2013. Exogenous application of hydrogen sulfide donor sodium hydrosulfide enhanced multiple abiotic stress tolerance in bermudagrass (*Cynodon dactylon* (L). Pers.). Plant Physiol. Biochem. 71, 226–234.

Shi, H., Ye, T., Han, N., Bian, H., Liu, X., Chan, Z., 2015. Hydrogen sulfide regulates abiotic stress tolerance and biotic stress resistance in *Arabidopsis*. J. Integr. Plant Biol. 57 (7), 628–640.

Signorelli, S., Corpas, F.J., Borsani, O., Barroso, J.B., Monza, J., 2013. Water stress induces a differential and spatially distributed nitro-oxidative stress response in roots and leaves of *Lotus japonicus*. Plant Sci. 201, 137–146.

Singh, V.P., Singh, S., Kumar, J., Prasad, S.M., 2015. Hydrogen sulfide alleviates toxic effects of arsenate in pea seedlings through up-regulation of the ascorbate–glutathione cycle: possible involvement of nitric oxide. J. Plant Physiol. 181, 20–29.

Sun, J., Wang, R., Zhang, X., Yu, Y., Zhao, R., Li, Z., et al., 2013. Hydrogen sulfide alleviates cadmium toxicity through regulations of cadmium transport across the plasma and vacuolar membranes in *Populus euphratica* cells. Plant Physiol. Biochem. 65, 67–74.

Suzuki, N., Katano, K., 2018. Coordination between ROS regulatory systems and other pathways under heat stress and pathogen attack. Front. Plant Sci. 9, 490.

Tang, R.J., Zhao, F.G., Garcia, V.J., Kleist, T.J., Yang, L., Zhang, H.X., et al., 2015. Tonoplast CBL–CIPK calcium signaling network regulates magnesium homeostasis in Arabidopsis. Proc Natl Acad Sci 112 (10), 3134–3139.

Tian, B., Zhang, Y., Jin, Z., Liu, Z., Pei, Y., 2017. Role of hydrogen sulfide in the methyl jasmonate response to cadmium stress in foxtail millet. Front. Biosci. 22, 530–538.

Tracy, F.E., Gilliham, M., Dodd, A.N., Webb, A.A., Tester, M., 2008. NaCl-induced changes in cytosolic free Ca^{2+} in *Arabidopsis thaliana* are heterogeneous and modified by external ionic composition. Plant, Cell Environ. 31 (8), 1063–1073.

Tuteja, N., Sopory, S.K., 2008. Chemical signaling under abiotic stress environment in plants. Plant Signal. Behav. 3 (8), 525–536.

Valivand, M., Amooaghaie, R., Ahadi, A., 2019. Seed priming with H_2S and Ca^{2+} trigger signal memory that induces cross-adaptation against nickel stress in zucchini seedlings. Plant Physiol. Biochem. 143, 286–298.

Virdi, A.S., Singh, S., Singh, P., 2015. Abiotic stress responses in plants: roles of calmodulin-regulated proteins. Front. Plant Sci. 6, 809.

Wang, R., 2012. Physiological implications of hydrogen sulfide: a whiff exploration that blossomed. Physiol Rev. 92 (2), 791–896.

Wang, B.L., Shi, L., Li, Y.X., Zhang, W.H., 2010. Boron toxicity is alleviated by hydrogen sulfide in cucumber (*Cucumis sativus* L.) seedlings. Planta 231 (6), 1301–1309.

Wang, Y., Ye, X.Y., Qiu, X.M., Li, Z.G., 2019. Methylglyoxal triggers the heat tolerance in maize seedlings by driving AsA-GSH cycle and reactive oxygen species-/methylglyoxal-scavenging system. Plant Physiol. Biochem. 138, 91–99.

Whiteman, M., Armstrong, J.S., Chu, S.H., Jia-Ling, S., Wong, B.S., Cheung, N.S., et al., 2004. The novel neuromodulater hydrogen sulfide: an endogenous peroxynitrite 'scavenger'? J. Neurochem. 90, 765–768.

Wilson, L.G., Bressan, R.A., Filner, P., 1978. Light-dependent emission of hydrogen sulfide from plants. Plant Physiol. 61 (2), 184–189.

Xiong, T.C., Ronzier, E., Sanchez, F., Corratgé-Faillie, C., Mazars, C., Thibaud, J.B., 2014. Imaging long distance propagating calcium signals in intact plant leaves with the BRET-based GFP-aequorin reporter. Front. Plant Sci. 5, 43.

Yang, Y., Guo, Y., 2018. Elucidating the molecular mechanisms mediating plant salt-stress responses. N. Phytol 217 (2), 523–539.

Yang, H., Mu, J., Chen, L., Feng, J., Hu, J., Li, L., et al., 2015. S-nitrosylation positively regulates ascorbate peroxidase activity during plant stress responses. Plant Physiol. 167 (4), 1604–1615.

Yu, L.X., Zhang, C.J., Shang, H.Q., Wang, X.F., Min, W.E.I., Yang, F.J., et al., 2013. Exogenous hydrogen sulfide enhanced antioxidant capacity, amylase activities and salt tolerance of cucumber hypocotyls and radicles. J. Integr. Agric 12 (3), 445–456.

Yuan, P., Yang, T., Poovaiah, B.W., 2018. Calcium signaling-mediated plant response to cold stress. Int. J. Mol. Sci. 19, 3896–3906.

Yun, B.W., Feechan, A., Yin, M., Saidi, N.B., Le Bihan, T., Yu, M., et al., 2011. S-nitrosylation of NADPH oxidase regulates cell death in plant immunity. Nature 478 (7368), 264–268.

Zeng, H., Xu, L., Singh, A., Wang, H., Du, L., Poovaiah, B.W., 2015. Involvement of calmodulin and calmodulin-like proteins in plant responses to abiotic stresses. Front. Plant Sci. 6, 600.

Zhang, H., Hu, L.Y., Hu, K.D., He, Y.D., Wang, S.H., Luo, J.P., 2008. Hydrogen sulfide promotes wheat seed germination and alleviates oxidative damage against copper stress. J. Integr. Plant Biol. 50 (12), 1518–1529.

Zhang, H., Tang, J., Liu, X.P., Wang, Y., Yu, W., Peng, W.Y., et al., 2009. Hydrogen sulfide promotes root organogenesis in *Ipomoea batatas*, *Salix matsudana* and *Glycine max*. J. Integr. Plant Biol. 51 (12), 1086–1094.

Zhang, H., Jiao, H., Jiang, C.X., Wang, S.H., Wei, Z.J., Luo, J.P., et al., 2010a. Hydrogen sulfide protects soybean seedlings against drought-induced oxidative stress. Acta Physiol Plant 32 (5), 849–857.

Zhang, H., Tan, Z.Q., Hu, L.Y., Wang, S.H., Luo, J.P., Jones, R.L., 2010b. Hydrogen sulfide alleviates aluminum toxicity in germinating wheat seedlings. J. Integr. Plant Biol. 52 (6), 556–567.

Zhang, H., Dou, W., Jiang, C.X., Wei, Z.J., Liu, J., Jones, R.L., 2010c. Hydrogen sulfide stimulates β-amylase activity during early stages of wheat grain germination. Plant Signal. Behav. 5 (8), 1031–1033.

Zhang, H., Wang, M.J., Hu, L.Y., Wang, S.H., Hu, K.D., Bao, L.J., et al., 2010d. Hydrogen sulfide promotes wheat seed germination under osmotic stress. Russ J. Plant Physiol. 57 (4), 532–539.

Zhang, H., Hu, S.L., Zhang, Z.J., Hu, L.Y., Jiang, C.X., Wei, Z.J., et al., 2011. Hydrogen sulfide acts as a regulator of flower senescence in plants. Postharvest Biol. Technol. 60 (3), 251–257.

Zhang, L., Pei, Y., Wang, H., Jin, Z., Liu, Z., Qiao, Z., et al., 2015. Hydrogen sulfide alleviates cadmium-induced cell death through restraining ROS accumulation in roots of *Brassica rapa* L. ssp. pekinensis. Oxid. Med. Cell. Longev. 2015.

Zhu, J.K., 2016. Abiotic stress signaling and responses in plants. Cell 167, 313–324.

Chapter 17

Hydrogen sulfide homeostasis in plants: An overview

Sunil K. Gupta[1,2,*], Naina Marwa[2], Ashutosh K. Pandey[3], Yun-Bing Zhang[1] and Jiao-Lin Zhang[1]
[1]*CAS Key Laboratory of Tropical Forest Ecology, Xishuangbanna Tropical Botanical Garden, Chinese Academy of Sciences, Yunnan, P.R. China,* [2]*Plant Ecology & Climate Change Science Division, CSIR-National Botanical Research Institute, Rana Pratap Marg, Lucknow, India,* [3]*School of Agricultural Sciences, Sharda University, Greater Noida, India*
**Corresponding author.*

17.1 Introduction

In plants, hydrogen sulfide (H_2S) has been proven to play an essential role in a variety of growth and developmental processes, for instance, adventitious root and lateral root formation (Zhang et al., 2009a). H_2S, a colorless gas with a strong odor of rotten eggs, has conventionally been considered as a phytotoxin, with a harmful impact on plant growth and development (Huo et al., 2018). H_2S was considered as merely a toxic by-product of cell metabolism, but recently its role as a novel gaseous signal molecule was reviewed (Li et al., 2016; Huo et al., 2018). H_2S contributes in seed germination to plant growth and development, as well as in the attainment of stress tolerance in plants. The exogenous/endogenous H_2S is responsible for developing tolerance to both biotic and abiotic stresses (Bloem et al., 2004; Shi et al., 2015; Mostofa et al., 2015). In different developmental stages of plant growth, plants do produce H_2S in the cytosol through enzymatic mechanisms, particularly desulfhydrases. It is also suggested that H_2S alleviates the negative effect of metals, for example, aluminum (Al) stress in wheat species (Zhang et al., 2008a) and barley species (Dawood et al., 2012). H_2S is proposed to be used in many metal stresses of Al and boron (Zhang et al., 2010a; Wang et al., 2010). Studies identifying the impacts of H_2S on metal toxicity in plants and the physiology of plants have been reported. There are many reports providing proof that the substantial stress from various metals in various crops is coped with via exogenous H_2S application (Wang et al., 2010; Zhang et al., 2010b; Ali et al., 2014a).

Hydrogen Sulfide in Plant Biology. DOI: https://doi.org/10.1016/B978-0-323-85862-5.00017-8

Over the years, sulfur has been documented as a vital constituent for plant growth. Sulfur is a factor of various biochemically essential amalgams, including cysteine, methionine, glutathione (GSH), various sulfolipids, and cofactors such as thiamine, vitamin B, biotin, lipoic acid, and coenzyme A (Trust and Fry, 1992). A significant proportion of total sulfur in the plant cell is usually in organic forms such as amino acids and sulfolipids; however, inorganic sulfur dominates when only excess sulfur is available (Allen and Raven, 1987). In contrasting to carbon and nitrogen isotopes, the use of stable sulfur isotopes in biochemical and physiological investigations are underrated. In the present chapter, we have discussed the potential role of H_2S that could contribute to the detrimental effects of its exposure to plant growth. Thus this chapter emphasizes the postulated role of H_2S as a priming molecule involved in the regulation of biochemical and physiological processes in plants.

17.1.1 Source of H_2S in plants

For H_2S to influence plant cells it must be present and in an optimum concentration. However, its low levels within the plant do not definitely mean that it will have a decisive role. There are two sources of H_2S: the environment and its presence intercellularly.

17.1.1.1 Environmental sources

Natural and human-made both H_2S can be included as environmental sources. Aiuppa et al. (2005) suggested volcanic discharge as one of the natural environmental sources of H_2S. Interestingly, the study indicated that the plants growing downwind from a volcanic crater would potentially be exposed to substantial levels of H_2S; to survive such plants must either be adaptive or resistant. However, environmental H_2S sources may also be contributed by coastal marine sediments (Hansen et al., 1978) or from anoxic soils, for example, marshland soils (Morse et al., 1987). Such releases affect both the aerial parts of the indigenous plants as well as their roots. SO_4^{2-}-reducing bacteria such as *Desulfovibriao* are present in waterlogged soils and marshes produce the sulfur-containing compound sulfite (Postgate, 1959).

Anthropogenic activities like sewage/waste treatment plants (Zhang et al., 2008b), various agricultural undertakings (Aneja et al., 2008), and geothermal industries (Bacci et al., 2000) also add to the environmental H_2S levels.

17.1.1.2 Intracellular sources

Intracellular sources include the production of H_2S by enzymatic mechanisms. Sulfide chemistry is intimately interwoven with the appearance of life on Earth. However, plant's intracellular sources include the release of H_2S by an effective mechanism of enzyme, desulfhydrases. Leon et al. (2002)

reported the presence of cysteine desulfhydrase in *Arabidopsis* plastids. Other plant groups have found an analogous enzyme which is situated in mitochondria (Riemenschneider et al., 2005). For a reaction to advance a required proportion of reactants and threshold energy is required, likewise, for H_2S to exhibit its role in plant cells an optimum concentration is needed. There are also other enzymes that have been proposed as having the options to synthesize H_2S as well. These enzymes are, for instance, d-cysteine desulfhydrase (Papenbrock et al., 2008), which synthesizes pyruvate, ammonia, and H_2S, and an enzyme engaged with cyanide metabolism, b-cyanoalanine synthase, which converts cysteine and cyanide to b-cyanoalanine and H_2S (Garcia et al., 2010).

17.1.2 Role of H_2S in plants

The biochemistry of H_2S and its possible intercellular crosstalk have been reviewed (Kolluru et al., 2013) and it intervenes with the physiological processes. Earlier it was thought that H_2S was merely a phytotoxic compound but later its role as a signaling molecule has been explored. Here, in this chapter, we are discussing its cytotoxic, signaling, and substrate roles in plants and cells.

17.1.2.1 As toxin

Hydrogen sulfide has conventionally been thought of as a phytotoxin, having harmful effects on plant growth and survival. Previous studies suggested H_2S to be a toxic reactive intermediate of cellular metabolism owing to its strong affinity for Fe^{2+}-constituting proteins such as cytochrome oxidase, hemoglobin, and myoglobin, which possibly had been the prime cause of the mass annihilation of species in the Permian (Lisjak et al., 2013). One of the effects seen, for example, is the inhibition of cytochrome oxidase in the mitochondria (see Dorman et al., 2002), where a level higher than 30 ppm caused significant inhibition in some tissues. In 1975, for instance, Joshi et al. (1975) showed that H_2S repressed oxygen release from young seedlings of six rice varieties. It was also reported that nutrient absorption was also reduced in some varieties of rice. H_2S has also been reported to impede oxygen release from rice cultivars and the absorption of phosphorus, a macronutrient (Li, 2013; Hancock and Whiteman, 2014; Calderwood and Kopriva, 2014). The persistence treatment of H_2S in many plant species and variety was performed across a variety of ranges (Thompson and Kats, 1978). Treatment of 3000 ppb of H_2S in some of the plants (for instance, medicago, grapes, sugar beets, pine, and fir) produces harmful injuries to leaves, defoliation, and reduced plant growth. These studies show H_2S behaved as a toxin in different plants species.

17.1.2.2 As substrate

However, exogenous application of H_2S resulted in the discharge of its donor as NaHS and morpholin-4-ium 4-methoxyphenyl (morpholino) phosphinodithioate (GYY4137), displaying significantly constructive effects on seed germination (Li and He, 2015; Wojtyla et al., 2016), plant organogenesis and development (Lin et al., 2012; Fang et al., 2014), and the optimum control of senescence (Zhang et al., 2011). These activities resulted in the procurement of tolerance for stress, for instance, salt (Christou et al., 2013), heavy metals (HMs) (Chen et al., 2013), drought (Christou et al., 2013), and cold resilience (Fu et al., 2013).

17.1.2.3 As signaling molecule

Reactive intermediates have been identified to be influential in the regulation of many cell signaling mechanisms, in both animals and plants (Hancock and Whiteman, 2014). Based on early research on nitric oxide (NO) (Palmer et al., 1987), it was soon recognized that various other reactive short-lived intermediates may participate in such processes. Wang (2002, 2003) proposed the addition of H_2S, beside NO and carbon monoxide (CO), as the third gasotransmitter. H_2S has been displayed to regulate a variety of biological reactions and to be made in response to various stresses. H_2S is established to function as a gaseous signaling biomolecule at low concentration like CO and NO in plants, and in addition it has been revealed that plants can vigorously produce endogenous H_2S under biotic or abiotic stress environments (Li, 2013; Calderwood and Kopriva, 2014; Yamasaki and Cohen, 2016).

The increase in production of endogenous H_2S levels within a plant cell is due to triggering environmental factors, like HM, heat/cold stress, salt, drought conditions, and pathogenic infection (Li, 2013; Calderwood and Kopriva, 2014).

17.2 H_2S effects on plant growth and development

H_2S, at low concentration, can enhance plant growth and development, leading to a sharp increase in global crop production and biofuel production (Dooley et al., 2013). Interestingly, recently researchers have concentrated on the physiological capacity of H_2S in plants. It was described that H_2S supported seed germination in wheat, diminished reactive oxygen species (ROS) damage because of copper toxicity, maintained chlorophyll scarcity, and regulated ROS loss due to osmotic stress in sweet potato leaves (Yang et al., 2008; Zhang et al., 2009b). Moreover, boron salinity, and aluminum metal stress were minimized by H_2S application (Li et al., 2010; Kong et al., 2010; Wang et al., 2012; Chen et al., 2013). Likewise, H_2S at a low level enhances embryonic root development of *Pisum sativum* and regulates flower senescence (Li et al., 2010; Zhang et al., 2011). Sun and Luo (2014) reported that exogenous application of NaHS on cucumber (*Cucumis sativus*) significantly improved hypocotyl and radicle lengths along with the soluble sugar content. NaHS usage also amplified the enzymatic activities of α-amylase, β-amylase, superoxide dismutase (SOD), and peroxidases (POD) and reduced the malondialdehyde (MDA) accumulation.

Hu et al. (2014) reported the significant slowing down of the ripening rate and senescence of mulberry fruit and reduced respiratory intensity and anthocyanin content due to NaHS treatment. Similarly, reduced respiration was also observed in green leafy vegetables/herbs, sweet Italian basil (*Ocimum basilicum*), and green curly kale (*Brassica oleracea var. sabellica*) due to H_2S, which also inhibited ethylene production, chlorophyll loss, various antioxidant factors, and ion leakage, indicating the role of H_2S in delaying senescence (Al Ubeed et al., 2017). Exogenous application of H_2S improves the root morphology, which might be due to an increase in ascorbate peroxidase (APX) activity in the roots in *Brassica napus* L. (Li et al., 2012). Moreover, Dawood et al. (2012) demonstrated that NaHS released H_2S, which mitigated the toxic effect of Al and improved the root morphology in barley.

17.3 H_2S effects on physiological homeostasis

Hydrogen sulfide is a signaling molecule and emerged as a regulator of variety of physiological events in plants (Garcia-Mata and Lamattina, 2010; Ali et al., 2014a). The key physiological processes such as root organogenesis (Zhang et al., 2009c), photosynthesis (Chen et al., 2011), seed germination (Dooley et al., 2013), and stomatal apertures (Lisjak et al., 2011a) are regulated by H_2S. As discussed above, H_2S plays a dual role; a cytotoxic at high concentrations and a signaling molecule at low levels, H_2S homeostasis in the plant is essential to utilize its physiological functions comprising cross-adaptation stimulation.

17.3.1 On gas exchange parameter

17.3.1.1 On photosynthesis

Previous reports have shown the exogenous application of H_2S induces its amalgamation into organic sulfur compounds, causing a rise in soluble thiols, comprising GSH and cysteine (Schütz et al., 1991). Chen et al. (2011) also confirmed a significant increase in nonprotein thiols, GSH, and cysteine content in *S. oleracea* after the optimal concentration application of NaHS. Miginiac-Maslow et al. (2000) described the significance of thiol redox adaptation in photosynthetic electron transfer. The ferredoxin/thioredoxin scheme in chloroplasts is essential for the regulation of CO_2 absorption in oxygenic photosynthesis (Buchanan, 1991). The ferredoxin/thioredoxin scheme, comprising ferredoxin, ferredoxin–thioredoxin reductase (FTR), and thioredoxin, is a common mechanism of light-mediated enzyme regulation (Buchanan, 1991). The photons reduce ferredoxin by photosynthetic electron transfer at the level of PSI. It reduces FTR, creating a dithiol which reduces the disulfide link of thioredoxin. H_2S is generally linked with pollutants like arsenic, boron, and mercury (Arnoórsson, 2004); however, it was reported to have synergistic effects on lichens (Loppi et al., 1998).

In addition, the natural oxidation of H_2S results in the formation of SO_2, a toxin pollutant that exceptionally restricts the growth of lichens (Gries et al., 1995; Deltoro et al., 1999). Significant variations in biomass and species composition of epiphytic lichen populations have been recognized nearby geothermal fields or geothermal springs discharging H_2S (Tretiach et al., 1999). Bertuzzi and Tretiach (2013) observed the maximum photochemical quantum produced which was strongly associated to the efficacy of PSII. The sharp, abrupt decline of *Fv/Fm* was detected in all the four studied lichens and could be inferred to be a direct influence of H_2S on PSII.

The ubiquitous enzyme Ribulose 1,5-bisphosphate carboxylase-oxygenase (Rubisco) assimilates atmospheric carbon dioxide within the Calvin–Benson cycle that is utilized by most photosynthetic organisms. Rubisco is a crucial enzyme regarding photosynthetic carbon fixation in plants, and the amount of stimulated Rubisco is thoroughly associated to Lsp, CE, and the rate of photosynthetic carbon assimilation (Lin and Hsu, 2004). Contrary to the studies reporting the negative effects of H_2S, Chen et al. (2011) observed a significant intensification in Rubisco activity in *Spinacia oleracea* seedlings treated with 100 μM NaHS, thus proposing an enhancement in photosynthesis. Additionally, H_2S, as a signaling molecule, can improve photosynthetic performance by promoting biogenesis of chloroplasts and increasing the ability of CO_2 fixation (Mostofa et al., 2015).

17.3.1.2 On stomatal conductance

H_2S has been found to perform a critical role in the plant's stomatal regulation. Alternatively, there is controversy that H_2S can influence or stimulate stomatal movement, and currently two distinct theories are considered regading H_2S' control of stomatal movements. Studies have suggested that H_2S can control stomatal movements at low levels, for instance, in guard cells, abscisic acid persuaded NO decreased in concentration by H2S donor, hence leads to affects stomatal aperture opening (Lisjak et al., 2010, 2011b; Duan et al., 2015). Moreover, some studies have also revealed that H_2S stabilizes the oxidation produced by H_2O_2, thus initiating the higher activity of antioxidant enzymes (Zhang et al., 2010b). ABA induces the generation of NO, which is directly correlated to the production of H_2O_2 (Bright et al., 2006). Thus it suggested that the probable negative effect of H_2S on H_2O_2 facilitated stomatal closure induced by ABA. The stomatal closure movement is induced by ATP-binding channels and their transporters are present in guard cells (Garcia-Mata and Lamattina, 2010). A few reports suggested that endogenous H_2S participates in the stomatal closure movement incited by ethylene, and L-cysteine desulfhydryl or D-cysteine desulfhydryl, with the catalytic activity enhancing the breakdown of L-cysteine desulfhydryl or D-cysteine desulfhydryl supplemented by abscisic acid (Hou et al., 2013).

17.3.1.3 On respiration

Mitochondria in higher plants constitute two respiratory pathways, a cyanide-sensitive cytochrome pathway (CP) and a cyanide-insensitive alternative pathway (AP). Respiration of AP relates to the respiratory chain by an additional terminal oxidase-alternative oxidase (AOX). Whereas AOX is part of the branched respiratory electron transport chain. AOX accepts electrons from the ubiquinone pool and reduces oxygen to water (Millenaar and Lambers, 2003). In general, the AP may dampen the generation of ROS during periods of rapid respiration (Maxwell et al., 1999). Therefore, H_2S may play an important role in regulating the AOX transcription mainly through the ROS-independent pathway, and AOX can dampen the generation of ROS.

17.3.1.4 On relative water content

H_2S has been shown to act as signaling molecules in numerous physiological processes, and play significant roles in plant cellular developments and mediate responses to both biotic and abiotic stresses in plants (Khan et al., 2017). Mostofa et al. (2015) reported an increase in relative water content (RWC) with H_2S supplements under Cd stress in rice. Importantly, the enhancement of RWC might lead to an enhancement in stomatal conductance and transpiration in H_2S-treated Cd-stressed rice plants which in turn contributes to the improvement of photosynthesis (Bharwana et al., 2014; Duan et al., 2015).

17.3.2 H_2S effects on mineral nutrition

Extensive efforts are ongoing to improve plants' ability to thrive under biotic and abiotic stresses. In this framework, the implementation of the probable roles of small biomolecules, for example, phytohormones and signaling molecules may be regarded as a potent means for transforming plants' compliance against adverse effects caused from dynamic global conditions. In plants H_2S has recently gained a reputation as being linked with adaptive reactions against environmental stressors. Studies showed that the application of a H_2S donor, for example, sodium bisulfide (NaHS), provided substantial tolerance to a range of abiotic stresses, such as drought, heat, HMs, and osmotic stresses (Christou et al., 2013; Shi et al., 2013; Ali et al., 2014b; Mostofa et al., 2014). Mostofa et al. (2015) reported that H_2S significantly limited Na^+ uptake, while enhancing the absorption of K^+, thus confining the augmented Na^+/K^+, indicating this may be done through the conservation of plasma membrane integrity, as was detected in *Arabidopsis* and alfalfa (Li et al., 2013; Lai et al., 2014). Likewise, H_2S has proven to have contributed to enhancing the photosynthetic performance of rice plants by transporting Mg, Fe, Zn, and Mn ions as well as by helping the synthesis of chloroplasts and increasing the skill of CO_2 fixation, as observed in *Spinacea oleracea* (Chen et al., 2011).

The involvement of H_2S in plant sulfur metabolism has been reported in the literature. The S supply may support plants to survive with Fe deficiency (Astolfi et al., 2010, 2012). Astolfi et al. (2010) revealed that barley showed a positive correlation between the nutritional status of S and its ability to manage with Fe deficiency. Furthermore, one of the primary responses to Fe deficit in plants (strategy II) is the release of phytosiderophores in the rhizosphere in order to chelate and solubilize Fe^{3+} (Astolfi et al., 2010, 2012). Phytosiderophores, a nicotianamine derivate produced from three molecules of S-adenosyl-methionine, demonstrates another potential connection between Fe and S metabolism (Chen et al., 2015). Under sulfur-deficit conditions, the release of phytosiderophores is concentrated (Astolfi et al., 2010, 2012); conversely, the question is whether H_2S acts as a sulfur complex or a signaling molecule to play a crucial role in the reaction to Fe scarcity in plants.

17.4 H_2S homeostasis during metal toxicity

HMs and metalloids, for instance, chromium (Cr), copper (Cu), zinc (Zn), arsenic (As), etc. are highly phytotoxic for plant growth, development, and reproduction. In plants, during their life cycle, they require more than 15 different HMs as nutrients for growth and development, but the excess of these HMs led plants to absorb them forcefully (Sharma and Chakraverty, 2013). HM stress prompts unrestrained ROS bursts leading to the impairment of antioxidant defense mechanism by the process of lipid peroxidation and the oxidation of proteins and nucleic acids. H_2S acts as a reductive constituent to directly cope ROS, such as peroxynitrite and hypochlorite (Whiteman et al., 2004), superoxide (Mitsuhashi et al., 2006), and hydrogen peroxide (H_2O_2) (Geng et al., 2004). This whole process results in the distortion of the protein moiety and nucleic acid structure (Sandhi et al., 2017).

The role of H_2S in tolerating metal toxicity was reported in several studies. Exogenous application of NaHS in pea seedlings resulted in efficient mitigation of As-mediated injuries, due to the maintenance of a high Cys level (Singh et al., 2015). Phytochelatins (PCs) are formed due to excess Cys moieties inside the cell and maybe helped to scavenge the toxic As within the cells (Banerjee and Roychoudhury, 2018c). Dawood et al. (2012) reported a decrease in Al uptake and the abundant ATPase and photosynthetic performance assisted in the alleviating role of H_2S in barley seedlings under Al phytotoxicity. Through the greater abundance of antioxidant enzymes and the activation of tonoplast Cd^{2+}/H^{+} antiporters, H_2S improved Cd phytotoxicity in *Populus euphratica* cells (Sun et al., 2013). In some other experiments, the pretreatment of NaHS to wheat seed resulted in better germination under Cr stress (Zhang et al., 2010c). In another study, citrate secretion was significantly reversed by exogenous application of NaHS to *Hordeum vulgare* seedlings under high concentrations of aluminum (Al^{3+}) (Chen et al., 2013).

So, the above reports suggest an exogenous application of H_2S has a crucial role during metal stress, and plant homeostasis is maintained. H_2S minimizes the absorption and transport of various metals, reduces the accretion of metals in plants, and improves optimum homeostasis for stress tolerance of plants to metal stress.

17.5 Hydrogen sulfide effects and enzymatic homeostasis in plants

17.5.1 On membrane degradation

MDA produced as a product of lipid peroxidation is an indicator of membrane damage caused due to overproduction of ROS under stress conditions (Liu et al., 2013; Shafiq et al., 2015). Exogenous application of NaHS (H_2S donor) has been reported to reduce the MDA content in several plant species exposed to varying environmental stress, for example, in pepper plants under Zn stress (Liu et al., 2016), mulberry with NaHS treatment (Hu et al., 2014), wheat under drought stress (Ma et al., 2016), and sweet potato under osmotic stress (Zhang et al., 2009b). Previous studies suggest a positive role for H_2S application in membrane integrity (Liu et al., 2016; Mir et al., 2015).

17.5.2 On reactive oxygen species defense and homeostasis

Environmental stresses can disturb the balance between generation and the scavenging of ROS in intracellular levels (Apel and Hirt, 2004). This imbalance of ROS may lead to various growth, physiological, biochemical, and crop productivity deteriorations. Plants have evolved various strategies to ameliorate oxidative toxicity by the induction of antioxidative defense responses against oxidative stress and upregulation to combat the adverse effects of ROS (Mullineaux and Baker, 2010).

H_2O_2 and the superoxide radical are major indicators for oxidative stress-induced damages. Shi et al. (2013) reported that exogenous application of H_2S on Bermuda grass under salt, osmotic, and cold stress showed oxidative protection, as indicated by significant lower concentrations of H_2O_2 and superoxide radical compared with nontreated plants.

The concentration of superoxide radical was reduced in the mulberry fruits with NaHS application as compared to the nontreated fruits in the plants (Hu et al., 2014). H_2S has been reported to have a protective role against oxidative stress and is considered to be a powerful reducing agent (Dawood et al., 2012; Li et al., 2012; Wang et al., 2012).

17.5.3 Antioxidant and antioxidant enzyme activity

Plants use their antioxidative defense system to alleviate the damage induced due to oxidative damage. The antioxidative defense system includes both enzymatic SOD, catalase (CAT), APX, dehydroascorbate reductase (DHAR), glutathione reductase (GR), and nonenzymatic antioxidants (ascorbate and GSH) (Gill and Tuteja, 2010).

Previous studies indicated the involvement of H_2S-mediated enhancement in the antioxidative defense mechanisms to reduce the negative effects of oxidative stress (Liu et al., 2016). For example, in mulberry fruit different antioxidative enzymes, SOD, CAT, and POD, showed enhanced activity with NaHS treatment (Hu et al., 2014). Indicating that the increased SOD enzyme activity can more efficiently dismutate the superoxide radical, and the enhanced activities of CAT and POD for efficient elimination of hydrogen peroxide radical (H_2O_2) (Mittler, 2002).

Similar trends of enhanced antioxidative defense were reported by Zhang et al. (2009b) in sweet potato under osmotic stress with H_2S-treated plants This indicated that the ameliorative effect of H_2S against osmotic stress was mediated mainly by the upregulation of the antioxidative defense, that is, higher levels of SOD, CAT, and APX and decreased concentrations of lipoxygenase and concentrations of H_2O_2 and MDA content.

GSH and ascorbate has been considered to be a potent antioxidant scavenging ROS and in maintaining the intracellular redox status. A higher concentration of GSH and ascorbate prevents cells from entering into programmed cell death or necrosis (Schafer and Buettner, 2001; Noctor et al., 2018). Under stress conditions plants face oxidative stress due to the increased production of ROS, leading to disturbed homeostasis. Higher concentrations of oxidized GSH and ascorbate indicate some of the important molecules such as NADH/NADPH are harder to maintain, resulting in the plant cells entering programmed cell death or necrosis (Schafer and Buettner, 2001). Shan et al. (2011) reported that water-stressed wheat seedlings treated with NaHS showed higher GSH (reduced) and ascorbate (reduced) concentrations.

Yu et al. (2013) suggested that H_2S-induced an increase in the level of antioxidative enzymes, such as SOD, CAT, APX, GR, GPX, and DHAR, while there was a decrease in the H_2O_2 concentration and MDA content under salt stressed in cucumber seedlings.

Likewise H_2S-mediated activation of the antioxidative defense system has been reported in different plant species, for example, *C. sativus* (Yu et al., 2013; Sun and Luo, 2014), *Fragaria* × *ananassa* (Christou et al., 2013), *Medicago sativa* (Wang et al., 2012; Lai et al., 2014), *Nicotiana tabacum* (Da Silva et al., 2017), *Oryza sativa* (Mostofa et al., 2015), and *Triticum aestivum* (Khan et al., 2017). This indicates that the H_2S treatment can have a significant protection against oxidative stress, especially when the plants face different environmental stresses.

17.6 Hydrogen sulfide and molecular homeostasis in plants

As discussed above, H_2S acts as a signaling molecule in plants, or plants simply respond and adapt to the presence of H_2S, and there must be a molecular mechanism involved in the whole process. If H_2S is extracellularly produced, it should be at the last stage of a pathway during which related cells can respond to it by a special mechanism, and this response may lead to a signal transduction cascade with an appropriate effect. However, if H_2S is in the mid-reaction pathway, then its formation can be an integral part of the same reaction pathway, and the concentration of H_2S may be minimized for further later effects.

To explore molecular homeostasis, H_2S assimilation in plants must be studied. The sulfur assimilation pathway in plants is divided into two reaction steps: sulfate reduction and cysteine synthesis. There are specific sulfate transporters (SULTRs; Buchner et al., 2004) which are responsible for sulfate transport (from soil to the plants) and their reduction in plants. Inside the plants, ATP sulfurylase (ATPS) in the initial step stimulates sulfate and the formation of adenosine-50-phosphosulfate (APS) occurs. Now, two enzymes APS reductase (APR) and sulfite reductase (SiR) by performing two-step reaction and reduces APS. (Fig. 17.1). Then the formation of cysteine takes place by the enzyme *O*-acetyl-(thiol)-serine lyase (OAS-TL) through reacting sulfide into O-acetylserine (OAS), and then cysteine acts as a sulfur source to all those organic molecules which contain sulfur in a reduced form, for instance, some proteins, GSH, vitamins and cofactors (Takahashi et al., 2011). Then serine delivers amino acid precursor for cysteine, and OAS formation.

Simply, in plants, numerous pyridoxal-5'-phosphate (PLP)-dependent D/L-cysteine desulfhydrases (D/L-CDs) are primarily responsible for H_2S generation (Cheng et al., 2013) and the removal of H_2S is mainly by O-acetylserine lyase (Lisjak et al., 2013).

17.6.1 The molecular mechanism behind H_2S homeostasis

In plants H_2S homeostasis is an essential source for mechanisms to withstand plant stress (Romero et al., 2014; Mostofa et al., 2015; Dai et al., 2017; Tian et al., 2017). Despite the wealth of recent publications suggesting new physiological roles for H_2S, its molecular mechanism of action is still not clear. As we know, H_2S has a dual role in the cell: one as a cytotoxin (at higher concentration), and the other as a signaling molecule (at optimum concentration).

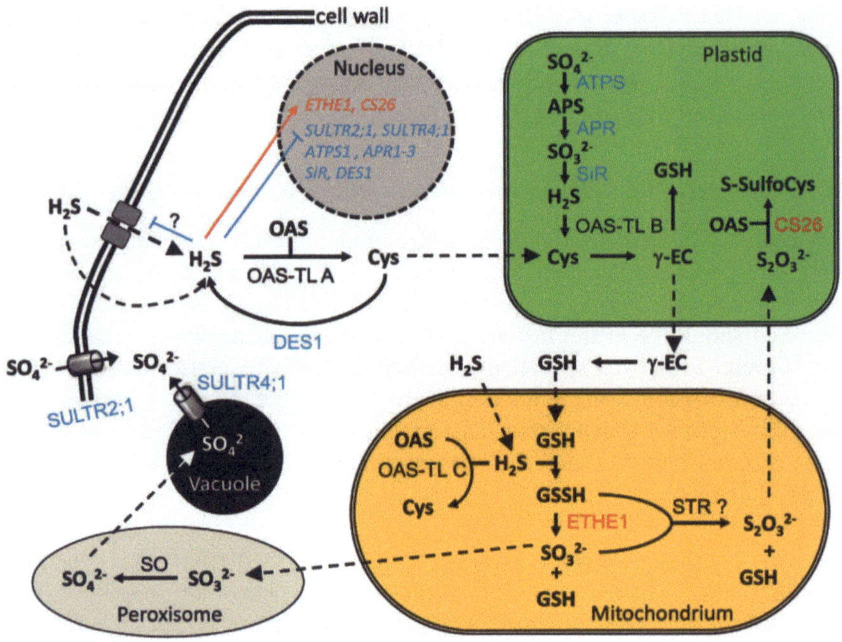

FIGURE 17.1 Schematic representation of the consequences related to H_2S exposure in *Arabidopsis* leaf cells. APR, adenosine APS reductase; APS, adenosine-5-phosphosulfate; ATPS, ATP sulfurylase; CS26, Sulfocysteine synthase; Cys, cysteine; DES1, Cys desulfhydrase 1; ETHE1, ethylmalonic encephalopathy protein 1; GSH, glutathione; GSSH, glutathione persulfide; OAS, O-acetylserine; OAS-TL, OAS(thiol)lyase; SIR, sulfite reductase; SULTR, sulfate transporter; SO, sulfite oxidase; STR, sulfurtransferase. *Adapted from Birke, H., De Kok, L.J., Wirtz, M., Hell, R., 2014. The role of compartment-specific cysteine synthesis for sulfur homeostasis during H_2S exposure in Arabidopsis. Plant Cell Physiol. 56(2), 358–367.*

H_2S homeostasis in plant cells is essential for it to exert its physiological and molecular functions. In plant cells there are many metabolic pathways to regulate H_2S homeostasis, like other signal molecules, that is, H_2O_2 and NO. H_2S homeostasis is closely regulated by LCD, DCD, SiR, cyanoalanine synthase (CAS), and cysteine synthase (CS Li (2013, 2015); Fig. 17.2). The syntheses of H_2S, amine, and pyruvate are carried out by the enzymatic activity of L-CD/D-CD through the breakdown of L-/ D-cysteine. Ferrodoxin donates its electrons for the reduction of sulfite by the enzyme SiR. Cysteine may produce H_2S by the catalytic activity of CAS in the presence of hydrogen cyanide. Cysteine synthase, namely OAS-TL, may react with H_2S by the *O*-acetyl-L-serine to produce cysteine, and vice versa a reaction may produce H_2S (Li, 2013, 2015; Fig. 17.2). In most plants H_2S formation is performed by L-CD or D-CD enzymes, which react to different stresses and enhance the stress tolerance. The remaining H_2S may be released to atmosphere (Li, 2013; Calderwood and Kopriva, 2014; Hancock and Whiteman, 2014).

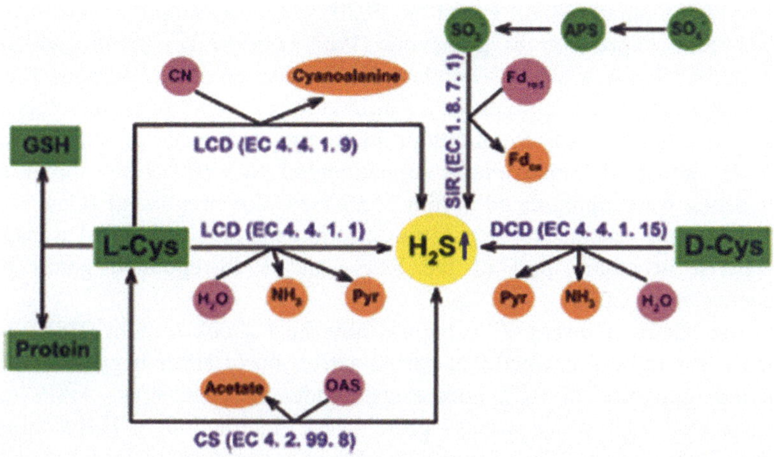

FIGURE 17.2 Schematic representation of hydrogen sulfide (H_2S) homeostasis in plant cells. LCD, L-cysteine desulfhydrase; DCD, D-cysteine desulfhydrase; SiR, sulfite reductase; CAS, cyanoalanine synthase; and CS, cysteine synthase. *Adapted from Li, 2015. Analysis of some enzymes activities of hydrogen sulfide metabolism in plants,* Methods Enzy mol. *555, 253–269.*

17.6.2 Biotechnological application of H_2S in plants

The signaling properties of hydrogen sulfide (H_2S), which is intracellularly generated in plants, are mainly observed during persulfidation, a posttranslational protein modification (PTM) that affects redox-sensitive cysteine residues. Some studies show exogenous application of H_2S decreases the damage caused by abiotic stress in plants and enhances physiological features such as seed germination, root development, and postharvest preservation of vegetables (Corpas et al., 2019; Huo et al., 2018; Deng et al., 2009). Up to now, exogenous applications have been carried out by using some other supplementary chemicals (polysulfide compounds: sodium hydrosulfide (NaHS) and inorganic sodium polysulfides (Na_2Sn) such as Na_2S_2, Na_2S_3, and Na_2S_4) capable of transporting H_2S inside plants (Corpas and Palma, 2020). These supporting polysulfide compounds are generally used by making aqueous solutions to deliver H_2S inside plants, but moderate pH corresponding to pKa should be maintained (Gun et al., 2004). In plant studies the NaHS is exogenously mixed into hydroponic solutions and in vitro growth media or sometimes a foliar spray can also be used on plants. Exogenous treatment with H_2S enhances nonenzymatic antioxidant systems which ultimately prevent ROS overproduction and regulate oxidative damage during abiotic stress.

17.6.3 H_2S-mediated regulation of gene and protein expression

Although the exact mechanisms involved in molecular regulation remain unknown, H_2S has been revealed to regulate gene expression to some extent

(Pandey and Gautam, 2020; Lin et al., 2020). Exogenous applications of H_2S prompt gene expression in grapevine (*Vitis vinifera* L.) plants, which is involved in the synthesis of secondary metabolites as well as various abiotic resistance defensive compounds that ultimately optimize plant development (Ma and Yang, 2018). In addition, microarray analysis of differentially expressed genes of tomato plants supplemented with NaHS has shown that 5349 genes were upregulated, while 5536 were downregulated (Guo et al., 2018). Defense responsive genes, transcription factors, and signal transduction are a necessary part of the ROS- and RNS-regulated genes pool (Antoniou et al., 2016).

In the "defense response" category, nine heat shock protein (HSP) were reported. It can be suggested that partial activation of HSPs triggered NaHS-mediated regulation of ROS homeostasis under different stress. HSPs must play a crucial role during protein−protein (P-P) interaction, as HSPs helps in protein folding (molecular chaperones); for instance, protein folding support in the arrangement of legitimate protein shape and avoidance of undesirable protein accumulation. HSPs have an important role in coping with abiotic stress which has been proved by many shreds of evidence (Zhang et al., 2015), for instance, drought, cold, and high salinity. Heat shock and high salinity are also responsible for induced MsHSP17.7 (Li et al., 2016). Overexpression of RcHSP17.8 enhanced tolerance to drought, osmotic, salt, and heat stresses (Jiang et al., 2009).

In the "signal transduction" category, mitogen-activated protein kinase (MAPK) and phospholipase C and D were mostly affected due to exogenous H_2S application. Guo et al. (2018) reported five genes with increased expression due to H_2S in tomato roots. The MAPK cascade is an evolutionarily conserved signal transduction module involved in transducing extracellular signals to the nucleus for appropriate cellular adjustment. MAPKK kinase (MAPKKK), a MAPK kinase (MAPKK), and a MAP kinase (MAPK) are three essential components of MAPK cascade. These three components are usually connected by the event of phosphorylation (Li et al., 2017).

Many studies have shown that transcription factors play important roles in regulating plant responses to the environment. Plants exposed to salt stress commonly react by inducing several stress response networks, a process that involves the action and interaction of many transcription factors. Some common cereals (wheat and maize) treated with ethylenediurea under tropospheric ozone conditions showed positive response for major regulatory proteins (Gupta et al., 2018, 2020). Guo et al. (2018) reported some transcription factors-related genes, for instance, MADS-box transcription factor, WRKY transcription factor, TCP transcription factor, were more abundant in the NaHS + nitrate, as compared with the nitrate stress.

17.6.3.1 H_2S-induced protein persulfidation under stress

H_2S can induce PTM of Cys residues in macromolecular protein complexes; this phenomenon is considered to be Cys persulfidation. S-nitrosation, S-sulfenylation, and S-gluthionylation are oxidative PTM of Cys residues. Persulfides formation is generally catalyzed by two compounds, namely, cysteine desulfurases and sulfurtransferases. About 5% of the proteins are persulfidated in the proteome of *Arabidopsis thaliana* (Aroca et al., 2017).

17.7 Conclusion and future prospects

Scientific research on H_2S suggests its role in various metabolic pathways regulating the homeostasis of plants under abiotic and biotic stresses. As a gasotransmitter, hydrogen sulfide inhibits the absorption and accumulation of metals in plants and improves the antioxidant capacity of plants to cope with metal stress. H_2S confers abiotic and biotic stress tolerance by optimizing the stress-responsive gene expressions, metabolic homeostasis, and ROS balance. This chapter contributes to a better understanding of the H_2S homeostasis at the physiological, biochemical, and molecular level. The exogenous application of H_2S might have a better protective role in plants, and it was explained by the improvement of gas exchange parameters, that is, photosynthesis, stomatal conductance, RWC, and plant respiration. The antioxidant capacity, the reduction of metal accumulation, and the regulation of protein expression might also prove H_2S homeostasis in plants which ultimately can provide a useful strategy to protect against the different environmental challenges (metal stress, drought, salt stress, cold stress, heat stress) encountered by plants during their growth and developmental stages.

Acknowledgments

The authors are grateful to the Director, Xishuangbanna Tropical Botanical Garden (XTBG), Chinese Academy of Science (CAS), China for their kind cooperation. SKG is thankful to XTBG, CAS for postdoctoral fellowship. NM is grateful to DST (Department of Science and Technology, India) for senior research fellowship. AKP is thankful to H.S Gaur, Dean, School of Agricultural Sciences for guidance and support.

References

Aiuppa, A., Inguaggiato, S., McGonigle, A.J.S., O'Dwyer, M., Oppenheimer, C., Padgett, M.J., et al., 2005. H_2S fluxes from Mt. Etna, Stromboli, and Vulcano (Italy) and implications for the sulfur budget at volcanoes. Geochim. Cosmochim. Acta 69 (7), 1861–1871.

Al Ubeed, H.M.S., Wills, R.B.H., Bowyer, M.C., Vuong, Q.V., Golding, J.B., 2017. Interaction of exogenous hydrogen sulphide and ethylene on senescence of green leafy vegetables. Postharvest Biol. Technol. 133, 81–87.

Ali, B., Gill, R.A., Yang, S., Gill, M.B., Ali, S., Rafiq, M.T., et al., 2014a. Hydrogen sulfide alleviates cadmium-induced morpho-physiological and ultrastructural changes in *Brassica napus*. Ecotoxicol. Environ. Saf. 110, 197–207.

Ali, B., Song, W.J., Hu, W.Z., Luo, X.N., Gill, R.A., Wang, J., et al., 2014b. Hydrogen sulfide alleviates lead-induced photosynthetic and ultrastructural changes in oilseed rape. Ecotoxicol. Environ. Saf. 102, 25–33.

Allen, S., Raven, J.A., 1987. Intracellular pH regulation in *Ricinus communis* grown with ammonium or nitrate as N source: the role of long-distance transport. J. Exp. Bot. 38 (4), 580–596.

Aneja, V.P., Schlesinger, W.H., Erisman, J.W., 2008. Farming pollution. Nat. Geosci. 1 (7), 409–411.

Antoniou, C., Savvides, A., Christou, A., Fotopoulos, V., 2016. Unravelling chemical priming machinery in plants: the role of reactive oxygen-nitrogen-sulfur species in abiotic stress tolerance enhancement. Curr. Opin. Plant Biol. 33, 101–107.

Apel, K., Hirt, H., 2004. Reactive oxygen species: metabolism, oxidative stress, and signal transduction. Annu. Rev. Plant Biol. 55, 373–399.

Arnórsson, S., 2004. Environmental impact of geothermal energy utilization. In: Gieré, R., Stille, P. (Eds.), Energy, Waste and the Environment: A Geochemical Perspective. Geological Society, London, pp. 297–336.

Aroca, A., Benito, J.M., Gotor, C., Romero, L.C., 2017. Persulfidation proteome reveals the regulation of protein function by hydrogen sulfide in diverse biological processes in *Arabidopsis*. J. Exp. Bot. 68, 4915–4927.

Astolfi, S., Zuchi, S., Hubberten, H.M., Pinton, R., Hoefgen, R., 2010. Supply of sulphur to S-deficient young barley seedlings restores their capability to cope with iron shortage. J. Exp. Bot. 61 (3), 799–806.

Astolfi, S., Zuchi, S., Neumann, G., Cesco, S., di Toppi, L.S., Pinton, R., 2012. Response of barley plants to Fe deficiency and Cd contamination as affected by S starvation. J. Exp. Bot. 63 (3), 1241–1250.

Bacci, E., Gaggi, C., Lanzillotti, E., Ferrozzi, S., Valli, L., 2000. Geothermal power plants at Mt. Amiata (Tuscany–Italy): mercury and hydrogen sulphide deposition revealed by vegetation. Chemosphere 40 (8), 907–911.

Banerjee, A., Roychoudhury, A., 2018. Strigolactones: multi-level regulation of biosynthesis and diverse responses in plant abiotic stresses. Acta Physiol. Plant 40, 86.

Bertuzzi, S., Tretiach, M., 2013. Hydrogen sulphide inhibits PSII of lichen photobionts. Lichenologist 45 (1), 101–113.

Bharwana, S.A., et al., 2014. Hydrogen sulfide ameliorates lead-induced morphological, photosynthetic, oxidative damages and biochemical changes in cotton. Environ. Sci. Pollut. Res. 21, 717–731.

Birke, H., De Kok, L.J., Wirtz, M., Hell, R., 2014. The role of compartment-specific cysteine synthesis for sulfur homeostasis during H_2S exposure in *Arabidopsis*. Plant, Cell Physiol. 56 (2), 358–367.

Bloem, E., Riemenschneider, A., Volker, J., Papenbrock, J., Schmidt, A., Salac, I., et al., 2004. Sulphur supply and infection with *Pyrenopeziza brassica* influence L cysteine desulfhydrase activity in *Brassica napus* L. J. Exp. Bot. 55, 2305–2312. Available from: 10.1093/jxb/erh236.

Bright, J., Desikan, R., Hancock, J.T., Weir, I.S., Neill, S.J., 2006. ABA-induced NO generation and stomatal closure in *Arabidopsis* are dependent on H_2O_2 synthesis. v J. 45 (1), 113–122.

Buchanan, B.B., 1991. Regulation of CO_2 assimilation in oxygenic photosynthesis: the ferredoxin/thioredoxin system: perspective on its discovery, present status, and future development. Arch. Biochem. Biophys. 288 (1), 1–9.

Buchner, P., Stuiver, C.E.E., Westerman, S., Wirtz, M., Hell, R., Hawkesford, M.J., et al., 2004. Regulation of sulfate uptake and expression of sulfate transporter genes in *Brassica oleracea* as affected by atmospheric H_2S and pedospheric sulfate nutrition. Plant Physiol. 136, 3396–3408.

Calderwood, A., Kopriva, S., 2014. Hydrogen sulfide in plants: from dissipation of excess sulfur to signaling molecule. Nitric Oxide 41, 72–78.

Chen, J., Wu, F.H., Wang, W.H., Zheng, C.J., Lin, G.H., Dong, X.J., et al., 2011. Hydrogen sulfide enhances photosynthesis through promoting chloroplast biogenesis, photosynthetic enzyme expression, and thiol redox modification in *Spinacia oleracea* seedlings. J. Exp. Bot. 62 (13), 4481–4493.

Chen, J., Wang, W.H., Wu, F.H., You, C.Y., Liu, T.W., Dong, X.J., et al., 2013. Hydrogen sulfide alleviates aluminum toxicity in barley seedlings. Plant Soil 362 (1-2), 301–318.

Chen, J., Wu, F.H., Shang, Y.T., Wang, W.H., Hu, W.J., Simon, M., et al., 2015. Hydrogen sulphide improves adaptation of *Zea mays* seedlings to iron deficiency. J. Exp. Bot. 66 (21), 6605–6622.

Cheng, W., Zhang, L., Jiao, C., Su, M., Yang, T., Zhou, L., et al., 2013. Hydrogen sulfide alleviates hypoxia-induced root tip death in *Pisum sativum*. Plant Physiol. Biochem. 70, 278–286.

Christou, A., Manganaris, G.A., Papadopoulos, I., Fotopoulos, V., 2013. Hydrogen sulfide induces systemic tolerance to salinity and non-ionic osmotic stress in strawberry plants through modification of reactive species biosynthesis and transcriptional regulation of multiple defense pathways. J. Exp. Bot. 64 (7), 1953–1966.

Corpas, F.J., González-Gordo, S., Cañas, A., Palma, J.M., 2019. Nitric oxide and hydrogen sulfide in plants: which comes first? J. Exp. Bot. 70, 4391–4404.

Corpas, F.J., Palma, J.M., 2020. H_2S signaling in plants and applications in agriculture. J. Adv. Res. 24, 131–137.

Da Silva, C.J., Fontes, E.P.B., Modolo, L.V., 2017. Salinity-induced accumulation of endogenous H2S and NO is associated with modulation of the antioxidant and redox defense systems in Nicotiana tabacum L. cv. Havana. Plant Sci. 256, 148–159.

Dai, C., Cui, W., Pan, J., Xie, Y., Wang, J., Shen, W., 2017. Proteomic analysis provides insights into the molecular bases of hydrogen gas-induced cadmium resistance in *Medicago sativa*. J. Proteom. 152, 109–120. Available from: 10.1016/j.jprot.2016.10.013.

Dawood, M., Cao, F., Jahangir, M.M., Zhang, G.P., Wu, F., 2012. Alleviation of aluminum toxicity by hydrogen sulfide is related to elevated ATPase and suppressed aluminium uptake and oxidative stress in barley. J. Hazard. Mater. 210, 121–128.

Deltoro, V.I., Gimeno, C., Calatayud, A., Barreno, E., 1999. Effects of SO_2 fumigations on photosynthetic CO_2 gas exchange, chlorophyll a fluorescence emission and antioxidant enzymes in the lichens *Evernia prunastri* and *Ramalina farinacea*. Physiol. Plant 105 (4), 648–654.

Deng, Y., Wang, C., Wang, N., Wei, L., Li, W., Yao, Y., et al., 2009. Roles of small-molecule compounds in plant adventitious root development. Biomolecules 9 (9), E420.

Dooley, F.D., Nair, S.P., Ward, P.D., 2013. Increased growth and germination success in plants following hydrogen sulfide administration. PLoS One 8, 62048.

Dorman, D.C., Moulin, F.J., McManus, B.E., Mahle, K.C., James, R.A., Struve, M.F., 2002. Cytochrome oxidase inhibition induced by acute hydrogen sulfide inhalation: correlation with tissue sulfide concentrations in the rat brain, liver, lung, and nasal epithelium. Toxicol. Sci. 65, 18–25.

Duan, B., Ma, Y., Jiang, M., Yang, F., Ni, L., Lu, W., 2015. Improvement of photosynthesis in rice (*Oryza sativa* L.) as a result of an increase in stomatal aperture and density by exogenous hydrogen sulfide treatment. Plant Growth Regul. 75 (1), 33–44.

Fang, H., Jing, T., Liu, Z., Zhang, L., Jin, Z., Pei, Y., 2014. Hydrogen sulfide interacts with calcium signaling to enhance the chromium tolerance in *Setaria italica*. Cell Calcium 56 (6), 472–481.

Fu, P., Wang, W., Hou, L., Liu, X., 2013. Hydrogen sulfide is involved in the chilling stress response in *Vitis vinifera* L. Acta Soc. Bot. Pol. 82 (4). Available from: 10.5586/asbp.2013.031.

Garcia, I., Castellano, J.M., Vioque, B., Solano, R., Gotor, C., Romero, L.C., 2010. Mitochondrial β-cyanoalanine synthase is essential for root hair formation in *Arabidopsis thaliana*. Plant Cell 22, 3268–3279.

Garcia-Mata, C., Lamattina, L., 2010. Hydrogen sulphide, a novel gasotransmitter involved in guard cell signaling. N. Phytol. 188, 97–984.

Geng, B., Chang, L., Pan, C., Qi, Y., Zhao, J., Pang, Y., et al., 2004. Endogenous hydrogen sulfide regulation of myocardial injury induced by isoprotenol. Biochem. Biophys. Res. Commun. 318, 756–763.

Gill, S.S., Tuteja, N., 2010. Reactive oxygen species and antioxidant machinery in abiotic stress tolerance in crop plants. Plant physiol. Biochem. 48 (12), 909–930.

Gries, C., Sanz, M.J., Nash, T.H. III, 1995. Effect of SO_2 fumigation on CO_2 gas exchange, chlorophyll fluorescence and chlorophyll degradation in different lichen species from western North America. Cryptogamic Botany, *AGRIS* **5**(3), 2013, 239–246.

Gun, J., Modestov, A., Kamyshny, A., Ryzkov, D., Gitis, V., Goifman, A., et al., 2004. Electrospray ionization mass spectrometric analysis of aqueous polysulfide solutions. Microchim. Acta 46, 229.

Guo, Z., Liang, Y., Yan, J., Yang, E., Li, K., Xu, H., 2018. Physiological response and transcription profiling analysis reveals the role of H2S in alleviating excess nitrate stress tolerance in tomato roots. Plant Physiol. Biochem. 124, 59–69.

Gupta, S.K., Sharma, M., Majumder, B., Maurya, V.K., Lohani, M., Deeba, F., et al., 2018. Impact of ethylene diurea (EDU) on growth, yield and proteome of two winter wheat varieties grown under high ambient ozone phytotoxicity. Chemosphere 196, 161–173.

Gupta, S.K., Sharma, M., Majumder, B., Maurya, V.K., Deeba, F., Zhang, J.L., et al., 2020. Effects of ethylenediurea (EDU) on regulatory proteins in two maize (*Zea mays* L.) varieties under high tropospheric ozone phytotoxicity. Plant Physiol. Biochem. 154, 675–688. Available from: https://doi.org/10.1016/j.plaphy.2020.05.037.

Hancock, J.T., Whiteman, M., 2014. Hydrogen sulfide and cell signaling: team player or referee? Plant Physiol. Biochem. 78, 37–42.

Hansen, M.H., Ingvorsen, K., Jøgensen, B.B., 1978. Mechanisms of hydrogen sulfide release from coastal marine sediments to the atmosphere. Limnol. Oceanogr. 23 (1), 68–76.

Hou, Z., Wang, L., Liu, J., Hou, L., Liu, X., 2013. Hydrogen sulfide regulates ethylene-induced stomatal closure in *Arabidopsis thaliana*. J. Integr. Plant Biol. 55 (3), 277–289.

Huo, J., Huang, D., Zhang, J., Fang, H., Wang, B., Wang, C., et al., 2018. Hydrogen sulfide: a gaseous molecule in postharvest freshness. Front. Plant Sci. 9, 1172.

Hu, H., Shen, W., Li, P., 2014. Effects of hydrogen sulphide on quality and antioxidant capacity of mulberry fruit. Int. J. Food Sci. 49 (2), 399–409.

Jiang, C., Xu, J., Zhang, H., Zhang, X., Shi, J., Li, M., Ming, F., 2009. A cytosolic class I small heat shock protein, RcHSP17.8, of Rosa chinensis confers resistance to a variety of stresses to Escherichia coli, yeast and Arabidopsis thaliana. Plant Cell Environ 32, 1046–1059.

Joshi, M.M., Ibrahium, I.K.A., Hollis, J.P., 1975. Hydrogen sulphide: effects on the physiology of rice plants and relation to straighthead disease. Phytophatology 65, 1165–1170.

Khan, M.N., Mobin, M., Abbas, Z.K., Siddiqui, M.H., 2017. Nitric oxide induced synthesis of hydrogen sulfide alleviates osmotic stress in wheat seedlings through sustaining antioxidant enzymes, osmolyte accumulation and cysteine homeostasis. Nitric Oxide 68, 91–102.

Kolluru, G.K., Shen, X., Bir, S.C., Kevil, C.G., 2013. Hydrogen sulfide chemical biology: pathophysiological roles and detection. Nitric oxide 35, 5–20.

Kong, W.W., Zhang, L.P., Guo, K., Liu, Z.P., Yang, Z.M., 2010. Carbon monoxide improves adaptation of *Arabidopsis* to iron deficiency. Plant Biotec. J. 8 (1), 88–99.

Lai, D., Mao, Y., Zhou, H., Li, F., Wu, M., Zhang, J., et al., 2014. Endogenous hydrogen sulfide enhances salt tolerance by coupling the reestablishment of redox homeostasis and preventing salt-induced K + loss in seedlings of *Medicago sativa*. Plant Sci. 225, 117–129.

Leon, S., Touraine, B., Briat, J.F., Lobréaux, S., 2002. The AtNFS2 gene from *Arabidopsis thaliana* encodes a NifS-like plastidial cysteine desulphurase. Biochem. J. 366 (2), 557–564.

Li, Z.G., 2013. Hydrogen sulfide: a multifunctional gaseous molecule in plants. Russ. J. Plant Physiol. 60 (6), 733–740.

Li, Z.G., He, Q.Q., 2015. Hydrogen peroxide might be a downstream signal molecule of hydrogen sulfide in seed germination of mung bean (*Vigna radiata*). Biologia 70 (6), 753–759.

Li, D., Xiao, Z., Liu, L., Wang, J., Song, G., Bi, Y., 2010. Effects of exogenous hydrogen sulfide (H_2S) on the root tip and root border cells of *Pisum sativum*. Chin. Bull. Bot. 45, 354–362.

Li, L., Wang, Y., Shen, W., 2012. Roles of hydrogen sulfide and nitric oxide in the alleviation of cadmium-induced oxidative damage in alfalfa seed ling roots. Biometals 25, 617–631.

Li, H., Ding, Y., Shi, Y., Zhang, X., Zhang, S., Gong, Z., Yang, S., 2017. MPK3- and MPK6-Mediated ICE1 Phosphorylation Negatively Regulates ICE1 Stability and Freezing Tolerance in Arabidopsis. Dev. Cell 43, 630–642.

Li, Z.G., Yang, S.Z., Long, W.B., Yang, G.X., Shen, Z.Z., 2013. Hydrogen sulphide may be a novel downstream signal molecule in nitric oxide-induced heat tolerance of maize (*Zea mays* L.) seedlings. Plant Cell Environ. 36 (8), 1564–1572.

Li, Z.G., Min, X., Zhou, Z.H., 2016. Hydrogen sulfide: a signal molecule in plant cross-adaptation. Front. Plant Sci. 7, 1621.

Lin, M.J., Hsu, B.D., 2004. Photosynthetic plasticity of phalaenopsis in response to different light environments. J. Plant Physiol. 161 (11), 1259–1268.

Lin, Y.T., Li, M.Y., Cui, W.T., Lu, W., Shen, W.B., 2012. Haem oxygenase-1 is involved in hydrogen sulfide-induced cucumber adventitious root formation. J. Plant Growth Regul. 31 (4), 519–528.

Lin, X., Yang, R., Dou, Y., Zhang, W., Du, H., Zhu, L., et al., 2020. Transcriptome analysis reveals delaying of the ripening and cell wall degradation of kiwifruit by hydrogen sulfide. J Sci Food Agric. Available from: 10.1002/jsfa.10260.

Lisjak, M., Srivastava, N., Teklic, T., Civale, L., Lewandowski, K., Wilson, I., et al., 2010. A novel hydrogen sulfide donor causes stomatal opening and reduces nitric oxide accumulation. Plant Physiol. Biochem. 48 (12), 931–935.

Lisjak, M., et al., 2011a. Hydrogen sulfide effects on stomatal apertures. Plant Signal. Behav. 6, 1–3.

Lisjak, M., Teklić, T., Wilson, I.D., Wood, M., Whiteman, M., Hancock, J.T., 2011b. Hydrogen sulfide effects on stomatal apertures. Plant Signal. Behav. 6 (10), 1444–1446.

Lisjak, M., Teklic, T., Wilson, I.D., Whiteman, M., Hancock, J.T., 2013. Hydrogen sulfide: environmental factor or signaling molecule? Plant Cell Environ. 36 (9), 1607–1616.

Liu, J., Zhu, Y., Du, G., Zhou, J., Chen, J., 2013. Response of *Saccharomyces cerevisiae* to D-limonene-induced oxidative stress. Appl. Microbiol. Biotechnol. 97 (14), 6467–6475.

Liu, X., Chen, J., Wang, G.H., Wang, W.H., Shen, Z.J., Luo, M.R., et al., 2016. Hydrogen sulfide alleviates zinc toxicity by reducing zinc uptake and regulating genes expression of antioxidative enzymes and metallothioneins in roots of the cadmium/zinc hyperaccumulator *Solanum nigrum* L. Plant Soil 400 (1–2), 177–192.

Loppi, S., Cenni, E., Bussotti, F., Ferretti, M., 1998. Biomonitoring of geothermal air pollution by epiphytic lichens and forest trees. Chemosphere 36 (4–5), 1079–1082.

Ma, Q., Yang, J., 2018. Transcriptome profiling and identification of functional genes involved in H_2S response in grapevine tissue cultured plantlets. Genes. Genom. 40 (12), 1287–1300.

Ma, D., Ding, H., Wang, C., Qin, H., Han, Q., Hou, J., et al., 2016. Alleviation of drought stress by hydrogen sulfide is partially related to the abscisic acid signaling pathway in wheat. PLoS One 11 (9), 0163082.

Maxwell, D.P., Wang, Y., McIntosh, L., 1999. The alternative oxidase lowers mitochondrial reactive oxygen production in plant cells. Proc. Natl. Acad. Sci. USA 96, 8271–8276.

Miginiac-Maslow, M., Johansson, K., Ruelland, E., Issakidis-Bourguet, E., Schepens, I., Goyer, A., et al., 2000. Light-activation of NADP-malate dehydrogenase: a highly controlled process for an optimized function. Physiol. Plant 110 (3), 322–329.

Millenaar, F.F., Lambers, H., 2003. The alternative oxidase: in vivo regulation and function. Plant Biol. 5, 2–15.

Mir, B.A., Khan, T.A., Fariduddin, Q., 2015. 24-epibrassinolide and spermidine modulate photosynthesis and antioxidant systems in *Vigna radiata* under salt and zinc stress. Int. J. Adv. Res. 3, 592–608.

Mitsuhashi, H., Yamashita, S., Ikeuchi, H., Kuroiwa, T., Kaneko, Y., Kiromura, K., et al., 2006. Oxidative stress-dependent conversion of hydrogen sulfide to sulfite by activated neurophils. Shock 24, 529.

Mittler, R., 2002. Oxidative stress, antioxidants and stress tolerance. Trends Plant Sci. 7 (9), 405–410.

Morse, J.W., Millero, F.J., Cornwell, J.C., Rickard, D., 1987. The chemistry of the hydrogen sulfide and iron sulfide systems in natural waters. Earth Sci. Rev. 24 (1), 1–42.

Mostofa, M.G., Yoshida, N., Fujita, M., 2014. Spermidine pretreatment enhances heat tolerance in rice seedlings through modulating antioxidative and glyoxalase systems. Plant Growth Regul. 73 (1), 31–44.

Mostofa, M.G., Saegusa, D., Fujita, M., Tran, L.S.P., 2015. Hydrogen sulfide regulates salt tolerance in rice by maintaining Na + /K + balance, mineral homeostasis and oxidative metabolism under excessive salt stress. Front. Plant Sci. 6, 1055.

Mullineaux, P.M., Baker, N.R., 2010. Oxidative stress: antagonistic signaling for acclimation or cell death? Plant Physiol. 154 (2), 521–525.

Noctor, G., Reichheld, J.P., Foyer, C.H., 2018. ROS-related redox regulation and signaling in plants, Seminars in Cell & Developmental Biology, Vol. 80. Academic Press, pp. 3–12.

Palmer, R.M., Ferrige, A.G., Moncada, S., 1987. Nitric oxide release accounts for the biological activity of endothelium-derived relaxing factor. Nature 327 (6122), 524–526.

Pandey, A.K., Gautam, A., 2020. Stress responsive gene regulation in relation to hydrogen sulfide in plants under abiotic stress. Physiol. Plant . Available from: 10.1111/ppl.13064.

Papenbrock, J., Riemenschneider, A., Kamp, A., Schulz-Vogt, H.N., Schmidt, A., 2008. Characterisation of cysteine-degrading and H2S-releasing enzymes of higher plants—from the field to the test tube and back. Plant Biol. 9, 582–588.

Postgate, J., 1959. Sulphate reduction by bacteria. Ann. Rev. Microbiol. 13 (1), 505–520.
Riemenschneider, A., Wegele, R., Schmidt, A., Papenbrock, J., 2005. Isolation and characterization of ad-cysteine desulfhydrase protein from *Arabidopsis thaliana*. FEBS J. 272 (5), 1291–1304.
Romero, L.C., Aroca, M.Á., Laureano-Marín, A.M., Moreno, I., García, I., Gotor, C., 2014. Cysteine and cysteine-related signaling pathways in *Arabidopsis thaliana*. Mol. Plant 7, 264–276.
Sandhi, A., Greger, M., Landberg, T., et al., 2017. Arsenic concentrations in local aromatic and high-yielding hybrid rice cultivars and the potential health risk: a study in an arsenic hotspot. Environ. Monit. Assess. 189, 184.
Schafer, F.Q., Buettner, G.R., 2001. Redox environment of the cell as viewed through the redox state of the glutathione disulfide/glutathione couple. Free. Radic. Biol. Med. 30 (11), 1191–1212.
Schütz, B., De Kok, L.J., Rennenberg, H., 1991. Thiol accumulation and cysteine desulfhydrase activity in H2S-fumigated leaves and leaf homogenates of cucurbit plants. Plant Cell Physiol. 32 (5), 733–736.
Shafiq, S., Akram, N.A., Ashraf, M., 2015. Does exogenously applied trehalose alter oxidative defense system in the edible part of radish (*Raphanus sativus* L.) under water-deficit conditions? Sci. Hortic. 185, 68–75.
Shan, C.J., Zhang, S.L., Li, D.F., Zhao, Y.Z., Tian, X.L., Zhao, X.L., et al., 2011. Effects of exogenous hydrogen sulfide on the ascorbate and glutathione metabolism in wheat seedlings leaves under water stress. Acta Physiol. Plant 33 (6), 2533.
Sharma, J., Chakraverty, N. 2013. Mechanism of plant tolerance in response to heavy metals. In: Rout G., Das A. (eds) Molecular Stress Physiology of Plants. Springer, India. https://doi.org/10.1007/978-81-322-0807-5_12
Shi, H., Ye, T., Chan, Z., 2013. Exogenous application of hydrogen sulfide donor sodium hydrosulfide enhanced multiple abiotic stress tolerance in bermudagrass (*Cynodon dactylon* L. Pers.). Plant Physiol. Biochem. 71, 226–234.
Shi, H., Ye, T., Han, N., Bian, H., Liu, X., Chan, Z., 2015. Hydrogen sulfide regulates abiotic stress tolerance and biotic stress resistance in *Arabidopsis*. J. Integr. Plant Biol. 57 (7), 628–640. Available from: 10.1111/jipb.12302.
Singh, V.P., Singh, S., Kumar, J., Prasad, S.M., 2015. Hydrogen sulfide alleviates toxic effects of arsenate in pea seedlings through up-regulation ofthe ascorbate–glutathione cycle: possible involvement of nitricoxide. J. Plant Physiol. 181, 20–29. Available from: https://doi.org/10.1016/j.jplph.2015.03.015.
Sun, Y.D., Luo, W.R., 2014. Effects of exogenous hydrogen sulphide on seed germination and seedling growth of cucumber (*Cucumis sativus*) under sodium bicarbonate stress. Seed Sci. Tech. 42 (2), 126–131.
Sun, J., Wang, R., Zhang, X., Yu, Y., Zhao, R., Li, Z., et al., 2013. Hydrogen sulfide alleviates cadmium toxicity through regulations of cadmium transport across the plasma and vacuolar membranes in *Populus euphratica* cells. Plant Physiol. Biochem. 65, 67–74.
Takahashi, H., Kopriva, S., Giordano, M., Saito, K., Hell, R., 2011. Sulfur assimilation in photosynthetic organisms: molecular functions and regulations of transporters and assimilatory enzymes. Annu. Rev. Plant Biol. 62, 157–184.
Thompson, C.R., Kats, G., 1978. Effects of continuous hydrogen sulphide fumigation on crop and forest plants. Environ. Sci. Tech. 12, 550–553.
Tian, B., Zhang, Y., Jin, Z., Liu, Z., Pei, Y., 2017. Role of hydrogen sulfide in the methyl jasmonate response to cadmium stress in foxtail millet. Front. Biosci. 22, 530–538. Available from: 10.2741/4500.

Tretiach, M., Monaci, F.A.B.R.I.Z.I.O., Baruffo, L., Bargagli, R., 1999. Effetti dell'H2S sul contenuto di elementi in tracce nel lichene epifita *Parmelia sulcata*. Boll. Della Soc Adriatica di Sci 78, 365–387.

Trust, B.A., Fry, B., 1992. Stable sulphur isotopes in plants: a review. Plant Cell Environ. 15 (9), 1105–1110.

Wang, R., 2003. The gasotransmitter role of hydrogen sulfide. Antioxid. Redox Signal. 5 (4), 493–501.

Wang, R.U.I., 2002. Two's company, three's a crowd: can H_2S be the third endogenous gaseous transmitter? FASEB J. 16 (13), 1792–1798.

Wang, B.L., Shi, L., Li, Y.X., Zhang, W.H., 2010. Boron toxicity is alleviated by hydrogen sulfide in cucumber (*Cucumis sativus* L.) seedlings. Planta 231, 1301–1309.

Wang, Y., Li, L., Cui, W., Xu, S., Shen, W., Wang, R., 2012. Hydrogen sulfide enhances alfalfa (*Medicago sativa*) tolerance against salinity during seed germination by nitric oxide pathway. Plant Soil 351 (1-2), 107–119.

Whiteman, M., Armstrong, J.S., Chu, S.H., Jia-Ling, S., Wong, B.S., Cheung, N.S., et al., 2004. The novel neuromodulater hydrogen sulfide: an endogenous peroxynitrite 'scavenger'? J. Neurochem. 90, 765–768.

Wojtyla, Ł., Lechowska, K., Kubala, S., Garnczarska, M., 2016. Different modes of hydrogen peroxide action during seed germination. Front. Plant Sci. 7, 66.

Yamasaki, H., Cohen, M.F., 2016. Biological consilience of hydrogen sulfide and nitric oxide in plants: gases of primordial earth linking plant, microbial and animal physiologies. Nitric Oxide 55, 91–100.

Yang, G., Wu, L., Jiang, B., Yang, W., Qi, J., Cao, K., et al., 2008. H_2S as a physiologic vasorelaxant: hypertension in mice with deletion of cystathionine γ-lyase. Science 322 (5901), 587–590.

Yu, L.X., Zhang, C.J., Shang, H.Q., Wang, X.F., Min, W.E.I., Yang, F.J., et al., 2013. Exogenous hydrogen sulfide enhanced antioxidant capacity, amylase activities and salt tolerance of cucumber hypocotyls and radicles. J. Integr. Agr. 12 (3), 445–456.

Zhang, L., De Schryver, P., De Gusseme, B., De Muynck, W., Boon, N., Verstraete, W., 2008a. Chemical and biological technologies for hydrogen sulfide emission control in sewer systems: a review. Water Res. 42 (1-2), 1–12.

Zhang, W.F., Zhang, F., Raziuddin, R., Gong, H.J., Yang, Z.M., Lu, L., et al., 2008b. Effects of 5-amino levulinic acid on oil seed rape seedling growth under herbicide toxicity stress. J. Plant Growth Regul. 27, 159–169.

Zhang, H., et al., 2009a. Hydrogen sulfide promotes root organogenesis in *Ipomoea batatas*, *Salix matsudana* and *Glycine max*. J. Integr. Plant Biol. 51, 1086–1094.

Zhang, H., Tang, J., Liu, X.P., Wang, Y., Yu, W., Peng, W.Y., et al., 2009b. Hydrogen sulfide promotes root organogenesis in *Ipomoea batatas, Salix matsudana* and *Glycine max*. J. Integr. Plant Biol. 51, 1086–1094.

Zhang, H., Ye, Y.K., Wang, S.H., Luo, J.P., Tang, J., Ma, D.F., 2009c. Hydrogen sulfide counteracts chlorophyll loss in sweet potato seedling leaves and alleviates oxidative damage against osmotic stress. Plant Growth Regul. 58 (3), 243–250.

Zhang, H., Tan, Z.Q., Hu, L.Y., Wang, S.H., Luo, J.P., Jones, R.L., 2010a. Hydrogen sulfide alleviates aluminium toxicity in germinating wheat seedlings. J. Integr. Plant Biol. 52, 556–567.

Zhang, H., Jiao, H., Jiang, C.X., Wang, S.H., Wei, Z.J., Luo, J.P., et al., 2010b. Hydrogen sulfide protects soybean seedlings against drought-induced oxidative stress. Acta Physiol. Plant 32 (5), 849–857.

Zhang, H., Hu, L.Y., Li, P., Hu, K.D., Jiang, C.X., Luo, J.P., 2010c. Hydrogen sulfide alleviated chromium toxicity in wheat. Biol. Plant (Prague) 54, 743–747.

Zhang, L., Pei, Y., Wang, H., Jin, Z., Liu, Z., Qiao, Z., et al., 2015. Hydrogen sulfide alleviates cadmium-induced cell death through restraining ROS accumulation in roots of Brassica rapa L.s sp. pekinensis. Oxid. Med. Cell Longev 1–11.

Zhang, H., Hu, S.L., Zhang, Z.J., Hu, L.Y., Jiang, C.X., Wei, Z.J., et al., 2011. Hydrogen sulfide acts as a regulator of flower senescence in plants. Postharv. Biol. Tech. 60 (3), 251–257.

Index

Note: Page numbers followed by "*f*" and "*t*" refer to figures and tables, respectively.

A

AAC synthase, 27–28
Abiotic stress(es), 1, 55–56, 104, 113, 156–161
  ameliorative role of phytohormones in plants under, 284–287
  crosstalk of H_2S with phytohormones under, 287–291
  H_2S
    in abiotic stress tolerance, 118–124
    in ionic and osmotic homeostasis in plants, 66–68
    metabolism, 114–116, 115*f*
    and plant growth regulators, 124–125
    in plants, 270–283
    and ROS crosstalk during abiotic stress in plants, 205–208
  heat stress and chilling, 157–161
  heavy metal stress, 157
  interaction of H_2S and physiological responses in plants, 68–69
  protein persulfidation induced by H_2S due to oxidative stress, 116–118
  salt and drought stress, 157
  signaling molecules in plants during, 305–309
ABREBs/ABA binding factors (ABFs), 285–286
Abscisic acid (ABA), 10–11, 90–91, 105, 120, 134, 155, 232–234, 257, 267–268
  crosstalk between hydrogen sulfide and, 246–247
  H_2S and abscisic acid interaction, 290
Abscission, 215–217
ACC oxidase, 27–28
Acid invertase (AI), 41–42
Actin-binding proteins (ABPs), 107–108, 288
Actinidia deliciosa. *See* Kiwifruit (*Actinidia deliciosa*)
Active oxygen species (AOS), 259–260
Adenosine 5'-phosphosulfate (APS), 88–90, 114–115
Adenosine 5'-phosphosulfate reductase (APR), 114–116, 204, 351
Adenosine triphosphate (ATP), 3–4, 58–59
Adonis amurensis, 230
Adonis pseudoamurensis, 230
Agency for Toxic Substances and Disease Registry (ATSDR), 276–281
Agriculture, 303
Alternative oxidase (AOX), 347
Alternative pathway (AP), 347
Aluminum (Al), 87–88, 319, 341
1-aminocyclopropane-1-carboxylate synthase (ACS1), 37–39, 289
1-aminocyclopropane-1-carboxylic acid oxidase (ACC), 37–39
1-aminocyclopropane-1-carboxylic acid oxidase (ACO_1), 37–39, 137
Ammonia (NH_3), 88–90, 115–116
Antioxidant system, 65–66, 233
Antioxidative defense mechanism, H_2S and, 206
Antioxidative enzymes, 203, 261
Antioxidative system, impact of H_2S on, 218–219
APETALA2-type transcription factor (AP2-type transcription factor), 284–285
Arabidopsis, 11–12, 107–108
  A. thaliana, 88–90, 119–120, 153, 257–258
  H_2S, 56–58
Arabidopsis K^+ Transporter 1 (AKT1), 311
Arabidopsis response regulators (ARRs), 316–317
Arsenic (As), 87–88, 348
AsA-glutathione (AsA-GSH), 11–12

Ascorbate peroxidase (APX), 5–6, 29–30, 92, 94–95, 116–117, 135, 153, 215, 231, 275–276
Ascorbate to dehydroascorbate ratio (AsA/DHA), 249
Ascorbic acid (AsA), 5–6, 56–58, 95, 120, 231
Aspergillus niger, 35–36
ATP sulfurylase (ATPS), 88–90, 351
ATP-binding cassette transporters (ABC transporters), 257
  multidrug resistant protein 5, 137
Autophagy, 207–208
Autophagy-related ubiquitin-like 18a (Atg18a), 92, 118
Auxin, 107–108, 284
  crosstalk between hydrogen sulfide and, 248–249
  H_2S and auxin interaction, 288–289
Auxin Signaling F Box Protein1 (AFB1), 248–249

B

Banana, 249–250
Basic-leucine zipper protein (bZIP), 306–307
β-cyanoalanine synthase (CAS), 58–59, 88–90, 160–161
Betaine aldehyde dehydrogenase (BADH), 121–122
Biotic stress, 104, 161. *See also* Abiotic stress (es)
Bipolaris sorokiniana, 245
Blumeria graminis, 245
Brassica napus, 153
Brassica oleracea var. sabellica. *See* Green curly kale (*Brassica oleracea var. sabellica*)
Brassinosteroids (BRs), 284

C

C-repeat binding factor (CBF), 122–123, 246–247
Ca^{2+}-dependent protein kinases (CDPKs), 305
  CDPK3, 185–186
Ca^{2+}/CaM-regulated receptor-like kinases 1 (CRLK1), 316–317
Cadmium (Cd), 87–88, 319
Calcineurin B-like proteins (CBLs), 305
Calcium, 305
Calcium signaling (Ca^{2+} signaling), 186, 304
  Ca^{2+}/CaM signaling regulates H_2S production and homeostasis, 185–186
  during different types of abiotic stress tolerance in plants, 309–328
  during temperature stress, 311–319
Calcium-dependent protein kinases (CDPKs), 59–60, 171–172
Calmodulin-like proteins (CMLs), 305, 315–316
Calmodulins (CaMs), 305, 313–316
Calvin–Benson cycle, 346
Carbon monoxide (CO), 55–56, 103, 105–107, 114, 140–142, 213–214, 243, 304
Catalase (CAT), 5–6, 29–30, 56–58, 92, 120, 135, 153–154, 203, 215, 231
Catharanthus roseus, 230
Cell membrane integrity, 35
Cell wall-associated proteins, 105
Cellular reactions, 105
Chilling, 157–161
  chilling-induced effects on plants, 3–7
    chilling stress and postharvest quality of fresh fruits and vegetables, 7
    gaseous exchange and photosynthesis, 4–5
    growth and development, 4
    oxidative stress and osmotic balance, 5–6
    seed germination, 3–4
    water and nutrient relations, 5
    yield attributes, 6
  stress, 2
    in plants, 3
    and postharvest quality of fresh fruits and vegetables, 7
Chilling injury (CI), 7, 39–40
Chlorophyll, 217
Chloroplasts, 123–124, 181–182, 217
Chromium (Cr), 87–88, 319, 348
Citrus aurantium plants, 120
Climacteric fruits, 27–28
Cold responsive 15A (COR15A), 122–123
Cold responsive 15B (COR15B), 122–123
Cold stress, 104, 311–312
  protective function of hydrogen sulfide under, 66
Colletotrichum musae, 43
Copper (Cu), 87–88, 319, 348
Crop plants, 219
Crude oil processing industries, 134

Cucurbitacin C (CuC), 122–123
Cyanoalanine synthase (CAS), 11–12, 114–115, 138–139, 150–151, 171–172, 203, 205, 268, 273–274
Cyclic guanosine monophosphate (cGMP), 188
Cynodon dactylon, 56–58, 61–62
CysSSCys metabolism, 116
Cystathionine beta synthase (CBS), 116
Cysteine (Cys), 91–92, 95–96, 150–151, 171–172
Cysteine synthase (CS), 58–59, 114–115, 203, 268, 273–274
Cysteine-persulfide (Cys-SSH), 105–107
Cytochrome c oxidase (CcO), 152
Cytochrome pathway (CP), 347
Cytokinins (CK), 284
Cytoskeleton, 190

D

D-cysteine desulfhydrase (DCD), 11–12, 29–30, 88–90, 114–115, 150–151, 171–172, 203, 268
D-cysteine desulfhydrase 1 (DCD1), 246–247
D-cysteine desulfhydrase 2 (DCD2), 246–247
D-lactate dehydrogenase (D-LDH), 306
D/L-cysteine desulfhydrases (D/L-CDs), 351
Defense mechanism, 202–203
Dehydration-responsive element binding-type protein (DREB protein), 246–247, 306–307
Dehydroascorbate (DHA), 291
Dehydroascorbate reductase (DHAR), 46, 234–235, 270–273
Dendrobium officinale, 232
  H_2S, 215–217
Desulfovibriao, 342
Dihydrogen sulfide, 114
Dinitrososulfites, 135–136
Diosphyros seedlings, 231
Disease-related genes, 243–244
Disulfide-S-oxides (RS(O) 2SR), 55–56
DNA cytosine methylation, 176
Drought, 55–56, 229
Drought stress, 157, 322–325
  ameliorative roles of H_2S during, 119–120
  Ca^{2+} signaling in plants under, 323–325
  H_2S
    effect of H_2S on plants under drought conditions, 322–323
    H_2S-induced adaptation in plants against, 324*t*
    impact on antioxidative system, 234–235, 235*t*
    impact on oxidative stress, 234
    impact on photosynthesis, 233–234
    role in plant biology, 231–233
    role in plant productivity, 235–236
    role in plants, 233–235
  impact on
    growth and photosynthesis, 230–231
    oxidative stress, 231
    plants, 230–231
Drought/water/osmotic stress, 275–276

E

Effector-triggered immunity (ETI), 244–245
Electrolyte leakage (EL), 121–122
Endogenous H_2S, 104
Enhanced disease susceptibility genes (EDS genes), 248
Environmental sources, 342
Environmental stress
  endogenous production of H_2S under, 203
  H_2S and, 104–105
Environmental stressors, 303
Enzymatic reactions, 151–152
Epigenetic(s), 174–176
  DNA cytosine methylation, 176
  feedback loop, 176–179
  H_2S-mediated changes, 175*t*
  histone modification, 174–175
  noncoding ribonucleic acid regulation, 176
Escherichia coli, 88–90
Ethylene, 27–28, 34, 40–41
  crosstalk between hydrogen sulfide and, 249–250
  H_2S and ethylene interaction, 289
Ethylene diamine tetra acetic acid (EDTA), 313–314
Ethylene receptor (ETR), 289
Ethylene response sensor1;2 (ERS1, ERS2), 289
Exogenous H_2S, 2, 56–58, 104

F

F-actin, 107–108
Ferredoxin–thioredoxin reductase (FTR), 345–346
Free radicals, 55–56
Fruit ripening, 156
  H_2S, NO, and melatonin in regulation of, 28–29

Fruit ripening (*Continued*)
crosstalk between H_2S, melatonin and NO, 29
crosstalk of H_2S with NO during ripening, 28
crosstalk of melatonin with NO, 28–29
H_2S in, 29–30
color changes, 30–31
delaying ripening and conserving quality during postharvest, 30–36
H_2S, ethylene production, and respiratory metabolism, 34–35
H_2S, membrane integrity, and lipid peroxidation, 35
and postharvest chilling injury, 31
and postharvest diseases, 35–36
and postharvest fruit softening, 36
and postharvest oxidative stress, 31–34
mechanism, 27–28
melatonin and, 42–46
action mechanism, 46
chilling injury, and disease reduction, 43–45
postharvest fruit quality, 42–43
postharvest senescence, 43
NO and, 36–42
action mechanism, 40–41
biochemical quality conservation, 39
chilling injury, 39–40
intracellular energy, and sugar metabolism, 41–42
postharvest diseases, 40
postharvest fruit quality, 37
postharvest senescence reduction, 37–39
Fusarium culmorum, 245
Fusarium head blight disease (FHB disease), 245

G

Gas exchange, 8–9
Gaseous exchange, 4–5
Gasotransmitters, 55–56, 149–150, 213–214
Gene regulation by H_2S in plants, 173
Ca^{2+}, Ca^{2+}/CaM, H_2S, and, 185–186
crosstalk between
H_2O_2 and H_2S, 189
NO and H_2S, 187–188
H_2S
cytoskeleton, and gene regulation, 190
and epigenetics, 174–176
H_2S-mediated mechanisms of transcription reprogramming, 190–192
hormones, epigenetics, and transcriptional landscape, 179–181
MAPKs, signal transduction, and transcriptional control, 186–187
and microRNA, and epigenetic feedback loop, 176–179
mitochondria, chloroplast, epigenetics, and gene regulation, 181–182
and redox-based gene regulation, 189–190
and transcription factors, 182–185
multilayered transcriptional regulation, 173–174
Gibberellic acid (GA), 105–107, 125, 140–142
crosstalk between hydrogen sulfide and, 250
H_2S and, 290–291
Gibberellins, 90–91, 284
Glutamate–cysteine ligase (GCL), 61, 206
Glutamine synthetase (GS), 92
Glutamylcysteine synthetase (*GSH*1), 95
Glutathione (GSH), 56–58, 95, 114, 171–172, 204, 218–219, 231, 261, 342
Glutathione 5-transferase φ-2 (STF2), 286
Glutathione peroxidase (GPX), 11–12, 231
Glutathione persulfide (GSSH), 116
Glutathione reductase (GR), 11–12, 31–34, 95, 120, 153, 270–273
Glutathione synthase (GS), 61, 206
Glutathione-S-transferase (GST), 215, 231
Glyceraldehyde 3-phosphate dehydrogenase (GAPDH), 92, 153–154
Glycine betaine (GB), 317–319
Glycine max, 120, 232, 257–258
Glyoxalase I (GLYI), 306
Glyoxalase II (GLYII), 306
Glyoxalase III (GLYIII), 306
Green curly kale (*Brassica oleracea var. sabellica*), 345
Guaiacol peroxidase (GPX), 94–95, 215

H

Haloxylon aphyllum, 230–231
Haloxylon persicum, 230–231
Heat shock protein (HSP), 14–15, 189, 311–312, 354

Heat stress, 157–161, 276. *See also* Drought stress
in plants, 7
tolerance, 1–2
Heat stress pretreatment (HSP), 65–66
Heavy metals (HM), 55–56, 87–88, 105, 315
H_2S signaling triggered by heavy metal stress, 92–93
stress, 157
ameliorative roles of H_2S during, 123–124
H_2S and Ca^{2+} signaling during, 319–322
H_2S-induced adaptation in plants against, 320*t*
Heme oxygenase-1 (HO-1), 290–291
Heme oxygenase-1/carbon monoxide (HO-1/CO), 91, 246
High-temperature
high-temperature-induced effects on plants, 7–10
gas exchange and photosynthesis, 8–9
nutrient and water relations, 9
osmotic balance and oxidative stress, 9
pollen production and pollination, 9–10
seed germination and plant growth, 7–8
yield attributes, 10
in plants, 7
Histone deacetylases (HDACs), 174
Histone modification, 174–175
Hydrogen gas (H_2), 55–56
Hydrogen peroxide (H_2O_2), 55–56, 88, 92, 104–107, 124, 152–153, 171–172, 270, 304
Hydrogen sulfide (H_2S), 2, 25, 55–56, 88, 103, 113, 133–134, 149, 171–172, 201–202, 213–214, 243, 257, 267–268, 303, 341
in acceleration of antioxidative mechanism in plants during abiotic stress condition, 61–66
biological significance, 139–140
bioprotective nature of hydrogen sulfide on plant thermotolerance, 65–66
biosynthesis in plant cells under adverse environmental conditions, 58–59
biotechnological application, 353
and chilling stress in plants, 11–14
crosstalk of H_2S with
NO during ripening, 28
other molecules during temperature stress, 10–11
with other signals, 140–142
during different types of abiotic stress tolerance in plants, 309–328
effects on
antioxidant and antioxidant enzyme activity, 350–351
and enzymatic homeostasis in plants, 349–351
gas exchange parameter, 345–347
membrane degradation, 349
mineral nutrition, 347–348
photosynthesis, 345–346
physiological homeostasis, 345–348
plant growth and development, 344–345
reactive oxygen species defense and homeostasis, 349–350
relative water content, 347
respiration, 347
stomatal conductance, 346–347
as emergent signaling molecule in plants, 59–61
and environmental stress, 104–105
and fruit ripening, 29–36
generation of H_2S in plants, 268–270
H_2S in plants under abiotic stress, 270–283
H_2S-induced high-temperature/heat tolerance, 15–17, 16*t*
H_2S-induced low-temperature/chilling tolerance, 12–14, 13*t*
H_2S-induced protein persulfidation under stress, 355
H_2S-induced regulation of temperature stress in plants, 10
H_2S-mediated regulation of gene and protein expression, 353–355
and high-temperature stress in plants, 14–17
homeostasis during metal toxicity, 348–349
improves salinity tolerance in plants, 62–63
interaction with other signaling molecule to enhance antioxidative defense mechanism, 61
in ionic and osmotic homeostasis in plants under various abiotic stresses, 66–68
metabolism, 114–116, 115*f*
in plants, 204
in mitigation of drought stress in plants, 62
and molecular homeostasis in plants, 351–355
molecular mechanism behind H_2S homeostasis, 351–352

Hydrogen sulfide (H_2S) (*Continued*)
in plants, 88–91, 202–205, 343–344
as signaling molecule, 344
as substrate, 344
as toxin, 343
and production of lateral roots, 107–108
protective function
of hydrogen sulfide under cold stress, 66
in response to heavy metal and metalloid toxicity, 63–65
on quality of different fruits during storage, 32*t*
regulation of chilling stress responsive genes, 11–12
regulation of high-temperature responsive genes, 14–15
signaling, 105–107
source of H_2S in plants, 342–343
synthesis in plants, 137–139
during temperature stress, 3–17, 311–319
Hydroxyl radicals (OH–), 55–56, 104
Hypotaurine (HT), 249

I

I-aminocyclopropane-1 carboxylate oxidase 1 (ACO1), 289
Indole acetic acid (IAA), 107–108, 140–142, 248, 289
Inducer of C-repeat binding factor expression 1 (ICE1), 122–123
Induction of CBF expression levels (ICE1), 284–285
Intracellular sources, 342–343
Ion homeostasis, 59–60
Ipomoea aquatica, 134–135
Ipomoea batatas, 232, 257–258
Iron-regulated transporter (*IRT*), 93–94
Iron-regulated transporter like protein (*ZIP*), 93–94

J

Jasmonic acid (JA), 140–142, 249, 284
crosstalk between hydrogen sulfide and, 249
H_2S and, 291
Jatropha curcas, 140–142

K

Kandelia obovata, 135
Kiwifruit (*Actinidia deliciosa*), 232

L

L-cysteine (L-Cys), 150–151
L-cysteine desulfhydrase (L-DES), 88–90, 122, 150–151, 160, 246–247, 260–261
L-DES1, 203
L-cysteine desulfhydrase (L-DES), 203
L-tryptophan, 42
Lamiophlomis rotata, 11–12
Lateral roots production, 107–108
L–cysteine desulfhydrase (LCD), 10–11, 29–30, 105, 114–115, 138–139, 171–172, 185–186, 203, 205, 246–247, 268–269, 313–314
Lead (Pb), 87–88, 319
Leaf senescence, 156
Lipoxygenase (LOX), 29–30, 94–95, 234–235

M

Malondialdehyde (MDA), 11–12, 31, 94–95, 114, 135, 215, 229–230, 260–261, 270, 313–314
MAPK kinase (MAPKK), 354
Medicago sativa seedlings, 120–121
Melatonin, 26–27
crosstalk of melatonin with NO, 28–29
and fruit ripening, 42–46
on quality of different fruits during storage, 44*t*
Metal tolerance protein (MTP), 291
Metal(s), 87–88
accumulation, 87–88
nitrosylation, 152
stress, 87–88, 276–283
H_2S in plants, 88–91
H_2S signaling triggered by heavy metal stress, 92–93
H_2S-induced metal stress tolerance, 93–96
H_2S-induced protein persulfidation during oxidative stress, 91–92
Metallothioneins (MTs), 87–88, 319
Methane (CH_4), 55–56
Methionine (Met), 171–172
Methionine adenosyl transferase1 (MAT1), 40–41
Methyl jasmonate (MeJA), 93–94, 124, 284–285, 306–307
Methyl salicylate, 36–37
Methylglyoxal (MG), 122, 303, 306

MicroRNAs (miRNAs), 93, 96, 176–179, 319–321
control H_2S biosynthesis, 178
epigenetic machinery controls gene transcriptions of, 177–178
H_2S modulates, 179
histone and DNA modifications, 177
MIR167, 248
MIR393, 248
MIR396, 248
MIR398, 248
Mineral nutrition, H_2S effects on, 347–348
Mitochondria, 181–182
Mitogen-activated protein kinase (MAPK), 11–12, 104, 171–172, 186–188, 286, 306–307, 316–317, 354
MPK6, 289
Mitogen-activated protein kinase (MEK1/2), 249, 291
Monodehydroascorbate reductase (MDHAR), 153, 275–276
Morpholin-4-ium 4-methoxypheny l (morpholino) phosphinodithioate (GYY4137), 114, 133–134
Multilayered transcriptional regulation, 173–174

N

Naphthalene acetic acid (NAA), 289
Natural resistance associated macrophage protein (NRAMP), 93–94, 291
Nematodes, 244
Neutral invertase (NI), 41–42
Next-generation sequencing (NGS), 173–174
Nicotiana tabacum, 122
Nicotinamide adenine dinucleotide (NAD), 182
Nitrate reductase (NR), 124, 151–152
Nitric oxide (NO), 10–11, 25–26, 55–56, 88, 103, 105–107, 114, 125, 135, 140, 149, 171–172, 213–214, 243, 304
crosstalk between H_2S and NO in plants under stress conditions, 156–161
crosstalk of melatonin with, 28–29
exogenous application, 25–26
and fruit ripening, 36–42
and H_2S and role in plants at physiological conditions, 154–156
impact on quality of different fruits during storage, 38*t*
signaling pathways, 152–154
reaction with metals, 152
reaction with thiols, 153–154
reactions with oxidants, 152–153
similarities and differences between H_2S and, 149–154
metabolic processes, 150–152
physicochemical aspects, 149–150
Nitrogen oxide, 55–56
Nitropersulfides, 135–136
Nonclimacteric fruits, 27–28
Noncoding ribonucleic acid regulation, 176
Nonenzymatic reactions, 151–152
Nutrient and water relations, 9

O

O-acetylserine (OAS), 88–90, 115–116, 138–139, 150–151, 206
O-acetylserine thiol lyase (OAS-TL), 11–12, 88–90, 138–139, 150–151, 160–161, 203
Ocimum basilicum. See Sweet Italian basil (*Ocimum basilicum*)
Osmolytes, 56–58
Osmoregulation, 6
Osmotic adjustment, 9
Osmotic balance, 5–6, 9
Osmotic modification, 6, 9
Osmotic stress, 275–276
Oxidative burst, 309
Oxidative stress, 5–6, 9, 56–58, 218
H_2S-induced protein persulfidation during, 91–92
protein persulfidation induced by H_2S due to, 116–118
impact of salinity on, 215

P

Pathogen
effect of H_2S on plant morphology, 245–246
H_2S-plant hormone crosstalk under pathogen attack, 246–250
impact on plant physiology, 244–245
Pathogenesis-related proteins (PR proteins), 247
Pathological process, 304
Pattern-triggered immunity (PTI), 245
Pectin methyl esterase (PME), 28
Penicillium digitatum, 46
Penicillium italicum, 35–36
Peroxidase (POD), 30–31, 120, 135, 231

Peroxynitrite (ONOO−), 55–56, 153
Persulfidation, 91–92, 172–173
Phenylalanine ammonia lyase (PAL), 11–12
Phlox paniculata, 134
Phosphoenolpyruvate carboxylase (PEPC), 217–218
Phospholipase C and D, 354
Photoprotective responses, 259
Photosynthesis, 4–5, 8–9, 217–218, 313–314
  PS-II, 8–9
  impact of salinity on, 215
Physiological drought, 62
Phytoalexin deficient genes (PAD genes), 248
Phytochelatins (PCs), 87–88, 123–124, 319, 348–349
Phytochelatins synthase 1 (*PCS*1), 123–124
Phytochelatins synthase 2 (*PCS*2), 123–124
Phytohormones
  ameliorative role in plants under abiotic stress, 284–287
  crosstalk
    between H_2S and, 291
    of H_2S with phytohormones under abiotic stresses, 287–291
  H_2S contributes to modulation of biosynthesis, signaling, and/or functions of, 180–181
  in H_2S production and homeostasis, 179–180
Plant growth regulators (PGRs), 124–125, 287
Plants, 1
  generation of H_2S in, 202–203
  growth, 1–2, 7–8
  H_2S
    metabolism, 204
    in plant biology, 215–217
  hormones, 245
  signaling, 105–107
Polar auxin transport (PAT), 288
Pollen production, 9–10
Pollination, 9–10
Polygalacturonase (PG), 25
Polyphenol oxidase (PPO), 29–30, 231
Polysulfides, 135–136
Populus euphratica, 250
Postharvest
  chilling injury, 31
  oxidative stress, 31–34
  senescence reduction, 37–39
Post-translational modifications (PTMs), 91–92, 120, 204
Proline, 9, 104–105
Proline dehydrogenase (ProDH), 313–314
Proline oxidase (POX), 231
Propargylglycine (PAG), 313–314
Protein arginine methyltransferases (PRMTs), 308–309
Protein modification (PTM), 353
Protein persulfidation induced by H_2S due to oxidative stress, 116–118
Protein–protein interaction (P-P interaction), 354
Proteins, 305
Pseudomonas putida, 88–90
Puccinia graminis, 245
Puccinia striiformis, 245
Putrescine (Put), 260–261
Pyridoxal-5'-phosphate (PLP), 351
Pyrroline-5-carboxylate synthetase (P5CS), 313–314

R

RAB18. *See* Responsive to ABA (RAB18)
Radiation
  antistress role of H_2S, 261–262
  effect, 258–260
  H_2S helps in amelioration of radiation stress by interacting with other compounds, 260–261
  H_2S triggered by UV radiation, 261
  plant's response against, 260
  stress, 260–261
Rat brain (RAB), 246–247
Reactive carbonyl species, 306–307
Reactive carboxyl species (RCS), 303
Reactive nitrogen species (RNS), 11–12, 28–29, 92, 149, 303, 308–309
Reactive oxygen species (ROS), 3–6, 28, 88, 113, 135, 152–153, 171–172, 202, 215, 229, 244–245, 303, 307–308
  cooperative interactions of H_2S and reactive oxygen species in plants, 208
  H_2S and, 207–208
Reactive oxygen/nitrogen species (ROS/RNS), 171–172
Reactive sulfur species (RSS), 133–134, 149, 303, 305–306
Receptor-like kinase (RLK), 306–307
Redox-based gene regulation, 189–190
Relative water content (RWC), 119–120, 347
Responsive to ABA (RAB18), 306–307

Responsive to dehydration (RD29B), 306–307
Responsive to dehydration/dessication (RD), 246–247
Rhizoctonia solani, 244–245
Ribulose 1, 5-bisphosphate carboxylase-oxygenase (Rubisco), 68–69, 217–218, 346
RNA-mediated interference (RNAi), 176–177
Root organogenesis, 155

S

S-adenosyl methionine (SAM), 40–41, 176, 182
S-glutathionylation (–SSG), 207
S-nitrosoglutathione reductase (GSNOR), 66, 160–161
S-nitrosothiols (SNOs), 135, 154
S-nitrosylation (–SNO), 154, 207
S-sulfenylation (–SHO), 207
S-sulfhydration, 91–92, 95–96, 116, 136–137
SA. *See* Salicylic acid (SA)
Salicylic acid (SA), 104–105, 122, 138–139, 247, 267–268
  crosstalk between hydrogen sulfide and, 247–248
Salinity, 55–56, 270
Salinity stress, 213
  ameliorative roles of H_2S during, 120–121
  H_2S
    in plant biology, 215–217
    in plant productivity, 219
    in plants, 217–219
  impact on plants, 214–215
    growth, 214
    oxidative stress, 215
    photosynthesis, 215
Salix matsudana, 232, 257–258
Salt overly sensitive 1 (SOS1), 310
Salt stress, 157, 213, 270–275, 309–311
  Ca^{2+} signaling during, 310–311
  effect of H_2S in plants under, 271*t*
  H_2S signaling during, 309–310
  effect of salt stress on antioxidative system, 216*t*
Secondary messengers, 303
Seed dormancy/germination, 155
Seed germination, 3–4, 7–8
Senescence, 37–39
Serine acetyltransferase (SAT), 88–90
Signal(ing)
  of hydrogen sulfide, 105–107
  molecule, 10–11, 88, 135–137, 231–232, 303
    importance of H_2S in plants as, 204–205
  pathways, 206
  transduction, 56–58, 186–187
Small interfering RNAs (siRNAs), 176
Sodium hydrosulfide (NaHS), 29–30, 62, 91, 114, 121–122, 124, 133–135, 155, 288, 353
Sodium nitroprusside (SNP), 36–37, 140–142, 157, 160
Sodium polysulfides (Na_2Sn), 353
Softening of fruits, 27
Soil salinity, 62–63
Solanum lycopersicum, 125, 289
Solar UV radiation, 258
Sophora davidii, 230–231
Spinacia oleracea, 135, 232
Stomatal closure, 155–156, 249–250
Storage at low temperatures, 161
Stress, 229
  tolerance, 88–90
Sucrose phosphate synthase (SPS), 41–42
Sucrose synthase (SS), 41–42
Sulfane, 114, 150
Sulfane. *See* Hydrogen sulfide (H_2S)
Sulfate transporter 3;1 (SULTR3;1), 290
Sulfate transporters (SULTRs), 351
Sulfenic acid (RSOH), 55–56, 92, 118
Sulfide, 135–137
Sulfinic acid (RSO_2H), 92, 118
Sulfite reductase (SiR), 88–90, 114–115, 150–151, 171–172, 203–205, 268, 351
Sulfonic acid (RSO_3H), 92, 118
Sulfur, 58–59, 152, 204, 342
  metabolism, 88–90
  species, 103
Sulfur-comprising defense compounds (SDCs), 204
Superoxide (O_2–), 55–56
Superoxide dismutase (SOD), 5–6, 31–34, 56–58, 94–95, 120, 135, 215, 231
Superoxide ions (O_2^-), 104
Sweet Italian basil (*Ocimum basilicum*), 345

T

TaEXPB23, 284–285
Temperature stress, 1, 3–17
  ameliorative roles of H_2S during, 121–123
  chilling stress in plants, 3
  chilling-induced effects on plants, 3–7
  crosstalk of H_2S with other molecules during temperature stress, 10–11
  H_2S
    and Ca^{2+} signaling during, 311–319
    and chilling stress in plants, 11–14
    H_2S-induced regulation of temperature stress in plants, 10
  high-temperature or heat stress in plants, 7
  high-temperature-induced effects on plants, 7–10
  interplay between H_2S and calcium signaling during plant response to, 317–319
Thiyl radicals (RS), 55–56
Tolerance mechanism, 113
Total soluble solids (TSS), 39
Toxic gas, 114
Toxic metals, 123–124
Toxins, 245
Transcription factors, 182–185, 183*t*
Transcription reprogramming, H_2S-mediated mechanisms of, 190–192
Transcriptional control, 186–187
Transpiration, 62
Transport Inhibitor Response1 (*TIR*1), 248–249
Trehalose, 14–15
Triticum aestivum, 61–62

U

Ultraviolet (UV)
  absorbing compounds, 258–259
  H_2S triggered by UV radiation, 261
  UV B radiation, 258–260
  UV-A, 259

V

Vegetative growth, 1
Vicia faba, 10–11, 140–142
Vitis vinifera, 11–12, 66, 314–315
VvCBF3 gene, 11–12
  VvCBF3I gene, 122–123
VvICE1 gene, 11–12, 122–123

W

Water
  deficit conditions, 234–235
  and nutrient relations, 5
  stress, 275–276
    H_2S under, 277*t*
Wheat, 245
Wild-type plants (WT plants), 247
WRKY transcription factors, 286

Y

Yield attributes, 6, 10

Z

Zea mays, 65–66
Zinc (Zn), 87–88, 319, 348
Zinc-regulated transporter (*ZRT*), 93–94

Kolibri 360 GmbH
Ferdinand-Jühlke-Straße 7
99095 Erfurt, Deutschland
produktsicherheit@kolibri360.de

Ilán Bizberg

Four Worlds of the Welfare State in Latin America

palgrave
macmillan

Ilán Bizberg
CEI
College of Mexico
Distrito Federal, Mexico

ISBN 978-3-031-44419-7 ISBN 978-3-031-44420-3 (eBook)
https://doi.org/10.1007/978-3-031-44420-3

Cover illustration: Maram_shutterstock.com

This Palgrave Macmillan imprint is published by the registered company Springer Nature Switzerland AG
The registered company address is: Gewerbestrasse 11, 6330 Cham, Switzerland

Paper in this product is recyclable.

Para Andrea e Iván por las alegrías e inquietudes
que han llenado mi vida.
Para Leticia que me ha hecho feliz.
A la memoria de mi padre y de mi madre.

Acknowledgments

I am extremely grateful to many people and institutions without whom this book would have never seen the light. First, thanks to the very generous institution where I have been able to work with total liberty and support for so many years: El Colegio de Mexico, and to all my colleagues at the Center for International Studies. I would like to recognize the CONACYT for its very generous financial support under the program SEP-Conacyt (project 220018), which has allowed me to travel to most of the countries of Latin America that I analyze, as well as to several international conferences where I have been able to present my ideas and get valuable feedback, and learn about other countries outside the continent from other researchers. I want to thank the Humboldt Foundation for the Research Award I received in 2020, in the middle of the pandemic, that permitted me to spend a sabbatical year in Germany in 2022, first at the University of the Ruhr in Bochum and then at the Lateinamerika Institut of the Freie Universität in Berlin. Special thanks go to my colleague in Bochum, Ludger Pries, and to Barbara Fritz in Berlin, who have discussed some ideas that have found their place in this work. To Solène Morvant-Roux for having read and discussed the final chapters.

I am especially grateful to my colleagues and friends of the research group we have formed to study the diversity of capitalisms in Latin America: Alberto Aziz, Graciela Bensusán, Robert Boyer, Jaime Marques Pereira, Bruno Théret, and Enrique Valencia.

I want to acknowledge the invaluable assistance of Mauricio Jiménez Hernández, who made most of the graphs, for his collaboration in the historical chapters and our fruitful discussions on different topics of this work. I would also like to thank Ana Inés Fernández Ayala and Hugo López Araiza for their effort to make this book more readable.

I am equally thankful to my students at El Colegio de México and the University of Bochum, who reacted to my presentations with questions and comments that helped to configure a more coherent work.

I wouldn't have been able to write this book without the support of my wonderful family, to whom I dedicate it.

Contents

1 **Introduction: The Welfare State as a Social Compromise** 1
The Welfare State as Protection of Society 2
The Welfare State as a Mediation Between the Different Societal Orders 6
The Narrative of This Book 14
References 20

2 **Four Worlds of Latin American Welfare States** 23
References 48

3 **The Political Economy of the Statist and Socio-Corporatist Welfare States in Latin America** 51
The Socio-corporatist Welfare Regimes (Uruguay and Argentina) 53
Uruguay 54
Argentina 63
The Statist or State-Dominated Welfare Regimes 72
Costa Rica 72
Brazil 78
References 84

4 **The Political Economy of the Commodified, the Familiarist, and the Hybrid Welfare States** 89
The Commodified Welfare State: Chile 91

The Familiarist Welfare States: Peru, Ecuador, and Bolivia 100
Two Hybrids: The Mexican and the Colombian Welfare States 112
References 123

5 The Health Systems: Their Structure and Performance Facing the Pandemic 127
Socio-Corporatist Systems: Uruguay and Argentina 131
Statist Systems: Costa Rica and Brazil 136
Commodified Systems: Chile 138
Familiarist Systems: Ecuador, Bolivia, Peru 140
The Mexican and Colombian Hybrids 144
The Health Regimes Facing the Pandemic 146
Appendix 159
References 161

6 The Welfare State and the Wage Relation 165
Public Social Investment and Coverage 169
The Turn to Social Assistance 180
Non-contributory Pensions 182
Assistance Programs 184
References 202

7 The Welfare State and Gender 207
The Types of Gendered Welfare States in Latin America 214
Employment and Labor Market Situation 214
Programs that Promote Separate Gender Roles 220
Programs that Encourage Women to Engage in an Individual Career or Carer Activity 225
Movements, Political Rights, and Institutions 230
References 239

8 Conclusions 243
References 253

References 255

Index 271

Acronyms

AFAM-PE	Asignaciones Familiares-Plan de Equidad
AFJP	Administradoras de Fondos de Jubilaciones y Pensiones
AFP	Administradoras de Fondos de Pensiones
AMLO	Andrés Manuel López Obrador
ANPOS	Asamblea Nacional Permanente de Organizaciones Sindicales
APRA	Alianza Popular Revolucionaria Americana
ASSE	Administración de Servicios de Salud del Estado
AUC	Autodefensas Unidas de Colombia
AUGE	Plan de Acceso Universal a Garantías Explícitas
BID	Banco Interamericano de Desarrollo
BPC	Beneficio de Prestação Continuada
CAP	Caixa de Aposentadoria e Pensóes
CCSS	Caja Costaricense de Seguridad Social
CEISAL	Centro Internacional de Estudios en Salud
CEPAL	Comisión Económica para América Latina y el Caribe
CGT	Confederación General de Trabajadores
CIESU	Centro de Informaciones y Estudios del Uruguay
CLACSO	Consejo Latinoamericano de Ciencias Sociales
CLAS	Comités Locales de Administración de Salud
CNT	Central Única de Trabajadores
COA	Confederación Obrera Argentina
COB	Central Obrera Boliviana
CONCAMIN	Confederación de Cámaras Industriales de los Estados Unidos Mexicanos

CONCANACO	Confederación de Cámaras Nacionales de Comercio, Servicios y Turismo
CONEVAL	Consejo Nacional de Evaluación de la Política de Desarrollo Social
CORA	Confederación Obrera Regional Argentina
CROC	Confederación Revolucionaria de Obreros y Campesinos
CROM	Confederación Regional Obrera Mexicana
CSTB	Confederación Sindical de Trabajadores de Bolivia
CTA	Central de Trabajadores de la Argentina
CTCh	Confederación de Trabajadores de Chile
CTM	Confederación de Trabajadores de México
CTP	Confederación de Trabajadores del Perú
CUT	Central Única dos Trabalhadores
DIF	National System for the Integral Development of the Family
EBAIS	Equipos Básicos de Atención Integral en Salud
ECLAC	Economic Commission for Latin America and the Caribbean
EPS	Entidades Promotoras de Salud
FARC	Fuerzas Armadas Revolucionarias de Colombia
FCEA-UDELAR	Facultad de Ciencias Económicas y de Administración – Universidad de la República
FODESAF	Fondo de Desarrollo Social y Asignaciones Familiares
FONASA	Fondo Nacional de Salud
FRS	French Regulation School
FSP	Fondo de Solidaridad Pensional
FSTMB	Federación Sindical de Trabajadores Mineros de Bolivia
GDP	Gross Domestic Product
IAMC	Instituciones de Asistencia Médica Colectiva
IAP	Institutos de Aposentadoria e Pensões
ICSS	Instituto Colombiano de Seguros Sociales
IESS	Instituto Ecuatoriano de Seguridad Social
ILO	International Labour Organization
IMAS	Instituto Mixto de Ayuda Social
IMF	International Monetary Fund
IMSS	Instituto Mexicano del Seguro Social
INAMPS	Instituto Nacional de Assistencia Medica da Previdencia Social
INAMU	Instituto Nacional de las Mujeres
INEGI	Instituto Nacional de Estadística y Geografía
INP	Instituto Nacional de Previsión
INPS	Instituto Nacional da Previdência Social
INSABI	Instituto de Salud para el Bienestar
IPS	Instituciones Privadas de Salud

ISAPRE	Instituciones de Salud Previsional
ISI	Import Substitution Industrialization
ISSSTE	Instituto de Seguridad y Servicios Sociales de los Trabajadores del Estado
LACEA	Latin American and Caribbean Economic Association
MAS	Movimiento al Socialismo
MEP	Ministerio de Educación Pública
MIES	Ministerio de Inclusión Económica y Social
MINSA	Ministerio de Salud
MNR	Movimiento Nacionalista Revolucionario
MSP	Ministerio de Salud Pública - Ministry of Public Health
NGO	Non-Governmental Organization
NHS	National Health Services
OCHISAP	Observatorio Chileno de Salud Pública
OECD	Organización para la Cooperación y el Desarrollo Económicos
OIT	Organización Internacional del Trabajo
OMS	Organización Mundial de la Salud
ONU	Organización de las Naciones Unidas
OPS	Organización Panamericana de la Salud
PANES	Plan de Atención Nacional a la Emergencia Social
PIT-CNI	Plenario intersindical de Trabajadores-Convención Nacional de Trabajadores
PNUD	Programa de las Naciones Unidas para el Desarrollo
PRI	Partido Revolucionario Institucional
PSD	Partido Social Democratico
PT	Partido dos Trabalhadores
PTB	Partido Trabalhista Brasileiro
SAR	Sistema de Ahorro para el Retiro
SEG	Seguro Escolar Gratuito
SGSSS	Sistema General de Seguridad Social en Salud
SMI	Seguro Materno Infantil
SNIS	Sistema Nacional Integrado de Salud
SNS	Servicio Nacional de Salud
SNTMMSRM	Sindicato Nacional de Trabajadores Mineros Metalúrgicos y Similares de la República Mexicana
SOCPRO	Social Protection Department, International Labour Office
SP	Seguro Popular
SPAM	Seguro Público del Adulto Mayor
SS	Secretaría de Salud
SSC	Seguro Social Campesino
SSO	Seguro Social Obligatorio
SSS	Servicio de Seguridad Social

SUMI	Seguro Universal Materno Infantil
SUS	Sistema Único de Saúde
TUS	Tarjeta Uruguay Social
UGT	Unión General de Trabajadores
UN	United Nations
UNESCO	United Nations Educational Scientific and Cultural Organization
UNICEF	United Nations Children's Fund
USA	Unión Sindical Argentina
WB	World Bank

List of Figures

Fig. 5.1	Current health expenditure (% of GDP)	130
Fig. 5.2	Health System Structure	132
Fig. 5.3	Per capita health expenditure (in current PPP)	133
Fig. 5.4	Government health expenditure (% of current health expenditure)	134
Fig. 5.5	Private health expenditure (% of current health expenditure)	134
Fig. 5.6	Out-of-pocket health expenditure (% of total)	135
Fig. 5.7	Excess Deaths (size of the circles) according to infrastructure and performance	149
Fig. 5.8	Number of Doctors per 1000 inhabitants vs. excess deaths	153
Fig. 5.9	Total health expenditure (% GDP) vs. excess deaths	154
Fig. 5.10	Private current expenditure (% total health expenditure) vs. excess deaths	155
Fig. 5.11	Out-of-pocket expenditure vs. excess deaths	156
Fig. 5.12	GDP variation vs excess deaths	159
Fig. 6.1	Public social expenditure (% GDP, 2021)	170
Fig. 6.2	Public Social Expenditure (% GDP: 2000–2021)	170
Fig. 6.3	Public spending on pensions (% GDP, 2017)	172
Fig. 6.4	Public health expenditure (% GDP, 2020)	175
Fig. 6.5	Government health expenditure (% GDP)	176
Fig. 6.6	Coverage by contributory pensions and by contributory and public health services of the total employed population (%, 2020)	177
Fig. 6.7	Informality rate, 2021	179

Fig. 6.8	Average net pension replacement rates	180
Fig. 6.9	Incidence of poverty among the population over 65 years of age (% 2020)	181
Fig. 6.10	Wage relation Index	201
Fig. 7.1	Labor participation rate, 2020	215
Fig. 7.2	Wage equity for similar work	216
Fig. 7.3	Participation of women in labor income	217
Fig. 7.4	Share of informal employment (urban sectors) (%)	218
Fig. 7.5	Unemployment rate (2020)	218
Fig. 7.6	Women with transfers as sole resource and with no income (% of total. 2020)	223
Fig. 7.7	Feminization of poverty index, 2020	224
Fig. 7.8	Gross enrollment rate in pre-primary education programs (3–5 years of age), (% of coverage, 2020)	228
Fig. 7.9	Time spent on non remunerated care or domestic tasks (hours per week)	231
Fig. 7.10	Women labor and gender policies indexes	238

List of Tables

Table 2.1	Welfare State typologies	44
Table 2.2	Typology of the Latin American welfare states	46
Table 5.1	Infrastructure, Performance, and Excess Deaths	148
Table 5.2	Correlations with excess mortality	151
Table 6.1	Wage relation of the different ideal types of welfare states	167
Table 6.2	Coverage and expenditure of non-contributive pensions (2021)	185
Table 6.3	Assistance programs in Latin America (2019)	186
Table 6.4	Poverty rates (CEPAL estimates) 2021	190
Table 7.1	Typology of gendered welfare states	212
Table 7.2	Class hours and workday extension programs in Latin America	229
Table 7.3	Maternal and Paternal leaves	230
Table 7.4	Institutions in charge of the social rights of women	233
Table 7.5	National plans for equality, political representation and parity	235

CHAPTER 1

Introduction: The Welfare State as a Social Compromise

It is crucial to define how one understands what a welfare state is, and its primary social significance before we begin discussing the particular situation of Latin America. As we will discuss in this first chapter, it is crucial to understand if the welfare state can be considered from a purely institutional perspective or if it is necessary to include a discussion of the social protection system as the result of a social compromise derived from a conflict between different social actors. This is not a mere semantic question but modifies how we understand and analyze of social protection.

It is quite surprising that most studies on the welfare state do discuss its foundations, main principles, or essence if we may use such a word. We should begin with a fundamental question: what is the social significance of the welfare state? And if we answer as Polanyi did, that the welfare state is how society defends itself, we have then to analyze not only the specific institutions that allow society to do so but also how the societies we are describing are constituted. Which are the actors that dominate the social relations of that society, and what type of social compromise have they achieved?

I. Bizberg, *Four Worlds of the Welfare State in Latin America*, https://doi.org/10.1007/978-3-031-44420-3_1

The Welfare State as Protection of Society

As Polanyi has stated, the welfare state is a necessity of modern capitalist societies to counter the consequences of the disembeddedness of the economy from society generated by capitalism. Social protection in modern capitalist societies is how society protects itself against the rupture it risks because it has disjointed the economic order from the other orders that constitute society. This happened when labor became merchandise bought and sold in a self-regulated market. "The submission of the labor force to market laws generates the need for a set of social security institutions to partially weld the social bonds that capitalist market forces have dissolved. The autonomization of the economic order from the political, social, and domestic order to which it was linked [in feudal times] and the correlative insulation of the domestic order from both the social and the political have dissolved the old social bonds" (Théret 1998).

The main idea of Polanyi, when he states that the welfare state is a way society protects itself from the market, is fundamental to analyzing how this process comes about and who implements it. This author and others who followed his footsteps, found that the fact that capitalism converted labor into fictitious merchandise, implied that society had to secure the life of those who could not find work and thus lacked the means for survival. Capitalism converted labor into fictitious merchandise implying that all human beings who fail to transfigure their labor into merchandise will die. The fiction resides in that labor, something inherent to human beings, a part of their essence, is transformed into something that can be exchanged as if it were an object. If society does not want to cope with the consequences of this fact, it has to take care of those who fail to transform their labor (for any reason) into merchandise.

Polanyi analyzes how, against their liberal ideology, the English capitalist class had to implement the Poor Laws and a system of salary compensation to ensure that entire sectors of society would not fall into complete abandonment and that people living in poverty were, in a particular manner, reinserted into society or taken care of, so that they did not endanger the whole of society. While economic dynamics create winners and losers and generate inequality and tensions between different sectors, social protection systems are a way to reduce these tensions, which, if left unchecked, would destroy the social order. In more affluent societies, the welfare state has served to weaken class struggle, made capitalistic exploitation more acceptable, and thus undermined the worker's social representatives.

This implies, partly, a moral obligation but mainly a question regarding the cohesion of society. Any society would succumb if many of its members were obliged to sell their labor force but could not find buyers. This is the profound reason for the existence of the social protection systems and the basis of the first Poor Laws that were established in England, which retained the poor in churches and later in State establishments, where they received the means of subsistence in exchange for work; this prevented them from becoming a disruptive element, especially if their number was very significant. Polanyi describes another system that preserved society from disruption: the Speenhamland system, established to ensure that all individuals received the minimum necessary to survive. "The society of the eighteenth century resisted unconsciously, all seeking to convert it into a mere appendix of the market. No market economy was conceivable, which did not entail a labor market, but the creation of such a market, in particular in a rural civilization such as that of England, did not demand anything less than the massive destruction of the traditional edifice of society. The economic advantages of a free labor market could not compensate for the social destruction that it provoked. The introduction of a new type of regulation became necessary, one that would protect labor, but, this time, against the functioning of the market itself" (Polanyi 1944: 113). The modern welfare state, which should be more accurately (and neutrally) called a **social protection system,**[1] is a result of this intent to set limits to the market to preserve the life and the essential well-being of the individuals who are, for some reason or another, unable to work. If no limits are set to the market, society as a whole could not survive.

Esping-Andersen's concept of demercantilization is a prolongation of this idea: since all social protection systems limit the domination of labor by the market, one can characterize the different welfare states by the degree to which they extract or detach women and men from the market. "Marx and Polanyi and, more recently, Lindblom (1977) are correct in arguing that [the market] is freedom behind prison walls, and hence fictitious. Workers are not commodities like others because they must

[1] Although we actually prefer the concept of Social Protection System because it defines better what the so-called welfare system is since it does not qualify it—as it is not always certain that we can call its effect "welfare"—in this book, I will also use the concept of Welfare System due to its ample use in the literature; I will use both terms interchangeably.

survive and reproduce both themselves and the society they live in [...] Decommodification is, therefore, a process with multiple roots. As Polanyi argued, [...] even at the apex of laissez-faire precapitalistic residues of communalism persisted, and novel mechanisms of protection emerged" (Esping-Andersen 1990: 37).[2]

Finally, this essential function of the welfare system lies behind Bruno Théret's idea that the social protection system stems from **a life debt**, a debt that all of us individuals living in a society have towards each other, which concretizes in debt that the State (as representative of the whole of society) has towards each of its citizens. According to Théret (1995), each member of society has a life debt towards the other members due to their mere existence in a collectivity. Théret defines life debt as ".. the principle of social reproduction [that exists] because any group, any society, is confronted with the mortality of its members, and to be sustainable, must face the need to ensure its perpetuity beyond their inevitable disappearance. This permanence requires the transmission of life debts between generations and the maintenance of the overall 'life capital' of society, which is ensured by the sequence, throughout individual life cycles, of a set of transactions, real and ritualized symbolism linking humans to each other and superhuman beings — ancestors, deities, spirits, nation, country (on the occasion of births, initiations, marriages, funerals, murders, sacrifices, etc.)" (Théret 1995).

As John H. Weiss has written, the poor relief measures that were implemented during the French Third Republic were partly based on the idea of solidarity, defended by a group called the "Solidarists", who supported the idea that social protection of a sector of society that is being marginalized has to arise from a type of social contract between all members of society. They contrasted charity from this contract funded on solidarity in these terms: "...charity results in a unilateral moral subordination of the recipient to the benefactor, whereas Solidarity remains a bilateral quasi-contract of mutual association, a reciprocal collaboration in social progress in which the benefactor and the recipient are both co-participants" (Charles Brunot, cited by Weiss 1983: 58).[3]

[2] In fact, we could consider that nowadays, we are experiencing a similar situation, but regarding nature. The marketization, the exploitation, of nature has reached such levels that our physical (rather than social) existence, as we are part of nature, is threatened.

[3] Origins of the French Welfare State: Poor Relief in the Third Republic, 1871–1914 Author(s): John H. Weiss

Théret refers to this concept of life debt to analyze the characteristics of economic exchange and the creation of money, which "...is the medium which gives a measurable and quantified form to this set of constitutive social relations of society through mediation: social interdependencies which take the form of obligations and reciprocal rights between members of society and between them and the collective organizations representing the whole of society, which are translated into debts and claims" (Théret 1995). One can also consider the exchange of rights in the social protection system, such as pay-as-you-go pension systems, paying contributions to the health system that may benefit others more than oneself, or even progressive taxation as such a debt. Social protection is also a relation between debt/claims between generations, as the young pay for the old, the healthy pay for the sick, and the employed pay for the unemployed. This debt/claims concern not the individuals as such, but as these relationships are socialized, they concern society as a whole "because the question of social protection arises in any society at a fundamental structural level, that of the material conditions for the reproduction of the primary 'natural resource': population and its 'life capital'" (Théret 1995).

Life debt, the self-preservation of society, may be assumed by social associations, organizations, or by the State in the representation of the social common good or as a last resource, by the families, which are the primary social associations. Again, Esping-Andersen's concept of decommodification goes in this direction when it considers that the social protection system is a manner in which society, by acting upon itself, extricates itself from the domination of the economy. For both authors, the ideal welfare system that would wholly pay the debt of life and be totally decommodifying would be a universal, unified, generous, and entirely equitable system. Some systems come close to this ideal type, and others are very far from it. The more universal and generous social security systems are both more decommodifying, in the sense of Esping-Andersen, and more befitting of a conception of the rights of the citizens to be kept safe from the adversities of life, guaranteeing their life debt.

Source: French Historical Studies, Spring, 1983, Vol. 13, No. 1 (Spring, 1983), pp. 47–78 Published by: Duke University Press.

The Welfare State as a Mediation Between the Different Societal Orders

Once we understand the principle that establishes the welfare system, we can analyze how it is constructed, as it is from the multiple interrelations between the dimensions of familiar, economic, social, and political life (Théret 2011). "The welfare State mediates between these three new autonomous spheres of social practice—the capitalist corporation, the nuclear family, and the administrative State—to reinstitute some societal cohesion. To work in that way, a social protection system has to be structured by three different macro-economic relations of "economic consubstantiality", "political alliance", and "domestic protection"" (Théret 1998).

The family is the first societal order cared for by the social protection system. Social protection systems have their most direct relationship with individuals and families, protecting them at different stages of life. To partially reconstitute social cohesion, the welfare state allocates funding and care services to infants, the sick, the old, the unemployed, and low income families and purveys educational services. It may also provide housing and certain other specific goods; the state or social corporations such as trade unions or mutualists need to take care of these needs as they have been disembodied from the social and domestic realms since the advent of capitalism (Polanyi 1944).

The principle behind this activity is, as Titmuss has written, "...the welfare state is not only a manner in which society protects itself but also shields individuals that need protection ... All collectively provided services are deliberately designed to meet certain socially recognized needs; they are manifestations, first of societies' will to survive as an organic whole and secondly, of the expressed wish of all the people to assist the survival of some people" (Titmuss 2018: po. 624). As Titmuss wrote, the primary function of the welfare state is to "...detach the needs of the individual from the needs of the family" (Titmuss 2018: pos. 354). If the welfare system does not take care of the elderly, the sick, and the unemployed, they have to be cared for by their families. When no structure socializes, makes public these needs, and society does not take care of them (at least partly), they are familiarized. This situation is also "culturally" determined, as in some societies, "...need and behavior were still conceived of in terms of the individual rather than the family or the work-group" (ibid.: pos332).

The welfare state rests on the essential recognition that certain situations of dependency are not the individual's responsibility, and thus should not be taken care of by the families but by society. "States of dependency arise for the vast majority of the population whenever they are not in a position to 'earn life' for themselves and their families [...] In industrialized societies, there are many causes of dependency; they may be natural dependencies such as in childhood, extreme old age, and child bearing. They may be caused by physical and psychological ill-health and incapacity [...], or they may be wholly or predominantly determined by social and cultural factors. These, it may be said, are the 'man-made' dependencies. Apart from injury, disease and innate incapacity, they now constitute the major source of instability in the satisfaction of basic needs. They include unemployment and under-employment, protective and preventive legislation, compulsory retirement from work, and the delayed entry of young people into the labor market ..." (ibid., 2018: pos 684). Certainly, human-made dependencies should be considered society's responsibility, but even natural dependencies may also be regarded as not the sole responsibility of individuals and families. This is to counter the fact that, to a great degree, natural dependencies have been considered the *natural* (naturalized) responsibility of families, especially of women. Contrary to this idea, the more advanced welfare states have socialized these dependencies, as discussed in this book.

The second order shaped by the social protection system is politics, as the social security system may serve to legitimize the State, a political regime, or even a specific government. Privileges, thus, may be oriented towards both a limited sector of the population and the most strategic in economic, and political terms, as it was at the genesis of most of the systems—for example, the army, the police, the State functionaries, then the workers, in a typical corporatist scheme. During the "Fordist" period in Latin America, known as import substitution industrialization, the expansion of social protection of the population integrated into salaried work legitimized the corporatist state. Nowadays, in a situation in Latin America, where a large part of the population is either poor or extremely poor, and almost half of the population is occupied in informal activities, the expansion of an assistance-oriented/minimal security model legitimizes numerous governments of the continent—a model that administers poverty, but does not resolve the ever-growing inequalities, informalization, and poverty (Lautier 2009).

Governments may also attempt to legitimize themselves by promoting a more universalistic scheme to address the population as a whole, reducing inequalities. This has happened first in not-so-unequal societies, where the middle class is majoritarian, and an interclass alliance is possible, as has been the case of the north-European countries where red/green alliances have been at the source of the expansion of the welfare state. In the case of Latin America, this has been the case of Uruguay and Costa Rica in the past. And more recently, during the commodity boom, the States that we have characterized as socio-developmentist (Bizberg 2019) were actively searching to formalize workers and raise the minimum wages, increased the social security expenditure, expanded the contributory and non-contributory pensions system, and attempted to fuse the different sub-systems of the social security system into a more coherent whole. They also tried to extend the health system's coverage, either by universalization or by focalization. All these measures function as a way of legitimation, but, as we will discuss immediately, they are also due to an economic rationale.

The economic order is also affected by the social protection system—by way of short- and long-term **economic** effects. While the Speenhamland system endangered the economy, the modern social protection mechanisms tend to stimulate it. Haggard and Kaufman (2008), in their book on the welfare systems in Latin America, Asia, and Central and Eastern Europe, have argued that export-oriented economic systems tend to have a less generous, assistance-oriented social protection system that complements low salaries to raise the competitiveness of the economy regarding the international markets—as was the case of the Eastern Asian countries in the 50s–70s and of China until the mid-2010s, before they turned their economy from exports to the internal market. This has also been the case of Mexico since it adopted its economic model of international outsourcing at the end of the eighties. On the contrary, an economy oriented towards the internal market, such as the import substitution model in Latin America and the Fordist systems of Europe and the United States during the "glorious thirty years," has a more generous social protection system. Such a system rests on the idea that spending on education, health, and pensions increases internal demand as individuals have to spend less on the services allocated by the State or need to save less for old age, and thus, have more resources to spend in what is produced by the domestic industry. When the government implements a more universal and efficient health system, whether public or a mixture of public and private, people need to save or spend less in case of a disease. As old age, health risks, and unemployment are socialized, individuals can direct more resources to consumption. The same can be said of governmental spending on education and housing. Some authors have even affirmed that social protection schemes, such as universal unemployment security, pensions, and health, are functional for entrepreneurs because social risk is socialized among the whole of society, as it compensates differences in terms of risk, unhealthy conditions, labor instability, and between sectors (Mares 2001).

The expansion of the welfare system—the increase in social spending, especially in education and health—has a medium- and long-term effect, as it increases human capital, eventually increasing productivity. One of the most significant differences between how countries in Latin America and those in Asia have tried to develop is the investment in health and education in the latter. Thus, social protection policies can have a short-term impact on demand through pensions, unemployment compensations, health investment, and expenditure, and as has been shown by the example of the Asian countries, a medium- and long-term impact on the economy and on the possibility of a country to develop, through upgrading, technological change, and productivity growth.

The social order intervenes more markedly in the construction of the welfare state. The welfare state is a social construction, as we have already mentioned, to protect society from its own processes, in sum, from its invention of capitalism, from the disengagement of the economy, from its familial, social, and political grounds, and from the fictitious conversion of labor and nature into merchandises; it is a societal construction to protect itself from itself. On the other hand, every social protection system is the result of a specific social compromise; it does not result spontaneously but is the result of the settlement of the conflict between the different social sectors that constitute a society for the wealth this society creates through its collective work; it is thus a (temporary, susceptible to change) social arrangement to a redistributive conflict.

Social protection is one of human beings's creations that have to be distributed. As Walzer (1983) has written, ..."human society is a distributive community... we come together to share, divide, and exchange ... but we also come together to make the things that are shared, divided, and exchanged". But this process of distribution neither occurs spontaneously nor equitably, "...different political arrangements enforce, and different ideologies justify diverse distributions of membership: power, honor, ritual eminence, divine grace, kinship and love, knowledge, wealth, physical security, work and leisure, rewards and punishments, and a host of goods more narrowly and materially conceived: food, shelter, clothing, transportation, medical care, commodities of every sort ... and a multiplicity of distributive procedures, agents, and criteria matches this multiplicity of goods" (Walzer 1983: 1). Distribution, then, implies power, conflict, and settlements, as different social groups have different ideas of how distribution should occur, on what basis, and which is the most just manner of doing so. Conflict, then, is as natural to human society as distribution. All of what society creates has to be distributed. Honors, power, love, and, of course wealth (Walzer 1983: 1).

On the other hand, as has been analyzed by most authors (Titmuss, Esping-Andersen, Korpi, Theret, Martinez Franzoni, Barba, Valencia), while a social security system may be signficantly extended and cover the majority of the population, it may nevertheless be structured in different social protection systems and increase inequality. In addition, a universal system that allocates a minimum to all the population does not reduce inequality or ensures social cohesion. Finally, a system that only takes care of the poor because it supposes that all the other sectors can take care of themselves, and where, in addition, being poor is considered a fault, and the persons living in poverty have actually to prove their poverty and accept aid as charity, does not fulfill the mission of easing social tensions; in fact, it probably increases them (Esping-Andersen 1990).

Even more fundamentally, the welfare state is a central element of what we may call a *social compromise* between different sectors of society regarding how to distribute part of the wealth created; in any society, there is a central conflict around distribution. On the other hand, as we already discussed, capitalism converted labor into a fictitious commodity, and by doing so, it extricated work and the economy from the societal order, society was obliged to protect its members and itself to prevent its disolution. Nonetheless, this did not occur spontaneously but through conflict between different sectors of society, which has led to other, more or less durable, social compromises throughout history.

The welfare state is a central part of this social pact that results from the distributive conflict and society's need to protect itself. The State's social and economic action does not depend on its mere will, nor does it respond to specific contexts, but instead to an institutionalized commitment (Boyer 2004). The wage relationship is this institutionalized commitment, which includes the social protection system and the regulation of the labor market and wages (Boyer 2004). The management regulation of the labor market and of wages and the conformation of the welfare state are the ways in which social conflict is institutionalized. More precisely, its structure and transformations depend on the outcome of the conflict between the different social actors: the working class/unions, employers, local/regional powers, and the national state (Théret 2005).

We use the concept of the of social contract as a social arrangement, a less totalizing conception than that of Rousseau, yet large enough to englobe economy, society, and politics. While Rousseau conceived the social contract as the foundation of any society in its relation with the State, we refer to what the French regulation school calls "*compromis social*", which is more than a mere short-term arrangement but less than Rousseau's foundational, totalizing concept of social contract, and yet large enough to englobe economy, society, and politics. *Compromis social* is an agreement defining a settlement (partly tacit, partly explicit) that includes how a specific society organizes production and distributes the wealth it produces. The social compromise emerges from a particular relationship between the social sectors, the State, and what the regulation school calls the dominant social coalition (Amable 2005; Boyer 2015; Bizberg 2019). It, therefore, depends on the position that the State, the popular sectors, and capital occupy.

One of the clearest examples of a social pact is Fordism, which lasted from the end of the Second World War to the mid-seventies and was in function in the United States and Europe. In general terms, it was based on the distribution of the gains of economic growth to the workers through salaries and social protection measures (pensions, health, unemployment insurance), defined via collective (labor-employers) negotiations and social investment by the state. It was a social pact between the workers (represented by trade unions) and the employers with State intervention, including direct wage increases and indirect social security benefits, pensions, health, paid vacations, etc. It generated a virtuous socio-economic circle, where enterprises produced goods through Taylorist and mass production innovation systems, that continuously boosted the production of basic and durable consumer products, a result of continuous productivity increases (through both innovations in labor organization and the inclusion of labor-saving technology), which was followed by a constant rise in salaries and social protection investment. In this manner, as society became wealthier and capital increased its profits, workers improved their well-being. The fact that the State guaranteed pensions and invested in the health system and education implied that workers had to save less to counter the risks of old age and disease, and for their children's education, they thus had more resources to spend. This allowed them to acquire the products that were being produced by the factories where they worked. On the other hand, as salaries and contributions increased continuously, making labor more expensive, employers were forced to invent new ways to increase productivity (innovating in the organization of work and technology), a factor that led to further increases of productivity and to a decrease in the price of unity of labor. The mechanism that permitted this virtuous cycle was not the autoregulated market but rather the cycles of conflict between labor and capital and their temporary agreements. The State intervened to regulate this conflict, force the parties to come to agreements, and channel resources, coming from general taxes and contributions by workers and employers to invest them in the welfare state. This social agreement depended on more or less closed national economies, with limited exports of manufactures; it was broken once the national industries began to export massively, and competition between countries began disrupting the national social Fordist compromise. The end of the this arrangement led to increased exports, the liberalization of the national markets, the opening and deregulation of the national markets, and the degradation of the welfare states (Aglietta 1997; Boyer 2015).

In Latin America, the redistributive pact has gone through different stages; before the Second World War, there existed an oligarchic pact between the government and the most strategic socio-political sectors (the army, the police, State functionaries), basically based on "from above" concessions of pension and health funds to a very restricted part of the population. In the second stage, after the war, a semi-fordist compromise between the State, industrialists, middle classes, and workers, served to back the import substitution industrialization effort by implementing wage-led growth that expanded and protected the internal market. This pact led to the expansion of the welfare state to include an increasing part of the population; in the countries where industrialization advanced the most, this effort led to the inclusion of almost two-thirds of the population in pension and health systems; in other countries that did not develop as much, only a small part of the population was included, which did not differ much from the situation of the first oligarchic stage of the most developed countries of the continent. A third stage, after the "lost decade" of the eighties, when most countries abandoned the import substitution strategy, liberalized, and opened their internal markets to imports and capitals, increasing informality and poverty, the pact changed in nature: unions and other social organizations were marginalized due to de-industrialization and repression, and, in most countries, a coalition was established by the State with national and foreign capital, and the middle classes that profited from the new economic profit-led growth model. In this compromise, increasing number of people living in poverty were "included" in a subordinate position through ever-growing assistance programs, conditional monetary transfers, and non-contributive pensions for the old. A fourth stage, in the year 2000, due to the high demand for commodities and their increased prices, basically caused by the enormous growth of China and the expansion of financial resources due to the monetary expansion of the developed countries, some Latin American countries intended a neo-developmentist growth model, partly molded on the previous import substitution industrialization, although now favoring both import substitution and commodity exports. The fact that these countries, basically Brazil and Argentina, were also actively expanding minimum salaries and their welfare systems, both the contributory and non-contributory schemes, as well as assistance, in addition to their health services as a way to increase the internal market, led to name this model as socio-developmentist. Some other countries that were fundamentaly exporting commodities, but were governed by an alliance of leftist parties with popular organizations and movements, basically Bolivia and Ecuador, used some of the resources coming from the exports of commodities to increase wages and expand their social protection system. Finally, some other countries continued with the liberal, profit-led model and the extension of assistance policies for the poor to preserve social peace; this was the case of Chile, Mexico, Colombia, and Peru (Bizberg 2019).

However, if one looks more precisely at the compromises that have been established in practically all countries and have led to the development of the different types of welfare states, even the most developed ones (the social-democratic ones of Northern Europe), the social arrangement is between and for the benefit of men, who have held most of the political posts and the leading positions of the social organizations in the past. Although this situation is evolving slowly, the pact, seen from a gendered perspective, defines a division between the productive and the reproductive work, between work that is remunerated and work that is not paid, and women generally occupy this latter type of work. "Gender also structures the division between well-paid, predominantly masculine professional and industrial activities, and activities that relate to the "pink collars" and domestic services, which are worse paid and predominantly feminine. This results in an economic structure that creates sexual forms of exploitation, marginalization, and destitution" (Fraser 2005: 26).

It is very telling in this respect that although Esping-Andersen and other authors have considered the social-democratic countries welfare states as equally advanced, seen from a gendered perspective, there are substantial differences within this group of countries, and coincidences between countries situated in this category, with others that were considered "inferior" by this author—the corporatist ones. This is the whole argument of Sainsbury's (1999) work, which we will discuss in the last chapter of this book, as it questions the entire argumentation of the previous chapters.

In some countries there has been an attempt to improve the position of women by what Fraser calls corrective redistribution, which "…draws on affirmative action, trying to guarantee women an equal part of the existing employment and formation…" the nature of gender inequality has not changed (Fraser 2005: 37). These corrections do not abolish gender differentiation, but reconfigure and reinforce it, as they do not "…reach the profound level in which the political economy is sexualized … [they fail] to combat the discriminatory behaviors, it does not attack the sexual division between remunerated work and non-remunerated work, nor the sexual division of remunerated work" (Fraser 2005: 37).

The Narrative of This Book

In Chapter 2, we discuss the perspectives of the studies of the welfare states to propose our own. On the one hand, there is a long tradition of studies that try to find a general, universal rule to explain the implementation and expansion of a generous welfare state; on the other, there are studies based on the consideration that the welfare state evolves in different manners according to no general rule, but to a specific historical process. From this second perspective, derive national studies and the effort to simplify complexity by constructing typologies to cluster similar social protection systems to analyze the situation of Latin America. In this book, we discuss various typologies and defend the idea that there is room for a different typology that adds elements to increase our comprehension of the situation of the welfare states in the continent.

In contrast to the usual typologies, we define the types of welfare state according to a non-normative, unique principle that is neither based on a characteristic of the welfare state, such as segmentation or dualism, nor as a continuum between two models, nor on an ideal model from which the other types diverge. We consider all social protection regimes a specific combination of State, market, society, and family and define our typology based on the dominant actor that administers and distributes social protection. Our typology interrelates the state, civil society, the market, and the domestic sphere: one is statist, another one is based on the market, another on the family, and finally, the last one on civil society.

In Chapter 3, we begin discussing the historical trajectory of our ten countries. We divide the analysis into two chapters due to the length of the theme. We begin with the socio-corporatist and statist welfare states. In the cases of Uruguay and Argentina, social actors have been central in shaping the welfare state. In the case of these two countries, a large part of the health system is directly controlled by social actors, the *mutualistas,* and the unions, although the State has been central in expanding both the health and the pension systems. Although the analysts consider that the Uruguayan welfare state was a creation of the state as social actors were weak, nonetheless, at least in the case of health, the State allocated services to the population that the *mutualistas* had already created; thus, in a sense, it responded to a demand that was already present, and had been created by civil society. The State created the Argentinian welfare state to divide, coopt, and prevent the autonomous organization of the labor movement. However, the unions were empowered during the Peronist era and were given the administration of the health services. While Perón ruled the country, they were subordinate to the State, but after the military coup of 1955, they gained autonomy through their opposition and negotiation with the successive governments and managed to continue administering the health system.

In both Brazil and Costa Rica, the statist character of the welfare state has prevailed until the present, although it is party privatized, especially in the case of Brazil. In this country, unionism, basically an administrative rather than a political actor, was much weaker than in Argentina or even Mexico; it was the State that created and centralized both the pension and the health structures. The present health system, the SUS, built in the wake of democratization, in the process that created the Constitution of 1988, is a centralized, statist health structure that the private health services have nevertheless eroded. Costa Rica never lived through a corporatist moment, nor were social actors strong the welfare state was a direct creation of the State and has maintained itself as one of the most extensive, equitable, and efficient in Latin America.

Chapter 4, discusses the commodified and familiarist welfare regimes, Chile for the first and Bolivia, Peru, and Ecuador for the second types. We analyze the socio-political conformations and dynamics that have given rise to and preserved the welfare states that depend most on the market and individuals and their families, but also on the market, as the latter acquire health services from private providers and pharmacies. The social protection system of Chile is in many respects the opposite of the Brazilian one, which is a State-led welfare system that is being commodified. It is the conception of the transformation that occurred during the military regime and the critical reforms the democratic governments have implemented. Although it is still the most commodified, the reforms it has undergone, especially the one of 2008 on health, have shifted the Chilean welfare state somewhat away from the market; pensions are still totally based on capitalization.

In contrast, the familiarist social protection regimes emerged in a very ad hoc manner, slowly and dispersedly, since the beginning of the twentieth century. Their trajectory was promoted neither by social organizations or a political party nor by the intention of the State to coopt the labor movement or any other social sector. The fact that the industrialization process only advanced limitedly in Peru, Ecuador, and Bolivia meant that the labor movement was also frail. Nonetheless, in all these three countries, cores of strong labor organizations existed, especially in extractive industries and in government, which were allotted certain privileges regarding pensions and health when they protested. They also received these advantages when a new government arrived that needed political support or after an authoritarian downturn or democratic reestablishment. This resulted in a reduced and atomized welfare state, where most of the population lacking organizational and pressure capacity was excluded. This situation was partly modified during the commodity boom in some countries.

We also analyze the historical trajectories of two welfare states that are hybrids of various ideal types. In the case of Mexico, about half of the population has access to pensions and health, basically State-controlled, while the other half are in a situation such as in the familiarist social protection systems, where it depends on their resources (monetary and human) and those of their families. Although the Colombian is by many respects a typical familiarist welfare state, since the Constitution of 1991, it offers health services to almost the totality of the population with both a pure contributory and a subsidized contributory system.

In Chapter 5, we go from a diachronic to a synchronic analysis of the Latin American welfare states. First, we analyze the health system, which we consider crucial as, in contrast to the pension system, we can see all four social actors acting in a common field. Regarding pensions, few, if any, funds are union-managed, as most are administered either by the State or by banks; in the health system, clinics and hospitals are contracted or administered by unions, social associations, the State, or private capital.

Of our ten countries, Uruguay and Costa Rica have come closest to the goal of universality. In addition, during the first decade and a half of the twenty-first century, several countries that saw the arrival to power of coalitions that included the popular classes expanded their health structures by taking advantage of the influx of resources due to the commodity boom and financial expansionism of the developed countries. The end of the commodity boom and the arrival of governmental coalitions dominated by financial and commodity exports interests put an end to the slow transformations of the health systems in countries such as Bolivia and Ecuador. It also strongly affected the efforts undertaken by Brazil in trying to consolidate one of Latin America's most ambitious health structures, which, like Costa Rica's, is State-run. Chile, a country that suffered significant deleterious transformations of its welfare state during the dictatorship, succeeded in rebounding an almost universal health scheme and regulating its (previously unfettered) private health sector. On the other hand, Colombia has virtually succeeded in covering its entire population in a mandatory contributory public service by subsidizing the inclusion of its lower income population. With a similar model, the *Seguro Popular*, Mexico continuously left around 20% of its population without coverage due to its voluntary and un-subsidized affiliation for all but the poor population.

The pandemic the world endured between 2020 and 2022 strained all the health structures of the continent, an exceptional situation that allows us to compare their performance and resilience. The fact that some countries weathered the crisis better than others was due, in part, to the condition of the health services themselves but also to the approach that the different governments adopted to deal with the epidemic: some governments initially denied the severity of the disease (Brazil, Mexico), others reacted late (Ecuador, Bolivia), while still others did everything they could to face the crisis from the very start (or even before, as soon as the pandemic hit the European countries), like Costa Rica, Uruguay, and Argentina.

In Chapter 6, we will analyze the welfare state as the result of a social compromise in the distributive conflict concerning the many creations of human society, including wealth and security. We will introduce the concept of wage relation, central to the French Regulation School, which is, in many respects, the translation of the social pact each society has come to define to limit the redistributive conflict. The specificities of the wage relation are either negotiated directly between the unions and the employers or defined through State intervention. Its conditions closely depend on the organization, strength, and capacity of action of the labor movement, the employer's organizations and their cooperative or confrontational relation. Each social compromise translates into a specific wage relation. Wage-led social pacts are those where the social actors have a significant force and/or an alliance with the State, while, on the contrary, profit-led social pacts are those where the popular classes are weak, and the coalition is sealed between the state and capital.

In most Latin American countries, minimum salaries are fixed by commissions, where the State has a decisive role. In addition, the state is also responsible for regulating the labor market, and intervenes directly in the collection of the contributions paid by workers and employers to finance the welfare state. It finances part of it through taxes, allocates some programs to poor workers, spends in education, health, and pensions, invests in clinics and hospitals, and thus defines part of the characteristics of the social security system. While during ISI, the social compromise had certain similarities to Fordism, although limited to a sector of society, state officials, and the industrial workers of the large enterprises, nowadays, the social pact has been radically modified. As the neo-liberal economic model created a host of people living in poverty and informal workers, the new social pact had to include them, although in a subordinate relation. Depending on the weight and organizational capacity of the salaried sectors, the compromise comprises both: the poor households and the salaried, or mainly the poor ones. This determines how far the turn to assistance has gone, whether assistance has substituted other forms of social protection, or whether assistance is complementary to the welfare state based on rights.

However, this chapter is not limited to this theoretical discussion but gives data on how the social pact is defined in each of the four types of welfare states we have defined in this book and draws a balance of which countries of our sample have a more advanced social protection system.

In Chapter 7, we discuss what has been left in silence in the previous chapters, namely the situation of women, half of the continent's population that does not take complete, or even close, advantage of its benefits. This is because the welfare state is oriented towards the productive sector of society and disregards its reproductive part. Nowadays, it is evident that the welfare state is both gender structured and contributes to the reproduction of a gendered labor market division. That is the reason why we end the book by trying to justify the perspective we have adopted, while at the same time point out how future studies of the welfare states of the continent should be oriented if they intend to go beyond this perspective.

In this chapter, we retake the typology of Diane Sainsbury and Nancy Fraser of gendered welfare states to analyze to what extent the Latin American countries are applying "woman-friendly" social policies and whether their particular welfare systems can be considered as coming close to one of the types. The typology has included all the dimensions that are fundamental to characterize the welfare regimes from a gender perspective; nonetheless, it is clear that in the case of Latin America, some of them are more significant than others and that some are totally absent: for example, assistance to low income families is more critical than in the developed countries, and some, like taxation, are practically inexistent.

We analyze the specific policies towards women, whether they allocate resources to compensate for their care activity, facilitate their integration into the labor market, or ease their possibility of choosing between a job or the moment when to pursue a career and care. We will discuss how the different governments have tried to implement some policies that favor women. However, they are basically assistance oriented rather than turned towards the increase of freedom.

References

Aglietta, M. (1997), "Postface", *Régulation et crises du capitalism*, Paris, Odile Jacob.

Amable, Bruno (2005), *Les cinq capitalismes*, Paris, Editions du Seuil.

Bizberg, Ilán (2019), *Diversity of Capitalisms in Latin America*, London, Palgrave Macmillan.

Boyer, Robert (2004), "New Growth Regimes, But Still Institutional Diversity", *Socio-Economic Review*, 2 (1): 1–32.

Boyer, Robert (2015), *Économie politique des capitalismes, Théorie de la régulation et des crises*, Paris, La Découverte.

Esping-Andersen, Gøsta (1990), *The Three Worlds of Welfare Capitalism*, Princeton, New Jersey, Princeton University Press.

Fraser, Nancy (2005), *Qu'est-ce que la justice sociale ? Reconnaissance et redistribution*, Paris, La Découverte.

Haggard, Stephan and Robert R. Kaufman (2008), *Development, Democracy, and Welfare States*, Princeton, Princeton University Press.

Lautier, B. (2009), "Gouvernement moral des pauvres et dépolitisation des politiques publiques en Amérique Latine", In N. Borgeaud-Garciandía, B. Lautier, R. Peñafiel, and A. Tizziani (ed.), *Penser le politique en Amérique Latine*, Paris, Karthala, 19–36.

Lindblom, C. (1977), *Politics and Markets: The World's Political-Economic Systems*, New York, Basic Books.

Mares, Isabela (2001), "Firms and the Welfare State: When, Why, and How Does Social Policy Matter to Employers?", In P. Hall and D. Soskice (ed.), *Varieties of Capitalism: The Institutional Foundations of Comparative Advantage*, Oxford University Press, 184–212.

Polanyi, K. (1944), *The Great Transformation*, New York, Farrar & Rinehart (cited from the French translation of 1983: Paris: Editions Gallimard).

Sainsbury, Diane (ed.) (1999), *Gender and Welfare State Regimes*, Oxford University Press.

Théret, Bruno (1995), *L'État, la finance et le social. Souveraineté nationale et construction européenne*, Paris, La Découverte.

——— (1998), "L'État-providence à l'épreuve des comparaisons internationals", In Philippe Auvergnon, Philippe Martin, Patrick Rozenblatt et Michèle Tallard (eds.), *L'État à l'épreuve du social*, París, Editions Syllepses.

——— (2011), "The Place of National Systems of Social Protection and Political Representation in Socio-Economic Regulation: A Morphogenetic Structuralist View on Institutional Change in Comparative Perspective with Special References to Japan and France", *Evolutionary Institutional Economic Review*, 7 (2): 177–208.

Titmuss, R. (2018), *Essays on the Welfare State (Reissue)*, Bristol University Press.

Walzer, Michael (1983), "Spheres of Justice: A Defence of Pluralism and Equality," *Philosophy*, 59 (229): 413–415.

Weiss, J. H. (1983), "Origins of the French Welfare State: Poor Relief in the Third Republic, 1871–1914". In *French Historical Studies*, Duke University Press, 13 (1): 47–78.

CHAPTER 2

Four Worlds of Latin American Welfare States

There are two ways of studying welfare states, and both have been adopted in Latin America. On the one hand, there is a long tradition of studies that try to find a general, universal rule to explain the implementation and expansion of a generous welfare state (Titmuss, Korpi, Haggard and Kaufman, Pribble, Garay, Huber, and Stephens). Other authors (Esping-Andersen, Théret, Sainsbury, Barba, Valencia, Martinez Franzoni, Filgueira) have considered that since in social sciences there are no universal rules and welfare systems evolve in different manners according to no general rule, but to a specific historical process, one should instead trace the trajectory of each protection system.

a. *Two Different Approaches to Study the Welfare State*

While the first type of study tries to find a unique series of events or forces behind the evolution of the welfare state, the second tries to find certain historical and structural similarities and differences in groups or clusters of countries. The first point of view intends to arrive at a general law, whereas the second elaborates a typology to group a certain number of cases that share certain characteristics. Most of the latest books on the welfare state in Latin America, published in the United States, follow the first perspective. In contrast, most of the studies done by Latin American

I. Bizberg, *Four Worlds of the Welfare State in Latin America*,
https://doi.org/10.1007/978-3-031-44420-3_2

researchers follow the second. The second group of researchers' familiarity with their object of study is probably why they are more hesitant to propose general explanations for such particular situations. Although in this book we subscribe to this second perspective, we nevertheless recur to some concepts of the first group of researchers to signal that even though there are no general laws, because there are too many exceptions to the rule, there are indeed certain conclusions that have been well defined by most studies. One of them is that the existence of a strong labor movement and a party linked to it has promoted the expansion and deepening of the welfare state and that the contrary is also the case, as has been signaled by Korpi, Huber and Stephens, and Pribble, where a strong labor movement is lacking or where there is no structural or institutional link between labor and party, the welfare state is less developed. On the other hand, in some situations, the advance of the social protection system is motivated by the State to either preempt the organization of the workers to demand its implementation or extension or divide them by privileging some sectors more than others, as has been signaled by Esping-Andersen, and Haggard and Kaufman.

One of the most ambitious and encompassing books written from the first perspective is that of Haggard and Kaufman (2008), which includes three continents and most of the developing world. It proposes two general hypotheses: the **democratic**, according to which countries where democracy has existed for an extended period tend to have more significant and more generous social protection coverage. The idea is that democracy facilitates citizen pressure for better social benefits, and social policies become a way of attracting supporters and voters. This position is very close to the one advanced by Walter Korpi, who considers that "well-developed welfare states and coordinated market economies have emerged in countries with strong left parties in long-term cabinet participation, or in countries with state-corporatist institutional traditions and confessional parties in intensive competition with left parties" (Korpi 2006: 1). The second, a top-down hypothesis, based on **legitimacy**, suggests that some social protection systems were put into effect in oligarchic regimes when the governing coalition was forced to integrate the popular sectors, in general, the workers; it thus used social policies as a legitimation mechanism to ensure their political affiliation. This type of trajectory gave rise to less extensive and highly heterogeneous social protection systems insofar as to better control the beneficiaries of the policies, they were segmented; those favored were in the most strategic sectors, generally situated in the industrial and urban activities, as in Brazil, Argentina, and Mexico. In contrast, the democratic trajectory (in Uruguay, Costa Rica, Chile, and Venezuela) gave rise to a more extensive and homogeneous scheme, which included rural sectors (Haggard and Kaufman 2008: 1–24).

In their book of 2001, Huber and Stephens propose a hypothesis similar to the democratic one of Haggard and Kaufman, although from a power resource perspective; rather than deriving the evolution of the welfare state from democracy, they consider the "...importance of the **strength** of the labor movements and affiliated parties in building generous, universalistic, and redistributive welfare states" (Huber and Stephens 2001: pos. 191). This means that the welfare state developed where either labor organizations forced governments to implement or expand social protection or motivated oligarchic governments to apply these measures as a preemptive policy to demobilize or prevent the organization of labor. This position is also different from the legitimizing hypothesis of Haggard and Kaufman, which focuses on a "top-down" alliance between the State and popular sectors, as Korpi considers the preemptive action of the State a reaction to the initiation of workers' organization and mobilization which threatens its power: "...early social insurance legislation ... can be partly seen as reactive responses to the Worker Question and early socialist party organization" (Korpi 2006: 9), thus a response of the government to a social mobilization challenge.

The book by Jennifer Pribble (2013) aligns with the **democratic** hypothesis, although she adds various conditions to the character and the organization of the political parties in power. She considers that although the literature argues that in European countries, the existence of democracy has been crucial to explain the expansion of the welfare state, in Latin America, an additional requisite has been the arrival of leftist parties to government. Nonetheless, this requisite is not sufficient as the ideological character of the governing party has to be accompanied by a specific organizational structure and the linkage mechanism employed by parties: the "strength of the tie between elites and base organizations, in combination with the party's ideology and the predominant linkage mechanism employed to appeal to the core constituency" are fundamental (Pribble 2013: 12). "This focus on overall party character, which I define as the combination of ideology, internal organization and external linkage mechanism, is innovative and bridges two bodies of political science research... internal organization of political parties, particularly the strength of the tie between elites and base organizations, in combination with the parties' ideology and the predominant linkage mechanism employed to appeal to the core constituency, has a profound effect on the timing and content of policy reform" (Pribble 2013: 12). She adds a third variable, electoral competition, which shapes social policy reforms. "In a setting where electoral competition is intense, parties face incentives to carry out reforms that appeal to a broad sector of society… left-leaning parties that face intense competition from the right might be pushed to enact relatively centrist reforms in order to attract moderate voters. Similarly, when parties of the right face intense competition from the left, they may carry out reforms that expand State involvement in an effort to appeal to a wider range of voters" (Ibid.: 13).

The work by Candelaria Garay (2016) also accepts the democratic hypothesis, adding the power resource perspective. However she shifts the focus from the usual actors that the welfare state has priviledged: the army, public officials, and the workers from industrial and extractive sectors, to the outsiders: the poor and marginal populations. This focus is a fundamental complement to most of the studies on the welfare state that concentrate on the "insiders", those that are well organized and pose a threat to governments if they are not coopted or taken into consideration, a perspective that overlooks a large part of the population (in fact, in most countries, the majority) of Latin American countries; and also the fact that most of the social policies in the last thirty years have focused on the poor instead of the previous "insiders".[1] She tries to answer why the outsiders have increased their share of the social policies (although, as is well known, the resources allocated to them are much smaller than those for the insiders) and why the governments have decided "to launch social policies for outsiders and for variation in the resulting policies" (Garay 2016: 8). She considers three causes that she deems insufficient: (1) the existence of resources coming from the commodity boom; (2) the arrival of left-wing coalitions into government; and (3) the diffusion of policy models, promoted by national (NGOs and national governments—these

[1] This question is similar to the one we will be asking at the end of this book on the genderization of the welfare state, as these theories and typologies leave out the reproductive work of women and the fact that they benefit much less from the welfare state than men due to their lower salaries, the fact that they occupy part time jobs and work in the informal part of the economy more often than men, especially in Latin America.

latter for clientelistic and electoral motives)and international organizations, that exert pressure to governments that have not implemented them (Garay 2016: 8).

Her explanation is electoral/mobilizational: "Incumbents expand social policy when faced with high levels of electoral competition for the vote of outsiders and/or with large-scale social mobilization by coalitions of social movements and labor unions that push for social policy expansion through the use of protest, institutional channels, or alliances with the governing party" (Garay 2016: 18). She also argues that when the pressure for expansion comes from above, that is from electoral competition between parties, the expansion of the social policies is more limited and restrictive, while, when the government responds to pressures from below, the resulting social policy models are more inclusive (Ibid., 2016: 19–20).

Although many studies that pretend to arrive at general conclusions on the evolution of the welfare states are very insightful and provide very precious information, one can always find exceptions to their findings and situations that cannot be explained. For example, the democratic hypothesis cannot explain why there are currently such fundamental differences between countries of the Latin American continent when all have embraced democracy for decades: some nations have an almost universal welfare state (Uruguay, Costa Rica), while others have a significantly reduced and assistance-oriented one (Peru, Bolivia, Ecuador, and Mexico). Outside of Latin America, we have two clear examples of this same paradox, in addition to the strong labor movement factor: both Great Britain and the USA (especially the latter), two democratic countries in which there was a strong labor movement during the period when the social protection system was implemented, did not develop a generous and universal decommodifying social protection system such as the one existing in Northern Europe. In the USA, after the New Deal, the advancement of a more extended welfare state was halted because the workers' unions were constrained to act within the Democratic Party, where there was a tie between liberals and conservatives (the Southern Democrats) who did not bet on an expansion of the welfare state (Krugman 2007 111–143). For its part, the British welfare state, despite being universal and widespread, did not upgrade to the generosity of social-democratic welfare states, although the Labor Party governed at many moments and the labor movement was powerful and closely linked to it, as there was never an alliance between the labor sectors, the middle classes, and the farmers as in Northern Europe (Esping-Andersen, 1990). Thus, one cannot consider that the mere existence of a democratic institutional context, the length of democratic rule, and the will of governments to obtain legitimation can be regarded as enough to explain the expansion of the welfare state.

These two examples back Esping-Andersen's main idea, based on a conscientious study of the history of the welfare states of the developed countries, that the origins of the social protection system have not been a specific context such as democracy or lack of legitimacy, nor a single social actor like a labor party or a workers movement, but coalitions between various social actors. "It is a historical fact that welfare state construction has depended on political coalition building. The structure of coalitions is much more decisive than are the power resources of any single class" (Esping-Andersen 1990: 30). Nevertheless, not even this can be considered a decisive causal force, but the task of research in the field is "to identify salient interaction-effects" (Esping-Andersen 1990: 29).

Both Pribble and Garay analyze more recent developments, once all countries of the continent had democracies. Both adduce significant additional arguments to the democratic thesis. Pribble considers that the idea that democracy helps explain the extension of the welfare state is not sufficient but that the ideology of the government is the fundamental real explanatory element. But even this is not enough—the leftist parties must entertain strong links with popular organizations. Thus, the fundamental element that explains a country's expansion of the social protection system is the existence of popular movements and parties linked to those interests. She also marks a difference between more or less inclusive welfare states, depending on whether the pressure for change comes from below or from the government. This has been the case of the Lula and first Dilma governments in Brazil, the Kirchner and the Fernández de Kirchner governments in Argentina, and those of Ecuador under Correa and Bolivia under Evo Morales. However, this argument fails to explain the expansion of the welfare state in Colombia and Chile, for example, as there was a significant expansion of health services during the second decade of the twenty-first century. While in Chile, it was undoubtedly two leftist governments, the one of Lagos and then two of Bachelet, that promoted significant changes in the health scheme, although their party, *Concertación de Partidos por la Democracia*, had practically no link with the existing social movements; it is well known that the Chilean labor movement has been structurally weakened since the promulgation of the labor law of the dictatorship in 1979 and the violent repression of the Pinochet junta. On the other hand, in both of her administrations, Bachelet faced a substantial student movement. In the case of Colombia, until Petro arrived at the presidency in 2022, the left did not govern the country; however, since 1991, under rightist governments, there was a significant expansion of the health system, which reached almost universal coverage and meant, as we will see below, that Colombia was able to weather rather successfully the coronavirus pandemic.

The book of Candelaria Garay innovates in focusing on the "outsiders" rather than on the usual beneficiaries of the welfare state: the army, the public officials, and the formal sectors. In fact, her study is not on the welfare state as such but on social policies oriented to the people living in poverty. She focuses on the marginalized: the poor households and the peasants, who are generally not or weakly organized. She considers that the expansion of social policies oriented to this population results from a competitive electoral arrangement, where the incumbents face a significant threat on the part of parties that are gaining sympathies and votes from outsiders. In addition, she describes how, when the outsiders link themselves to organized social or labor movements, the result is an even more inclusive welfare state. This can explain the expansion of all the assistance programs during the presidencies of Lula da Silva: the Bolsa Familia, the expansion of the SUS, the housing program: "Minha casa minha vida", among others; as is well known, during the two presidencies of Lula, and the first of Dilma, the relation between the labor unions and the numerous social movements that pursued the democratic transition and actively intervened in the writing of the 1988 Constitution had the effect of exerting pressure on a government they considered their own. In the case of Argentina, it was not only the fact that Kirchner and Fernández de Kirchner came from the Peronist movement but also that the crisis that burst after the Menem presidency due to the collapse of the convertibility monetary model gave rise to an impressive outburst of social movements, one of the most important being the unemployed *piquetero* movements. Garay explains the expansion of ample policies oriented to the "outsiders" because these movements managed to ally with the labor unions, specifically with the CTA.

In contrast, in Mexico and Chile, assistance programs for outsiders have been a strategy of the parties in government, certainly under electoral pressure from other parties, but without significant social mobilization. Garay considers that the absence of social movements has meant that in both Chile and Mexico, assistance policies are more restrictive in contrast to the ones of Argentina and Brazil; this is somewhat questionable. This may have been right until the reforms of Lagos and Bachelet, as these two governments modified the Chilean system, especially regarding health, in a much more inclusive direction. This is also the case in a country like Colombia, where a health structure giving services to a sizeable non-contributive population has expanded to become practically universal without social mobilization and, without significant political challenges to the ruling party. The same can be said of the system of Uruguay, which includes both formal and informal workers, insiders, and outsiders, which is the most extensive and egalitarian in the Latin American context.

Notwithstanding these relative exceptions, the ample empirical data of Garay clearly illustrates how the poor population has become a crucial political element that all governments have to consider, but also how social policies are designed to administer rather than solving their fate, as Bruno Lautier signaled since the eighties (Lautier 2007).

In addition to the various hypotheses relating the welfare state to democracy, another one links its character to a **development model**. The protected, inward-oriented development model that was implemented in Latin America from the 40s to the 80s, and again in some countries in South America during the commodity super cycle, could be more easily accommodated with social policies directed towards the urban and working classes, which were the sectors on which the model was politically—as well as economically, since they constituted the internal market that favored the development of the domestic demand—based; the peasant and the urban poor populations were relegated. The counterexample was the low level of social protection in the Southeast Asian countries, which were outward oriented and depended on exporting manufactures with high labor content, which risked losing competitiveness with higher costs and less flexibility in social protection due to denser union organization. Instead, these countries' governments invested more in long-term human capital, especially in education (Haggard and Kaufmann 2008: 1–24). Huber and Stephens also consider, as a possible explanation, the relation between the welfare state and the mode of production; they "argue that patterns of relationships among enterprises, banks, labor and the government that favor the provision of capital to enterprises on preferential terms, that provide for training and retraining of a highly qualified labor force that support efforts in R&D and that facilitate wage restraint are the essential infrastructure on which a generous welfare state can be built and sustained" (Ibid., pos. 223). On the contrary, "dependence on export competitiveness requires some mechanism for keeping wage costs under control" (pos. 227). Finally, these authors consider that production modes imply "a clear connection to the impact of the international economy on the construction and maintenance of welfare states... both its economic and political dimensions" as well as the effects of globalization and deregulation for financial capital in the case of retrenchment (pos. 223–7).

Nonetheless, it is essential to consider that the two models defined by these authors do not exhaust the possible diversity of capitalisms, as they cannot be reduced based on their characterization as an open or a protected economy or an internal or external market-based one. For example, in the case of Latin American countries, we have argued that two other capitalisms exist, both dependent on the export of commodities that we have called rentier. Although both are outward looking, one expands its welfare state and increases wages, while the other holds back both. The reason behind this difference is a strong link, or pressure, from social movements on the government in the first case and a weak one in the second (Bizberg 2019). In addition, the Asian capitalist model does not exist as such since, in this area of the world, there is at least as much diversity as in Latin America. Additionally, export-oriented economies, such as Korea, Taiwan, and even Japan, have considerably stopped their strategy of halting the rise of salaries and social investment. However, they never shifted their economies towards the internal market (Harada and Toyama, in Boyer, Uemura and Isogai 2011).

Considering both the strong and weak points of the analyses we have been discussing up to now, in this book, we will argue that the expansion of the welfare state emerges from multiple rarely unique sources. On the one hand, the existence of a strong labor movement, or of popular movements in general, forces governments to implement or expand the welfare state. However, although we privilege the hypothesis of union strength and autonomy together with their ability to construct coalitions that exert pressure to extend social protections in much the same way as they support industrialization processes, such as import substitution industrialization. It is also possible, as we have seen in Europe and also in Latin America, most notably in Mexico, Argentina, and Brazil, that governments implement social protection measures to preempt the emergence of an autonomous labor movement and try to co-opt it. In countries where social movements are weak, like Peru, Colombia, and Ecuador, following social or political dislocations (military coups, democratic transitions), governments may increase social protection rights as a compensation and legitimation mechanism. The decree of a new constitution that includes new social rights, like in Brazil in 1988 and Colombia in 1991, may also be a moment of social protection expansion as the written text forces even non-progressive governments to comply, at least, to a certain degree. We also agree that in situations with significant electoral competition, for most parties, expanding social protection may be a way to attract voters. Finally, welfare state expansion may be compatible with specific economic models. An export-oriented capitalism in a developing country generally requires lower salaries and lower social protection, as the latter means higher costs for enterprises. In contrast, a model of import substitution capitalism requires, on the contrary, the expansion of the internal market and thus an increase of both wages and social protection rights; this marks the contrast between Mexico during the last 30 years and Brazil and Argentina during the presidencies of Lula da Silva, and Kirchner and Fernández de Kirchner during the recent commodity super cycle (Haggard and Kaufman 2008; Bizberg 2019).

In contrast to the studies that pretend to find general explanations, most of the analyses undertaken by scholars in Latin America have a less ambitious (and less uncertain) goal and have followed Esping-Andersen's indication that "the hope of finding a single powerful causal force must be abandoned" (1990: 29). This is most probably because they consider that the countries of the continent are so distinct that any intention to conceive a general rule does not do justice to the situations these nations have experienced. Thus, they do not intend to find defining elements that explain the development of the welfare state in general or the variables that would determine the evolution of a more generous, extensive, and universal social protection system. Still, they instead try to define the main characteristics of the welfare states that exist in Latin America, their differences and similarities, or, as Martinez Franzoni and Sanche-Ancochea propose, their architecture. This is why many of the studies of the area tend to focus on a single welfare state, study it in detail, in its particularities, while others try to reduce the complexity of a case-by-case analysis and focus on the similarities or differences of groups of countries in order to construct typologies.

One of the first typologies for the Latin American countries was that of Carmelo Mesa-Lago, based on the periodization of the advent of the welfare states, a fact that, according to the author had significant consequences on the characteristics of these systems. Mesa-Lago differentiated between what he called the pioneer countries, which adopted social protection before the Second World War: Uruguay, Argentina, Chile, Cuba, and Brazil, and those that followed; he also added Costa Rica to this first group, although this country developed its social protection system after the Second World War. The countries of the first group implemented retirement programs in the 1920s and 1930s, and like other pioneering countries in Europe, originally "protected the best organized occupational groups [...] evolved in a fragmented way, [...] leading to deep stratification [...]. These systems "gradually incorporated larger groups [...] reaching [at their highest level] 70% or more, [becoming] virtually universal if non-contributory programs [where] the poor are taken into account" (Mesa-Lago 1994: 17). The second group of countries (Bolivia, Colombia, Costa Rica, Ecuador, Mexico, Panama, Paraguay, Peru, and Venezuela) implemented social protection programs after the Second World War. "In these countries, there was a central institution in charge of covering the whole population, but at the beginning, its action was limited to the capital and the big cities, [...] the coverage ... [reached its highest point, between] 13 and 60%. Specific programs protected the most powerful groups [...] [although] fewer special funds were created than in the pioneer countries, so the degree of stratification was lesser" (Mesa-Lago 1994: 17). In fact, although Costa Rica belongs to this second group of countries, it managed to reach the level of coverage of the first group, a country which according to Evelyn Huber has the closest to a social-democratic social protection scheme as defined by Esping-Andersen; it is more universal, less segmented, and more equitable (also see Martinez Franzoni and Sánchez Ancochea 2016). Although our description refers to retirement programs, it more or less coincides with that of the health systems, as according to Mesa-Lago, in 1990, the countries of the pioneer group had almost universal health coverage, between 74 and 96%, while the others covered between 16 and 58%, and less (Mesa-Lago, 2008: 7).

More recent authors have tried to construct a more qualitative typology, inspired on Esping-Andersen. Most of them have stressed that in many Latin American countries, the State was the leading actor that implemented and expanded the welfare state, in a similar manner to that described by this author for nineteenth-century Germany. They have thus considered that the corporatist-statist or Bismarkian welfare regime model applies to many of the continent's countries. Indeed, this ideal type is easily transposed to countries such as Argentina, Brazil, and Mexico in the thirties, when the state, dominated by national-popular political movements, allocated pension rights, and in some cases health services, to government officials and the workers in the best-organized and most strategic economic branches. A process analogous to what Esping-Andersen has called a conservative modernization in countries such as Germany occurred in these countries, where the State "sought to achieve two simultaneous results [...] to consolidate divisions among wage-earners by legislating distinct programs for different class and status groups [...] and tie the loyalties of the individual directly to the monarchy or the central State authority" (199: 24). Conservative modernization was also a manner to prevent the emergence of independent unions that had begun to organize: "The corporatist status-differentiated model springs mainly from the old guild tradition. The neo-absolutist autocrats, such as Bismarck, saw in this tradition a means to combat the rising labor movements" (Esping-Andersen 1990:24), in addition to the effort of integrating the German nation that had emerged as a result of the war of 1870 against France (Crouch et al. 1993). The conservative-corporatist model allocated privileges to public officials and the workers in the most strategic economic sectors throughout the German Empire, starting with the functionaries and the army. It assigned generous privileges to these sectors to link them in a dependent relation to the State while allocating less generous ones to other sectors; in this way, it divided unionism through the granting of privileges in a segmented manner.

On the other hand, this model is also considered a conservative response to capitalism, a reaction of the traditional ruling circles (Bismarck and the governments of the Third Republic in France) to the development of this economic transformation, more specifically, to the commodification of labor). This political reaction to an economic process tried to delay the erosion of the old institutions: the church, the guilds, and the family; up to a certain degree, the social protection programs oriented to distinct social groups were a way in which the guilds were maintained, while the conservative character of this "modernization" extends up to the present, as the German welfare state is a system "that supports a 'male breadwinner' model, belittles the social rights of women" (Sainsbury 1999: 78). According to Esping-Andersen, "the corporatist regimes are also typically shaped by the church, and hence strongly committed to the preservation of traditional familyhood. Social insurance typically excludes non-working wives, and family benefits encourage motherhood. Daycare and similar family services are conspicuously underdeveloped; the principle of 'subsidiarity' emphasizes that the State will only interfere when the family's capacity to service its members is exhausted" (Esping-Andersen 1990: 27).

In Latin America, Carlos Barba, Enrique Valencia, Fernando Filgueira, Martinez Franzoni, and Sánchez Ancochea are probably the researchers who have conceived the most elaborate typologies. Although they do not coincide wholly with the typology of Esping-Andersen, they are inspired by it and adopt the first of the types defined by this author: the corporatist one, which is a socio-political regime established both for political and for socio-economic reasons, intended to coopt and divide the most active and strategic social actors by offering them privileged socio-economic conditions. This type, named universalistic by Barba and segmented universalism by Filgueira, coincides most closely with the existing welfare state of Uruguay, Argentina, and Chile (before the reforms of Pinochet). The argumentation of Filgueira is that one of the most significant characteristics of the social protection systems of the Latin American countries included in this model is that they comprise a large majority of their population in pension and health schemes, as well as give education to the totality of the population at the primary and secondary level, under a segmented model. Using Esping-Andersen's central concept, what characterizes these regimes is that they "offered extended levels of 'decommodification' both in the provision of services outside the market and in the provision of monetary benefits for various situations of job inability" (Filgueira, w/d). In addition, according to Barba, "like the conservative European regimes, the axis of the Latin American paradigm was essentially political, since it conceived the welfare state in a corporatist form intended to preserve status differences and privileged the breadwinner regime, in which the family and particularly women were responsible for allocating welfare services, while adult men played the role of providers and right holders for the rest of the family through formal employment [...] that only incorporated members of the social coalition which promoted the industrialization project, particularly the labor organizations and the middle classes, forged by the modern sectors of the economy and the state, but left aside the rest of society" (Barba 2004: 15[2]). According to this author, while in Europe, this corporatist welfare regime expanded to include most of the population and evolved from the conservatism and authoritarianism of its beginnings towards a neo-corporatist model based on the coordination of social actors: trade unions and employer associations in the case of Latin America this corporatist type continued to evolve towards an "exclusionary and non-democratic social security regime that did not allow an expansive process of rights and social citizenship, but rather of clientelist relations" which profited the officials of the public sector and the workers of the most modern and strategic enterprises, while excluding the workers of the informal sector and the peasants, leading "to high levels of inequality in income distribution, coverage of social protection, and quality of social services" (Ibid).

Both Filgueira and Barba depart from Esping-Andersen's definitions for their other two types of welfare state. As is well known, Esping-Andersen defines a liberal welfare regime that some authors have named residual, characteristic of the Anglo-Saxon countries, which is based on minimum assistance for the people living in poverty. Although it existed as a prototype since the beginning of the seventeenth century, with the English Poor Laws, the contemporary regime was established during the crisis of 1929 under the New Deal, as well as in Great Britain after the Second World War, based on the report of 1942 of William Beveridge. In contrast to the political rationale of the corporatist regime, the liberal is basically moved by a socio-economic rationale: guaranteeing the cohesion of society, oriented to assure citizens with minimum resources so they can survive if they lose their jobs or once they have finished their working cycle; it ensures that people who are near or in the poverty trap are rescued from it. As discussed by Esping-Andersen, this regime is very commodifying since individuals depend on the market if they want to get more than the minimum, and also because to get that minimum, they have to prove they are poor, thus rejoining the idea of the deserving poor in the Poor Laws (Esping-Andersen 1990: 26–9).

The third model defined by Esping-Andersen is the social democratic. This model also points to social cohesion but through universality and equity. Benefits are allocated based on citizenship; they do not depend

[2] https://repositorio.cepal.org/bitstream/handle/11362/6080/S047561.pdf?sequence=1.

on the individuals's job situation, nor are they assigned only to the poor households, but they intend to equalize all of the members of society. The social-democratic model of the northern European countries originates in a similar approach to the Beveridgean regime. Still, unlike the way it was implemented in the Anglo-Saxon countries, based on a minimum to be acquired by all citizens, and where in this manner, the benefits given by the State are meager because they are set based on minimum wages, in the Scandinavian countries the benefits were adjusted towards the income of the middle classes. The political economy for this trajectory divergence is that the welfare state in the Scandinavian countries was expanded based on an alliance between the popular and the middle classes once the latter was integrated into the universal scheme. In the liberal regimes, the middle classes remain outside and receive a minimum from social protection and add to it market solutions: private or group savings in the case of pensions, private health insurance, etc. Corporatist regimes integrate the middle classes (professionals, civil servants), through their own corporations in a segmented and differentiated manner[3] (Esping-Andersen 1990: 26–9).

These two latter types, the liberal model and the social-democratic model, have not been adopted by the typologies of Latin America. Concerning Andersen's social-democratic model, Barba considers that none of the Latin American welfare states have had the continuity of those of the industrialized countries. On the other hand, except in some countries, basically Argentina and, to a certain extent, Mexico, due to its lack of autonomy from the state, the working class has not been a pressure factor demanding the expansion of the welfare state, a central feature of the North European states. He also mentions differences in the level of industrialization and income, which have been crucial: while in the developed countries "salarization" has been practically universal, in Latin American countries, salaried sectors have been accompanied by

[3] Esping-Andersen also makes differences between the universal ans statist British health system and the North American one, which is basically private. In the latter, individuals must purchase private insurance, either individually or through their companies, while the public system is geared towards people with low incomes (Medicaid) and the elderly (Medicare). The Obama reform universalized private insurance by making them accessible to all: either through the obligation of insurers to offer "group" insurance to individuals and/or giving tax credits to individuals who require them to purchase insurance and sanctioning those who do not have one despite these facilities, although it still excludes illegal migrants, around 11 million people (Martin, S., in Bizberg and Martin 2012).

large sectors of informal workers in both the urban and the rural zones (Barba 2004: 14).

One can also consider that a liberal-residual welfare regime has never existed in Latin America because after the nineteenth century, during the period of the export of commodities, and until the 1990s, most of the Latin American countries have seldom been liberal. During the period of ISI, the State was interventionist, protectionist, and, in fact, mercantilist. Only after the crisis of the eighties, the lost decade, liberalism, or neo-liberalism, began to impose itself on the continent. The welfare state was instituted and developed in most of these countries during the interventionist period; it began to be eroded in the direction of a residual model only since the 1990s, so in most countries, it is hybrid. However, although one could consider it to be neo-liberal at present, it cannot be regarded as *residual* as in the developed countries since the poor households are, in effect, a residual part of society, while the people with low incomes in Latin America are a large proportion of the population, and in some cases the majority.

Barba, Valencia, and Filgueira propose alternative types of welfare regimes besides to the corporatist one: a dual and an exclusionary welfare regime. The dual regimes that basically correspond to Brazil and Mexico are "built on stratified forms of labor insertion that coexist with care and informal mechanisms to attend to the well-being of those who lack formal employment, associated with clientelistic and patrimonial practices". They "generated a great institutional segmentation, stratified coverage in social security and health, as well as a great dualism in the rights of those who participated in the formal economy and those who did not... caregiving was considered a family task because it was naturalized as a female responsibility, and in countries with a larger indigenous or Afro-descendant population, the trend was to strengthen the reproduction of the social inequalities inherited from the colonial period and gender inequalities" (Barba and Valencia 2021: 2). The exclusionary regime exists in the poorer countries of the continent: Ecuador, Bolivia, and Central America. In this model, as its name indicates, only a very reduced part of the population has the right to pensions and health: the army, the police, some of the functionaries, and industrial workers of the most strategic sectors (oil, gas, electricity), while the great majority does not have access to these services and has to rely on their families or on the health service market, when old or sick (Filgueira, op. cit.).

The contribution of Martinez Franzoni to this discussion is crucial as she **explicitly** introduces the family, and more specifically, women, in addition to the State and the social actors, as one of the central defining actors of the welfare state.[4] In fact, the exclusionary regime can be defined "positively" as one where the family, and more precisely women, protect the individuals in substitution of the State and the different social organizations. She defines a specific **informal-familiarist** welfare regime that exists in the continent's poor countries. Although in all of Latin America, the family has a fundamental role in compensating the lack of a comprehensive social protection system, even in the most advanced ones, it continues to fail to ensure women equivalent conditions to those of men in terms of job opportunities, wages, and share of household tasks, in a "truly family-based welfare regime [where] the population largely relies on family arrangements, as social public policies are inadequate or non-existent in countries like El Salvador or Bolivia" (Martinez Franzoni et al. 2009, and Martínez Franzoni 2008). She defines two other types of regime, the **statist-productivist (or focalized)** one, where "informal arrangements interact with public policy that either emphasizes targeted policy or focuses [public policy]"—a regime that exists in very liberal contexts such as Chile, with a "more universal yet stratified social protection regime", which she names **protectionist (or stratified)**, that also exists in Costa Rica, Uruguay, Argentina, and Brazil (Martinez Franzoni et al. 2009, and Martínez Franzoni 2008). In a more recent book, Martinez Franzoni and Sánchez-Ancochea rename this latter regime **segmented corporatism,** adopting, in fact, the definition of Barba and Filgueira. On the other hand, they consider that Costa Rica is advancing towards a system similar to the one of the social-democratic countries, which they call **egalitarian universalism** (Martinez Franzoni and Sánchez Ancochea 2016).

Nevertheless, the most significant contribution of Martinez Franzoni is that she includes gender as a dimension that has the potential to radically modify the typology of the Latin American continent, as it has occurred in the case of developed countries. One cannot understand the welfare state if one ignores gender, both because gender fundamentally modifies our vision of the welfare state and because the welfare state shapes the

[4] In his recent articles, Barba has also introduced the family as a significant variable.

unequal structure of society without which capitalism would be impossible, where women contribute with care and reproductive labor, while men are engaged in productive labor. And, even more fundamentally, while productive labor is considered work, is retributed, and is the source of privileges that go beyond men's working phase, reproductive labor is a task that "**naturally**" corresponds to women and is thus non-retributed. In the last chapter of this book, we will discuss in more detail the contributions of this author, where we question the whole architecture of the theoretical propositions we will define in this chapter and which will guide the next ones.

There is finally another interesting typology that of Cecchini and Martinez (2011), where the models of welfare regimes are defined based on the concept of citizen's rights to social protection in the sense of T.H. Marshall. They define a continuum from **citizen rights** to **social protection** to **assistance,** which they describe as the two dominant types in the present period characterized by an economic model based on competitivity—a continuum resembling the one of Esping-Andersen centered on decommodification. Their conceptualization is based on how social protection is allocated and whom it benefits. They define four types, while they consider that only the last two of them are pertinent to describe the present situation: (1) protection based on formal employment; (2) protection due to emergencies (like unemployment); (3) assistance and access to promotion (based on the idea of increasing human capital to extract individuals from poverty and prevent its generational transmission); and (4) social protection as a right, linked to citizenship (2011: 38). These authors consider that in the twenty-first century, once the countries of Latin America have abandoned both ISI and the radical neo-liberalism, to adopt what they call, a systemic competitivity model, only the last two types dominate. They thus base their typology on the conception of Marshall, who, as is well known, considers a historical evolution of the acquisition of rights by humanity: from civil to political rights, and then to social rights. In Latin America, there are countries where the majority of the population has no rights. However, they get assistance from their governments, while others have advanced in constructing their social protection system enough to consider they are near to guaranteeing social citizenship to their population.

b. *A typology based on the societal orders*

Although the different typologies that we have been describing are very valuable and have been applied in enlightening manners in numerous books and articles to explain the situation of social protection in Latin America, in this book, we will defend the idea that there is room for a different typology that adds elements to increase our comprehension of the situation of the welfare states in the continent. A typology has to be internally coherent, meaning that all of the ideal types must be defined based on of one single basic principle so that each type is as exclusive as possible, that there are no overlaps between types that reduce their explanatory capacity; if a typology is constructed on various principles, then overlaps are inevitable. The concepts that should be used to conceive the ideal types do not arise from a theory but are historically based; they are founded on a historical analysis of the actors that intervene in the construction of the welfare states. We should remember that we are constructing ideal types that do not coincide with the existing welfare states because these are hybrids that combine (in different proportions) the ideal types. Nevertheless, some actual social protection systems approach an ideal more than others. However, to retain their explanatory capacity, the ideal types *per se* must be as differentiated as possible, and they should not be hybrids.

We have analyzed three different perspectives to classify the Latin American welfare systems; the first one approaches it from the point of view of the concept of segmentation (corporatism, dualism, exclusion); the second one from the perspective of familiarization/focalization, and the third one from the perspective of citizenship. Nonetheless, none of these perspectives is totally coherent; Barba, Valencia, and Filgueira include a type external to the conceptual frame, that of dualism, which is basically descriptive. There is thus an overlap between corporatism and dualism, as corporatism can also be described as dualist since there is a formal segmented sector and an informal one. While the dualist regimes exclude part of the population—the informal sector and the peasants—the segmented regimes can also be described as dualistic, as the State covers the remaining population outside the corporatist sector in a deficient social protection scheme. In the case of the typology of Martinez Franzoni, one can see how there is an overlap between the informal-familiarist type and the focalized welfare states. While it is evident that the family (especially women) carries the weight of caring for those persons that the market leaves behind, nonetheless, this population is generally attended by focalized social policies implemented by the state. In contrast, the focalized states are supposed to prioritize relying on these mechanisms of focalization.

On the other hand, most typologies are normative, as they consider one of the types as the best. Indeed, the typology of Esping-Andersen also falls into this problem when he defines the social-democratic model as the best based on his decommodification principle. The most common way of (falsely) comparing countries is to postulate one of them as the ideal and compare the others against it. In other cases, several types of regimes are considered, placing them in a hierarchy that goes from best to worst. This is a derivative of the first type, in which an ideal type is postulated. In both cases, it is a form of classification that starts by finding the similarities and differences that can be found empirically to classify the systems and name one of them as the most suitable; one country is considered as the model: Sweden, Japan, and the United States, and is then used as the reference against which the others are contrasted. This procedure is not a comparison but a description based on how far each country is from the ideal model.

My third criticism touches even the most noteworthy typologies, such as the one of Esping-Andersen. In fact, he defines an ideal model based on the concept of decommodification, which states that social-democratic welfare states centered on the State (universal, egalitarian, and generous) are on the top of a hierarchy that goes from more to less decommodifying. Next come corporatist-segmented regimes based on wage earners, civil society, or trade unions. And finally, there is the residual liberal welfare state, where the market predominates and social security is residual. Most of the typologies that have been elaborated on the social protection regimes of our continent consider that universalization is the goal towards which all of them should advance. However, universalization should not be considered only regarding coverage but must also be understood as equity and generosity (Martinez Franzoni and Sanchez-Ancochea, 2018). Nonetheless, according to these authors, from a perspective very similar to that of Andersen, this goal can **only** be achieved in regimes where the State finances (through taxes) and offers health services and pensions, the best being a statist one. The argument in the case of health services is that to the extent that they are integrated at the national level, they prevent the entry of "external" providers (whether social or private) which introduce disparities and make the system less equitable. The same is true regarding pensions, the best one is a single scheme, under the control of the state, based on citizenship, which homogenizes the conditions. This means that a more segmented structure, in which not-for-profit mutualist associations intervene (cooperatives of doctors, communities, unions, NGOs, etc.) or even for-profit enterprises, is less desirable, even if it achieves coverage of the population and if it is equitable, both are external and inevitably tend to introduce segmentation and inequality (Ibid.). For example, although the book by Cecchini and Martínez identifies the actors we will be talking about next: the family, society, the market, and the State as providers of social security, they consider that the "complete social responsibility [...] corresponds fundamentally to the State as social protection is conceived as a right and not as a privilege" (2011: 21). The book by Martinez Franzoni and Sanchez-Ancochea compares Costa Rica and Uruguay, and concludes that the first is more universal than the second, basically due to its being state based.

The existence of other actors in the provision of health introduces complexity and is thus a potential source of inequalities. As we will discuss further, one of the consequences of the private sector's intervention in the health structure of both Brazil and Chile was its inferior performance during the pandemic. Nonetheless, these deficiencies may be solved through governmental regulation and surveillance by social actors, groups of beneficiaries and providers, unions, and mutualists rather than canceling all intervention of any actor other than the state.

On the other hand, this position does not consider the evident risks of having the State as the exclusive provider of all social protection. The fact that in some countries, like France and most of Latin America, where social protection is wholly dominated by the State, or Great Britain, where the state dominates the health services, has had significant consequences at a time of retrenchment and declining social protection since the 1990s was a direct result of "statization" (Théret 1998). Concentration of the social protection system in the hands of the State has been functional to the process of its downgrading. This operation has been much more difficult where social protection relies on social organizations. The more state-dependent social protection regimes have more successfully substituted social rights by means-testing and assistance. As we will see in the case of Latin America, during the last thirty years, in most countries, we have lived through the privatization of pensions on one hand, which went from segmented "pay-as-you-go" schemes to individual capitalization schemes, complemented by non-contributive programs allocated by

Table 2.1 Welfare State typologies

Esping Andersen	*Martínez franzoni*	*Filgueira/ Barba Solano*	*Martinez Franzon and Sánchez-Ancochea*	*Ceccini and Martinez*
Liberal/Residual: US, Canada, Great Britain	Informal: central America, Ecuador, Peru, Colombia	Exclusionary: central America, Ecuador, Peru, Colombia		Assistance
Conservative corporatist: Germany	Focalized: Argentina y Chile	Dual: Brazil, Mexico	Segmented Corporatist: Uruguay	Intermediary
Socio-Democratic: Sweden	Stratified: Brazil, Costa Rica, Mexico, Uruguay	Stratified universalist: Argentina, Uruguay, Chile	Egalitarian universalist: Costa Rica	Citizenship

Source Own elaboration, based on various works of the authors, cited in the references

the state, which offer minimal pensions, or programs guaranteeing a minimum to those workers who have contributed to an individual fund but did not manage to save enough to get a minimum allowance. On the other hand, centralized State health structures, which are inefficient and underfinanced, are substituting civil society-based health schemes in some countries with the argument that they are segmented, which they are effectively (Table 2.1).

In contrast to these typologies, we define the types of welfare state according to a non-normative, unique **principle**, neither based on characteristics such as segmentation or dualism, nor as a continuum between two models or built upon an ideal model from which the other types diverge (in both the developed and underdeveloped world, there exist various types of social protection systems that have been developed historically and are equally capable of protecting society). We adopt the methodology proposed by Bruno Théret, who defines certain parameters through a historical analysis of the countries that one wants to compare, to induce a principle that becomes the theoretical concept that will allow us to deduce the different types. Some of these types may be coherent and stable, others not. We would then study how the different countries approach these types. As we already said, none of the countries coincide with any specific type; some are nearer to one of them, others are further away, and some are "perfect" hybrids of two or more types.

Unlike the perspectives we have been discussing, which begin with an empirical analysis of the characteristics of social protection regimes in order to define "ideal forms", Théret proposes a typology based on different combinations of the societal orders or spheres that define social protection regimes. These orders are determined on the basis of Polanyi's central idea that "The national social protection systems act as a mediation between these three autonomous spheres of social practices—**the capitalist corporation, the nuclear family and the administrative State**—that participate to the reinstitution of societal cohesion" (Polanyi 1944, cited by Théret 2011: 181). Théret adds a fourth sphere: civil society, that is the social actors that intervene in the materialization of the social protection schemes, which in some countries administer and allocate these services directly; as in the case of health, the Argentinian *Obras Sociales* and the Uruguayan *mutualistas*.

We will thus ground our typology on the dominant **actor** that administers and allocates social protection. Although social protection systems are all a specific combination of state, market, society, and family, our ideal types will be defined by a dominant actor in each. In this manner, we will end up with a typology of protection systems that interrelates the state, civil society, the market, and the domestic sphere: one is statist, another one is based on the market, another on the family, and finally, the last one on civil society.

This scheme coincides, in several of its types, with those elaborated by other Latin American authors, but rather than being focused on how the welfare state allocates resources, it is focused on the dominant actor. It does not propose a model that is considered the best, but different types that may be more or less coherent with these countries' economic, social, and political institutions (Table 2.2).

We define four types of welfare state, or four worlds of welfare state as Esping-Andersen called them. The first one is the statist regime, where the State is the dominant actor in the social protection system, as it is this actor that recollects, administers, and allocates pensions and health services. In this case, society is a secondary actor, although there may be some organizations or institutions, some of them stemming from the origins of the social security system, that occupy themselves on these services: they may be civil society associations, like the medical profession, unions, or even sectors of the state, that have maintained their funds

Table 2.2 Typology of the Latin American welfare states

Type/ Societal Order Hierarchy	*Statist*	*Socio-corporatist*	*Commodified*	*Familiarist*
State	Dominant	Complementary	Subsidiary/ regulatory	Marginal/ strategic sectors
Society	Secondary	Dominant	Marginal	Marginal
Market	Residual	Residual	Dominant	Complementary
Domestic	Dependent on coverage and efficacy	Dependent on coverage and efficacy	Complementary	Dominant
Country that approaches the ideal type	Costa Rica	Argentina, Uruguay	Chile	Perú, Bolivia

Source Own elaboration

and services: the army, the police, for example. Depending on the efficacy of the services that the State apportions, families and the market are marginal. However, some systems allow a particular share of pensions and health services to be allotted by the private sector (for-profit or non-profit; the distinction is crucial), and in some schemes, the government turns to private health services as complementary.

The second one, named in the same way as many Latin American scientist have done, is the corporatist or **socio-corporatist regime**. This regime is defined by the distinct, in some cases majoritarian administration of the welfare state by society, be it unions or civil society associations. The model is the German or French systems, where society assures or participates in the entire social protection system. Although, in many countries in Latin America, this participation does not cover more than 50% of the total population, and the State has to complement the welfare system to assure the rest of the population, the regime's character is totally different from the statist. In general, in this type, the domestic sphere and the market play a much smaller role due to the "demonstration" effect of social action.

The commodified regime is one where the market is dominant, setting the pace, and the State. However it may assure a more significant proportion of the population; nevertheless, is in subsidiary relation to the market. Private enterprises insure the individuals: banks in the case of pensions and insurance companies and private services in the case of health. When these market services fail, individuals cannot pay for their health insurance or do not get a decent income from their private funds, the State comes in. The domestic order, that is the families are complementary to the market and the public protection system. The social order, society, is marginal.

Finally, in the familiarist regime, the domestic order, the families, and especially women, are central and dominant; they have to allot a signficant portion of the resources, monetary and care-wise, to the young, the sick, the unemployed, and the elders. The State allocates basic assistance, a minimum to survive. Society is present as a charity, churches, and ONGs. Families must resort to the market to acquire what they need when a family member is sick; thus, the market plays a complementary but significant role, much higher than in statist and socio-corporatist regimes.[5]

While the first three social protection regimes are coherent and can reach universal coverage, it is clear that the fourth type, the familiarist, is not a coherent in principle because it leaves the entire burden of social protection on the family; it is a type in permanent crisis, due in part to the fact that governmental assistance can neither compensate the deficiencies of the market, nor of the family, and to the fact that the families of the continent are undergoing signficant changes due to cultural transformations in their character, due to the changing role of women, migration, violence, etc.

We are nonetheless left with three coherent types of welfare regimes that emerged from a non-hierarchical typological construction, which, if we look at certain quantitative studies done on the welfare systems in our continent and other parts of the world can be as "protective". In fact, two quantitative studies that try to evaluate which of the Latin American welfare states are the most comprehensive[6] coincide that, in effect, they are of various types; Uruguay, Chile, Costa Rica, Brazil, and Argentina all

[5] Although the Japanese welfare model has been sometimes defined as familiarist, it is paternalistic and actually rests on a social actor: the companies (it has also been defined as "companyist"). As is well known, although it rests on an economic actor, the companies, it does depend on the market, but on a paternalistic relationship of the enterprises towards its employees (Lechevallier 2015).

[6] Although probably this concept does not coincide with the idea of universality of Martinez Franzoni and Sánchez Ancochea 2016, because it does not include equality (together with coverage and generosity) as one of its dimensions, it may still be close to it.

seem to be equally protective. Although one may be skeptical of specific inclusions in the group of the most comprehensive welfare states, since none of them is a universal system in the conception of Martinez Franzoni and Sánchez Ancochea 2016, the critical idea of these studies is that they conclude that all these countries, with different welfare regimes, are equally comprehensive (Ocampo and Gómez Arteaga 2016. Cruz Martinez 2021).

Another study, this time of health systems in European countries, concludes how very different types of health systems can achieve the same level of performance (in terms of coverage, equality, and sufficiency, which amount to an indicator very close to that defined by Martinez Franzoni and Sánchez-Ancochea as universalism). While in the British health system, the State is both single payer and provider, healthcare is primarily private in the Dutch and Swiss systems. A segmented health system (the German one) is basically administered and provided by social organizations and associations; it is based on negotiations between companies and unions, while the State collects contributions and is provided by private insurance and clinics (both for profit and non-profit), under contract by both social actors. In the private health systems of the Netherlands and Switzerland, individuals contract either individual or collective insurance with private companies (by way of their employment), and services are offered by private hospitals, clinics, and doctors. Although the systems of these two countries are private, they are strictly regulated regarding coverage, prices, co-payments, the cost of medicines, etc. The State imposes the obligation to be insured by imposing fines to those who are not, while public collectivities negotiate coverage for individuals who do not have the resources to pay for insurance (including refugees and illegal migrants) with private insurance companies and offer subsidies; they then get the same quality of services than the wage earners (Schneider et al. 2017).

In the following chapters, we will analyze the different welfare states of the ten countries we have included in this book from this perspective.

References

Barba Solano, Carlos (2004), "Régimen de bienestar y reforma social en México", CEPAL, *Serie Políticas Sociales*, No. 92.

——— (2021), "El régimen de bienestar mexicano: inercias, transformaciones y desafíos", *serie Estudios y Perspectivas-Sede Subregional de la CEPAL en México*,

191, Ciudad de México, Comisión Económica para América Latina y el Caribe (CEPAL).
Bizberg, Ilán and Scott B. Martin. (2012), *El Estado de Bienestar ante la Globalización, el caso de Norteamérica*, México, El Colegio de México.
Bizberg, Ilán (2019), *Diversity of Capitalisms in Latin America*, London, Palgrave Macmillan.
Ceccini, Simone and Rodrigo Martínez (2011), "Protección social inclusiva en América Latina: una mirada integral, un enfoque de derechos", *Libros de la CEPAL*, 111.
Crouch, Colin (1993), *Industrial Relations and European State Traditions*, Oxford, Clarendon Press.
Cruz-Martínez, Gibrán (2021), "Mapping Welfare State Development in (post) Neoliberal Latin America", *Social Indicators Research*, 157: 175–201.
Esping-Andersen, Gøsta (1990), *The Three Worlds of Welfare Capitalism, Princeton*, Nueva Jersey, Princeton University Press.
Garay, Candelaria (2016), *Social Policy Expansion in Latin America*, New York, Cambridge University Press.
Haggard, Stephan and Robert R. Kaufman (2008), *Development, Democracy, and Welfare States*, Princeton, Princeton University Press.
Harada, Y. and H. Tohyama (2011), "Asian Capitalisms: Institutional Configurations and Firm Heterogeneity", In R. Boyer, H. Uemura and A. Isogai, *Diversity and Transformations of Asian Capitalisms*, London and New York, Routledge, 243–263.
Huber E. and J. D. Stephens (2001), *Development and Crisis of the Welfare State: Parties and Policies in Global Markets*, University of Chicago Press.
Krugman, Paul (2007), *L'Amerique que nous voulons*, París, Flammarion.
Korpi, Walter (2006), "Power Resources and Employer-Centered Approaches in Explanations of Welfare States and Varieties of Capitalism: Protagonists, Consenters, and Antagonists", *World Politics*, 58 (2): 167–206.
Lautier, B. (2007), "Les politiques sociales au Brésil durant le gouvernment de Lula: aumône d'Etat ou droits sociaux?", *Problèmes d'Amérique latine*, 63: 51–76.
Lechevalier, Sébastien (2015), "L'hétérogénéité croissante des entreprises au sein du capitalisme japonais: Faits, causes, conséquences, et implications", In Akinori Isogai, Hiroyasu Uemura y Robert Boyer (dir.), *Capitalismes asiatiques: Diversité et transformations*, Presses Universitaires de Rennes.
Mesa-Lago, Carmelo (1994), *Changing Social Security in Latin America. Towards Alleviating the Social Costs of Economic Reform*, Boulder and London, Lynne Rienner Publishers.
Martínez Franzoni, Juliana, Maxine Molyneux and Diego Sánchez-Ancochea (2009), "Latin American Capitalism: Economic and Social Policy in Transition", *Economy and Society*, 38 (1): 1–16.

——— (2008), *¿Arañando bienestar? Trabajo remunerado, protección social y familias en America Central*, Buenos Aires, Consejo Latinoamericano de Ciencias Sociales- CLACSO.

——— (2016), *The Quest for Universal Social Policy in the South: Actors, Ideas and Architectures*, Cambridge: Cambridge University Press.

——— (2018), "Undoing Segmentation? Latin American Health Care Policy During the Economic Boom", *Social Policy & Administration*, Special number on Latin America Press, 52 (6): 1181–1200.

Mesa-Lago, Carmelo (2008), *Reassembling Social Security. A Survey of Pensions and Health Care Reforms in Latin America*, Oxford, Oxford University Press.

Ocampo, José Antonio and Natalie Gómez-Arteaga (2016), "Social Protection Systems in Latin America: An Assessment", *Extension of Social Security series*, 52, Geneva, International Labour Office, Social Protection Department (SOCPRO).

Pribble, Jennifer (2013), *Welfare and Party Politics in Latin America*, Cambridge University Press.

Polanyi, K. (1944), *The Great Transformation*. New York, Farrar & Rinehart (cited from the french translation of 1983: Paris: Editions Gallimard).

Sainsbury, Diane (ed.) (1999), *Gender and Welfare State Regimes*, Oxford University Press.

Schneider E., D. Sarnak, D. Squires, A. Sharnav and M. Doty (2017), *Mirror, mirror 2017: International Comparison Reflects Flaws and Opportunities for Better Health Care*, New York, Commonwealth Fund.

Théret, Bruno (1998), "L'État-providence à l'épreuve des comparaisons internationals", In Philippe Auvergnon, Philippe Martin, Patrick Rozenblatt et Michèle Tallard (eds.), *L'État à l'épreuve du social*, París, Editions Syllepses.

——— (2011), "The Place of National Systems of Social Protection and Political Representation in Socio-Economic Regulation: A Morphogenetic Structuralist View on Institutional Change in Comparative Perspective with Special References to Japan and France", *Evolutionary Institutional Economic Review*, 7 (2): 177–208.

CHAPTER 3

The Political Economy of the Statist and Socio-Corporatist Welfare States in Latin America

As we already discussed in the first chapter, we will use the concept of the social contract in the sense of a social compromise, what the French regulation school calls *compromis social,* a less totalizing conception than that of Rousseau's foundational contract but more than a mere short-term arrangement; a type of social pact that includes how a particular society organizes production and distributes the wealth it creates, and in this last sense defines how the social protection system is configured. The social compromise emerges from a specific relationship between the social sectors and the State; hence, it depends on the positions that the State, the popular sectors, and capital occupy.

Based on different combinations of class coalitions and through the pressure of the social actors that fought for their expansion, most of the European regimes moved away from statism to socio-corporatism. The

---

With the collaboration of Mauricio Jiménez Hernández

I. Bizberg, *Four Worlds of the Welfare State in Latin America*,
https://doi.org/10.1007/978-3-031-44420-3_3

Bismarkian welfare state in continental European countries changed radically; it lost its original conservative statist character when it expanded to incorporate most of the labor sectors and the middle classes to be supported by a broad social coalition, becoming socio-corporatist. In the case of the northern European countries, a Beveridgean system not only accomplished expanding universally through a "red-green" alliance between working-class sectors and small agricultural landowners, but since the end of the Second World War, under the push of the social-democratic parties, it incorporated the middle classes and expanded its benefits adopting the expectations of the latter. Although the Beveridgean regimes in most Anglo-Saxon countries developed, they failed to become as generous as the corporatist, as the middle classes chose a private way of protecting themselves, and they lacked the support of the most powerful unions and employers that opted for a social protection through contractual negotiations. While the pension system remained at a minimal level in the United States, Great Britain, and Canada, the latter countries followed a different trajectory in their health system, which expanded universally under the direction of the State.

In contrast, the fact that in most Latin American countries the politico-legitimating role of social protection policies was not only foundational but continues to be relevant today implies that, in many nations of the continent, the State maintains its primordial role. On the other hand, in most cases, social actors have either been too weak, were controlled by the State, or have adopted a radical stance, rejecting any negotiation with the government or with political intermediaries. The cases of Uruguay and Argentina are exceptions, as we will promptly discuss, as social actors have been central in shaping the welfare state. In the case of these two countries, a large part of the health system is directly controlled by social actors, the *mutualistas,* and the unions, although the State was central to expanding both the health and the pension systems. In the case of Uruguay, although the analysts consider that the Uruguayan welfare system was a creation of the State as social actors were weak, and nonetheless, for example, in the case of the health system, the government allocated the population services that already existed in the *mutualistas*;

thus, in a sense, the State responded to a demand that was already present and had been created by civil society.

The government prompted Argentinian welfare state to divide, coopt, and prevent the autonomous organization of the labor movement; however, the unions achieved greater autonomy since, during the Peronist era, they were empowered and were given the health system to administer and finance. And while when Perón ruled the country they were subordinate to the State, after the military coup of 1955 when the *Justicialista* party was illegalized, they strengthened their autonomy through their opposition and negotiation with the governments of the next 20 years; they also managed to maintain their social power, by continuing to control the health system. In contrast, in Mexico, unionism continued to be subordinated to the State due to the capacity of the PRI to maintain itself in power, while in Brazil, unionism was much weaker as an actor—it was fundamentally an administrative rather than a political actor—it was loosely rooted in the workplace and to the workers, and continued to be heteronomous until the beginning of the democratic transition in the late seventies. Thus, in both countries, the statist character of the welfare state has prevailed until the present. The same can be said of the trajectory of Costa Rica, which never lived a corporatist moment and where the welfare state was a direct creation of government.

The Socio-corporatist Welfare Regimes (Uruguay and Argentina)

Among the countries we are discussing, there are two in which, although both pensions and health services extended as a process of conservative modernization, their trajectory led them to depend on civil society increasingly. In fact, in the case of the health system of Uruguay, the process was even the opposite: since the mid-nineteenth century, health services were first implemented by civil society associations, the mutualists, while the State expanded them to the rest of the population in the third decade of the twentieth century. In Argentina, the unions created the *Obras Sociales*, which administer the health funds and give services to the workers of the sectors they represent, and nowadays to all who want to adhere. In both Uruguay and Argentina, civil society has resisted the frequent attempts to privatize or centralize these services in the hands of the State.

Uruguay

One of the most prominent analysts of the Uruguayan social policies considers that "In the 1950s, Uruguay was known as a pioneer in the development of the welfare state [...] due to the ... implementation, in the second decade of this century, of a series of measures that created a social protection system designed to be provident and to anticipate demands" (Rial 1984, cited by Filgueira 1995: 1). One of the precedents is the very early enactment, in 1829, by President Rondeau, of the payment of a pension to the families of soldiers that died or were maimed during independence (Mesa-Lago 1978: 70–1). As usual, in most welfare systems, the establishment of pensions has always been the initiative of the State, typically insuring first the army and the police, then the civil servants, educators, and later still the blue-collar workers of the most strategic public enterprises, and finally the private sector workers (Mesa-Lago 1978: 264).

More decisively, during the first presidency of José Battle (1903–1907), the *Fondo de Pensión al Servicio Civil* was created to regularize the few pension funds that had been previously created (for police officers, firefighters, teachers, and civil servants) and expand coverage and increase benefits (Martinez Franzoni, and Sanchez-Ancochea 2016: 10). In 1919, the government created the Retirement and Pension Fund for Public Services Employees and Workers, which protected railway, telegraph, telephone, water, and gas workers—a fund that included not only old-age pensions, but also disabilities, unemployment, and death (Ruezga Barba 2009: 209). In 1925, the Retirement and Pension Fund for Employees of Banking Institutions and Stock Exchange was created, which covered all bank employees, both private and official (Mesa-Lago 1978: 74). Furthermore, during the twenties, the *Caja de Jubilaciones y Pensiones de Empleados y Obreros de Servicios Públicos* expanded services to the workers of the mutualist health institutions, to dock and electricity workers, and to the private companies' workers, hospitals, and political parties (Ruezga Barba 2009: 218). In 1928, a new law allowed many workers in the private sector to join this fund (García Repetto 2011: 24).

During the first two decades of the twentieth century, the Battlista (1904–1930) reforms, intended to "build a model country", as he wrote, permanently shaped Uruguay: in the economic sphere, he nationalized, bringing the industry under the control and protection of the State; in the social sphere, he reinforced the labor movement, buttressed the social protection and the labor market, as well as assistance for the poor; in the rural areas, he progressively eliminated the livestock *latifundios*, promoted small farmers, and endorsed an equilibrium between agriculture and cattle raising; in the fiscal sphere, he increased taxes to the rich and reduced consumer taxes; he supported education, the defense of national/cosmopolitan identity, the secular State, and the emancipation of women; and finally, promoted the politicization of society, direct democracy, and a collective executive power (Caetano 2019: 134). In fact, Battle implemented an ample State-directed social-democratic program, which the Uruguayans prefer to call Republican, that has marked the character of his country.

Once reelected (1911–1915), after a sojourn in Europe, José Battle y Ordoñez intended to implement what was an advanced welfare system for his time; nevertheless, "the Battlista elite needed to soften the most radical aspects of the program [as]...the dominant sectors in control of the export production imposed limits on the construction of the [welfare] model" (Filgueira 1995: 10). However, the result, according to Mesa-Lago, was that by the end of the 1930s, Uruguay already had a comprehensive social security structure (Mesa-Lago 2008: 4). One example, Battle even succeeded in implementing the *Ley Madre*, an *avantgarde* (for the moment) program that allowed women working in industry and commerce to leave their jobs to raise their children, although it also had a conservative and natalist character, as they intended to reinforce the family (Mesa-Lago 1978: 74).

The initiatives of the two presidencies of Battle have led Filgueira to affirm that the model imposed in Uruguay was a statist one, as "...it was not society that pressured the political system and the State to deliver social programs. Nor did the State see the need to regulate an emerging society with new subjects and actors (Uruguay was a rural society, not an industrializing one as in Germany, for example). Rather, the political elites, through the State, sought to create an alternative to the traditional society" (Filgueira 1995: 10). On the other hand, "...the absence of a robust civil society and economic development to create pressure for the State to transform its social dimension contributed to the development of a particular type of welfare state, one that was statist, anticipatory, and paternalistic. This style strongly influenced the Uruguayan political culture, which also became statist" (Ibid: 11). According to historian Gerardo Caetano, the importance of the State in Uruguay also derived from a weak oligarchy and capitalist class, the absence of dominant classes (Caetano 2019: 134).

Nonetheless, the regime that emerged from Battlism was more Republican than State oriented; the political parties became the center of the political and social life of the country, becoming, in many cases in a clientelist manner, the "...main intermediaries between the demands of a civil society lacking strong corporations and a public space defined almost monopolistically by the State." However, the fact that an incipient associationism that the political parties mediated had emerged during the second half of the nineteenth century, implied a "precocious densification of a political society", which gave birth to the two parties that dominated most of the political life of the country: the *Blancos* and the *Colorados,* but "to the detriment of a weak and segmented civil society" (Caetano 2019: 132–33).

Furthermore, "...in those years there were numerous labor movements demanding greater pension coverage, or else the creation of new funds and the pressure to increase these funds through consumption taxes and not by worker wage contributions ... the Retirement and Pension Fund for Employees of Banking Institutions and the Stock Exchange (1925) was founded, which covered all bank employees, both private and official" (Mesa-Lago 1978: 74). One of the most signficant of these movements was the anarchist Congress of the Worker's Federation (Filgueira 1995: 11). As statist as Battlism was, it never tried to take control of the trade unions that it had "...helped create, expand, and institutionalize. In fact, in contrast to the populist experiences of other countries such as Argentina and Brazil [and Mexico] in the 40s and 50s, the Uruguayan State eventually became the product of society and of its organizations; these were not subordinate to State directives nor political parties" (Filgueira 1995: 12).

This is especially true in the case of the mutualists: the citizens searching for a way to guarantee health services. Although since the first Battlista presidency, priority was given to public action in most of the spheres of the economic, political, and social life, if one directs one's gaze towards health services, one can see how it was civil society that, since the beginning of the mid-eighteen century, and without the intervention of the State, created many mutualists that gave access to health to a significant proportion of Uruguayans, that had to depend on neither the market nor the public services.

These mutual aid associations were created during the mid-nineteenth century by the successive migrant waves and then by other religious, professional, and workers' groups and even political parties. They began as mutual aid societies that, as in Europe and other countries of Latin America, were created as savings funds to assist those who lost their jobs, got sick, grew old, or died. They prevailed as health insurance in the case of Uruguay (Martinez 2007: 46). Some of them were the *Sociedad Francesa de Socorros* created in 1852, *the Asociación Española*[1] in 1853, *La Fraternidad* in 1866, the *Círculo Napolitano* (1880), the *Círculo Católico de Obreros* (1885), the *Sociedad Fraternidad Unida* (1879), the Círculo Católico de Paso Molino (1900), the *Casa de Galicia* and the *Mutualista del Partido Nacional (1917)*, the *Mutualista del Partido Colorado*, the *Médica Uruguaya* and *España Mutualista (1931)*, and the *Centro Asistencial del Sindicato Médico (1935)*. Some workers' organizations started medical insurance funds, which were self-administered tripartite with the workers, employers, and public presence. These associations were formed as autonomous, non-State public entities (Ruezga Barba 2009: 211). The expansion of these associations was spectacular: in the capital, Montevideo, membership in these associations increased from 2.4% of the population in 1885 to 61.5% in 1930 (Martínez 2007: 8; Barran 1992, quoted by Martinez Franzoni and Sanchez-Ancochea 2018: 86). Although the mutualists offered their services exclusively to their members at the outset, they then opened their services to whoever paid their fees (Filgueira 1995: 19).

Although the Battlista government did increase public intervention in health, building new hospitals and providing health services, "...the State's role in the provision of health services remained overshadowed by the ... mutual aid societies (IMCs)" (Setaro 2005; cited by Pribble 2013: 43). Battle never tried to limit the *mutuales*; in fact, during the 1930s "the medical profession created their own association, the Medical Union, and laid the foundation for private, nonprofit medical institutions, to provide health services to the middle sectors, adding their services to the preexisting mutualist organizations" (Filgueira 1995: 14). Thus, "by the 1940s, these *mutuales*[2] had consolidated themselves as the primary

[1] This mutualist continues to offer health services and has 187,473 members (Ministry of Health 2020).

[2] These *mutuales* can be defined as democratic civil associations of non-profit private law that promote equality among their members based on a principle of intergenerational

provider of health care for Uruguay's middle class" (Filgueira 1995, cited by Ibid.: 45). Battlismo and the "European patterns of the middle of the 30 s" that he introduced in Uruguay resulted in the grouping of the existing public health care-related institutions in the Ministry of Public Health. "The new framework tried to meet the health needs of low-income sectors and to implement a general policy of prevention" (Filgueira 1995: 14).

In March 1933, Gabriel Terra took power through a military coup and akin to his authoritarian character and in a similar way to European fascism and other Latin American dictators, sought to establish a corporatist model of relations with civil society, especially with the unions (Filgueira 1995: 15). Although he failed in his attempt, he promoted the centralization and concentration of the pension funds, which were merged into the Institute of Retirement and Pensions of Uruguay and canceled their supervision by the active and pensioned workers (Garcia Repetto 2011: 28). In compensation, he extended the coverage of pensions: the Retirement and Pension Fund for Industry, Commerce and Public Services was created in January 1934 (Ruezga Barba 2009: 210).

The 1940s were also years of expansion for the Uruguayan pension scheme: that year, the fund mentioned above incorporated the workers of State-owned companies; a year later, it allowed the affiliation of independent workers, and in 1942, it expanded to include domestic workers. The coverage of the Uruguayan pensions advanced even further when, in 1943, rural workers gained access to it (Ruezga Barba 2009: 210). However, by 1947, the administrative and organizational problems resulting from the centralization of the Terra period had shown their limits. The institutional and financial autonomy of the Savings Banks was proposed (García Repetto 2011: 35), with the result that the following year, the Retirement and Pensions Institute was disaggregated into three different pension funds: the Civil Service and Education fund, the Industry, Commerce, Public Services, and the Rural Fund, and the Domestic Service and Old Age Pensions (Ruezga Barba 2009: 211).

solidarity. Such associations would be supervised in terms of medical, administrative, and financial assistance. The mutualists are directed by a Board of Directors, the control of the Fiscal Commission, and in some of them (e.g., the Spanish Association and Casa de Galicia), there exists a Representative Assembly. The positions are honorary, and the representatives are elected by direct vote (Martinez 2007: 9, 13–14).

Public health spread slower than pensions. "Until the sixties, the wide coverage of the State pension system did not compare with the scant State support for health protection. Only a weak network of public hospitals in Montevideo and clinics in the province provided limited medical care. At the same time, hospitals and private practices began to grow and increase their prices due to the better quality of service and high demand, and medicines become more expensive" (Mesa-Lago 1978: 77). Facing this situation, the government began to secure health benefits for industrial workers and low-level commerce and service workers while still favoring the middle sectors without affecting industrial, financial, and rural producers' interests. These initiatives were supported by the middle classes imbued with a republican and egalitarian political culture that legitimized the seemingly paradoxical demand for "privileges for all" (Figueira 1995: 22). One could consider that the social compromise was being extended from the middle classes to the popular classes and the State. Prosperity from exports through the late 1940s and early 1950s (Figueira 1995: 22) enabled the sheltering of the popular sectors by the hospitals and clinics of the Ministry of Public Health and by the university hospital, which opened in the late 1940s (Filgueira 1995: 19). In a sense, at least in the case of health, the State expanded its services as a reaction to the initiatives of civil society. So, fundamentally, when, beginning in the sixties, public services were developed to take care of the population not covered by the *mutuales*, the State was, to a certain degree, following a trend set by civil society, extending the services that the different associations had created as democratic, non-profit organizations, establishing common funds to attend illness and accommodate or find work for its members, and later on, in the twentieth century, for other members of society.

The Constitution of 1967 created the Social Security Bank (*Banco de Previsión Social*), which did not function as a bank but as the institution that centralized the pensions and retirement funds. Although it was intended to unify the system, it did not achieve this end, as many sectors did not join: both in the public sector (military, state banks) and in the private sector (universities, private banks' employees, notaries). And although it was projected to function as a quadripartite institution, including the workers' representatives, it was finally State managed (Filgueira 1995: 25). A few years later, a new civic-military regime resumed the centralization of pensions. In 1979, the 9th Institutional Act eliminated the autonomy of the BPS, which managed the different existing pension funds and replaced it with a single fund. A process of control that the first democratic government reversed under Julio María Sanguinetti after thirteen years of dictatorship. In 1986, his government reinstated it as an autonomous entity, and five different services were created to meet specific needs: retirement and old-age benefits, asset benefits, tax advice and collection, administration and services, and health, all under the Ministry of Labor and Social Security (García Repetto 2011: 45).

In the nineties, as in other countries of the Southern Cone, Uruguay entered a liberal (or neo-liberal) phase under President Luis Alberto Lacalle and the second presidency of Julio Sanguinetti (1990–1999), with the imposition of what has been named the Washington Consensus. Nonetheless, due to the country's republican and social-democratic tradition, the social protection reforms were not as radical as those of its neighbors. In the economic sphere, the State retreated from the economy, which opened and was financialized; in fact, the project was to transform Uruguay (a country renowned for its stability) into the leading financial center for Latin America—something that was achieved up to a certain extent.

The consequences of this liberal turn on pensions had nothing in common with the privatization that Chile (in 1981) and Mexico (in 1995) undertook. As in other countries of the continent, the government created an individual capitalization pillar, complementary to pay-as-you-go. In this manner, the system ended up with three different pillars: a public system of intergenerational solidarity, a mandatory individual savings regime in charge of public or private entities, and the personal decision to inject voluntary savings into the capitalization pillar (Ruezga Barba 2009: 215).

In contrast to these minor transformations of pensions, the labor market was flexibilized and unionemployer negotiations were reformed in a significant manner "countering the collective negotiations that had been in place for 50 years" since the *Ley de Consejos de Salarios* of 1943, except for the civic-military dictatorship that was initiated in 1973. Collective agreements by branch were weakened: the outcomes of negotiations between the unions and the entrepreneurial organizations would not be automatically applicable to all the enterprises of the branch but only to whom was affiliated to the unions; an immediate direct effect of this situation (which added up to the brutal repression of the previous dictatorship) was the weakening of the labor movement and its main unique central union: the PIT-CNI (*Plenario Intersindical de Trabajadores-Convención Nacional de Trabajadores*) that ended up representing little more than 10% of the economic active population, "...a historically low proportion, especially since the unification of the labor movement of the first half of the seventies" (Caetano, G. 2019: 240). Nevertheless, this situation of the labor movement did not prevent the *Frente Amplio*, where the labor movement was one of the principal partners, from winning the election of the year 2000.

Other reforms were undertaken in education, state structure and the constitution during the nineties and beginning of the 2000s. Conversely, the organization of health was left almost untouched; it continued to operate at the two levels we have discussed: the mutualists together with the non-profit institutions and the public sector. The health sector was not only divided into the diverse mutualists and nine autonomous health insurers for workers of various sectors, but while the latter was accessible to a large part of the middle class, the public services covered the poorest sectors. This meant an unequal fragmenting of health care; in fact, the private, non-profit organization concentrated three times more resources per person than the public one[3] (Haggard and Kaufman 1978; Martínez Franzoni and Sánchez-Ancochea 2016: 10).

To give a solution to this central problem, at different moments, the State has tried to unify the health structure; both the military in the seventies and the democratic government of the *Frente Amplio*, in the 2000s did so with different methods and ends. The State Health Services Administration (ASSE) was created in 1975 as a decentralized institution charged with ensuring universal coverage of medical care and disability benefits, as well as preventive medicine and rehabilitation. The contributions to this scheme were to be collected through the aforementioned *Banco de Previsión Social* to be transferred to the institution chosen by the individuals: both public and the mutualists (Martínez 2007: 48). In this manner, the State would become the sole payer of the expenses of the individuals to their health purveyors (Ruezga Barba 2009: 214). In addition, a National Resources Fund was instituted to pay for catastrophic or high-impact illnesses such as organ transplants or heart surgery—a fund that sought to assure high-cost interventions for the people with low

[3] This is why these authors consider that, even today, although the Uruguayan system is one of the protection systems with the greatest coverage among Latin American countries, its fragmentation has prevented universalizing its benefits.

incomes as well as to mitigate the finances of small mutualists (Martínez Franzoni and Sánchez-Ancochea 2016: 87). A further attempt to reduce such differences was the standardization of the service packages offered with shared protocols among health service providers (Martínez Franzoni and Sánchez-Ancochea 2016: 52, 60). Nonetheless, none of these modifications solved the problem of the disparities between both schemes (Martínez Franzoni and Sánchez-Ancochea 2016: 87–88).

Significant changes were again intended after the economic crisis caused by the contagion of the 2001 Argentinian collapse. The reforms were also projected to remedy the decline in the quality of public services and the growing dependence of mutualists on governmental subsidies. In 2008, the *Sistema Nacional Integrado de Salud* (SNIS) was created, which included compulsory insurance, including children and adolescents that had not been covered before. In addition, the National Health Fund (FONASA) was created to channel all the contributions to health. The BPS administers this fund and currently transfers resources to providers on a per capita quantity based on age and risk conditions. In this manner, "...FONASA pays similar amounts to the mutual societies and to the public provider—thus narrowing the historical inequality of the system… [in addition] Contributions were made more progressive by differentiating monthly fees according to income levels and number of children—fees range between 3 to 8%". In this manner, between 2007 and 2008, 500,000 new beneficiaries joined the SNAIS, and although the mutual societies prevailed and were even strengthened financially, they now care for more people with fewer resources (Martínez Franzoni and Sánchez-Ancochea 2016: 88–89).

The idea behind this last reform was to universalize and establish a unique structure that would concentrate all providers (the public and the mutuals); nonetheless, it encountered strong resistance from the medical professions and civil society (mutualists and unions) (Fuentes y Midaglia 2018). What the reform did achieve, which is central to advancing the homogenization of the conditions of both health schemes, is the transformation of the structure of financing, "...a universal and solidary pooled fund was established, oriented to solve the enormous inequalities between the contributions of individuals, families, and the state" (Olekser 2015: 16–17). The *Frente amplio* instituted a sole payer: the FONASA, designed to allocate resources to providers considering risk, age, and sex. Therefore, although health services are still segmented, they are financed by one single payer and are regulated by the State, and while the cooperatives and mutualists were preserved as providers, the private-for-profit organizations were excluded; only those that were in the scheme at the time of the reform were left (Olesker 2015: 16–17). This means that while civil society and the State intervene in the health organization, the market is intended to disappear progressively. While this is effectively the case, health is nearer to a socio-corporatist regime than to a statist one, both because of its trajectory and because of the presence of the mutualists.

Argentina

If one focuses on the pensions, the origins of the Argentinean welfare state are very similar to those of Uruguay and, in fact, other pioneer pension schemes in Latin America; the government began allocating pensions to the army and public officials and then extended benefits to the rest of society, going from the most to the least strategic economic sectors. The first pensions for old age or disability were approved for high-level public servants through the *Montepío de Justicia y Real Hacienda* in 1810, immediately after independence. Thenceforth, during the governments of Bartolomé Mitre (1862–1868) and Domingo Faustino Sarmiento (1868–1874), these benefits were extended to all public servants (Mesa-Lago 1978: 16). In 1877, President Nicolás Avellaneda (1874–1880) introduced the first pensions financed by the State, benefiting judges, and his successor, general Julio A. Roca (1880–1886, 1898–1904), extended protection to public school teachers in 1885 (Mesa-Lago 1978: 161). In 1904, the National Retirement and Civil Pensions Fund offered pensions to workers of autonomous bodies and public companies (Schulthess and Demarco 1993: 24).

Under the *Unión Cívica Radical* government of Hipólito Yrigoyen, the middle classes were the main objective of public policies since this sector was its primary support. In 1921, a fund for old age, survivors, and disabilities was established for personnel of various productive areas controlled by foreign companies, such as trams, telephones, telegraphs, gas, electricity, buses, and radio, as well as workers of clinics and hospitals. And in 1922, a pension fund was created for employees of private and public banks, insurance companies, and savings funds. During the second presidency of Yrigoyen, marked by the crisis of 1929, his government tried to extend pensions to marine merchants, merchants, industrial workers, printers, and journalists, but the law was abolished before it came into force. Instead, a military coup deposed him in 1930.

In the case of health, the evolution of the Argentinian services were not, as the those of Uruguay, a result of the spontaneous action of civil society. Although unions established the first associations for health care, their expansion occurred during Peronism, in the process of concession/negotiation with the government under Juan Domingo Perón (since he was Minister of Labor of the military government), intended to coopt the relatively strong labor movement that was dominated by socialist and anarchist ideologies. The first association aimed to ensure the health of its members was, like the Uruguayan, linked to a sector of migrants, in this case, the French Mutual Aid Society (1854). Nonetheless, immediately afterward, the labor mutualists emerged; in 1856, the Shoemakers Society, in 1857, the Typographical Society of Buenos Aires, and in 1858, the *Unione e Benevolenza*. In 1919, the Railway Union—the main union of the time—established mandatory contributions by the railway employers (Garay, O. E. 2017).

In this manner, union organization was much more dominant in the case of Argentina, even before Peronism, although this movement gave it a significant drive. Unions had organized and were active since the end of the nineteenth century: in 1887, railroad and printing workers formed the first union in Argentina; in 1909 the Argentine Regional Workers' Confederation (CORA) was constituted by the merger of the *Unión General de Trabajadores* (UGT, created by the socialists in 1903) and some other unions; in 1896, there were 26 strikes by railroad workers, builders, sailors, and printing workers. Until 1910, the dominant anarcho-syndicalist ideology of the workers' organizations prevented any connection with the government. This type of unionism declined in the 30s and practically disappeared when a more conciliatory labor movement emerged and was promoted by Perón in order to consolidate labor as one of his main political supports; in 1930, the very important *Confederación General de Trabajadores* CGT emerged from the fusion of the *Unión Sindical Argentina* (USA) and the *Confederación Obrera Argentina* (COA) (Mesa-Lago 1978: 162–163).

The existence of a strong labor movement and its capacity to exert pressure on successive governments was decisive for the construction of the welfare state that began being implemented during the first two decades of the twentieth century, which, up to a certain degree was consolidated in its contemporary characteristics under Peronism. However, in contrast to the Uruguayan trajectory, where Battle intended to implement his scheme "from above" and his successor failed in his attempt at corporatist control, in Argentina, Brazil and Mexico, corporatism prospered. As a result, the construction of the Argentinian welfare state was a process of labor movement pressure, repression/negotiation, and concession, a prevalent socio-corporatist regime that has survived until the present, while the others we have mentioned have declined.

Although not related to the welfare state, the first such event of pressure/negotiation/concession characteristic of corporatism was the response of President Roque Sáenz Peña (1910–1914) to social discontent with the universalization, compulsion, and confidentiality of the vote and voter registration (Mesa-Lago 1978: 162). This author considers the strategy of social security of the government of Perón as the core of his policy to control the labor movement and coopt its support for his party; this, in addition to outlawing the communist and the anarcho-syndicalists, intervening in the socialist unions, limiting the right to strike, and especially imposing the legal requirement of formal recognition of unions on the part of the Ministry of Labor to be able to represent workers and engage in collective negotiations. Thus, he converted the *Confederación General del Trabajo* (CGT), created in 1930, into a unified labor union that backed his government (Mesa-Lago 1978: 164).

Until the government of Juan Domingo Perón, "the Argentine social security system provided pensions for a small segment of the labor force [...] he considerably enlarged the pension system of the military, granted life insurance to civil servants, and introduced healthcare for those employed in the legislature, the judiciary, banking and insurance, railroads, and the merchant marine" (Mesa-Lago 1978: 164). During the first Peronist government, the welfare state acquired most of its modern characteristics: with the backing of the Ministry of Labor, unions in different sectors of the economy created funds that incorporated the contributions of both workers and employers. "The old employers' funds from the beginning of the century, made up of company contributions based on a percentage of salary, were extended as a model [...] emulating the experience of the Railway Fund", the main union of the time (Garay O. E. 2017).

While, to a certain degree, social protection was a concession from the government, it was happening in a context where the mere presence of numerous autonomous and powerful unions represented pressure by a social actor. The unions were strong enough to threaten the government when Perón arrived at the Ministry of Labor, under the military government of the Revolution of 1943; he was obliged to expand social protection if he wanted to accomplish a popular national alliance. Once Perón was ousted from the presidency in 1955, as it was in opposition, the labor movement became more autonomous. It continued to exert pressure on successive governments (even the military ones), with the result that towards the end of the 1950s, the coverage of pensions was 70% of the workforce (Haggard and Kaufman 2008: 95). It was also during the Peronist government that the health services expanded, in what is known as the *Obras Sociales*, which, according to our perspective where we give particular importance to the structure of the health system, to a great extent determine the character of the welfare state in this country.

As early as 1944, Peronism created the Social Service Commission to promote the development of social services for each labor sector (Alonso 2000, p. 66); they were established and controlled by the unions as they were founded on the existing workers' mutual aid associations. The tendency of the Argentinian unions to directly allocate services to workers is something that "…involved a strong redistribution of power towards the unions" (Alonso 2000, p. 66). There were two modalities: those of the large unions that provided services directly through their hospitals and those of the small and medium-sized ones that provided them through contracts with universities, associations, and private institutions (Alonso 2000, 67). After the fall of Peronism, unions continued to be in charge of medical services, and the military withdrew their support to public health institutions, with their consequent degradation. Towards the end of the 1960s (before the expansion of the 1970s), only 26.7% of the population of the metropolitan area of Buenos Aires received medical services from the *Obras sociales* (Alonso 2000, 67).

In the following decades, in spite of the attempts to centralize pensions in the hands of the State during the successive military and civic governments, social protection was not reduced, due to the mobilization capacity of the labor movement. The military junta that demoted Perón sought the unification and standardization of the system and increased coverage and benefits to broad sectors of workers and employees to undermine the Peronist bases. Next, the civic government of Arturo Frondizi (1958–1962) also tried to unify the welfare and assistance institutions under the National Institute of Social Security. Finally, during the military government of 1966–1973, the mandatory nature of employee and employer contributions for all salaried workers was established. According to Garay, this latter reform, concerning "the mandatory contribution for all wage earners and their control of the *Obras Sociales*, strengthened the role of the unions, as they gave the union organization a continuous economic flow" (Garay O.E. 2017).

In 1967, the military government ended the National Institute of Social Protection, which had the representation of unions and employers, and centralized pensions directly through the Ministry of Social Welfare—a measure that, on the one hand, concentrated a crucial financial source on the government and, on the other, weakened the unions. In 1970, the government of General Juan Carlos Organía reduced the number of *Bancos de Ahorro* from thirteen to three, covering: industry, commerce, and civil activities; state and public services; and the self-employed (Schulthess and Demarco 1993: 24). On the other hand, the military issued a decree that considerably expanded health coverage, taxed employers, and strengthened the control of the unions on the *Obras Sociales*, which can be interpreted as a compromise between taking power away from the unions in the case of pensions, in exchange of preserving and even strengthening their power over the health services. The coverage of the *Obras Sociales* had a very considerable expansion in 1970 with a law which established a compulsory employer contribution of 2% of the payroll, 1% by the workers, in addition to the fact that it included retirees and pensioners who contributed 2% of their salaries. Coverage thus increased from 3.5 million people in 1967 to 16 million after itsdecree (Alonso 2000, 67).

The explanation for the need of a compromise with labor is related to the broader social context of the country. The expansion of social protest in what became known as the *Cordobazo* had weakened the government, and the concessions to trade unionism were a way of defusing the political and social conflict. As Alonso says, this shows that "...the corporatist character of the Argentine social protection resulted more from the ability of the unions to penetrate the State than from the State's ability to coopt the social sectors vertically" (Ibid.: 70). And in fact, in contrast to Chile, Argentina's dictatorship faced a stronger resistance from trade unionism, which prevented it from implementing the structural reforms to the welfare state that Pinochet applied in Chile (Berins Collier 1999, Munck 2004, Papadoupulos, op. cit.: 118). This was because, as Stepan has written, there is a significant contrast between both Brazil, Chile, and Argentina; while in the first two, an authoritarian regime was established, in Argentina, there existed an authoritarian "situation" (Stepan 1985) mainly due to the powerful presence of trade unionism.

Consequently, it was not during the military government that the most critical changes in national social protection took place, but "paradoxically" during the *Justicialista* government of Carlos Menem (1989–1999). Although he had arrived in power due to the traditional support of the working class for his party, basically the CGT, the labor base of the party, had been weakening due to the massive growth of the poor households during the economic crisis and the hyperinflation of the eighties, which had become a significant support of *menemismo* (Sidicaro 2003–2004). Menem passed significant liberalization and privatization measures: in 1993, the pension scheme was reformed, creating a mixed structure that included both individual capitalization and intergenerational distribution; Menem (and Alfonsín before him) tried to disconnect the health services, the *Obras Sociales* from the control of the unions, and concentrate them in the hands of the State, with the aim, in the logic of the liberal reforms, of increasing the weight of the private sector, which in Argentina was around 10%, compared to the *Obras Sociales*, which accounted for more than 55%. However, neither Alfonsín nor Menem achieved their aim, which is why the health services have remained in the hands of the unions until today. Most authors consider that this was the result of a negotiation between the Menem government and the unions where the CGT accepted the privatization of public companies and the flexibilization of the labor market in exchange for preserving the health services in union hands (Munck 2004: 11, Murillo 2000).

The crisis of 2001–2002 had a profound impact on both the economic model followed during the Menem years and social protection. From the beginning of the 2000s, until around 2015, both Argentina and Brazil had transformed their capitalistic model from liberalism to what I have called (following some Brazilian authors) a socio-developmentist type of capitalism (Bizberg 2019). The liberal model applied in Argentina through the nineties, based on "convertibility", that had fixed the parity of the peso to the dollar, fell into a profound crisis because only pesos that were backed by dollars in the central bank's reserves could be issued, which, in turn, implied that Argentine's economic policy was determined by the monetary policies of the United States and depended on foreign currency inflow from exports of Argentine goods and other productive and speculative investments. Although the currency board scheme worked well for a few years, the model was maintained for too long, even though in the late 1990s and early 2000s, the country was plunged into a deep economic depression, resulting in a massive increase of unemployment.

Economic, labor, and welfare policies underwent a 180-degree turn after the financial crisis and the arrival of Néstor Kirchner (2003–2007) to power (Bensusán 2010, Palomino 2000, Alonso 2000). This resulted from the unprecedented boom in social movements (*piqueteros*, neighborhood assemblies, factory occupations, among others) and the rebirth of the Peronist labor movement due to the most profound economic and political crisis that the country had lived through since the thirties. The governments that resulted from the 2003 elections were forced to find support in all these different social groups to get over this multifaceted crisis. The most significant measure that marked the social policies of the next government, that of Fernández de Kirchner (2207–2015), was the renationalization, in the midst of the subprime crisis in 2008, of the pension funds, which had been partially privatized during the Menem presidency. The Fernández de Kirchner government unified pensions under a State-controlled pay-as-you-go pension scheme, eliminating the capitalization pillar administered by the *Administradoras de Fondos de Jubilaciones y Pensiones* (AFJP) (CEPAL 2010: 8–9). This measure was not only a reaction to the economic crisis but had been prepared by various previous reforms: the first one went into effect in 2005, allowing early retirement, with reduced benefits, for the unemployed who had contributed a minimum of 30 years, although they had not completed the 35 years required for retirement. The second, from 2007, allowed all workers of retirement age to receive benefits even if they had not contributed the required 35 years; these workers would have to continue paying their contributions. These two reforms reversed the decline in coverage of the elderly population, which had fallen by approximately 10% in the previous decade and allowed coverage to return to 77% in 2007. Nevertheless, the benefits were meager; the average was a proportion of 33% of the average salary (Huber and Stephens 2012: 187).

In the case of the health services, although the Menem government did not succeed in modifying them substantially, as it did not achieve to withdraw the *Obras Sociales* from the control of the unions, he did issue a decree freeing workers from the obligation to join their union's *Obra Social*, thus allowing them to choose the one they considered most convenient. Nonetheless, this law did not result in a large migration from one health service to another, even less towards the private sector. On the other hand, the Kirchner government extended the coverage of health services (Mesa-Lago 2009: 15). In this manner, although in the case of pensions, unions ended up losing control of the resources after so many attempts to centralize and homogenize them since the first Peronism, they managed to maintain the control of the *Obras Sociales*.

Whereas most of the assistance programs in the rest of the continent were merely monetary distribution transfers, the Argentinian government conceived of a program, *Trabajar*, that focused on labor rather than on assistance; it consisted of temporary jobs focused on community needs such as the construction or repair of roads, clinics, schools, etc. (Weitz-Schapiro 2006). This program[4] was a response to the *piquetero* movement that had surfaced before the deep economic crisis of the end of convertibility, contesting the effect of the liberal measures of the Menem government. It began with protests against the closure of public companies in the country's south (Neuquén, Salta and Jujuy), which, in many cases, constituted the only local source of employment. These closures generated street blockades that often led to real revolts of the inhabitants of small towns: the *puebladas*. *Piquetero* actions increased with the impressive growth of unemployment that spawned between 1990 and 2001: from 15% in 1992 to 40% just before the crisis broke out. In those years, and primarily during the De la Rua government, the movements spread to the rest of the country, especially to the Buenos Aires area, where the results of the neo-liberal economic model were even more severe and affected private sector companies (Svampa and Pereyra 2005). The crisis of unemployment and factory closures also led to a broad movement of hundreds of factory occupations of the plants that had been (or threatened to be) closed by their owners and where workers organized to keep them running.

When Néstor Kirchner came to power in 2003, primarily buoyed by the Peronist trade unionism and the movements that emerged in the crisis, especially the *piqueteros*, he was forced to considerably expand social policies to alleviate the adverse conditions of the bases of these movements. Membership in *piquetero* organizations increased from 1,000 in 1997 to more than 200,000 in 2004 (Franceschelli and Ronconi 2005:15). As of 2004, roadblocks decreased for three main reasons: the change in policy and the attributions in the social programs that were conceded to their organizations. The most important *piqueteros* organization, the 125,000-member *Tierra y Vivienda* Federation, led by

[4] It was later substituted by the program *Jefes y Jefas de Hogar*, that was a more traditional monetary transfer's program, and then by the universal *Asignación Universal por Hijo para protección social*, that did not demand any labor retribution.

Luis D'Elía and incorporated into the *Confederación de Trabajadores de Argentina*[5], managed 75,000 temporary work programs at any one time. In this way, it became a critical support for the Kirchner government, as well as for that of Fernández de Kirchner.

Both the *piquetero* movement and the *Trabajar* program were innovative responses to the organization of the unemployed. As is well known, the unemployed are challenging to organize as they have lost contact with other workers and a collective milieu. However, in the case of the *piqueteros,* they maintained, in the first place, strong links with all the other inhabitants of the locality where the source of labor, resources, and collective life had disappeared, but, in the second place, they had a strong trade union tradition and Peronist subculture that imbued their social and political cultures. Unemployment, together with the fact that there was no policy to compensate for the loss of employment in a society in which joining the union meant so much (not only pensions, but health services through the *Obras Sociales*, and other benefits like hotels in touristic places, etc.), gave rise to a movement that developed outside of the structures of the *Justicialista* party, and expanded exponentially following the 2001 crisis (Swampa and Pereira 2004). Surely enough, this movement and its policy response constitute a reaction against neo-liberalism (Silva 2009). Still, the profound significance of these actions was that they were not only reactive, but they were fundamentally proactive actions demanding the human right to decent work and bearing witness to the importance of civil society in the configuration of the welfare state in Argentina.

The power that the labor movement has shown that it upholds in Argentina has prompted many authors to consider the relationship between the labor movement and the State during the 15 years of the Kirchner and Fernández governments as neo-corporatist (Palomino and Trajtemberg 2006; Etchemendy and Collier 2007; Bensusán 2016). Even the assistance program *Trabajar* that the governments of Duhalde and Néstor Kircher implemented bears witness to the importance of organized labor in this country. Although the labor movement has lost power in the last decade, it is still the most significant social actor in the country, and,

[5] An organization that not only associates workers, but also unemployed and other social movements; it held an autonomous position *vis-à-vis* the CGT, although it never received formal registration.

concerning social policy, it still controls a substantial part of the health services.

The Statist or State-Dominated Welfare Regimes

The Brazilian and Costa Rican cases are the nearest examples of a welfare regime led by the State. Nevertheless, one crucial difference between both is that since the Costa Rican welfare state was a late-comer, it was never as segmented as the socio-corporatist ones, and consequently, it was the State that implemented and oversees social protection since the forties. On the other hand, the Brazilian welfare state developed in a segmented manner under the *Estado Novo* as a fundamental component in the effort of the authoritarian government of Getulio Vargas to control civil society. The Costa Rican welfare state was born as a statist regime, centralized and unified, and developed in this manner until the 1980s, when, although it maintained its character, it began to be undermined. The Costa Rican State managed to include in one unique institution, the *Caja Costarricense de Seguridad Social* (CCSS) or just *Caja*, pensions, health[6], as well as assistance to the poor households and non-contributory pensions; it could hence integrate the impoverished population at once to assistance and health. In the Brazilian case, the military centralized the numerous pension funds managed by the unions and expanded pensions to the peasants to get their support. Then, with the return of democracy, the universal *Sistema Único de Saúde* (SUS) was promoted by civil society but managed and financed by the State through contributions and taxes, although provided by public and private hospitals and clinics.

Costa Rica

There are forerunners to the creation of a unique State institution centralizing all of the features of social protection in this country: the law that created the National Insurance Bank, in charge of the Labor Accident Insurance of 1924, and the Ministry of Health and Social Protection in 1927, an institution in charge of hospitals, hospices, maternity homes, and children's shelters (Miranda Gutiérrez 1993), that was established with

[6] The Mexican IMSS inspired this model, but in contrast to Mexico, whereas the IMSS has never included more than 30% of the population, in Costa Rica, the Caja covers almost all the population.

"...a conscious political decision ... of the intervention and responsibility by the Costa Rican State, for the technical and financial management of public health" (Ruezga Barba 2009:122). Nevertheless, the definite event that determined what Martinez Franzoni and Sanchez-Ancochea call the architecture of the welfare state in this country was the establishment of the *Caja* by the Social Security Law that implemented a compulsory social security institution that included both pensions and health, under President Rafaél Ángel Calderón Guardia (1940–1944), among other progressive social reforms, most notably the Labor Code (Ruezga Barba 2009: 122). This law established the foundations for the gradual expansion of the system to its universalization (Martínez Franzoni and Sánchez-Ancochea 2016: 121–122).

Martinez Franzoni and Sanchez-Ancochea consider that social security was, at least partly, Calderon Guardia's personal project. However, other elements must be taken into consideration, such as the increase of the communist party electoral support, the Catholic ideology of social protection as part of the construction of class collaboration and social peace, as well as the contemporary ideas on social policy coming from Europe, most notably Beveridgean England. To implement these reforms, the president had the support of a group of experts who had studied in Europe in the interwar period, as well as the aforementioned Catholic Church and the communists (Martínez Franzoni y Sánchez-Ancochea 2016: 109–111). What was crucial at this first moment was that neither civil society, the political parties (except the threat that the communist party represented), nor the business sector played an important role in promoting it, nor were there organized actors rejecting it. The only attempt to veto the creation of a unified welfare state came from some workers' sectors who had previous pensions, such as the employees of the municipal government of the capital, the public bank, and customs; nothing compared to the situation of the countries of the pioneers of the Southern Cone, who faced numerous groups with privileges (Martínez Franzoni y Sánchez-Ancochea 2016: 116–117).

One of the social groups that had been most defiant to a previous attempt to eliminate the wage ceiling was the doctors, who attended the non-affiliated (in 1944, they had organized their Medical Union) since they considered that this act violated the principle of medical freedom. A three-day strike in 1945 paralyzed most health services, and the *Caja* backed down (Martínez Franzoni and Sánchez-Ancochea 2016: 120). This conflict helped the government perceive the risks of a public health deal including private providers, who showed themselves as a powerful pressure group, and increased fees and prices of medicines without control. This led the *Caja* to establish the Polyclinic Hospital, which by 1952 had reached 200 beds, and to hire more doctors: between 1950 and 1955, the CCSS hired two-thirds of the newly registered doctors in the country (Martínez Franzoni and Sánchez-Ancochea 2016: 119). A few decades later, the *Caja* became the sole health provider.

The expansion of the public health system was significant in the 60s: while in 1960, only 15.4% of the population was shielded (Mesa-Lago 1994: 95), it increased to 60% by the end of the decade (Rosenberg 1979 in Martínez Franzoni y Sánchez-Ancochea 2016: 128). Although the Costa Rican social protection had increased its insured population base significantly, there was a large gap between the health services allotted to the salaried workers and those for the people living in poverty and the self-employed; in addition, the latter lacked access to pensions, the health services were underfunded, and many higher income workers were still uninsured. Under the presidency of José Figueres (1970–1974), who won the election on a "war against misery" slogan, social security became mandatory for all salaried workers, and the wage ceiling was eliminated, which meant that the workers with higher wages were obliged to enter the system, and their contributions would be calculated based on their total salary, all of which increased the resources of the *Caja* considerably (Martínez Franzoni y Sánchez-Ancochea 2016: 125–126).

It was in 1973 when all of the health service programs and infrastructure were transferred to the Costa Rican Social Security Fund and transformed the *Caja* into the sole payer and provider of health services, in the mode of the British NHS.[7] It is important to emphasize that "...the transfer of all hospitals to the CCSS forced the institution to take responsibility for the poor" (Martínez Franzoni and Sánchez-Ancochea 2016: 138), and forced the government to implement a rural health program that entrusted the State with preventive and primary health care in the smallest localities (Martínez Franzoni and Sánchez-Ancochea 2016: 139): 200 health centers serving 600,000 people were created (Haggard and Kaufman 2008: 92). The Rural Health Program, which sought to reach

[7] We will discuss below how beginning in the 90 s, and especially during the years 2000, an outside option was developed, and the private sector began allocating increasing health acts.

the population with health posts in areas without services and with house-to-house visits, was crucial to give decent health services to the poor households (Ruezga Barba 2009: 124). "In comparison to most other Latin American societies, these efforts to reach into rural areas and poor neighborhoods were distinctive and had a positive impact on inequality, poverty, and infant mortality" (McGuire 2001a, 7–10, cited by Haggard and Kaufman 1978: 92).

All this implied that from that moment on, the middle classes, the salaried workers, the independent workers, and the low income households[8] would get health services in the same institutions, as they all would be subject to compulsory insurance. In this manner, the best payed would contribute to both the health and the pay-as-you-go schemes, transferring resources to the poor (Santos Basso 2005: 192). To ensure that the incorporation of the vulnerable population to social insurance was intended to be on equal terms as the rest of the population, and if the middle classes were to be prevented from migrating to private health (which started happening in the 2000s), the system had to provide the latter the level of services they demanded. Although eliminating the wage ceiling had given the *Caja* a significant quantity of resources, the *Caja's* finance would still be strongly strained, and the government would need to increase its resources significantly.

The government created "...a path-breaking social assistance fund, the *Fondo de Desarrollo Social y Asignaciones Familiares* (FODESAF), which generated a new income stream, coming from payroll taxes … made funding progressive as contributions came from the non-poor population" (Martínez Franzoni and Sánchez-Ancochea 2016: 147 and 123). In addition, this new fund granted non-contributive pensions, school meals, technical training, and preventive and primary medical assistance. The fact that the resources of the fund fed the CCSS facilitated equal access to services for the rural and most marginalized sectors without fragmenting the system, thus preventing the creation of a particular program of pensions and health oriented to the impoverished households, as has been the case in all the rest of Latin America (Martínez Franzoni and Sánchez-Ancochea 2016: 125–126).

This process of centralization was not achieved without opposition, although the transfer of all hospitals to the CCSS meant better

[8] These sectors would enter the integrated system with their contributions assumed by the State.

salaries for most doctors. On the other hand, while the possibility of contracting out services from the public to the private sector was barred, doctors would not be forced to serve in the public sector. However, the College of Physicians demanded the freedom to attend to patients outside public health (Martinez Franzoni and Sanchez-Ancochea 2016: 134–5). There was also opposition from business people who considered that the unification would imply a high fiscal burden that would affect the economy and end up hurting the quality of the health organization. In contrast, most unions[9] supported the reform (Martínez Franzoni and Sánchez-Ancochea 2016: 134), and as the army was eradicated after the 1948 civil war, one of the sectors that has been set apart in all of the Latin American countries was non-existent.

As a result of all these changes, from 1960 to 1980, spending as a percentage of GDP for social protection quintupled, going from 2 to 9% of GDP (Mesa-Lago 1985: 43), it became the fourth highest in Latin America, approaching the pioneer countries; in the health sector, it was the highest in Latin America with 6.8% of GDP (Mesa-Lago 1985: 58–59). More specifically, health coverage reached 90% of the population (Ruezga Barba 2009: 124) "...virtually the entire population was covered, between two-thirds and three-fifths by social security and the rest by social assistance and public health programs" (Mesa-Lago 1985: 45). In consequence, "mortality rates in 1980 were among the lowest in Latin America and life expectancy was among the highest". For Martinez Franzoni and Sanchez-Ancochea, "In terms of equity, the elimination of the wage ceiling increased the contribution of high-income groups. In terms of generosity, integration of hospitals into a single institution secured services for the whole population. The rural health program also helped reach the scattered population living far away from dispensaries and more complex medical facilities [and refer acute and chronically ill patients to the CCSS clinics and hospitals (Casas and Vargas 1980: 269) cited by Ibid.: 141)]. In terms of quality, additional resources contributed to more investment and thus better services for all" (Martínez Franzoni and Sánchez-Ancochea 2016: 149).

During the 90 s, the rural health program was further reinforced by the creation of the *Equipos Básicos de Atención Integral en Salud* (EBAIS),

[9] It is interesting to note the difference between the situation of Costa Rica and that of the socio-corporatist countries (Uruguay and Argentina); in the former, unions do not have specific pension funds nor health services.

which consisted of countrywide health posts composed of a team of doctors, nurses, and technicians per 1,000 households. This program reduced the proportion of inadequate access to health from 22% of the population in 1994 to 13% in 2000 (Martínez Franzoni and Sánchez-Ancochea 2016: 149). "This was an important improvement in terms of access to care, as the Health Ministry's primary care program had not achieved universality. Today, 94% of the population has access to a primary health care post and, while not functioning optimally yet, a referral network from the primary health care teams to the less and more specialized hospitals in the regional and national centers of the country is in place … Consequently, while out-of-pocket payments are substantial, no catastrophic health expenditure occurs in Costa Rica" (Vargas and Muiser 2013: 7). And the fact that this program was financed by the FODESAF, whose resources came from payroll taxes "...made funding progressive as contributions came from the non-poor population" (Martínez Franzoni and Sánchez-Ancochea 2016: 158).

In this manner, the health services weathered the "lost decade" of the eighties and the neo-liberal backlash of the nineties relatively unchanged, with the consequence that "Costa Rica's policy architecture has remained relatively stable in terms of eligibility, entitlements, and service provision … Moreover, additional reforms during this period further unified the policy architecture: primary care services expanded nationwide and health care and pension insurance for the self-employed became mandatory" (Martínez Franzoni and Sánchez-Ancochea 2016: 151–2).

In addition, although there has been a modification of pensions, this country avoided the more radical reforms imposed in Chile and Mexico, where the pay-as-you-go was eliminated, but also the solution of Brazil and Argentina, where there is no capitalization pillar. Costa Rica did establish an individual capitalization pillar administered by the private banks; in 1995, the government established the Private Regime of Complementary Pensions (Ruezga Barba 2009: 126). In the year 2000, it made the subscription of self-employed workers to the public pillar compulsory and added the possibility of voluntary and distinct contributions to the mandatory individual capitalization pillar; two pension funds stayed out of the *Caja:* that of the teachers and the Judicial Power (Ruezga Barba 2009: 126–127).

Brazil

The Brazilian welfare state developed before the Second World War with a segmented corporatist scheme; it began covering the armed forces, continued with the civil servants, and then with the most strategic economic sectors. Nevertheless, unlike Uruguay or Argentina, there were practically no initiatives coming "from below", from civil society to implement health services. Contrary to Argentina, unions were not as organized nor active before the State, under the presidency of Getulio Vargas, stimulated their emergence and incorporated them into the political regime. In this manner, although the unions managed to control the pensions that Vargas' government created, they functioned as an instrument of governmental cooptation and subordination rather than empowerment.[10] Accordingly, the Brazilian State had, since the beginning, a more robust intervention in pensions and health than what we have seen for Argentina and Uruguay; both schemes were finally centralized and put under the lead of the State during the military dictatorship in the case of pensions, and during the first democratic government of the post-transition, in the case of health.

The authoritarian Vargas government, which ruled from 1930 to 1945 and again from 1950 to 1954, "established a structured relationship between the State and the working class ... by coopting discrete segments of the laboring classes into a dependent position within a formalized corporatist structure. The structure divided labor horizontally even as it incorporated sectors of labor vertically into an asymmetric pattern of relationships with the State in which control flowed from the top down" (Malloy 1979: 61). The strategy used by the Brazilian government, the *Estado Novo*, was, in a similar way as the Bismarkian conservative modernization, intended as a way of control of the labor movement, preemptively fragmenting its organizations by channeling distinct welfare state benefits to the strategic sectors of the urban population.

Although the retirement and pension fund for railway workers (*Caixa de Aposentadoria e Pensões*, CAP) dates from 1923 (Ruezga Barba 2009: 86), it was during the government of Vargas that most pension funds were created: in 1933, one of the first *Institutos de Aposentadoria e Pensões (IAP),* organized nationally for a specific labor sector, was created (J.

[10] A similar situation occurred in Mexico at about the same time, as discussed in the next chapter.

Malloy 1979: 67). In the following year, the institute of bank workers and that of industrial workers were founded. By 1937, there were already 137 funds (Ruezga Barba 2009: 86). In this way, the structure of relations between the government and labor during *varguismo* expanded and consolidated in a segmented social protection system that provided basic protection to all organized sectors of the urban working class, but excluded unorganized groups and the vast rural sector—a manner of consolidating power in a few labor leaders that would be loyal to the government: the so-called *pelegos*. "The result was a bewildering array of laws and operating rules and a variety of protection and funding patterns" (Malloy 1979: 69).

With the election of Kubitschek, who arrived at the presidency supported by both parties created by Vargas: the *Partido Trabalhista Brasileiro* (PTB) and the *Partido Social Democratico* (PSD), the unions increased their control of pensions. Union representation was enhanced with the *Lei Orgánica* of 1960, to such degree that Malloy affirms that the funds "...were penetrated by the unions and turned into an important source of power to labor and political leaders identified with organized labor [...], inadvertently empowering labor through a relatively independent potential source of political power" (Malloy 1979: 73).

The military junta that took power in 1964 modified the system radically. During the authoritarian regime, the political interests of the military coincided with the technocrats who sought to reform social protection to simplify and standardize it. The technocrats did not have the power to promote these changes under a democratic regime; the opportunity to do so appeared when their interests aligned with the will of the authoritarian government to reduce the power of popular organizations (Malloy 1979: 124).

The number of pension funds began to decrease beginning with the democratic governments that followed the *Estado Novo*: down to 35 in 1959, under Kubitschek, to 6 in 1960, under Goulart, and then, under the military, pension funds were united in one sole State-controlled institute, as a way of weakening the unions (Malloy 1979: 102). The unification process concluded with the creation of the *Instituto Nacional da Previdência Social* (INPS), which unified all the IAPs into a single fund, and extended social protection to the entire rural sector and later on (1972) incorporated domestic workers (Malloy 1979: 129). The creation, in 1966, of the *Fundo de Garantia do Tempo de Serviço* (FGTS), with the contribution by employers of 8% of the workers' salary, helped flexibilize the labor market, as it permitted to fire workers (and labor leaders) before ten years of service without protection from the union, but with an income saved in the unemployment fund (Ruezga Barba 2009: 86; Malloy 1979: 126). However, "...beyond achieving the long-sought goal of unification, the creation of the INPS had a normative and political significance [...] for the first time since 1923, the concept of class representation in the administration of social protection had been eliminated, thereby unseating the main source of power of organized labor and placing this central political resource under firm State control" (Malloy 1979: 124). All this was achieved with the previous repression of the labor leaders and their replacement by government officials in the main unions.

Before military rule, the various pension funds financed the health services for their members in the private market. This meant that each institute left health services to the end as it had to cover all other obligations, something that had as a consequence that the funds with fewer members and better paid had superior health services, while the most dangerous and unhealthy professions received the worst—a situation aggravated by the fact that the best health facilities were in the urban areas of the Southeast of the country (Malloy 1979: 110); the non-salaried and the poor had to resort to charity or the market. And although "the Social Security Organization Law, passed in 1960, standardized the benefits provided by the different IAPs... they remained as isolated structures, serving their clients separately" (Lobato and Burlandy 2000: 82).

Although from the fifties to the end of the sixties, the leading health provider was the private sector, the Ministry of Health was responsible for preventive health and chronic diseases in hospitals, health centers, and primary health units at the three levels of government. When, in 1966, the military unified the social protection system, health care underwent "strong centralization at the federal level combined with policies designed to strengthen the private sector and aimed at increasing economic growth" (Lobato and Burlandy 2000: 79). It adopted the characteristic of a "dichotomy of institutions within the health care system; the growth in coverage through private provision of health care; incomplete coverage; and regressive financing" (Lobato and Burlandy 2000: 83). While the public sector provided health coverage through profit and non-profit contracts on fee-for-service payment, with no control over the kind of medical care delivered, private health companies provided services contracted by individuals or companies (Lobato and Burlandy 2000: 84).

From the 1970s on, health coverage was extended through both the social protection linked to contributions and through the expansion of emergency care to cover the whole population. "This provoked an unprecedented increase in the demand for services. The *Instituto Nacional de Assistencia Medica da Previdencia Social* (INAMPS) increasingly contracted with third parties [...] [and] gave the private sector a progressively more important role in service provision. As a result, the publicly owned network shrank and deteriorated; while 27% of all hospital beds were public, 73% belonged to the private sector" (Lobato and Burlandy 2000: 85).

One of the principal limitations of this system was that "...the progressive extension of coverage carried out during the 1970s was not followed by a change in funding. The resources collected by the Ministry of Social Security and Social Assistance were expected to cover not only retirements and pensions, which resisted cuts because their amounts were predetermined, but also medical care, which [were not assigned] a specific budget". While the health coverage expanded, "...the contributions were based on payroll, and tended to decrease", with the consequence that public (both social security and State) medical-care expenses were significantly reduced during recessions (as during the 1980s) (Lobato and Burlandy 2000: 85). This resulted in the health system crisis of the eighties, which "included the proliferation of expensive medical care without a corresponding change in the method of financing; a method of paying the private sector that stimulated an increase in expensive specialized procedures, as well as fraud; difficulty in controlling finances because of the disorganized structure of the system; deterioration in the quality of services..." (Ibid., 86).

It was during the crisis of the eighties when, in parallel to the democratic transition, the project of universal public health materialized. What eventually became the Sistema Único de Saúde SUS with the Constitution of 1988, promoted by a group of health functionaries called the *sanitaristas*, had as its goal a free, single, universal, but decentralized health structure, both at the State and at the municipal levels (Lobato and Burlandy 2000: 86). The unified decentralized health system proceeded with the transfer of health units and human resources from the Federal Government to the State governments. "Provision of health services had to become the responsibility of municipal governments, aided financially by the federal government and the States. However, all services had to operate within a unified system" (Santos 1997, cited by Lobato and Burlandy 2000: 89–90).

The creation of the SUS, partially based on the British universal scheme (NHS), rapidly expanded the offer of free health services: by 2003, 79% of the population attended these services regularly, and 57% of all medical acts were financed by the State, 26% by private insurance plans—at least partially—and 15% by the patients themselves. Regarding hospitalization, the SUS funded 68% of the events, while the private plans financed 24% (Lautier 2007: 56–57).

In addition to the expansion of health, the rest of the Brazilian welfare state weathered quite successfully the height of the neo-liberal reforms in Latin America; it contrasts significantly with the rest of the countries in the region insofar as, in the first place, during this decade it did not reduce spending on health and education as most other countries in the continent (Lautier, B. 2007, 53). It did not abandon pay-as-you-go pensions; or even institute a mixed model. In fact, in the context of democratization in which civil society played a dominant role, and intervened significantly in the drafting of the 1988 Constitution, one of the primary ordinances of this constitution was the universalization of education, the constitutional declaration of social protection as a citizen right and as an obligation of the State. The constitution has exerted pressure on successive governments to expand their social policies, although some of them did not advance at the pace set by the laws and more recently have even reversed track.

There were surely attempts to implement radical liberal measures; however, the fact that democratization preceded the reforms aroused an active resistance against these attempts. Almost immediately after the promulgation of the Constitution, the government of Collor de Melo tried to modify the pension system but faced strong opposition in Congress, especially from the PT, and the civil servants of the ministries of social protection themselves. Unlike the rest of the Latin American countries, where the former officials of the social ministries were displaced or even carried out the liberal reforms themselves, in Brazil, they stayed in place as they were deeply committed to the principles of public and universal social protection regime. The functionaries of the social protection institutions in Brazil and those of the economics ministry, who were committed to the socio-developmental agenda, argued that any policy that led to the reduction of contributions would make the system unsustainable. This group of experts had strong support from the CUT and the Labor Party (PT). Although the next government, that of Cardoso, also tried to implement liberal albeit less radical measures, it was equally unable to get them through Congress due to the opposition of the PT (De Melo).

The government of Lula implemented a reform to pensions that neither privatized them nor implemented a capitalization pillar, but in a similar way as has been done in Europe, adopted parametric changes: it imposed taxes on pensions if they exceeded a certain level, applied a ceiling to the allowances and raised the minimum retirement age to 65 years: the government proposed a limit of ten times the minimum wage but had to accept a higher level for specific categories such as the judiciary. But clearly, the reform maintained the principle of a public and intergenerational solidarity pension scheme.[11]

Nevertheless, in the case of health, the increasing intervention of the private interests that had characterized the trajectory of the health scheme until the democratic transition, began to fragment the structure; something that has been weakening the universality of the Unified Health System (SUS). This is what has led some analysts to argue that the agenda established by the 1988 Constitution has been "deconstructed" and that its main objectives, including the homogenization of social protection benefits, the expansion of coverage, the establishment of minimum levels of benefits, the introduction of mechanisms aimed at preserving the real value of benefits over time, and more diverse and equitable financing reforms, have only been partially achieved.

Since the nineties, at the very beginning of the implementation of the SUS, critics such as Vianna and Texeira (1997) considered that it was an evolution towards a "perverse Americanization" of health. Despite the fact that public health has been able to insure the majority of the population and is managed by a powerful State apparatus, its services and benefits are of very low quality and have focused on providing services for the low income households, while the workers with stable purchasing power and the middle classes have been induced to opt for private or semi-private protection, in particular company-level collective medical services (Vianna and Teixeira 1997: 264–267). They mentioned how private protection was inceasing, especially through schemes managed by insurance companies and medical cooperatives, which had grown at

[11] Veja on line, 25/04/2003.

rates of 56 and 46%, respectively, between 1980 and 1989, while group medicine, offered by the enterprises to their workers, had grown by 15%. This means that, as in the United States, health programs had become a central object of negotiation between employers and employees (Vianna and Texeira 1997: 266–267).

In this way, although the institutional reform was intended to implement a Brazilian NHS, Brazil has become one of the countries in the region with the greatest participation of the private sector in health. The financing of the services is disconnected from the health services themselves; municipal authorities buy services from public and private providers, while consumers choose between the different providers, who are then reimbursed (Gutierrez Arriola 2002: 65). In addition, the Brazilian welfare state has continued being highly dependent on the sole contributions to the social protection system, which has a reduced capacity to cover the new benefits and new beneficiaries established by the constitution. This is so much more crucial as the decentralization process increased municipal responsibilities to finance and provide health services (Draibe 1997: 223–4). All of this has accelerated the deterioration of the quality and credibility of the SUS (Viana and Texeira 1997: 265).

References

Alonso, Guillermo (2000), *Política y Seguridad Social en la Argentina de los '90*, Buenos Aires-Madrid, Niño y Dávila Editores.

Antía, F., M. Castillo, G. Fuentes, and C. Midaglia (2018), "La renovación del sistema de protección uruguayo", *Revista Uruguaya De Ciencia Política*, 22 (2): 145–185.

Berins Collier, R. (1999), *Paths Toward Democracy: The Working Class and Elites in Western Europe and South America*, New York, Cambridge University Press.

Bizberg, Ilán (2019), *Diversity of Capitalisms in Latin America*, London, Palgrave Macmillan.

Bensusán, G. (2010). Políticas laborales en América Latina: orientaciones, alcances y resultados. *Ponencia presentada en CEISAL*. Toulouse, 2 de julio.

——— (2016). Organizing workers in Argentina, Brazil, Chile and Mexico: The Authoritarian-Corporatist Legacy and Old Institutional Designs in a New Context", *Theoretical Inquiries in Law*, 17 (1), 131–161.

Caetano, Gerardo (2019), *Historia mínima de Uruguay*, México, El Colegio de México.

CEPAL (2010), *La reacción de los gobiernos de las Américas frente a la crisis internacional: una presentación sintética de las medidas de política anunciadas hasta el 31 de diciembre de 2009*, Santiago de Chile, United Nations.

Draibe, S. M. (1997). "A política social na américa latina: o que ensinam as experiências recentes de reformas?", In E. Diniz and S. de Azevedo (eds.), *Reforma do Estado e Democracia no Brasil*, Brasilia, UnBenap.

Etchemendy, S. and R. B. Collier (2007), "Down But Not Out: Union Resurgence and Segmented Neocorporatism in Argentina (2003–2007)", *Politics & Society*, 7 (3): 363–401.

Filgueira, Fernando (1995), "A Century of Social Welfare in Uruguay", *Democracy and Social Policy Series*, Working Paper #5 - Spring.

Franceschelli, I., and L. Ronconi (2005). "Clientelism, Workfare and the Emergence of the *piqueteros* in Argentina", LACEA Conference, http://depot.gdnet.org/gdnshare/pdf2/gdn_library/annual_conferences/seventh_annual_conference/Franceschelli_Parallel_1_4.pdf. Accessed 15 May 2018.

Garay, Oscar Ernesto (2017), "Obras Sociales: Antecedentes", *Diccionario Enciclopédico de la Legislación Sanitaria Argentina*, https://salud.gob.ar/dels/entradas/obras-sociales-antecedentes#:~:text=Las%20obras%20sociales%20tomaron%20un,con%20la%20revoluci%C3%B3n%20de%201943.

Garcia-Subirats, I., et al. (2014), "Determinantes del uso de distintos niveles asistenciales en el Sistema General de Seguridad Social en Salud y Sistema Único de Salud en Colombia y Brasil", *Gaceta Sanitaria*, 28 (6): 480–488.

García Repetto, Ulises (2011), *El financiamiento de la seguridad social en el Uruguay (1896–2008): una aproximación a su análisis en el largo plazo*, Uruguay, Universidad de la República, consultado el 26 de mayo de 2020 en: https://www.colibri.udelar.edu.uy/jspui/bitstream/20.500.12008/4207/5/dt-09-11.pdf.

Gutierrez Arriola, A. (2002), *México dentro de las reformas a los sistemas de salud y seguridad social en América Latina*, México, Siglo XXI.

Haggard, Stephan and Robert R. Kaufman (2008), *Development, Democracy, and Welfare States*, Princeton, Princeton University Press.

Huber E. and J. D. Stephens (2012). *Democracy and the Left*, Chicago and London, The University of Chicago Press.

Lautier, B. (2007), "Les politiques sociales au Brésil durant le gouvernment de Lula: aumône d'Etat ou droits sociaux?", *Problèmes d'Amérique latine*, 63: 51–76.

Lobato, L. and L. Burlandy (2000), *The Context and Process of Health Care Reform in Brazil. In Reshaping Health Care in Latin America*, Ottawa, International Development Research Centre.

Martínez Franzoni, Juliana, and Diego Sánchez-Ancochea (2016), *The Quest for Universal Social Policy in the South: Actors, Ideas and Architectures*, Cambridge: Cambridge University Press.

Martínez Franzoni, Juliana (2007), *Regímenes de bienestar en América Latina*, Madrid, Fundación Carolina.
——— (2018), "Undoing Segmentation? Latin American Health Care Policy During the Economic Boom", *Social Policy & Administration*, Special number on Latin America Press, 52 (6): 1181–1200.
Malloy, James. (1979), *The Politics of Social Security in Brazil*, Pittsburgh, University of Pittsburgh Press.
Mesa-Lago, Carmelo (1978), *Social Security in Latin America. Pressure Groups, Stratification and Inequality*, Pittsburgh, University of Pittsburgh Press.
——— (1985), *El Desarrollo de la seguridad social en América Latina*, Santiago de Chile, Organización de las Naciones Unidas.
——— (1994), *Changing Social Security in Latin America. Towards Alleviating the Social Costs of Economic Reform*, Boulder and London, Lynne Rienner Publishers.
——— (2008), *Reassembling Social Security. A Survey of Pensions and Health Care Reforms in Latin America*, Oxford, Oxford University Press.
Miranda Gutiérrez, Guido (1993), "La salud en la historia de Costa Rica", en *La seguridad social en Costa Rica*, México, Conferencia Interamericana de Seguridad Social.
Munck, R. (2004), "Introduction", *Latin American Perspectives*, 31 (4): 3–20.
Murillo, M. V. (2000), "From Populism to Neoliberalism: Labor Unions and Market Reforms in Latin America", *World Politics*, 52 (2): 135–174.
Olesker, Daniel (2015), "Igualdad, universalidad y accesibilidad: economía política y estado actual de la política pública de salud", In Miguel Fernández Galeano, et al., *Economía, política y economía política para el Acceso y la cobertura Universal de Salud en Uruguay*, Montevideo, Organización Panamericana de la Salud.
Palomino, H. and D. Trajtemberg (2006), Una nueva dinámica de las relaciones laborales y la negociación colectiva en la Argentina. *Revista de Trabajo*, 2 (3): 47–68.
Palomino, H. (2000), "Los sindicatos en la Argentina contemporánea", *Nueva Sociedad*, 169: 121–134.
Pribble, Jennifer (2013), *Welfare and Party Politics in Latin America*, Cambridge University Press.
Ruezga Barba, Antonio (1994), *La Seguridad Social en Colombia*, México D.F., Conferencia Interamericana de Seguridad Social.
Ruezga Barba, Antonio (2009), *Seguridad social. Una visión latinoamericana, México*, Centro Interamericano de Estudios de Seguridad Social.
Santos, L. (1997), "O poder regulamentador do estado sobre as açôes e os serviços de saúde", in Fleury, S.M. (ed.), Saúde e democracia: a luta do CEBES, Sao Paulo, Lemos Editorial, 241–280.

Santos Basso, Óscar Luis (2005), "Costa Rica" en *El nuevo derecho de las pensiones en América Latina*, México, Instituto de Investigaciones Jurídicas, Centro Interamericano de Estudios de Seguridad Social: 187–253.

Schulthess, Walter and Gustavo Demarco (1993), *Argentina: Evolución del Sistema Nacional de Previsión Social y propuesta de reforma*, Santiago, CEPAL/PNUD.

Silva, E. (2009), Challenging Neoliberalism in Latin America, Cambridge, Cambridge University Press.

Stepan, A. (1985), "State Power in the Southern Cone of Latin America", In P. B. Evans, D. Rueschemeyer, and T. Skocpol (eds.), *Bringing the State Back In*. Cambridge, Cambridge University Press.

Svampa, M., & Pereyra, S. (2005), "La experiencia piquetera", *Medio Ambiente y Urbanización*, 61 (1): 97–112.

Vargas, Juan Rafael and Jorine Muiser (2013). "Promoting Universal Financial Protection: A Policy Analysis of Universal Health Coverage in Costa Rica (1940–2000)", *Health Research Policy and Systems*, 11–28, http://www.health-policy-systems.com/content/11/1/28.

Werneck Vianna, Maria Lúcia Teixeira and M. L. Texeira (1997). "A americanização (perversa) da seguridade social no Brasil", In Eli Diniz and Sergio de Azevedo (ed.), *Reforma do estado e democracia no Brasil: dilemas e perspectivas*, Brasília, Editora Universidade de Brasília, 295–348.

Weitz-Schapiro, Rebecca (2006), "Partisanship and Protest: The Politics of Workfare Distribution in Argentina", *Latin American Research Review*, 41 (3): 122–148.

CHAPTER 4

The Political Economy of the Commodified, the Familiarist, and the Hybrid Welfare States

In this chapter, we will discuss the socio-political conformations and dynamics that have given rise and maintained the welfare states that depend most on the market and those that rely on individuals and their families, but, evidently, also on the market, as individuals and families acquire health services from private providers and pharmacies. The case of Chile is in many respects the opposite of the Brazilian one, which as we have seen in the last chapter, is a State-led welfare state that is being commodified as the market is taking over more and more space in health services. Before the important reforms that the democratic governments have implemented to the Chilean welfare state, it was the most commodified of the continent; however, these reforms, especially the reform on health of 2008, have shifted the Chilean social protection system somewhat away from the market, although pensions are still totally based on capitalization.

The case of Chile's social protection can be understood by the *coup d' etat* of September 1973, where the military imposed what has been defined as a pure neo-liberal experiment, radically dismantling what was the most akin to a socio-democratic welfare state in Latin America.

With the collaboration of Mauricio Jiménez Hernández.

I. Bizberg, *Four Worlds of the Welfare State in Latin America*,
https://doi.org/10.1007/978-3-031-44420-3_4

Before the coup, both pensions and health were promoted and expanded owing to the relation between a strong labor movement and the Socialist Party, a vibrant democratic party system, as well as the ideology of the Christian Democratic Party. This welfare state was dismantled in favor of a commodified social protection scheme, where pensions were fully capitalized and administered by private banks, and where health (and education) depended greatly on the private for-profit sector, mainly because the government channeled a considerable part of workers' contributions to this sector, did not exert (until 2008) a regulatory function, and retrieved those who could not continue affording private health services or who were expelled by it for different reasons; the role it was assigned was absolutely subsidiary.

Familiarist social protection systems emerged in an *ad hoc* manner, slowly and dispersedly since the beginning of the twentieth century. Their trajectory was neither promoted by social organizations linked to a political party, nor by the intention of the State to coopt the labor movement or any other social sector. The fact that the industrialization process was frail in Peru, Ecuador, and Bolivia meant the emergence of a small and weak labor movement. Nonetheless, in all these three countries there existed cores of solid labor organizations: in mining, oil, dockers, governmental institutions, and primary education, which were allotted certain special conditions that the rest of the population did not have concerning pensions and health when they protested and struck, pressuring the government. They also received these advantages when a new government arrived that needed political support (something that happened very often) after a downturn (be it a military coup or a return to a civil government). This resulted in a reduced and atomized welfare state, which some authors call exclusionary, that excluded the majority of the population that lacked organizational and pressure capacity. This situation was partly modified during the most recent commodity boom, especially in those countries where a redistributive coalition governed, for example in Ecuador and Bolivia, which tried to expand social protection by allocating non-contributive pensions and enlarging public health services.

We will also analyze the historical trajectories of two welfare states that are hybrids of various ideal types. We must remind the reader that particular countries do not coincide with the ideal types we have defined as a manner to reduce complexity and differentiate between nations in a way that makes sense, that, to a large degree, all countries are hybrids. But whereas most countries approach an ideal type to a certain point, although they vary at different moments of their historical trajectory, other countries are almost at an equal distance from two or more ideal types, and are thus "real" hybrids; this is the case of Mexico and Colombia, as we will argue below. Half of the Mexican population has access to a scheme that allocates both pensions and health, basically state-controlled, although formally administered by a committee composed of representatives of employers and unions. The other half, the informal workers, neither contribute nor receive pensions. However, once they reach 65, they receive a non-contributory retirement, currently equivalent to 40% of the minimum salary. On the other hand, this population formally has access to a public health service. In reality, this system does not have the capacity to give them regular health services but only emergency and, imperfectly, chronic disease care. In consequence, this population basically depends on their own resources (monetary and human) and those of their families; the huge out-of-pocket expenses bear witness to this situation. In the case of Colombia, although a typical familarist welfare state until recently, after the constitution of 1991, it has achieved offering health services to almost the totality of the population with both a wholly and a subsidized contributory scheme; the dramatic decrease of out-of-pocket expenses is a clear indicator of this situation (see Chapter 5).

The Commodified Welfare State: Chile

It is important to remember that Chile was one of the pioneer welfare states in Latin America; in fact, it was the one that resembled most the European social democratic countries. First, the relationship between the labor movement and the Chilean Socialist Party was similar to that of the European social-democratic conformations. There were other countries on the continent where the labor movement had a close relation with political parties, Peronism with the *Partido Justicialista* and the Mexican CTM with the PRI, but none had the autonomy from and the influence of the Chilean trade unions upon the party. This is one of the reasons why the Chilean welfare state was, before the military dictatorship, one of the most extensive, equitable, and generous in Latin America. A system that "...developed gradually and in a stratified manner during the first 70 years of the twentieth century, through State concessions to powerful pressure groups that flourished in a climate of pluralist democracy. At the beginning of the 1970s, Chilean social protection covered all social risks (old age disability, death, professional risks, common illness, maternity, unemployment, and family allowances), and population coverage exceeded two-thirds, the second highest in the region. Due to the long democratic period, strong unionism and political mobilization produced … the 'massification of privilege', since as the lower social strata gained power, they exerted pressure on the State to approximate some of the generous conditions enjoyed by privileged groups" (Mesa-Lago 1985: 101).

According to Mesa-Lago, the presidency of Arturo Alessandri Palma (1920–1925) marks the beginning of significant reforms in the field. Chilean social protection originated in the context of a long wave of mining protests in the north of the country, the dock workers of Valparaíso, and the urban middle classes of Santiago in the first two decades of the twentieth century, which impelled Arturo Alessandri to power from 1920 to 1925. Social pressure forced his campaign and then his government to present a social legislative package to a conservative-dominated Congress, which finally approved it, under pressure from the army, who had threatened a coup if the laws were not passed. During this presidency, trade unions were recognized, the Ministry of Labor and Social Security was created, and protections for advanced age, illness, and work accidents were instituted (Mesa-Lago 1978: 24–25). Three laws constituted the base of Chilean social protection: the employer's responsibility for work accidents and work-related illnesses; a compulsory insurance fund that guaranteed medical care, hospitalization, medicines, and old-age and disabled pensions; and finally, a mandatory corporate savings program to pay retirement pensions and life insurance for white-collar workers (Mesa-Lago 1978: 25). Under this same presidential period, the most significant pension funds were instituted: the Social Security Service (SSS) that protected the blue-collar workers (who also received health services) and the Private Employees Fund, as well as the Fund of Public Employees and Journalists. During the next 50 years, many other pension funds were established for different groups of workers (Mesa-Lago 1985: 101). Regarding health, "…the sickness-maternity program for workers evolved into the National Health Service (SNS), which integrated various services and would eventually incorporate the homeless. The separation between workers and employees in the realm of pensions was reproduced with the creation of the National

Employee Medical Service, which covered private and public employees" (Mesa-Lago 1985: 101). The Constitution of 1925 consolidated the social gains obtained in this period.

In his second presidency (1932–1938), Alessandri leaned more towards the middle classes' interests than labor. However, in the last two years of his term, he was compelled to respond to the pressure of the labor class; a significant indicator of the strength that the labor movement had acquired during the great depression was that the *Confederación de Trabajadores de Chile* (CTCh) was founded in December 1936. The presidency was obliged to expand social protection regarding family allowances and unemployment insurance as well as access to preventive medicine and maternity insurance (Mesa-Lago 1978: 26).

During the 1930s and 1940s, pensions and health coverage increased in correspondence with the electoral competition between the center-left and the center-right. Another great impulse occurred in the 1960s and early 1970s consequent with the expansion of the electorate (women won voting rights, and the secret ballot was imposed in the countryside), which gave rise to an intense competition between the recently created Christian Democracy and the socialist-communist coalition led by Allende. Both during the Frei and Allende governments, the welfare state was greatly expanded (Haggard and Kaufman 2008: 89–90).

At the beginning of the 70s, Chilean social protection was, together with the Uruguayan, one of the most advanced on the continent. It protected more than 70% of the population (being practically universal considering health care and pensions), and "...granted one of the most generous benefit packages and conditions, and stood out in the quality and levels of its health services" (Mesa-Lago 1985: 99). The system covered benefits for old age, maternity, seniority, death, survival, unemployment, and work accidents (Ruezga Barba 2009: 100).

Nevertheless, the fragmentation of the welfare state caused deep inequalities between the various social classes. As in most other countries in Latin America, successive governments had deliberately stratified social protection to fragment the labor movement (Haggard and Kaufman 2008: 89–90). Although around 90% of the affiliated population belonged to the Social Security Service: the Private Employees Fund and the Public Employees Fund, an additional 30 pension funds existed, which for Mesa-Lago, represented a "bureaucratic maze" (Mesa-Lago 1985: 102). In the early 1970s, this legislative and bureaucratic maze "comprised hundreds of institutions and programs: 31 old-age pensions,

30 seniority pensions, 30 disability pensions, 35 for sickness – maternity funds, 55 social assistance programs, and dozens more for family allowances and unemployment. Each institution had notable differences in legislation, administration, financing, and benefits. There were at least half a dozen public bodies with functions of supervision or collaboration, but there was no effective coordination. The principles of unity, uniformity, and solidarity did not work in the system. Still, due to democratic plurality, there was a tripartite representation in the administration of the institutions of the private sector" (Mesa-Lago 1985: 101). In addition, "In the early 1970s, the Chilean social security network cost set a continental record [17% of GDP], surpassed only by the most developed countries in Europe. It was alleged that this high cost harmed employment, but the system generated a slightly progressive impact on distribution. The system was also criticized for its required heavy state subsidy" (Mesa-Lago 1985: 101).

However, after the bloody military coup of September 1973 and the implementation of the purest model of liberal orthodoxy, as Ffrench-Davis calls it, the old welfare state was dismantled. According to this author, it was the "...main case of the modern application of orthodoxy due to its purity, depth, and extension of its coverage" (Ffrench-Davis 2008: 80). This is why Chile was the first country in Latin America to transform its social protection system radically and why it did so in such a profound way. The lack of resistance on the part of the affected sectors was due to the brutal repression of the Pinochet dictatorship; the political character that had characterized Chilean unionism, which had been its strength in democracy, became its weakness in the face of repression. In fact, unlike the Mexican and Argentinian unionism, Chilean unionism was much more unified and centralized, both organizationally and politically. It was not divided into sectors or specialties but grouped all the workers of a specific productive sector, without considering the professional qualification or the type of work they performed (Zapata 1993: 64). In addition, Chilean trade unionism was highly politicized; Chileans were able to act in politics before they were given social rights (Zapata 1993: 68). Hence, the political representation of the workers nearly predominated over the union; the Chilean unionists were concurrently leaders of the political parties of the left (socialist and communist). In this way, when the military regime banned and decapitated the political parties, it also decapitated unionism; this prevented the labor leaders from retreating to the union sphere and from exercising any resistance to government measures, as the Peronist or the Polish *Solidarnosc* unionists did when facing dictatorship.

Thus, the Chilean dictatorship achieved a total transformation its social protection system. The pension reform was carried out in 1981, even before the economic crisis that affected the continent and caused the implementation of neoliberal economic reforms in the rest of its nations. It was completed by the "Chicago Boys", a group of Chilean economists educated at the University of Chicago, who succeeded in their reforms owing to the convergence of their criticism of the existing pension scheme with the political interests of the military. They also had the support of local financial sectors, who saw lucrative opportunities in privatizatizing the pension funds. They also had the support of both the IMF and the World Bank, who wanted to promote the reorientation of Latin American social policy towards economic objectives: the development of the domestic financial markets and the constitution of an internal source of capital to complement external sources. This turn also meant the transfer of the debate on social policy reforms from experts in the field to the ministries of finance and local and international private financial institutions (Laurell 1997; Turcotte 2012).

In this way, Chile moved from pay-as-you-go administered by sectoral funds (there existed 35 such funds)[1] to an individual capitalization scheme managed by private banks. This reform eliminated the segmentation of the pension system, although it did not eliminate the privileges of some sectors, such as the army. In the new scheme, the government and employer contributions were eliminated and replaced by a mandatory deduction of 10% of the monthly salary; the employers would only allocate 0.85% for occupational accident insurance; in addition, the contributions of 2 to 3% for health were eliminated in 1988 (Marcel and Arenas de Mesa 1991: 13). The funds are managed by private entities regulated by the government, the *Administradoras de Fondos de Pensiones* (AFP's); the State guarantees a minimum return on accumulated savings and ensures total capital, in addition to a minimum pension (Graham 1998: 48–49).

[1] http://www.afphabitat.cl/servicios/preguntas-frecuentes/informacion-economica/beneficios-del-sistema-de-afp/en-que-consistia-el-antiguo-sistema-de-pensiones#:~:text=%C2%BFEn%20qu%C3%A9%20consist%C3%ADa%20el%20Antiguo,com%C3%BAn%20con%20aportes%20del%20Estado.

The most tangible impact of this reform was that it reduced coverage and condemned many workers to receive only the minimum pension (Graham 1998: 48–49). Mesa-Lago calculated the decline in work-force coverage from 76% in 1979 to 57% in 1982. However, according to this author, the decline in the 1980s was due more to the consequences of the economic crisis: unemployment, the expansion of the informal sector, and evasion of contributions, rather than the transformation of the system (Mesa-Lago 1994: 117), once the crisis was over, and even though employment had grown and informality decreased, at the beginning of the years 2000 there was still no real improvement: the coverage of pensions was around 64% of employed workers, which is high for Latin America, but was still far below what it was in the late 1970s (Mesa-Lago 2001: 1313).

On the other hand, although one of the justifications for privatization was that competition between different pension funds would increase the system's efficiency and decrease administrative costs, this was not the case in Chile or the countries that followed its example. Administrative costs are generally very high and have to be fully covered by the workers: 2.5–2.7% in Chile, Uruguay, and Bolivia; 3.0%-3.8% in Peru, Colombia, and Argentina; and 4.13% in Mexico (Mesa-Lago 2001: 1315). Furthermore, although the reforms were intended to reduce fiscal subsidies, this has not happened either; in fact, they have increased considerably: in Chile, the budgetary costs of the reform grew from 3.8 to 6.1% of GDP between 1981 and 2000, a reduction to 3.3% of GDP has been projected for 2040, 60 years after the reform (Mesa-Lago 2001: 1318).

Equally radical modifications were carried out in the health[2] sector. The military replaced the universal and unified health scheme, the *Sistema Nacional de Salud*, with a decentralized public service that gives services to the population that cannot afford private insurance. It proceeded to a decentralization process in which the control over the establishments, equipment, and personnel was transferred to the municipalities. It also allowed that the services that covered the middle and upper classes were privatized. The result was that private insurance companies ended up insuring individuals of working age with low-risk jobs. In contrast, people living in poverty, those who have to leave the private sector because they lose their jobs or cannot continue disbursing the co-payments of

[2] And in education, which was also privatized, with even more success than health.

chronic disease, and those with greater risk—for whom the cost of insurance is prohibitive—depend on the overcharged public service: the *Fondo Nacional de Salud* (Fonasa) (Gutierrez Arriola 2002: 65). On the other hand, concurrent to the growth of the private sector, the quality of public health services deteriorated as they were deprived of the contribution of the groups with higher income levels. The ISAPREs mainly insure young, high-income individuals: whereas in 2020 they insured 21.4 of the total population between 30 and 44 years old, they only included 9.2% of those over 60. In contrast FONASA, insured 70% of those between 30 and 44 and 83.7% of the total population over 60. This means that many of the insured in the ISAPREs are expelled when they age and are more likely to get sick (https://datasocial.ministeriodesarrollosocial.gob.cl/fichaIndicador/736/1). In some cases, individuals are so ill that, having to incur too expensive treatment, they are forced to emigrate to the public sector as they cannot cover the costs of the co-payments. In this manner, the State became truly subsidiary, in the commonsensical significance of the term, since it had to care for the individuals who were no longer profitable for private insurers.

Although the democratic governments preserved the economic model imposed by the dictatorship, partly due to its success in assuring continuous economic growth while with high inequality and principally because it was the base of a political consensus that had allowed a relatively smooth transition to democracy, they attempted to correct the most acute injustices of social protection and slightly modified the industrial relations system. The Lagos government (2000–2006) established a State-financed supplement to guarantee a minimum pension, equivalent to a minimum wage, to all pensioners in case they could not capitalize enough to ensure that amount. On the other hand, it expanded the public health institutions to cover just over 70% of the population and obliged the ISAPREs to include the most common diseases in their insurance plans (Mesa-Lago 2009: 13).

Even after these reforms, the Ministry of Social Development calculated that more than 33% of the total employed did not contribute to pension funds and that half of those who contributed would be unable to finance a minimum pension at the end of their labor life and would not qualify for the guaranteed minimum pension (Berstein et al. 2005 and Arenas de Mesa et al. 2006 in Robles 2011: 33).

The Lagos government instituted a minimum salary pension for workers who did not reach this level through capitalization. In 2008, another significant pension reform, realized by the *Concertación* government, modified the scheme considerably. Another critical reform was effected during the Bachelet presidency (2006–2010), which, although it did not transform the AFP's pensions, it determined that "...the disbursement of the Disability and Survivorship Insurance cease to be associated with the AFPs and is now paid by the employer ... the fixed commission charged by the AFPs is eliminated and competition between the pension services is promoted as a way aimed at lowering affiliates' premiums" (Arenas de Mesa 2010 cited by Robles 2011: 34). In addition, in this same administration, the universalization of a social, non-contributory pension for the people living in poverty was implemented, and access to paid pensions for the elderly was made more flexible (Mesa-Lago 2009: 15–16).

Although these reforms were in line with what the more progressive democratic governments that governed this country since 2000 (Lagos and Bachelet) had been doing, the measure was undoubtedly a deepening of the reformist trend. During the global crisis of 2008–2009, the government extended unemployment insurance (the Solidarity Unemployment Fund) to include workers who did not have a permanent contract. The second government of Bachelet (2014–2018) also implemented changes to the Pinochet labor law, especially in what concerned the prior possibility of replacing striking workers both internally and externally and the facility that the employers previously had to dismiss their employees without justification,[3] although it did not modify a crucial aspect: the decentralization of collective bargaining at the company level.[4]

The democratic governments, especially those of Lagos and Bachelet, also considerably modified the health system. They extended the coverage of public services and imposed regulations on the private ISAPREs (Mesa-Lago 2009: 13). In 2004, the Lagos government approved the so-called *Plan de Acceso Universal a Garantías Explícitas* (AUGE), which outlines certain diseases that both public and private health institutions must prioritize, which were defined in function of their incidence among the Chilean population. This meant that they should be excluded from waiting lists and be integrally paid for by the Fonasa; it began with 40 ailments and

[3] https://www.dt.gob.cl/portal/1626/w3-propertyvalue-165849.html#recuadros_articulo_17659_26.

[4] It established the possibility of negotiations between various trade unions of the same branch and the employers, if the employers accepted them.

has gradually increased to 87 by the end of 2022. The plan also imposed a limit on co-payments which could not exceed 20% of the total hospitalization costs (Robles Farías 2013: 28; Becerril-Montekio et al. 2011a, 2011b; Cechini et al. 2014). It is a fact that the last three socialist governments have, on the one hand, introduced measures to regulate the private health sector and on the other, have considerably strengthened public care.

Some analysts consider that AUGE "...substantially modifies access to health benefits by the citizenry as a whole and is structured under a rights-based approach" (Robles 2011: 41). Nevertheless, the fact that the public sector does not have the capacity to take care of all the cases defined by the law and that they have to be treated in the private sector, means that the public sector has to pay a considerable amount of resources to the ISAPRES that have to come from public finances, because "...the opposition to the Lagos government rejected the creation of a common fund to finance the AUGE between the public and private sectors, the ISAPRES had to create a payment scheme different from the one used by the State. This eventually led to all ISAPRE users paying an additional fee for AUGE...for every $100 that the ISAPRES collects for AUGE, $47 correspond to profits" (Manuel Inostroza, health superintendent of the Lagos government, cited by Universidad Andres Bello 2015). This has meant the growth of the public hospitals' debt and the increase of the private health and out-of-pocket expenses, and the fact that those who do not have a pathology covered by AUGE have to enter a waiting list (Ibidem.).

In this way, although the democratic governments that ruled Chile after the year 2000 effectively eliminated the most unjust elements of the social reforms of the dictatorship (Riesco 2009), it seems clear that Chile has not abandoned its liberal model in which economic rationality prevails over the social rationale and the commodifying character of its social protection system: pensions are capitalized, and health depends very strongly on private health services. This is one of the reasons why, at the end of the year 2019, part of the population, especially the young, manifested with such force against the commodification of education, pensions, and health, a model that has preserved the inequalities of a country that has been growing continuously in the last 40 years but has not implemented a manner of distributing the acquired wealth.

The Familiarist Welfare States: Peru, Ecuador, and Bolivia

The familiarist welfare regime has been defined as a limited system of protection that covers only the most strategic socio-economic sectors (civil servants, the army, the police, the strongly unionized and conflictual economic branches) but excludes the majority of the population, who have to deal individually or depend on their families, mostly its female members, to deal with old age and health; in these countries, even when public protection formally covers the whole of the population, it does not do so in reality, forcing families to resort to the private health market.

The first phases of the historical trajectory of the welfare state of these countries are not very different from those of the other Latin American countries we have been describing: the army, the police, and eventually the civil servants were the first to receive pension rights and health services guaranteed by the State (in the case of these countries, between the second and third decades of the twentieth century). Nonetheless, after this first wave of expansion of social protection, the rest of the population acquired rights in a very disparate manner, only when their civil society organizations or unions were strong enough (in organizational or mobilizational terms) to extricate from their governments a pension or a health scheme, or when a recently democratized or authoritarian government required legitimation. Thus, there was no general allocation of rights to whole sectors of the population in orde to coopt them within a statist corporatist scheme, as in Brazil, Mexico, or Argentina, or within a socio-corporatist compromise, as in Costa Rica and Uruguay. Also, no well-organized union movement linked to a political party as the socialist and communist parties in Chile, or political parties confronted in strong political competition, such as the Chilean Christian democrats, existed in these countries.

In the case of the three countries we have considered as familiarist, the historical trajectory of their social protection expanded in an ad hoc manner depending on the strength of a specific social actor and the political necessities of a particular government. In Bolivia, the miners organized the very powerful *Confederación de Obreros Bolivianos* in the early twenties of the twentieth century. Then they acquired a central role in the 1952 revolution. In Peru, it was also the miners, dockers, and teachers whose unions won social rights between the end of the Second World War and the seventies. Finally, in Ecuador, Jiménez Barbosa (2017) considers that the most significant gains have been the consequence of democratization after 1978, the pressure of the movements of indigenous, labor, and student organizations against neo-liberalism at the end of the 90s, and again at the end of the first decade of the twenty-first century, under the government of Rafael Correa (2007–2017) (Jiménez Barbosa et al. 2017: 127). Nonetheless, the fact that social movements and/or the political parties were not solid or willing enough to impose the extension of social rights beyond the sectors that had acquired them through their pressure or as a result of the government's drive to gain legitimacy resulted in a significant part of the population being left unprotected, and having to depend on their means in these three countries.

As we have already seen in the other countries of Latin America, the governments of these three countries began allocating social protection for disability, old age, and death to the most strategic socio-political sectors. In Peru, before the reforms of the Leguía government in the 1920s, the only sectors that had pension rights were the military and the civil servants, and this since 1850 (Mesa-Lago 1978: 114). In Bolivia, public officials got retirement rights in 1831 (Ruezga Barba 2009: 86). Ecuador was, in many senses, a "laggard" as the first pension fund benefiting teachers was created in 1923, although a few years later, it advanced faster than the other two as in 1928 it extended these rights to civil servants, banking workers, and military (Ruezga Barba 2009: 138). In the twentieth century, other sectors of the extractive industries began acquiring retirement rights; but while during the nineteenth century, these rights were a concession from the government, in the twentieth century, and in the case of the blue-collar workers, these rights were acquired due to social and political pressures. In Bolivia and Peru, these rights were linked to the fact that since the beginning of the century, different working-class sectors had begun organizing and striking.

Although the extent of Bolivian industrialization was minimal, the artisans of La Paz founded the first union of the country in 1908: the *Federación Obrera de La Paz*, and a few years later, the anarchist *Federación Obrera Internacional* (Ponce 1969: 8). In 1918, the *Federación Obrera del Trabajo* was instituted, with the affiliation of miners and railroad workers. Since 1919, miners had been demanding the eight-hour working day, which was only allocated to the workers of the miners of the Huanuni. The first general strike that mobilized railroad and tramway workers and chauffeurs happened in 1922 in La Paz.

The following year, on the 1st of May, the government triggered the first massacre of miners of a peaceful demonstration (Ponce 1969: 10).

The first significant social protection measure they extracted did not have to do with pensions but with labor accidents, something coherent in consideration of the economic sectors involved, especially mining and railroads. In 1924, a law covering workers' professional accidents and deaths was instituted, and a compulsory savings fund for workers was established (Ruezga Barba 2009: 86). Nonetheless, it was not until the end of the disastrous *Guerra del Chaco* that affected the economy very profoundly, and led to significant social and political turmoil, that the welfare state expanded. Although in 1925, the miners, together with the urban sectors, took over the *Federación* and abandoned its anarchist character to embrace Marxism, it was in reality after the Chaco War that, according to Ponce, the ideological phase of unionism ended and syndicalism began (Ponce 1969: 9–11). In fact, on the 10th of May 1936, an alliance of workers, students, and the indigenous population began a general strike that caused the fall of the incumbent government and the arrival of the military, who proclaimed a "socialist military revolution". This government decreed a new Constitution which included several essential articles in terms of social protection, both regarding public health and labor (Mendizábal Lozano 2002: 166); in 1938, the Ministry of Health and Labor was founded, and a year later, the government decreed a Labor Law (Ponce 1969: 11).

It was also after the Chaco War, in 1935, that the *Caja de Seguro y Ahorro Obrero* was instituted, which held the funds for accidents and illnesses, and workers' savings (Ruezga Barba 2009). The fact that this was a response to the pressure exerted by the workers' movement is evident from the opening paragraph of the law: "...The implementation of the Workers' Insurance and Savings Fund is of real urgency, as an institution designed to eliminate reasons for conflict between employers and workers and simplify the procedure for the disbursement of compensations, broadening the guarantees to workers and setting standards for managing the mandatory workers' savings". In addition, the fund would be managed "by a Board of Directors with representatives of the Government, two representatives of the large producers, one representative of the small producers, *and a representative of the workers* ... appointed by the supreme government from among the mining workers' leaders of the country".[5] The contributors to this fund would be the mining companies, the workers, and the State. The long-term idea was that the benefits of the fund would be eventually extended to all workers.

In Peru, according to Mesa-Lago, the first significant social security law was decreed by president Leguía during his first presidency (1908–1912), protecting the worst-paid sector of the industrial workers of private enterprises against work-related accidents. This law followed the first strikes of the anarcho-syndicalist sugar, mining, and port workers unions. A year before this law, Leguía had expanded the pensions to the Air Force. During his second term (1919–1930), Leguía extended life insurance to white-collar workers, "...initiating the tradition that set this group apart from the blue-collar workers" (Mesa-Lago 1978: 115). His second term also saw the rise of the first health service measures, oriented towards miners and dockers, two sectors with strong unions. This law established the obligation of enterprises with more than 1000 workers to provide free medical assistance and establish health facilities when located far from the few public hospitals that existed at that time (Mesa-Lago 1978: 114–115).

However, the modern Peruvian welfare state dates from the 1930s, under president Benavides (1933–1939). The context that determined the establishment of social protection was the creation of numerous unions, a strong wave of strikes, and the foundation of two of the principal leftist parties of Perú, organized by two of the most significant social leaders of Latin America: José Carlos Mariátegui, founder of the Peruvian Socialist Party, and Raul Haya de la Torre initiator, in 1924, of the APRA, *Alianza Popular Revolucionaria Americana*, the most signficant Peruvian party of the next 40 years. The significant mobilization that these events wrought led to the resignation of Leguía and a military coup, with the consequent repression of the labor movement. When Benavides arrived at the presidency, he allowed for the reorganization of the unions, legalized the APRA, and amnestied its militants; however, a year later, the APRA was again banned and the unions curbed.

Nevertheless, this context led to the creation, in 1935, of the Ministry of Public Health, Labor and Social Welfare, a significant acknowledgment of the importance that the social question and labor had acquired. A year later, the *Seguro Social Obligatorio (SSO)*, which covered the totality

[5] https://www.derechoteca.com/gacetabolivia/decreto-supremo-22-05-1935-3-del-22-mayo-1935/.

of the blue-collar workers, was created. The SSO allocated health and maternity care, pensions for old age and disability, subsidies for temporary incapacity, and funeral aid for blue-collar workers, self-employed, and domestic workers; it also expanded the old civil-servant legislation of 1850 to include most public employees (Mesa-Lago 1978: 116). While this law represented a measure for all workers, the Benavides government also implemented measures that divided the labor movement: it allocated "a retirement fund for the Callao longshoremen in 1934, safety regulations for miners, and compulsory mutual insurance funds for teachers and judiciary employees, in 1936" (Mesa-Lago 1978: 116).

In Ecuador, as in the other two countries, at the beginning of the twentieth century, only the military had protection against disability, old age, and death, financed by the State without intermediation or funds. More universal measures concerning protection against professional risks for the entire working force were instituted in 1921; they included compensation, retirement funds for disabled workers, and medical assistance (Ruezga Barba 2009: 137). In 1928, a pension fund was created to extend coverage to all public and banking workers, it was open to anyone who contributed voluntarily. These latter reforms were achieved under the *Revolución Juliana*, a coup that was led, in July 1925 by young officers of the army that represented the popular sectors that had suffered the repression that had ended the first general strike a few years before. The Junta also created public health leagues and organized the National Sanitary Service, but most importantly it created the Ministry of Social Welfare, Labor, Agriculture, Charity, Health, Hygiene, Statistics and Immigration and Colonization (Paz y Miño Cepeda 2002: 47). During the decade of the 30s, Ecuador, as most countries in Latin America, went through a deep economic crisis; in the case of this country, it was mainly due to the dive of the price of its main export: cacao. Important labor mobilizations followed, which led to a new round of institution creation (Ycaza 1991). In 1935, the Compulsory Social Security Law was decreed, and the *Instituto Nacional de Previsión* (INP) was created to coordinate the application of compulsory social security (Ruezga Barba 2009: 139). In 1937, the Private Employees and Industrial Workers Fund was instituted (Ruezga Barba 2009: 138).

The next very significant period for the welfare states of our three countries occurred after the Second World War. In the case of Bolivia, the period that preceded the Revolution of 1952 was full of turmoil and noteworthy measures in the social protection system. The military government of Gualberto Villarroel (1943–1946) was characterized by a significant organizational activity of the unions, sustained by the government. It was during this government that the *Federación Sindical de Trabajadores Mineros de Bolivia* (FSTMB) was founded in the mining district of Huanuni (Ponce 1969: 14). It was also in this same period that the first indigenous organization saw the light: the First Bolivian Indigenous Congress was organized in La Paz in 1946. However, the end of this experience was tragic, as the Confederation of Bolivian Workers (CSTB) contributed to the overthrow of the president, then following a series of teacher's strikes that led to numerous disorders, President Villarroel was assassinated (Ponce 1969: 14–15).

In the years that preceded the Revolution of 1952, a series of events were to many degrees its forerunners. First, in 1947, the Catavi mine, owned by the weathiest man in Bolivia, Simón Patiño, dismissed hundreds of workers. That gave way to a strike, then a lockout by the enterprise, the arrest of several mining leaders who were then expelled to Chile, and the kidnapping and assassination of some functionaries of the principal mines of the country by the miners. All of this led to a general strike in 1950, which was strongly repressed and signaled the beginning of a civil war that extended to many of the principal cities of the country: Cochabamba, Potosí, Sucre, and Santa Cruz. In addition, in the parliamentary elections that were held in 1950, the gains of the main leftist party were not recognized by the government, a situation that eventually led to a general rebellion that ended with the overthrow of the government and the arrival of the MNR to government (Ponce 1969: 16–17).

It was in these turbulent times before the Revolution that some of the most significant transformations of social protection occurred: at the end of 1949, the *Instituto Boliviano de Seguridad Social* was created, which formally implemented the universality and obligatory character of social security in Bolivia, which was to include all public, private, white- and blue-collar workers, and be financed in a tripartite manner (Pérez Patón 1951: 29). In addition, in 1951 the labor risk law from 1924 was updated and merged with the mandatory social security law so that maternity, sickness and disability, old age and death were also recognized as social risks and not exclusively the professional ones (Ruezga Barba 2009: 87). It is possible to argue that these laws were decreed as a manner of trying to appease the labor and the political mobilization but failed to do so.

Once the revolution triumphed, "...the revolutionary measures of the government of the MNR affected the life, motivations, and objectives of

Bolivian trade unionism. Unionism became one of the pillars on which the regime rested ... [a relationship] of control and mutual support between the government and workers, which will last from 1952 to 1956" (Ponce 1969: 20).[6] In the last year of this first period of the close relationship between the government and the Central Obrera Boliviana (COB), which represented mainly the miners but united most workers of Bolivia, the government promulgated the Social Security Code conceived to regulate the health funds, and according to Mendizábal Lozano, inaugurated a new phase of the health services organization by the Ministry of Health, which eventually led to the constitutional reform of 1961, when the country first established the right to health (Mendizábal Lozano 2002: 208–209).

The military, who ruled from 1965 to 1982, did not undertake any fundamental changes to the pension or the health schemes. Only after democratization did things begin moving again, but in contradictory directions. Under the government of Paz Estenssoro, an important health program was launched: the *Plan Global de Salud*, which pretended to extend primary health services all over the country through regionalization and social participation via local and district councils (Mendizábal Lozano 2002: 340). A few years later, under the impulse of the transformative *Ley de Participación Popular* of 1994, decreed during the government of Sánchez de Lozada, that resulted in very significant social and political changes, both in terms of organization of local social organizations, the decentralization of decisions and resource obtainment at the local level in a country that had been centralized, but where the State was very weak and did not reach most of the localities. In this context, health services were also decentralized and began receiving considerably more resources, which grew from 2.71% of GDP to 3.98% after the law (Mendizábal Lozano 2002: 373).

[6] In fact, in contrast with Peru, Ecuador, and Colombia, both the military government of Villarroel and especially that of the MNR, intended to establish a corporatist relation with the labor movement (as in Mexico, Argentina, and Brazil). The military tried to make unionization obligatory and promoted the organization of the *Asamblea Nacional Permanente de Organizaciones Sindicales* (ANPOS), which was meant to contain all unions; however, their attempt failed, as most of the unions decided to maintain their independence. In contrast, the government that emerged from the revolution of 1952 did succeed in incorporating the most important miners' unions in the COB, and in including it as a member of government (Ponce 1969: 20).

Regarding pensions, we find a contrary evolution under the democratic transition and the lost decade: the privatization of pension schemes constructed from the beginning of the twentiethcentury. In 1997, also under Sánchez de Losada, in the neo-liberal wave of the nineties that swept Latin America, Bolivia passed from pay-as-you-go to capitalization. At the same time, the government instituted a solidary wage: the *Bono Solidario* (Bonosol) for those who had neither contributed nor saved enough. This non-contributive pension program was greatly expanded during the presidency of Evo Morales, being at the same time the most extended of the non-contributive pension programs in the continent and the least generous (see Table 6.3 in Chapter 6).

In Peru, a significant wave of labor organization occurred under the presidency of Prado y Ugarteche (1939–1945), which gave birth to a new labor organization: the *Confederación de Trabajadores del Perú* (CTP). According to Mesa-Lago, health insurance was expanded during this presidency by constructing six hospitals in Lima and other cities. While on the one hand, the government enforced pensions for old age, it continued with the segmented allocation of allowances for the miners, construction workers, dockers, and drivers, who were incorporated into the *Seguro Social Obligatorio* (Mesa-Lago 1978: 117). The next government, that of Bustamante Rivero sustained by the APRA, which was legalized but not allowed to contend in the elections, saw a record number of union foundations and a significant expansion of the welfare state for all blue-collar workers as well as for the best-paid white-collar workers that backed the party (Mesa-Lago 1978: 117).

During the government of Manuel Prado (1956–1962), again with the support of APRA several unions, the welfare state once again expanded, responding to the pressure of specific sectors, this time coming from the National Association of Civil Servants, with the creation of a pension fund for public employees in 1960. With the example of the civil servants, the CTP pushed for a pension fund for blue-collar workers and obtained one in 1961. By the middle of the sixties, under the presidency of Belaunde Terry, the multiplicity of the pension funds began to be perceived as a problem; his government founded the National Council of Social Security to try to unify them (Mesa-Lago 1978: 120). Although this government did not achieve this goal, the next government partly attained unification after the military coup of Velasco Alvarado in 1968. Although it was a leftist and nationalist military government that undertook a nationalization program and an agrarian reform, it tried to weaken the labor movement, most specifically the CTP, to "...assume control over the social security institutions and initiate radical reforms" (Mesa-Lago 1978: 121). The military terminated the participation of workers and employers in the administrative councils of social security funds (ibid.: 122), and in 1979, the different pension funds were unified into a unique one, and the differences between employee and worker insurance were eliminated (Ruezga Barba 2009: 196). However, important exceptions, such as civil servants, military personnel, and police officers were preserved (Mesa-Lago 1978: 123).

In the case of health, the significant reforms of Peru included fees for services, means-testing, a basic package of health services, and decentralization of the administration of local health clinics (Ewig 2006, cited by Noy 2017: 10). In 1994, *Salud Básica Para Todos* was introduced: a basic package of health services with the focus on extending services to poor and remote areas was implemented and the administration of health services was decentralized via the *Comités Locales de Administración de Salud* (CLAS), which aimed to expand coverage and improve the efficiency via community participation (Noy 2017: 10). "In the late 1990s two different schemes, also run by the Peruvian Ministry of Health, the *Seguro Materno Infantil* (SMI) and the *Seguro Escolar Gratuito* (SEG) targeted pregnant women, mothers and children, and children in public schools, respectively, who did not have other forms of public or private insurance, though such means-testing was not typically enforced … targeting has therefore utilized to extend access for certain groups of people" (Noy 2017: 10). In 2001, these two programs were united and the *Seguro Integral de Salud* was created (Parodi 2005, cited by Noy 2017: 11), "...this was a response to an estimate that approximately 58% of the population did not have health insurance, and relied entirely on the Ministry of Health" (Alcade-Rabanal et al. 2011, p. S248 cited by Noy 2017: 11).

It is noteworthy that the most significant changes in the Ecuadorian health scheme have always been coupled with political and social changes: after the democratic transition at the end of the 70s, with the protests against neo-liberalism in the nineties, and following the emergence of a civic movement during the first five years of the new millennium that resulted in the presidential election of Rafael Correa (2007–2017) and the promulgation of a new Constitution (Jiménez Barbosa et al. 2017: 127).

According to Ruezga Barba, although the Ministry of Public Health was created in 1967, the delay in social security coverage in Ecuador is noteworthy as it was not until 1971 that the universalization of social security was formally specified in the *Código de Seguridad Social*. Even then, this formal measure was opposed by the sectors, that were already affiliated, and by the employers, "...the first opposed the redistribution of benefits, and the second claimed increased production costs; as a result, the code was canceled three months after its entry in force" (Ruezga Barba 2009: 141). Things began to change with the democratic transition and the activation of the labor movement at the beginning of the 80s. There was a first national strike in May 1981, four others during that year and the next, and then again in 1982 and 1983, which, according to Ospina, "...marked the highest level of worker mobilizations ever" (2009: 24, cited by Herrera, 2015).

During the 1980s, labor movements coincided with the "lost decade", resulting in reduced public investment in health. In 1984, the regulation that fixed the prices of medicines was annulled. "Also in this decade, the universalist discourse began to be dismantled, and governmental action began to focus on the groups considered 'most vulnerable': children and mothers" (Velasco 1997: 28, cited by Minteguiaga 2013: 48). Likewise, the proposals for "decentralization" gained strength. In health, an attempt was made to "regionalize" the Ministry of Public Health services, transferring responsibilities to subnational levels (Minteguiaga 2013: 48).

Nevertheless, it was only in the 1990s that social protection was expanded after a new wave of indigenous, labor, and student movements. In 1994, the free maternity law was enacted. A few years later, the 1998 constitution included a specific section on health and established a tripartite scheme with the participation of the State, companies, and workers under principles of universality, equity, and solidarity. In 2001, the Law on Social Security was promulgated. In 2006, the Universal Health Insurance Program was created, being promoted by a loan of 90 million dollars from the Inter-American Development Bank, which also allowed for the financing of a Catastrophic Illness Coverage Program to attend to high-cost cases. This program was implemented by the Ministry of Economic and Social Inclusion (MIES) and the Ministry of Public Health (MSP) (Jiménez Barbosa et al. 2017: 130).

The 2008 Constitution of Ecuador, under the leftist government of Rafael Correa (2007–2017), the so-called "*Revolución Ciudadana*", recognizes the protection of health as a human right and of the State as responsible for guaranteeing it with universal and free services (Chang Campos 2017: 453). Its articles consider health integrally and the need for essential services and recognize social security as an inalienable right, in addition to the principle of citizen participation and transparency in the management of social protection (Jiménez Barbosa et al. 2017: 127). Although in this period, coverage effectively increased through voluntary contributions and the obligation of affiliating all workers in a dependency relationship as well as all domestic workers, nonetheless "...no progress was made in a comprehensive reform of the system ... to universalize the services ... Social security only covers affiliates and does not extend to family groups. Nor is the formal sector of the economy fully covered" (Minteguiaga 2013: 55). In addition, there is no single national health system, although "...progress was made in the expansion of the primary services, free medicines and home care in poor neighborhoods: the consultation rate of the Ministry of Public Health per 100,000 inhabitants, went from 95 in 2006 to 243 in 2010" (Source: MSP/SENPLADES 2011, cited by [Minteguiaga 2013: 53]).

As we have already mentioned, the situation of these three countries changed with the arrival of leftist governments in Bolivia and Ecuador (the *Movimiento al Socialismo* (MAS) in Bolivia and the *Movimiento de Revolución Ciudadana* in Ecuador), but especially with the increased resources that came due to the commodity supercycle, of which all these countries benefited as they export mainly raw materials. Even Peru and Colombia, where the left did not arrive to power, also expanded social protection with the resources they obtained from the surge of the demand and prices of their exports. Nonetheless, in all three countries (as well as in Colombia and Mexico), the expansion of both the pension and health systems has been achieved through targeting rather than universalization and under a minimal assistance model. Even in Bolivia and Ecuador, where beginning with the leftist governments of Evo Morales and Rafael Correa, a universal scheme was implemented; they only allocate minimal non-contributive pensions to older people, help poor children or uniparental (basically women-led) families and formally[7] (although not really) offering the right of health care to all.

Both the historical trajectory and the way social protection has been expanded in the last 20 years explain why, at present, a significant percentage of the population of these three countries receive minimal protection or are even excluded from social protection. The fact that only a minority of the population benefits from the welfare state is because, in these countries, there exists a vast informal sector (ranging from 50 to 60% of the employed population). In the case of health, while contributive social security funds cover the formal sector of the population, most of the population receives services through a limited and underfinanced public health sector. On the other hand, part of this population, especially in the rural areas, needs more coverage, as the health centers are concentrated in urban areas.

One of the reasons for the low expansion of the welfare state in these countries is the small portion of the population that is covered by contributory social protection schemes that require formalization, union organization, etc. This population exerts a very strong pressure on the public pension and health systems that have to shield the great majority of the population and do not succeed in doing so. A sharp contrast with the countries where, as discussed above, the coverage of pensions and health is extensive (Uruguay and Argentina): where more than half of the population is included. In the three countries we are discussing, social protection comprises less than a quarter of its citizens: in Peru 26%; in Bolivia 22%; and in Ecuador 22%. In addition, private health insurance is also minimal due to the few people who can afford it; the highest is Peru, with 10%, while in Ecuador it is 3%, and in Bolivia, it is 1.5%. This is in apparent contrast with Chile, where private insurance includes 16% of the population, or Brazil, where it covers around 23%. In this manner, the great majority of the population has to depend on the public sector,

[7] The amount of financing the health system has received is in no way sufficient to cover all the population formally included: although formally an ample sector (or even the totality of the population) has a right to these services, in reality, for different reasons (type of care needed, overcrowding of health services, lack of knowledge of their rights, etc.) not all can profit from them; we will see how this translates into a great percentage of the population lacking health services. On the other hand, even formally the system does not include the total population due to the fact that there are certain conditions that not all citizens comply.

which, since it is very deficient, must rely on families, as 59% of the population in Bolivia does not have health insurance, 56% in Ecuador, and 17% in Peru (see Fig. 5.3).

Two Hybrids: The Mexican and the Colombian Welfare States

While the welfare states described in the last two chapters point towards a unique type, some are equidistant to two types. This is the case with both the Mexican and the Colombian social protection systems. In the Mexican case, the State has a significant role in the contributive pension and health schemes, through the IMSS and ISSSTE (covering private and public workers, respectively). In contrast, its non-contributive pension and health programs are very ungenerous and do not cover the rest of the population entirely, leaving most of the spending of this population under the responsibility of the individuals and their families; it is in this respect, a system that is a hybrid between a statist and a familiarist social protection regime. In the case of Colombia, until the constitution of 1991, the welfare state of this country was fundamentally familiarist. Still, starting with its promulgation, social protection expanded considerably, especially concerning health, to cover practically all Colombians, although, in contrast, pensions still cover a significantly reduced percentage of the population. Health services are directly assumed by the State in its hospitals and clinics and partly by private insurance companies in private hospitals and clinics, although subsidized. In this manner, we consider this system a hybrid between a statist and a commodified welfare regime.

Let's start by describing the historical trajectory of **the Mexican welfare state**. The Mexican social protection system not only expanded through State intervention to coopt or control the popular sectors, but at some point in history, especially after the worker movements of the mid-1950s, when salaries suffered due to high inflation, the welfare state expanded as a response to these movements. The State that emerged from the 1910–1917 revolution established a national-popular alliance with workers and peasants in the 1930s, which was made possible thanks to the emergence and mobilization of many unions and peasant organizations. This led to increased strikes and land invasions that resulted in thousands of collective contracts and an extensive agrarian reform decreed by the government of Cárdenas (1934–1939), profoundly affecting the interests of capitalists and landlords.

The situation changed radically with the beginning of the Second World War, the arrival of a more conservative government, and the economic boom the country was facing due to the increase in exports to countries at war as the start of an import substitution strategy. The government needed to demobilize the labor organizations empowered during the previous administration to profit from this situation. One of the ways how this was intended to be achieved was by creating, in 1943, the *Instituto Mexicano del Seguro Social* (IMSS), which was assumed to provide a pension and health services for workers in the formal economy, mainly in urban areas, to the exception of the few sectors that already had their own pension schemes (Gordon 1999: 52–53). At the war, the government of Ávila Camacho convinced the workers and employers to sign an explicit developmentist pact.

While the character of the IMSS was universalist in its vocation, as it progressively pretended to include all workers in the formal sector, it never expanded to include the less organized sectors of the people living in poverty and peasants. On the contrary, the system was segmented further to responf to another wave of worker mobilizations. In 1958–1959, workers and peasants reacted with a wave of strikes and land invasions to 12 years of unpopular economic policies, high inflation, wage freezes, and, in 1956, a sharp devaluation. Unlike many other Latin American countries, this mobilization did not lead to a military coup, as happened, for example, in Brazil in 1964. On the contrary, the Mexican political regime managed to strengthen its political alliance with the workers and peasants, although in an even more segmented and corporatist manner, by allocating special conditions for the most strategic labor sectors: oil, railway, electricity, and sugar workers (Gordon 1999: 56). The government of López Mateos (1956–1962) also created the *Instituto de Seguridad y Servicios Sociales de los Trabajadores del Estado* (ISSSTE), to attend the public employers, which included the members of the largest union of the country, the teachers' union. In this way, the government focused on the most strategic economic sectors, with the better-organized unions that had been most active during the movement, to detach them from the rest of the working class. While the official discourse maintained that the Mexican government was strengthening the pact of the 1930s with the popular sectors, it was, in reality, redirecting part of its resources toward the most militant social groups. Consequently, social protection moved away definitively from the more universalist character that the IMSS had inaugurated and adopted a more clearly corporatist model.

As in most of Latin America, social protection was profoundly modified after the crisis of the 1980s and the "lost decade". Nonetheless, there were profound differences between countries depending on the character of the political regime that implemented the reforms and the ability of civil and political society to challenge them. In Mexico, as in Chile, the reforms were implemented by authoritarian regimes, which could not only impose a liberal economic model and considerably reduce social protection but also weaken the social actors durably, especially unionism. In fact, in both Chile and Mexico, the reforms were more radical, mainly because economic liberalization had preceded democratization. On the contrary, to the extent that both in Argentina and Brazil, democratization preceded economic reforms, resistance from civil society was greater, which implied that the scope of the reforms was not so radical, and social actors themselves became strengthened.

Even though Mexico totally transformed its pension scheme following the Chilean model, going from pay-as-you-go to pure capitalization, the process was much longer because although the country was living under an authoritarian regime, it was not an exclusive one like the military dictatorships of the Southern Cone, but an inclusive one, where the Mexican government had to deal with the unions that were part of the power structure. The liberalization process took more than 20 years: the opening of the economy occurred with the entry of Mexico to the GATT in 1986, the privatizations occurred primarily in the six-year term of Salinas (1988–1994), while the transformation of the pensions was carried out in 1995 for the IMSS and in 2007 for the ISSSTE.

The transformation of the pension system required, first, the weakening of the corporatist social actors that constituted the most important base of political support of the regime of the PRI. This occurred, in part, mechanically, with the reduction of the public administration and the privatization of public companies. In addition, the democratization process also diminished the strength and influence as unions, as well as popular and peasant organizations that were under the control of the State, became increasingly less effective as an electoral tool. Finally, the pressure of globalization and the adoption of the international outsourcing economic model imposed flexibility on the external (at the level of the national economy) and internal (at the plant level) labor markets. Finally, the fact that the modifications to the pension plans were going to concern workers who would be hired after the law came into force also played a fundamental role, as those already on the job were going to be able to choose between staying in the old scheme or joining the new model. All these elements resulted in low social resistance to the modifications.

The government implemented the first pension reform in 1992; it added an individual capitalization pillar, the Retirement Savings System (SAR). The law that radically reformed pensions in the Mexican spirit dates from 1995, when the new scheme, based exclusively on individual savings and capitalization, was applied to private sector workers affiliated to the IMSS, representing around 80% of the total. The new scheme excluded government employees (affiliated to the ISSSTE) and workers in other sectors such as oil and electricity, the army, the navy, and the employees of the IMSS. However, in contrast to the Chilean model, both employers and the government would continue to contribute: employer contributions went from 13 to 12% of the salaries, while the government increased its participation from 0.4% to between 1.4% and 5.9%, depending on the wage level. Workers' contributions remained at 2.1% (Gutierrez Arriola, op. cit.: 53). The government guaranteed a minimum pension at the same level as before: 100% of the minimum wage.

In the case of public health, the rationale was not the Chilean one, as privatization was not be feasible in the Mexican case; the modifications tried to alleviate two of the most critical problems of the Mexican health organization: its low coverage and the impressive out-of-pocket spending, more than 50% of total health spending before the reform (see Fig. 5.6). At the beginning of President Fox's term (2000–2006), it was accepted that more than half the population did not have access to any health service. However, those who did not contribute formally had access to the public health structure of the *Secretaría de Salud*, in practical terms they only had access to services in emergencies. In addition, it is common for patients who attend public clinics and hospitals but have to pay for medicines (sometimes even those prescribed during their stay in the hospital) and other things such as gauze, blankets, etc. The disinvestment in the eighties and nineties deteriorated these services and those of the formal workers in the IMSS and ISSSTE once the financial resources that represented the pension contributions of the workers who migrated to capitalization disappeared. That is why many people opt for low-cost private health services for minor ailments instead of waiting in clinics and losing a day of work.

To try to alleviate these deficiencies, the Mexican government decentralized health services in 1995; from that moment on, the State and municipal authorities operated the public health services while the federal government, in principle, regulated and monitored the services. On the other hand, the IMSS, the ISSSTE, and the SSA established a cost recovery arrangement in which patients payid a fee according to their income, with no charge for the poorest (Laurell, op. cit.). In the same reform, the state implemented a plan to extend coverage by voluntarily incorporating uninsured workers and families into the IMSS; they would have to pay a small contribution, while those earning less than three minimum wages would be exempted. It was thought this would lead informal sector workers to join (Gutierrez Arriola, op. cit.: 188–189). This program did not have the expected results since the voluntary integration of workers of the informal sector did not occur.

In 2003, Fox government launched the *Seguro Popular*, which once again aimed to achieve universal health care. It was a decentralized organization where the federal states would provide health services, financed equally by both the federal government and the state government (and, in principle, also by contributions of the individuals that would join voluntarily, although they would foreseeably be minimal), based on the number of affiliates. Those persons with an income under the lowest three deciles would pay nothing, while those beyond would be charged a small fee. It was a voluntary program based on providing a series (a package) of medical services to its members; as there was no obligation to do so, like in Colombia, it could never include the totality of the population.

It was believed that this form of integration, where the individuals would have a formal affiliation (with an identification), would give them a way to demand a right to health in contrast to the previous situation when they had, in principle, access to this system but no recognized rights. In principle, workers could go to any health center run by the state and the federal governments, and even with private ones that had signed an agreement with the program. Although this program aimed to extend health coverage to the entire population not insured by the contributory schemes, it was never able to attract the vast majority of informal sector workers who would have had to contribute (Bizberg and Martin 2012). Although public spending and infrastructure had not grown at the same rate a few years after the creation of the *Seguro Popular*, the lack of access to health services was considerably reduced: from 38.4% in 2008 to 16.9% in 2018 (CONEVAL 2019, cited by Barba Solano 2021: 97). The *Seguro Popular* gave an official document to its beneficiaries so that they could prove their affiliation and have access to the medical centers of the Ministry of Health. A Universal Catalog of Health Services was created, which defined the illnesses that would be covered. In 2010, the catalog included a list of 275 medical interventions (Valencia Lomelí et al. 2013: 29). In 2015, the catalog included 284 interventions (Flamand and Moreno Jaimes 2015).

Nevertheless, the Seguro Popular did not solve the segmentation and inequality of the Mexican Health organization; on the contrary, to a certain degree, it reinforced it. In 2010, 50% of affiliates to the contributory schemes were concentrated in the three deciles with the highest income (only 17.5% of affiliates belonged to the four poorest deciles), 63.8% of the members of the *Seguro Popular* belonged to the poorest deciles (Barba Solano 2021: 97). On the other hand, while the IMSS and ISSSTE reached 64.5% coverage of all the diagnoses and interventions, the SP only 12.9%, that is, coverage five times lower than the former. On the other hand, although regional inequalities in health coverage were reduced, they did not disappear: although the *Seguro Popular* reduced the distance between itself and the contributory schemes in the medium-level regions, in the welthier regions there was a lesser lack of health services done by this program, while in the poorer areas, the opposite occurred (Barba Solano 2021: 98). Furthermore, since it only offers a package of health services, the services it provided were considerably more limited than those financed by contributory social security schemes (Durán Arenas 2011, cited by Valencia Lomelí et al. 2013: 29). On the other hand, the program introduced a new breach amid the population which had no other choice than using these services, as the affiliates of the SP would be privileged in terms of attention and medicines, against those who were not affiliated (Brachet-Márquez 2007: 346). As we will discuss next, Colombia, which adopted the same program with some modifications, solved this division by making the affiliation compulsory and subsidizing those who could not pay.

Some other programs were introduced by the following governments, such as the Health Insurance for a New Generation, Safe Pregnancy Insurance, and the Catastrophic Expense Protection Fund. The first program formally guaranteed complete and lifetime health services for all children born on or after December 1, 2006. The Fund for catastrophic expenses covered 40 treatments corresponding to eight severe and expensive diseases: However, "...not all necessities have been met since they depend on the available resources defined according to the budget set by the Fund's technical committee" (Valencia Lomelí et al. 2013: 28–29).

The government of AMLO that took office in 2018 canceled the *Seguro Popular*, spent the resources of the Catastrophic fund during the pandemic, and substituted both by the *Instituto de Salud para el Bienestar* (Insabi), which undoubtedly was supposed to be universal and which is credited to attend all the non-formal population, but as we will see in Chapter 6, it did not increase the public resources for health suffciently. The Insabi was founded in January 2020, just a few months before the pandemic and before all of its organization was defined, which in reality regressed us to the situation before the creation of the Seguro Popular, where all formally had a right to health services. Still, in reality, they did not have access to them.

We have, in consequence, defined Mexico's social protection system as a hybrid between statism and familiarism. It is statist, or State-corporatist, as although the unions and employers are represented in both the IMSS and the ISSSTE, the government makes most of the decisions. In the IMSS, the governing board consists of 30 members, among which are the secretary of labor, of health, of the ministry of finance, the main employer associations (*Concamin* and *Concanaco*), and the labor representatives of the main confederations (CTM, CROM, CROC, SNTMMSRM),[8]and in the ISSSTE, the governing organ consists of ten representatives of different government institutions and nine members of the various unions of government employees and workers.[9] While the statist/corporatist social protection sector covers less than half of the Mexican population, the other half has to deal with care, old age, and health based on its resources. Although there existed a non-contributory retirement program and a public health scheme (*Seguro Popular* from 2003 to 2019, and now *INSABI*), neither the SP nor the *Insabi* covers all the population that is not incorporated by contributions, the cancellation of the SP has worsened things, around 28% of the Mexicans say they have no health insurance, while before this it was around 20%. On the other hand, while

[8] https://www.imss.gob.mx/conoce-al-imss/estructura.

[9] https://www.gob.mx/issste/acciones-y-programas/h-junta-directiva-del-issste.

social pensions are universal, they are very low (40% of the minimum wage with the latest increases).

In the Colombian case, due to the complex socio-political situation of this country, where violence has been prevalent since the assassination of the liberal presidential candidate, Eliéser Gaitán, in the election of 1948, most of the innovations concerning the welfare state came from governmental initiatives rather than from pressure from civil society. Nonetheless, one of the first of these actions was that of president Alfonso López Pumarejo (1934–1938), who implemented a series of social and economic reforms under his program of "*Revolución en Marcha*", which, according to Uribe Gómez (2009: 89–90), was a response to labor mobilization of the 20s. This president propped the birth of the *Confederación de Trabajadores de Colombia* in 1936, and then in 1938, the *Ministerio de Trabajo, Higiene y Previsión Social*. However, according to Ruezga Barba (1994: 13–14), the various attempts to implement pension funds, health services, and even a labor code from the beginning of the twentieth century until the end of the Second World War failed, although many of them had received the support of the government.

In this manner, the first fund of social protection that was created in this country was in 1945, under a conservative party government, under the name of *Caja Nacional de Previsión Social*, while the following year, the Ministry of Labor and the *Instituto Colombiano de Seguros Sociales* (ICSS) was created, all of which benefited the unions and professional organizations but excluded the peasants, the independent workers, and the indigenous population (Uribe Gómez 2009: 93). However, its real impact was somewhat limited and health services were precarious. At this time, there were only 45,627 employees affiliated with the social protection structure, with a total of 60,000 direct beneficiaries (Ruezga Barba 1994: 20).

Social protection bagan to take shape with the creation of the Ministry of Labor and the ICSS. The ICSS was intended to provide coverage to workers for professional risks or illness, and maternity, disability, old age, and death; it was also supposed to provide medical, surgical, and dental care (Uribe Gómez 2009: 111–113). In 1949, with the implementation of compulsory social security for specific sectors, the care service for non-occupational illnesses and maternity began (Ruezga Barba 1994: 32). Nevertheless, in response to the mobilization of the medical profession, social protection was segmented with the creation of the *Caja Nacional de Previsión* for the employees of the public sector (Uribe Gómez 2009: 94).

The limited ensuing development of the welfare state in this country was due first to what in Colombia is named "*La Violencia*", which was, in fact, a civil war between the two main political parties, the liberal and the conservative, that had dominated the political life of the country since the nineteenth century, which were socially profoundly rooted, especially in the countryside, in the villages and small cities; each of these was factually dominated by one or the other party and fought against each other. The detonation of "the violence" was the assassination of Eliéser Gaitán, the liberal candidate for the presidency. The conflict lasted ten years until the political actors found a way out with a pact between the two parties. This inaugurated the Frente Nacional period, in which both parties would alternate in power and divide all the governmental posts among them. This pact, which lasted until the mid-80s, resulted in a political system where clientelism became the "...principal way in which citizens and government related" (Uribe Gómez 2009: 97).

According to Zuñiga Romero, the context of armed conflict has been a fundamental factor in explaining the weakness of trade unionism, as with the emergence of the Self-Defense Forces of Colombia (AUC) and the Revolutionary Armed Forces of Colombia (FARC), "...the mere suspicion of attempting to create a union was sufficient for ordering a homicide"; 2,515 trade unionists murdered in twenty-one years, which makes Colombia the country with the highest rate of violations of unionists rights (Zúñiga Romero 2012: 196). The violence generated by guerrilla and the drug cartels led to an unfavorable context for social activity in general and unionism in particular; the maelstrom created by the war against guerrilla and drug lords made social mobilization impractical; social movements were either coopted by the two parties of the pact or repressed in the context of the war against the guerrilla and the drug cartels. In the 1960s, union density reached its highest level: a mere 12.5% (Zúñiga Romero 2012: 192).

The weakness of unions and other non-violent social organizations relevant to the struggle around formulating of social protection policies defined how social protection evolved, basically through governmental initiatives, in tune with the international directives of the OMS-OPS (Uribe Gómez 2009: 244). During the *Frente Nacional* period, any implementation of improvements in the distribution of public services was regionally politicized by the various political actors as patronage meant for the exercise of power (González 1997: 188).

In 1975, a very limited *Sistema Nacional de Salud* was implemented, with three pillars: a public one for the poor households, a contributory one for formal workers, and a private one; while the coverage of public health was 40%, the contributory 20%, and the private 11%, 30% of the population was left without health services by the end of the 80s (Uribe Gómez 2009: 111). In 1990, the health system was again reformed, to decentralize it in administrative and financial terms, prioritizing primary and preventive care (Uribe Gómez 2009: 125).

Nonetheless, the 1991 Constitution marked a significant transformation in the country's social and political life, including health. "As it did in Brazil, the assembly provided a platform for popular and left-oriented groups previously marginalized from political power. The new constitution also incorporated broad commitments to the provision of universal health coverage, education and other forms of social protection" (Haggard and Kaufman 2008: 293). The Constitution included a specific article on social security that defined: "...social security is a public service of a mandatory nature that will be provided under the direction, coordination and control of the State, subject to the principles of efficiency, universality and solidarity ... All inhabitants are guaranteed the inalienable right to social security" (Cardona et al. 1999: 26).

In 1993, the *Sistema General de Seguridad Social en Salud* (SGSSS) was created, which transformed the Colombian health structure into a public compulsory system based on contributions from employees and employers of the formal sectors of the economy (to which resources of direct taxes are added) and a subsidized sector based on individual contributions and public subsidies for the sectors of the population working informally, or lacking sufficient resources. The SGSSS compels individuals to enroll in the *Entidades Promotoras de Salud* (EPS), which administers contributions and offers defined health plans delivered by either public or private health institutions (IPS) (Bernal and Barbosa 2015: 43–45).

The State concentrates the contributions of the workers (12.5% of their income), those of the companies, and of the government itself in the *Fondo de Solidaridad y Garantía;* in addition, 1.5% of the salary of members of the contributory scheme goes to the subsidiary, in addition to the public contribution to this second sector (Giedion and Villar Uribe 2009: 853): the result is that 42.8% of the total corresponds to the contributory scheme and 48.5% to the subsidized (Guerrero et al. 2011:

150). The State has, in this manner, become the single payer. The only sector excluded from this scheme is the 4.9% of the population, which has its own services: teachers, army, police, public universities, and oil workers.

Under the government of César Gaviria (1990–1994), the government implemented a modest liberalizing measure to the pension system, with the creation of a capitalization pillar, while at the same time committing itself to a "sweeping reorganization and expansion of access to health insurance and provision" (Haggard and Kaufman 2008: 293).

In 1994, all citizens were compulsorily incorporated into the pension system consisting of two different retirement schemes: the *Régimen Solidario de Prima Media con Prestación Definida* (the pay-as-you-go scheme) and the *Régimen de Ahorro Individual con Solidaridad* (the individual capitalization scheme) (Cardona et al. 1999: 26). Starting in 1995, the rates of the contributory scheme increased: both employers and workers contributed 24.5% of their base income (12% to health and 12.5% to pensions). Employees contributed proportionally, with 1/3 to health and 1/4 to pensions, while the employer, public or private, had to contribute the rest. One innovation concerning other pension funds was that all workers with a base income over four minimum wages had to contribute one percentage point to a Pension Solidarity Fund (FSP), allocated to the pensions of workers with lower incomes (Tassara 2015: 342).

Based on our discussion, the Colombian system is similar to the other familiarist ones. However, the 1991 constitution pushed a reform of health that defined a significant deviation of its trajectory from the other familiarist systems. In effect, in 1993, the *Sistema General de Seguridad Social en Salud* (SGSSS) was founded to regulate public health and guarantee access to the right to health defined in the constitution; its goals were to assure equity, universality, quality, free choice, solidarity, and integrity. On the one hand, it had liberal components: the decentralization of the provision of services, and the expansion of the role of the private insurers and providers (Haggard and Kaufman 2008: 294), and separated functions between insurers and providers; on the other hand, it targeted fiscal spending on the poorest population (Molina and Rodríguez 2007: 92), "...provided low-income families with an opportunity to enroll in a subsidized system of healthcare organizations ... coverage increased from just over 20% in 1993 to 57% in 1997 and was especially rapid in rural areas and among the very poor" (Haggard and Kaufman 2008: 294).

In this manner, although, as we will discuss more in detail in Chapter 5, the health services implemented in Colombia since 1993 did not resolve the inequality inherent in a health structure formed by two different sectors, nor did it improve significantly the quality of health care, it was successful in increasing considerably the coverage of the health system, to attain a level similar to those of the most comprehensive systems of the continent; this had a very significant result, the reduction of the out-of-pocket expenses of the Colombians. This latter indicator proves that coverage was not only formal but also real in many respects. And this is what allows us to define the Colombian welfare state as a hybrid between a familiarist social protection regime, especially concerning pensions and the position of the assistance programs in the architecture of its welfare state, and a statist regime if one considers health.

References

Alcade-Rabanal, J. E., O. Lazo-González, and G. Nigenda (2011), "Sistema de salud de Perú", *Revista Salud Pública de México*, 53 (2): 243–254.

Arenas de Mesa, Alberto (2020), *Los sistemas de pensiones en América Latina: institucionalidad, gasto público y sostenibilidad financiera en tiempos del COVID-19*, Santiago, CEPAL.

Arenas de Mesa, Alberto, María Claudia Llanes, and Fidel Miranda Bravo (2006), "Protección social efectiva, calidad de la cobertura, equidad y efectos distributivos del sistema de pensiones en Chile", *Serie Financiamiento del Desarrollo*, 172, Santiago, CEPAL.

Barba Solano, Carlos (2021), "El régimen de bienestar mexicano: inercias, transformaciones y desafíos", *serie Estudios y Perspectivas-Sede Subregional de la CEPAL en México*, 191, Ciudad de México, Comisión Económica para América Latina y el Caribe (CEPAL).

Becerril-Montekio, Victor M., et al. (2011a). Sistema de salud de Brasil, *Salud Pública de México*, 53 (2): 120–131.

——— (2011b), "Sistema de salud de Chile", *Salud Pública de México*, 53 (2): 132–142.

Bernal, O. and S. Barbosa (2015), "La nueva reforma a la salud en Colombia: el derecho, el aseguramiento y el sistema de salud", *Revista Salud Pública de México*, 57 (5): 433–440.

Bizberg, Ilán and Scott B. Martin. (2012), *El Estado de Bienestar ante la Globalización, el caso de Norteamérica*, México, El Colegio de México.

Brachet-Márquez, V. (2007), "La reforma de los sistemas de salud y previsión social en México, 1982–2000", In V. Brachet-Márquez (ed.), *Salud pública y regímenes de pensiones en la era neoliberal: Argentina, Brasil, Chile y México*, México, El Colegio de México.

Cardona, Álvaro, et al. (1999), *Impacto de la reforma de la seguridad social sobre la organización institucional y la prestación de los servicios de salud en Colombia*, Medellín, Universidad de Antioquia.

Chang Campos, Caroline (2017), "Evolución del sistema de salud de Ecuador. Buenas prácticas y desafíos en su construcción en la última década 2005–2014", *Anales de la Facultad de Medicina,* 78 (4), octubre-diciembre: 452–460.

Ffrench-Davis, R. (2008), Chile entre Neoliberalismo y Crecimiento con Equidad. Reformas y Políticas Económicas desde 1973, Santiago de Chile, J.C. Sáez Editor.

Flamand, Laura and Carlos Moreno-Jaimes (2015), Seguro Popular y Federalismo. Un Análisis de Política Pública, México, Centro de Investigación y Docencia Económicas.

Giedion, Ursula and Manuela Villar Uribe (2009), "Colombia's Universal Health Insurance System", *Health Affairs*, 28 (3): 853–863.

González, Fernan E. (1997), *Para leer la política: ensayos de historia política colombiana*, Bogotá, Cinep/Programa por la paz.

Gordon, R. Sara (1999), "Del universalismo estratificado a los programas focalizados", In Martha Schteingart (ed.), *Políticas sociales para los pobres en América Latina*, México, Editorial Porrúa.

Graham, Carol (1998), *Private Markets for Public Good. Raising the Stakes in Economic Reform*, Washington, DC, Brookings Institution Press.

Guerrero, Ramiro, Ana Isabel Gallego, Victor Becerril-Montekio, and Johanna Vazquez (2011), "Sistema de salud de Colombia", *Salud Pública de México,* 53 (2): 144–155.

Gutierrez Arriola, A. (2002), *México dentro de las reformas a los sistemas de salud y seguridad social en América Latina*, México, Siglo XXI.

Haggard, Stephan and Robert R. Kaufman (2008), *Development, Democracy, and Welfare States*, Princeton, Princeton University Press.

Herrera, S. Gonzalo (2015), "Situación, estrategia y contexto de los sindicatos en el Ecuador ", *Documentos de Trabajo*, 51, Buenos Aires, CLACSO.

Jiménez Barbosa, Wilson Giovanni, et al. (2017), "Transformaciones del Sistema de Salud Ecuatoriano", *Revista Universidad y salud*, 19 (1): 126–139.

Laurell, Asa Cristina (1997), *La reforma contra la salud y la seguridad social*, México, Era-FES.

Marcel, Mario and Alberto Arenas de Mesa (1991), "Reformas a la seguridad social en Chile", *series de Monografías*, 5, Washington, DC, Banco Interamericano de Desarrollo.

Mendizábal Lozano, Gregorio (2002), *Historia de la Salud Pública en Bolivia. De las Juntas de sanidad a los Directorios locales de salud*, La Paz, Organización Panamericana de la Salud/Organización Mundial de la Salud, https://elagoraasociacioncivil.files.wordpress.com/2015/09/historia-de-la-salud-pc3bablica-en-bolivia.pdf.

Mesa-Lago, Carmelo (1978), *Social Security in Latin America. Pressure Groups, Stratification and Inequality*, Pittsburgh, University of Pittsburgh Press.

——— (1985), *El Desarrollo de la seguridad social en América Latina*, Santiago de Chile, Organización de las Naciones Unidas.

——— (1994), *Changing Social Security in Latin America. Towards Alleviating the Social Costs of Economic Reform*, Boulder and London, Lynne Rienner Publishers.

——— (2001), "Privatización del sistema de pensiones en América Latina", In Rolando Franco (coord.), *Sociología del desarrollo, políticas sociales y democracia*, México, Siglo XXI-CEPAL, 218–237.

——— (2009), *Efectos de la crisis global sobre la seguridad social de salud y pensiones en América Latina y el Caribe y recomendaciones de políticas. Serie Políticas Sociales*, Santiago de Chile, CEPAL.

Minteguiaga, A. (2013), "Política y políticas sociales en el ecuador reciente: dificultades asociadas a la salida del ciclo neoliberal", *Revista De Ciencias Sociales*, 135–136: 45–58.

Molina, Gloria and Claudia Milena Rodríguez (2007), "Participación pública y privada en la provisión de servicios de salud, Colombia 2006", *Investigación y educación en Enfermería*, Medellín, 25 (1), marzo: 90–99.

Noy, Shiri (2017), "Healthy Targets? World Bank Projects and Targeted Health Programs and Policies in Costa Rica, Argentina, and Peru, 1980–2005", *Oxford Development Studies*, 46 (2): 164–183.

Paz y Miño Cepeda, J. J. (2002), *Revolución Juliana. Nación, Ejército y burocracia*, Quito, Ediciones Abya-Yala.

Pérez Patón, Roberto (1951), *El seguro social en Bolivia*, La Paz, Editorial Centenario.

Ponce, G. Jaime (1969), "El sindicalismo boliviano: resumen histórico y perspectivas actuales", *Desarrollo económico*, 9 (33) , abril-junio: 3–31.

Riesco, M. (2009), "El modelo social chileno comienza a cambiar", *Revista Internacional del Trabajo*, 128 (3): 311–330.

Robles Farías, Claudia (2013). *Sistemas de protección social en América Latina y el Caribe, Chile*. Santiago, Cepal, 52 pages.

Robles, Claudia (2011), *El sistema de protección social de Chile: Una mirada desde la igualdad*.

Ruezga Barba, Antonio (1994), *La Seguridad Social en Colombia*, México D.F., Conferencia Interamericana de Seguridad Social.

——— (2009), *Seguridad social. Una visión latinoamericana*, México, Centro Interamericano de Estudios de Seguridad Social.

Tassara, Carlo (2015), "Políticas públicas de protección social y lucha contra la pobreza en Colombia: logros y desafíos", *Papel Político*, 2 (20): 233–351.

Turcotte, Sylvain F. (2012), "Reforma a los fondos de pensiones públicos e integración económica", In Ilán Bizberg and Scott Martin (eds.), *El Estado de bienestar ante la globalización, el caso de Norteamérica,* México, El Colegio de México, 257–286.

Universidad Andrés Bello. El impacto del Auge a 10 años de su creación, *https://www.ispandresbello.cl/el-impacto-del-auge-a-10-anos-de-su-creacion/* ,https://www.ispandresbello.cl/el-impacto-del-auge-a-10-anos-de-su-cre acion/ (2015).

Uribe Gómez, Mónica María (2009), *La contienda por las reformas del sistema de salud en Colombia (1990–2006)*, tesis, El Colegio de México.

Valencia Lomelí, Enrique, Darcy Tetreault Weber, and David Foust Rodríguez (2013), *Sistemas de protección social en América Latina y el Caribe: México*, Santiago de Chile, CEPAL.

Velasco, Margarita (1997), "El reto de la salud para Todos", *Pobreza y Capital Humano*, Quito, Secretaría Técnica del Frente Social, 23–58.

Ycaza, Patricio (1991), *Historia del Movimiento Obrero*, Quito, CEDIME.

Zapata, Francisco (1993), *Autonomía y subordinación en el sindicalismo latinoamericano*, México, Fondo de Cultura Económica-El Colegio de México.

Zúñiga Romero, Marjorie (2012), "El declive del sindicalismo en Colombia y sus consecuencias frente al conflicto colectivo", *Revista de Derecho*, julio: 189–213.

CHAPTER 5

The Health Systems: Their Structure and Performance Facing the Pandemic

We put special emphasis on the health system since we consider not only that it has attracted less attention in most studies on the welfare state, except for the one by Martínez Franzoni and Sánchez-Ancochea (2016), but also because the way health services are structured and organized tells us more clearly both about the differences between countries and between the different types of welfare regimes. This is because health services combine the action of the State, the social actors, the market, and the family in a more complex and evocative manner. It is also less evident that the private sector is a negative factor, as in the case of pensions, it can first be non-profit, offer specific services that the State does not provide, or be contracted and strictly regulated by the government. It is also important to note that in most cases in Latin America, practically no social actors intervene in the pension system, whereas they actively participate in health services.

None of the Latin American countries has achieved a fully universal, egalitarian, and efficient health system, although Uruguay and Costa Rica have probably come closest to this goal. Nevertheless, during the first decade and a half of the twenty-first century, several countries which saw the arrival to power of coalitions that included the popular classes tried to develop their health structures by taking advantage of the commodity boom and the influx of financial resources from developed countries, mainly from the United States.

I. Bizberg, *Four Worlds of the Welfare State in Latin America*, https://doi.org/10.1007/978-3-031-44420-3_5

With the commodity boom of the first decade and a half of the 2000s, several countries where redistributive social coalitions arrived to power expanded the social protection structure towards universalizing health services. The end of the commodity boom and the arrival of governmental coalitions dominated by financial and commodity export interests put an end to the slow transformations of health services in countries such as Bolivia and Ecuador. It also strongly affected the efforts undertaken by Brazil in trying to consolidate one of Latin America's most ambitious health structures, which, like Costa Rica's, is State-run. It is quite true that, even before this economic and political reversal, the SUS (created in 1988) could not achieve its egalitarian and universal objectives, due to the lack of public investment and the absence of regulation of the private health sector. This situation led to increased personal spending not only among the rich and the middle classes but also among workers and civil servants, individually or through collective contracts. Likewise, Argentina's near universal health scheme was seriously de-financed during the Menem administration (in the nineties). This evolution resulted in significant inequalities between the different health sectors in these two countries: in Brazil, between the private and the SUS, and regionally, in Argentina, among the *Obras Sociales*, and between them and public health.

Chile, a country that suffered significant deleterious transformations of its welfare state during the dictatorship, has rebounded to an almost universal health scheme and regulated its (previously unfettered) private health sector. However, while only a reduced percentage of the population can afford the latter (16%), the public sector continues to be much less generously funded. On the other hand, Colombia has almost succeeded in covering its entire population with a mandatory contributory public service by subsidizing the inclusion of its poorer population. This situation has decreased the importance of private health services (Guerrero et al. 2011). This contrasts with Brazil, which has experienced the contrary. Mexico, with a similar model to the Colombian one (the *Seguro Popular*, until 2020), continuously left around 20% of its population without coverage due to its voluntary and un-subsidized affiliation for all but the impoverished population. The INSABI[1] that substituted this scheme left even more people excluded from health services, around

[1] This program has been cancelled as well, due to its inability to cover all of the non-affiliated population and its inability to assure the provision of medicines to this

30%. Except for Chile and Colombia, most countries regulate poorly the private health services.

The pandemic the world endured between 2020 and 2022 strained all the health structures of the continent, an exceptional situation that allows us to compare their performance and resilience. The fact that some countries weathered the crisis better than others was due, in part, to the condition of the health services themselves but also to the approach that the different governments adopted to deal with the epidemic: some governments initially denied the severity of the disease (Brazil, Mexico), others reacted late (Ecuador, Bolivia), while still others did everything they could to face the crisis from the very start (or even before, as soon as the pandemic hit the European countries), like Costa Rica, Uruguay, and Argentina.

In this chapter, we will analyze the latest changes in the health systems, their main characteristics in terms of coverage, structure, investment, infrastructure, historical performance, and their performance during the pandemic.

Figure 5.1 shows that health spending follows the same trend described in the next chapter concerning the differences between the welfare states of two blocks of countries: Argentina, Brazil, and Uruguay, and the rest. They coincide with what Carmelo Mesa-Lago considered the pioneer countries that adopted social protection before the Second World War (Uruguay, Argentina, Chile, Cuba, and Brazil; one must add Costa Rica, which developed after the war but caught up with this group) and those that came afterward. Chile is an interesting case as it underwent a substantial transformation during the dictatorship and reversed its trajectory to join the countries that spent the most.

But, while the amount spent by different countries may be equivalent, there are significant differences regarding the source of spending: government, private capital, or families, whether public health is managed by the State or by civil society organizations (unions or mutualists), and finally, the coverage of each of the subsystems. To begin with, it is helpful to analyze both the population coverage of the regime and each health sector, as well as the characteristics of the dominant one. In Fig. 5.3, we can see that the three so-called pioneer countries are, again, those

population, at the end of April 2024. It was substituted by what existed before the Seguro Popular, the IMSS-Solidaridad, now called IMSS-Bienestar.

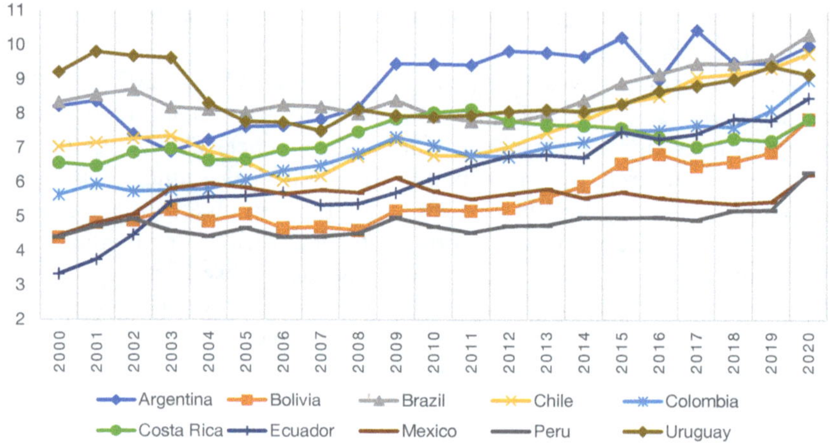

Fig. 5.1 Current health expenditure (% of GDP) (*Source* Own elaboration based on: Global Health Expenditure Database, WHO)

that shield almost their entire populations. Nevertheless, there are significant differences regarding the structure of each. Some countries have a majoritarian State-supported public sector: Brazil, Chile, and Costa Rica. In others, such as Uruguay and Argentina, health services are based on contributions managed by civil society.

There are obvious exceptions: on the one hand, two countries: Costa Rica and Colombia, have almost achieved universal coverage. As mentioned, Costa Rica joined the leading pack following the reforms undertaken after the 1940s. Colombia's universal coverage is based on a contributory and a subsidized health structure administered by the State; this latter one is, nevertheless, less generous than the contributory one; even so, one of the significant specificities of this country's health scheme is that it covers migrants since 2018.[2] On the other hand, Chile also seems to have succeeded in covering the whole of its population with public health services, despite the importance of the private sector. We will see further that, although Chile spends as much per capita as Uruguay and

[2] https://www.minsalud.gov.co/Paginas/Colombia-llego-al-aseguramiento-universal-en-salud-al-alcanzar-el-99.6.aspx#:~:text=Para%20la%20vigencia%202022%20a,y%20se%20incrementar%C3%A1%20%24300.000%20millones.

more than Brazil and Argentina, the private expenditures are very considerable. This is also the case in Brazil, despite a public service sector (the SUS) with universal coverage.

However, it is not enough to consider expenditure, it is necessary to analyze the structure and specific characteristics of each system. It is evident that coverage universality can be achieved through different combinations of public, social, and private services. We will discuss how these combinations, especially their hierarchy, determine the regime's character: whether the State, civil society, the market, or families are predominant, following the typology we have developed.

In Fig. 5.2, we can see that Uruguay, Argentina, Chile, and Brazil have higher health protection, even though their systems have followed different trajectories and coverage is obtained by diverging combinations of the State, social organizations, and the private sector. Two countries (Argentina and Uruguay) have preserved a regime dominated by social organizations (unions and mutualists) while developing a State sector to cover those not included in a contributive scheme. Only one country, Costa Rica, has succeeded in approaching these four countries with a single, State-run, centralized health model. With the 1988 Constitution, Brazil moved away from its corporatist / segmented trajectory and, like Costa Rica, founded a centralized public scheme (the SUS), leaving more and more room for the private sector over the years. Finally, even if the military dictatorship in Chile tried to impose a health structure dominated by the private sector, since the democratic transition, it has been improving its public service, subjecting private health to rising regulation.

Socio-Corporatist Systems: Uruguay and Argentina

The health systems of Uruguay and Argentina remain highly segmented, as from their beginnings: a large part of the population (56% in Uruguay and 65% in Argentina) is insured in a scheme based on contributions and health benefits administered by social actors—unions in Argentina and mutualists in Uruguay. There is, in addition, a sector that is governed and provided by the State and financed by taxes; private health is more limited.

The Uruguayan and the Costa Rican health systems are the only ones in Latin America that resemble a European or a Canadian one, a socio-democratic health care scheme; in Uruguay, the State is the sole payer and financially covers 95% of the population (Aran and Laca 2011).

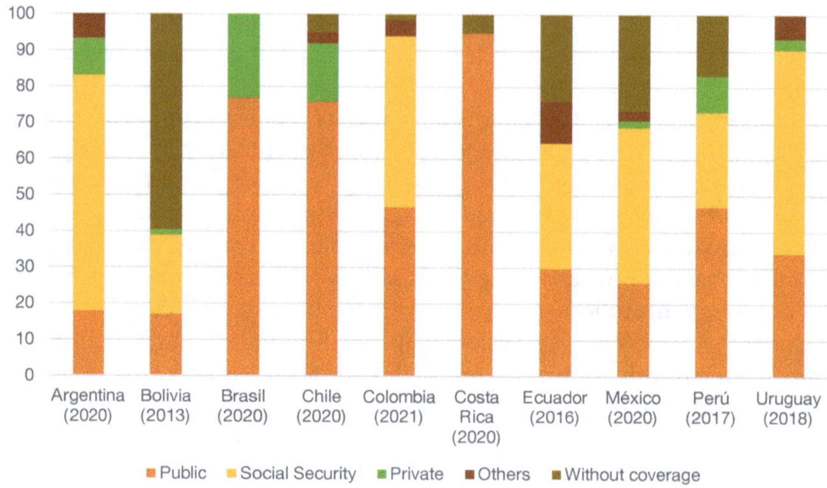

Fig. 5.2 Health System Structure (*Sources* Own elaboration based on: ARG: https://www.estadisticaciudad.gob.ar/si/genero/principal-indicador?indicador=tip_cob_sal; BRA: https://www.commonwealthfund.org/international-health-policy-center/countries/brazil; CHI: https://www.trade.gov/knowledge-product/chile-healthcare; COL: https://www.minsalud.gov.co/proteccionsocial/Paginas/cifras-aseguramiento-salud.aspx; COS: https://www.binasss.sa.cr/ojssalud/index.php/gestion/article/view/177/318; ECU: https://www.pbs.org/frontlineworld/rough/2007/06/ecuador_healthlinks.html; http://ilo.org/wcmsp5/groups/public/---americas/---ro-lima/---sro-lima/documents/publication/wcms_510914.pdf; MEX: https://www.inegi.org.mx/temas/derechohabiencia/; PER: https://rpmesp.ins.gob.pe/index.php/rpmesp/article/view/3998; URU: https://www.gub.uy/ministerio-salud-publica/tematica/cobertura-poblacional)

Even if the mutualists manage health care, the sole payer is the State, through FONASA, which concentrates both taxes and contributions from employers and employees and distributes them to the various providers: both to the health services administered directly by the State (*Administración de Servicios de Salud del Estado*—ASSE), which covers 34% of the most modest population and to the *Instituciones de Asistencia Médica Colectiva* (IAMC), which are private non-profit organizations that provide comprehensive health care to 56% of Uruguayans (Aran and Laca 2011: 269).

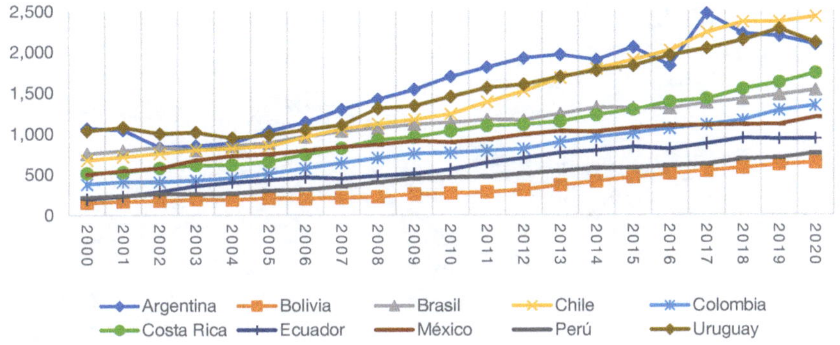

Fig. 5.3 Per capita health expenditure (in current PPP) (*Source* Own elaboration based on World Bank Open Data)

However, it is a stratified system, as there are differences between the different mutualists, but more importantly, there are those who cannot pay for the health services of these institutions and have to opt for public services; there is another group that is affiliated to the mutualists and in addition reimburses for certain special conditions. Finally, some can pay in full for private medical care, a small percentage: 5% (Ruezga Barba 2009: 214). Thus, although "...a nationwide system of medical prevention and services exists, it is administered in a fragmented way, responding to the interests of a divided medical corporation and to the interests of groups of members who seek to preserve those institutions with the highest standards of care" (Filgueira 1995: 40).

In this manner, even though the Uruguayan health services can be considered universal due to the population they cover and socio-corporatist due to the weight of the mutualists, they are fragmented due to differences in quality between providers. They are undoubtedly unequal, to a certain degree, as public hospitals receive around 25% of the total health budget, while mutualist associations receive 75% of the budget, including State subsidies (Martínez Franzoni and Sánchez-Ancochea 2016: 87–88).

However, the existence of a single payer that covers most health services reduces the segmentation of services. As can be seen in the following figures, in recent years, the per capita health expenditure in this country has risen sharply from 1000 to 2300 dollars (Fig. 5.3), but this mainly concerns the share of public services, which has gone from

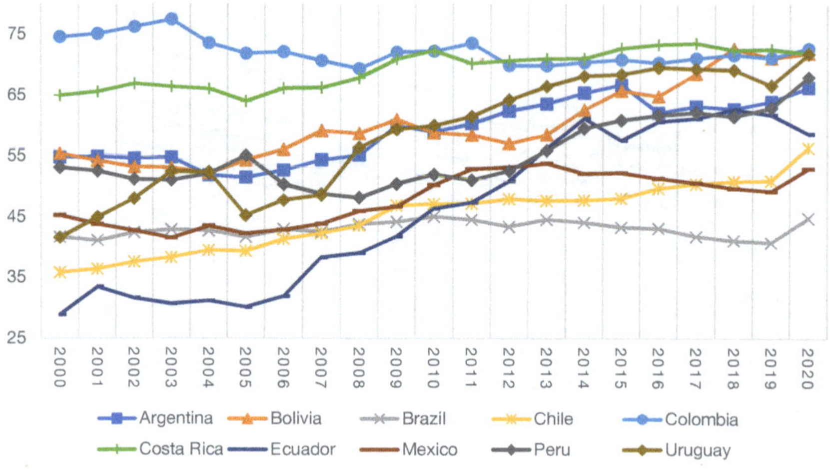

Fig. 5.4 Government health expenditure (% of current health expenditure) (*Source* Own elaboration based on World Bank Open Data)

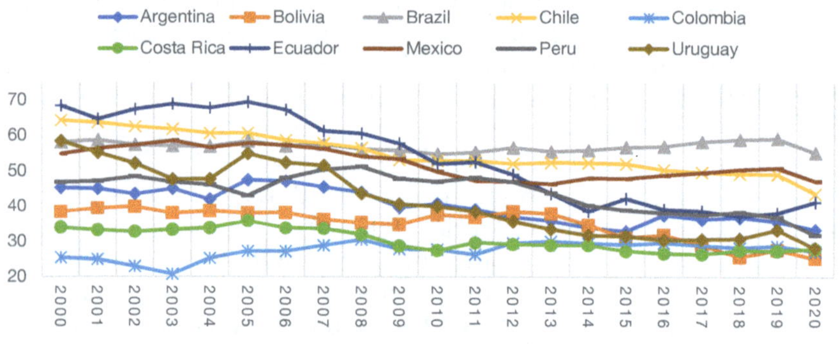

Fig. 5.5 Private health expenditure (% of current health expenditure) (*Source* Own elaboration based on World Bank Open Data)

representing 41% to 73% of the total (Fig. 5.4), since the private part of expenditure has decreased from 58 to 33% of the total (Fig. 5.5) and out-of-pocket expenses have also been considerably reduced (15%) (Fig. 5.6); they are, along with those of Colombia, the lowest of the continent; the out-of-pocket expenditure of Uruguay is comparable to that of the average of the European Union countries (16.6%).

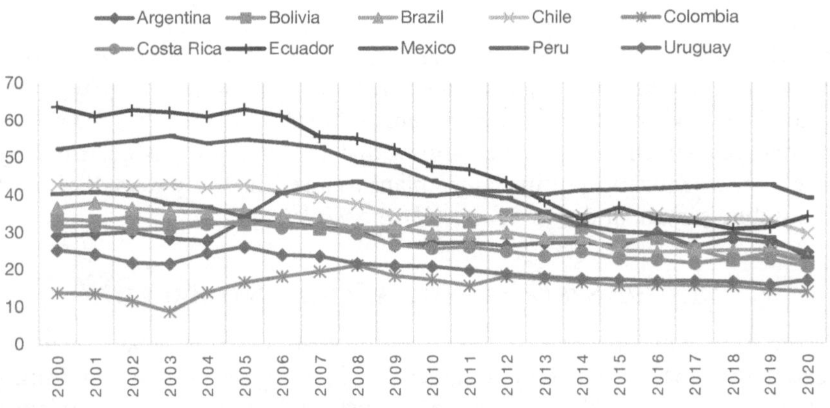

Fig. 5.6 Out-of-pocket health expenditure (% of total) (*Source* Own elaboration on the basis of World Bank Open Data)

Argentina's system is also based on civil society and is similarly segmented. The *Obras Sociales* are administered by trade unions and financed by contributions from workers and employers (8% of their salary) in full autonomy from the State; they cover 42% of the population. If we add the national contributory sector (8%) and the provincial *Obras Sociales* (14%), the social sector represents 64% of the total (Repetto and Potenza 2011; in Fig. 5.2 it is 65%). The remainder, 25%, receive care in public hospitals or through private insurance (Belló and Beccerril-Montekio 2011; currently, it is 28%). Argentina has a high level of per capita spending, which strongly increased since 2003 even though it was reduced in the last years of Macri's government.[3] Although the total and per capita expenditures are similar to Uruguay's, the share of private and out-of-pocket spending levels are higher, around 40% and 25%, respectively. During the last 20 years, Argentina has gone through two contrary processes: during the first Kirchner and Fernández de Kirchner periods, the health coverage of both the social and the public services was considerably extended, and the latter improved (Cecchini et al. 2013: 35). Due to the reduction of the price of commodities that occurred around 2012, since 2016, during the Fernández de Kirchner presidency, and then in the

[3] We have information until the year 2019, due to the deep crisis that Argentina is going through at present, it probably is even lower.

Macri government, the proportion of governmental health expenditure has decreased in about 5%. As a result, private spending has increased, reaching 38% of the total, somewhat higher than in Uruguay; even though out-of-pocket expenditure remained stable (28%), it is 10 points higher than in Uruguay, which shows that the increase has been mainly due to the acquisition of insurance plans as a result of the degradation of the public service and that of the *Obras Sociales*.

Statist Systems: Costa Rica and Brazil

Costa Rica comes nearest to a State-led health regime, as it has a centralized and unified health service. Since 1973, all the various public infrastructures and financial resources were transferred to the *Caja Costaricence de Seguro Social* (CCSS). Unlike Brazil and other statist systems similar to the NHS, the CCSS is not funded primarily by taxes but with contributions from employers and workers (Sáenz et al. 2011). On the other hand, non-formal employment individuals are obliged to join this fund, their contributions being paid by the government. Finally, the self-employed can participate voluntarily (Santos Basso 2005: 192). In 1998, the hospital de-concentration law brought about a transformation in the allocation of resources: from one based on their historical trajectory to one according to the future needs calculated upon established agreements (Martínez Franzoni and Sánchez-Ancochea 2016: 170). This law resulted in fiscal and administrative autonomy not accompanied by effective coordination and encouraged the duplication of doctors, spending on advisers, and corruption (ibid., 163–164). This has had the effect of decreasing the quality of the services, the retreat of users from the public sector, and the reinforcement of the private sector, and is a challenge to the universalist institutional architecture of Costa Rican health protection (Martínez Franzoni and Sánchez-Ancochea 2016: 167–170). The health services have suffered a relative retrenchment caused by the expansion of "the private health services and the spread of conflicting dual practices among CCSS doctors", which, according to Martinez Franzoni and Sánchez-Ancochea, have had as their consequence that "policy outputs are less universal today that they were two or three decades ago, primarily due to lessening quality and growing inequality in service provision" (Martínez Franzoni and Sánchez-Ancochea 2016: 151–152). On the other hand, in contrast with the countries that profited from the commodity boom, the international subcontracting economy of Costa Rica grew less quickly and encountered financial problems that contributed to the erosion of its integrated and efficient health services and the expansion of private care. For example, whereas in 2003, 36% of Costa Ricans felt that hospitals had not been modernized, in 2007, this percentage had risen to 74% (Poltronieri 2011).

Despite all this, the entire Costa Rican population can access medical care (Cercone and Pacheco Jiménez 2008), and private services have not become dominant. In fact, Costa Rica has some of the lowest private (27%) and out-of-pocket expenses (22%) in the continent, comparable to those of Uruguay.

Universal and State-led coverage of health services in Brazil was enshrined in the 1988 constitution, the result of a democratization process involving a very intense mobilization of civil society organizations and movements, among them the *sanitaristas* who promoted a very advanced public health project. The *Sistema Único de Saúde* (SUS) is a universal, unified health system modeled on the British NHS and financed by taxes and contributions at the three federal levels: the federation, the states, and the municipalities. It formally includes the entire population and includes all diseases and conditions (in contrast to the former Mexican Seguro Popular and the Chilean FONASA, which define a limited list of diseases and procedures) and has aimed to equalize access to care. However, it has not achieved to do so, as we will see below. In addition, the SUS does not charge any co-payments, can direct patients to private hospitals or clinics, and is monitored by more than 5000 municipal health councils, in which more than 100,000 citizens participated in the first decade of the twenty-first century (Becerril-Montekio et al. 2011a).

Nevertheless, it is noteworthy that even though the expenditure per capita has increased considerably in the last 30 years, private and out-of-pocket expenditures have grown even faster, reaching the vast figure of 60% of the total for the first and 24% for the second. The expansion of private expenses is linked to problems inherent to the SUS: basically, quality of services and waiting times (Becerril-Montekio et al. 2011a). While the Brazilian is a State-led universal scheme, and it was thus supposed to be more homogeneous compared to segmented structures, the fact that the share of private participation is so high, unlike the cases of Uruguay and Argentina, for example, explains an increased segmentation among private/State lines. After 20 years of existence, public healthcare facilities grew from 22,000 in 1981 to 75,000 in 2009, and hospital beds in the public sector increased from 22 to 35 percent of the total; nevertheless, in 2013, 50% of hospital beds were private (Gragnolati et al. 2013: 2–3). Somewhat more recent data for 2015, place the percentage of public hospital beds at 70%, and while 75% of Brazilian citizens rely solely on SUS, 23% have private medical/hospital insurance (commonwealthfund.org.). In addition, the share of private healthcare plans charging co-payments has grown from 22 to 52%; amid private hospitals, 38% were non-profit, and 63% were for-profit (commonwealthfund.org).

A study based on a poll comparing Brazil with Colombia, displays the same trends. Regarding the type of services used, 74% of those interviewed used the SUS exclusively, while 19.4% attended private centers; people who used both were 6.6%. Regarding primary services, 80% went to the public ones, while 18.7% used the private ones. Concerning specialized attention, the percentage using public facilities went down to 63%. Finally, in the case of emergencies, the percentage rises to 87.4% for public and 11.5% for private (García-Subirats et al. 2014: 482). In this manner, the Brazilian health services based on the SUS have allowed private health services to occupy an ever-growing place, with a tendency to commodification.

Commodified Systems: Chile

As we have already discussed, the reforms undertaken under the dictatorship promoted the decentralization and privatization of health. However, the military project was that private insurance would become majoritarian; this never happened, as it is too expensive for most Chileans (Becerril-Montekio et al. 2011b). The private health sector includes mainly the middle and upper classes: 44% of ISAPRE beneficiaries belong to the wealthiest quintile, while those in the lowest quintiles are beneficiaries mostly of FONASA; in the 4th quintile it is already only 17% who benefit from FONASA (Ministry of Health, 2010a, p. 195, cited by Robles 2011: 42–43). Consequently, in 2020, 76% of the population was covered by public sector services (FONASA), 16% by private insurers, the ISAPREs, 3% by other services, and 5% without coverage (see Fig. 5.2). Any individual can be treated by the public sector while enrolled in an ISAPRE; in fact, many use the public health sector for emergencies, hospitalizations, and intensive care treatments. Whereas the private companies insure individuals of working age with low-risk jobs, the rest of the population is assisted by the public institutions dedicated to providing health services to those who do not have private insurance, those who have left the private sector because they lose their jobs or are too sick to pay for the increasing co-payments, and those with greater risk jobs that have to pay dues they cannot afford (Gutierrez Arriola 2002: 65), or when they stop being able to pay the annual fee for the insurance that increases annually or once they retire; in consequence, 27.4% of FONASA contributors are over 60 years of age, but only 3.3% of ISAPRE contributors are of that age (Gattini 2018: 9). In fact, the public structure "picks up" the individuals who are no longer profitable for private insurers.

Both sectors receive resources from the FONASA, a fund that collects the contribution of 7% of the salary of formal workers in the private and public sectors, as well as those who contribute independently; the private sector is, in addition, paid for directly by its affiliates; in addition, taxes complete its deficit. There is, thus, a financial imbalance between the two schemes that benefits the private one. While 41% of the population beneficiary of FONASA were contributors (the rest are subsidized individuals or beneficiaries), in the ISAPRES, 50% are contributors (data for 2011; Gattini 2018: 9). FONASA's resources from contributions amounted to only 37.3% of its total, so from the rest, 58.5% corresponded to taxes and 4.3% of co-payments. In contrast, in the ISAPRES, more than half (52.3%) originates in contributions, 19.6% in an additional voluntary contribution, and 26.1% in co-payments (Gattini 2018: 9). Another effect is that the public sector has been deprived of the contribution of groups with higher income levels, with the result that the quality of public health services has deteriorated.

Although an improvement has been observed over time, inequality between both sectors is evident regarding the use of resources: if, in 2001, the per capita spending in the private sector was 2.7 times that of the public, this proportion was still 1.6 times in 2009. "The reduction of this gap may be due to the incorporation of the health reform, preliminary results of the reform show that the AUGE Plan has been used by 3.2 million people (almost 20% of the population), 75% by people belonging to groups A and B of FONASA[4]" (Robles 2011: 48).

[4] "For those who receive care within the public service—which include those who decide to do so and those who receive a State assistance subsidy and are not contributors—the care modality was established around two options: those who do not have resources have access to a predefined modality (Institutional Attention Modality) in which they are attended in establishments of the public structure, while those who are contributors

It is for all these reasons, and notwithstanding the modifications the Chilean health system has undergone since the democratic transition, that we consider it a commodified scheme; the State is fundamentally subsidiary not only because it takes care of those that the private services drive out, but because it generously subsides them.

Familiarist Systems: Ecuador, Bolivia, Peru

We have called the systems of Peru, Bolivia, and Ecuador familiarist because most of the population is without coverage. Thus individuals and families face many expenses, not due to private insurance but out-of-pocket payments. Even in a hybrid system like the one of Mexico, private insurance coverage is quite limited, around 2.5% of the population. In contrast, in countries where this type of insurance is considerable, it is around 16% for Chile and 23% for Brazil. The familiarist character is defined by the fact that the primary source of health expenditure comes from families and is paid directly to doctors and private clinics or for purchasing medicines, as very often these are neither provided by social security nor public institutions.

Although the health services of Peru, Ecuador, and Bolivia considerably expanded during the commodity boom, they have merely extended the previously existing focused policies to provide minimal health services to the population that needs them.[5] Thus they have left behind a many people, as can be seen in Fig. 5.2: Bolivia, 59%; Peru, 17%; and Ecuador, 24%. These three countries, together with Mexico, are the ones that spend less per capita of our sample: Bolivia and Peru spend less than 500 dollars, and Ecuador spends just above this amount (Fig. 5.3). The health structures in these countries depend very heavily on private

(and their charges), can choose between a network of private providers that are part of FONASA (Free Choice Modality). Depending on the income of the people, four groups of beneficiaries are identified for the first modality: those who live in a situation of indigence or have a minimum income receive free care in the system (sections A and B), while in the following two sections of income, free primary care is received, but 25% (group C) and 50% (group D) of the value of more complex benefits in the public structure must be paid. Under the second care modality, the system covers the cost of 50% of most benefits" (Robles 2011: 43).

[5] "In Peru targeting has therefore been utilized to extend access for certain groups of people, which because of its limited health system, by definition means increased coverage" (Noy 2017).

spending. However, in contrast to Brazil and Chile, for example, the spending on private services has less to do with the acquisition of insurance than with direct spending by individuals and families: private spending in Ecuador is 38% and 31% out-of-pocket. In Peru, private spending is 37%, and out-of-pocket expenses are 28%. The small amount of general spending in Bolivia explains the low level of private and out-of-pocket expenses; in addition, traditional medicine is very considerable in this country; in fact, 87% of private spending on health is out-of-pocket (Ledo and Soria 2011: 114).

In the case of Peru, the main reforms to the health scheme have intended to extend rather than to improve the quality of the services: the reforms of the 1990s included means-testing and a basic package of health services, together with decentralization, "two different schemes ... the *Seguro Materno Infantil* (SMI) and the *Seguro Escolar Gratuito* (SEG) targeted pregnant women, mothers, and children attending public schools" (Noy 2017: 10).

The reason why a high percentage of Peruvians lack health coverage is that the leading institution, the *Sistema Integral de Salud*, is, like the extinct Mexican *Seguro Popular*, voluntary; most of the population that is not affiliated consider that it is not worth to do so for different reasons: waiting time, quality, the fact that the affiliated still have to pay for the services, the idea that they are young and do not need health insurance, etc. If, in principle, all Peruvians have access to the public health services offered by the Ministry of Health (MINSA), they must pay a recovery fee which "is variable, and subject to the discretion of the organizations"; in fact, the recovery fee represents a significant part of the total resources of the Ministry: 14% (Alcalde-Rabanal et al. 2011: 274).[6] Under these conditions, Peruvians prefer to go to a private doctor or clinic rather than wait for their turn in public clinics, losing a day's work and, ultimately, having to pay the fee and buy their medicines. In most countries of the continent, there are pharmacies where a doctor prescribes drugs or treatment after an "on the spot" cheap or even free consultation when the medicines are purchased at the same dispensary. This accounts for the high out-of-pocket expenses in this country: 31%.

On the other hand, the health system of this country has significant administrative liabilities; there is "...little coordination among the three

[6] "79% of MINSA's financing in 2007 came from general taxes, 15% from recovery quotas, and 6% from donations and transfers" (Alcalde-Rabanal et al. 2011: 248).

subsectors of public, private, and social security, and … few separations of functions in terms of financing, purchasing, delivery, and regulation, within the health system" (Noy 2017: 4). There is an enormous concentration of services in the capital, Lima: 53.2% of doctors, 40.2% of nurses, 44.2% of dentists, and 41.5% of nursing technicians and assistants (Alcalde-Rabanal et al. 2011: 250). Finally, an essential amount of the budget of the ministry goes to current spending and little on investment: "In 2007, around 37% of the MINSA budget was dedicated to spending on personnel and social obligations (holidays, compensation, gratuities, and family allowances, among others), 6.2% on pension obligations (social security health payments and pensions) and 46% on goods and services. Investment spending and other capital spending were below 11%" (Alcalde-Rabanal et al. 2011: 249).

While Bolivia and Ecuador sought to universalize health care through public sector expansion during the commodity boom, they failed. In Bolivia and Ecuador, there exists a health protection scheme that covers formal workers with the best wage and working conditions, a small private sector that protects the middle and upper classes, and a public sector that is, in principle, open to the rest of the population, but which leaves a considerable percentage of the population without assistance.

While spending (especially public expenditure) on health as a proportion of GDP has been increasing steadily since the mid-1990s in Bolivia and exceeds those of Peru and Mexico, in terms of per capita expenditure it does not. While the social protection sector consists of eight health funds and two mixed funds that cover formal workers and their families (22% of the population, 28% according to Fig. 5.2) (Ledo and Soria 2011: 112), public health consists of the *Seguro Universal Materno Infantil* (SUMI), implemented in 2003, which includes all children under five and all women of childbearing age, and the *Seguro Público del Adulto Mayor* (SPAM); together, both cover 17% of the population (14.2, according to Fig. 5.2) (Ledo and Soria 2011: 114). Young people and the productive age population that are not beneficiaries of the public insurance scheme or social security must resort to private services to meet their health needs (Ledo and Soria 2011: 114). Although the possibility of voluntarily joining the social security structure exists for the self-employed, as we have seen in the case of Mexico, Peru, and Ecuador, few self-employed, mostly informal workers, see incentives to do so.

In addition, according to an evaluation of maternal and child insurance in Bolivia, the SUMI "…has not adequately taken care of the poorest population ... it has mainly benefited the population with enough means to pay for health services" (Arce 2013: 24). In fact, "According to the 2000 Household Survey, one of the main reasons for the unattended health demand is the lack of economic means to cover the rates of public institutions" (Arce 2013: 23). "The SUMI has also shown different results according to geographic area and income group: in urban areas, it mainly favors the poorest population and in rural areas, the richest groups" (Arce 2013: 24).

Consequently, a study comparing catastrophic expenses in eight Latin American countries places uninsured Bolivian households as the most susceptible to present these types of costs. Another study, with data from 2002, indicates that 13% of Bolivian households presented catastrophic health expenses that year (Ledo and Soria 2011: 114).

Finally, Ecuador is similar to Bolivia. In the 1998 Constitution, health was defined as a universal right. In 2006, the government created the universal health insurance scheme and increased taxes on foreign oil companies, raising the royalty on profits from 50 to 90% (Conaghan 2011) to significantly increase spending on social programs, as one can see especially from 2008 to 2015 (Fig. 5.1). In fact, with this increase, this country spends more on health relative to GDP than Mexico, Peru, Bolivia, and even Costa Rica and Colombia, landing just below the most developed systems of the continent (see Fig. 5.1).

The contributive health institution, the *Instituto Ecuatoriano de Seguridad Social* (IESS), formally covers around 35% of the population. Nevertheless, as in other familiarist systems, due to the high level of informality, the IESS and the *Seguro Social Campesino* (SSC), do not include all of the formal sector (Minteguiaga 2012: 55).

On the other hand, while the public sector, the *Ministerio de Salud Pública* (MSP), formally protects the rest of the population, in reality, it offers health services to 30% of Ecuadorians; however, significant progress has been made through the expansion of primary care service, free medicines, and home care in poor neighborhoods (Minteguiaga 2012: 53). The public non-contributive sector of the MSP offers the broadest provision of services in the country: 47% of outpatient and hospital units (Lucio et al. 2011: 183). In 2008, the Ecuadorian health system had 21,575 beds (10,273 private and 11,302 public), a ratio of 1.6 beds per 1,000 inhabitants (Lucio et al. 2011: 183). In 2007, the Catastrophic Illness Coverage Program was implemented, with 60 million dollars, to finance the treatment of high-cost diseases in selected cases (Lucio et al. 2011: 182).

Nonetheless, the goal of universal coverage has yet to be achieved since 24% of Ecuadorians still lack these services. Per capita spending is low, at 900 dollars. Then, while private and out-of-pocket spending have decreased continuously since 2006, they are still very high: 38% and 30%, respectively, just below Mexico and Chile and just below Brazil in private health expenditure, the highest of our sample of countries.

The Mexican and Colombian Hybrids

With the creation of the *Sistema General de Seguridad Social en Salud* (SGSSS) in 1993, the Colombian health system became a public compulsory scheme in which almost 100% of Colombians are covered by 968 public and 4,565 private health institutes (EPS) (Bernal and Barbosa 2015: 434). However, it excludes some treatments and illnesses, and health institutions charge user fees (Bernal and Barbosa 2015: 435).

Although in 2012, the mandatory health plans of the two sectors were harmonized in order to try to end the existing inequity between them, the Colombian system still suffers from inequalities between the population covered by the subsidized services and those covered by the contributory services, which are related to the differences between the benefit packages and their level of funding (García-Subirats et al. 2014). In addition, while the EPSs of the subsidized sector send its adherents to public health providers, those of the contributive sector send them to private providers; nevertheless, the law mandates that private EPSs provide services to at least 60% of public patients. There are also significant differences between what each EPS covers, although a minimum is defined by the mandatory health plan (Bernal and Barbosa 2015; Guerrero et al. 2011). In 2008, the law was amended to standardize both sectors and forbid private providers from refusing health care to patients from the public sector. García-Subirats' study shows the weight of the public structure in allocating health services: 92.2% of those interviewed used only the public SGSSS and 5.6% the private services, while the rest used both. More specifically, 93.3% of Colombians attended public services procedures, and 5% attended private ones for primary health. In the case of specialized attention, the trend is similar as 93.5% attended public health centers. Finally, in the case of emergencies, 95% of patients did so in the public sector (García-Subirats et al. 2014: 482).

Nevertheless, even though Colombia has an almost unified system and a low percentage of private health (28%, similar to Uruguay and Costa Rica) and of out-of-pocket spending (15%, lower than Uruguay), when it comes to per capita spending, it is in the bottom tier. This situation has as its consequence, as will be seen in Sect. 5.6 of this chapter, a significant lack of health infrastructure and one of the lowest historical performances, comparable to those of Peru, Bolivia, and Ecuador, which we have designated as familiarist, while, as we will also discuss, the performance of Colombia during the pandemic was much better.

The Mexican system is also a hybrid, now between a social security sector (IMSS, ISSSTE, and the two smaller contributory services of Pemex, the navy, and the army), a public health institution (Secretaría de Salud/INSABI[7]), and a private sector that serves both the wealthiest segment of the population, but also the poor and the middle classes, due to the deficiencies of the other two structures. The two major contributory health institutions: the IMSS (private workers) and the ISSSTE (civil servants) are overburdened and underfunded. The public organization, under the direction of the federal and state Secretaries of Health, provides services to the population that is not affiliated with any contributory scheme (formally, 52% of the population is affiliated, but in reality, the percentage is only 26%, as another 26% lacked coverage[8]), has the same shortcomings, but aggravated due to the proportion of the population that it covers.

In 2003, the government of Fox launched the *Seguro Popular*, which once again aimed to achieve universal health care; nevertheless, as it was a voluntary program based on providing a series (a package) of medical services to its members, it never managed to include all the population that had no social health insurance; part of the explanation is that there was no obligation to do so, as it does in Colombia. On the other hand, the financial resources allocated to this institution did not follow the impressive expansion the program was representing. In Fig. 5.1, we can see that spending increased considerably the first year of the program,

[7] Now, after the end of Insabi, the IMSS-Bienestar.

[8] With the end of the *Seguro Popular*, the situation worsened with Insabi, as the percentage of people affirming they did not have health insurance increased from 16% in 2018 to 28% in 2021 according to Coneval. https://www.coneval.org.mx/Medicion/MP/Documents/MMP_2018_2020/Notas_pobreza_2020/Nota_tecnica_sobre_la_carencia_por_acceso_a_los_servicios_de_salud_2018_2020.pdf.

by 1% of the GDP, but stagnated in the following years, receding in the last years of Peña Nieto (2012–2018) almost to the level it had before its creation. Even in terms of beds, the situation was paradoxical: according to ECLAC, in 2000, there were 0.8 hospital beds for every 1,000 inhabitants in Mexico, while in 2010, the proportion had dropped to 0.7 (ECLAC 2011). However, per capita health expenditure has been increasing steadily since 2000, although it is in the group of the familiarist countries (Fig. 5.3). While the proportion of public spending began rising in 2003, reaching a high in 2014, it started descending since then. The first data on out-of-pocket spending was, at first, reduced marginally from 51.9 to 49% (*Reforma*, 08/28/12: 2). However, since its inception, these expenses went from a high of 55% of the total to 42% in 2019.

In addition, not all diseases and acts were considered in the *Seguro Popular*; only 284 were. At least until 2018, a severe exclusion of the SP was that it did not include dialysis, which is essential for the survival of thousands of Mexicans who suffer from serious cases of diabetes. This disease is adopting the characteristics of a national epidemic (El Universal, 26/12/2017). Until 2016, this institution did not cover heart attacks in people over 60 years of age, strokes, multiple sclerosis, or lung cancer (OECD 2016: 13).

In any case, the current government of López Obrador replaced the Seguro Popular with a universal, centralized system, the *Instituto de Salud para el Bienestar* (INSABI), barely three months before the onset of the pandemic; it was then in its infancy in terms of organization and was fundamentally underfunded. The Insabi also abandoned the contributory principle, which gave those included formal rights. As we have mentioned, the result of substituting the certainly imperfect *Seguro Popular* was decreased coverage.

The Health Regimes Facing the Pandemic

The health crisis that befell the world in December 2019 and Latin America head-on in March 2020 was a terrible health tragedy that has claimed millions of lives. It is also an opportunity to measure the performance of the various health regimes of the continent. Even though the number of contagions would be an excellent indicator of the country's ability to control the epidemic, the fact that this indicator depends on the number of tests each country completes and that there are significant differences in this respect makes it unreliable. The number of deaths each country records is equally untrustworthy, as countries differ in the way and efficacy of (voluntarily or unintentionally) counting deceases. The number of deaths in excess from the last year compared to previous years is thus the most reliable indicator.

To assess the causes of the divergences among countries, we will refer both to the qualitative characteristics of the health systems, analyzed in the previous sections of this chapter, and to quantitative indicators, and then, briefly, to the political responses of the government.

In Table 5.1 and the following Fig. 5.7, we have assessed the situation of the different countries regarding health infrastructure and performance before the pandemic. Concerning infrastructure, Uruguay, Argentina, Chile, Brazil, and Costa Rica were the best endowed, with few dissimilarities. However, as we have seen, these five countries have three different types of regimes. On the other hand, in terms of performance, two countries with high infrastructure indexes: Brazil and Argentina, are below the high performers.

In general terms, the pandemic was a humanitarian disaster of significant proportions in the entire continent, except for two countries: Uruguay and Costa Rica. With data for October 2021, once the pandemic had peaked and was descending, we have divided our countries into three groups: Peru, Ecuador, Bolivia, and Mexico which had more than 400 deaths per 100,000 inhabitants. There is an intermediate group formed by Brazil, Argentina, Colombia, and Chile, with between 200 and 355 deaths. Finally, two countries had the lowest rates: Costa Rica and Uruguay, which had less than 200 deaths per 100,000 inhabitants.

The correlation between infrastructure, past performance,[9] and the number of deaths is very significant. In Fig. 5.7, we can see that the two countries which stand out in infrastructure and performance also had the fewest deaths during the pandemic. On the opposite side of the spectrum, the countries with the weakest infrastructure and the lowest performance have had the most deaths: Bolivia, Ecuador, Peru, and Mexico.

We have already mentioned how the two countries that stand out from the rest, Costa Rica and Uruguay, have the most solid health systems, even if they are very different in character and structure: one

[9] Past performance in defined by variables that measure the efficiency of the health system in terms of cancer preparedness, vaccination, among others; the complete list is in the Appendix.

Table 5.1 Infrastructure, Performance, and Excess Deaths

Country	*Infrastructure Index*	*Performance Index*	*Excess deaths/100,000 inhabitants*	*Correlations*	*Infrastructure*	*Performance*	*Deaths*
ARG	72.59	79.40	254	Infrastructure	1	0.62	−0.504
BOL	22.87	44.49	416	Performance	0.62	1	−0.506
BRA	58.00	76.34	355	Deaths	−0.504	−0.506	1
CHI	42.42	87.65	231				
COL	39.93	74.73	326				
COS	42.13	84.54	129				
ECU	33.33	67.21	435				
MEX	39.39	78.19	479				
PER	31.38	70.08	707				
URU	67.98	85.25	174				

Source See the Appendix at the end of the chapter

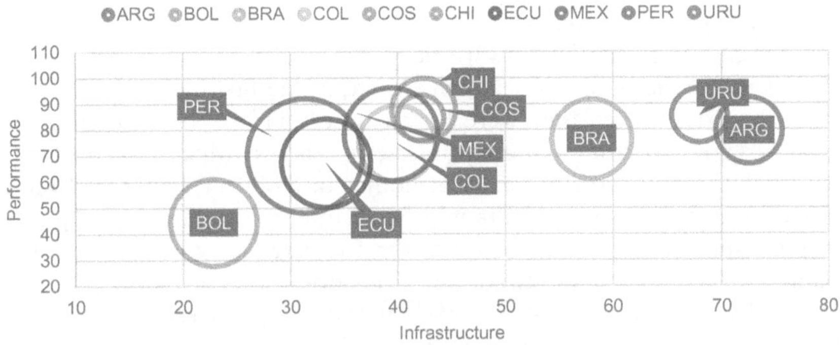

Fig. 5.7 Excess Deaths (size of the circles) according to infrastructure and performance (*Source* See the Appendix at the end of the chapter. The size of the circles indicates the number of deaths per 100,000 inhabitants)

is State-governed, and the other is socio-corporatist. In addition, it is also interesting that Costa Rica, with less infrastructure, achieves a similar performance in confronting the pandemic as Uruguay, something which could result from less segmentation when compared to the latter. They are indeed small countries, but Bolivia and Ecuador are also relatively small and performed poorly.

The poorest performance against the pandemic was found in the three countries we have named familiarist: Peru, Bolivia, and Ecuador, as well as Mexico, which we have defined as a hybrid between statism and familiarism. In all four, a considerable part of the population is without coverage, total health expenditure is low, and out-of-pocket is high; all variables closely related to excess deaths during the pandemic, as shown in Figs. 5.9 and 5.11.

The fact that families depend on their means makes them quite helpless when facing health problems, and more understandably during a pandemic. The fact that health relies so heavily on out-of-pocket expenses means that people hesitate before going to the doctor or the hospital and do so only when their condition worsens, which in the case of the Covid-19 pandemic resulted in many of the sick contacting the clinics when it was too late, once they were too seriously ill, or dying at home. We have learned that for Covid, a delay in the onset of treatment before the autoimmune shock begins and the individual lacks oxygen is fundamental in preventing aggravation, which can lead to death. In almost all the countries of the continent, we saw panic purchases of oxygen tanks, which testifies to the fact that many patients tried, in the first place, to get by at home, and only when they were very seriously ill did they start looking for a hospital. In some countries, most notably in Mexico, almost half of Covid-related deaths occurred at home. In addition, between 40 and 50% of Mexican Covid patients who were interned in any of the public hospitals died, while in the private sector, the percentage was around 20% (Sánchez-Talanquer et al. 2020: 28). This does not only show that care provided in public hospitals is deficient but also, at least partially, that many patients arrived once they were too sick.

While the extreme cases can be quite clearly explained, the situations of Argentina, Chile, Brazil, and Colombia require further analysis. The answer to why these countries did not fare as well as expected in relation to the indexes of infrastructure and past performance that are similar to those countries that fared better has to do with recent trends in terms of public investment in health, but especially with the structure of the system (basically its dependence on out-of-pocket expenditure and the sources of inequality).

Argentina, a country with an infrastructure and a performance of health services (before the pandemic) among the highest in the continent, did not perform as would correspond to its situation. Whereas Uruguay and Argentina are very similar in infrastructure and performance, health expenditure as a percentage of GDP, and number of doctors, Argentina did not perform as well in the face of the pandemic. We have mentioned how, especially during the Macri government (Dec. 2015–Dec. 2019), the social and public health sectors deteriorated due to the reduction of public investment; this can be seen in Figs. 5.1 and 5.4, but especially in the statistics that indicate an increase in private health expenditure, which went from a historical low in 2015 (33%), to 38% in 2018 (Fig. 5.5), even though out-of-pocket expenses remained stable. We can see in Table 5.2 that there is a robust correlation between the level of total expenditure on health (−0.71), the government expenditure as a percentage of GDP (−0.77) and out-of-pocket expenditure (0.44), and deaths during the pandemic; negative in the first two cases, and positive in the third. So, the fact that public health expenditure was reduced in the first place and that out-of-pocket spending in this country is higher than in the other countries, which are similar in terms of infrastructure and performance may be part of the explanation.

Table 5.2 Correlations with excess mortality

Variable	*Correlation with Excess deaths per 100,000 inhabitants*
Number of beds per thousand inhabitants	−0.31
Average number of intensive care beds per hundred thousand inhabitants	−0.42
Number of doctors per thousand inhabitants	−0.71
Number of nurses per thousand inhabitants	−0.14
Ventilators per hundred thousand inhabitants	−0.29
Total expenditure on health per capita, PPP	−0.69
Total expenditure on health as a % of GDP	−0.71
Private expenditure on health as % of total health expenditure	0.28
Government expenditure on health as % of total health expenditure	−0.27
Government expenditure on health as % of gross domestic product (GDP)	−0.77
Out-of-pocket expenditure 2017	0.44
GDP growth 2020	−0.80
Percentage of GDP spent on trying to mitigate the economic effect of the pandemic	0.47

Source See Appendix at the end of the chapter

Another part of the explanation is the structure of the Argentinian health system. But structure may mean many things. First, its most basic and obvious meaning is the difference between having or not having health coverage, which determines how frequently people make doctor appointments (Ballesteros 2014). Most of the articles we have consulted consider that income level is one of the main factors that explain inequality in Latin America, as most systems have a pro-rich and pro-educated bias (both coincide most of the time[10]). Nonetheless, this fact does not allow us to distinguish between the Latin American countries, as they all fall into the same situation.

[10] And, in fact, one could also include in the pro-rich and pro-educated bias of the health system, such a bias towards a basic health condition. Many studies have related poorer and less educated populations (for example: FIEL 2007) and the effects of bad health conditions as a risk in case of Covid has been thoroughly proven.

As we have analyzed, the Argentinian health organization is highly segmented, based on the *Obras Sociales* and a public sector. This segmentation is the source of inequalities both in funding and in the quality of services, first amid the various *Obras Sociales* managed by the unions and among them, the public and private services. Some of the *Obras Sociales* fare better than others, and these are superior, in general, to the public services. On the other hand, there are substantial regional differences between the capital and the suburbs of Buenos Aires (the *conurbado*), and between the capital and the provinces (Palacios et al. 2020). Financial and infrastructure resources are highly concentrated in Buenos Aires, Córdoba, and Rosario. For example, there are 10.2 doctors and 7.3 beds per 1,000 inhabitants in the capital, while in the province of Misiones, there are 1.2 and 1.1, respectively. This explains why the capital performed much better (around 30% fewer deaths) than the *conurbado*. Whereas the number of beds is not as strongly correlated with the prevention of fatalities (−0.31 for regular beds, −0.42 for intensive care beds), the number of doctors is highly correlated (−0.71) (Palacios et al. 2020). Another study that centers on cervical cancer also signals geographic inequality: "...a disease symbolic of health inequality; primarily a cancer of the poor, socially vulnerable women, even though there are simple and relatively low-cost means of preventing it" (Arrossi et al. 2008*:* 50). This study found that "...in addition to the rich/educated/labor active bias, there was a geographic one, as the mortality rate in the province of Jujuy is almost four times higher than that of the province of Buenos Aires. Women from the Northwest and Northeast regions (the least developed areas of the country) were almost three times more likely to be underusers of pap screening [...] [and] have the country's highest cervical cancer mortality" (Arrossi et al. 2008: 56).

Brazil, with a comparatively well-equipped and efficient health organization (Table 5.1), also displayed poor performance in the face of the health crisis. On the one hand, even if other infrastructure factors are high, it has a significant shortage of doctors akin to the familiarist systems (Fig. 5.8). This may be part of the explanation, given the strong correlation between number of doctors and deaths. A complementary elucidation is the large proportion of private expenditure, particularly out-of-pocket; Brazil is among the countries with the highest level of infrastructure and performance, the one with the utmost out-of-pocket spending: around 25% (Fig. 5.6). While the studies we consulted coincided with the fact that, like in Argentina, there are significant differences

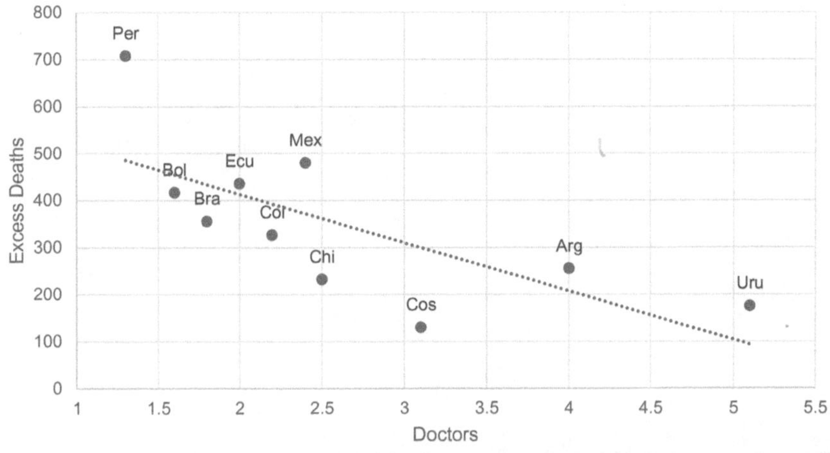

Fig. 5.8 Number of Doctors per 1000 inhabitants vs. excess deaths (*Source* See Appendix)

concerning health access between the rural and urban areas and amid poorer regions of the country (North and Northeast) and that these characteristics reinforce each other (Figueireido Santos 2018), some other studies stress the fact that one of the primary sources of inequality lies between those who are privately insured (25% of the population) and those who rely only on the SUS (Barbosa and Cookson 2019; García-Subirats et al. 2014). So, while in Argentina, the less effective management of the crisis (compared with Uruguay and Costa Rica) was probably partly due to the segmentation of the health organization and, especially (as Uruguay is also a segmented health system) the rural/urban and regional inequalities (that coincide with levels of income), in Brazil, in addition to these same rural/urban and regional differences, the SUS has been sapped by private insurance (Barbosa and Cookson, op. cit.).

But it is quite probable that another crucial factor to explain the ruinous management of the health crisis is the total irresponsibility of the federal government headed by president Bolsonaro, who never took it seriously, was against the use of masks and vaccines, and made coordination at the national-level inoperative (in a manner that is comparable to what occurred in the United States under Trump); nonetheless, the governments of some states, as Sao Paulo and Rio, managed to cope

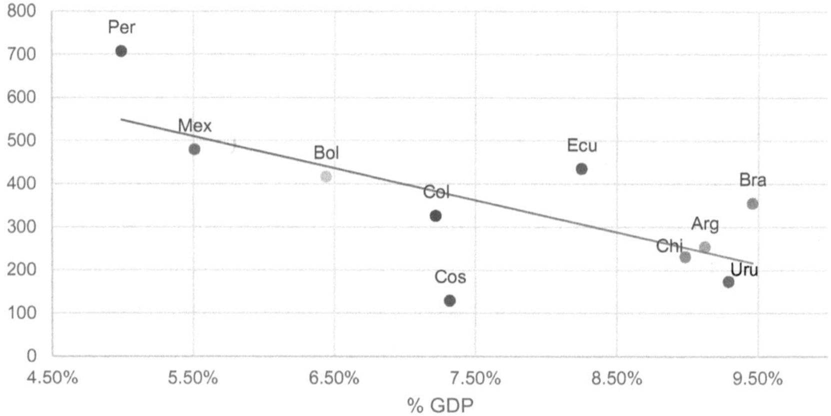

Fig. 5.9 Total health expenditure (% GDP) vs. excess deaths (*Source* See Appendix)

better. While federal State structures that operate effectively (Canada and Germany) addressed the pandemic better than many centralized countries (England and France), federal structures where there is no coordination between the federation and the states showed poor performance (the United States and Brazil).

Chile is another case that deserves to be discussed at greater length to understand why it performed less effectively than expected concerning its infrastructure and past performance. It is a system that has been reformed by social-democratic governments over the past 20 years and has achieved universal coverage. It has a mid-range health infrastructure but is top in our performance index. It also has one of the highest levels of total expenditure in relation to GDP (Fig. 5.2) and the highest per capita spending (with Uruguay) (Fig. 5.4). The reason why this country failed to measure up to our benchmark countries during the pandemic is similar to that we defined for the case of Brazil. Although Chile is a centralized country and does not have the extreme regional inequalities of Brazil, it is comparable in the high level of private and, especially, out-of-pocket spending (Figs. 5.10 and 5.11), among the highest in the continent. These variables have a solid correlation with deceases during the pandemic. Like all other health systems in Latin America (with their differences), the Chilean is lopsided towards the wealthy, and the best predictor of health service usage is education and employment status. However, a particular feature of the Chilean structure is that it also discriminates by type of health disbursement between those who have private insurance (and thus also spend more on out-of-pocket in a privatized health system such as the Chilean, where insurance fees and co-payments are very high) and those who depend only on the public health services (Núñez and Chi 2013; Vásquez et al. 2013). In fact, as Vásquez et al. write, "...while 79% of the total population was in FONASA, and public insurance covered a greater share (87.5%) of the population aged 65 or older, ... ISAPRES beneficiaries, who accounted for 13% of the total population, were disproportionately rich, with 45% belonging to the top income quintile".

Finally, Colombia is also an "outlier", but in the contrary direction from the three countries we have just discussed—it has not fared as poorly as one would have expected if one considers that it has one of the lowest infrastructure and performance levels of the continent, and a relatively low total health expenditure (similar to Ecuador, Bolivia, and Peru), but did not have the level of deceases of these countries or of Mexico (Table 5.1 and Fig. 5.9). This may be explained by the fact that its health system has

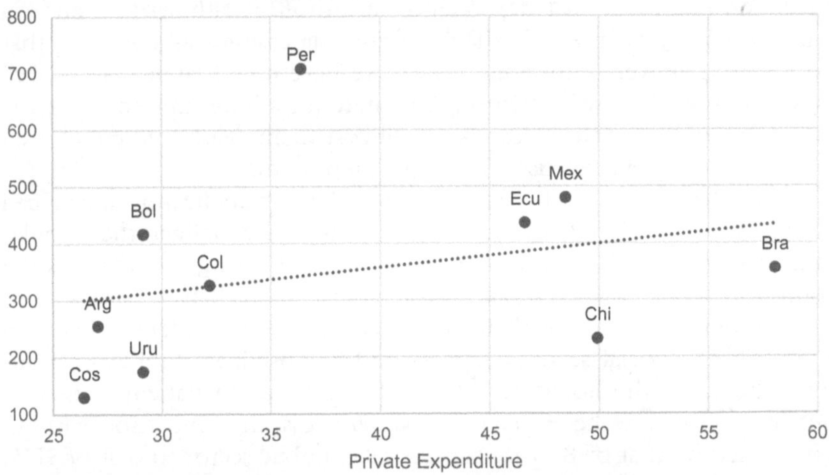

Fig. 5.10 Private current expenditure (% total health expenditure) vs. excess deaths (*Source* See Appendix)

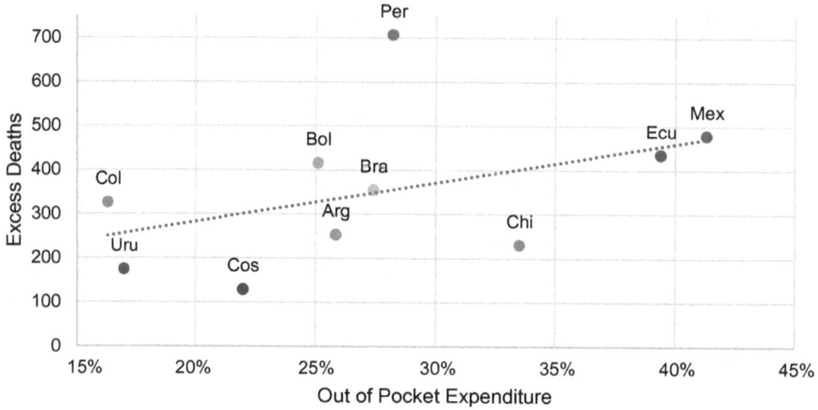

Fig. 5.11 Out-of-pocket expenditure vs. excess deaths (*Source* See Appendix)

practically covered its entire population, which has translated into a very low level of private and especially out-of-pocket spending (Figs. 5.5, 5.6, 5.10, and 5.11); both are crucial variables for defining excess mortality. From the discussion on Chile and Brazil, we have learned about the effects on inequality that dependence on private health services and on out-of-pocket fees entail. The latter is also the case for all countries that do not have universal coverage, which we have called familiarist, as well as Mexico; in all these cases, people hesitate for a long time to go to the doctor or to the hospital, because it will cost them dearly and could even lead to catastrophic expenses that could ruin them.

As we have previously discussed, the Colombian health system can be considered a single payer scheme, as the State obliges the population to enroll in either a private or a public health plan, concentrates the contributions of all workers, and subsidizes the individuals who lack sufficient resources to pay for their plan. It is, however segmented between a contributive social security scheme and a subsidized public contributive scheme, both allot different benefits and attend patients either in public or in private health centers. Nonetheless, in comparison with an akin structure, that of Brazil, with a parallel public sector to that of SUS, the latter suffers from very high usage of emergency services and "...a certain weakness in treating chronic diseases in primary attention [...] as compared to Colombia" (García-Subirats et al. 2014). This is most probably due to long-waiting periods in Brazil, which increases the disparity between the public and private sectors. Another study shows that the 1993 reform in Colombia increased the use of preventive and curative outpatient services, a 22% increase in the supply of drugs, and, even more importantly, a 53% reduction in out-of-pocket expenses for outpatients and 50% for inpatients (Ruiz Gómez et al. 2013). In this way, we can risk concluding that the Colombian health system, admittedly segmented and having an income bias like the rest of Latin American organizations, had a better performance in the face of the pandemic than similar countries concerning infrastructure and past performance due to reduced private health expenditure and, more importantly (and with a higher correlation index), low out-of-pocket outlays.

To conclude, the most significant indicators to explain the number of deaths are those we could have expected. While the number of doctors per capita and the percentage of total health expenditure in relation to GDP have a negative correlation, the level of private and out-of-pocket expenditures have a positive correlation. Countries with the fewest doctors and lowest public spending—Peru, Bolivia, Ecuador, and Mexico—had a high number of deaths per capita; on the contrary, the countries with the most doctors and health expenditure—Costa Rica and Uruguay—had the fewest.

Nonetheless, we also found specificities regarding of the type of health regime. Those who were the best able to fight the epidemic have been the ones of the two countries that are the most universal, regardless of whether they are socio-corporatist or State-led (Uruguay and Costa Rica). Although they have different types of health systems, both Uruguay's and Costa Rica's coped relatively well with the health crisis. Whereas Uruguay is segmented, Costa Rica has a unified State scheme. It should be noted that Costa Rica had a very high performance before the emergency and also in the face of the pandemic, with less infrastructure; this reinforces the idea that a centralized regime can spend less to be efficient.

The historical performance of health services based on the percentage of children vaccinated, maternal death, cancer preparedness index, etc. does not always match with infrastructure. However, the correlations between the two are quite high (0.59). In fact, there are significant disparities in a few countries, namely Chile, Mexico, and Costa Rica. Chile has a predominantly market-dominated, historically efficient health system (index = 88) but a lower infrastructure index. Faced with the crisis, it was in the middle group in terms of number of deaths in proportion to the population, which is most probably the consequence of the lack of infrastructure for all Chileans (while the private sector is very well provided, the public one is much less so) and of the fact that out-of-pocket health expenditure is very high. Mexico has one of the worst performances in the face of the epidemic despite a relatively high historical performance; what played a decisive role in this case, as in the Chilean case, was the lack of infrastructure and a very high out-of-pocket expense (the highest of all the countries considered). The fact that health depends so heavily on out-of-pocket costs means that people hesitate before going to the doctor or the hospital, and thus, they only go when they are already very sick.

With the highest infrastructure, historical performance and lower out-of-pocket expenses, Argentina performed worse than Uruguay and Costa Rica. This can be explained on the one hand by the public investment reduction in health since the end of the commodity boom, but above all by the inequities of a highly segmented scheme that concentrates resources and infrastructure in the capital and some other big cities. Brazil also has a reasonably high level of infrastructure and a high level of past performance, but was strongly hit by the epidemic. We can also consider that this is the consequence reducing health spending since 2015, but above all, due to the inequity between private and public care.

Colombia is another exception: its health system has one of the poorest infrastructures and the lowest historical performances in the continent in terms of our indexes; only Ecuador, Bolivia, and Peru have weaker infrastructure and worse performance. Nevertheless, in the face of the epidemic, it has fared better than these three countries and even Mexico and Brazil. This may be explained by a health structure covering almost the entire population, which results in a very low level of private expenditure, but above all, out-of-pocket expenses, we have discussed how the latter has had a deleterious effect during the pandemic.

Finally, in the cases of Mexico and Brazil, we also have to account for a situation that is difficult to quantify: the irresponsible posture of their leaders in the face of the epidemic. The Mexican president did not take the health crisis seriously at the beginning. Even when the situation worsened, and the government began to advise social distancing (without imposing strict containment and without providing support to the informal or formal sectors to achieve this), it did not recommend the widespread use of face masks; in fact, the Mexican president appeared masked in public in very few occasions. The case of Brazilian President

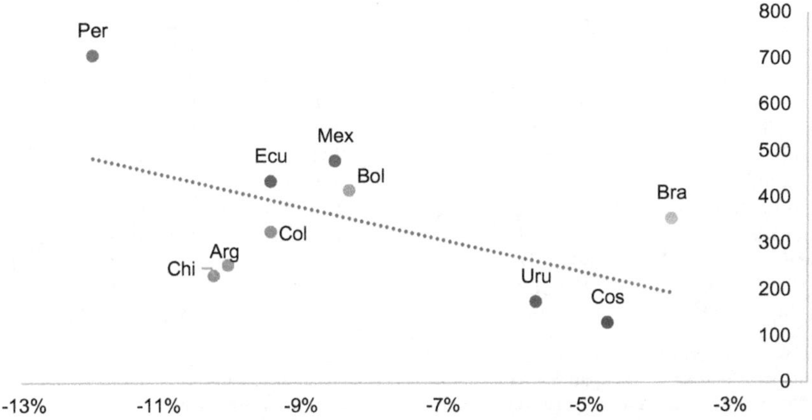

Fig. 5.12 GDP variation vs excess deaths (*Source* See appendix and ECLAC [2021] for GDP)

Bolsonaro is much worse, as he not only avoided advise on masking and social distancing, but sabotaged these measures when taken by state governments, hindered the purchase of vaccines and boycotted vaccination. In fact, like most other leaders in Latin America, they did not call for strict confinement since, in the continent, the informal sector is so large that these measures would be impossible to implement. Still, both AMLO and Bolsonaro clearly (at the beginning and specific moments in the case of the Mexican president, always in the case of the Brazilian) prioritized the economy rather than life itself. This gamble failed, as shown in Fig. 5.12, where the economies of the countries with more deaths have also fallen sharply. In the case of Bolivia, Peru, and Ecuador, rather than irresponsible governments, it is political instability that partly explains their performance. Argentina, Chile, and Colombia imposed stricter confinement measures than other countries of the continent, which led to a profound fall in the economy but probably saved lives.

Appendix

Sources of Tables 5.1 and 5.2 and the following figures.

Country	*Excess mortality*	*Excess mortality per*	
Argentina*	115,416	254	March/20–7/Oct/21
Bolivia	48,378	416	Jan/20–July/21
Brazil	746,715	355	Jan/20–Aug/21
Chile	43,834	231	Jan/20–20/Sept/21
Colombia	163,956	326	Jan/20–5/July/21
Costa Rica**	6,584	129	March/20–7/Oct/21
Ecuador	75,559	435	Jan/20–20/Sept/21
Mexico	603,507	479	Jan/20–30/Aug/21
Peru	229,863	707	Jan/20–20/Sept/21
Uruguay***	6,062	174	March/20–7/Oct/21

https://ourworldindata.org/excess-mortality-covid
*https://datosmacro.expansion.com/otros/coronavirus/argentina
**https://datosmacro.expansion.com/otros/coronavirus/Costa-Rica
***https://datosmacro.expansion.com/otros/coronavirus/Uruguay

The indexes:

They are both constructed as a simple average of the percentages of the variables considered for each index. The variables not expressed in percentages are converted into these by a simple rule of three. The highest value for each variable is rounded up to set the 100% benchmark, and each country's percentage is then defined with respect to it.

The following variables are used for each of the indexes:

Infrastructure: Number of beds per thousand inhabitants 2017*; 2. Average number of intensive care beds per 100,000 inhabitants*; Number of doctors per thousand inhabitants (2017)*; Number of nurses per thousand inhabitants*; Radiotherapy units per million inhabitants (latest year available)*; Global Health Security Index 2019**; Mechanical ventilators per hundred thousand inhabitants***.

Performance: Percentage of the population vaccinated per year*; Percentage of women examined at least four times during pregnancy*; Maternal mortality rate per 100,000 births*; Mortality in children under five per 1000 births*; Cancer Preparedness Index****.

*OECD, World Bank Group (2020), "Health at a Glance: Latin America and the Caribbean 2020", OECD Publishing, París, https://www.oecd-ilibrary.org/sites/6089164f-en/index.html?itemId=/content/publication/6089164f-en;

**2019 Global Health Index, https://www.ghsindex.org/; ***Global Health Intelligence (2020), "Infografia sobre el nivel de preparación de hospitales latinoamericanos ante coronavirus"; ****World Cancer Initiative (2020), "Preparación para el abordaje del cáncer en América Latina", https://www.iccpportal.org/system/files/resources/PREPARACION_PARA_ABORDAJE_CANCER_AMERICA_LATINA.pdf.

References

Alcalde-Rabanal, J. E., O. Lazo-González, and G. Nigenda (2011), "Sistema de salud de Perú", *Revista Salud Pública de México*, 53 (2): 243–254.

Aran, Daniel and Hernán Laca (2011). "Sistema de salud de Uruguay", *Revista Salud Pública de México*, 53 (2): 265–274.

Arce, Javier Monterrey (2013), *Sistemas de protección social en América Latina y el Caribe: Estado Plurinacional de Bolivia*, Santiago, CEPAL.

Arrossi, Silvina, et al. (2008), "Social Inequality in Pap Smear Coverage: Identifying Under-Users of Cervical Cancer Screening in Argentina", *Reproductive Health Matters*, 16 (32): 50–58.

Ballesteros, Matías (2014), "Un análisis sobre las desigualdades en el acceso a los servicios de salud en la población adulta urbana de Argentina a partir de datos secundarios", *Documento de Jóvenes Invetigadores*, 41, Buenos Aires, Instituto de Investigaciones Gino Germani, Facultad de Ciencias Sociales, UBA.

Barbosa, Estela Capelas and Richard Cookson (2019) "Multiple Inequity in Health Care: An Example from Brazil", *Social Science & Medicine*, 228: 1–8.

Becerril-Montekio, Victor M., et al. (2011a). Sistema de salud de Brasil, *Salud Pública de México*, 53 (2): 120–131.

——— (2011b), "Sistema de salud de Chile", *Salud Pública de México*, 53 (2): 132–142.

Belló, Mariana and Victor M. Beccerril-Montekio (2011), "Sistema de Salud de Argentina", *Revista Salud Pública de México*, 53 (2): 96–108.

Bernal, O. and S. Barbosa (2015), "La nueva reforma a la salud en Colombia: el derecho, el aseguramiento y el sistema de salud", *Revista Salud Pública de México*, 57 (5): 433–440.

Cecchini S., F. Filgueira and C. Robles. (2013). *Sistemas de protección social en América Latina y el Caribe. Una perspectiva comparada*. Santiago de Chile, CEPAL.

Cercone, James and José Pacheco Jiménez (2008), "Costa Rica: 'Good Practice' in Expanding Health Care Coverage", In Pablo Gottret, George J. Schieber and Hugh Waters (eds.), *Good Practices in Health Financing: Lessons from*

Reforms in Low- and Middle Income Countries, Washington, DC, The World Bank.
Conaghan, Catherine (2011), "Ecuador. Rafael Correa and the Citizen's Revolution", In Steven Levitsky and Kenneth Roberts, *The Resurgence of the Latin American Left*, Baltimore, The Johns Hopkins University Press: 260–282.
ECLAC (2011), *Statistical Yearbook for Latin America and the Caribbean*, Santiago de Chile: United Nations.
——— (2021), *2020 Preliminary Overview of the Economies of Latin America and the Caribbean*, Santiago, United Nations.
Figueireido Santos, J. A. (2018), "Classe Social, território e desigualdade de saúde no Brasil", *Saúde Soc.*, 27 (2): 556–572.
Filgueira, Fernando (1995), "A Century of Social Welfare in Uruguay", *Democracy and Social Policy Series*, Working Paper #5 - Spring.
Fundación de Investigaciones Económicas Latinoamericanas (FIEL) (2007), *La Desigualdad en la Salud*, Buenos Aires, Fundación de Investigaciones Económicas latinoamericanas.
García-Subirats, I., et al. (2014), "Determinantes del uso de distintos niveles asistenciales en el Sistema General de Seguridad Social en Salud y Sistema Único de Salud en Colombia y Brasil", *Gaceta Sanitaria*, 28 (6): 480–488.
Gattini, Cesar (2018), "El sistema de salud en Chile", Santiago: OCHISAP Observatorio Chileno de Salud Pública (OCHISAP), Escuela de Salud Pública U. de Chile. Disponible en: http://www.ochisap.cl.
Gragnolati, Michele, Magnus Lindelow, and Bernard Couttolenc (2013), *Twenty Years of Health System Reform in Brazil: An Assessment of the Sistema Único de Saúde*, Directions in Development, Washington, DC, World Bank.
Guerrero, Ramiro, Ana Isabel Gallego, Victor Becerril-Montekio, and Johanna Vázquez (2011), "Sistema de salud de Colombia", *Salud Pública de México*, 53 (2): 144–155.
Gutierrez Arriola, A. (2002), *México dentro de las reformas a los sistemas de salud y seguridad social en América Latina*, México, Siglo XXI.
Ledo, C. and R. Soria (2011), "Sistema de Salud de Bolivia", *Revista Salud Pública de México*, 53(2): 109–119.
Lucio R., N. Villacré, and R. Henríquez (2011), "Sistema de salud de Ecuador", *Salud Publica*, 53(supl 2): 177–187.
Martínez Franzoni, Juliana, and Diego Sánchez-Ancochea (2016), *The Quest for Universal Social Policy in the South: Actors, Ideas and Architectures*, Cambridge: Cambridge University Press.
Minteguiaga, A. (2012), "Política y políticas sociales en el ecuador reciente: dificultades asociadas a la salida del ciclo neoliberal", *Revista De Ciencias Sociales*, 135–136: 45–58.

Noy, Shiri (2017), “Healthy Targets? World Bank Projects and Targeted Health Programs and Policies in Costa Rica, Argentina, and Peru, 1980–2005”, *Oxford Development Studies*, 46 (2): 164–183.

Núñez, Alicia and Chunhuei Chi (2013). Equity in Health Utilization in Chile, *International Journal for Equity in Health*, 12 (58): 1–16.

OECD (2016), *Estudios de la OCDE sobre los Sistemas de Salud: México*, OECD.

Palacios, Alfredo, Natalia Espinola, and Carlos Rojas-Roque (2020). “Need and Inequality in the use of Health Care Services in a Fragmented and Decentralized Health System: Evidence for Argentina”, *International Journal for Equity in Health*, 19 (67): 1–14.

Poltronieri, Jorge (2011), *Proyecto de Investigación Estructuras de la Opinión Pública, Comunicado de prensa, Encuesta de opinión pública XXIII – Panorama global*, San José, Costa Rica, CIRMA/UCR, 45 pages.

Repetto, Fabián and Fernanda Potenza dal Masetto (2011), Protección social en la Argentina, *Serie Politicas Sociales*, Santiago de Chile, *CEPAL*. no. 174, 68 pages.

Robles, Claudia (2011), *El sistema de protección social de Chile: Una mirada desde la igualdad*.

Ruezga Barba, Antonio (2009), *Seguridad social. Una visión latinoamericana*, México, Centro Interamericano de Estudios de Seguridad Social.

Ruiz Gómez, Fernando, Teana Zapata Jaramillo, and Liz Garavito Beltrán (2013), “Colombian health care system: results on equity for five health dimensions, 2003–2008”, *Revista Panamericana de Salud Pública*, 33 (2): 107–115.

Sáenz, María del Rosario, et al. (2011). Sistema de salud de Costa Rica, *Salud Publica*, 53 (2), 156–167.

Sánchez-Talanquer, Mariano, et al. (2020). *La respuesta de México al COVID: estudio de caso*, Institute for Global Health Sciences-UCSF, Santiago.

Santos Basso, Óscar Luis (2005), “Costa Rica” en *El nuevo derecho de las pensiones en América Latina*, México, Instituto de Investigaciones Jurídicas, Centro Interamericano de Estudios de Seguridad Social: 187–253.

Vásquez, Felipe, Guillermo Paraje, and Manuel Estay (2013), “Income-Related Inequality in Health and Health Care Utilization in Chile, 2000–2009”, *Revista Panamericana de Salud Pública*, 33 (2): 98–106.

CHAPTER 6

The Welfare State and the Wage Relation

In Chapter 1, we defined the welfare state as the result of a social compromise in the conflict concerning the distribution of one of the many creations of human society, in this case social protection. It should then become evident that the history of the welfare states of both the developed and underdeveloped countries is such a compromise. In the developed countries, Fordism was a concord where productivity increases were linked to salary raises and to the expansion of the social protection system. There existed a tacit agreement between capital and labor to the effect that the State used contributions and taxes to improve workers' lives in order not only to secure social peace but also to increase the production, productivity, and quality of the manufactures. Another social compromise, the Japanese, has been agreed between the enterprises and their workers, where social peace between unions and employers was guaranteed through continuous, although moderate, salary hikes and social protection that the enterprises themselves allotted: lifelong labor, health security, and pensions; the two latter with the support of the State (Lechevalier 2015).

Although many of the historical accounts of the implementation of the welfare state in Germany, at the time of Bismarck, as well as in Brazil, Mexico, and Argentina in the thirties and forties have been described as a concession on the part of the government to workers to coopt them politically and prevent their autonomous organization, it is clear that in

I. Bizberg, *Four Worlds of the Welfare State in Latin America*,
https://doi.org/10.1007/978-3-031-44420-3_6

its most profound sense, it involved a pact, accepted by both government and labor and not a unilateral event. The workers consented to the social policies implemented by the government, sealing a compromise that was questioned periodically by labor to demand better conditions. As has been exceptionally analyzed by Cardoso and Falleto (1969), import substitution industrialization was based upon a social alliance between the State, workers, and national industrialists, and the welfare state was a fundamental component of this alliance. The same can be said of the socio-democratic welfare regimes, which were founded on a pact between workers and farmers, a red-green coalition as Esping-Andersen has called it, which was sealed and administered by the State, the latter becoming the guarantee of the compromise as the agent that collects the contributions and taxes that fund the health, pension, and unemployment insurance.

While during import substitution industrialization in Latin America, the social compromise had certain similarities to Fordism, although limited to a sector of society, in some countries this sector was more extensive than in others, as we have seen in the historical reconstruction of Chapters 3 and 4, at present the social compromise has been totally modified in most countries. As the neoliberal economic model created a host of poor and informal workers (see Table 6.1)—in some countries, they represent around half of the population—the new social compromise had to include them, although in a subordinate relation. Depending on the formal sectors' power and organizational capacity and the extension of the informal, the compromise encompasses both: the formal and the informal, or mainly the people living in poverty. This determines how far the turn to assistance has gone, whether assistance has substituted other forms of social protection, or whether assistance is complementary to the welfare state based on rights; we will address this issue in the second section of this chapter.

Because the Fordist mode of capitalism depended fundamentally on the scope of "salarization", i.e., the formalization of workers, and as in Latin America, this process never included the majority of the working population because the informal sector was, then as now, very large, Fordism in this continent was at best incomplete. While most countries in the continent attempted to implement wage-led growth during the ISI period, and some of them, again, during the commodities super cycle, other countries gave up implementing a redistributive social contract after the crisis of the eighties. The "lost decade" that ensued ended the social pact implicit in

Table 6.1 Wage relation of the different ideal types of welfare states

	Socio-corporatist	*Statist*	*Commodified*	*Familiarist*
Public social Expenditure	High	High	Low	Low
Health	High coverage	High coverage	Low coverage	Low coverage
Pensions	High coverage	High coverage	Low	Low
Non-contributory pensions	Low	Low	High	High
Assistance programs	Complementary to social rights	Complementary to social rights	Central to social protection system	Central to social protection system
Informality	Low	Low	High	High
Wage policy	High minimum salaries	High minimum salaries	Minimum salaries linked to productivity	Low minimum salaries
Labor market policy	Formalization	Formalization	Limited active policies to reduce informality	Limited active policies to reduce informality

Source Own elaboration

the import substitution model existing in many countries between the State, the popular classes, and the entrepreneurs. This pact sometimes referred to as partial or limited Fordism, because it was limited to a sector of the country's population: the workers of the formal sector and the government functionaries, the national entrepreneurs and the State, were abandoned in Mexico, Chile, Colombia, and Peru. In other countries of the continent, the military regimes had already broken the pact before the eighties: in Brazil since 1964, in Chile in 1973, and in Argentina in 1975.

The abandonment of import substitution industrialization in the wake of the debt crisis of the eighties and the democratization process that ensued in most Latin American countries at almost the same time gave rise to different social compromises whose character depended on the force of the social organizations and their influence on the democratic transition. In some cases, for example Brazil, Argentina, and Uruguay, the democratic transition occurred before the imposition of the liberal economic model; in others, for example in Mexico and Chile, it happened after, while in the first, social and political organizations impeded the radical imposition of neoliberalism, in the second, the absence of opposition led to a more far-reaching liberal model. And then, during the first and a half decade of the years 2000, during the commodities super cycle, the application of a wage-led or a profit-led economic model and a redistributive or a wage-repressive social model depended on the force and organizing capacity of civil society and the character of the parties in power (Bizberg 2019).

It is now helpful to introduce the concept of the wage relation, central to the French Regulation School (FRS), which can be interpreted as the translation of the social compromise each society has come to define to limit the redistributive conflict. The concept of **wage relation** concerns not only salaries but "...the totality of the juridical and institutional conditions that determine the use of labor as well as the existence of the workers" (Boyer 2015). It thus comprises also "indirect" forms of determining the existence of workers, such as the conditions in which labor is negotiated in the labor market and the social security system (health, pensions, unemployment benefits, etc.).[1] The specificities of the wage relation are either negotiated directly between the unions and the employers or defined through State intervention. Its conditions closely depend on the organization, strength, and capacity of action of the labor movement, the employer's organizations, and their cooperative or confrontational relation (Crouch 1993). On the other hand, in most countries, the minimum salaries are fixed by commissions where the State has a decisive role. In addition, the State is co-responsible for regulating the labor market. It intervenes directly in collecting the contributions paid by workers and employers to finance the social protection system. It finances part of it through taxes; it allocates some programs to poor workers; it spends on education, health, and pensions; it invests in clinics and hospitals, and thus defines part of the characteristics of the welfare state. Each social compromise translates into a specific wage relation.

[1] It is evident that a country's tax structure must be considered as forming part of the wage relation, although in a "passive" way; the level of taxes allows the State to redistribute wealth and spend more or less to protect its population. And then, the level of taxes is also a compromise between capital and labor and between those that earn more and less in a society. We have to think of the consensus of the different social sectors to pay very high taxes in the countries of Northern Europe because they get much in return from the State and they preserve society from inequality, in contrast to the reluctance to pay taxes in such unequal countries like those of Latin America.

In Table 6.1, we set the **theoretical** characteristics of how the different types of social protection systems we discussed in the previous chapters translate into wage relations. We will then analyze the empirical correspondence based on expenditure, social protection coverage, and wage data.

Public Social Investment and Coverage

Figure 6.1 clearly shows that the proportion of public resources dedicated to social policies and health is very significant in three countries we have cataloged as having a socio-corporatist and a statist welfare state: Argentina, Brazil, and Uruguay. The rest of the countries have all lagged well behind. The cases of Argentina and Brazil are interesting because of the very significant growth of social expenditure from the beginning of the year 2000 until the mid-2010s, something clearly related to the fact that the social actors had a considerable capacity to exert pressure on their respective governments, and by the model of development they implemented, a wage-led growth regime that required the increase of salaries and a hike of investment in social policies, which was facilitated by the boom of the price of commodities that both countries produced.

In contrast, the countries where the State spends less in relation to GDP are the familiarist: Ecuador, Peru, and Bolivia. This is also the case of the countries we have designated as hybrids: Mexico and Colombia, although Mexico is the country that spends the least. In terms of progression in the last 21 years, we can see that Peru and Mexico are the two countries where this expenditure progressed more slowly, something we can understand if we remember that both countries have been pursuing a liberal exporting model, Peru a liberal rentier one, while Mexico, an international subcontracting model, both based on a profit-led growth and a coalition between the State and capital. In contrast, Ecuador and Bolivia have been increasing their expenditure on the basis of their redistributive rentier model based on an alliance between the State and the popular sectors (Bizberg 2019) (Fig. 6.2).

The cases of Costa Rica, Chile, and Colombia occupy the mid-range of social expenditure. We have defined Chile as approaching a commodified welfare regime, and we can see that social spending until 2018 coincides with this characterization; expenditure only increased after the massive mobilization at the end of 2019 and during the pandemic almost to reach the most generous countries. Although we do not know if this tendency

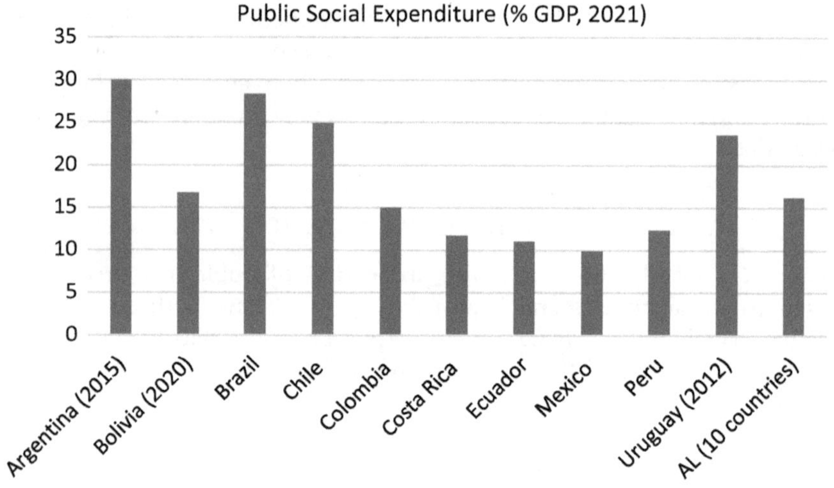

Fig. 6.1 Public social expenditure (% GDP, 2021) (*Source* Own elaboration based on Cepalstat data)

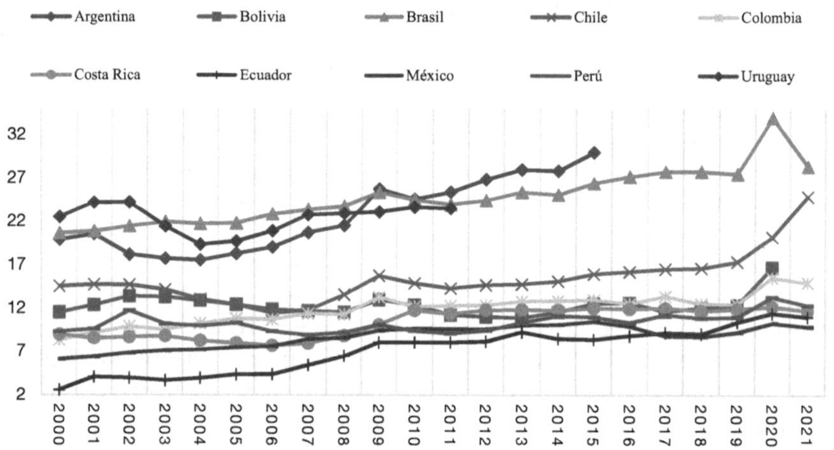

Fig. 6.2 Public Social Expenditure (% GDP: 2000–2021) (*Source* Own elaboration based on Cepalstat data)

will continue, we can suppose so with the government of Boric. If one takes into consideration that, in the last two decades, under three left-leaning governments, the most extreme characteristics of liberalism in the social protection scheme were alleviated and that the growth of its economy has been constant during the last thirty years, it is not understandable why it is still lagging behind the countries that spend the most; one has to refer here to the fact that social actors are weak and that, up to the present government, it continued to follow a liberal rentier model, and a profit-led growth, where the priority is given to profits rather than wages.

As we have seen in Chapter 5, although Colombia has a relatively limited welfare state, since the promulgation of the 1991 Constitution, it has significantly expanded its health services. Although, as we have discussed in the previous chapter, it continues to have an inegalitarian welfare state, where many citizens get low-quality health care, it has been able to reach almost total coverage since it has implemented the obligation to affiliate, subsidizing the population that has no resources to do so, and in addition it has spent more than the familiarist welfare states; both measures have allowed this country to reduce considerably out-of-pocket expenditures.

If one looks now at the amount spent in more specific sectors of the welfare state, first in contributory pensions, the same pattern appears (see Fig. 6.3). The socio-corporatist and statist regimes, including Costa Rica (at a lower level), spend more—an expected situation since the pension scheme is more inclusive; the State both completes the deficits of the contributory private schemes (the AFPs) and allocates pensions to its employees. The case of Chile, where pensions are based on capitalization, and the State only allocates a minimum salary to those who do not save enough to reach this wage level, the public spending is meager, at the level of the familiarist welfare schemes. Mexico, where pensions were also privatized in the 90s, is at the same level of Chile. The significant difference is that while the percentage of coverage in Chile is 65%, it is half that size in Mexico.

Public spending on the health system is more equilibrated; the corporatist regime spends the most, together with Colombia, which has established, as we discussed in the preceding chapter, compulsory health affiliation with State subsidies and legislation that obliges a homogenization of both the contributory and the subsidized schemes, both of which impulse higher spending by government.

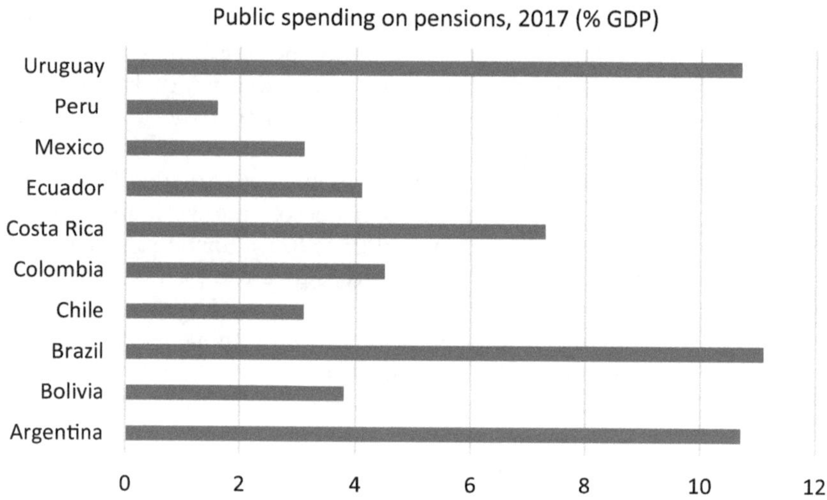

Fig. 6.3 Public spending on pensions (% GDP, 2017) (*Source* Alberto Arenas de Mesa 2020)

Familiarist social protection systems have emerged slowly and dispersed since the beginning of the twentieth century. Their trajectory was neither promoted by social organizations or political parties nor by the intention of the state to coopt the labor movement. This resulted from a weak industrialization process in Peru, Ecuador, and Bolivia. However, in all three, there existed nuclei of solid labor organizations: in mining, oil, governmental institutions, primary education, and dockers, which, when they protested and striked, pressuring the government to allocate them special conditions. They also received advantages when a new government (democratic or authoritarian) arrived that needed political support. This resulted in a limited and atomized welfare state. Lately, with the growth of informality and poverty, these countries expanded non-contributive pensions and enlarging their public health scheme to dispense minimum services.

We also analyzed the historical trajectories of two hybrid welfare states: Mexico and Colombia. Half of the Mexican population has access to a structure that allocates both pensions and health, which is state-controlled, the other half, the informal and the poor workers, receive assistance and non-contributory pensions. On the other hand, this population formally has access to public health services, but, in reality, does not get regular health services but only emergency and, to a certain degree, chronic disease care. Consequently, half of the population depends on their resources, those of their families, and the care services of its female members. In the case of Colombia, after the constitution of 1991, it has achieved offering health services to almost the totality of the population with both a pure and a subsidized contributory scheme, although significant differences between both structures still exist.

The pandemic the world suffered in 2020 and 2022 put to the test the different health systems of Latin America and, thus, the type of welfare regime the diverse countries were nearest to. Although the number of deaths each country endured depended greatly on many variables because, like a war or a great catastrophe, the pandemic was a **total, multifaceted event**. Although we do not pretend to find the principal cause that defined the proportion of deaths because it was a unique occurrence that befell all countries simultaneously, it can point to the characteristics and tell us something about which health schemes are more or less efficient in controlling such a dire situation. First, the number of doctors per capita and the percentage of total health expenditure to GDP have a negative correlation with deaths. In contrast, the level of private and out-of-pocket spending have a positive correlation.

The health systems of Uruguay and Costa Rica, the most universal, had the best performance tackling the crisis; they are both the countries with most doctors and health expenditure. A difference between both is that although Costa Rica has as a good a record of past performance as Uruguay, it has less infrastructure; however, in the face of the pandemic, it was as effective. This reinforces the idea that a centralized scheme can spend less to be efficient.

The social protection systems that did worst in the pandemic were the familiarist, those of Peru, Bolivia, Ecuador, and hybrid Mexico; this is due to the lack of infrastructure, the fewest doctors, and the lowest public spending, but especially by the fact that these countries have the highest level of out-of-pocket expenses, which meant that the sick waited until the last moment to attend the clinics and the hospitals, either because they knew there was no place, because the hospitals were not capable of tackling the number of sick, or because they had the previous experience that any attendance to a health service, even if it is public costs them a large sum.

The other countries did not do as well as expected, if one looks at their infrastructure and past performance, or did better, as in the case of Colombia. The reason why Argentina, a socio-corporatist system with the highest level of infrastructure and historical performance, performed worse than Uruguay and Costa Rica can be explained above all by the insufficiency of a socio-corporatist system, partly its corporatist segmentation and partly to geographical disparities: the centralization of the best health services in Buebos Aires.

Another statist health structure strongly hit by the epidemic, Brazil, also has a relatively high infrastructure level and good past performance. We can conclude that its relatively poor performance during the pandemic was, above all, due to the inequity that exists between private and public care, thus a result of the privatization of a statist system, as well as to geographical disparities resulting from a federal State structure that has both advantages and disadvantages.

The fact that the private health sector is so determinant in Chile is why it performed less effectively than expected regarding its infrastructure and past performance. It is a scheme that has succeeded in achieving universal coverage, has a mid-range health infrastructure and is on the top in our performance index. Nonetheless, the reason why this country failed to measure up to our benchmark countries during the pandemic is similar to what we defined for the case of Brazil. Although Chile is a centralized country and does not have the extreme regional inequalities of Brazil, it is comparable in the high level of private and, especially, out-of-pocket spending (Figs. 5.10 and 5.11 of the previous chapter), among the highest in the continent. These are variables have a solid correlation with deceases during the pandemic.

Colombia did better than expected in the pandemic. Although its health configuration is one of the continent's poorest concerning infrastructure and past performance, it has managed to built a health structure covering almost the entire population. This results in a low level of private expenditure, but above all, out-of-pocket expenses; we have analyzed how the latter had a deleterious effect during the pandemic (Fig. 6.4).

Figure 6.5, shows the impressive increase in Colombia's health spending since 2000 from 4% to 7%, surpassing Costa Rica and Brazil. The low amount of public health spending in Brazil and its stagnation since the year 2000; however, it increased facing the pandemic (as most other countries), confirms the privatization of the health organization of this country which we analyzed in Chapter 5. Chile doubled its expenses

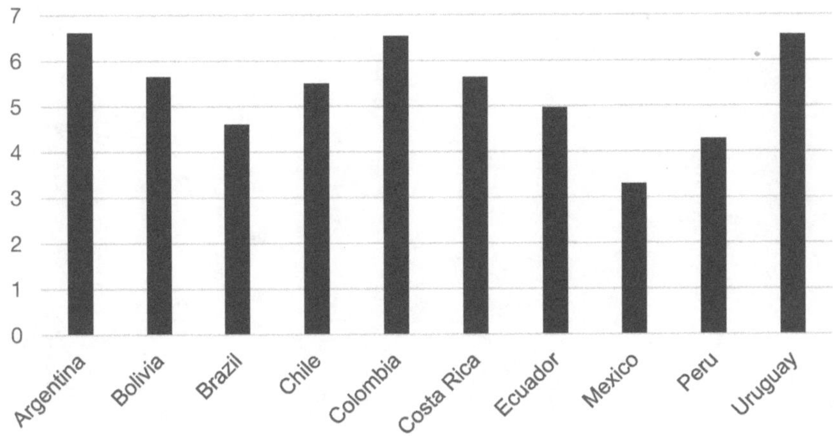

Fig. 6.4 Public health expenditure (% GDP, 2020) (*Source* Elaborated based on Global Health Expenditure Database)

from the level of Mexico (3.3%) (where this country stagnated) to the level of Argentina and Uruguay (6.5%).

Finally, Costa Rica, a country with a statist welfare state, which allows it to do more with less, as one can conclude with the analysis of its impressive performance during the pandemic, with less health infrastructure than the most efficient countries, and even of some of the least efficient ones during the health crisis. Although this country spends less on health than Uruguay and Argentina, it has been able to structure one of the most universal health systems in the continent, both in terms of coverage, equity, and relative generosity, probably due to its centralized and unified character, as Martinez Franzoni and Sánchez Ancochea have demonstrated.

As Esping-Andersen has written, although how much a country spends in the different segments that comprise the social protection system is an indicator of its breadth and generosity, how it is spent is even more critical as it reflects the characteristics of the welfare state and which sectors benefit more.

Coverage is an indicator that **structurally** distinguishes the different types of welfare regimes. In general terms, familiarist welfare states have a lower coverage of their population in pensions and health than the other types. In Fig. 6.6, we can see that Uruguay, Costa Rica, Brazil, Chile, and

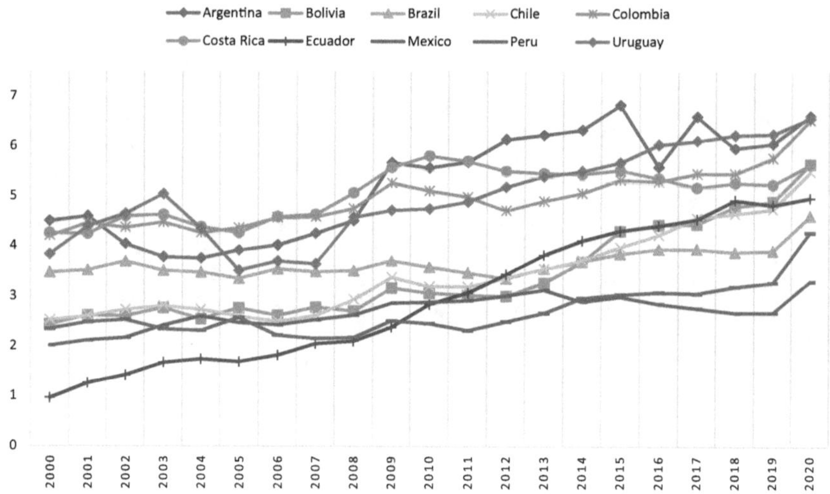

Fig. 6.5 Government health expenditure (% GDP) (*Source* Elaborated based on Global Health Expenditure Database)

Argentina have the highest levels of both pension and health coverage, while Colombia, Ecuador, Mexico, and Peru, have much lower ones; Colombia has low coverage in pensions but joins the countries that assure almost total coverage in health services. As we have seen in the previous chapter, the Brazilian health organization has been increasingly privatized, and the Chilean one also depends significantly on private expenditure.

It is evident that while pension affiliation coincides inversely with informality—in fact, it is one of the ways how it is measured—(Fig. 6.3) health coverage may be higher even with an elevated informality rate, as is the case of Colombia, as the public health scheme can achieve in including relatively efficiently most of the population of a country, even the informal sectors. In fact, this was one of the handicaps of the Mexican health program, the *Seguro Popular,* which automatically inscribed the poorer population that received governmental assistance but demanded a contribution from individuals of the informal sector that earned more than the lowest three income deciles, something that almost totally inhibited their inclusion. This handicap was resolved by the Colombian system, structurally similar to the Mexican scheme, by making affiliation mandatory and subsidizing the contributions for those without resources.

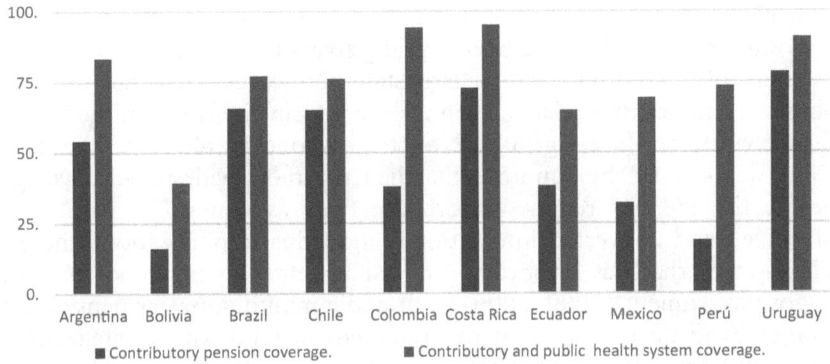

Fig. 6.6 Coverage by contributory pensions and by contributory and public health services of the total employed population (%, 2020) (*Source* Own elaboration based on Cepalstat and ILO [2021], *Panorama de la Protección Social en América Latina y el Caribe: Avances y retrocesos durante la pandemia*, octubre; ARG: https://www.estadisticaciudad.gob.ar/si/genero/principal-indicador?indicador=tip_cob_sal; BRA: https://www.commonwealthfund.org/international-health-policy-center/countries/brazil; CHI: https://www.trade.gov/knowledge-product/chile-healthcare; COL: https://www.minsalud.gov.co/proteccionsocial/Paginas/cifras-aseguramiento-salud.aspx; COS: https://www.binasss.sa.cr/ojssalud/index.php/gestion/article/view/177/318; ECU: https://www.pbs.org/frontlineworld/rough/2007/06/ecuador_healthlinks.html; http://ilo.org/wcmsp5/groups/public/---americas/---ro-lima/---sro-lima/documents/publication/wcms_510914.pdf; MEX: https://www.inegi.org.mx/temas/derechohabiencia/; PER: https://rpmesp.ins.gob.pe/index.php/rpmesp/article/view/3998; URU: https://www.gub.uy/ministerio-salud-publica/tematica/cobertura-poblacional)

In the cases of Uruguay, Brazil, and Argentina, the relationship between public social expenditure, coverage, and informality is conclusive: they are countries that spend much on social protection, have extensive coverage of both pensions and health, and have relatively lower informality rates (particularly Uruguay). However, nowadays, the case of Argentina is less conclusive, as this country has lately reached a high level of informality due to the crisis that followed the commodities super cycle, especially the disastrous economic management of the Macri and Fernández presidency.

Chile is another country that stands out as an exception, but in the opposite direction; we have been arguing that this country is the clearest example of a commodified welfare state. However, coverage of both pensions and health is relatively high, since informality is one of the lowest on the continent (see Fig. 6.7). Nonetheless, one should remember that Chile was one of the countries that had the most widespread coverage before the military regime, before the "pay as you go" scheme was privatized and converted into a purely individual capitalization scheme. This system that has drastically reduced benefits, so much so that the Lagos government (2000–2006) instituted a minimum salary pension for workers who do not arrive at this replacement level with capitalization. Thus, as we will see below, while pension coverage is very high in Chile, it allocates a low level of pensions because it depends on capitalization, having, consequently, a high percentage of poor older Chileans.

Mexico, a hybrid, although it does not have the lowest social protection coverage, is much behind the larger countries of the continent and some of the smallest ones, such as Costa Rica, Ecuador, and Colombia. This is basically due to the large proportion of informal workers in this country, which has to do with the fact that the economic model of international outsourcing in such a large country as Mexico, which competes in the global economy based on low salaries and low labor costs, generates informality and poverty.

In Fig. 6.8, shows some additional characteristics of the different types of welfare regimes concerning pensions. Here, we have to read this graph in two different manners. The data seems quite paradoxical because we have socio-corporatist and familiarist welfare regimes among the countries with the highest replacement rates. The countries with the highest replacement rates are the ones with socio-corporatist and statist welfare regimes: Uruguay, Argentina, Costa Rica, and Brazil, quite behind, but we also have two familiarist welfare states: Peru and Ecuador. This may be explained by the fact that the small percentage of people who have pensions in these two countries get, in effect, relatively generous pensions corresponding to their previous wages. Still, they are a small minority of the population. The other familiarist welfare states get low replacement rates, and so does Chile due to its capitalization model, which reaffirms its characteristic as a commodified regime, and Mexico, which also adopted this same pension scheme.

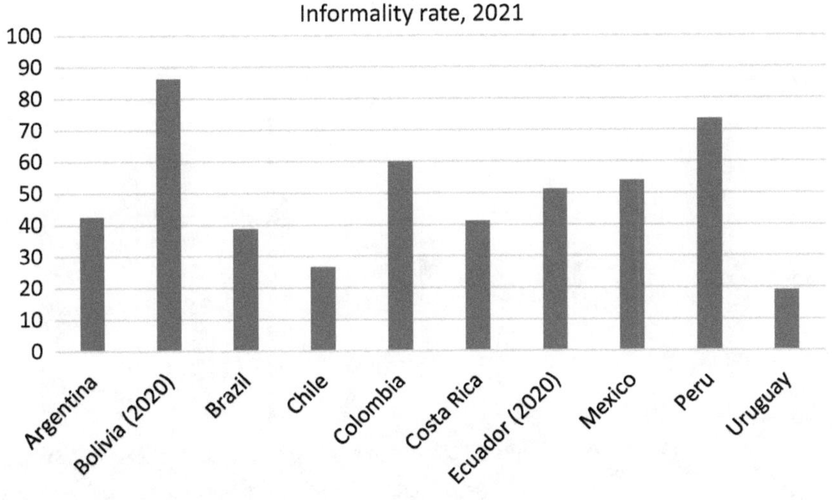

Fig. 6.7 Informality rate, 2021 (*Source* OIT [2021], *Panorama laboral 2021: América Latina y el Caribe*, Lima; Instituto de Estudios Avanzados para el Desarrollo [2022], "Informalidad, principal característica del empleo en Bolivia", https://www.inesad.edu.bo/2022/05/04/informalidad-la-principal-caracteristica-del-empleo-en-bolivia/; Dirección General de Estadística de Colombia, https://www.dane.gov.co/index.php/estadisticas-por-tema/mercado-laboral/empleo-informal-y-seguridad-social#:~:text=Para%20el%20total%20nacional%2C%20en,anterior%20(59%2C9%25); OIT, https://www.ilo.org/lima/sala-de-prensa/WCMS_780991/lang--es/index.htm)

The consequence of a small proportion of the population receiving pensions in the familiarist social protection systems and reduced replacement rates in countries like Mexico,[2] Colombia is the proportion of older persons in poverty, as we can see in Fig. 6.9. While in Uruguay, Argentina,[3] Brazil and Chile the proportion of older people in poverty is much lower.

[2] In the case of Mexico, this situation has probably changed as the government of AMLO has expanded and significantly raised the amount of the non-contributory pensions. However, they are still low, as they cover 40% of the minimum salary.

[3] The situation of Brazil, but especially Argentina, has changed dramatically in the last four years, so these data have to be considered as a historical trend.

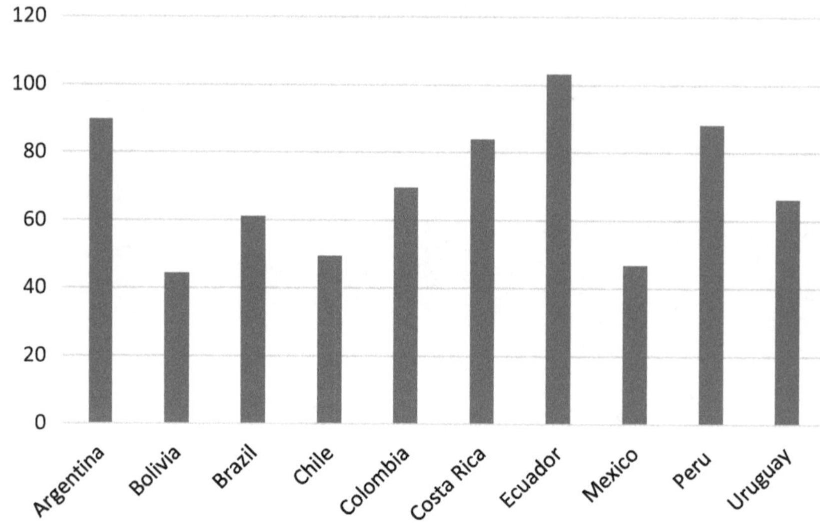

Fig. 6.8 Average net pension replacement rates *Source* (BID 2015)

The Turn to Social Assistance

In the last pages, we have seen how one of the most fundamental differences between the socio-corporatist and statist welfare regimes and the commodified and familiarist ones is the coverage and benefits of pensions and health. Another difference, although subtler and, as we will discuss further, is the relation between concessions and rights, between charity and entitlements; especially in the case of the assistance programs, both non-contributory pensions and conditional cash transfers that have thriven in Latin America since the 90's.

According to Esping-Andersen, one of the most crucial distinctions between the socio-democratic and corporatist welfare regimes on the one hand and the liberal one on the other is the fact that in the first two, assistance is complementary and articulated with the contributory or universal tax-funded social protection systems, which are central in these regimes. In contrast, in the liberal welfare regime, assistance is the main feature, as it substitutes contributory schemes. In a profound sense, citizenship is replaced by needs rights by aid. We go from a regime based on the debt the State has on behalf of its citizens to a debt its citizens have

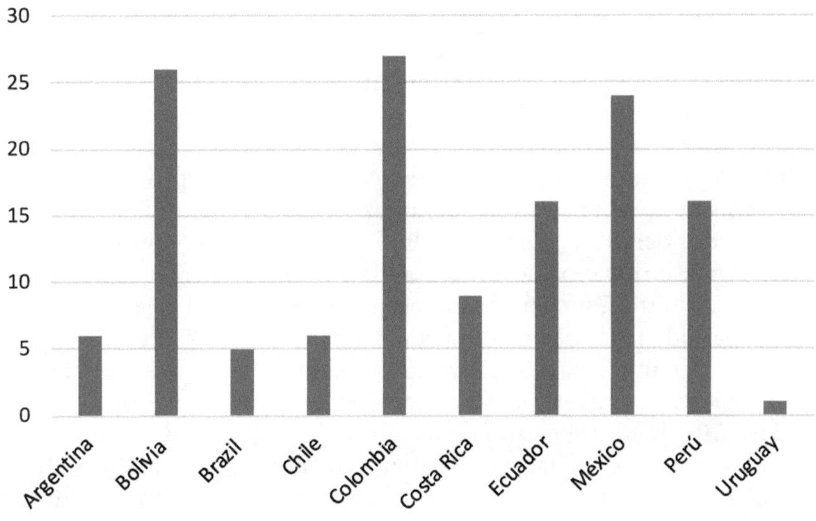

Fig. 6.9 Incidence of poverty among the population over 65 years of age (% 2020) (*Source*: Base on CEPAL, 2022 https://repositorio.cepal.org/server/api/core/bitstreams/43a39b21-edc7-478e-9085-348efae44cfa/content, P. 74)

towards the State, which is the feature that characterizes assistance when it is a central feature of a regime (Morvant-Roux and Carmona 2019).

According to Esping-Andersen and Théret, we have already discussed how universal measures, defined based on citizenship, can be considered a way of decommodification and debt that the State owes its citizens; they are a citizen right. On the contrary, conditional cash transfers, as well as any other focalized assistance, are the degree zero of social protection, a residual concession of the State, allocated once the person has proven that she/he is poor, has a disability, is an impoverished mother or father with a child, etc. These allocations are the opposite of an obligation of the State versus its citizens; they are the modern version of the oppressive Poor Laws implemented in England and other countries that considered being poor as an individual liability rather than a fault of society relative to its members.

As has been amply discussed in the literature, one of the most radical changes that occurred in most of the Latin American countries during the eighties and nineties was the shift from a tendential (although very slow in most countries) incorporation of citizens to a public structure generating social rights based on employment, on salarization, to a model where rights were substituted by assistance, where public intervention in social protection renounced to give rights to its population, but intended to implement programs oriented to particular social categories. Since the beginning of the neoliberal period during the end of the eighties, the ambition of most governments has no longer been to expand the guaranteed rights to social protection but, as taking care of one's future has been basically considered a private issue, the State would only focus on assisting those who are not capable of integrating into the market. In all Latin American countries, income-based aid played a marginal role throughout the ISI period; assistance to the people living in poverty was residual, as it was seen a temporary phenomenon that economic growth would solve. While this was probably an actual project in Mexico and Brazil in the 30s and 50s, it soon became rhetoric used solely to legitimize a policy that sought to consolidate a corporatist relationship with the reduced core of salaried workers who enjoyed not only formal employment and wage stability but also access to the social protection system. Whereas during the "Fordist" (or semi-fordist) epoch in Latin America, of import substitution industrialization, the expansion of social protection oriented to the growing population that was being integrated into salaried work legitimized a fundamentally corporatist system since the "lost decade" the people with low incomes have become a sector to be sought as political support through clientelist mechanisms (Lautier 2004).

Non-contributory Pensions

Non-contributory pensions may be considered assistance if they are focalized on older persons living in poverty and those who do not get the equivalent of a minimum salary in countries where pensions are based on capitalization. They can be considered as entitlements if they are universal and are no longer based on a program but are sustained by legislation. Nonetheless, in most countries (except in Mexico), they are not universal and are far from actually representing a substitute for a contributory scheme because the allowances are much lower than what persons earned when they were active; in Mexico, this is also true. Thus, in most of Latin America they can be considered a downgraded substitute of a contributory scheme.

Latin American countries have tried to expand pension coverage in two radically different ways. In some countries, we have seen an intent to formalize a more significant part of the population through active measures such as tax subsidies, higher surveillance of informal workers, restrictions on temporary work and subcontracting, and unionization of service employees and domestic workers (Brazil, Argentina, Uruguay, and Chile). In all these cases, one can consider an intent to consolidate them as entitlements. In most countries, however, we have seen a trend to merely expand non-contributory pensions to the entire population over a certain age (Argentina, Chile, Costa Rica, Bolivia, and Mexico) (Cecchini et al. 2013: 34).

One would expect (see Table 6.1) countries with lower contributory pension affiliation distribute more non-contributory allowances; a situation that may be interpreted in terms of generosity but which can also be understood as a manner of substituting rights by assistance. Familiarist social protection systems turn more resolutely towards this tendency; the most palpable examples are Bolivia and Mexico, where affiliation to pensions is very low (30% or lower), while non-contributory pensions are practically universal. And, in fact, in Mexico, the government of López Obrador has universalized old-age pensions (by inscribing them in the constitution) and significantly raised the amount, universalizing assistance rather than entitlements.[4] Nevertheless, we can also see that in these two countries, where although social pensions are almost universal, the allotted amount is far below the minimum salary, something that explains the large proportion of poor old population. The fact that there is a higher proportion of poor senior adults in the countries with more expanded non-contributory pensions is an indicator of the fact that this kind of support, and, in general, assistance to the poor, does not combat poverty,

[4] On the other hand, this government has converted assistance into financial products, something which promises to have profound consequences on the social relation between the individuals and the State, as has been analyzed in the case of Mexico by Solène Morvant-Roux, 2019 and for Brazil by Lena Lavinas; a theme which opens up an array of interpretations of the social protection systems in Latin America which we have not been able to analyze in this book. Morvant-Roux considers that this transformation converts the debt of the State vis a vis its citizens to a debt of the citizens to the State, as this latter may impose an obligation on the part of the recipients of assistance to spend it in only in a specific manner, which can be easily controlled by electronic means. The conditionality of most of the assistance programs was already this kind of a transformation, and financialization is a deepening of this evolution.

but administers it for economic and political reasons, as Bruno Lautier analyzed.[5]

On the contrary, the countries with the most extended contributory pension coverage. Argentina, Uruguay, and Costa Rica have the lowest proportion of the population on non-contributory allowances, an indicator that assistance has not substituted social rights. However, the middle ground is more ambiguous: Peru, Colombia, and Ecuador, that have low coverage of contributory pensions, also have low non-contributory coverage; the first two are the countries that spend the least in these programs. On the other hand, Brazil, which has a high proportion of its citizens affiliated to pension schemes, has a very high proportion on non-contributive allowances: the BPC (Beneficio de Prestação Continuada) covers 4,700,000 elderly or disabled people over the age of 65 and who live in a home with an income of less than a quarter of the minimum wage (Lautier 2007: 60–62 and https://dds.cepal.org/bpsnc/ptc), a sign of complementarity between social rights and assistance.

The amount of what the countries spend on social pensions is also telling and contrasts with coverage: the countries that allocate a low proportion of these kinds of pensions are the ones that spend most (proportionally to the persons it covers), as their marker is the level of remuneration of the contributory allowances or at least the minimum salary. In Table 6.2, one can see the disparity between what non-contributory pensions in the socio-corporatist (Uruguay and Argentina) and statist (Brazil) welfare states allot in relation to the minimum wage, in contrast to non-contributive pensions in commodified and familiarist welfare states: the first represents between 70 and 100% of the minimum salary (except in Costa Rica), while the second embodies less than 50%. The case of Chile is also telling; a commodified social protection system, with high contributive pension coverage, results in a significant percentage of poor older people, as neither the capitalization model guarantees a decent pension and the replacement rate of the non-contributive pensions is meager.

Assistance Programs

As in the familiarist welfare states, public social expenditure is low, social rights are limited, so we would expect that assistance becomes dominant: in fact, this indicator works partially: Whereas in Bolivia, the programs

[5] A study by Yaschine (2015), in the case of Mexico demonstrated that the very extensive poor relief programs applied in Mexico had no effect in reducing poverty.

Table 6.2 Coverage and expenditure of non-contributive pensions (2021)

	Non-contributory coverage (65 years and over) (%)	*Public expenditure on non-contributory old-age and disability pensions (% GDP)*	**Percentage of minimum salary*
Argentina	4.3	0.86*	100
Bolivia	90.5	1.52	15
Brazil	40.5	1.76	100
Chile	17.0	0.52	50
Colombia	21.3	0.16	9
Costa Rica	14.7	0.45	28
Ecuador	26.2	0.59	25
Mexico	93.5	0.59	45
Peru	20.3	0.11	27
Uruguay	3.5	0.50	71

Source CEPAL, Base de datos de programas de protección social no contributiva en América Latina y el Caribe https://dds.cepal.org/bpsnc/ilp

* It includes both: *Pensión Universal para el Adulto Mayor* and the *Programa de pensiones no contrib utivas*

Bono Juancito Pinto and *Bono Madre Niño-Niña Juana Azurduy* attain 55% of the population, in Colombia, the program—*Más Familias en Acción*, reaches 16% of the Colombians, in Ecuador, the programs *Bono de Desarrollo Humano* and *Desnutrición Cero* include 19% of the population, finally in Peru, the program *Juntos* covers 10% of Peruvians. In Mexico, a hybrid, program *Progresa* included 24% of the population. Under the presidency of Lopez Obrador, the resources for assistance programs have not increased but have been concentrated in two main programs, the old-age pensions and the scholarships for children for primary and secondary school, almost reaching the level of Bolivia.

Nonetheless, assistance is also vastly extended in the socio-corporatist and statist models: in Brazil, *Bolsa Familia* includes 25% of the Brazilians. In Argentina, the program *Asignación Universal por Hijo para Protección Social* covers 24% of the population, and in Uruguay, *Asignaciones familiares–Plan Equidad and Tarjeta Uruguay Social* benefits 24% of the population (Table 6.3).

Thus, neither funding nor coverage of the population of the assistance programs distinguishes between welfare regimes as well as non-contributory pensions. Nevertheless, if we analyze some of the characteristics of the programs, we will be able to distinguish some differences between the regimes, however subtle. There are two countries where, up to the present, rights have been substituted by assistance, especially in regarding the pension scheme, as in both Chile and Mexico, they have

Table 6.3 Assistance programs in Latin America (2019)

Country	*Main programs*	*Focused population*	*Expenditure (% GDP)*	*Coverage of total population*	*Transfer Amount (Maximum) US Dollars*
Argentina	*Asignación Universal por Hijo para protección social*	Minor child of an unemployed family of an informal worker; workers in private homes; self-employed	0.63	9.95	71.96
Bolivia	*a.Bono Juancito Pinto* *b. Bono Madre Niño-Niña Juana Azurduy*	a. Children and adolescents under 21 years old attending formal school b. Pregnant and lactating women with children under 2 years	a. 0.16 b. 0.07	a. 50.42 b. 4.75	29 (per year)
Brazil	*a. Bolsa Familia* b.*Brasil auxilio*	a. Families with per capita income Families with per capita income below the poverty line b. Families in poverty and extreme poverty; only if they include pregnant women, nursing mothers, or children and adolescents up to 21 years of age	a. 0.45 (2019) b. n. d	a. 25.2 b. n. d	a. 91.3 (per family) b. 20 to 40 US dol./person

Country	*Main programs*	*Focused population*	*Expenditure (% GDP)*	*Coverage of total population*	*Transfer Amount (Maximum) US Dollars*
Chile	*Chile Seguridades y Oportunidades/Ingreso ético familiar*	Families in extreme poverty, with children between 6 and 18 years of age	0.02	4.38	72
Colombia	*Más Familias en Acción*	Vulnerable and poor families. Indigenous families with children under 18 years of age	0.14	16.61	54.7
Costa Rica	*Avancemos*	Families with economic difficulties, with children between 12 and 25 years old to continue sending them to secondary school	0.19	4.03	68.1
Ecuador	a. *Bono de Desarrollo Humano (madres)* b. *Desnutrición Cero*	a. Families in extreme poverty with children under 16 years of age b. Families with children under one year of age and pregnant women	a. 0.61 b. 0.01 (2015)	a. 14.37 b. 4.15 (2015)	a. 150 b. 5 (2015)
Mexico	*a.Prospera (2018)* *b.Becas Benito Juárez*	a. Poor families b. Scholarships for students from poor families, for primary and secondary schooling	a. 0.35 b. 0.33	a. 24.8 b. 48.8	a. 128 (per family) b. 40*

(continued)

Table 6.3 (continued)

Country	*Main programs*	*Focused population*	*Expenditure (% GDP)*	*Coverage of total population*	*Transfer Amount (Maximum) US Dollars*
Peru	*Juntos*	Families in extreme poverty with pregnant women, widowed parents or mothers, or children under 19 years of age	0.10	10.08	59.9
Uruguay	*a. Asignaciones familiares–Plan Equidad* *b.Tarjeta Uruguay Social*	a. Poor families with children under 18 years old b. Men and Women with children under three years of age, poor pregnant women, transgender, homeless people, women victims of violence and/or human trafficking, people with chronic diseases and in extreme destitution	a. 0.30 b. 13 (2018)	a. 10.79 b. 12.91 (2019)	a. 65.5 b. 161.8 (2019)

Source CEPAL, Base de datos de programas de protección social no contributiva en América Latina y el Caribe https://dds.cepal.org/bpsnc/ilp

* (For the 3 last years of high school) -escuela preparatoria- the amount allocated is around 160 U.S dollars

been privatized, and the government only assures a minimum allowance for those who fail in saving enough in their capitalization schemes.

Chile was the first country where the neoliberal economic model was implemented; the Pinochet dictatorship imposed a market-oriented social protection system after harshly repressing the political parties, unions, and social organizations that could have shown any resistance. One of the fundamental aspects of this model was social policy targeting, which, with the privatization of pension funds and the commodification of health and education, became a model of neoliberal social policy for the continent (and beyond) in the 1990s. The more fundamental idea behind this model was the depolitization of the State, as was overtly defended a few years later by the Argentinian *junta* (O'Donnell 1979, pp. 291–294, cited by Buchanan 1987: 349). The idea of the Argentinian military (as conceptualized by O'Donnell) was that the State had been conquered by the popular sectors, basically by the *peronista* unions, and that the coup had as its central mission to rid it from the influence of these actors that strayed economic and social policies that should be neutral.

The democratic governments in Chile maintained this shift from social rights towards assistance. Although these governments have certainly corrected some of the most inequitable elements of the social reforms of the dictatorship, they have upheld the market orientation of its social protection system, in which economic rationality prevails over the social (Riesco 2009). On the one hand, the privatization of the pension funds, which was central to dismantle the previous contributory and rights-based scheme, was not questioned, although the present government of Boric has announced their reform and a return to "pay as you go". On the other hand, the democratic governments expanded assistance by including sectors of society beyond those in extreme poverty: poor women, young people, indigenous people, the older adults, and the people with disabilities (Papadopoulos 1999: 115–116), although they spend a minimal amount on these and give coverage to a small proportion of their population; something we could relate to the low proportion of informality (26%) as well as of poverty (14%), relative to the rest of the countries in the continent (see Table 6.4).

The Bachelet 2008 reform flexibilized access to contributory pensions for the old (Mesa-Lago 2009: 15–16) and set a minimum retirement (the minimum wage) for workers unable to capitalize enough resources during their active life. In addition, the government extended unemployment insurance to include those workers without a permanent contract.

Table 6.4 Poverty rates (CEPAL estimates) 2021

Country	*Poverty rate, 2021*	*Extreme poverty rate, 2021*	*Gini index, 2021*
Argentina	27.9	3.7	0.392
Bolivia	29.0	9.9	0.418
Brazil	24.3	8.3	0.537
Chile	14.2*	4.5*	0.475
Colombia	35.4	15	0.528
Costa Rica	17.3	3.7	0.501
Ecuador	28.5	7.6	0.466
Mexico	37.4*	9.2*	0.452*
Peru	19.3	4.2	0.426
Uruguay	4.8	0.1	0.402

Source Own elaboration based on CEPAL data. *2020

It also implemented an exceptional measure even for the more advanced economies: fiscal exemptions for companies that maintain and qualify their workers (CEPAL 2010: 17). This country, like all others of the continent, has also created several programs for labor inclusion, such as the *Fondo al Trabajo de la Mujer*, *Formación para el Trabajo*, and *Desarrollo de Competencias Laborales*, which in the case of the first includes 390,000 women (https://dds.cepal.org/bpsnc/ptc).

Although Mexico's welfare state is a hybrid between familiarism and statism, the neoliberal model of assistance implemented since the mid 90s put in place one of the world's first conditional monetary transfer programs called *Progresa*, the leading Mexican direct cash transfers' program (named differently in the successive governments, going from *Progresa* to *Prospera*) became a model for the rest of Latin America. The stated goal of the direct cash transfers to households living in extreme poverty was to help the poor households overcome some of their deprivations and to avoid the generational reproduction of poverty by increasing the human capital of the young (Valencia 2008). Resources were received directly by the mother, who was responsible for ensuring that the children assisted the school and the health centers. It is clear that this program was not a right but that it was part of a government policy based on a sophisticated focalization mechanism (Valencia 2008). Compared to resources oriented to pensions and health, which amounted to about 5% of GDP at the time, assistance-oriented resources were meager, less than 0.5%. Yet, they were very extended as they included more than 5 million families, around 25% of the total population (https://dds.cepal.org/bpsnc/ptc). Notwithstanding, a significant part of the people living in poverty were excluded: single men, families without children, persons who live in localities that are not considered poor, or where there are no schools or health centers (Lautier 2007: 65). On the other hand, the corporatist welfare structure began to be dismantled: in 1995, the private workers' pension scheme, of the IMSS, was privatized; the public workers' pensions of the ISSSTE suffered the same fate in 2005, they both went from pay-as-you go to an individual capitalization scheme.

The present government of López Obrador promised a total transformation of the social protection system, which it rightly considered to be focused on assistance rather than on rights; the government canceled both the principal conditional transfers program, *Prospera*, and the public health program, *Seguro Popular*. It then concentrated all the assistance programs in the *Secretaría del Bienestar* and health in the *Instituto de Salud para el Bienestar* (INSABI),[6] expanded their range, and significantly raised some programs' amounts. However, this has not signified the end of the conditional monetary transfers to allocate rights; on the contrary, it has reinforced them. The only program defined as a right is the universal pension for adults over 65, who received 4,800 pesos every two months in 2023, approximately 120 dollars a month. This program obtains half of the total social budget and it is not only given to the poor households, but to middle and higher classes also receive it without any taxation, so it is regressive. On the other hand, there has been a budget reduction of other programs dedicated to the impoverished households because most of the resources have been channeled to the universal pension, but also due to the austerity measures that the government has implemented, and because there has not been a fiscal reform.[7] Other important programs that also affect the people living in poverty and other vulnerable populations have been reduced, such as those dedicated to decreasing violence against women, searching for missing persons, and maintaining shelters for women under threat of violence (Jaramillo-Molina 2022). Even the budget for health has suffered, there was a

[6] This program was a failure; it was canceled by the same government that implemented it at the beginning of 2023.

[7] As Máximo Jaramillo-Molina (2022) has written, the total that the government of AMLO spends on the poor is lower than what the past governments spent from 2009 to 2016. It spent in 2021 4.7% of GDP, while the peak occurred during the previous Peña Nieto presidency, with 5.3%.

significant increase in 2020 due to the pandemic, but in 2021, it reverted to the level of 2019.

Prospera was substituted by a similar conditional cash transfer program, through scholarships allocated to children in all schooling levels. However, the only one that is universal is the one given to youngsters attending "preparatory" school (three years after primary school), with 1680 pesos each two months (about 42 dollars a month). There exists a scholarship for primary school children of low income families with the same amount, a program for single mothers and fathers for children between 1 and 4 years old, 1600 pesos bimonthly, children with a handicap receive 3,600 bimonthly, while elderly persons with a handicap receive 2800 pesos.[8]

Two other social programs that are not just monetary transfers but are meant to promote productive activities and technical formation of the young are *Sembrando Vida* and *Jóvenes Construyendo el Futuro,* respectively. The first one allocates 5000 pesos monthly for four years to peasants with at least 2.5 hectares of uncultivated land to plant trees. This program has been criticized for three different reasons: first to be included in the program, some peasants have cut the trees that existed in their lot or modified the traditional usage of their land[9]; this led to significant deforestation, the opposite of what the program was expected to generate. On the other hand, it is a program that only guarantees support for four years, which is too short to assure the growth of trees and there is no guarantee that the program will continue in the next government,[10] which would thus unintentionally convert the program into a mere monetary transfer. Finally, the continuity and success of this program are further endangered by the fact that it is individualized, particular peasants subscribe to it, and all existing social and communal organizations are marginalized. In addition, the indigenous communities are not taken into consideration; a situation that not only endangers its result but creates tensions within the communities, especially the indigenous ones (Bertoli 2020; Bertoli et al. 2023).

[8] https://apoyoseconomicos.org/.

[9] https://consumidoresorganicos.org/2019/12/02/sembrando-vida-ya-causa-danos-ambientales-Y-sociales-en-campeche/ (cited by Bertoli, Max A. 2020: 76).

[10] As is usual in Mexico, programs are sexennial; they only last as long as the government that implements them.

The other program *Jóvenes Construyendo el Futuro,* is a massive (777,000 in 2021) one-year contract directed to young who neither study nor work to become apprentices in an enterprise: the pay is 3,800 pesos. It is criticized because there is no follow-up; it is ignored if the young get a job after that year if what they actually do in the position increases their qualification, etc. As in the previous tree-planting program, it is not clear if it is not simply another way to distribute money, and on the other hand, if it is given with no political bias, as one does not know how the beneficiaries are chosen, or even by whom; so, it may be more prone to be used for clientelist purposes than the previous neoliberal focalized conditional monetary transfers. Finally, the government of AMLO has implemented a labor reform that may mean, in the medium to long term, the empowerment of unions and pressure to formalize and acquire social rights. It also reformed the law to restrict subcontracting workers, which may mean more formal jobs. Depending on what happens with the Mexican unions and the actual implementation of the law on subcontracting, we may see some progress toward social rights. However, neither in this country nor in Chile have the pension schemes, or the situation of the health services been modified, which are central for a welfare state based on rights.

While in all of Latin American countries focalized assistance programs were considerably extended since the 90's, due to the considerable growth of poverty since the "lost decade" of the eighties and the implementation of the neoliberal economic model, the main argument we are advancing in this chapter is that whereas in the commodified and familiarist welfare regimes; assistance was implemented as a **substitute** to all other types of social protection, as it was accompanied by the dismantling of the contributory social protection schemes, in the social-corporatist and statist welfare regimes they are still, more or less, **complementary** to a social protection regime based on rights. In Uruguay and Costa Rica, as well as in Brazil (under Lula and the first government of Dilma) and Argentina (under Kirchner and Fernández de Kirchner) assistance programs such as *Bolsa Familia* and the *Asignación Universal por Hijo*, respectively, were accompanied by significant strategies to formalize informal workers and increase minimum wages in order to, as Laura Vecinday has written to "... contribute to improve the income of [...] the recipients who are inserted in the world of work more than the provision of assistance" (Vecinday 2017: 55–56). In these four countries, pay-as-you-go pension schemes were maintained and, as we will now discuss, assistance was accompanied by labor integration or formation consistent programs. Nonetheless, as we will be able to see in what follows that, in no country of the continent, there is a clear example of an effective complementarity between assistance and rights, as assistance is still gaining ground in the region.

Uruguay is a good example of the complementarity between assistance and the welfare state, where the first does not substitute the second. When this country's economy suffered deeply by the profound crisis that occurred in Argentina in 2001–2002 due to the close link between both countries, the response of the government that arrived to power in 2005, the leftist *Frente Amplio*, with a very close relation to the leading union organization of the country, the *Central Única de Trabajadores*—CNT, the first reaction was the expansion of the assistance program, the *Plan de Atención Nacional a la Emergencia Social* (PANES), to protect the increasing number of unemployed and of the social groups that had been impoverished (Vecinday 2017: 55). This program was supplemented with the *Tarjeta Uruguay Social (TUS)* and the *Asignaciones Familiares-Plan de Equidad (AFAM-PE)*,[11] which covered " ...more than 80% of households with children in the lowest income decile [...] while in 1990 coverage did not exceed 20%. The available studies indicate that PANES, AFAM-PE, and TUS contributed substantially expanding the social protection network, the reducing indigence and inequality, and to a lesser extent, income poverty" (FCEA.UDELAR 2022).[12]

Nonetheless, these actions to deal with the social emergency did not intend to transform the social protection system, nor did they situate assistance at the center of the welfare state. Antía et al. consider that assistance was, in effect, a complement to the "...universally oriented protection package, whether through employment or strategic social policies" (Antía et al. 2018). Neither was it isolated from the other programs; in fact, it was accompanied by the Equity Plan launched in 2007 in the Social Reform Action Plan (MIDES 2011, cited by Vecinday 2017: 65). The Equity Plan was based on structural components such as the tax and health reform and the restructuring of the employment, housing,

[11] "The AFAM-PE program was created to alleviate poverty and promote school attendance, especially in the secondary cycle. The TUS, meanwhile, was conceived to promote the consumption of food and basic necessities in households in situations of extreme socioeconomic vulnerability" (Antía et al., 2018: 8).

[12] https://fcea.udelar.edu.uy/blog/6007-13-05-2022-que-sabemos-sobre-los-programas-de-transferencias-no-contributivas-en-uruguay-una-sintesis-de-resultados-de-investigacion-disponibles-sobre-el-panes-afam-pe-Y-tus.html.

and education policies. Two objectives [were] pursued by the reforms: "...a) influence social equity through changes in the structural components and b) consolidate and extend assistance, articulating these benefits with those of a universal nature" (Vecinday 2017: 65). In addition, according to Raczynski (1995: 12) ..."initiatives to overcome poverty are not limited to assistance policy, [as] ...in addition to emergency employment [the government effected]: (i) sectoral social policies: investment in basic education and health services, and policies and subsidies for housing, community equipment, and health infrastructure; (ii) policies that support the productive capacity of poor sectors ... (iii) policies to support social organization and training in poor sectors ...; (iv) labor and remuneration policies" (cited by Vecinday 2017: 56).

In addition, there are several of labor insertion programs, such as *Uruguay Trabaja*, which is oriented to labor insertion of long-term unemployed people in households in situations of socioeconomic vulnerability. This program benefits around 3000 people a year who are trained to permit them into reintegrate the labor market and "...carry out transitory jobs of public value for 30 hours a week and for a period of up to 8 months, during which a subsidy called "Support for Labor Insertion" of 2.35 BPC[13] is received" (Extracted from http://www.mides.gub.uy/uruguay-trabaja)" (Vecinday 2017: 62).

But even more important, the government actively promoted the formalization of employment through "... regulations that seek to include and set employment conditions of some categories of workers with a high incidence of informality, such as domestic workers or subcontracted workers ... the expansion of the *monotributo* scheme that sought to facilitate the formalization process of small companies, which were excluded from social security" (Lanzilotta 2009, cited by Antía et al. 2018: 8). These measures, along with an intensification of labor inspection, resulted in a reduction of informality from "35% in 2006 to 27% in 2012" (Antía et al. 2018: 8).

In Costa Rica, the complementarity between universal policies and assistance to the poor "... allowed the country to reap, in the long term, important achievements in terms of human development ... of a social protection system developed in Costa Rica in the mid-twentieth century based on universality and recognition of citizen rights. It promoted

[13] It is an index (*Base de Prestaciones y Contribuciones*) that represents about one-fourth of a minimum salary, so 2.35 BPCs are about 60% of the minimum salary.

universal social policies, which fixed minimum welfare standards or "common floor" for all citizens. The preceding was later complemented with the development of selective policies to combat poverty, assistance, and social promotion, for the most vulnerable groups" (Román 2010: 53). And, in fact, the *Instituto Mixto de Ayuda Social* (IMAS), that assists the poor, does not limit itself to monetary transfers, but " ...works in three main areas: social assistance, which includes meeting the basic needs of families, food security, emergency care, childhood;... credit and training to families seeking to develop productive ideas ... Community development promotes actions aimed at improving housing, public infrastructure and other benefits granted without spending (such as easing limitations to build)..." (Román 2010: 27).

One of the most significant programs at the intersection of universality based on citizens' rights and assistance is the Avancemos program, a scholarship oriented not only to low income families but to all adolescents who have economic difficulties to continue schooling. In principle, it is not fundamentally different from the Mexican scholarships allocated by the government of AMLO; however, *Avancemos* is focused on adolescents between 12 and 22 years and covers a more reduced population, only 4% of the total population, against 48% of Mexico, thus with a more significant amount of resources per capita: 0.16% of GDP compared to the *Benito Juárez* scholarships (0.33%) (Table 6.3 and Román, 2010: 68). Although the program has a particular educational purpose, it is inscribed in the IMAS, an institution that allocates assistance, rather than in the Ministry of Education, something which, according to Román, means this program is facing a quandary, of either reinforcing assistance and a clientelist approach or being a proper complement to the policies for the universalization of secondary education. "Until now, and due in large part to the scale of *Avancemos*, a logic closer to assistance has operated, which is limited to the delivery of transfers, with weak monitoring of compliance with conditionalities and a total disassociation of the educational offer of the MEP [Ministerio de Educación Pública]" (Román 2010: 55).

In the first place, Brazil did not transition from a pay-as-you-go pension scheme to capitalization; it did not even implement a private capitalization pillar to complement the public pensions, as did most countries before eliminating the latter. Also, the health structure, the *Sistema Único de Saude* (SUS), is universal, based on citizenship; the result of a significant intervention of civil society organizations, in this case, the *sanitaristas*, who not only played an essential role in the drafting of the 1988

Constitution, which established universal social protection as a right, but oversaw and in many cases ran the program. This group of experts had strong support from the leading trade union organization (CUT) and the Partido dos Trabalhadores (PT) (De Melo 1997: 295–296). On the other hand, the governments of Lula and Dilma implemented very significant programs to formalize informal workers through fiscal and work inspection measures. Finally, Brazil was the first major country in the world to institute (since 2004) a basic citizenship income, the *Renda Básica de Cidadania*, which was intended to cover all Brazilians by 2008, replacing all other assistance and minimum pensions. Although it has not been implemented, it is a right and a fundamental reference in the Constitution (Lautier 2007: 54).

Since the end of the eighties, under the first democratic government, the official discourse maintained that assistance programs should be considered as part of a general policy based on the rights of citizens and that they should be integrated into social protection systems (Távares Soáres 1999: 36). Although Collor de Melo intended far-reaching neoliberal changes to social protection, Brazil's social policy remained relatively protected. Brazilian targeted policies, primarily implemented locally, continued being generally more universalist and unconditional (Valladares 1999).

In 1995, the Cardoso government implemented the *Comunidad Solidaria*. This anti-poverty program allocated assistance to low-income groups, including food aid, primary education, and child health, and employment and income generation schemes. The PT governments increased the amount allocated to the poorest categories, doubled the pensions for rural workers and scholarships for children, and implemented the "zero hunger" program. During Lula's second term, practically all the assistance programs were grouped in the *Bolsa Família* and expanded in resources and coverage, reaching almost 60 million people in 2021 (https://dds.cepal.org/bpsnc/ptc). All of these, although targeted, have a universal character: funds are not allocated according to poverty indices, as is the case in most countries, and which excludes a large part of poor people, nor are they tied to conditions that the beneficiaries must comply with, as was the case in the Mexican *Progresa*, for example (Valladares 1999: 62–67). In fact, *Bolsa Familia* is not a conditional transfer, as it does not depend on families having children under 18 years (as is the case of *Prospera* in Mexico and other similar programs in the continent), but it was a universal allocation for people with low incomes (Cecchini et al. 2013: 35; BID 2016). On the contrary, the program implemented by Bolsonaro required the presence of children, and it was established at the same time that all other parts of the welfare state were being attacked.

The 2001–2002 crisis marked a significant turn in virtually every aspect of Argentina's social and economic policies concerning the previous Menem government (Bensusán 2010; Palomino and Trajtemberg 2006; Alonso and Di Costa 2011).[14] In the social protection domain, the government of Néstor Kirchner expanded the program *Trabajar*, similar to *Uruguay Trabaja*, but created in the mid-1990s in response to the *piquetero* movement that revendicated the right to work to be employed, rather than to be assisted; it consisted in temporary jobs focused on community work such as the construction or repair of roads, clinics, schools, etc.; instead of monetary transfers with the conditionalities of the next generation of programs, here the contribution on the part of the recipient was labor (Weitz-Schapiro 2006). In 2016, the *Plan Argentina Trabaja* included 531,000 persons, with a budget of 0.14 of GDP (https://dds.cepal.org/bpsnc/ilp).

The program *Jefes y Jefas de Hogar Desocupados* also had as its central idea the fight against poverty through temporary jobs. In exchange for an allowance, the beneficiaries had an obligation to participate either in community tasks, such as the construction or maintenance of infrastructure, social services (defined by NGOs, religious organizations, or local governments), or their requalification or continuation of their formal studies, or finally the launching of micro-enterprises (Kostzer 2008). Contrary to most assistance programs, which are paternalistic, and the assisted lack the sense of dignity that a job can give them, and their human capital is degraded from long-term unemployment, this program was unorthodox and directed to the unemployed and not to the poor and served to reinsert individuals to labor; the central demand of the piquetero movement that forced the government to implement it was the "right to work" (Kostzer 2008). This program reached a high just after the 2001 crisis, under the Duhalde presidency, with almost 2,000,000 beneficiaries (Delamata 2008: 134). It concerned 11% of the active population and contributed to decrease unemployment by 2.5%. In addition, with the support of the peronist unions, this government implemented policies to extend the coverage of health services through the *Obra Social*

[14] The Macri government made a 180-degree turn in the other direction.

of retired workers. It reduced the population without health services (Mesa-Lago 2009: 15).

Equally important to all these differences in the assistance programs of Uruguay, Costa Rica, Brazil and Argentina are the active labor policies tending to formalize workers are fundamental to increase coverage to the social protection system: the governments of Kirchner, Lula and Dilma fostered a public policy aimed at the formalization of workers based on tax incentives and stricter work inspection, (Maurizio 2014: 7; Berg 2011), as well as actions to empower labor unions. In fact, during the commodity boom, both Brazil and, to a lesser degree, Argentina implemented a socio-developmentist capitalist model (Bizberg 2019) based on a wage-led growth strategy that required the growth of the internal demand which was set at the center of the new model of development (Stockhammer 2011). This is why these countries were actively searching to formalize workers, significantly increase the minimum wages (even above mean salaries), and boost social security expenditure to reach a universal scheme by expanding contributory and non-contributory pensions. They also actively endeavored to develop the health structure, mainly by universalization, and by attempting to unify the different sectors of the social security structure into a more coherent whole. In addition, there was an intensive program of housing for people living in poverty, financed wholly or partly by the State as well as increased access to loans by the popular sectors, which permitted them to acquire durable consumer products.

We can thus conclude from this discussion that the substitution of the rights to social protection by assistance oriented social policies has been a way to legitimize governments that administer their poor population by assigning them a minimum to survive in a society that generates ever-growing inequalities, informalization, and poverty (Lautier 2004). It has also become a mechanism for obtaining votes in electoral competitions, as Garay (2016) has analyzed.

In more fundamental terms, regarding our discussion on the social compromise, the one existing between the State, unions, middle classes, and industrialists of the import substitution industrialization model has been replaced by a different pact, one between the State, large multinationals, financial interests and the reduced sectors of workers, and middle classes that have profited from the opening of the economy, and of profit-led growth (Bizberg 2019; Bresser-Pereira 2024). People living in poverty, an ever-larger proportion of the population generated by the end of the expansion of salarization, growing informality, and higher inequality, have, in a certain sense, become part of the social pact, although in a subordinate position and especially as socio-political counterpart to organized labor and civil society. This has meant the substitution of a welfare state based on the expansion of the contributory pension and health schemes by the extension of minimum assistance and primary health; even the liberal idea of increasing human capital and of hindering the process of intergenerational poverty has been abandoned, aid has become a manner of gaining the vote of the deprived populations.

Finally, we want to conclude more generally, analyzing which regimes, considered in their totality, are more extensive, generous, and with better outcomes. Thus, we have elaborated an index bringing together most of the data we have collected and presented in the different tables and figures concerning pensions and health. Figure 6.10 shows that the social compromise/wage relation in the socio-corporatist and statist welfare states ranks higher than in the other two. As expected, the familiarist social protection systems and Mexico are the worst rated. Chile deserves special mention as, with a commodified social protection scheme, it is below Argentina, Uruguay, and Brazil, even lower than Costa Rica and near Colombia.[15]

[15] These last conclusions regarding the rating of the different countries on a quantitative scale are comparable to the ones of two articles that pretend to rank the social protection systems of Latin America, but for two significant exceptions, those of Chile and Colombia. The article by José Antonio Ocampo and Natalie Gómez-Arteaga (2016), based on nine variables, basically structural ones (universality, coverage, and social spending), finds that Chile is one of the top systems, just below Uruguay, while Colombia comes much lower, in the intermediate level. In the other quantitative article by Gibrán Cruz-Martínez (2021), based on 15 variables, Chile also comes second, and Colombia is much behind. However, this article includes not only structural indicators but also outcomes and quality, and in the index based on the quality indicators, Chile descends from the second to the eighth level, and Colombia ascends from the eighth to the fifth, approaching our ranking. This may explain why our index, where almost half of the indicators are related to quality, comes closer to Cruz-Martinez's when measuring the quality of the different welfare systems in Latin America.

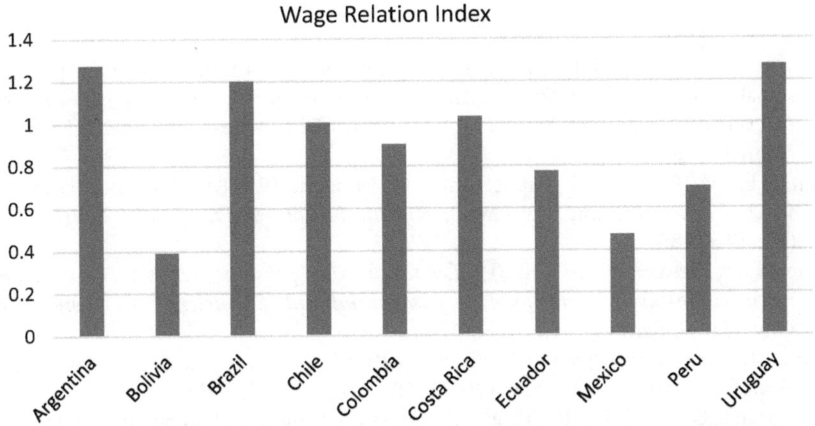

Fig. 6.10 Wage relation Index (*Source* Own elaboration, based on the following normalized variables: Public Social Expenditure, Public spending on pensions (% of GDP), Current health expenditure (% of GDP), Coverage by any health sub-system (public or social; % of the total population), Coverage by contributory pensions (% of the total population), Incidence of extreme and moderate poverty among the population over 60, Out-of-pocket health expenditure (% of total), Average net pension replacement rates, Infrastructure index, Performance index (see Appendix in Chapter 5). To ensure a comprehensive evaluation, the data for each indicator have been normalized by rescaling them using their respective maximum and minimum values

Ocampo and Gomez Arteaga	*Cruz-Martinez (global index)*	*Cruz Martinez (quality or performance index)*	*Bizberg*
Uruguay	Uruguay	Uruguay	Uruguay
Chile	Chile	Argentina	Argentina
Costa Rica	Brazil	Brazil	Brazil
Argentina	Argentina	Costa Rica	Costa Rica
Brazil	Costa Rica	Colombia	Chile
Colombia	Mexico	Bolivia	Colombia
Peru	Bolivia	Peru	Ecuador
Mexico	Colombia	Chile	Peru
Ecuador	Ecuador	Ecuador	Mexico
Bolivia	Peru	Mexico	Bolivia

References

Alonso, G. and V. Di Costa (2011), "Cambios y continuidades en la política social argentina, 2003–2010", In M. Uribe Gómez (coord.), *Los vaivenes de las políticas sociales en Argentina, Colombia, Chile, México y Uruguay*, México, Miguel Ángel Porrúa.

Antía, F., M. Castillo, G. Fuentes, and C. Midaglia (2018), "La renovación del sistema de protección uruguayo", *Revista Uruguaya De Ciencia Política*, 22 (2): 145–185.

Arenas de Mesa, Alberto (2020), *Los sistemas de pensiones en América Latina: institucionalidad, gasto público y sostenibilidad financiera en tiempos del COVID-19*, Santiago, CEPAL.

Banco Interamericano para el Desarrollo (2015), *El panorama de las pensiones en América Latina y el Caribe*, OCDE/Banco Mundial/BID.

Bensusán, G. (2010). Políticas laborales en América Latina: orientaciones, alcances y resultados. *Ponencia presentada en CEISAL.* Toulouse, 2 de julio.

Berg, J. (2011), "Laws or Luck? Understanding Rising Formality in Brazil in the 2000s", In S. Lee & D. McCann, *Regulating for Decent Work: New Directions in Labour Market Regulations.* Geneva and Basingstoke, ILO and Palgrave Macmillan.

Bertoli, Max A. (2020), Sembrando Vida: l'organisation communautaire au service du développement national?, thesis, Faculté des Sciences de la Société/ Institut de Démographie et de Socioéconomie.

Bertoli, Max, Solène Morvant Roux, and Ilan Bizberg (2023), "Le rôle de la Société civile dans le déploiement des projets économiques et politiques alternatifs: l'expérience mexicaine de la 4T", *Revue de la Régulation*, 33.

BID (2016), Síntesis-del-programa-Bolsa-Familia-en-Brasil, Banco Interamericano de Desarrollo, https://publications.iadb.org/publications/spanish/viewer/S%C3%ADntesis-del-programa-Bolsa-Familia-en-Brasil.pdf.

Bizberg, Ilán (2019), *Diversity of Capitalisms in Latin America*, London, Palgrave Macmillan.

Boyer, Robert (2015), *Économie politique des capitalismes, Théorie de la régulation et des crises*, Paris, La Découverte.

Bresser-Pereira, Luiz Carlos (2024), *The Rise and Fall of Neoliberal Rentier Capitalism. The political economy of New Developmentalism in the 20th and 21st centuries*, Oxford University Press.

Buchanan, Paul G. (1987), "The Varied Faces of Domination: State Terror, Economic Policy, and Social Rupture During the Argentine 'Proceso' 1976–81", *American Journal of Political Science*, 31 (2), May: 336–382.

Cardoso F. H. and E. Falleto (1969), *Dependencia y desarrollo en América Latina*, México, Siglo XII Editores.

Cecchini S., F. Filgueira and C. Robles. (2013). *Sistemas de protección social en América Latina y el Caribe. Una perspectiva comparada*. Santiago de Chile, CEPAL.

CEPAL (2010), *La reacción de los gobiernos de las Américas frente a la crisis internacional: una presentación sintética de las medidas de política anunciadas hasta el 31 de diciembre de 2009*, Santiago de Chile, United Nations.

Crouch, Colin (1993), *Industrial Relations and European State Traditions*, Oxford, Clarendon Press.

Cruz-Martínez, Gibrán (2021), "Mapping Welfare State Development in (post) Neoliberal Latin America", *Social Indicators Research*, 157: 175–201.

De Melo, M. A. (1997), "As reformas constitucionais e a previdência social 1993–1996", In E. Diniz and S. de Azevedo (eds.), *Reforma do estado e democracia no Brasil*, Brasília, Editora Universidade de Brasília.

Delamata, G. (2008), "Luchas sociales, gobierno y Estado durante la presidencia de Néstor Kirchner", In *Desempenho de Governos Progressistas no Cone Sul: Agendas Alternativas ao Neoliberalismo,* Rio de Janeiro, Edicoes IUPERJ: 128–146.

FCEA, Universidad de la República de Uruguay (2022), "¿Qué sabemos sobre los programas de transferencias no contributivas en Uruguay? Una síntesis de resultados de investigación disponibles sobre el PANES, AFAM-PE y TUS", *Universidad de la República de Uruguay,* 13 May, https://fcea.udelar.edu.uy/blog/6007-13-05-2022-que-sabemos-sobre-los-programas-de-transferencias-no-contributivas-en-uruguay-una-sintesis-de-resultados-de-investigacion-disponibles-sobre-el-panes-afam-pe-y-tus.html.

Garay, Candelaria (2016), *Social Policy Expansion in Latin America*, New York, Cambridge University Press.

Instituto de Estudios Avanzados para el Desarrollo (2022), "Informalidad, principal característica del empleo en Bolivia", https://www.inesad.edu.bo/2022/05/04/informalidad-la-principal-caracteristica-del-empleo-en-bolivia/.

Jaramillo-Molina, Máximo Ernesto (2022), "Primero los menos pobres: Política social, desigualdad y pobreza durante el sexenio de López Obrador", *Análisis Plural*, 1.

Kostzer, D. (2008), "Argentina: A Case Study on the Plan Jefes y Jefas de Hogar Desocupados, or the Employment Road to Economic Recovery", *Working Paper No. 534*. Annandale-on-Hudson, New York, The Levy Economics Institute of Bard College.

Lautier, B. (2004), Las políticas sociales en México y Brasil: seguridad social, asistencia, ausencia. In C. Alba and Ilan Bizberg (eds.), *Democracia y Globalización en México y Brasil*, México, Colegio de México.

——— (2007), "Les politiques sociales au Brésil durant le gouvernment de Lula: aumône d'Etat ou droits sociaux?", *Problèmes d'Amérique latine*, 63: 51–76.

Lechevalier, Sébastien (2015), "L'hétérogénéité croissante des entreprises au sein du capitalisme japonais: Faits, causes, conséquences, et implications", In Akinori Isogai, Hiroyasu Uemura y Robert Boyer (dir.), *Capitalismes asiatiques: Diversité et transformations*, Presses Universitaires de Rennes.

Maurizio, R. (2014), "Labour Formalization and Declining Inequality in Argentina and Brazil in the 2000s: A Dynamic Approach", *ILO Research Paper*, 9, Geneva.

Mesa-Lago, Carmelo (2009), *Efectos de la crisis global sobre la seguridad social de salud y pensiones en América Latina y el Caribe y recomendaciones de políticas. Serie Políticas Sociales*, Santiago de Chile, CEPAL.

Morvant-Roux, Solène and Mariana Carmona (2019). "La financiarisation des politiques sociales au Mexique: la laborieuse conversion des transferts monétaires en produits financiers", *Mondes en développement*, 2019/4 (n° 188).

Ocampo, José Antonio and Natalie Gómez-Arteaga (2016), "Social Protection Systems in Latin America: An Assessment", *Extension of Social Security series*, 52, Geneva, International Labour Office, Social Protection Department (SOCPRO).

OIT (2021), *Panorama de la Protección Social en América Latina y el Caribe: Avances y retrocesos durante la pandemia*, octubre.

Palomino, H. and D. Trajtemberg (2006), Una nueva dinámica de las relaciones laborales y la negociación colectiva en la Argentina. *Revista de Trabajo*, 2 (3): 47–68.

Papadopoulos, J. (1999), Igualdad, democracia y políticas sociales. In M. Schteingart (ed.), *Políticas sociales para los pobres en América Latina*, México, Miguel Ángel Porrúa.

Raczynski, D. (1995), "Focalizacion de programas sociales: Lecciones de la experiencia chilena", In Pizarro, C., Raczynski, D. and Vial, J., Eds., Politicas economicas y sociales en el Chile democratico, CIEPLAN-UNICEF, Santiago.

Riesco, M. (2009), "El modelo social chileno comienza a cambiar", *Revista Internacional del Trabajo*, 128 (3): 311–330.

Román, Isabel (2010), "Sustentabilidad de los programas de transferencias condicionadas: la experiencia del Instituto Mixto de Ayuda Social y 'Avancemos' en Costa Rica", *Serie Políticas Sociales*, 160, Santiago de Chile, CEPAL.

Stockhammer, E. (2011), "Crecimiento basado en los salarios: introducción", *Boletín Internacional de Investigación Sindical*, 3 (2): 183–208.

Távares Soáres, L. (1999), "Política Social y pobreza. Brasil en el contexto latinoamericano", In M. Schteingart (ed.), *Políticas sociales para los pobres en América Latina*, México, Miguel Angel Porrua.

Valencia Lomelí, E. (2008), "Las transferencias monetarias condicionadas como política social en América latina. Un balance: Aportes, límites y debate", *Annual review of Sociology*, 34: 499–524.

Valladares, L. (1999), "Programas sociales para los pobres en Brasil. Los casos de Río de Janeiro y Belo Horizonte", In M. Schteingart (ed.), *Políticas sociales para los pobres en América Latina*, Mexico, Miguel Angel Porrua.

Vecinday, L. (2017), "Alteraciones institucionales y tecnológicas de la política de asistencia social en el Uruguay del siglo XXI: ¿hacia un nuevo modelo de gestión social de la pobreza?", *Fronteras*, 10: 55–67.

Weitz-Schapiro, Rebecca (2006), "Partisanship and Protest: The Politics of Workfare Distribution in Argentina", *Latin American Research Review*, 41 (3): 122–148.

Yaschine, Iliana (2015), *¿Oportunidades? Política social y movilidad intergeneracional en México*, El Colegio de México.

CHAPTER 7

The Welfare State and Gender

From a gendered point of view, what we have been discussing in this book is that it does not explain much about the welfare state, as it has left in silence the situation of women, of half the population of the continent, that does not take fully, or even closely, advantage of its benefits. This is because the welfare state is oriented toward the productive sector of society and ignores its reproductive part, although both are constitutive of capitalism (Federici, 2019). In addition, gender "...structures the division between professional and industrial activities, well paid, predominantly masculine, and activities that relate to the 'pink collars' and domestic services, worse paid, predominantly feminine. This results in an economic structure that creates sexual forms of exploitation, marginalization, and destitution" (Fraser, 2005: 26). This means that the social protection system is gender-structured and contributes to the reproduction of a gendered labor market division. Although this book has as its primary objective the discussion of the welfare states of Latin America from a general perspective, it falls into this same pitfall. For this reason, we would like to end by pointing out how future studies of the social protection systems of the continent should be oriented if they intend to go beyond this trap.

At present, an ample literature discusses how most studies on the welfare state are gender-blind. The same goes for the different typologies, as they are all centered on men as the primary providers of the

I. Bizberg, *Four Worlds of the Welfare State in Latin America*,
https://doi.org/10.1007/978-3-031-44420-3_7

household, and thus, as the main beneficiaries of social protection. From a gendered perspective, the social compromise established in almost all countries, even the most developed ones, is a pact between men, who have occupied most of the political, organizational, and associational positions until recently. Consequently, the welfare state defines a division between productive and reproductive labor, between remunerated and non-remunerated work that is not remunerated, and between considered and de-considered jobs. As in all countries, not only do the social protection systems of Latin America have an impact on gender, but they are central to defining the socio-economic situation of women; the welfare states and social policies, in general, operate to preserve the patriarchal type of capitalism; one that is blind to the fact that reproductive labor is concentrated on women (Sainsbury, 1999).

In consequence, the typologies are "...falsely universalizing (implicitly masculinist) analytic frames undergirded almost all comparative studies of welfare states [...] Feminists, from the 1970s on, continually pointed out how the gendered underpinnings of systems of social provision and the specific situations of women were occluded in mainstream analyses" (Orloff, 2009: 319). The central argument of various comparative books (Sainsbury, 1999; O'Connor et al. 1999; Daly, 2020) has been that while certain countries group together when considered from the traditional typologies defined by Esping-Andersen and others, from a gendered perspective there are substantial differences among them. There are sometimes more coincidences with some countries situated outside a specific type.

We must begin our discussion with the affirmation that how the welfare state was constructed to make capitalism viable was to exclude women from the productive sphere and relegate them to the private domain through the reinforcement of the family, the naturalization of care, and child upbringing. Capitalism requires the alienation of the producers from the means of production the products of their work, and, even more fundamentally, from work itself, but it also separates the productive processes from the reproductive processes. This is achieved through both physical separation and their realization by distinct subjects: men usually perform the productive tasks, while women do the reproductive ones, and while the first are paid and salaried, the second are not paid for and taken for granted. To legitimize this development, it had to be accompanied and justified by a cultural process through "...the devalorization and invisibilization of domestic work and its naturalization as the 'work of women'" (Federici, 2019: 20).

However, thanks to the feminist movement and the rise of women to positions of power, "women-friendly" policies have been increasingly implemented in most countries. Although this has been more clearly the case in the more developed countries, they have also been implemented in Latin America. However, the differences between countries may not be so significant. This point of view will be more and more relevant since the feminist movement in Latin America has gained importance during the last decades. It has compelled many of the countries of the continent to open their political systems to the representation of women, to the adoption of women's rights (divorce, abortion, etc.), as well as to the LGBTQ + rights. The pace of this process will probably intensify as women progressively occupy the political and the public sphere and modify the posture of their governments, as we have seen lately in Chile and Colombia, which have adopted parity in politics and an increasingly gendered perspective. At the same time, there is always a possibility of regressions, as in Brazil and Mexico, but also outside Latin America.

As Orloff has written, taking into consideration gender as a foundation of the welfare state is fundamental not only to characterize the social protection system but as "...gendered identities and agency—including orientations to family and employment—are not pre-political, or 'natural'" (Orloff, 2009: 322), it is crucial to start by analyzing how gender is defined in our societies, and how both the productive sphere, as well as the welfare sphere, shape the place of women in society. This is because "...welfare provision, alongside other political and social institutions, is involved in shaping gendered divisions of labor and the preferences, needs, and desires that sustain it" (Orloff, 2009: 322).

This not only means that "gender matters" (Orloff, 2009: 318) but that due to the invisibilization of the complementarity between production and reproduction, it matters even more when we analyze the social protection system; "gender is (in part) constituted by systems of social provision and regulation and, in turn, shapes them" (Orloff, 2009: 318). This is why, although we have carried out a comparative analysis, we cannot stop short of trying to gender it. Although to be consequent with our intent of analyzing the continent's welfare states, we should have begun from this perspective. We wanted to engage in the discussion with the work done by our predecessors, and we thus had to start from the most usual perspective. On the other hand, it was not until relatively recently that studies in Latin America have gendered the subject, most notably by Angulo Salazar and Morvant-Roux, 2018; Molyneux, 2002, 2006 and 2007; Martinez Franzoni, 2008; Blofield and Martinez Franzoni, 2014; Rossel, 2013; Tromben and Podestá, 2019; Llanos and Roza, 2015; Donoso and Valdés, 2007; Lexartza Artza et al., 2019; Altamirano, 2024.

According to Sainsbury, there are two ways in which we can introduce gender into the discussion of the welfare state, the first is "...to build gender into mainstream theoretical frameworks [...] by reconstructing the core ideas and the analytical categories so that they include gender", the second "holds that the mainstream theories are fundamentally lacking, and new frameworks and models must be reformulated [...] which is against a strategy that merely adds on women to frameworks where the male is the norm" (Sainsbury, 1999: 2). According to this author, "Ann Orloff's framework (1993) represents the most systematic effort to build gender into the welfare state regime concept. First, rather than taking the nexus between the market and the state, Orloff refashions this dimension of variation as state, market, and family relations. This modification recognizes the family and women's unpaid work as a variation in social provision. Her second dimension is the pattern of gender stratification produced by entitlements. She distinguishes between gender differentiation and gender inequality. Gender differentiation in entitlements occurs when claims to benefits are based on the traditional division of labor between the sexes. Men receive benefits as family providers and workers, while women claim benefits as wives and mothers. Gender inequality refers to differences in the benefit levels of women and men, often affected by gender differentiation. Benefits tied to participation in the workforce are usually more generous than benefits claimed on the basis of wifely or motherly labor [...] Orloff proposes access to paid work as her third dimension. The fourth dimension of her framework is the capacity to form and maintain an autonomous household" (Sainsbury, 1999: 3).

In contrast to this intent to include gender into the traditional typologies, Sainsbury considers that while the conventional typologies describe one part of the picture, basically one of the formal, male breadwinners, a typology that centers on the role of women as workers and carers has to be radically different. Her strategy is "...to recast the generalizations of the feminist critique into dimensions of variation, constructing two contrasting ideal types: the male-breadwinner and the individual models" (Sainsbury, 1999: 3). She also "...emphasizes the importance of gender and familial ideologies as a key variation. Her framework additionally highlights whether social rights are familiarized or individualized" (Sainsbury, 1999: 4).

The main idea of Sainsbury is that because the welfare states are based on a specific conception of the pretended role of women in society, "...gender policy regimes can be distinguished on the basis of ideologies that describe actual or preferred relations between women and men, principles of entitlement, and policy constructions" (Sainsbury, 1999: 79). She considers that in order "...to overcome these shortcomings we need to focus on how gender is constructed in policies—and regularities in the patterning of gender relations in legislation" (Sainsbury, 1999: 77). She defines three types of social protection systems, depending on the role which is assigned to women, through cultural/ideological models, which translate into specific manners of organizing the economy and society: "The first one, **the male-breadwinner regime** is characterized by a gender ideology of male privilege based on a strict division of labor among the sexes. It assigns different tasks and obligations to women and men, resulting in unequal entitlements in social provision". The second one is the ideology of **the separate gender roles,** which "...emphasizes the significance of differences between the sexes, and social rights are allocated on the basis of these differences ... Women may receive benefits for caring responsibilities irrespective of their marital status. However, the principle of maintenance privileges men on the labor market, and employment and wage policies are oriented to the male worker" (Sainsbury, 1999: 77–9).

The first type of welfare regime ignores the role of women as workers or carers; it is a male breadwinner type, which corresponds to most of the actual systems and the traditional typologies. The second, the separate gender roles regime, explicitly acknowledges the specific tasks that, in most societies, are assigned to women and tries to compensate and equalize the retribution of these tasks to those of men in the productive economy. Social policies in this type tend to favor monetary transfers to women doing care work, either when they do so for their own children or elders. This second ideal type, which Fraser calls the *caregiver parity model*, keeps care work in the family context but seeks to reevaluate it through public assistance, such as caregiver allowances. "It preserves the gendered division of caregiving and breadwinning, but seeks to make gender difference costless and to upgrade women to the status of citizen caregivers" (Borchorst and Siim, 2008: 211).

Table 7.1 Typology of gendered welfare states

	*Male Breadwinner**	*Separate gender roles*/Caregiver Parity***	*Individual earner*/ Universal breadwinner***	*Universal Caregiver Parity***
Life Pattern	Man	Man and woman	Man and woman	Woman
Employment	Priority to men	Priority to men	Aimed at both	Aimed at both
Labor Market	Rigid	Rigid	Flexible	Flexible
Taxation	Familial	Familial	Individual	Individual
Pensions	Contributory	Contributory	Taxpayer	Taxpayer
Caring Work	Unpaid	Paid at home	Paid in and out home	Aimed at both
Entitlements	Assistance	Assistance	Citizenship	Citizenship
Public Services	Scarce	Scarce	Ample	Ample
Parental Leave	Mother	Mother	Mother and father	Mother and father

Source Own elaboration based on Sainsbury (1999) *and Fraser (1994)**

The third type of welfare regime is one that supports women to pursue an individual career, independently of their care obligations or will to pursue them, empowering them to pursue a life independent of their husbands or partners; the model life pattern is that of man, and it seeks to "...turn women into citizen workers like men by, among other things, moving care work from the family into the market and the State" (Borchorst and Siim, 2008: 211). "The ideology of **the individual earner–carer regime** differs fundamentally from the other two in that the preferred relations between women and men are shared roles and obligations, leading to equal rights. Both sexes have entitlements as earners and carers, and policies are structured to enable women to become workers and men to become caregivers". "Social rights and tax obligations are attached to the individual rather than the family" (Sainsbury, 1999: 77–9) (Table 7.1).

We include a fourth type, which Nancy Fraser calls the **universal caregiver model**, defined by an equivalent situation between men and women, not only concerning an individual career **but also regarding care**. This regime is patterned on the life of women and does not "...aim to make women's lives the same as that of men's, but to 'make difference costless'" (Fraser, 1994: 606). Both men and women should dedicate time to care, which ought to be alternated among the couple. In this model, "...many (although not all) women [and men] will follow the current U.S. female practice of alternating spells of full-time employment, spells of full-time care work with part-time employment ... To this end, several major new programs are necessary. One is a program of care-giver allowances to compensate childbearing, child-raising, housework, and other forms of socially necessary domestic labor; the allowances must be sufficiently generous at the full-time rate to support a family, hence equivalent to a breadwinner wage. Also required is a program of work-place reforms, these must facilitate the possibility of combining supported care work with part-time employment and of transitioning between different life states ... Finally ... there must be programs to ensure continuity of all the basic social welfare benefits, including health, unemployment, disability, and retirement insurance" (Fraser, 1994: 606). This regime "...aims at removing gendered segregation by making women's life patterns the norm for both women and men" (Borchorst and Siim, 2008: 211).

In this regime, women should be able to conjugate an individual career and care for their children and elders; women should be able to be "...daughters, daughters in law, and mothers while earning their own living in the paid labor market..." (Hansen, Dahl, and Horn, 2022: 445), and so should men. The question regarding the welfare state is what incentives it should allot to allow women/men to discard (temporarily or totally) the labor of care and to facilitate the equal distribution between men and women of this task that has been traditionally assigned exclusively to women, through an ideological process of naturalization. How does the social protection system reconsider care as a social task rather than an individual one while allowing both men and women to spend care time? This implies creating social structures that help care for children, the sick, and the elder allowing families to depend on the public sphere for part of these tasks and, on the other hand, paying an equivalent amount for care as for a job and considering the years spent on care as part of the time accumulated for pensions for both men and women. This means considering reproductive as equivalent to productive tasks and that care should no longer be the unique responsibility of women, men, or families, an individual question, to become a social one. This configuration would allow both men and women to spend time on care voluntarily and not in a compulsory manner.

The Types of Gendered Welfare States in Latin America

In the following pages, I will analyze to what extent the Latin American countries are applying "woman-friendly" social policies and whether the welfare systems of the different countries can be considered as coming close to one of the four types based on the various dimensions of Table 7.1. The typology has included all the fundamental dimensions to characterize the welfare regimes from a gender perspective. Nonetheless, it is clear that in the case of Latin America, some of them are more significant than others; for example, assistance to low income families is more important than in developed countries, and some, like taxation, are practically inexistent, as the imposition system does not make any difference between families where a man or woman, or both work.

Following this theoretical/typological discussion, we can consider that most Latin American countries have a male breadwinner regime and, to some extent lean towards separate gender roles. Thus, the relationship between all the nations of the continent with the other two types of gendered welfare state lies in the future: some countries are moderately tending to one of the two types. In these two types of regimes, where all the Latin American systems are located, there are two ways of implementing "women-friendly" social policies: one oriented towards the individual earner, the other towards care. One can also analyze to what extent the social protection system facilitates women's careers and makes life easier for women who need to both have a job and care labor, by making public care institutions available. Another significant measure is facilitating and promoting men caring for children and elders, for example, allocating both men and women unexchangeable, paternal/maternal, and elder care leaves. However, we are very far from this situation when one considers that maternal leave in most of the continent is 14 weeks, while paternal leave is just a few days; the imbalance is impressive.

Employment and Labor Market Situation

Before discussing any of the specific policies towards women, whether towards allocating resources to compensate for their care activity, facilitating their integration into the labor market, or easing their possibility of choosing the moment to pursue a career and care, we must analyze the situation in which women are in the different countries. From Fig. 7.1, we can see how two countries, Mexico and Peru, have the lowest participation rates of women in the labor market, around 35%, while three very different countries, Uruguay, Ecuador, and Bolivia, have the highest level of participation, around 50%. While Uruguay has one of the most advanced welfare systems, which we have characterized as socio-corporatist, Bolivia and Ecuador are, on the contrary, two of the most limited and least advanced welfare states. This apparent paradox can be explained by the fact that while social/cultural advancement in Uruguay explains more extensive labor participation of women; it is rather economic needs that pushe women to work in the poorest countries.

In no country in the continent (nor, in fact, in the world) is there wage equality between men and women; however, the gap is most significant in the two countries with the lowest participation, high informality, and poverty: Peru and Mexico. It is also high in Bolivia, although participation is high, which may be a sign that it is more related to the general situation of the labor market in this country, with high informality and poverty, as well as to the general condition of the social protection system we have been discussing in the previous chapters of this book. In the cases of Uruguay and Costa Rica, there is a coincidence between high labor

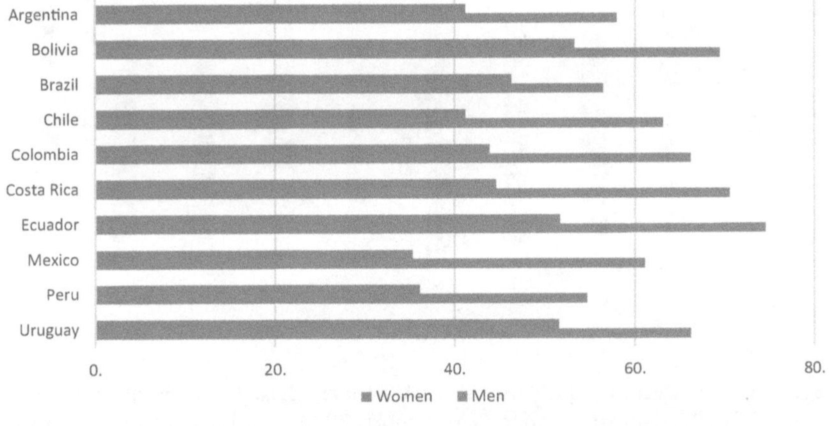

Fig. 7.1 Labor participation rate, 2020 (*Source* Based on CEPAL, 2021 "La autonomía económica de…)

participation and a more universal and generous welfare state, as well as relatively low informality (Fig. 7.2).

Another indicator of the wage gap is women's contribution to each country's total labor income. This is a better indicator of the wage gap, as it is an indicator that incorporates both the actual situation of women in the labor market, their involvement, and their earnings. If women were equally included in the formal labor market and earned the same amount as men, the percentage of women's participation in the labor income would be 50%. But this is not the case for the world as a whole; before the pandemic, the amount gained by women was much lower than that of men: 34%. In the case of Latin America, there are some significant differences between countries. Once again, the South Cone countries, which have a more extensive and, except for Chile, a more generous welfare state, have women earning more than in the rest of the countries of the continent, nearing 40% of men's wages, while in the other countries they are between 30 and 35% (Fig. 7.3).

One explanation for this difference between women's and men's wages is that the former are more prone to be occupied in part-time or informal jobs (Espejo, 2022: 27–8). "This is because women require the flexibility

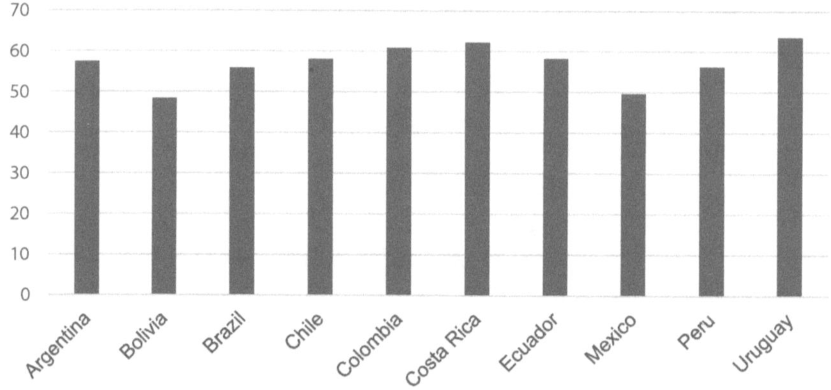

Fig. 7.2 Wage equity for similar work (*Source* Based on World Economic Forum (2022), Global Gender Gap Report 2022, 13 July, https://www.weforum.org/reports/global-gender-gap-report-2022/?DAG=3&gclid=CjwKCAjwm4ukBhAuEiwA0zQxk81ZEG7refMr5SKQ8lWNMf9OlAzq2gR-J2KAswpMoPVIB0XsXtR62xoCn-oQAvD_BwE)

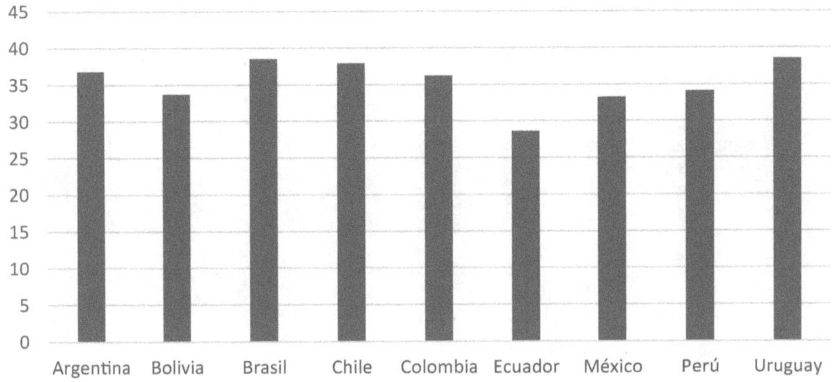

Fig. 7.3 Participation of women in labor income (*Source* World inequality lab database)

of the informal sector to be able to deal with work at home [and thus] ... the chances of being paid or unemployed are different for men and women" (Espejo, 2022: 29). Figure 7.4 is, to a certain extent, proof of this fact: in the first place, in most countries (except Uruguay) the percentage of women in the informal economy is greater than that of men. Where informality of women is higher is again in the countries where participation is lower and social protection systems are the weakest: Peru and Mexico.

In general, women have a higher percentage of unemployment than men. The gaps in unemployment rates between women and men coincide with the rates of informality. The fact that countries with the strongest welfare state have higher unemployment rates than the countries where the structure of the social protection system is weak can be explained by the fact that there is a better register of the rate itself because of ampler social protection; in Colombia and Costa Rica, unemployment is higher than in Peru, Mexico, Ecuador, and Bolivia because it is more carefully calculated. For all countries, the gap in unemployment is probably more significant than what is shown in the figure because in many cases, women are not looking for a job, although they would desire one as they do not have time, and thus are not considered unemployed. In the countries where the unemployment percentage of women is lower than that of men, the situation of women not looking for a job or working informaly is probably accentuated. It also follows that in most countries with

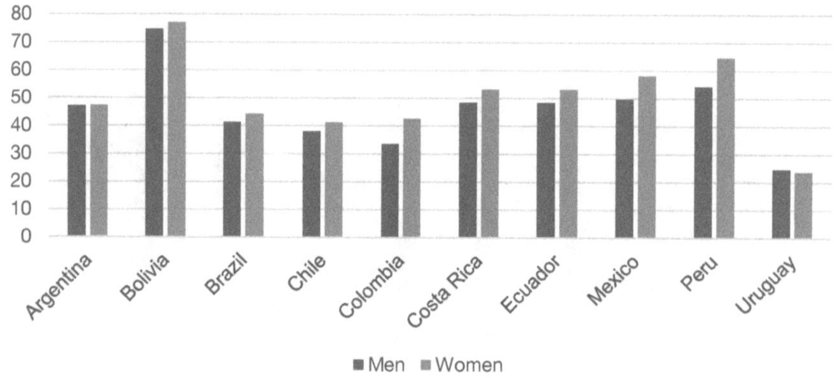

Fig. 7.4 Share of informal employment (urban sectors) (%) (*Source* Based on International Labour Organization (2018), Women and Men in the Informal Economy: a Statistical Picture, Geneve)

a significant inclusion of women in the labor market; the percentages of unemployment are higher in women that in men (Fig. 7.5).

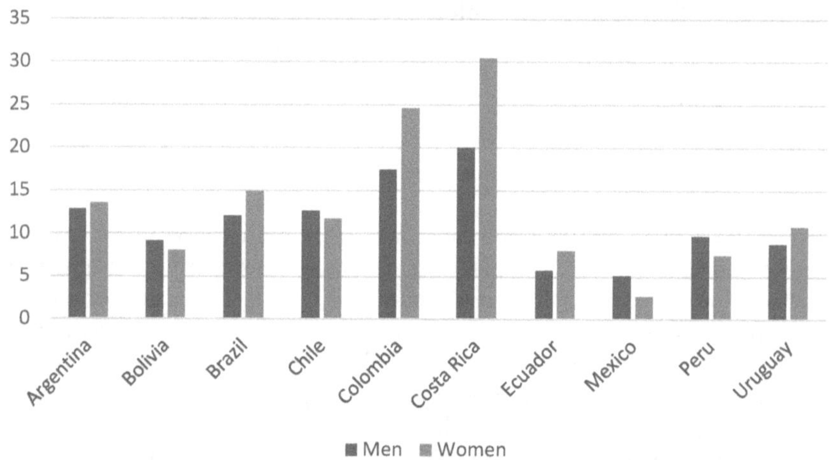

Fig. 7.5 Unemployment rate (2020) (*Source* Based on CEPAL, 2021, "La autonomía económica de...)

A society where women can be independent, where women can choose to pursue a job career or occupy themselves with care, and change from one to the other requires an earner–carer welfare regime. First, it requires a labor market where women have the same possibility of getting a job in the formal sectors and where the salaries of women and men are equal. We have been able to see that the situation of the labor market in all our countries does not seem to be set toward this goal; in fact, we have not registered any policies aimed at moving in that direction; the labor market measures we pointed at the beginning of this chapter defined by the authors we have cited, especially Diane Sainsbury and Nancy Fraser: that the labor market be aimed at both men and women equally, that it be flexible enough to allow women (and men) to take a leave from their jobs to care for their children or elders (or friends), that pensions be based on taxpayer money rather than contributions, and that entitlements be based on citizenship facilitates this latter choice, in addition to specific programs that we will discuss in the last section of this chapter.

According to Fraser, to improve the possibility for women to work, the governments have to stimulate a full employment policy because "...it has the potential to alter the previous imbalance between men's and women's labor-market status" (Sainsbury, 1999: 100). Nonetheless, this is not enough, but what is also needed are labor market reforms that go in the direction of allowing women (and men) to take leave from their jobs without losing them, nor their salaries (which would have to be balanced by the state), nor their privileges linked to seniority, such as the advancing within the internal hierarchy linked with time on the job, including pension rights. Flexibility to allow for "...exit and entry employment without losing security or seniority [...] as well as a program of retraining and job search for those not returning to their old jobs" (Fraser, 1994: 605). The same rights would be allocated to men.

None of the countries of Latin America are advancing towards this scenario. However, in the four countries that we have considered to have either socio-corporatist, statist or even commodified welfare states, the situation of women is more favorable than in the so-called familiarist ones, or Colombia and Mexico.

Next, we will discuss next how the different governments have tried to implement some policies that favor women. However, they are basically assistance-oriented rather than aimed toward increasing freedom. So, while a more women-friendly labor market increases the possibility of employment for women, the existence of public assistance specifically oriented toward women puts more pressure on them. While the first measures tend to undermine gender differences, the second ones strengthen them. Nonetheless, in Latin America, both allow women to live autonomously of their spouses, as independent workers in the first case and as mothers (especially when the mother is alone in raising a family) in the second.

Programs that Promote Separate Gender Roles

Assistance is the degree zero of the social protection system, not only in the case of the non-gendered welfare state but also in the gendered one. This is because assistance to women is allocated based on their being poor mothers, assisting them to take care of their children. In fact, never for taking care of their elders, which is still considered a natural duty of women. Assistance and most conditional monetary transfers demand something from women in the context of their families; they are thus maternalistic and perpetuate the separate roles of women, in contrast to that of men (Jenson and Nagels, 2016: 323). Focused entitlements, such as assistance programs oriented to low income families, as is the case of many of the programs that have been implemented in Latin America, such as *Oportunidades* in Mexico, *Bolsa Familia* in Brazil, the Argentinian *Asignación Universal por Hijo*, are all allocated to women as head of families, or as female members of poor households. Thus, they reproduce gender inequality, and in addition, since they compel women to fulfill specific tasks to continue receiving the aid, "...they depend upon women fulfilling their 'traditional' social roles and responsibilities […] by basing the program on normatively ascribed maternal responsibilities, in effect making transfers conditional on 'good motherhood'. Men are not incorporated in any serious way, and no effort is made to promote the principle that men and women might share responsibility for meeting project goals, let alone for taking an equal share in caring for their children. These programs un-ambiguously rest on normative assumptions concerning 'women's roles' so that the work women undertake in ensuring that children's needs are met, is taken for granted as something that mothers 'do'. The social relations of reproduction therefore remain unproblematized, with the work performed simply naturalized" (Molyneux, 2006: 438).

Most Latin American social policies oriented towards impoverished families (*Oportunidades*, *Bolsa Familia*) impose significant tasks on women; including when they are not the heads of family, as most of the programs impose conditions to receive the aid (Molyneux, 2007, Jenson and Nagels, 2016). In fact, "...several studies indicate that the dynamics of the transfer programs tend to compromise women's time and, therefore, constitute a strong disincentive for their entry into the labor market (Martínez Franzoni & Voreend, 2010, cited by Rossel, 2013: 29) or an "inactivity trap" (Rodríguez Enríquez, 2011, Ibidem). In more general terms, "...the approach tends to reinforce a maternalistic care view of children and to omit the consideration of women as workers or, at least, to give more relevance to their role as mothers rather than as workers" (Rossel, 2013: 29). M. Molineux and Angulo, as well as Morvant-Roux studied the Mexican program *Oportunidades,* which became a flagship for the rest of the programs for the people living in poverty in the continent and was promoted by the international aid agencies, most notably the World Bank, and argues that "...*Oportunidades* represents a novel combination of earlier maternalist social policy approaches with the conditional, co-responsibility models associated with the recent approaches to social welfare and poverty relief endorsed by international policy actors" (Molineux, 2006: 425).

According to Angulo and Morvant-Roux, the assistance programs impose, in an informal manner, additional tasks on women that do not correspond to their care responsibilities nor to their position as recipients of the transfer, far beyond taking the children to school and to clinics for vaccination and check-ups. They intrude even deeper in their control of women's time and intimacy, as they result "... in social dynamics where the borders of co-responsibility become quite weak and gives rise to the monitoring of various spheres of women's daily life and a discipline that touches the time, body and privacy of the holders" (Angulo and Morvant-Roux, 2018:92). In their interviews, these two authors found out that the beneficiaries of the transfers are compelled to help clean the streets and alleys, clinics, and schools in exchange for assistance, something that leads to the substitution of local agents to allocate certain services, and signifies that the debt of the State towards its citizens is transformed into its contrary: a debt of the citizens towards the State. On the other hand, as a result of the interaction between the women who receive the conditional transfers and the operating agents of the program, their intimacy is intruded by the fact that the agents are allowed to visit their homes to check if women are boiling the water, have latrines, take out their garbage, as they are responsible for the health of their children (Angulo and Morvant- Roux, 2018: 98).

In Latin America, most programs oriented to women are conditional and allocated to women as heads of low income families. However, they do not even fulfill to pay women to accomplish their tasks as carers. Although there is a certain level of empowerment of women who receive transfers that allows them to advance towards a separate gender roles model; they are not only maternalistic (Jenson and Nagels, 2016), but in most countries, the allocation is so low that it does not suffice to even come near the possibility of being able to lead an equivalent life to that of an independent earner. Many of them are forced to have a parallel activity, which can only be informal due to their lack of time. Conditional cash transfers certainly ease the economic condition of women, especially of poor indigenous women, and in that respect, "empower" them. Still, they do not make it easier to carry out an individual career. In addition, as we have seen, conditional programs impose on women an extra workload that is not shared by men, irrespective of them having a part-time or full-time job. Thus, one can consider that as they are based on the fact that women maintain their status as mothers or wives and do not reduce their dependence on their husbands responsibilities as heads of a family.

Figure 7.6 shows the proportion of women receiveing transfers and those without income in the countries we analyze. It is interesting to notice how in the countries we have seen have a more structured pension scheme (even though it was not one of the most generous, as in the case of Chile), women get fewer transfers, and at the same time, they are also the ones where fewer women have no income. The countries with more extensive and generous welfare systems and more women-inclusive labor markets are the ones where transfers cover a more reduced part of the population: Argentina (26.8%); Uruguay (16.3%); Chile (26.3%). It is also in these countries where there is a smaller proportion of women that do not have any income: Argentina (19.1%); Uruguay (12.9%); and Chile (18.4%).

On the other hand, where there is a lower structure of contributive pensions, higher informality, and worse paid jobs, there are more women receiving transfers and also not having any income; this is in the countries we identified as having a familiarist welfare system: in Bolivia, 46.5% of women receive some assistance; Colombia and Ecuador are also very high, above 40%; then come Peru and Mexico with around 35%.

Two countries are outliers, concerning what we have seen regarding their social protection systems: Brazil, where the proportion of women receiving economic aid is very high (38%), and Costa Rica (41%). While

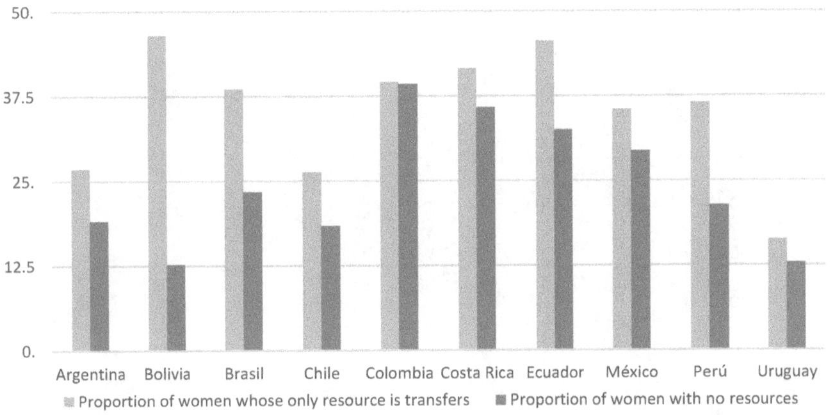

Fig. 7.6 Women with transfers as sole resource and with no income (% of total. 2020) (*Source* Cepalstat)

the percentage of women with no income in Brazil is relatively low, (23.4%), in Colombia and Costa Rica it is high, around 35%. The expansion of *Bolsa Familia* may explain the case of Brazil. In Colombia, Costa Rica, Ecuador and Mexico, women with no resources constitute a significant proportion of the total, they neither work nor receive any aid from the government and are thus absolutely dependent on their spouse or other family members.

In Fig. 7.7, we can see that poverty is higher among women than men, around 10% higher, something that is not exclusive to developing countries and that has to do with the fact that women have more tasks to accomplish than men, that impede their having formal or full-time jobs; they occupy more frequenlty informal jobs, and even when they have a formal job, they are worse paid. Less difference in poverty between men and women than expected may result from a higher poverty level in the country as a whole. The contrary is also true, which may be the reason why, in Uruguay, the country with the lowest level of poverty in Latin America, women's poverty is triple than that of the other countries.

From Table 6.2 (in the previous chapter) listing the different transfer programs oriented to families, that is, basically to women, we can see that the three countries that spend the most relative to GDP are Argentina, Brazil, and Uruguay. These also are the countries in which

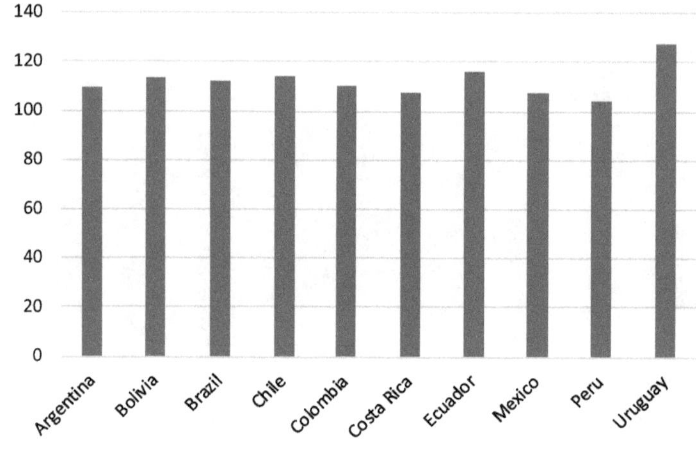

Fig. 7.7 Feminization of poverty index, 2020 (*Source* Cepalstat)

the programs are more generous. The population coverage in these three countries is similar, between 25 and 30%. In the case of Bolivia, the program covers more than half of the population but allocates a reduced amount. However, it is very significant for those who receive it in a low-income country. In other countries, such as in Mexico, *Oportunidades*[1], which was the model for most of the countries of Latin America, it benefited a similar proportion of families; they are all given to women, so one can consider that it is a women-oriented transfer, although as it is conditioned, and forces women to send their children to school, take them to the clinics, it is thus familiarist rather than women-oriented. In Mexico, the amount given was half that of the previous three countries; the same can be said of the program in Colombia.

In this manner, although theoretically and practically, we must consider that assistance programs are a familiarist type of policy that do not

[1] This program was canceled in 2019 by the new government that arrived in power and was substituted by scholarships given to primary school students (universal) and secondary and higher education scholarships oriented to impoverished households. In contrast to the previous program, these scholarships are not allocated directly to mothers, so they are left without income.

promote independence of women from their spouses and families; nevertheless, in Latin America, we should nuance this affirmation. As we saw before, few countries in the continent have a pro-woman labor market, so the situation of women in this respect is quite different from that of some of the more advanced welfare states. Thus, any resources women acquire are significant for their capacity to be more independent. Even programs that demand mothers send their children to school and take them regularly to medical examinations to survey their weight and health favor their autonomy in relation to other women without any financial resources. However, it is evident that they do not promote an independent life because they depend on having a family and children of a certain age.

Only a universal allocation of resources to all citizens without any conditionality on women or men could define a type of welfare state that stimulates either separate gender roles, or individual earner–carer trajectories of women. Sainsbury writes, "...when citizenship or residence is the basis of entitlement, unmarried and married persons have equal rights, and in marriage husband and wife enjoy the same rights. This basis of entitlement also neutralizes a gender differentiation in social rights because paid and unpaid work provides equal benefits" (Sainsbury, 1999: 79). In this way, the less the social protection system of a country is based on programs that are founded on the fact that women have to take care of their children, are lone mothers, live in a low-income family, the more equal the social protection system is, since it considers its citizens as equal, irrespective of gender and promotes equality between women and men.

Programs that Encourage Women to Engage in an Individual Career or Carer Activity

It is the State, on the one hand, and men in co-responsibility, on the other, who can fundamentally facilitate the possibility of women to engage in an individual career. "By assigning more care responsibilities to the State and fathers, these policies have the capacity to reduce gender inequalities in the burden of care" (Blofield and Martinez Franzoni, 2014: 111). According to these authors, "State co-responsibility policies that reconcile work and family entail not only defamiliarization through public provision or subsidies to the private offer of early childhood care and education services, but also, and very importantly, workdays compatible with the full-time working day, to allow mothers and fathers to use these services" (*Ibid*). But maybe, even more fundamentally, it entails co-responsibility of men with the tasks undertaken by women. As Esping-Andersen has written, it requires "...feminizing the male life cycle" (Esping-Andersen, 2009, p. 99, cited by Blofield and Martinez F., 2014: 111), which "generally entails sequential policies that allow reorganizing gender roles between mothers and fathers, without jeopardizing remuneration or job continuity" (*Ibid.*).

However, as we have seen, there have, in reality, been no reforms of the labor market and retirement structure to favor flexibility to allow women and men to decide the periods in their lives when they will work and when they choose to take care of their children or elders, there have certainly been some measures, although limited, that facilitate that women engage in an individual career: among them is the existence of ample public services for caring for early childhood care and pre-primary schooling, as well as, once in primary school, that school hours coincide as much as possible with working hours. Thus, the provision of public services is central for women desiring to pursue a job, as well as for women and/or men who want to combine childbearing or taking care of a family member with working full or part-time or taking leave from their jobs for some time and restarting a few months or years later.

One of the most essential services a State can and should guarantee is nursery and pre-primary schools. Concerning the existence of nursery schools for children under three years old, very few countries cover even a minimum percentage. We have data for some countries; of these, in 2020, the only two where the percentage is over 30% coverage were Brazil (37.5%) and Uruguay (39.7%). The other countries cover a minimal part of these needs, around 5% (in Argentina and Peru), and even lower for all the other countries. There exists a relation between higher coverage in early education and higher integration of women to the labor market on paid jobs: "Uruguay, Brazil, Chile and Ecuador, where more than 20% of children between the ages of 0 and 3 are enrolled in child care, participation of women is over 50%, while, on the contrary, in countries where the percentage of children registered is low, participation in the labor market is more limited (in Mexico[2], for example, less than 10% of children between 0 and 3 years old are registered)" (Altamirano, in press: 54).

Concerning children between 3 and 6 years old, coverage changes significantly. Uruguay and Brazil continue to be the most advanced countries in terms of coverage, while two other countries add up, Costa Rica and Peru, where almost 100% of children are taken care of at day centers. In contrast, in Argentina, Bolivia, and Chile, around 80% of children have a place in pre-primary schooling; finally, the most backward countries, Mexico (71%), Colombia, and far behind, Ecuador (56%) (Fig. 7.8).

Another significant provision facilitating women to pursue a career is the length of school workdays. Mexico has the shortest pre-primary days, from 8:00 am. to 12:30 pm., and a workday extension program, which will probably not be effective because it will depend on a consensus of parents, teachers, and school administrators. Brazil is next, with few working hours a year: 800 hours a year, but a national workday extension program; then Argentina and Colombia, none with enough hours to allow the person taking care of children to cover an entire workday. The countries where this latter is possible are Uruguay, Chile, Bolivia, and Costa Rica, all with around 8 hours of daily schooling (Table 7.2).

Another indicator of how much freedom women have acquired to combine a career with care, in this case of children, is how much time is given for maternal or paternal leave. Most countries in Latin America allot barely the minimum amount of time established by the ILO convention, which is 14 weeks for women and basically nothing for men. The difference between the countries is a few weeks. Those countries that allot more time are, once again, Brazil, Chile, Colombia, and Costa Rica. Although Uruguay only allocates women the minimum of 14 weeks for maternal leave, in this country, and in Chile, both parents can extend the leave for up to 12 weeks in Chile and 18 in Uruguay. This means these two countries have effectively supported women to pursue an autonomous career,

[2] In Mexico, there existed a program that, in 2012, had instituted 9,536 nurseries that cared for nearly 293,000 children (in addition to the official ones administered by the ISSSTE and IMSS), which were coordinated by a social assistance agency, the DIF (National System for the Integral Development of the Family, created in 1977) (Barba, 2021: 100). Melina Altamirano (2024) has found that the number of children concerned was between 327.854 and 452,880 in 2018 (p. 253). In 2022, the government of AMLO canceled this program by a direct allocation of 800 pesos (around 40 US dollars) per month to mothers with children under three years old.

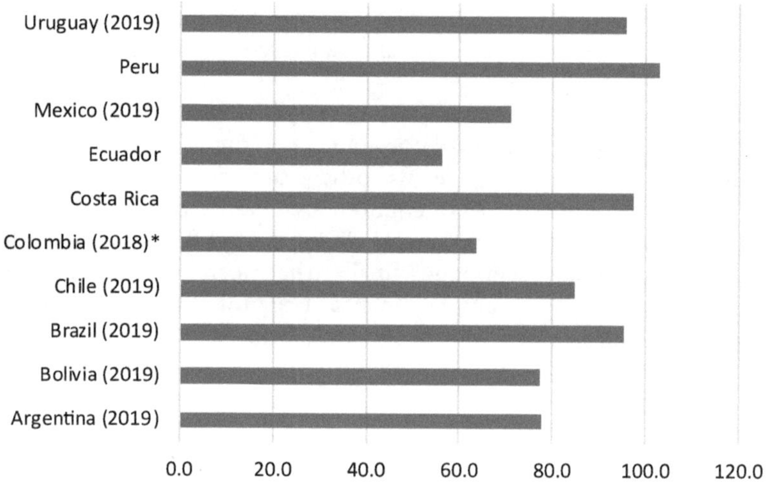

Fig. 7.8 Gross enrollment rate in pre-primary education programs (3–5 years of age), (% of coverage, 2020) (*Source* UNESCO, UNICEF, CEPAL (2022), *La encrucijada de la educación en América Latina y el Caribe*. Informe regional de monitoreo ODSS4-Educación 2030, ONU Source for Colombia: https://compite.com.co/wp-content/uploads/2020/11/CPC_INC_2020_2021_Educacion.pdf * For Colombia we have net preschool coverage, which somewhat underestimates the percentage.

while in the second place we have Brazil and Costa Rica and a bit behind Argentina (Table 7.3).

However, no country has actually established measures that enable men to become caregivers, which would be a condition of the **individual earner–carer regime** defined by Sainsbury. Uruguay and Chile are the countries that, although very far from this goal, have advanced the most; as in the case of Chile, there is the possibility that when the child is 18 weeks, and the mother's leave is over, she can transfer a few weeks to the father, and in Uruguay, it is the father that can take the dispensation. However, to allow men to pursue the same trajectory as women, it would be necessary that the laws regarding parental leave be universal, as well as those allowing men and women to take care of an old or sick person. We can measure inequality between men and women by regarding time spent on non-remunerated care and domestic tasks; in all of Latin

Table 7.2 Class hours and workday extension programs in Latin America

Country	*Hours per year + / Hours per week + +*	*Name of the program*	*Date of creation*
Argentina	n.d/25	*Jornada extendida en las escuelas primarias / Política Nacional para la Ampliación de la Jornada Escolar en el Nivel Primario*	2006/2011
Bolivia	n.d./40	No specific program	–
Brazil	800	*Programa Mais Educação / Novo Mais Educação*	2008/2019
Chile	1,015/38	*Jornada Escolar Completa Diurna*	1997
Colombia	950/25	*Programa Nacional de Día Escolar Completo / Jornada Única*	2014
Costa Rica	1,209	No program	–
Ecuador	n.d	No specific program	–
Mexico	760/22.5	*Programa Escuelas de Tiempo Completo**	2007*
Peru	n.d./30	*Jornada Escolar Completa*	2015
Uruguay	n.d./35	*Escuelas de tiempo completo*	1998

Sources Fernando Ceballos (2022), "Efectos de ampliar la jornada escolar de América Latina: una revisión bibliográfica", Revista Innova Educación, vol.4, núm.3, pp. 152–167; Leandro Bottinelli (2016), ***La ampliación de la jornada escolar en perspectiva. Estudio sobre la situación de la jornada extendida y completa en el nivel primario de Argentina,*** Paraguay, Ministerio de Educación y Deportes; Ministerio de Educación Nacional (2022), "Jornada Única", https://www.mineducacion.gov.co/portal/especiales/Jornada-unica/; + https://data.oecd.org/eduresource/teaching-hours.htm
+ + Different internet pages
* Ended by the government of AMLO in 2022; in the new program the assembly of each school (teachers, directives and parents) decides the length of the workday https://laescuelaesnuestra.sep.gob.mx/storage/recursos/2022/06/flTFKVEccH-06_Guia%20Horario%20Extendido.pdf

America, as in most of the rest of the world, women principally accomplish these tasks, but there are significant disparities between countries. The countries where the difference between the time spent by men and women is less are: Uruguay, Brazil, and Bolivia; in Mexico it is more extensive, followed closely by Ecuador and Costa Rica (Fig. 7.9).

Table 7.3 Maternal and Paternal leaves

Country	*Duration*	
	Maternal leave (weeks)	*Paternal leave (days)*
Argentina	13	2
Bolivia	13	3
Brazil	17	5
Chile	18 + 12*	5 days + 6 weeks**
Colombia	18	8
Costa Rica	17	0
Ecuador	12	10
Mexico	12	5
Peru	14	4
Uruguay	14 + 18*	13 days + 18 weeks**
Minimum duration established by Agreement 183 of the ILO	14	-

Source Varinia Tromben and Andrea Podestá, 2019: 34–35
* Uruguay and Chile are the only countries where fathers can request paternity leave with full rights immediately at the end of the maternity leave for up to 12 weeks in Chile and 18 weeks in Uruguay. In the latter, in addition, some of the parents receive a transfer on their salary. In both countries, the time extends to 24 weeks of the infant's life. They are the only countries that comply with recommendation no. 165 of the ILO
** Optional weeks in substitution of mothers[3]

Movements, Political Rights, and Institutions

We do not pretend to find a causal relation between the social rights of women and social and political participation, but we want to signal some affinities. It is interesting to mention how in the countries we have seen as the most advanced in terms of gender that is Uruguay, Brazil, Chile,

[3] "In Chile, female workers are entitled to twelve full weeks postnatal parental leave once the postnatal period (12 weeks) of maternity leave has ended until the boy or girl reaches 24 weeks of life. The Chilean parental postnatal leave has three modalities: (i) 12 extra weeks for mothers (with a subsidy of 100% of the salary and a maximum of US$3,100 per month); (ii) 18 extra weeks for the part-time mother (with a 50% subsidy and a maximum of $1,550 per month); and (iii) transfer a few weeks to the father (in the full-time mode you can transfer up to a maximum of 6 weeks and in a half-time, a maximum of 12 weeks). This way, once the child has completed 18 weeks, the worker can transfer a few weeks to the father. In Uruguay, the subsidy for newborn care consists of an economic benefit that replaces the salary that is granted to the mother or father (alternately or as they decide) once the maternity leave is over, which makes it possible to reduce half the working day until the child is six months old" (Varinia Tromben and Andrea Podestá, 2019: 34–35)".

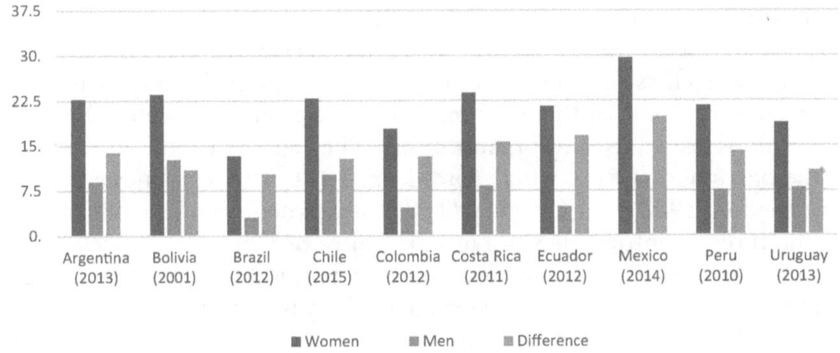

Fig. 7.9 Time spent on non remunerated care or domestic tasks (hours per week) (*Source* Bidegain and Calderón, 2018)

Argentina, and Costa Rica, it coincides with the strength of the feminist movements as well as with the presence of women in the institutional arena.

In the South Cone countries, which suffered dictatorships in the sixties and seventies, we saw the emergence of human rights organizations with significant participation of women, as in Argentina: *Madres de Plaza de Mayo* (1977) and *Abuelas de la Plaza de Mayo*; in Brazil: *Movimiento Femenino por la Amnistía* (1972); in Chile: *Agrupación de Mujeres Democráticas* (1973), *Agrupación de Familiares de Detenidos Desaparecidos* (1975) and *Agrupación de Familiares de Ejecutados Políticos* (1978); and in Uruguay: *Grupo de Madres y Familiares de Procesados por la Justicia Militar* (1982). In countries that went through civil wars or situations that resembled them, significant organizations where women were central also emerged: in Peru (*Coordinadora Nacional de Derechos Humanos*-1985) and Colombia (*Asociación de Familiares de Detenidos Desaparecidos*)[4] (Donoso and Valdés, 2007: 28).

[4] It is interesting to note that in the case of Mexico, where there were no military coups, it is since the recent human rights crisis, from the beginning of the Calderón government in 2006, that numerous women's organizations have been created to look for their disappeared family members, as well as young women organizations began to protest against the increase of violence against women and femicides. This situation will have an impact, as it happened in the other countries we are analyzing, of increasing the demand for other women's rights.

In the institutional arena, in Table 7.4, we have the date of creation of the different institutions that deal with policies oriented explicitly towards women, as well as their institutional hierarchy within the government of each country; some of them are ministries or secretaries, while others are institutions or counseling organizations. The oldest of these institutions is the one in Costa Rica, which dates from 1974. The next ones are those of Ecuador (1980) and Uruguay (1987). Regarding the formal hierarchy of the three countries, they are just Institutes or Councils. However, the INAMU of Costa Rica is presided by a woman, the Minister of the Condition of Women. The countries where a Ministry is in charge are Argentina, Brazil, Chile, and Peru; all date from the nineties. The more recent institution is that of Bolivia, which, although it is not a Ministry, seems to have a more comprehensive reach as its title calls for de-patriarchization rather than just equity. The Argentinian Ministry includes gender diversity. In the countries that seem to be more behind, the institutions charged with the social rights of women are or have been institutes: Mexico, Ecuador, Colombia (until 2023, when Gustavo Petro won the elections and named his vice-president in charge of a Ministry of Equity).

An interesting indicator of where women's rights point to is the practical trajectory that their governments define in order to reach political equality between men and women. According to Donoso and Valdez (2007), there are important differences between the scope of the national projects to reach parity. All the countries propose political participation as an instrument of equality; nonetheless, some are more proactive than others, some propose positive discrimination or quotas within the structure of government; this is notably the case of Colombia (*Acuerdo Nacional por la equidad entre mujeres y hombres- Politica nacional mujeres constructores de paz y desarrollo* (2003), Costa Rica (*Política Nacional para la igualdad y la equidad de género* (2007–2017), Ecuador (*Plan de igualdad de oportunidades* (2005–2009), Mexico (*Programa nacional para la igualdad de oportunidades y no discriminación conrta las mujeres* (2001–2006), Peru, (*Ley y plan nacional de igualdad de oportunidades entre mujeres y varones* (2006–2010), and Uruguay (2007–2011). In addition, some plans propose parity as an explicit instrument for reaching equity: Costa Rica, Ecuador, and Uruguay. And, finally, the plans of Costa Rica and Uruguay are the most ambitious, as they propose concrete steps to reach parity at different levels; something that obviously makes the whole difference (Donoso and Valdez, 2007: 44) (Table 7.5).

Table 7.4 Institutions in charge of the social rights of women

Country	*Institutions in charge of social rights of women*	*Start of activities*	*Previous institutions*
Argentina	Ministerio de las Mujeres, Género y Diversidad	2019	-1992: Consejo Nacional de la Mujer -2017: Instituto Nacional de las Mujeres
Bolivia	Servicio Plurinacional de la Mujer y de la Despatriarcalización	2019	None
Brazil	Ministerio de la Mujer, Familia y Derechos Humanos	2019	-1997: Secretaría de Derechos Humanos de la Presidencia de la República -2015: Ministerio de las Mujeres, de la Igualdad Racial y de los Derechos Humanos
Chile	Ministerio de la Mujer y la Equidad de Género	2016	-1991: Servicio Nacional de la Mujer y la Equidad de Género
Colombia	Ministerio de la Igualdad	2023	-1995: Dirección Nacional para la Equidad de las Mujeres -1999: Consejería Presidencial para la Equidad de la Mujer
Costa Rica	Instituto Nacional de las Mujeres (Presided by a woman Minister)	1998	-1974: Oficina de Programas para la Mujer y la Familia -1976: Centro Nacional para el Mejoramiento de la Mujer y la Familia -1986: Centro Nacional para el Desarrollo de la Mujer y la Familia
Ecuador	Consejo Nacional para la Igualdad de Género	2014	-1980: Oficina Nacional de la Mujer -1987: Dirección Nacional de la Mujer -1997: Consejo Nacional de las Mujeres
México	Instituto Nacional de las Mujeres	2001	-

(continued)

Table 7.4 (continued)

Country	*Institutions in charge of social rights of women*	*Start of activities*	*Previous institutions*
Perú	Ministerio de la Mujer y Poblaciones Vulnerables	2012	-1996: Ministerio de Promoción de la Mujer y del Desarrollo Humano -2002: Ministerio de la Mujer y Desarrollo Social
Uruguay	Instituto Nacional de las Mujeres	2005	-1987: Instituto de la Mujer -1992: Instituto Nacional de la Familia y la Mujer

Source Own elaboration

To conclude our discussion of the relationship between social protection and gender, we want to draw a balance of the situation of women in Latin America. In this case, we have elaborated two indexes; in one, we try to measure the more "women-friendly" labor markets, and in the second, the social support women get from the welfare state in different countries. Although none of our countries near a universal earner-carer or even less a universal caregiver parity regime, some have a more open labor market for women. In contrast, others allow them more independence through rather timid (in most of our countries, compared to the developed countries, especially the northern European ones) social policies that will enable them to pursue an independent career (Fig. 7.10).

The only country in which the two indexes coincide is Uruguay. We have already seen that it has the most advanced welfare system in Latin America. We can consider that Uruguay is the country that comes nearest to an individual earner-carer model, although still very far from it. Both Uruguay and Argentina have socio-corporatist social protection systems, so it is a bit surprising that, in the case of Argentina, the situation of women in the labor market is not as favorable as that of Uruguay, but this may be explained by the economic upturns and downturns this country has been going through for decades but also may be because its welfare state is structured around a male breadwinner model.

The two other countries where there is both a high index of women labor and of gender policies are Brazil and Chile. Brazil has a strong participation of women in the labor market, although it also has a significant wage difference and informality rate difference; thus, women are

Table 7.5 National plans for equality, political representation and parity

Country	*Latest action plan available*	*Proposes political participation as an instrument of equity*	*Proposes establishing an electoral quota law (in countries without quota laws)*	*It raises positive discrimination or quotas within the government structure*	*Explicitly raises parity as an instrument of equity*	*Set concrete goals or deadlines to achieve parity at various levels*
Argentina	Plan for equal opportunities between men and women in the world of work	X				
Bolivia	Plan for equal opportunities between men and women (2004–2006)	X				
Brazil	National policy plan for women (2004–2007)	X				
Chile	Equal opportunities plan (2000–2010) (evaluation of the first phase 2000–2005)	X	X			

(continued)

Table 7.5 (continued)

Country	*Latest action plan available*	*Proposes political participation as an instrument of equity*	*Proposes establishing an electoral quota law (in countries without quota laws)*	*It raises positive discrimination or quotas within the government structure*	*Explicitly raises parity as an instrument of equity*	*Set concrete goals or deadlines to achieve parity at various levels*
Colombia	Within the framework of the National Agreement for equality between women and men – National policy women builders of peace and development (2003)	X		X		
Costa Rica	National policy for gender equality and equity (2007–2017)	X		X	X	X
Ecuador	Equal opportunities plan (2005–2009)	X		X	X	
Mexico	National program for equal opportunities and non-discrimination against women – PROEQUIDAD (2001–2006)	X		X		

Country	*Latest action plan available*	*Proposes political participation as an instrument of equity*	*Proposes establishing an electoral quota law (in countries without quota laws)*	*It raises positive discrimination or quotas within the government structure*	*Explicitly raises parity as an instrument of equity*	*Set concrete goals or deadlines to achieve parity at various levels*
Peru	Law and National Plan for equal opportunities between women and men (2006–010)	X		X		
Uruguay	First national plan for equal opportunities and rights (2007–2011)	X	X	X	X	X

Source Asociación Latinoamericana de Organizaciones de Promoción al Desarrollo A.C. (2007), p. 44

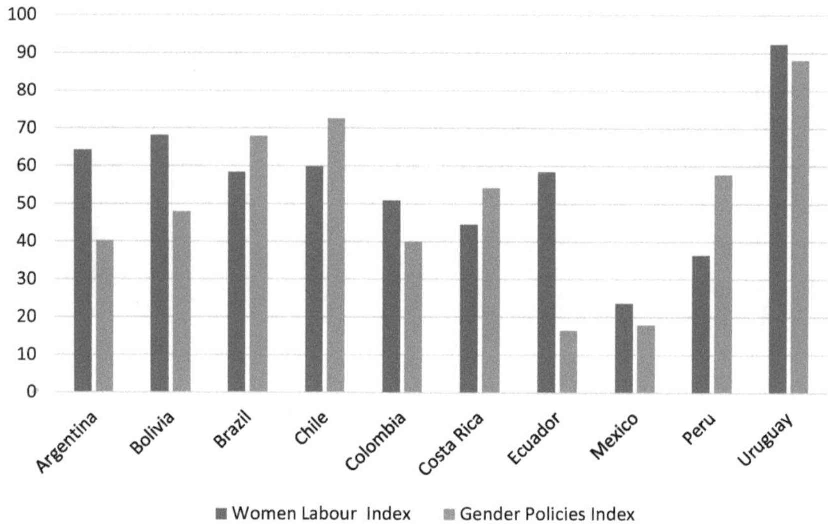

Fig. 7.10 Women labor and gender policies indexes (*Source* Own elaboration)

massively represented in the labor market but in a disadvantageous situation. Gender policies are, in this case, some of the most developed, defined by gross enrollment in pre-primary school, maternal leave, and proportion of time spent in care between men and women, although in this respect, the differences between countries are minimal as men almost do not participate in household activities in the continent.

In the case of Chile, the situation of women in the labor market is similar to that of Brazil concerning participation and wage differences between men and women, excluding the level of informality. This situation is coherent with the characteristics we have found in previous chapters on the commodified character of this country's social protection system. However, regarding public support for women's independence, Chile, Brazil and Uruguay, have the best situation. In fact, Chile, like Uruguay, has the most advanced maternal leave conditions and support for infant schooling.

The situation of Costa Rica is paradoxical with what we know about its welfare system. This country has one of the most universal health and pension schemes in Latin America; it can definitely be considered a very advanced welfare state; some authors have considered it the most progressive due to its homogeneity (it compares more favorably with Uruguay) as this latter is segmented while the Costa Rican is statist, unified, and thus more equitable. However, considering the condition of women, it is not among the highest, as we could have expected: it has a relatively high gender policies index but a rather low women labor index.

We have yet to mention three other countries: Bolivia and Ecuador, and to a certain degree also Colombia, that (surprisingly) have some of the highest woman-friendly labor market indexes, nearing or higher than that of Argentina. As we have seen in the previous chapters of this book, they have a significantly large poor population, 39%, 36%, and 50%, respectively (see Table 6.4 in the previous chapter), a high percentage of informality, and have what we have called familiaristic social protection regimes. This situation pushes women to enter the labor market massively. The difference between the three countries is that the public support index in Bolivia and Colombia is relatively high, while the one in Ecuador is the lowest, Bolivia has effectively customized its educational system to facilitate women to do both: participate in the labor market and take care of their children. Peru has done the same, although there is relatively low participation of women in the labor market; the high index of Peru in public support for women is basically due to the expansion of schooling for children between 3 and 6 years old, which is universal, and the existence of a school workday extension program. In contrast, concerning the policies pursuing autonomy, Bolivia needs to catch up in two indicators that are not in the index but which we have discussed: coverage of enrollment of children in nurseries and pre-primary schools and maternal leave, which is below the minimum duration of the ILO.

The two countries where the situation of women in the labor market is the worst are Mexico and Peru. Mexico is also a country with less public support for women's independence; it is a typical breadwinner model.

References

Altamirano, Melina (2024), *La política del cuidado infantil en México. Políticas públicas y ciudadanía en un régimen de bienestar fragmentado*, Mexico, El Colegio de México.

Angulo Salazar, L. del C. and S., Morvant-Roux, 2018. "Corresponsabilidades de las mujeres beneficiarias del programa Prospera en cuatro regiones de

México: entre la deuda moral y su disciplinamiento", In *Géneros, Revista de investigación y divulgación sobre los estudios de género*, no. 24.

Asociación Latinoamericana de Organizaciones de "Promoción al Desarrollo A.C. (2007), Participación política de las mujeres en América Latina. Informe Regional", Santiago de Chile.

Barba, C. (2021), *El Régimen de Bienestar Mexicano. Inercias, Transformaciones y Desafíos*, Santiago, CEPAL.

Bidegain, Nicole and Coral Calderón (comps.) (2018), *Los cuidados en América Latina y el Caribe*, Santiago de Chile, CEPAL.

Blofield, Merike and Juliana Martínez Franzoni (2014), "Trabajo, familia y cambios en la política pública en América Latina: equidad, maternalismo y corresponsabilidad", *Revista CEPAL*, 114.

Borchorst, Anette and Birte Siim (2008), "Woman-Friendly Policies and State Feminism. Theorizing Scandinavian Gender Equality", *Feminist Theory*, 9 (2): 207–224.

Bottinelli, Leandro (2016), *La ampliación de la jornada escolar en perspectiva. Estudio sobre la situación de la jornada extendida y completa en el nivel primario de Argentina*, Argentina, Dirección Nacional de Información y Estadística Educativa.

Ceballos, Fernando (2022), "Efectos de ampliar la jornada escolar de América Latina: una revisión bibliográfica", *Revista Innova Educación*, 4 (3): 152–167.

Comisión Económica para América Latina y el Caribe (CEPAL) (2021), "La autonomía económica de las mujeres en la recuperación sostenible y con igualdad", *Informe especial COVID-19*, 9, febrero.

Daly, Mary (2020), *Gender Inequality and Welfare States in Europe*, Cheltenham, Edward Elgar.

Donoso, O. Alina and E. Teresa Valdés (2007), "Participación política de las mujeres en América Latina. Informe Regional", Chile, Centro de Estudios para el Desarrollo de la Mujer (CEDEM).

Espejo, A. (2022), "Informalidad laboral en América Latina: propuesta metodológica para su identificación a nivel subnacional", *Documentos de Proyectos (LC/TS.2022/6)*, Santiago, Comisión Económica para América Latina y el Caribe (CEPAL).

Esping-Andersen, Gøsta (2009), *The Incomplete Revolution: Adapting to Women's New Roles*, Cambridge, The Polity Press.

Federici, Silvia (2019), *Le Capitalisme Patriarchal*, La Fabrique Éditions.

Fraser, Nancy (1994), "After the Family Wage. Gender Equity and the Welfare State", *Political Theory*, 22 (4): 591–618.

Fraser, Nancy (2005), *Qu'est-ce que la justice sociale ? Reconnaissance et redistribution*, Paris, La Découverte.

Jenson, Jane and Nora Nagels (2016). "Social Policy Instruments in Motion. Conditional Cash Transfers from Mexico to Peru", *Social Policy & Administration*, 52 (1): 323–342.

Hansen, L. L., H. M. Dahl, and L. Horn (2022), *A Care Crisis in the Nordic Welfare States? Care Work, Gender Equality and Welfare State Sustainability*, Bristol, Policy Press.

Lexartza Artza, Larraitz, María José Chaves Groh, Ana Carcedo Cabañas, and Amelia Sánchez (2019), *La brecha salarial entre hombres y mujeres en América Latina: En el camino hacia la igualdad salarial*, Lima, OIT/Oficina Regional para América Latina y el Caribe.

Llanos, Beatriz and Vivian Roza (2015), *Partidos políticos y paridad: Un desafío de la democracia en América Latina*, Washington, DC, Banco Interamericano de Desarrollo/Instituto Internacional para la Democracia y la Asistencia Electoral.

Martínez Franzoni, Juliana (2008), *¿Arañando bienestar? Trabajo remunerado, protección social y familias en America Central*, Buenos Aires, Consejo Latinoamericano de Ciencias Sociales- CLACSO.

Molyneux, Maxine (2002), "Gender and the Silences of Social Capital: Lessons from Latin America", *Development and Change*, 33 (2): 167–188.

——— (2006), "Mothers at the Service of the New Poverty Agenda: Progresa/ Oportunidades, Mexico's Conditional Transfer Programme", *Social Policy and Administration*, 40 (4): 425–449.

——— (2007), *Change and Continuity in Social Protection in Latin America. Mothers at the Service of the State?*, Switzerland, United Nations Research Institute for Social Development.

O'Connor, Julia S., Ann Shola Orloff, and Sheila Shaver (1999), *States, Markets, Families. Gender, Liberalism and Social Policy in Australia, Canada, Great Britain and the United States*. Cambridge, Cambridge University Press.

Orloff, Ann Shola (2009), "Gendering the Comparative Analysis of Welfare States: An Unfinished Agenda", *Sociological Theory*, 27 (3): 317–343.

Rodríguez Enríquez, Corina (2011), "Programas de transferencias condicionadas de ingreso e igualdad de género. ¿Por dónde anda América Latina?", *Serie Mujer y Desarrollo*, Santiago de Chile, CEPAL.

Rossel, Cecilia (2013), *Políticas para las familias en América Latina: Panorama de políticas de reducción de pobreza y conciliación entre trabajo-familia*, Santiago, CEPAL.

Sainsbury, Diane (ed.) (1999), *Gender and Welfare State Regimes*, Oxford University Press.

Tromben, Varinia and Andrea Podestá (2019), *Las prestaciones familiares públicas en América Latina*, Santiago de Chile, CEPAL.

WEF, World economic forum (2022), *Global Gender Gap Report 2022*, 13 July, https://es.weforum.org/publications/global-gender-gap-report-2022/.

CHAPTER 8

Conclusions

As we have been discussing in this book, the foremost purpose of the welfare state is the self-preservation of society, which, perceived from the point of view of the relation between the State and the population is the entitlement to the debt of the State towards its citizens. The function of social security became necessary when capitalism uprooted the economy from society, when, in terms of Polanyi, it disembodied the economy from the social and domestic realms and made individual lives totally dependent on the market. The social protection system intends to reconstitute social cohesion as far as possible. For Esping Andersen and Théret, the ideal goal of the welfare system is to reconstitute the original relation of man with the economy and society, to re-embed the economy in society. However, because capitalism does not allow the decommodification of the lives of individuals, the welfare state seeks to create a system that would pay the debt of life, and build a universal, unified, generous, and wholly equitable system. Most systems of Latin America are very far from this ideal, although some are closer to it than others.

But while this is the fundamental function of the social protection system, it is not something that appears spontaneously or results from a unilateral action of the State or a single social actor, but, as all social actions, results from a conflict and a temporary compromise that defines how all what society creates is distributed. We have cited Walzer's idea that society makes goods that must then be distributed and that how

I. Bizberg, *Four Worlds of the Welfare State in Latin America*,
https://doi.org/10.1007/978-3-031-44420-3_8

this distribution is achieved is defined by a conflict first and then agreement. This distribution is achieved by institutions created to deliver social goods, institutions that solidify settlements that are nevertheless temporary and are transformed when the social arrangement that created them is modified.

One of these distributive institutions is the welfare state, which allocates the different measures of protection offered by society to its members: unemployment benefits, health services, pensions, and education. Like all social institutions, it is administered and assigns its benefits through a combination of different social actors. Social protection can be assumed by social associations and organizations, by the State in the representation of the social common good, and in the last resource by the families or the market, in a combination of all these four actors.

The welfare state has an impact on four different orders: it has the most direct relationship with individuals and their **families,** protecting them at various stages of life; **politics** is involved as the social security system may serve to legitimize the State, a political regime, or even a specific government; the **economy** is affected with short- and long-term consequences: increased investment in the social protection system permits individuals to save less and consume more, and advances in human capital profits the rise of productivity, added value, and quality of goods in a virtuous cycle, which is the way Fordism functioned; in the present, minimum social protection systems sustain a profit-led economic model, adapted to attract capital in a globalized economy. Finally, every social protection system profits some sectors of society more than others; some are more universal and equitable, and some invest more in pensions, others in health, others still on education or favoring women to pursue an individual existence without relying on their family or partner. The different characteristics a welfare system assumes depend on the force of the diverse actors and their relation to the State, which is also an autonomous actor; in this respect, it is the result of a specific **social compromise**, of a settlement in the distributive conflict between the sectors that constitute a society for what it creates through collective work.

In the book, we propose a typology that describes the situation of social protection in Latin America, a typology that tries to be internally coherent, meaning that the ideal types must be defined based on a single fundamental concept so that each type is as exclusive as possible of the others that intends that there are no overlaps between types. The concepts that base the construction of our ideal types are based on a historical analysis of trajectories of the construction of the welfare states. The ideal types do not coincide with the existing welfare states; as these are hybrids that combine ideal types; the actual social protection systems approach the ideal types to different extents.

We base our typology on the dominant **actor** that administers and apportions social protection. Social protection systems are all a particular combination of State, society, market, and family. We have defined four types of welfare regimes, or four worlds of welfare states as Esping-Andersen has called them: a **statist regime**, where the state is the dominant actor, as the actor that recollects, administers, and allocates pensions and health services. Civil society is a secondary actor, although there may be some organizations or institutions, some of them stemming from the genesis of the social security system, that are active; depending on the efficacy of the services that the State assigns, families are marginal, and so is the market. The **socio-corporatist regime** is defined by the explicit, in some cases majoritarian, administration of the welfare state by society, be it unions or civil society associations; the State complements this regime, and the family and the market usually play a much smaller role. The **commodified regime** is one where the market is dominant, where it sets the pace, and the State, although it may assure a more extended proportion of the population, is in a subsidiary relation to the market; the civil society organizations are usually absent, and the family plays a significant role, complementing the market and State coverage. Finally, in the **familiarist regime**, the domestic order is central, and families have to allot the most considerable portion of the resources, monetary, and care-wise, to the young, the sick, the unemployed, and the elders. The State allocates basic, minimal assistance. Civil society is present only as charity by the churches and NGOs. The market plays a complementary but more prominent role than in the statist or the socio-corporatist regimes, as families have to resort to it to acquire what they need.

In the following chapters, we briefly described the historical trajectory of the welfare systems of Latin America based on this typology. We explained how, in the case of Uruguay, since the mid-nineteenth century, health services were implemented by civil society associations the mutualists, while the State extended them to the rest of the population in the third decade of the twentieth century. In Argentina, the unions created the *Obras Sociales*, which administer health funds and give services to the workers they represent and to all who want to affiliate. In both Uruguay and Argentina, civil society has resisted the frequent attempts by different governments to concentrate these services in the hands of the State or to privatize them.

The Brazilian and Costa Rican cases are the ones that most closely approach a **statist** welfare system, especially concerning the health system. The Costa Rican welfare state has been a centralized and unified social protection system since the forties, the *Caja Costarricense de Seguridad Social* allocates both pensions and health, as well as assistance to the poor and non-contributory pensions. This system was born as a unified statist system. In contrast, the Brazilian social protection system developed first in a segmented manner under the *Estado Novo* in the thirties, and only after the proclamation of the 1988 Constitution did it become a centralized, universal system. In the Brazilian case, the military centralized the numerous pension funds managed by the unions and extended pensions to the peasants. Then, with the return of democracy, the universal *Sistema Único de Saúde* (SUS) was created, administered, and financed by the State with contributions and taxes, although provided by public as well as private hospitals and clinics.

The circumstance of Chile's social protection system can only be understood by the *coup d'Etat* of September 1973, when the military dismantled what was the welfare state most akin to a socio-democratic one in the continent. Before the coup, pensions were comprehensive, and health services were universal and centralized by the State. It was replaced by a **commodified** social protection system, in which pensions were fully capitalized, and health and education depended heavily on the private for-profit sector. Until 2008, the State did not exert a regulatory function; its role was essentially subsidiary; however, since the governments of Lagos and Bachelet, Chile has been advancing towards a more regulated system, less dependent on the market, although pensions continue to be based on capitalization, and the health system is still strongly marketized. It is interesting to note that the opposite is happening in the Brazilian system; a State-led welfare regime is being commodified as the market is taking over more and more space, especially in health.

Familiarist social protection systems have emerged slowly and in a dispersed manner since the beginning of the twentieth century. Their development was neither promoted by social organizations or political parties, or by the State intending to coopt the labor movement as in Mexico, Brazil, and Argentina. The frail industrialization process in Peru, Ecuador, and Bolivia meant a weak labor movement. However, in all three, there existed nuclei of solid labor organizations: in mining, oil, government, primary education, and dockers, which when they protested, pressured the government to give them special conditions. They also received advantages when a new government (democratic or authoritarian) required political support. This resulted in a limited and atomized welfare state. Lately, with the growth of informality and poverty, these countries have expanded non-contributive pensions and enlarged their public health system to dispense minimum services.

We also analyzed the historical trajectories of two hybrid welfare states: Mexico and Colombia. Half of the Mexican population has access to a system that allocates both pensions and health, basically State-controlled; the other half, the informal workers and the people living in poverty, receive assistance and non-contributory pensions. On the other hand, this population formally has access to public health services, which do not apportion regular health services but only emergency ones and to a certain degree, chronic disease care. Consequently, half of the population depends on their resources, which must be bought in the market. In the case of Colombia, beginning with the constitution of 1991, it has achieved to offer health services to practically the totality of the population with both a wholly contributive and a subsidized system. However, significant differences between both systems still exist in terms of quality.

The pandemic the world suffered from 2020 to 2022 put to the test the health systems of Latin America, and notwithstanding the great tragedy it has been for millions of people, something which we are seriously taking into consideration when talking about deaths, has served us, researchers, to evaluate the resilience of the different types of welfare regimes we have defined; it was a single occurrence that befell all countries at the same time. Although the number of deaths each country endured depended greatly on many variables because, as with any **total** event, a war, or a great catastrophe, the pandemic was multifaceted. We, thus, do not pretend to unearth the leading cause that defined the proportion of deaths.

First, for all countries, the number of doctors per person and the percentage of total health expenditure in relation to GDP are negatively correlated with deaths. In contrast, the level of private and out-of-pocket spending is positive. The health systems of Uruguay and Costa Rica, the two most universal, regardless of whether they are socio-corporatist or State-led, had the best performance tackling the crisis; they are the countries with the most doctors and health expenditure. A difference between both is that although Costa Rica has as good a record of past performance as Uruguay and endured the pandemic effectively, it has less infrastructure; this reinforces the idea that a centralized system can spend less and be as efficient.

The social protection systems that did worst were the familiarist ones, those of Peru, Bolivia, Ecuador, and the hybrid Mexico; this is due to the lack of infrastructure, fewer doctors, and lowest public spending, but especially by the fact that these countries have the highest level of out-of-pocket expenses, which mean that the sick waited until the last moment to attend the clinics and hospitals, either because they knew there was no place, because the hospitals were not capable of tackling the number of sick, or because they had the previous experience that any attendance to a health service, even if it is public, is very costly due to lack of medicines and other materials.

The other countries we studied could have done better taking into consideration their infrastructure and past performance. In contrast, Colombia did better that expected. The reason why Argentina, a socio-corporatist system with the highest level of infrastructure and best historical performance, performed worse than Uruguay and Costa Rica can be explained on the one hand by the reduction of public investment in health since the end of the commodity boom, but above all by the inequities of a highly segmented scheme, that concentrates resources and infrastructure in the capital and some other big cities. In this case, we can blame the insufficiency of a socio-corporatist system partly on its corporatist segmentation and its geographical disparities, the centralization of the best health services in Buenos Aires.

A statist health system strongly hit by the epidemic, Brazil also has a relatively high infrastructure level and of past performance. We can also consider that this was the consequence of the reduction of health spending since 2015, but above all to the inequity between private and public care, thus a result of the privatization of a statist system, as well as to geographical disparities, due to a federal system that has both advantages and disadvantages.

The fact that the private sector is so determinant in Chile is why it performed less effectively than expected concerning its infrastructure and past performance. It is a system that has succeeded in achieving universal coverage and has a mid-range health infrastructure but is on the top of our performance index. However, the reason why this country failed to measure up to our benchmark countries is similar to that of Brazil.

Although Chile is a centralized country and does not have the extreme regional inequalities of Brazil, it is comparable in the high level of private and, especially out-of-pocket spending, which is among the highest in the continent. These variables have a robust correlation with deaths during the pandemic.

Colombia is another exception, but that contrasts with the last three cases in that it did better than expected. Its health system has one of the poorest infrastructures and the lowest past performances according to our indexes; only Ecuador, Bolivia, and Peru have a feebler infrastructure and worst performance. Its success may be explained by a health structure that covers almost the entire population, which ensues in a minor level of private expenditure, but above all, out-of-pocket expenses; we have discussed how the latter had a deleterious effect during the pandemic.

In Chapter 6 we try to conclude how the different welfare regimes represent a distinct wage relation, which defines a specific social compromise, as we have discussed. We have seen how one of the most fundamental differences between the socio-corporatist and statist welfare regimes and the commodified and familiarist ones is the level of coverage and benefits of the pension and health schemes. Another difference is that while the first two types of regimes tend towards social protection as a system of rights and entitlements, the second tends toward social security as assistance and concessions.

In most countries in Latin America, the pact between the State, unions, middle classes, and industrialists during the import substitution industrialization period and wage-led growth between the 1930s and the 1970s was substituted by a pact between the State, large multinationals, and the reduced sectors of workers and middle classes that have profited from the opening of the economy, and of profit-led growth. The people with low incomes, an ever-larger proportion of the population generated by the end of the expansion of salarization, growing informality, and higher inequality, have, in a certain sense, become part of the social pact, although in a subordinate position, as a weak "partner", and especially as a socio-political counterpart to organized labor and social movements. This has meant substituting a welfare state based on expanding the contributory pensions and health schemes by another extending minimum assistance and basic health care.

While in all of the Latin American countries, focalized assistance or universal minimal non-contributory programs have been considerably extended since the 1990s, due to the considerable growth of poverty since the implementation of the neo-liberal economic model, our main argument is that, in the commodified and familiarist welfare states, assistance has been accompanied by the dismantling of the contributory social protection schemes and implemented as a **substitute** to social rights; this has served to gain votes, as Garay (2016) has analyzed, and to legitimize governments that nonetheless administer the impoverished populations while the model generates ever-growing inequalities, informalization, and poverty (Lautier, 2004). In the social-corporatist and statist welfare states, assistance programs have not substituted the welfare system-based entitlements; assistance is still, more or less, **complementary** to a social protection system based on rights. This is evident in the manner in which some of the past governments of Brazil (Lula and Dilma 1), Argentina (Kirchner and Fernández de Kirchner), Uruguay (*Frente Amplio*), and Chile (Lagos and Bachelet) have tried to expand coverage of the pension system by formalization through active measures such as tax subsidies, higher scrutiny of informality in enterprises, restrictions on temporary work and subcontracting, unionization of service employees and domestic workers, intending to consolidate them as entitlements.

In most countries, however, we have seen a trend to expand non-contributory pensions, focalized on the impoverished older people, and those who do not attain a minimum at retirement in countries where the pension system is based on capitalization. In some countries, for example, in Mexico, they are universal and are even inscribed in the Constitution. However, they are far from representing a valid substitute to a contributory scheme because they assign a much inferior amount than that of active workers. In conclusion, although some countries have tried to complement assistance with rights, assistance programs continue to expand significantly in all countries, gaining increasingly more ground, accompanied by the financialization of the welfare state (Morvant-Roux, 2019).

Finally, we tried to conclude which of the regimes is more extensive and generous; we have thus elaborated an index combining most of the data concerning pensions, health, and assistance. In Figure 6.1, shows how the social compromise/wage relation in the socio-corporatist and statist welfare states ranks higher than in the other two. As expected, the worst ranked are the familiarist social protection systems, together with Mexico. Chile deserves special mention, as its commodified social protection scheme ranks lower than similar countries in the Southern

Cone and Costa Rica, and it is just above the level of Colombia in terms of extension, generosity, and performance.

The last chapter of the book is based on the corroboration of a series of feminist researchers to the effect that most of the studies and typologies of the welfare state are gender-blind and that most of the discussion has been centered on men as the leading providers of the household; that in consequence, the analyses have been "...falsely universalizing, implicitly patriarchal..." as, in fact, men have been the primary beneficiaries of the welfare state. Indeed, from a gendered perspective, most of the studies on the welfare state ignore that most of the Latin American countries, even the most advanced ones, are fundamentally male breadwinners' systems. However, as we discussed, some of them are leaning towards a separate gender roles regime, and in some, there are limited but real elements that point to an individual earner–carer regime.

From a gendered perspective, the social compromise that has been established in almost all countries, even the most developed ones, has been a pact between men, who have occupied most of the decision-making positions until very recently. If one wants to understand the welfare state from a gendered perspective, a different typology has to be adopted, like the one elaborated by Sainsbury, which defines three types: **the male breadwinner regime**, **the separate gender roles, and the individual earner–carer regime.** The first is based on male privilege, determined by a strict division of labor among the sexes. Men have permanent jobs and the entitlements that result from them, some extend to their families and spouses, who occupy themselves in care in priority. The second one "...emphasizes the significance of differences between the sexes, and social rights are allocated on the basis of these differences ..." (Sainsbury 1999: 78); men get benefits and have entitlements based on work, while women receive benefits and assistance for caring for their children. The third type of welfare state is based on equal obligations of employment and care between women and men, which leads to equal rights. "Both sexes have entitlements as earners and carers, and policies are structured to enable women to become workers and men to become caregivers". The cultural model is that of women (Sainsbury 1999: 79).

The separate gender roles regime is based on assistance to families or directly to women to assist them to dedicate time to care. All of the countries of Latin America have increasingly implemented these types of programs that are allocated to women of low-income families. In this manner, these programs empower women by dealing them their own resources in a separate gender roles scheme, although in a maternalistic role. However, as in most countries the provision is so low that it does not suffice to even come near to the possibility of allowing them to lead an equivalent life to that of an independent earner, many women are forced to have a parallel activity which, in many cases, can only be informal due to their lack of time. On the other hand, still more profoundly, in the case of Latin America, various authors (Molyneux 2007; Angulo and Morvant-Roux 2018) have found that these programs not only dispense compensation to women in order to make them responsible for the care of their children but impose additional responsibilities, both domestic and communal, as well as impinge on the intimacy regarding their use of time and body (Angulo and Morvant-Roux 2018: 92).

The countries with the highest positions in the "woman-friendly" labor market index are Uruguay, Bolivia, Argentina, and Ecuador. While Uruguay has one of the most advanced welfare systems in both general and gendered terms, Bolivia, Ecuador, and Peru have, on the contrary, two of the most limited social protection systems, which we have called familiarist, and some of the highest poverty rates; all of which explains why women are compelled to enter the labor market, even if they also have care responsibilities. Thus, although this indicator should indicate a favorable situation for women in separate gender roles, in the case of these latter countries it signifies a rather adverse situation of women. The case of Argentina is intriguing as it has, like Uruguay, a socio-corporatist welfare regime, a long tradition of labor organization, and a highly contested industrial relations system, but does not have a women's labor index as high as that of its neighbor, as one could have expected.

The situation of Costa Rica is inconsistent with what we have learned about its welfare system. This country has one of the Latin America's most universal health and pension systems; it can be considered a very advanced welfare state; some authors have considered it the most progressive due to its homogeneity and equity compared to the other best systems of Uruguay. However, in terms of the condition of women, it is in the middle echelon regarding women's labor conditions and gender policies. Nonetheless, the country, together with Uruguay, has established the most advanced projects for equality, political representation, and parity, a condition that may change the situation of women in this country in the coming years.

Finally, when we discussed social policies that favor the individual career of women, help women to deal with life and make it easier for women who need to do both work and care labor by making public care institutions available, we concluded that while most Latin American welfare states are nearer either to the male breadwinner or to the separate gender roles type, although some countries have implemented measures that make them moderately lean towards to the individual earner-carer regime, this is especially true of Uruguay and Chile; both have the most advanced maternal leave conditions, as well as institutions to support infant schooling. Nevertheless, in Latin America, no country has established measures that promote or even enable men to devote a more prominent role as caregivers, something which would demand that the possibility and responsibility of parental leave be shared equally by both men and women.

References

Angulo Salazar, L. del C. and S., Morvant-Roux, 2018. "Corresponsabilidades de las mujeres beneficiarias del programa Prospera en cuatro regiones de México: entre la deuda moral y su disciplinamiento", In *Géneros, Revista de investigación y divulgación sobre los estudios de género*, no. 24.

Garay, Candelaria (2016), *Social Policy Expansion in Latin America*, New York, Cambridge University Press.

Lautier, B. (2004), Las políticas sociales en México y Brasil: seguridad social, asistencia, ausencia. In C. Alba and Ilan Bizberg (eds.), *Democracia y Globalización en México y Brasil*, México, Colegio de México.

Molyneux, Maxine (2007), Change and Continuity in Social Protection in Latin America. Mothers at the Service of the State?, Switzerland, United Nations Research Institute for Social Development.

Morvant-Roux, Solène et Mariana Carmona (2019). "La financiarisation des politiques sociales au Mexique: la laborieuse conversion des transferts monétaires en produits financiers", *Mondes en développement*, 2019/4 (n° 188).

Sainsbury, Diane (ed.) (1999), *Gender and Welfare State Regimes*, Oxford University Press.

References

Aglietta, M. (1997), “Postface”, *Régulation et crises du capitalism*, Paris, Odile Jacob.

Alcade-Rabanal, J. E., O. Lazo-González, and G. Nigenda (2011), “Sistema de salud de Perú”, *Revista Salud Pública de México*, 53 (2): 243–254.

Alonso, G. and V. Di Costa (2011), “Cambios y continuidades en la política social argentina, 2003–2010”, In M. Uribe Gómez (coord.), *Los vaivenes de las políticas sociales en Argentina, Colombia, Chile, México y Uruguay*, México, Miguel Ángel Porrúa.

Altamirano, Melina (2024), *La política del cuidado infantil en México. Políticas públicas y ciudadanía en un régimen de bienestar fragmentado*, Mexico, El Colegio de México.

Amable, Bruno (2005), *Les cinq capitalismes*, Paris, Editions du Seuil.

Amaury, Max (2020), *Sembrando Vida: l'organisation communautaire au service du développement national?,* tésis, Faculté des Sciences de la Société / Institut de Démographie et de Socioéconomie.

Angulo Salazar, L. del C. and S., Morvant-Roux, (2018). “Corresponsabilidades de las mujeres beneficiarias del programa Prospera en cuatro regiones de México: entre la deuda moral y su disciplinamiento”, In *Géneros, Revista de investigación y divulgación sobre los estudios de género*, no. 24.

Antía, F., M. Castillo, G. Fuentes, and C. Midaglia (2018), “La renovación del sistema de protección uruguayo”, *Revista Uruguaya De Ciencia Política*, 22 (2): 145–185.

Aran, Daniel and Hernán Laca (2011). “Sistema de salud de Uruguay”, *Revista Salud Pública de México*, 53 (2): 265–274.

I. Bizberg, *Four Worlds of the Welfare State in Latin America*, https://doi.org/10.1007/978-3-031-44420-3

Aranco, Natalia, Marco Stampini, Pablo Ibarrarán, and Nadin Medellín (2018), *Panorama de envejecimiento y dependencia en América Latina y el Caribe*, Banco Interamericano de Desarrollo.

Arcidiácono, P., G. Gamallo, L. Pautassi, and M. Straschnoy (2014), "Brechas de bienestar en el acceso a las prestaciones sociales. Acerca de las asignaciones familiares y la asignación universal por hijo en Argentina", *Universidad y sociedad. Los desafíos de la investigación interdisciplinaria*, Eudeba – Universidad de Buenos Aires, Buenos Aires, pp. 201 a 232.

Arce Ramírez, Claudio Arturo (2020), "Financiamiento y cobertura del Seguro de Salud en Costa Rica: desafíos de un modelo exitoso", *Gestión en salud y seguridad social*, 1 (1): 12–20.

Arenas de Mesa, Alberto (2020), *Los sistemas de pensiones en América Latina: institucionalidad, gasto público y sostenibilidad financiera en tiempos del COVID-19*, Santiago, CEPAL.

Arenas de Mesa, Alberto, María Claudia Llanes, and Fidel Miranda Bravo (2006), "Protección social efectiva, calidad de la cobertura, equidad y efectos distributivos del sistema de pensiones en Chile", *Serie Financiamiento del Desarrollo*, 172, Santiago, CEPAL.

Arrossi, Silvina, et al. (2008), "Social Inequality in Pap Smear Coverage: Identifying Under-Users of Cervical Cancer Screening in Argentina", *Reproductive Health Matters*, 16 (32): 50–58.

Asociación Latinoamericana de Organizaciones de Promoción al Desarrollo A.C. (2007), Participación política de las mujeres en América Latina. Informe Regional, Santiago de Chile.

Aziz, Alberto and Ilan Bizberg (coord.) (2021), *Variedades de Capitalismo en Crisis*, México, El Colegio de México.

Ballesteros, Matías (2014), "Un análisis sobre las desigualdades en el acceso a los servicios de salud en la población adulta urbana de Argentina a partir de datos secundarios", *Documento de Jóvenes Invetigadores*, 41, Buenos Aires, Instituto de Investigaciones Gino Germani, Facultad de Ciencias Sociales, UBA.

Banco Interamericano para el Desarrollo (2015), *El panorama de las pensiones en América Latina y el Caribe*, OCDE/Banco Mundial/BID.

Barba, C. (2021), *El Régimen de Bienestar Mexicano. Inercias, Transformaciones y Desafíos*, Santiago, CEPAL.

Barba Solano, Carlos (2004), "Régimen de bienestar y reforma social en México", CEPAL, *Serie Políticas Sociales*, No. 92.

——— (2021), "El régimen de bienestar mexicano: inercias, transformaciones y desafíos", *serie Estudios y Perspectivas-Sede Subregional de la CEPAL en México*, 191, Ciudad de México, Comisión Económica para América Latina y el Caribe (CEPAL).

Barba Solano, Carlos and Enrique Valencia Lomelí (2015), "Diversos universalismos en las trayectorias de las políticas de salud en México y Brasil", manuscript, Universidad de Guadalajara, presented at the XXXIII international Congress of LASA.

Barbosa, Estela Capelas and Richard Cookson (2019) "Multiple Inequity in Health Care: An Example from Brazil", *Social Science & Medicine*, 228: 1–8.

Becerril-Montekio, Victor M., et al. (2011a). Sistema de salud de Brasil, *Salud Pública de México*, 53 (2): 120–131.

——— (2011b), "Sistema de salud de Chile", *Salud Pública de México*, 53 (2): 132–142.

Belló, Mariana and Victor M. Beccerril-Montekio (2011), "Sistema de Salud de Argentina", *Revista Salud Pública de México*, 53 (2): 96–108.

Bensusán, G. (2010). Políticas laborales en América Latina: orientaciones, alcances y resultados. *Ponencia presentada en CEISAL*. Toulouse, 2 de julio.

——— (2016). Organizing workers in Argentina, Brazil, Chile and Mexico: The Authoritarian-Corporatist Legacy and Old Institutional Designs in a New Context", *Theoretical Inquiries in Law*, 17 (1), 131–161.

Berg, J. (2011), "Laws or Luck? Understanding Rising Formality in Brazil in the 2000s", In S. Lee & D. McCann, *Regulating for Decent Work: New Directions in Labour Market Regulations*. Geneva and Basingstoke, ILO and Palgrave Macmillan.

Berins Collier, R. (1999), *Paths Toward Democracy: The Working Class and Elites in Western Europe and South America*, New York, Cambridge University Press.

Bernal, O. and S. Barbosa (2015), "La nueva reforma a la salud en Colombia: el derecho, el aseguramiento y el sistema de salud", *Revista Salud Pública de México*, 57 (5): 433–440.

Bertoli, Max A. (2020), Sembrando Vida : l'organisation communautaire au service du développement national?, thesis, Faculté des Sciences de la Société / Institut de Démographie et de Socioéconomie.

Bertoli, Max, Solène Morvant Roux, and Ilan Bizberg (2023), "Le rôle de la Société civile dans le déploiement des projets économiques et politiques alternatifs: l'expérience mexicaine de la 4T", *Revue de la Régulation*, 33.

BID (2016), Síntesis-del-programa-Bolsa-Familia-en-Brasil, Banco Interamericano de Desarrollo, https://publications.iadb.org/publications/spanish/viewer/S%C3%ADntesis-del-programa-Bolsa-Familia-en-Brasil.pdf.

Bidegain, Nicole and Coral Calderón (comps.) (2018), *Los cuidados en América Latina y el Caribe*, Santiago de Chile, CEPAL.

Bizberg, Ilán (2019), *Diversity of Capitalisms in Latin America*, London, Palgrave Macmillan.

Bizberg, Ilán and Scott B. Martin. (2012), *El Estado de Bienestar ante la Globalización, el caso de Norteamérica*, México, El Colegio de México.

Blofield, Merike and Juliana Martínez Franzoni (2014), "Trabajo, familia y cambios en la política pública en América Latina: equidad, maternalismo y corresponsabilidad", *Revista CEPAL*, 114.

Borchorst, Anette and Birte Siim (2008), "Woman-Friendly Policies and State Feminism. Theorizing Scandinavian Gender Equality", *Feminist Theory*, 9 (2): 207–224.

Bottinelli, Leandro (2016), *La ampliación de la jornada escolar en perspectiva. Estudio sobre la situación de la jornada extendida y completa en el nivel primario de Argentina*, Argentina, Dirección Nacional de Información y Estadística Educativa.

Boyer, Robert (2004), "New Growth Regimes, But Still Institutional Diversity", *Socio-Economic Review*, 2 (1): 1–32.

——— (2014), "Is More Equality Possible in Latin America? A Challenge in a World of Contrasted but Interdependent Inequality Regimes", *Working Paper Series*, 67, Berlin, desiguALdades.net International Research Network on Interdependent Inequalities in Latin America.net.

——— (2015), *Économie politique des capitalismes, Théorie de la régulation et des crises*, Paris, La Découverte.

Brachet-Márquez, V. (2007), "La reforma de los sistemas de salud y previsión social en México, 1982–2000", In V. Brachet-Márquez (ed.), *Salud pública y regímenes de pensiones en la era neoliberal: Argentina, Brasil, Chile y México*, México, El Colegio de México.

Bresser-Pereira, Luiz Carlos (2024), *The Rise and Fall of Neoliberal Rentier Capitalism. The political economy of New Developmentalism in the 20th and 21st centuries*, Oxford University Press.

Buchanan, Paul G. (1987), "The Varied Faces of Domination: State Terror, Economic Policy, and Social Rupture During the Argentine 'Proceso' 1976–81", *American Journal of Political Science*, 31 (2), May: 336–382.

Caetano, Gerardo (2019), *Historia mínima de Uruguay*, México, El Colegio de México.

Cardona, Álvaro, et al. (1999), *Impacto de la reforma de la seguridad social sobre la organización institucional y la prestación de los servicios de salud en Colombia*, Medellín, Universidad de Antioquia.

Cardoso F. H. and E. Faleto (1969), *Dependencia y desarrollo en América Latina*, México, Siglo XII Editores.

Ceballos, Fernando (2022), "Efectos de ampliar la jornada escolar de América Latina: una revisión bibliográfica", *Revista Innova Educación*, 4 (3): 152–167.

Cecchini S., F. Filgueira and C. Robles. (2013). *Sistemas de protección social en América Latina y el Caribe. Una perspectiva comparada*. Santiago de Chile, CEPAL.

Ceccini, Simone and Rodrigo Martínez (2011), "Protección social inclusiva en América Latina: una mirada integral, un enfoque de derechos", *Libros de la CEPAL*, 111.

CEPAL (2010), *La reacción de los gobiernos de las Américas frente a la crisis internacional: una presentación sintética de las medidas de política anunciadas hasta el 31 de diciembre de 2009*, Santiago de Chile, United Nations.

Cercone, James and José Pacheco Jiménez (2008), "Costa Rica: 'Good Practice' in Expanding Health Care Coverage", In Pablo Gottret, George J. Schieber and Hugh Waters (eds.), *Good Practices in Health Financing: Lessons from Reforms in Low- and Middle Income Countries*, Washington, DC, The World Bank.

Chang Campos, Caroline (2017), "Evolución del sistema de salud de Ecuador. Buenas prácticas y desafíos en su construcción en la última década 2005–2014", *Anales de la Facultad de Medicina,* 78 (4), octubre-diciembre: 452–460.

Comisión Económica para América Latina y el Caribe (CEPAL) (2021), "La autonomía económica de las mujeres en la recuperación sostenible y con igualdad", *Informe especial COVID-19*, 9, febrero.

Comisión Económica para América Latina y el Caribe (CEPAL) (2022), *Panorama social de América Latina 2021*, Santiago de Chile, CEPAL.

Conaghan, Catherine (2011), "Ecuador. Rafael Correa and the Citizen's Revolution", In Steven Levitsky and Kenneth Roberts, *The Resurgence of the Latin American Left*, Baltimore, The Johns Hopkins University Press: 260–282.

Coneval (2021), *Nota técnica sobre la carencia por acceso a los servicios de salud, 2018–2020*, México, Coneval, 5 de agosto.

Consejo Nacional de Políticas Sociales (2011), *La reforma social. Hacia una nueva matriz de protección social del Uruguay*, Presidencia de la República Oriental de Uruguay.

Crouch, Colin (1993), *Industrial Relations and European State Traditions*, Oxford, Clarendon Press.

Cruz-Martínez, Gibrán (2021), "Mapping Welfare State Development in (post) Neoliberal Latin America", *Social Indicators Research,* 157: 175–201.

Daly, Mary (2020), *Gender Inequality and Welfare States in Europe*, Cheltenham, Edward Elgar.

De Melo, M. A. (1997), "As reformas constitucionais e a previdência social 1993–1996", In E. Diniz and S. de Azevedo (eds.), *Reforma do estado e democracia no Brasil*, Brasília, Editora Universidade de Brasília.

Delamata, G. (2008), "Luchas sociales, gobierno y Estado durante la presidencia de Néstor Kirchner", In *Desempenho de Governos Progressistas no Cone Sul: Agendas Alternativas ao Neoliberalismo,* Rio de Janeiro, Edicoes IUPERJ: 128–146.

Donoso, O. Alina and E. Teresa Valdés (2007), "Participación política de las mujeres en América Latina. Informe Regional", Chile, Centro de Estudios para el Desarrollo de la Mujer (CEDEM).

Dowbor, L. (2009), "La crisis no es solo económica. Reflexiones globales y algunos apuntes sobre Brasil", *Nueva Sociedad*, 224: 179–200.

Draibe, S. M. (1997). "A política social na américa latina: o que ensinam as experiências recentes de reformas?", In E. Diniz and S. de Azevedo (eds.), *Reforma do Estado e Democracia no Brasil*, Brasilia, UnBenap.

ECLAC (2011), *Statistical Yearbook for Latin America and the Caribbean*, Santiago de Chile: United Nations.

——— (2021), *2020 Preliminary Overview of the Economies of Latin America and the Caribbean,* Santiago, United Nations.

Espejo, A. (2022), "Informalidad laboral en América Latina: propuesta metodológica para su identificación a nivel subnacional", *Documentos de Proyectos (LC/TS.2022/6)*, Santiago, Comisión Económica para América Latina y el Caribe (CEPAL).

Esping-Andersen, Gøsta (1990), *The Three Worlds of Welfare Capitalism*, Princeton, New Jersey, Princeton University Press.

——— (2009), *The Incomplete Revolution: Adapting to Women's New Roles*, Cambridge, The Polity Press.

Etchemendy, S. and R. B. Collier (2007), "Down But Not Out: Union Resurgence and Segmented Neocorporatism in Argentina (2003–2007)", *Politics & Society*, 7 (3): 363–401.

FCEA, Universidad de la República de Uruguay (2022), "¿Qué sabemos sobre los programas de transferencias no contributivas en Uruguay? Una síntesis de resultados de investigación disponibles sobre el PANES, AFAM-PE y TUS", *Universidad de la República de Uruguay,* 13 May, https://fcea.udelar.edu.uy/blog/6007-13-05-2022-que-sabemos-sobre-los-programas-de-transferencias-no-contributivas-en-uruguay-una-sintesis-de-resultados-de-investigacion-disponibles-sobre-el-panes-afam-pe-y-tus.html.

Federici, Silvia (2019), *Le Capitalisme Patriarchal*, La Fabrique Éditions.

Ffrench-Davis, R. (2008), *Chile entre el neoliberalismo y el crecimiento con equidad. Reformas y políticas económicas desde 1973*, Santiago, J. C. Saenz Editor.

Figueireido Santos, J. A. (2018), "Classe Social, território e desigualdade de saúde no Brasil", *Saúde Soc.*, 27 (2): 556–572.

Filgueira, Fernando (1995), "A Century of Social Welfare in Uruguay", *Democracy and Social Policy Series*, Working Paper #5 - Spring.

Franceschelli, I., and L. Ronconi (2005). "Clientelism, Workfare and the Emergence of the *piqueteros* in Argentina", LACEA Conference, http://depot.gdnet.org/gdnshare/pdf2/gdn_library/annual_conferences/seventh_annual_conference/Franceschelli_Parallel_1_4.pdf. Accessed 15 May 2018.

Fraser, Nancy (1994), "After the Family Wage. Gender Equity and the Welfare State", *Political Theory*, 22 (4): 591–618.

——— (2005), *Qu'est-ce que la justice sociale ? Reconnaissance et redistribution*, Paris, La Découverte.

Fundación de Investigaciones Económicas Latinoamericanas (FIEL) (2007), *La Desigualdad en la Salud*, Buenos Aires, Fundación de Investigaciones Económicas latinoamericanas.

Garay, Candelaria (2016), *Social Policy Expansion in Latin America*, New York, Cambridge University Press.

Garay, Oscar Ernesto (2017), "Obras Sociales: Antecedentes", *Diccionario Enciclopédico de la Legislación Sanitaria Argentina*, https://salud.gob.ar/dels/entradas/obras-sociales-antecedentes#:~:text=Las%20obras%20sociales%20tomaron%20un,con%20la%20revoluci%C3%B3n%20de%201943.

García Repetto, Ulises (2011), *El financiamiento de la seguridad social en el Uruguay (1896–2008): una aproximación a su análisis en el largo plazo*, Uruguay, Universidad de la República, consultado el 26 de mayo de 2020 en: https://www.colibri.udelar.edu.uy/jspui/bitstream/20.500.12008/4207/5/dt-09-11.pdf.

Garcia-Subirats, I., et al. (2014), "Determinantes del uso de distintos niveles asistenciales en el Sistema General de Seguridad Social en Salud y Sistema Único de Salud en Colombia y Brasil", *Gaceta Sanitaria*, 28 (6): 480–488.

Giedion, Ursula and Manuela Villar Uribe (2009), "Colombia's Universal Health Insurance System", *Health Affairs*, 28 (3): 853–863.

González, Fernan E. (1997), *Para leer la política: ensayos de historia política colombiana*, Bogotá, Cinep/Programa por la paz.

Gordon, R. Sara (1999), "Del universalismo estratificado a los programas focalizados", In Martha Schteingart (ed.), *Políticas sociales para los pobres en América Latina*, México, Editorial Porrúa.

Gragnolati, Michele, Magnus Lindelow, and Bernard Couttolenc (2013), *Twenty Years of Health System Reform in Brazil: An Assessment of the Sistema Único de Saúde*, Directions in Development, Washington, DC, World Bank.

Graham, Carol (1998), *Private Markets for Public Good. Raising the Stakes in Economic Reform*, Washington, DC, Brookings Institution Press.

Guerrero, Ramiro, Ana Isabel Gallego, Victor Becerril-Montekio, and Johanna Vazquez (2011), "Sistema de salud de Colombia", *Salud Pública de México*, 53 (2): 144–155.

Gutierrez Arriola, A. (2002), *México dentro de las reformas a los sistemas de salud y seguridad social en América Latina*, México, Siglo XXI.

Haggard, Stephan and Robert R. Kaufman (2008), *Development, Democracy, and Welfare States*, Princeton, Princeton University Press.

Hansen, L. L., H. M. Dahl, and L. Horn (2022), *A Care Crisis in the Nordic Welfare States? Care Work, Gender Equality and Welfare State Sustainability*, Bristol, Policy Press.

Harada, Y. and H. Tohyama (2011), "Asian Capitalisms: Institutional Configurations and Firm Heterogeneity", In R. Boyer, H. Uemura and A. Isogai, *Diversity and Transformations of Asian Capitalisms*, London and New York, Routledge, 243–263.

Herrera, S. Gonzalo (2015), "Situación, estrategia y contexto de los sindicatos en el Ecuador", *Documentos de Trabajo*, 51, CLACSO, Buenos Aires.

Huber E. and J. D. Stephens (2001), *Development and Crisis of the Welfare State: Parties and Policies in Global Markets*, University of Chicago Press.

——— (2012). *Democracy and the Left*, Chicago and London, The University of Chicago Press.

Instituto de Estudios Avanzados para el Desarrollo (2022), "Informalidad, principal característica del empleo en Bolivia", https://www.inesad.edu.bo/2022/05/04/informalidad-la-principal-caracteristica-del-empleo-en-bolivia/.

International Labor Organization (2018), *Women and Men in the Informal Economy: A Statistical Picture*, Geneva, ILO.

International Trade Administration (2021), "Chile-Healthcare", 16 January, https://www.trade.gov/knowledge-product/chile-healthcare.

Jaccoud, L. (2019). "Democracia, desarrollo y reconfiguraciones de la protección social en Brasil", in Alberto Aziz and Ilan Bizberg (coord.), *Variedades de Capitalismo en Crisis*, México, El Colegio de México.

Jaldín, Juan Pablo, et al. (2019), "Health Expenditure Based on Insurance Type and Age Prior to Implementation of Universal Health Coverage in Bolivia", *Gaceta Médica Boliviana*, 42 (2): 194–197.

Jaramillo-Molina, Máximo Ernesto (2022a), "Primero los menos pobres: Política social, desigualdad y pobreza durante el sexenio de López Obrador", *Análisis Plural*, 1.

Jaramillo-Molina, Máximo Ernesto (2022b), "Los más pobres al final: cuatro años de política social de la 4T", *Nexos*, Mayo.

Jenson, Jane and Nora Nagels (2016). "Social Policy Instruments in Motion. Conditional Cash Transfers from Mexico to Peru", *Social Policy & Administration*, 52 (1): 323–342.

Jiménez Barbosa, Wilson Giovanni, et al. (2017), "Transformaciones del Sistema de Salud Ecuatoriano", *Revista Universidad y salud*, 19 (1): 126–139.

Korpi, Walter (2006), "Power Resources and Employer-Centered Approaches in Explanations of Welfare States and Varieties of Capitalism: Protagonists, Consenters, and Antagonists", *World Politics*, 58 (2): 167–206.

Kostzer, D. (2008), "Argentina: A Case Study on the Plan Jefes y Jefas de Hogar Desocupados, or the Employment Road to Economic Recovery", *Working Paper No. 534*. Annandale-on-Hudson, New York, The Levy Economics Institute of Bard College.

Krugman, Paul (2007), *L'Amerique que nous voulons*, París, Flammarion.

Lanzilotta, Bibiana (2009), "El empleo por cuenta propia y la cobertura de seguridad social en Uruguay" In Fabio Bertranou, *Trabajadores independientes y protección social en América Latina,* Montevideo, OIT.

Larrañaga, O., D. Contreras, and G. Cabezas (2015), *Políticas contra la pobreza: de Chile Solidario al Ingreso Ético Familiar*, Santiago, PNUD.

Laurell, Asa Cristina (1997), *La reforma contra la salud y la seguridad social*, México, Era-FES.

Lautier, B. (2004), Las políticas sociales en México y Brasil: seguridad social, asistencia, ausencia. In C. Alba and Ilan Bizberg (eds.), *Democracia y Globalización en México y Brasil*, México, Colegio de México.

——— (2007), "Les politiques sociales au Brésil durant le gouvernment de Lula: aumône d'Etat ou droits sociaux?", *Problèmes d'Amérique latine*, 63: 51–76.

——— (2009), "Gouvernement moral des pauvres et dépolitisation des politiques publiques en Amérique Latine", In N. Borgeaud-Garciandía, B. Lautier, R. Peñafiel, and A. Tizziani (ed.), *Penser le politique en Amérique Latine*, Paris, Karthala, 19–36.

Lechevalier, Sébastien (2015), "L'hétérogénéité croissante des entreprises au sein du capitalisme japonais: Faits, causes, conséquences, et implications", In Akinori Isogai, Hiroyasu Uemura y Robert Boyer (dir.), *Capitalismes asiatiques: Diversité et transformations*, Presses Universitaires de Rennes.

Ledo, C. and R. Soria (2011), "Sistema de Salud de Bolivia", *Revista Salud Pública de México,* 53(2): 109–119.

Lexartza Artza, Larraitz, María José Chaves Groh, Ana Carcedo Cabañas, and Amelia Sánchez (2019), *La brecha salarial entre hombres y mujeres en América Latina: En el camino hacia la igualdad salarial*, Lima, OIT/Oficina Regional para América Latina y el Caribe.

Lindblom, C. (1977), *Politics and Markets: The World's Political-Economic Systems*, New York, Basic Books.

Llanos, Beatriz and Vivian Roza (2015), *Partidos políticos y paridad: Un desafío de la democracia en América Latina*, Washington, DC, Banco Interamericano de Desarrollo/Instituto Internacional para la Democracia y la Asistencia Electoral.

Lobato, L. and L. Burlandy (2000), *The Context and Process of Health Care Reform in Brazil. In Reshaping Health Care in Latin America*, Ottawa, International Development Research Centre.

Lucio R., N. Villacré, and R. Henríquez (2011), "Sistema de salud de Ecuador", *Salud Publica*, 53(supl 2): 177–187.

Malloy, James. (1979), *The Politics of Social Security in Brazil*, Pittsburgh, University of Pittsburgh Press.

Marcel, Mario and Alberto Arenas de Mesa (1991), "Reformas a la seguridad social en Chile", *series de Monografías*, 5, Washington, DC, Banco Interamericano de Desarrollo.

Marco Navarro, Flavia (2012), *El Bono Juancito Pinto del Estado Plurinacional de Bolivia Programas de transferencias monetarias e infancia*, Santiago: CEPAL.

Mares, Isabela (2001), "Firms and the Welfare State: When, Why, and How Does Social Policy Matter to Employers?", In P. Hall and D. Soskice (ed.), *Varieties of Capitalism: The Institutional Foundations of Comparative Advantage*, Oxford University Press, 184–212.

Martínez Franzoni, Juliana (2007), *Regímenes de bienestar en América Latina*, Madrid, Fundación Carolina.

——— (2008), *¿Arañando bienestar? Trabajo remunerado, protección social y familias en America Central*, Buenos Aires, Consejo Latinoamericano de Ciencias Sociales- CLACSO.

Martínez Franzoni, Juliana, and Diego Sánchez-Ancochea (2016), *The Quest for Universal Social Policy in the South: Actors, Ideas and Architectures*, Cambridge: Cambridge University Press.

——— (2018), "Undoing Segmentation? Latin American Health Care Policy During the Economic Boom", *Social Policy & Administration*, Special number on Latin America Press, 52 (6): 1181–1200.

Martínez Franzoni, Juliana, Maxine Molyneux and Diego Sánchez-Ancochea (2009), "Latin American Capitalism: Economic and Social Policy in Transition", *Economy and Society*, 38 (1): 1–16.

Maurizio, R. (2014), "Labour Formalization and Declining Inequality in Argentina and Brazil in the 2000s: A Dynamic Approach", *ILO Research Paper*, 9, Geneva.

Mendizábal Lozano, Gregorio (2002), *Historia de la Salud Pública en Bolivia. De las Juntas de sanidad a los Directorios locales de salud*, La Paz, Organización Panamericana de la Salud/Organización Mundial de la Salud, https://elagoraasociacioncivil.files.wordpress.com/2015/09/historia-de-la-salud-pc3bablica-en-bolivia.pdf.

Mesa-Lago, Carmelo (1978), *Social Security in Latin America. Pressure Groups, Stratification and Inequality*, Pittsburgh, University of Pittsburgh Press.

——— (1985), *El Desarrollo de la seguridad social en América Latina*, Santiago de Chile, Organización de las Naciones Unidas.

——— (1994), *Changing Social Security in Latin America. Towards Alleviating the Social Costs of Economic Reform*, Boulder and London, Lynne Rienner Publishers.

——— (2001), "Privatización del sistema de pensiones en América Latina", In Rolando Franco (coord.), *Sociología del desarrollo, políticas sociales y democracia*, México, Siglo XXI-CEPAL, 218–237.

——— (2008), *Reassembling Social Security. A Survey of Pensions and Health Care Reforms in Latin America*, Oxford, Oxford University Press.

——— (2008), "Un reto de Iberoamérica en el siglo XXI: La extensión de la cobertura de la seguridad social", *América Latina Hoy*, 48: 67–81.

——— (2009), *Efectos de la crisis global sobre la seguridad social de salud y pensiones en América Latina y el Caribe y recomendaciones de políticas. Serie Políticas Sociales*, Santiago de Chile, CEPAL.

Mezones-Holguín, Edward, et al. (2019), "Cobertura de aseguramiento en salud: el caso peruano desde la ley de aseguramiento universal", *Revista Peruana de Medicina Experimental y Salud Pública*, 36 (2): 196–206.

Ministerio de Salud (2020), *A tu servicio,* Uruguay, Recuperado de https://atuservicio.msp.gub.uy/.

Minteguiaga, A. (2013), "Política y políticas sociales en el ecuador reciente: dificultades asociadas a la salida del ciclo neoliberal", *Revista De Ciencias Sociales*, 135–136: 45–58.

Miranda Gutiérrez, Guido (1993), "La salud en la historia de Costa Rica", en *La seguridad social en Costa Rica*, México, Conferencia Interamericana de Seguridad Social.

Molina, Gloria and Claudia Milena Rodríguez (2007), "Participación pública y privada en la provisión de servicios de salud, Colombia 2006", *Investigación y educación en Enfermería,* Medellín, 25 (1), marzo: 90–99.

Molyneux, M. and Diego Sánchez-Ancochea (2009), Latin American Capitalism: Economic and Social Policy in Transition, *Economy and Society*, 38 (1) February.

Molyneux, Maxine (2002), "Gender and the Silences of Social Capital: Lessons from Latin America", *Development and Change*, 33 (2): 167–188.

——— (2006), "Mothers at the Service of the New Poverty Agenda: Progresa/Oportunidades, Mexico's Conditional Transfer Programme", *Social Policy and Administration*, 40 (4): 425–449.

——— (2007), *Change and Continuity in Social Protection in Latin America. Mothers at the Service of the State?,* Switzerland, United Nations Research Institute for Social Development.

Monterrey Arce, J. (2013), *Sistemas de protección social en América Latina y el Caribe: Estado Plurinacional de Bolivia*, Santiago, CEPAL.

Morvant-Roux, Solène et Mariana Carmona (2019). "La financiarisation des politiques sociales au Mexique: la laborieuse conversion des transferts monétaires en produits financiers", *Mondes en développement*, 2019/4 (n° 188).

Munck, R. (2004), "Introduction", *Latin American Perspectives*, 31 (4): 3–20.

Murillo, M. V. (2000), "From Populism to Neoliberalism: Labor Unions and Market Reforms in Latin America", *World Politics*, 52 (2): 135–174.

Noy, Shiri (2017), "Healthy Targets? World Bank Projects and Targeted Health Programs and Policies in Costa Rica, Argentina, and Peru, 1980–2005", *Oxford Development Studies*, 46 (2): 164–183.

Núñez, Alicia and Chunhuei Chi (2013). Equity in Health Utilization in Chile, *International Journal for Equity in Health*, 12 (58): 1–16.

Ocampo, José Antonio and Natalie Gómez-Arteaga (2016), "Social Protection Systems in Latin America: An Assessment", *Extension of Social Security series*, 52, Geneva, International Labour Office, Social Protection Department (SOCPRO).

O'Connor, Julia S., Ann Shola Orloff, and Sheila Shaver (1999). *States, Markets, Families. Gender, Liberalism and Social Policy in Australia, Canada, Great Britain and the United States.* Cambridge, Cambridge University Press.

OECD (2016), *Estudios de la OCDE sobre los Sistemas de Salud: México*, OECD.

OIT (2018), *Presente y futuro de la protección social en América Latina y el Caribe*, Lima.

——— (2021a), *Panorama de la Protección Social en América Latina y el Caribe: Avances y retrocesos durante la pandemia*, octubre.

——— (2021b), *Panorama laboral 2021: América Latina y el Caribe*, Lima.

Olesker, Daniel (2015), "Igualdad, universalidad y accesibilidad: economía política y estado actual de la política pública de salud", In Miguel Fernández Galeano, et al., *Economía, política y economía política para el Acceso y la cobertura Universal de Salud en Uruguay*, Montevideo, Organización Panamericana de la Salud.

Orloff, Ann Shola (2009), "Gendering the Comparative Analysis of Welfare States: An Unfinished Agenda", *Sociological Theory*, 27 (3): 317–343.

Palacios, Alfredo, Natalia Espinola, and Carlos Rojas-Roque (2020). "Need and Inequality in the use of Health Care Services in a Fragmented and Decentralized Health System: Evidence for Argentina", *International Journal for Equity in Health*, 19 (67): 1–14.

Palomino, H. and D. Trajtemberg (2006), Una nueva dinámica de las relaciones laborales y la negociación colectiva en la Argentina. *Revista de Trabajo*, 2 (3): 47–68.

Palomino, H. (2000), "Los sindicatos en la Argentina contemporánea", *Nueva Sociedad*, 169: 121–134.

Papadoupulos, J. (1999), Igualdad, democracia y políticas sociales. In M. Schteingart (ed.), *Políticas sociales para los pobres en América Latina*, México, Miguel Ángel Porrúa.

Paraguay, Ministerio de Educación y Deportes; Ministerio de Educación Nacional (2022), "Jornada Única", https://www.mineducacion.gov.co/portal/especiales/Jornada-unica/.

Paz y Miño Cepeda, J. J. (2002), *Revolución Juliana. Nación, Ejército y burocracia*, Quito, Ediciones Abya-Yala.

Pérez Patón, Roberto (1951), *El seguro social en Bolivia*, La Paz, Editorial Centenario.

Polanyi, K. (1944), *The Great Transformation.* New York, Farrar & Rinehart (cited from the french translation of 1983: Paris: Editions Gallimard).

Poltronieri, Jorge (2011), *Proyecto de Investigación Estructuras de la Opinión Pública, Comunicado de prensa, Encuesta de opinión pública XXIII – Panorama global,* San José, Costa Rica, CIRMA/UCR, 45 pages.

Ponce, G. Jaime (1969), "El sindicalismo boliviano: resumen histórico y perspectivas actuales", *Desarrollo económico,* 9 (33) , abril-junio: 3–31.

Pribble, Jennifer (2013), *Welfare and Party Politics in Latin America,* Cambridge University Press.

Raczynski, D. (1995), "Focalizacion de programas sociales: Lecciones de la experiencia chilena", In Pizarro, C., Raczynski, D. and Vial, J., Eds., Politicas economicas y sociales en el Chile democratico, CIEPLAN-UNICEF, Santiago.

Repetto, Fabián and Fernanda Potenza dal Masetto (2011). Protección social en la Argentina, *Serie Politicas Sociales,* Santiago de Chile, *CEPAL.* no. 174, 68 pages.

Rial, Juan (1984), "Nivel de vida y calidad de vidad de las clases subalternas de Montevideo al finalizar el siglo XIX e inicios del XX", *Informes #22,* Montevideo, CIESU/PISPAL.

Riesco, M. (2009), "El modelo social chileno comienza a cambiar", *Revista Internacional del Trabajo,* 128 (3): 311–330.

Robles Farías, Claudia (2013a). *Sistemas de protección social en América Latina y el Caribe, Chile.* Santiago, Cepal, 52 pages.

——— (2013b), "Los desafíos de la protección social en un país de renta alta: el caso chileno", *Revista Uruguaya de Ciencia Política,* 22: 94–112.

Robles, Claudia (2011), *El sistema de protección social de Chile: Una mirada desde la igualdad.*

Rodríguez Enríquez, Corina (2011), "Programas de transferencias condicionadas de ingreso e igualdad de género. ¿Por dónde anda América Latina?", *Serie Mujer y Desarrollo,* Santiago de Chile, CEPAL.

Román, Isabel (2010), "Sustentabilidad de los programas de transferencias condicionadas: la experiencia del Instituto Mixto de Ayuda Social y 'Avancemos' en Costa Rica", *Serie Políticas Sociales,* 160, Santiago de Chile, CEPAL.

Rossel, Cecilia (2013), *Políticas para las familias en América Latina: Panorama de políticas de reducción de pobreza y conciliación entre trabajo-familia,* Santiago, CEPAL.

Ruezga Barba, Antonio (1994), *La Seguridad Social en Colombia,* México D.F., Conferencia Interamericana de Seguridad Social.

——— (2009), *Seguridad social. Una visión latinoamericana,* México, Centro Interamericano de Estudios de Seguridad Social.

Ruiz Gómez Fernando, Teana Zapata Jaramillo, and Liz Garavito Beltrán (2013), "Colombian health care system: results on equity for five health dimensions, 2003–2008", *Revista Panamericana de Salud Pública,* 33 (2): 107–115.

Sáenz, María del Rosario, et al. (2011). Sistema de salud de Costa Rica, *Salud Publica,* 53 (2), 156–167.

Sainsbury, Diane (ed.) (1999), *Gender and Welfare State Regimes*, Oxford University Press.

Sánchez-Talanquer, Mariano, et al. (2020). *La respuesta de México al COVID: estudio de caso*, Institute for Global Health Sciences-UCSF, Santiago.

Santos Basso, Óscar Luis (2005), "Costa Rica" en *El nuevo derecho de las pensiones en América Latina*, México, Instituto de Investigaciones Jurídicas, Centro Interamericano de Estudios de Seguridad Social: 187–253.

Schneider E., D. Sarnak, D. Squires, A. Sharnav and M. Doty (2017), *Mirror, mirror 2017: International Comparison Reflects Flaws and Opportunities for Better Health Care*, New York, Commonwealth Fund.

Schulthess, Walter and Gustavo Demarco (1993), *Argentina: Evolución del Sistema Nacional de Previsión Social y propuesta de reforma*, Santiago, CEPAL/PNUD.

Stepan, A. (1985), "State Power in the Southern Cone of Latin America", In P. B. Evans, D. Rueschemeyer, and T. Skocpol (eds.), *Bringing the State Back In*. Cambridge, Cambridge University Press.

Stockhammer, E. (2011), "Crecimiento basado en los salarios: introducción", *Boletín Internacional de Investigación Sindical*, 3 (2): 183–208.

Svampa, M., & Pereyra, S. (2005), "La experiencia piquetera", *Medio Ambiente y Urbanización*, 61 (1): 97–112.

Tassara, Carlo (2015), "Políticas públicas de protección social y lucha contra la pobreza en Colombia: logros y desafíos", *Papel Político*, 2 (20): 233–351.

Tavares Soares, L. (1999), "Política Social y pobreza. Brasil en el contexto latinoamericano", In M. Schteingart (ed.), *Políticas sociales para los pobres en América Latina*, México, Miguel Angel Porrua.

Théret, Bruno (1995), *L'État, la finance et le social. Souveraineté nationale et construction européenne*, Paris, La Découverte.

——— (1998), "L'État-providence à l'épreuve des comparaisons internationales", In Philippe Auvergnon, Philippe Martin, Patrick Rozenblatt et Michèle Tallard (eds.), *L'État à l'épreuve du social*, París, Editions Syllepses.

——— (2011), "The Place of National Systems of Social Protection and Political Representation in Socio-Economic Regulation: A Morphogenetic Structuralist View on Institutional Change in Comparative Perspective with Special References to Japan and France", *Evolutionary Institutional Economic Review*, 7 (2): 177–208.

Tikkanen, Roosa, et al. (2020), *International Profiles of Health Care Systems*, The Commonwealth Fund, 228 pages, https://www.commonwealthfund.org/sites/default/files/2020-12/International_Profiles_of_Health_Care_Systems_Dec2020.pdf.

Titmuss, R. (2018), *Essays on the Welfare State (Reissue)*, Bristol University Press.

Tromben, Varinia and Andrea Podestá (2019), *Las prestaciones familiares públicas en América Latina*, Santiago de Chile, CEPAL.

Turcotte, Sylvain F. (2012), "Reforma a los fondos de pensiones públicos e integración económica", In Ilán Bizberg and Scott Martin (eds.), *El Estado de bienestar ante la globalización, el caso de Norteamérica,* México, El Colegio de México, 257–286.

Universidad Andrés Bello (2011), El impacto del Auge a 10 años de su creación, https://www.ispandresbello.cl/el-impacto-del-auge-a-10-anos-de-su-creacion/.

Uribe Gómez, Mónica María (2009), *La contienda por las reformas del sistema de salud en Colombia (1990–2006),* tesis, El Colegio de México.

Valencia Lomelí, E. (2008), "Las transferencias monetarias condicionadas como política social en América latina. Un balance: Aportes, límites y debate", *Annual review of Sociology*, 34: 499–524.

Valencia Lomelí, Enrique, Darcy Tetreault Weber, and David Foust Rodríguez (2013), *Sistemas de protección social en América Latina y el Caribe: México,* Santiago de Chile, CEPAL.

Valladares, L. (1999), "Programas sociales para los pobres en Brasil. Los casos de Río de Janeiro y Belo Horizonte", In M. Schteingart (ed.), *Políticas sociales para los pobres en América Latina*, Mexico, Miguel Angel Porrua.

Vargas, Juan Rafael and Jorine Muiser (2013). "Promoting Universal Financial Protection: A Policy Analysis of Universal Health Coverage in Costa Rica (1940–2000)", *Health Research Policy and Systems*, 11–28, http://www.health-policy-systems.com/content/11/1/28.

Vásquez, Felipe, Guillermo Paraje, and Manuel Estay (2013), "Income-Related Inequality in Health and Health Care Utilization in Chile, 2000–2009", *Revista Panamericana de Salud Pública*, 33 (2): 98–106.

Vecinday, L. (2017), "Alteraciones institucionales y tecnológicas de la política de asistencia social en el Uruguay del siglo XXI: ¿hacia un nuevo modelo de gestión social de la pobreza?", *Fronteras*, 10: 55–67.

Velasco, Margarita (1997), "El reto de la salud para Todos", *Pobreza y Capital Humano*, Quito, Secretaría Técnica del Frente Social, 23–58.

Walzer, Michael (1983), "Spheres of Justice: A Defence of Pluralism and Equality," *Philosophy*, 59 (229): 413–415.

WEF, World economic forum (2022), https://es.weforum.org/publications/global-gender-gap-report-2022/.

Weiss, J. H. (1983), "Origins of the French Welfare State: Poor Relief in the Third Republic, 1871–1914". In *French Historical Studies*, Duke University Press, 13 (1): 47–78.

Weitz-Schapiro, Rebecca (2006), "Partisanship and Protest: The Politics of Workfare Distribution in Argentina", *Latin American Research Review*, 41 (3): 122–148.

Werneck Vianna, Maria Lúcia Teixeira and M. L. Texeira (1997). "A americanização (perversa) da seguridade social no Brasil", In Eli Diniz and Sergio de Azevedo (ed.), *Reforma do estado e democracia no Brasil: dilemas e perspectivas*, Brasília, Editora Universidade de Brasília, 295–348.

World Bank/OECD (2015), *Panorama de las pensiones: América Latina y el Caribe*, Banco Mundial y Banco Interamericano para el Desarrollo.

Yaschine, Iliana (2015), *¿Oportunidades? Política social y movilidad intergeneracional en México*, El Colegio de México.

Ycaza, Patricio (1991), *Historia del Movimiento Obrero*, Quito, CEDIME.

Zapata, Francisco (1992), "Transición democrática y sindicalismo en Chile", *Foro Internacional*, 32 (5): 702–721.

——— (1993), *Autonomía y subordinación en el sindicalismo latinoamericano*, México, Fondo de Cultura Económica-El Colegio de México.

Zúñiga Romero, Marjorie (2012), "El declive del sindicalismo en Colombia y sus consecuencias frente al conflicto colectivo", *Revista de Derecho*, julio: 189–213.

Index

A

Abuelas de la Plaza de Mayo, 231

Acuerdo Nacional por la equidad entre mujeres y hombres- Politica nacional mujeres constructores de paz y desarrollo, 232

*Administración de Servicios de Salud del Estado (*ASSE), 132

Administradoras de Fondos de Jubilaciones y Pensiones (AFJP), 69

Administradoras de Fondos de Pensiones (AFP), 95

affirmative action, 14

Agrupación de Familiares de Detenidos Desaparecidos, 231

Agrupación de Familiares de Ejecutados Políticos, 231

Agrupación de Mujeres Democráticas, 231

Alessandri Palma, Arturo, 92

Alfonsín, Raúl, 68

*Alianza Popular Revolucionaria Americana (*APRA), 103, 107

Allende, Salvador, 93

anarcho-syndicalism, 64, 65, 103

Argentina, 13, 15, 18, 25, 28–31, 33–35, 37, 39, 47, 52, 53, 56, 63–65, 67, 68, 71, 76–78, 96, 100, 106, 111, 114, 129–131, 135, 137, 147, 150, 152, 159, 165, 167, 169, 174–177, 179, 183–185, 190, 193, 194, 198, 199, 222, 223, 226–235, 239, 245, 246, 248, 250, 252

Argentine Regional Workers' Confederation (CORA), 64

Asamblea Nacional Permanente de Organizaciones Sindicales (ANPOS), 106

Asignaciones Familiares-Plan de Equidad (AFAM-PE), 194

Asignación Universal por Hijo para Protección Social, 70, 185, 186

Asociación de Familiares de Detenidos Desaparecidos, 231

Asociación Española, 57

I. Bizberg, *Four Worlds of the Welfare State in Latin America*,
https://doi.org/10.1007/978-3-031-44420-3

assistance, 13, 18, 19, 29, 30, 36, 40, 43, 47, 55, 66, 70–72, 75, 76, 94, 103, 104, 110, 142, 166, 173, 176, 180–182, 184, 185, 189–191, 193–197, 199, 200, 211, 214, 219, 220, 224, 247, 249, 250
assistance for women, 19, 189, 211, 219–222, 225, 251
assistance-oriented model, 7, 8, 19, 27, 190, 219, 220
assistance welfare regime, 19, 180, 185, 193, 211, 214, 245, 249
Avancemos, 196
Avellaneda, Nicolás, 63
Ávila Camacho, Manuel, 113

B
Bachelet, Michelle, 28–30, 98, 189, 246, 250
Banco de Previsión Social, 59, 61
Bancos de Ahorro, 67
Barba, Carlos, 10, 23, 35–39, 41, 117
base organizations, 25
Battle y Ordoñez, José, 55
Belaúnde Terry, Fernando, 107
Beveridgean regime, 37, 52
Bismarck, Otto von, 34, 165
Bismarkian welfare state model, 34, 52
Blancos, 56
blue-collar, 54, 92, 101, 103–105, 107
Bolivia, 13, 16–18, 27, 28, 33, 38, 90, 96, 101, 104–107, 110, 111, 128, 129, 140, 142, 143, 145, 147, 149, 155, 157–159, 169, 173, 185, 215, 217, 222, 224, 227, 229, 232, 239, 246, 249, 252
Bolsa Familia, 29, 185, 186, 193, 197, 220, 223
Bolsonaro, Jair, 153, 159, 198
Bono de Desarrollo Humano, 185, 187
Bono Juancito Pinto, 185, 186
Bono Madre Niño-Niña Juana Azurduy, 185, 186
Bono Solidario (Bonosol), 107
Boric, Gabriel, 171, 189
Brazil, 13, 15, 17, 18, 25, 28, 30, 31, 33, 34, 38, 39, 43, 47, 53, 56, 67, 68, 77, 82, 84, 100, 106, 111, 113, 114, 121, 128–131, 136–138, 140, 141, 144, 147, 150, 152, 154, 156, 158, 160, 165, 167, 169, 174, 175, 177, 178, 182–186, 190, 193, 196, 197, 199, 209, 220, 223, 226, 227, 229–235, 238, 246, 248, 250
Buenos Aires, 64, 66, 70, 152, 248

C
Caetano, Gerardo, 55, 56, 61
Caixa de Aposentadoria e Pensões (CAP), 78
Caja Costarricense de Seguridad Social (CCSS), 72, 136, 246
Caja de Jubilaciones y Pensiones de Empleados y Obreros de Servicios Públicos, 54
Caja de Seguro y Ahorro Obrero, 102
Caja Nacional de Previsión Social, 119
Calderón Guardia, Rafael Ángel, 73
Cárdenas, Lázaro, 113
Cardoso, F.H. and Falleto, E., 166
caregiver parity regime, 211
Casa de Galicia, 57, 58
Catastrophic Expense Protection Fund, 117
Catastrophic Illness Coverage Program, 109, 143
Catavi mine, 105

Catholic Church, 73
Catholic ideology, 73
Central Obrera Boliviana (COB), 106
Central Única de Trabajadores (CNT), 194
Centro Asistencial del Sindicato Médico, 57
charity, 245
Chicago Boys, 95
Chile, 95, 96, 99, 100, 105, 111, 114, 128–130, 140, 141, 147, 150, 154, 156, 157, 159, 167, 169, 171, 174, 175, 178, 183, 184, 189, 193, 222, 226–228, 230, 231, 234, 238, 246, 248–250, 253
Christian Democratic Party, 90
Círculo Católico de Obreros, 57
Círculo Católico de Paso Molino, 57
Círculo Napolitano, 57
citizen rights, 195
Civil Service and Education fund, 58
clientelist relations, 36
Cochabamba, 105
Código de Seguridad Social, 109
Collor de Melo, 197
Colombia, 13, 17, 28, 29, 32, 91, 96, 110, 112, 116, 117, 120, 128–130, 138, 145, 147, 150, 155, 157, 158, 167, 169, 171, 173, 174, 176, 178, 185, 209, 219, 224, 227, 231, 232, 239, 247, 249, 251
Colorados, 56
Comités Locales de Administración de Salud(CLAS), 108
commodification, 34, 99, 138, 189
commodified Systems, 138
commodified welfare state, 178
commodity boom, 8, 16, 17, 26, 127, 128, 136, 140, 142, 158, 199, 248
commodity super cycle, 30, 32
communism, 65, 73, 93, 94, 100
compromis social, 11, 51
Concertación de Partidos por la Democracia, 28
conditioned monetary transfers, 220
Confederación de Cámaras Industriales de los Estados Unidos Mexicanos (Concamin), 118
Confederación de Cámaras Nacionales de Comercio de los Estados Unidos Mexicanos (Concanaco), 118
Confederación de Obreros Bolivianos, 100
Confederación de Trabajadores de Argentina, 71
Confederación de Trabajadores de Chile (CTCh), 93
Confederación de Trabajadores de Colombia, 119
Confederación de Trabajadores del Perú (CTP), 107
Confederación de Trabajadores de México (CTM), 91, 118
Confederación General del Trabajo (CGT), 65
Confederación General de Trabajadores (CGT), 64
Confederación Obrera Argentina (COA), 64
Confederación Regional Obrera Mexicana (CROM), 118
Confederación Revolucionaria de Obreros y Campesinos (CROC), 118
Confederation of Bolivian Workers (CSTB), 105
Consejo Nacional para la Igualdad de Género, 233
conservatist-corporatist model, 34
conservative modernization, 34, 53, 78

conservatives, 27
contributory pensions system, 8
contributory services, 144, 145
convertibility monetary model, 29
Coordinadora Nacional de Derechos Humanos, 231
Cordobazo, 67
corporatist state, 7
corporatist-statist welfare state, 169, 250
Correa, Rafael, 28, 101, 108–110
Costa Rica, 8, 15, 17, 18, 25, 27, 33, 39, 43, 47, 53, 72, 76, 77, 100, 127–131, 136, 137, 143, 145, 147, 153, 157, 158, 160, 169, 171, 173–175, 178, 183–185, 187, 190, 193, 195, 199, 217, 222, 227, 229–233, 236, 238, 247, 248, 252
coup d'Etat, 246

D
decommodification, 4, 5, 35, 40, 42, 181
De la Rúa, Fernanado, 70
demercantilization, 3
democratic hypothesis, 25–27
Democratic Party, 27, 90
democratic transition, 29, 32, 53, 81, 83, 107, 108, 131, 140, 167
Desnutrición Cero, 185
development model, 30
distributive community, 9
distributive conflict, 9, 10, 18, 244
domestic order, 2, 47, 245
Domestic Service and Old Age Pensions, 58

E
economic order, 2
Ecuador, 13, 16–18, 27, 28, 32, 33, 38, 90, 101, 104, 106, 109–111, 128, 129, 140, 142, 143, 145, 147, 149, 155, 157–159, 172, 173, 176, 178, 184, 185, 215, 217, 222, 226, 227, 229, 230, 232, 233, 236, 239, 246, 248, 249, 252
egalitarian universalism, 39
electoral competition, 26, 27, 32, 93, 199
elites, 25, 55
Entidades Promotoras de Salud (EPS), 121
Equipos Básicos de Atención Integral en Salud (EBAIS), 76
Equity Plan (EP), 194
Escuelas de tiempo completo, 229
España Mutualista, 57
Esping-Andersen, Gøsta, 3–5, 10, 14, 23, 24, 28, 32–37, 40, 42, 45, 175, 180, 181, 208, 226, 245
Estado Novo, 72, 78, 79, 246
exclusionary regime, 38, 39
export-oriented capitalism, 32
external market-based economy, 31

F
familiarist welfare state, 17, 171, 175, 178, 184, 250
Federación Obrera de La Paz, 101
Federación Obrera del Trabajo, 101
Federación Obrera Internacional, 101
Federación Sindical de Trabajadores Mineros de Bolivia (FSTMB), 105
Fernández de Kirchner, Cristina, 28, 29, 32, 69, 71, 135, 193, 250
Filgueira, Fernando, 23, 35, 36, 38, 39, 41, 54–59, 99, 133, 135, 183

Fondo de Desarrollo Social y Asignaciones Familiares (FODESAF), 75, 77
Fondo de Pensión al Servicio Civil, 54
Fondo de Solidaridad y Garantía, 121
Fondo Nacional de Salud (FONASA) [Chile], 62, 97, 132, 137–139, 155
Fordism, 11, 18, 165–167, 244
formalization of employment, 195
formal workers, 30, 115, 139, 142
Free Choice Modality, 140
Frei, Arturo, 93
French Mutual Aid Society, 64
French Regulation School (FRS), 168
French Third Republic, 4
Frente Amplio, 61, 62, 194, 250
Frente Nacional, 120
Frondizi, Arturo, 66
Fundo de Garantia do Tempo de Serviçio (FGTS), 79
Fund of Public Employees and Journalists, 92

G
Gaitán, Eliéser, 119, 120
Garay, Candelaria, 23, 26–30, 199, 250
gender, 19, 39, 207–211, 214, 220, 222, 225, 226, 230, 232, 234, 238, 251–253
gender differentiation, 14, 210, 225
gender inequality, 14, 210, 220
gender parity, 212
General Agreement on Tariffs and Trade (GATT), 114
generational transmission of poverty, 40
Grupo de Madres y Familiares de Procesados por la Justicia Militar, 231
Gualberto Villarroel, 105
Guerra del Chaco, 102
guilds, 34

H
Haggard, S. and Kaufman, R., 8, 23–25, 30, 32, 61, 66, 74, 75, 93, 121–123
Health Insurance for a New Generation, 117
health scheme, 17, 28, 35, 44, 62, 83, 96, 100, 108, 112, 118, 128, 130, 141, 172, 176, 200, 249
Huanuni, 101, 105
Huber, Evelyn, 23–25, 30, 33, 69
hybrid welfare state, 172, 247

I
ideology, 2, 25, 26, 64, 90, 211
Import Substitution Industrialization (ISI), 7, 12, 13, 31, 166, 167, 182, 199, 249
IMSS-Solidaridad/IMSS-Bienestar, 129
individual capitalization scheme, 43, 95, 122, 178, 191
individual earner–carer regime, 212, 225, 228, 251, 253
Industry, Commerce, Public Services, and the Rural Fund, 58
informal-familiarist welfare regime, 39
informality rate, 176, 177, 234
informal workers, 18, 30, 38, 91, 142, 166, 178, 183, 193, 247
Ingreso Ético Familiar-Subsidio de Seguridades y Oportunidades, 187
insiders, 26
Instituciones de Asistencia Médica Colectiva (IAMC), 132
Instituciones de Salud Previsional (ISAPRE), 97–99, 138, 139

Institute of Retirement and Pensions of Uruguay, 58
Institutional Attention Modality, 139
Instituto Boliviano de Seguridad Social, 105
Instituto Colombiano de Seguros Sociales (ICSS), 119
Instituto de Salud para el Bienestar (INSABI), 118, 146, 191
Instituto de Seguridad y Servicios Sociales de los Trabajadores del Estado (ISSSTE), 112–118, 145, 191
Instituto Ecuatoriano de Seguridad Social (IESS), 143
Instituto Mexicano del Seguro Social (IMSS), 112–118, 145, 191
Instituto Mixto de Ayuda Social (IMAS), 196
Instituto Nacional da Previdência Social (INPS), 79, 80
Instituto Nacional de Assistencia Medica da Previdencia Social (INAMPS), 81
Instituto Nacional de las Mujeres (Costa Rica), 233
Instituto Nacional de las Mujeres (México), 233
Instituto Nacional de las Mujeres (Uruguay), 234
Instituto Nacional de Previsión (INP), 104
Institutos de Aposentadoria e Pensões (IAP), 78
Inter-American Development Bank, 109
internal market-based economy, 8, 31
International Monetary Fund (IMF), 95
international subcontracting model, 169

J
Jefes y Jefas de Hogar Desocupados, 198
Jimenez Barbosa, Wilson Giovanni, 100
Jornada Escolar Completa, 229
Jornada Escolar Completa Diurna, 229
Jornada extendida en las escuelas primarias/Política Nacional para la Ampliación de la Jornada Escolar en el Nivel Primario, 229
Jóvenes Construyendo el Futuro, 192, 193
junta, 29, 66, 79, 104, 189
Juntos, 185
Justicialista party, 53, 71

K
Kirchner, Néstor, 28, 29, 32, 135, 193, 250
Kubitschek, Juscelino, 79

L
Labor, 2, 3, 9–12, 15, 16, 18, 19, 24, 25, 27–32, 34, 35, 38, 52, 53, 55, 56, 61, 64–71, 78–80, 90, 91, 93, 94, 97, 98, 101–105, 107, 109, 113, 114, 118, 119, 152, 165, 166, 168, 172, 178, 190, 193, 195, 198–200, 207, 209–211, 213–216, 218, 219, 221, 222, 225–227, 234, 238, 239, 246, 249, 251–253
Labor Accident Insurance, 72
labor insertion, 195
labor integration, 193
labor market, 3, 11, 19, 55, 60, 68, 80, 168, 195, 207, 211, 213–216, 218, 219, 221, 225, 226, 234, 238, 239, 252

labor movement, 15, 16, 18, 24, 27, 28, 31, 53, 54, 60, 64–66, 69, 71, 78, 90, 91, 93, 103, 104, 106, 108, 109, 168, 172, 246
labor participation rate, 215
Lacalle, Luis Alberto, 60
La Fraternidad, 57
laggard, 101
Lagos, Ricardo, 28, 30, 97–99, 178, 246, 250
La Paz, 101, 105
latifundios, 55
Lautier, Bruno, 7, 30, 82, 182, 184, 191, 197, 199, 250
La Violencia, 120
Law on Social Security, 109
legitimacy hypothesis, 24
Leguía, Augusto, 101, 103
Lei Orgánica, 79
level of poverty, 223
Ley de Consejos de Salarios, 60
Ley de Participación Popular of 1994, 106
Ley Madre, 55
Ley y plan nacional de igualdad de oportunidades entre mujeres y varones, 232
liberal commodity exporting model, 169
liberalism, 68, 171
liberal-residual welfare state, 38
liberal welfare regime, 36, 180
life capital, 4, 5
life debt, 4, 5
López Mateos, Adolfo, 113
López Obrador, Andrés Manuel (AMLO), 146, 183, 191
López Pumarejo, Alfonso, 119
Lula da Silva, Luiz Inácio, 29, 32

M

Madres de Plaza de Mayo, 231
male breadwinner regime, 214, 251
Mariátegui, José Carlos, 103
Marshall, T.H., 40
Martínez Franzoni, Juliana, 39, 61, 62, 73–77, 133, 136, 221
Marxism, 102
Más Familias en Acción, 185, 187
maternal leave, 214, 227, 238, 239, 253
means-testing, 108, 141
Médica Uruguaya, 57
Mendizábal Lozano, Gregorio, 106
Menem, Carlos, 29, 68–70, 128, 198
merchandise, 2, 9
Mesa-Lago, Carmelo, 33, 54–56, 59, 63–65, 69, 74, 76, 92–94, 96–98, 101, 103, 104, 107, 108, 129, 189, 199
Mexico, 8, 13, 15, 16, 18, 25, 27, 29, 31, 33, 34, 37, 38, 53, 56, 60, 65, 72, 77, 78, 91, 96, 100, 106, 110, 114, 128, 129, 140, 142–144, 146, 147, 149, 150, 155, 157, 158, 160, 165, 167, 169, 171–173, 175, 176, 178, 179, 182–185, 190, 192, 196, 197, 209, 215, 217, 219, 220, 222, 224, 227, 230–232, 236, 239, 246–248, 250
middle class, 8, 13, 28, 35, 37, 52, 59, 61, 83, 145, 199
military coup, 15, 32, 53, 63, 90, 94, 103, 107, 113
minimum security model, 7
minimum wage, 8, 37, 83, 97, 115, 116, 119, 122, 184, 189, 193, 199
Ministerio de la Igualdad, 233
Ministerio de la Mujer, Familia y Derechos Humanos, 233
Ministerio de la Mujer y la Equidad de Género, 233

Ministerio de la Mujer y Poblaciones Vulnerables, 234
Ministerio de las Mujeres, Género y Diversidad, 233
Ministerio de Salud (MINSA), 141, 142
Ministerio de Salud Pública (MSP), 143
Ministerio de Trabajo, Higiene y Previsión Social, 119
Ministry of Economic and Social Inclusion (MIES), 109
Ministry of Labor and Social Security, 60, 92
Ministry of Labor (Colombia), 119
Ministry of Public Health (Ecuador), 109
Ministry of Public Health, Labor and Social Welfare (Peru), 103
Ministry of Public Health (MSP), 109, 110, 143
Ministry of Social Security and Social Assistance (Brazil), 81
Ministry of Social Welfare, Labor, Agriculture, Charity, Health, Hygiene, Statistics and Immigration and Colonization (Peru), 104
Mitre, Bartolomé, 63
monotributo, 195
Montepío de Justicia and Real Hacienda, 63
Montevideo, 57, 59
Morales, Evo, 28, 107, 110
Movimiento al Socialismo (MAS), 110
Movimiento de Revolución Ciudadana, 110
Movimiento Femenino por la Amnistía, 231
mutuales, 57, 59
Mutualista delPartido Colorado, 57
Mutualista del Partido Nacional, 57
mutualists, 6, 43, 53, 56, 57, 61–64, 129, 131–133, 245

N

National Association of Civil Servants, 107
National Comprehensive Health System (SNAIS), 62
National Council of Social Security, 107
National Employee Medical Service, 93
National Health Fund, 62, 132, 137–139, 155
National Health Service (SNS) (Chile), 92
National Health Service (NHS), 74, 82, 137
National Institute of Social Protection, 67
National Insurance Bank, 72
National Resources Fund, 61
National Retirement and Civil Pensions Fund, 63
National Sanitary Service (Peru), 104
neo-corporatist model, 36
neo-liberalism, 38, 40, 71, 101, 108
New Deal, 27, 36
non-contributory pensions system, 8
non-governmental organizations (NGOs), 26, 43, 198

O

Obras Sociales, 45, 53, 66–71, 128, 135, 136, 152, 245
oligarchic regimes, 24
Oportunidades, 220, 221, 224
Organía, Juan Carlos, 67
out-of-pocket spending, 115, 144–146, 154, 156, 174, 249
outsiders, 26, 27, 29

outward-looking economy, 31

P

*Partido dos Trabalhadores (*PT), 82, 83, 197
Partido Revolucionario Institucional (PRI), 53, 91, 114
Partido Social Democratico (PSD), 79
Partido Trabalhista Brasileiro (PTB), 79
paternal leave, 214, 227, 230
Patiño, Simón, 105
pay-as-you-go pension system, 43, 60, 69, 75, 77, 82, 107, 114, 122, 193, 196
Paz Estenssorro, Víctor, 106
peasants, 29, 36, 41, 72, 112, 113, 119, 192, 246
pelegos, 79
pension affiliation, 176, 183
pension and health system coverage, 12, 110, 111, 176, 200, 249
pension scheme, 58, 63, 68, 83, 95, 113, 114, 171, 185, 191, 193, 222
Pension Solidarity Fund (FSP), 122
Peronism, 64, 66, 70, 91
peronist unions, 198
Perón, Juan Domingo, 15, 53, 64–66
Peru, 13, 16, 27, 32, 33, 90, 96, 100, 101, 103, 107, 108, 110, 111, 140–143, 147, 149, 157, 159, 167, 169, 172, 178, 184, 215, 222, 226, 227, 231, 232, 239, 246, 248, 249, 252
Petro, Gustavo, 29, 232
pink collar, 14, 207
Pinochet, Augusto, 29, 35, 67, 94, 98, 189
pioneer countries, 33, 76, 129
piquetero movement, 29, 70, 71, 198
Plan de Acceso Universal a Garantías Explícitas(AUGE), 98, 99, 139
Plan de Atención Nacional a la Emergencia Social (PANES), 194
Plan de Igualdad de Oportunidades, 232
Plan Global de Salud, 106
Plenario Intersindical de Trabajadores-Convención Nacional de Trabajadores (PIT-CNI), 61
Polanyi, Karl, 1–3, 6, 45, 243
political coalitions, 28
political dislocations, 32
political economy, 14, 37
political equality, 232
Política Nacional para la Igualdad y la Equidad de Género, 232
Poor Laws, 2, 3, 36, 181
poor relief measures, 4
Potosí, 105
Prado, Manuel, 107
Prado y Ugarteche, Manuel, 107
Pribble, Jennifer, 23–26, 28, 57
Private Employees and Industrial Workers Fund (Peru), 104
Private Employees Fund (Chile), 92
private health system, 48
private insurance, 37, 48, 82, 96, 108, 112, 135, 138–140, 153, 155
private spending, 141
productive labor, 40
profit-led social pacts, 18
Programa Escuelas de Tiempo Completo, 229
Programa Mais Educação/Novo Mais Educação, 229
Programa Nacional de Día Escolar Completo/Jornada Única, 229
Programa Nacional para la Igualdad de Oportunidades y no

Discriminación Contra las Mujeres, 232
progressive taxation, 5
Prospera, 187, 190, 191, 197
protected economy, 31
protectionist (or stratified) welfare regime, 39
Public Employees Fund, 93
public health system, 74, 247
public spending, 116, 146, 157, 171, 173, 248
puebladas, 70

Q
qualified labor force, 31

R
Railway Fund, 65
Régimen de Ahorro Individual con Solidaridad, 122
Régimen Solidario de Prima Media con Prestación Definida, 122
regulation of the labor market, 11, 18, 168
Renda Básica de Cidadania, 197
rentier capitalism, 31
reproductive labor, 40, 208
research and development (R&D), 31
residual welfare regime, 38
Retirement and Pension Fund for Industry, Commerce and Public Services, 58
retirement program, 33, 118
Retirement Savings System (SAR), 115
Revolución Ciudadana, 109
Revolución en Marcha, 119
Revolución Juliana, 104
Revolutionary Armed Forces of Colombia (FARC), 120
right to strike, 65
right to work, 198
Roca, Julio A., 63
Roussef, Dilma, 28, 29, 193, 197, 199, 250
Ruezga Barba, Antonio, 54, 57, 58, 60, 61, 73, 76–80, 93, 101, 102, 104, 105, 108, 109, 119, 133

S
Sáenz Peña, Roque, 65
Safe Pregnancy Insurance, 117
salarization, 37, 166, 182, 199, 249
Salinas de Gortari, Carlos, 114
Salud Básica Para Todos, 108
Sánchez-Ancochea, Diego, 39, 48, 61, 62, 73–77, 133, 136
Sánchez de Losada, Gonzalo, 107
Sanguinetti, Julio María, 60
sanitaristas, 81, 137, 196
Santa Cruz, 105
Sarmiento, Domingo Faustino, 63
school workdays, 227
Secretaría del Bienestar, 191
Secretaría de Salud, 115, 145
segmented corporatism, 39
segmented health system, 48, 153
segmented universalism, 35
Seguro Escolar Gratuito (SEG), 108, 141
Seguro Integral de Salud, 108
Seguro Materno Infantil (SMI), 108, 141
Seguro Popular, 17, 116–118, 128, 137, 141, 145, 146, 176, 191
Seguro Público del Adulto Mayor (SPAM), 142
Seguro Social Campesino (SSC), 143
Seguro Social Obligatorio (SSO), 103, 107
Seguro Universal Materno Infantil (SUMI), 142

Self-Defense Forces of Colombia (AUC), 120
Sembrando Vida, 192
separate gender roles regime, 211, 251
Servicio Plurinacional de la Mujer y de la Despatriarcalización, 233
Shoemakers Society, 64
Sindicato Nacional de Trabajadores Mineros, Metalúrgicos, Siderúrgicos y Similares de la República Mexicana (SNTMMSRM), 118
single payer pension system, 48, 122, 133, 156
Sistema de Ahorro para el Retiro (SAR), 115
Sistema General de Seguridad Social en Salud (SGSSS), 121, 122, 144
Sistema Integral de Salud, 141
Sistema Nacional de Salud, 96, 121
Sistema Nacional Integrado de Salud (SNIS), 62
Sistema Único de Saúde (SUS), 15, 29, 72, 81, 82, 84, 128, 131, 137, 138, 153, 156, 196, 246
social arrangement, 9, 11, 13
social cohesion, 6, 36
social compromise, 1, 9–11, 18, 51, 59, 165–168, 199, 208, 244, 249–251
social conflict, 67
social contract, 4, 11, 166
social corporations, 6
social democracy, 36
social democratic welfare regime, 40, 89, 131, 166, 180, 246
social expenditure, 167, 169, 177, 184
Socialist Party (Chile), 91
Socialist Party (Peru), 103
social order, 2, 9, 47
social protection, 1, 2, 4, 5, 7–9, 11, 27, 28, 30–37, 39–47, 51, 52, 54, 55, 65–68, 72–74, 76, 79–84, 89, 90, 92–95, 97, 99–105, 109–112, 118–121, 123, 128, 129, 142, 165, 166, 169, 171, 177, 181–183, 189, 193, 194, 197–199, 208, 209, 213, 217, 222, 225, 234, 239, 243–246, 248–250, 252
social protection rights, 32
social protection system, 1–10, 13, 14, 16, 17, 19, 24, 27, 28, 32, 33, 35, 39–41, 43–46, 51, 54, 79, 80, 84, 89, 94, 95, 99, 104, 112, 165, 168, 169, 172, 173, 175, 179, 180, 182–184, 189, 191, 194, 195, 197, 199, 207–209, 211, 213–215, 217, 220, 222, 225, 234, 243–246, 248, 250, 252
Social Reform Action Plan, 194
Social Security Code, 106
Social Security Fund (CCSS), 74
Social Security Organization Law, 80
Social Security Service (SSS), 92, 93
social security system, 5, 7, 8, 10, 18, 45, 65, 168, 244, 245
Social Service Commission, 66
social tensions, 10
Sociedad Francesa de Socorros, 57
Sociedad Fraternidad Unida, 57
societal orders, 6, 10, 41, 45
socio-corporatist systems, 131, 174, 248
socio-corporatist welfare state, 15, 219, 250
socio-developmentist state, 8
socio-political regime, 35
Solidarists, 4
solidarity, 4, 58, 60, 83, 94, 109, 121, 122

Solidarity Unemployment Fund, 98
Solidarność, 95
Southern Cone, 60, 73, 114, 251
Southern Democrats, 27
Speenhamland system, 3, 8
State Health Services Administration (ASSE), 61
state-led health system, 136
statist-productivist (or focalized) welfare regime, 39
statist systems, 136
statist welfare state, 15, 169, 175, 250
subsidized services, 144
Sucre, 105
Support for Labor Insertion, 195
syndicalism, 102

T

Tarjeta Uruguay Social (TUS), 185, 189, 194
Terra, Gabriel, 58
Théret, Bruno, 2, 4–6, 11, 23, 43–45, 181
Titmuss, Richard, 6, 10, 23
Trabajar, 70, 71, 198
trade unions, 6, 11, 36, 56, 71, 91, 92, 98, 135
traditional family, 35
Typographical Society of Buenos Aires, 64

U

underemployment, 7
unemployment, 7, 9, 11, 40, 54, 68, 70, 71, 80, 92–94, 96, 98, 166, 168, 189, 198, 213, 217, 244
unemployment rates, 217
Unión Cívica Radical, 63
Unione e Benevolenza, 64
Unión General de Trabajadores (UGT), 64
unionism, 15, 34, 53, 64, 67, 70, 92, 94, 102, 106, 120
Unión Sindical Argentina (USA), 64
universal breadwinner regime, 212
universal caregiver regime, 234
Universal Health Insurance Program, 109
universal health system, 175
universalistic scheme, 8
universalistic welfare state, 25
Uruguay, 8, 15, 17, 18, 25, 27, 30, 33, 35, 39, 43, 47, 52–55, 57, 58, 60, 63, 64, 76, 78, 96, 100, 111, 127, 129–131, 134, 136, 137, 145, 147, 149, 150, 153, 154, 157, 158, 160, 167, 169, 173–175, 177–179, 183–185, 189, 190, 194, 199, 215, 217, 222, 223, 226–232, 234, 238, 245, 247, 248, 250, 252, 253
Uruguay Trabaja, 195, 198

V

Valencia, Enrique, 10, 23, 38, 41, 117, 118, 190
Valparaíso, 92
Vargas, Getulio, 72, 76–79
varguismo, 79
Velasco Alvarado, Juan, 107

W

wage gap, 216
wage-led growth strategy, 199
wage-led social pacts, 18
wage relation, 18, 165, 167, 168, 201, 249, 250
wage relationship, 10
Walzer, Michael, 9, 10
Washington Consensus, 60
Weiss, John H., 4

welfare state, 1–3, 6–19, 23–39, 41, 42, 44–46, 48, 52–55, 63–67, 71–73, 78, 82, 84, 89–91, 93, 94, 100, 102–104, 107, 111, 112, 119, 120, 123, 127–129, 165–168, 171, 172, 175, 178, 184, 190, 193, 194, 198, 200, 207–217, 220, 225, 234, 239, 243–247, 249–253
white collar, 92, 103, 105, 107
William Beveridge, Chancellor of the Exchequer, 36
women's rights, 209, 232
World Bank, 95, 135, 221
World War II, 11, 12, 33, 36, 52, 78, 100, 104, 113, 119, 129

Y

Yrigoyen, Hipólito, 63

Z

"zero hunger" program, 197

SPRINGER NATURE

GPSR Compliance

The European Union's (EU) General Product Safety Regulation (GPSR) is a set of rules that requires consumer products to be safe and our obligations to ensure this.

If you have any concerns about our products, you can contact us on ProductSafety@springernature.com

In case Publisher is established outside the EU, the EU authorized representative is:

Springer Nature Customer Service Center GmbH
Europaplatz 3
69115 Heidelberg, Germany